1 MONTH OF
FREE
READING

at

www.ForgottenBooks.com

By purchasing this book you are eligible for one month membership to ForgottenBooks.com, giving you unlimited access to our entire collection of over 1,000,000 titles via our web site and mobile apps.

To claim your free month visit:

www.forgottenbooks.com/free949106

ISBN 978-0-260-45199-6
PIBN 10949106

This book is a reproduction of an important historical work. Forgotten Books uses
state-of-the-art technology to digitally reconstruct the work, preserving the original format
whilst repairing imperfections present in the aged copy. In rare cases, an imperfection in
the original, such as a blemish or missing page, may be replicated in our edition. We do,
however, repair the vast majority of imperfections successfully; any imperfections that
remain are intentionally left to preserve the state of such historical works.

BULLETINS AND CONVENTION PROCEEDINGS
of the
AMERICAN RAILWAY ENGINEERING ASSOCIATION
(Engineering Division, Association of American Railroads)

CONTENTS, VOLUME 63

* Because of practical considerations, the binder for the Bulletins of Vol. 63 is not designed to hold Part 2 of Bulletin 570, February 1962, which contains only prints of the new and revised trackwork plans presented for adoption at the 1962 Convention.

CONTENTS

Published by the American Railway Engineering Association, Monthly, January, February, March, November and December; Bi-Monthly, June-July, and September-October, at 2211 Fordem Avenue, Madison, Wis.; Editorial and Executive Offices, 59 Van Buren Street, Chicago 5, Ill.
Second class postage paid at Madison, Wis.
Accepted for mailing at special rate of postage for in Section 1103, Act of October 3, 1917, authorized on June 29, 1918.
Subscription $10 per annum.

Insulated Rail Joint Development and Research*

A. PRELIMINARY

The insulated rail joint presents a difficult design problem. The joint is a load-carrying member of the track, and for satisfactory service it must have flexural strength and stability comparable to the full rail section and still maintain a good degree of electrical insulation between the rail ends through wide temperature variations and in the presence of water and contaminants of many types. The joint, bolts, bushings and end posts are also subjected to the forces from temperature changes which put high compressive loads on the insulating materials, especially in welded track. Since all types of insulation available are of lesser physical strength and modulus of elasticity than steel, the insulated joints have been not only items of high first cost but also sources of high maintenance costs compared to the rest of the track.

A number of requests from interested railroad engineers and the later assignment of the subject to a subcommittee of Committee 4—Rail, led to consideration of the problem by the AAR research staff. A. L. Essman, chief signal engineer of the Chicago, Burlington & Quincy Railroad and past chairman of the AAR Signal Section, actively assisted in the work on the subject. The research work, designing, testing, and preparation of this report were in the direct charge of Randon Ferguson, electrical engineer of the research staff, under the general direction of G. M. Magee, director of engineering research. Manufacture of specimens and materials was done principally by the Gatke Company of Chicago in the early part of the test program and at present by the Johnson Rubber Company of Middlefield, Ohio, which is fabricating the AAR design of molded, bonded and armored joint under the trade name "VulcaBond." A number of railroads have cooperated by installing and observing service test specimens to supplement the laboratory tests in determining the effectiveness of the various features of the designs.

B. ANALYTICAL CONSIDERATIONS

1. Discussion of Design and Objectives

Previous laboratory and field investigations and theoretical analyses[1] concerning the mechanics of the action of the rail joint brought out the fact that very high compressive and abrasive forces are developed between the rails and the joint bars, especially at the top mid-length of the joint. Consideration of these forces made it evident that no insulation known was strong enough to resist these forces without major damage and that the insulation must be protected by the higher strength steel. The bonding of steel protection to the insulation seemed a logical way to best protect the insulation. The protective steel armor bonded to the insulation would accomplish the following important objectives:

---

* An extended abstract of Report ER-9 issued by the Research Department, AAR, incorporating some additional information developed subsequent to the publication of the original report. Copies of the original report can be obtained from the director of engineering research, AAR, at 3140 South Federal St., Chicago 16.

[1] Fifth and Sixth Progress Reports, Special Committee on Stresses in Railroad Track, American Railway Engineering Association, Vol. 31 (1930) and Vol. 35 (1934) respectively.

a. Direct abrasive slipping contact under high compression would be kept from the insulation and taken by the steel which is better able to withstand this type of stress.

b. The stiffness of the steel armor relative to that of the insulation would cause concentrated loadings, such as those from the rail ends, to be distributed over larger areas of insulation with resultant decreased insulation unit loading and stresses.

c. Confinement of the insulation by the steel armor and insulation adjacent to the load application points and the bonding will give the insulation greater ability to carry load and take large deflections. This has been demonstrated by research work done on concrete in triaxial loading[2] and is especially helpful for materials that are weaker in tensile strength than compressive strength. The restraint of the insulation to lateral movement under the compressive loading due to the bonding and nearby presence of insulation not under compressive loading corresponds in effect to a triaxial loading. It is apparent from this discussion that the points of application of bearing pressures must be kept at a distance from the edges of the insulation and steel armor to obtain the most benefit from this action in containment and load distribution on the insulation.

d. The bonding of the steel and insulation in the joint bars makes the flexural *strain* the same in the steel and insulation at the bond, but the *stress* or load is carried mostly by the higher ·strength steel because of its greater stiffness (higher modulus of elasticity). Conversely, this means that the insulation has a lower flexural stress which will increase its service life.

2. Mechanics of the Action of the Rail Joints in Track and Their Relation to Joint Design

The rail provides a continuous support for a wheel load or group of wheel loads that distributes the load over a considerable number of ties so that the tie reactions developed to support the wheel load are all of lesser amounts than the wheel loads, and the pressures on the ballast are correspondingly lessened. The reduction and equalizing of these ballast pressures are the basic requirements for optimum maintenance conditions. The ballast and subgrade act in essentially an elastic manner under most conditions so that the deflection is proportional to load. This action is expressed mathematically by the fundamental equation of a beam on a continuous elastic support.[3]

$$EI \frac{d^4y}{dx^4} = -uy$$

where E is the modulus of elasticity of steel, I the moment of inertia of the rail section, u the modulus of track support, y the deflection at any point, and x the position of that point with respect to the load.

When the rail ends are joined together by a rail joint, the vertical moment in the rail is zero at the rail ends and must be carried entirely by the joint bars at that point and shared by the rail and the bars at other points in the joint bar area. It is not feasible to have the two joint bars form a member as stiff as the full rail; also, there are relative movements and slippages due to lack of fit. So it follows that there is a discontinuity in

[2] F. E. Richart, Anton Brandtzaeg and R. L. Brown, A Study of Concrete Under Combined Compressive Stresses, Bulletin 185, Engineering Experiment Station, University of Illinois.
[3] First Progress Report, Joint Committee on Stresses in Railroad Track, Vol. 19 (1918) p. 872, Proceedings, American Railway Engineering Association.

the moment and deflection diagrams at the rail ends which increases the deflection at the mid-length of the joint relative to that obtained from the full rail. It is axiomatic that the closer the rail joint stiffness and deflections approach those of the full rail the more nearly will the maintenance required by the ties and ballast at the joint be comparable to that of the rest of the rail. So the design objective for any joint, insulated or unin-sulated, should be to approach the condition of full continuity as closely as possible.

If the lesser moment at the rail ends with the wheel at the rail ends is designated by M_1, that in the full rail by M_0, and a constant of proportionality by k (less than unity), the effect of the discontinuity may be calculated from the equation

$$M_1 = k M_0$$

and the relation given previously for the action of the rail on a continuous elastic support.[4] This solution is quite long and complicated and indicates double the usual deflection at the load point for $k = 0$ (with no joint bars at the rail ends). This result has been found to be only approximately true in the track, probably because the ballast action is not linear under highly localized variations of load. However, the analysis does indicate that flexible members used at the rail ends will greatly increase the tie loads and inevitably the maintenance requirements. Thus one criterion of a rail joint can well be its deflection compared to a full rail under the same moment, and this has been used in the following discussion as one comparison of the various insulated joints tested.

The fact that a joint is stiff and does take a given moment with a small deflection does not necessarily indicate that it is going to stand up well in track. The very fact that it does take a large moment means that it has high stresses, perhaps higher than a more flexible joint that shirks its share of the load by increasing the tie loads. Thus it is conceivable that the flexible joint may show up well in service tests if maintenance is such that the tie support is kept well up to the joint.

The fact that the joint should develop maximum moment requires that it be designed to do so with the least amount of stress, play and internal movements possible. Thus the load-carrying members should be of the most efficient sections, with best fit possible, and of the stiffest, strongest and toughest material. Since all insulations at present usable for this purpose are weaker and less stiff than steel, the maximum and most efficient use of steel must be made to obtain the strongest joint. As previously pointed out, the internal forces developed in the rail joint are quite high and so concentrated that any known insulation will be damaged at their points of application; so it is essential to protect the insulation as much as possible. The use of bonded armor and insulation seemed the best and most feasible method of accomplishing this protection. The armor also protects the insulation from the high abrasive forces resulting from relative slippage between the parts under these high compressive forces.

C. LABORATORY TESTS

Manufacturing facilities available to the program at the start required that the insulation be applied to the joint as a sandwich shim having inner and outer armor, with the inner armor having a gap at the center if the shims were the full length of the joint, or to use two shims on each side of one-half length. To eliminate rolling or forging a special bar section, a 115-lb RE bar was used for 132-lb RE rail, with the shim making up the difference in fishing height. Since the shim at this time could only be made in uniform thickness, the bars had to be slightly reformed to obtain the proper fit in the

[4] Fifth Progress Report, Special Committee on Stresses in Railroad Track, Vol. 31 (1930) p. 93, Proceedings, American Railway Engineering Association.

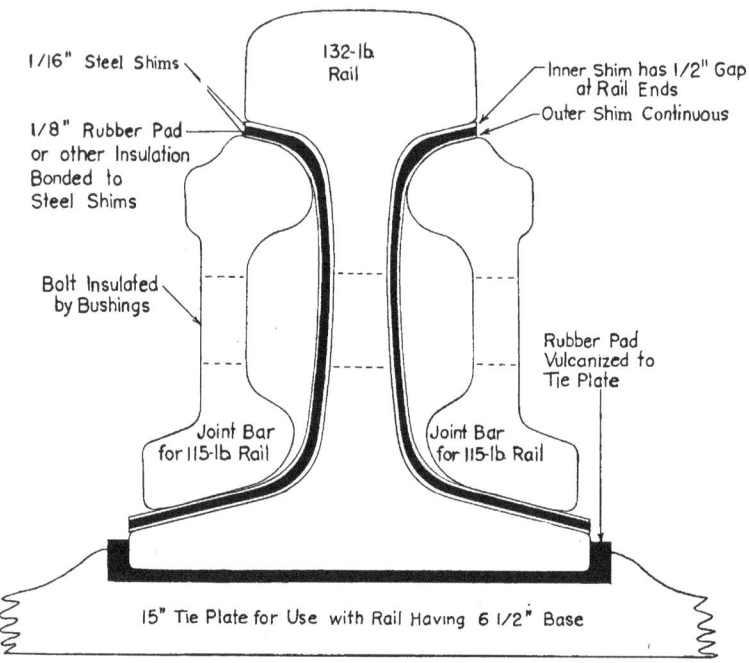

Fig. 1—Insulated joint with sandwich-type insulating shims.

fishing areas. The bars also required special drilling for the insulating thimbles and end post insertion. Fig. 1 shows a cross section of this type of joint, including an insulating tie plate obtained by molding insulation into a tie plate for a 155-lb PRR rail.

Considerable time elapsed in the various phases of the initial program because of difficulties in getting the parts processed. Reformed and drilled bars were finally obtained for a laboratory test in 1953. The laboratory test was made in a rolling-load testing machine with a 33-in stroke that applies a positive moment and 50 percent reversal in each cycle. The joints were loaded with a 44,400-lb load developing 400,000 in-lb positive moment and 200,000 in-lb negative moment. After various preliminary tests and trials a joint with reformed bars and sandwich shims was tested in the summer of 1953. It ran 303,200 cycles and failed by fracture of the joint bars. The joint bars were found to be very soft (160 BHN), which has been shown to be quite conducive to a short joint life in repeated loading.

Field tests of this type of joint construction for 129-lb TR rail were started on the CB&Q Railroad late in 1954, but the design was such that bearing pressure of the rail head was concentrated near the edges of the shims, which caused them to deteriorate quickly, and they were shortly thereafter removed from service. The insulation was squeezed out of the shim by the bearing pressures between bar and rail. While it still appeared that a sandwich shim of proper design and material could be an improvement, it was decided that molding and bonding the insulation and armor directly to the joint bar would make a much stronger, simpler unit. Consultation with people in the rubber industry indicated that this could be done. It offered the additional advantage that standard sections could be used and the required fit could be obtained by design of the mold. No reforming of the bars would be necessary. Testing of material samples was done to obtain the type of insulating material that would be best suited to the needs before proceeding with the fabrication.

1. Tests of the Molded, Bonded, Vulcanized and Armored Insulated Joints

The molded and bonded insulated joint was designed in Feb. 1957 (trade name VulcaBond) and the first joint was received for test in November 1958 from the Johnson Rubber Company. A standard 115-lb RE joint bar was used and the insulation and armor bonded and molded to it to fit the rail, each bar being an integral unit. A cross section of the joint is shown in Fig. 2. The original washer plate bushing assembly is also shown in this figure. This joint ran 502,000 cycles in the rolling-load testing machine with a load of 44,400 lb, the same load used in tests of 115-lb uninsulated bars. A view of a VulcaBond joint in the rolling-load testing machine is shown in Fig. 3. This life compares favorably with the life of the uninsulated bars.[6] The failure was by fatigue fracture of one bar which had a BHN of 201, a somewhat low value. As mentioned before, low hardness values have been found to contribute to short fatigue life.

Three other joint bars tested for fatigue life "ran out" to 2,000,000 cycles and were still in fair to good serviceable condition. This number of cycles is considered representative of the flexural demand on a joint during its normal service life. The deflections of this joint in the 36-in span were 0.03 to 0.04 in compared to a measured deflection of 0.024 in for a full rail in the same machine. The deflection of the joint compared to that of the rail is an indication of its efficiency as a flexural unit in the track. These three joints had various changes in design, such as easements at top mid-length, shot-peened bars (at mid-length) and an improved bushing, insulating washer plate assembly which was also molded to the joint bar for better strength and ease of assembly (see Fig. 5). Several other VulcaBond joints were tested for shorter periods to determine effect of some other factors. A 2,000,000-cycle run covers a period of 3 to 4 months, depending on the running time per day.

2. Other Insulated Joint Tests

Rolling-load tests were made on several other types of insulated joints, including the standard continuous armored and unarmored types. They are briefly described below:

a. A continuous armored joint for 132-lb RE rail with vulcanized fiber insulation. Deflections were 0.06 to 0.07 in. The test was discontinued at 293,000 cycles because of impact on machine, looseness of parts and deterioration of fiber mechanically. Insulation was still good electrically.

b. A continuous unarmored 129-lb TR rail joint with vinyl insulation ran 457,400 cycles and grounded on one end. The insulation was badly worn at top mid-length.

c. A continuous armored 115-lb joint with Fabco insulation ran 152,700 cycles. The insulation shorted out on one end and was badly worn. Deflection was ⅛ in.

[6] Proceedings, American Railway Engineering Association, Vol. 53 (1954) p. 878, Vol. 57 (1956) p. 818, Vol. 58 (1957) p. 1005.

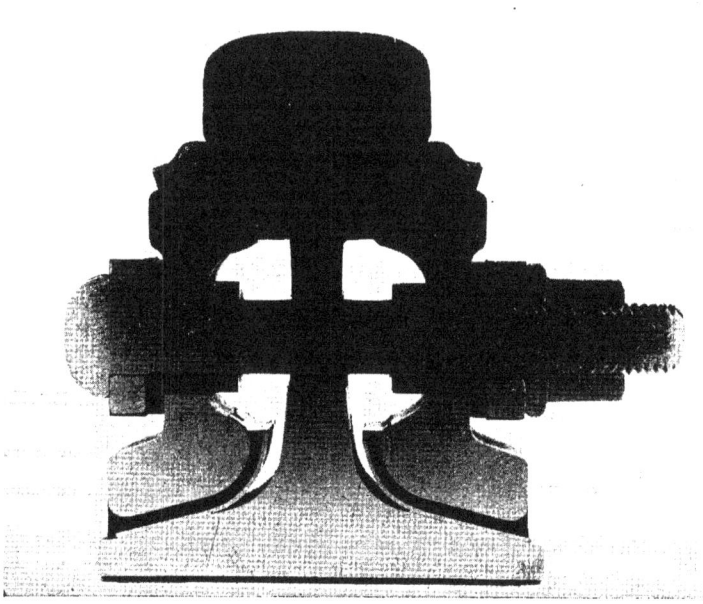

Fig. 2A—Cross section of first design of molded and bonded
VulcaBond insulated joint.

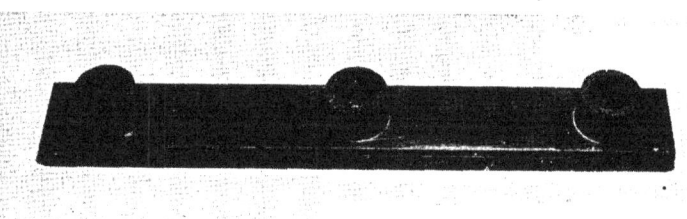

Fig. 2B—Washer plate-bushing of first design VulcaBond joint.

Fig. 3—**VulcaBond** insulated joint being tested in rolling-load machine.

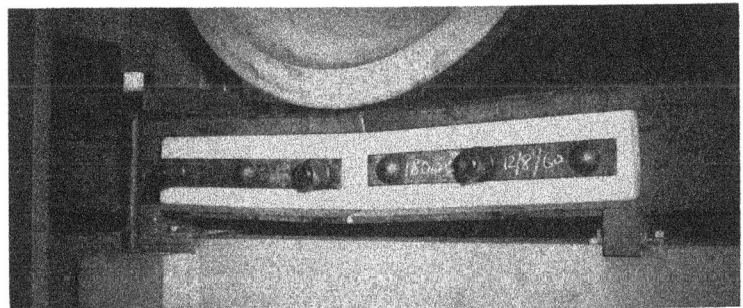

Fig. 4—**Plastic** joint under load in rolling-load testing machine.

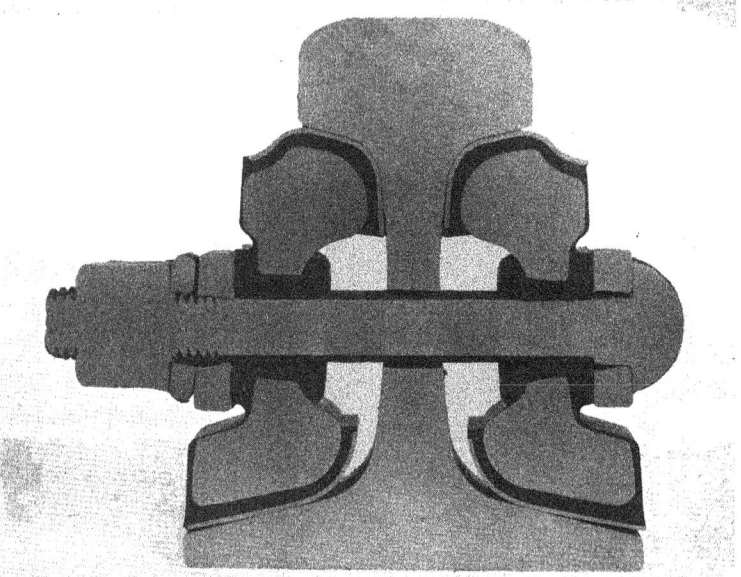

Fig. 5—Cross section of new design of VulcaBond insulated joint.

d. A continuous armored 115-lb joint with Fabreeka insulation shorted out on one end at 50,000 cycles and one bar broke at 199,700 cycles. The insulation was badly worn. Deflection was 0.07 in maximum.

e. Three 132-lb RE rail joints made by the Permali Company of laminated resin impregnated wood were tested. Each whole bar is composed of this material. Two joints failed under static loads of 55,000 lb and 27,500 lb; the third ran 17,500 cycles under 15,000-lb load. The bars appeared to fail from horizontal shear. The deflection was 0.16 in with the 15,000-lb load.

f. A joint with bars of nylon-type material for 129-lb TR rail gave a deflection of 0.43 in with 18,000-lb load. It was not tested because the large deflection did not permit running it in the rolling-load machine (see Fig. 4).

g. Four joints of fiber glass and plastic for 132-lb and 133-lb rail were tested and performed as follows: No. 1 ran 590,000 cycles at 22,000 lb (½ load). The test was stopped because of impact and large deflections. Bars had started to fail at middle. Deflections were 0.12 to 0.17 in. No. 2 (a little heavier bar) ran 35,700 cycles at ½ load and broke at 2,800 cycles at full load (44,400 lb). Deflections were 0.12 to 0.43 in. No. 3 ran 112,300 cycles under ½ load and failed by horizontal shear and transverse breaks. Deflections was 0.15 in. No. 4 ran 52,500 cycles under ¾ load and failed in a manner similar to No. 3. Deflection was 0.13 in.

D. SERVICE TESTS OF VULCABOND INSULATED JOINTS

Because of the excellent laboratory performance of the VulcaBond joint, service tests were installed to determine what difficulties might develop due to factors not present in the laboratory tests. Six joints were installed on four railroads. Two principal defects found were lack of flange clearance and weakness of the bushing washer plate assembly. Changes in the design overcame these difficulties apparently satisfactorily. Other joints have been installed on about a dozen railroads and their performance will be followed by the research staff and the manufacturer. Table 1 shows the location and results of an inspection after about one year of service of the original six joints and two joints on EJ&E welded track (see Fig. 6). No reports of difficulty have been received after two years of service except in the case of the original bars on the PRR where the worn condition of the rail head caused heavy flange contact, and bars of the later design were installed.

E. DISCUSSION OF RESULTS OF THE TESTS

The laboratory tests in the rolling-load machines give very good comparative data on the action of the joints in flexure and the fatigue life under flexure. The AAR joint (VulcaBond) had the smallest deflection, being 0.03 to 0.04 in compared to 0.024 in for the full rail. A standard armored continuous type was next lowest (0.06 in), with the plastic (Sustamid) joint the greatest, with 0.43 in under 40 percent load.

Three AAR joints were run to 2,000,000 cycles and still remained serviceable. Several others showed similar life in tests for effect of some of the design changes. A test of 2,000,000 cycles has been considered equivalent to the service life of uninsulated joints. The armored continuous joint was run 293,000 cycles and would have run longer, but the joint was giving considerable impact on the testing machine and the insulation and armor were breaking up. A continuous-type unarmored joint with vinyl insulation was run 404,000 cycles, but failed electrically and the insulation was considerably damaged. The others tested ran fewer cycles, had lesser loads, or both, and in one case, could not be tested in the rolling-load machine. The fiber-glass joints were the best of the plastic type tested but had considerably greater deflections than either the Vulcabond or continuous type and a lesser life in fatigue.

The rolling-load tests, of course, do not apply some of the forces and conditions found in service, such as longitudinal tension and compression forces, temperature variations, time effect and corrosion factors. The VulcaBond joints were placed in track on several railroads to determine the suitability of the design in regard to these other factors. Several weaknesses were soon evident, as previously mentioned. These were:

a. Lack of flange clearance, which was remedied by redesign of the section.

b. Breakage of the bushings with washer plate bushing assembly design shown in Fig. 4B. This was improved in the new section by molding the bushings and insulating plate into the bars with reinforcing rings on the inside face.

c. Damage to top armor from cocking of the bars in the fishing. This was eliminated by decreasing the height of the bars to get proper fit.

The molding of the bushings or thimbles into the VulcaBond bars has greatly strengthened the bushings, and it is believed that they will be able to last the serviceable life of the bar without replacement in jointed track. If the rolling-load tests are a valid criterion, this service life should be similar to that of an uninsulated joint. However,

Fig. 6—Two views of VulcaBond insulated joints in welded track.

TABLE 1—RESULTS OF INSPECTION OF VULCABOND SERVICE TEST JOINTS

Joints Inspected - September 14 and 15, 1960

Insulation Resistance in Megohms
Hardness - D Scale Durometer

Railroad Location Install. Date	Ins. Res. Min.	Ins. Res. Max.	Tie Plate Cond.	Flange Contact	Hardness EB*	Hardness WB	Armor	Remarks
Milwaukee Road M.P. 71.5 Install. 8/28/59	8	800	Good	Heavy	84	84	No Cracks Good	Most of thimbles broken. Top armor curling up from flange contact. End post poor condition. Replaced with new design bars.
Milwaukee Road M.P. 72.7 Oakwood Crossing Install. 8/28/59	50	200	Good	Slight	89	86	No Cracks Good	Six thimbles broken. End post armor partially loose.
Ill. Central M.P. 56.8 Kankakee, Ill. (at crossing) Install. 9/8/59	5	800	One Replaced	Moderate	80	75	No Cracks Some Indent. at Middle	Tie plate broken during installation replaced. End post armor loose on one side. Thimbles cracked, but still in place.
Ill. Central M.P. 56.8 Kankakee, Ill. (north of crossing) Install. 9/8/59	400	10,000	Good	Moderate	75	80	No Cracks Good	Top armor north end reversed end for end on one bar. Seven thimbles cracked, but still in place. End post fair shape.
Northern Pacific M.P. 75 + 437 Sauk Rapids, Minn. Install. 7/22/59	200	10,000	One Chipped	Slight	85	85	No Cracks Slight Indent.	End post fair. Five thimbles broken.
+ Pennsylvania RR M.P. 302.5 Derry, Pa. Install. 4/29/60	200	10,000	Good	None	--	--	No Cracks	Armor curling due to cocked bars. Outer bar worst. End post good. Thimbles very good. (Original bars were removed due to heavy flange contact and bond failure of an upper armor.)
** E.J. and E. 1/2 mile west of Spencer Crossing N.R. Install. 5/60	5	10,000	One Chipped	None	85	85	Excellent No cutting, cracks or curling	Inspected 11/25/60. End post badly battered by tread contact. New armored post put in. Bolts bent. Some bushings show cracks, but are still in tact. Under considerable tension with rail at 70° F.
** E.J. and E. 1/2 mile west of Spencer Crossing S.R. Install. 5/60	0	10,000	One Chipped	None	85	85	Excellent No cutting, cracks or curling	Inspected 11/25/60. End post badly battered by tread contact. New unarmored post put in. Chip out of one bushing each bar. One with crack. Some tension in joint with rail at 70° F.

* NB & SB in some cases

+ Note: These are new design bars with thimbles and washer plates molded on bar.
  Original bars damaged by heavy flange contact, a derailment and poor bond on one armor strip.

** Note: Welded track. These are new design joint bars with thimbles and washer plates molded on bar.

welded track puts much more severe strains on the bushings, and it is planned to strengthen them further to meet this requirement. The two joints under service tests in welded track are of the improved design, and their bushings show some evidence of damage, but were in serviceable condition after about 14 months of service.

It is possible, of course, to go back to the practice of having removable bushings if the life of the bushings is less than that of the joint bars. The bars themselves, when properly made and installed, show indications of a long service life. Tests are being made on other variations of design which will possibly be further improvements.

The Lateral and Longitudinal Distribution of Loading in Steel Railway Bridges

By W. W. Sanders, Jr., W. H. Munse, and N. M. Newmark*

A. INTRODUCTION

This report summarizes the results of the investigation of the lateral and longitudinal distribution of loading in steel railway bridges conducted at the University of Illinois under the sponsorship of AREA Committee 30.**

It has often been suggested that the present design specifications for the distribution of live load in railway bridge floors, although giving satisfactory designs, are generally too conservative. With this in mind, the study summarized herein was undertaken at the University of Illinois to develop a more realistic specification for the distribution of live load in the floor systems of steel railway bridges.

The first portion of the study consisted of the preparation of a review of the existing methods of analysis for bridge floor systems and related structures, and then the development of methods of analyses to accurately determine the distribution of wheel loads to bridge floor systems. The second objective of the investigation was to obtain a number of solutions of simulated bridge floors using these methods of analysis. These solutions include evaluations of the two general types of bridges now in use and are compared with the results of a number of actual bridge tests conducted during the last 10 years by the Association of American Railroads The third and final objective was to derive for design office use, a method of computing the live-load distribution which takes into account the principal factors affecting the distribution.

Two basic categories of bridge floor systems were studied in the investigation. The first floor system consisted of a number of transverse floorbeams supported by heavy longitudinal edge girders or trusses and the second consisted of a number of longitudinal beams connected by a series of transverse diaphragms. In addition to the study of the two general bridge types, results were obtained which indicated the effect on the behavior of the bridges of (1) size and spacing of the floorbeams or stringers, (2) type of floor covering, (3) diaphragms, (4) ballast, and (5) number of tracks.

B. DEVELOPMENT OF METHODS OF ANALYSIS

An analytical study of steel railway bridge floors offers numerous problems. In most cases, the bridges are indeterminate to a number of degrees. Also, the structural design is such that the variations are many, even in bridges of a given general type. Therefore, to study the various types of railway bridge floors in general terms it was necessary to idealize the structures. For this study, the idealized structure was assumed to be a gridwork of beams supporting a slab and/or plate, simply supported on two edges and free on the other two edges.

* Respectively, assistant professor of civil engineering, professor of civil engineering, and head, Department of Civil Engineering, University of Illinois.
** Copies of the complete report may be obtained from the director of engineering research, Association of American Railroads, 3140 S. Federal St., Chicago 16.

Although there are several mathematical procedures available which may be used to analyze this gridwork of beams, only two are used in the study of the behavior of the grid: (1) a moment distribution procedure developed by N. M. Newmark and presented in Bulletin 304 of the University of Illinois Engineering Experiment Station and (2) the orthotropic plate theory. A complete derivation of the necessary equations for the orthotropic plate analysis using the required boundary conditions is given in the complete report of this investigation, entitled "The Lateral and Longitudinal Distribution of Loading on Steel Railway Bridges", by W. W. Sanders, Jr. and W. H. Munse, Report No. ER-5, Association of American Railroads Research Department. The first of these two mathematical procedures assumes no interaction between the slab and the beams, whereas the second method assumes full composite action.

The effect of ballast was obtained by modifying the concentrated wheel load with the beam-on-elastic-foundation analysis. The general method of analysis used in the study of the diaphragms was developed by B. C. F. Wei (Vol. 85, No. ST5, *ASCE Proceedings*, May 1959). To take into account the effect of diaphragms, the structure is divided into two separate systems: (1) the diaphragms and (2) the bridge without the diaphragms, and the forces acting between the two systems determined. The problem was then reduced to finding moments in a bridge without diaphragms but subjected to the added diaphragm forces.

In the development of these methods of analysis for the various distribution problems in a steel railway bridge floor, numerous assumptions and empirical modifications have been used. In order to determine whether or not these methods of analysis will yield beam moments close to those which will be obtained in the actual structure, these procedures were used to analyze the results of a series of field tests on bridge structures under actual service loadings. These bridge tests were conducted by the Association of American Railroads and had previously been reported by AREA Committee 30. The correlations obtained between the results computed with the various mathematical procedures used in this study and the results obtained in the field tests demonstrate the effectiveness of the procedures to simulate the distribution of loads in actual bridges.

C. RESULTS OF INVESTIGATION

To obtain suitable design procedures for the different types of floor systems of steel railway bridges, it was necessary to study the effect of each of the variables affecting the behavior of the floor systems. Limits of values were selected for each of these variables based on a study of existing bridges. Numerous bridge floor systems were obtained by combining the various values of the variables which affect the floor system behavior. A number of solutions were then obtained using the numerical procedures developed previously.

The results of the numerical analyses are discussed on the basis of the five principal parameters which determine the behavior of a steel railway bridge floor: aspect ratio $\left(\dfrac{d}{a}\right)$, bridge stiffness factor (H), diaphragm stiffness ratio (r), depth of ballast (b), and location of the axle loads.

Where: $d =$ beam spacing

$a =$ beam span

$$H = \frac{E_b I_b}{aN}$$

$$N = \frac{E_c I_o}{1 - \mu^2}$$

$$r = \frac{E_d I_d}{E_b I_b}$$

E_b and E_c are the modulii of elasticity for the beam and slab respectively, I_b and I_c are the corresponding moment of inertia, and $E_d I_d$ is the stiffness of the diaphragms.

Each of these factors affects the behavior of the bridges with transverse floorbeams or bridges with longitudinal beams in a different manner. Therefore, the effect of the five factors with respect to each individual type of bridge are considered separately.

1. Bridges With Transverse Floorbeams With Slab

The behavior of a bridge with a reinforced concrete floor slab is affected most by the aspect ratio, d/a, and the bridge stiffness factor, H. A low value of H corresponds to a relatively stiff slab in comparison to the floorbeams. As H becomes larger, the beams become stiffer in relation to the slab, and the slab loses some of its effectiveness in distributing the load.

Since the aspect ratio and the bridge stiffness factor affect both the distribution of moment to the beams and the percentage of total moment in the beams, it is impossible to separate their effects. However, it should be noted that for a constant aspect ratio, an increase in bridge stiffness factor (i.e. stiffer beams) not only increases the percentage of total beam moment in the loaded beam but increases also the total beam moment itself.

The effect of ballast varies considerably, depending upon the relative stiffness of the slab and the beams. When the slab is very stiff in relation to the beams, a major part of the load distribution is obtained through the slab. However, as the stiffness of the beams increases the slab loses some of its distributing ability, and the ballast becomes more important as a distributing agent.

2. Bridges With Transverse Floorbeams Without Slab

Transverse floorbeam bridges without a reinforced concrete deck are generally constructed with only a steel or wrought iron deck plate covering the floorbeams, and are rarely constructed without a ballasted track. With this thin plate structure, the primary longitudinal distribution of live load to the floorbeams is by the ballast and the rail. Because of the negligible effect of the plate and the diaphragms, the distribution study for this bridge type was based completely on a beam-on-elastic-foundation analysis.

3. Bridges With Longitudinal Beams With Slab

The load distribution in bridges with longitudinal beams and with slabs, just as in the case of transverse floorbeam bridges with slabs, is a function of the aspect ratio (d/a) and the bridge stiffness (H). However, in bridges with longitudinal beams the transverse position of the track with respect to the beams is also important. In most longitudinal beam bridges the track or tracks are centered on the bridge. For a single-track bridge, therefore, the rail generally would be located 2.5 ft each side of the center line of the bridge. For a double-track bridge the track will generally be located approximately 7 ft each side of the bridge centerline. Then, as the beam spacing is changed the rail moves into a different position with respect to the individual beams. This variation in relative position must be kept in mind when considering the behavior of bridges with different beam spacings.

Because of the complex inter-relationship between the various factors which affect the behavior of this bridge type, specific statements cannot be made concerning these effects. However, it was found that the effect of the slab on the transverse distribution of the moments diminishes as the relative stiffness of the beams increases. Not only does more of the total beam moment go to the beams under the track, but the total beam moment itself increases. It was also noted that the maximum percentages of the total beam moment generally occurs in the beams nearest the center line of the track. However, if the beams are very stiff (high H) and the beam spacing is such that two of the beams lie near the rails, the maximum moment may occur in these two beams. In addition, the numerical results again indicated that the ballast has relatively small effect on the beam moments when a slab is used.

4. Bridges With Longitudinal Beams Without Slab

The purpose of this phase of the investigation was to obtain an indication of the behavior of longitudinal beam bridges with relatively heavy diaphragms but without a concrete deck slab. Numerical results have been obtained for various combinations of variables which are more likely to be used in designs of bridges of this type.

By a simplification of the various methods of analysis and by approximations of the numerical results showing the effect of the component parts of the floor system, tentative design procedures as developed at the University of Illinois have been presented for the four general types of steel railway bridges discussed above. These design procedures are currently being considered by AREA Committee 30 for possible inclusion in the AREA design specifications. In this brief summary, only the basic outline of the assignment has been presented. The complete details of the study, including comprehensive analytical results, are given in the final report, AAR Report ER-5, presented to AREA Committee 30.   ....

Rail Wear Tests on The St. Louis–San Francisco Railway

Introduction

Some maintenance people have questioned the effectiveness of locomotive flange or track lubricators in preventing rail wear on curves under heavy sanding conditions. The Reading Company has conducted tests which have shown that lubricated rails wear slightly more than dry rails under these conditions, presumably because the grease and sand mixture acted as a grinding compound. Their tests also showed a large reduction in wear using molydisulphide stick flange lubricators on their locomotives. Their results with track lubricators are contrary to those obtained in the AAR tests on the Bessemer & Lake Erie (AREA Proceedings, Vol. 56, page 269.)

Since this subject is of interest to many roads, tests were made in cooperation with the St. Louis–San Francisco Railway which had appropriate conditions for conducting such a study. B. H. Crosland, then chief engineer, and O. E. Fort, assistant chief engineer maintenance, arranged for the necessary facilities and assistance for conducting these tests.

Procedure

Two curves on the Frisco's single-track main line north of Birmingham, Ala., were selected for test purposes. They are slightly over a mile apart, have nearly identical wear conditions, have had very rapid rail wear, and are in a location of heavy sanding. Table 1 and Fig. 1 give the details of these curves. Curve C–717–36 to 718–12, which has had a slightly lower rate of wear in past service, was not lubricated. Curve C–719–26 to 718–35 had a "Meco" single rail lubricator using graphite grease at the far end of the two curves. Since the two curves are of opposite direction, there was no carry-over of lubricant from the high rail of the lubricated curve to the high rail of the dry curve.

The effectiveness of the lubrication was gaged by taking rail profiles (transverse head sections) at various intervals of time. These profiles show the amount of steel worn from the heads of the rail. Profiles of both high and low rails were taken at intervals of six poles along the circular portion of the dry curve and at intervals of four poles along the circular portion of the lubricated curve. Gage, elevation and curvature of the high rail were measured at each point where a profile was taken.

The test work by the AAR research staff was under the general direction of W. M. Keller, vice president—research, and G. M. Magee, director of engineering research. Randon Ferguson, electrical engineer, was in direct charge of the tests, assisted by Ralph Schinke, stress analyst, who prepared this report and others previously issued. G. E. Warfel, division engineer, Frisco, arranged for maintenance of the lubricator and obtained data on the gross traffic tonnage which operated over the test curves between profile measurements. H. V. West, Jr., Maintenance Equipment Company, and N. H. Dohrn, Railroad Products Division, American Brake Shoe Company, inspected and approved the lubricator installation after it was changed over from two-rail to one-rail operation on March 4, 1959.

SIXTH PROGRESS REPORT

Seven sets of profiles were taken of the rail undergoing these wear tests. The first and last sets of profiles of the two test curves are shown in Figs. 2, 3 and 4. These profiles show significantly less wear on the curve with the track lubricator than on the curve with no lubrication. The high rail of the curve with no lubrication, Fig. 2, was

17

changed out on March 8, 1960 after 14 months of service and 26,977,000 gross tons of traffic, since the wear had reached a point where it was economical to relay the rail. The high rail of the curve with the track lubricator, Fig. 3, has had about the same amount of wear after 27 months of service and 51,726,000 gross tons of traffic.

In order to test the effectiveness of molybdenum–disulphide as a rail lubricant, applications of moly in the stick form designed for locomotive flange use were made to the new high rail of the curve which was previously unlubricated. These applications were made from spring-loaded holders on a motor car two or three times a week, since the use of locomotive flange lubricators was not feasible because of the large number of locomotives operating over the test curves. These sticks had a very hard binder because they were designed to deposit the moly at a slow rate when in continuous contact with a locomotive wheel flange. Because of their hardness, not enough of the moly could be applied in this testing application effectively to cover the side of the head of the rail with the limited time and manpower available. Fig. 4 shows that the wear rate with this limited amount of stick application is less than the wear rate on this same rail when unlubricated, but greater than the wear rate on the curve with the lubricator. This difference between the moly-lubricated rail and the unlubricated rail was not apparent in the earlier profiles. The side of the head of the high rail had a rough abraded appearance which is typical of unlubricated rail. Rectangular blocks of moly of larger cross sections were substituted for the cylindrical sticks, but they did not appear to be any more effective. It would have been interesting to try a moly block with a soft crayon-like carrier which would have permitted coating the side of the head of the rail with a single pass of the motor car, but these were not available from the supplier.

These service tests show very definitely that track lubricators are effective to the extent of doubling the life of the outer rail in the presence of heavy sanding. The tests have been discontinued.

TABLE 1—TEST CURVE DATA—MAIN LINE NORTH OF BIRMINGHAM, ALA.,
SINGLE TRACK, 132 RE RAIL SECTION

Curve C–717–36 to 718–12

No lubrication
8°12′ curvature, left hand
3 in elevation
30 mph allowable speed
128°52′ central angle
1.05 percent grade, ascending southbound
New high rail laid January 21, 1959 and March 8, 1960
Old high rail gave 22 months service

Curve C–719–26 to 719–35

Lubricated by Meco single-rail lubricator at Mile Post C–720–0 (top of grade)
Graphite grease. Changed over from two-rail lubricator on March 4, 1959
8°30′ curvature, right hand
4 in elevation
35 mph allowable speed
73°57′ central angle
1.0 percent grade, ascending southbound
New high rail laid January 20, 1959
Old high rail gave 19 months service

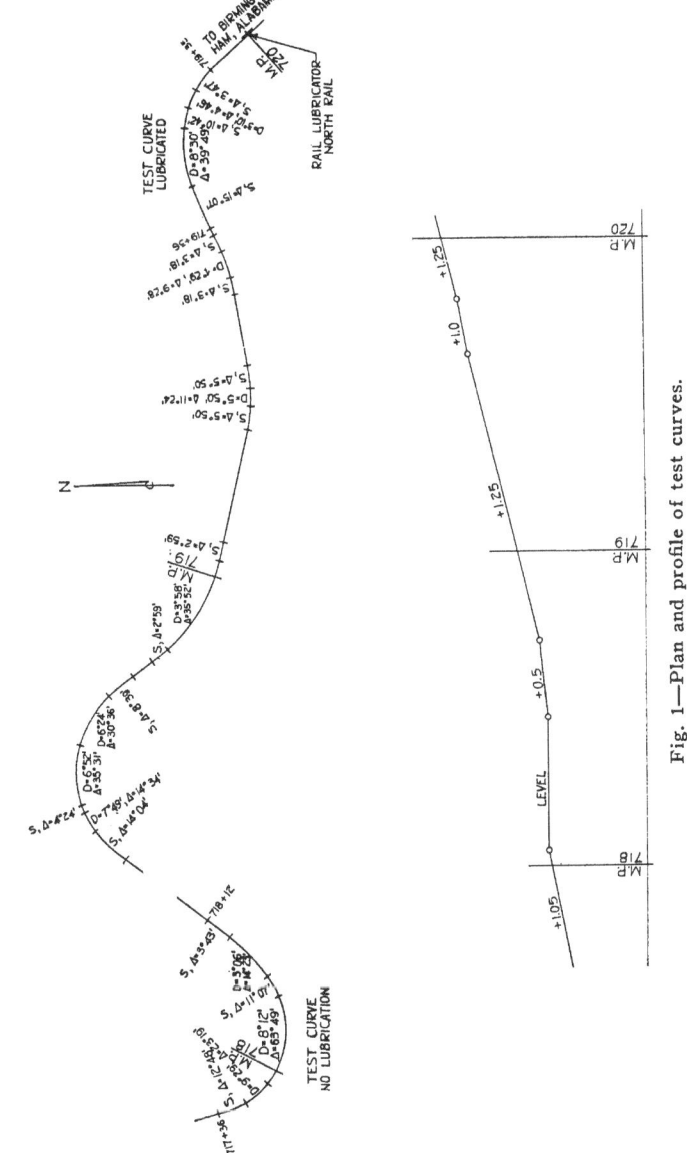

Fig. 1—Plan and profile of test curves.

$8^\circ 12'$ Curvature, 3 Inch Elevation – High Rail Laid January 21, 1959
First Profile March 5, 1959
Sixth Profile March 8, 1960 – High Rail Changed Out
Gross Traffic January 21, 1959 to March 8, 1960 – 26,977,000 tons

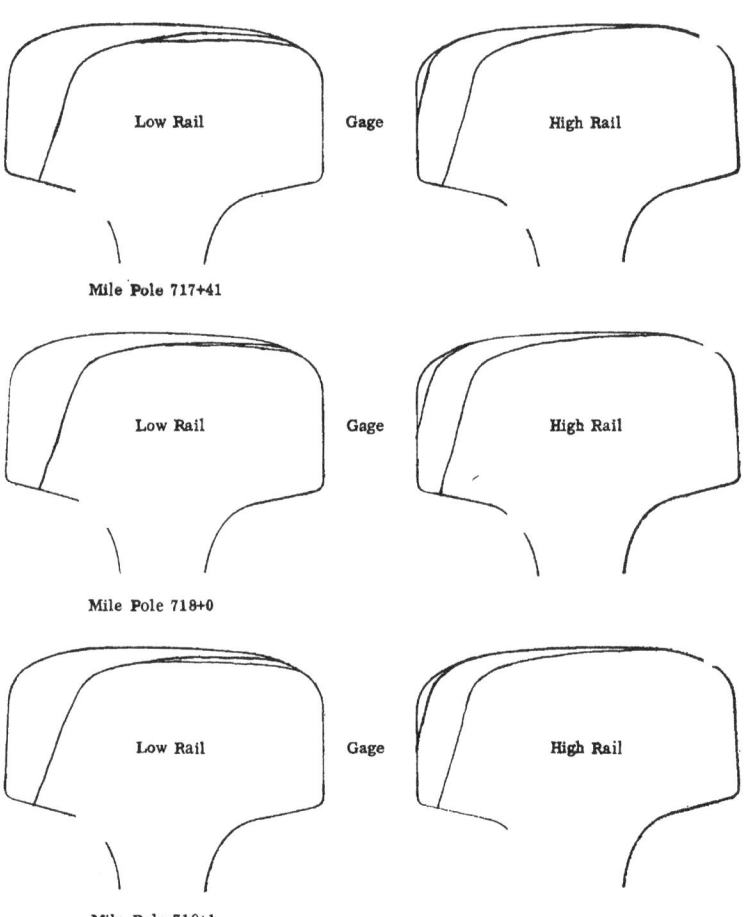

Mile Pole 717+41

Mile Pole 718+0

Mile Pole 718+1

Fig. 2—Rail profiles, Curve C–717–36 to 718–12, no lubrication.

8° 12' Curvature, 3 Inch Elevation - High Rail Laid January 21, 1959
First Profile March 5, 1959
Sixth Profile March 8, 1960 - High Rail Changed Out
Gross Traffic January 21, 1959 to March 8, 1960 - 26,977,000 tons

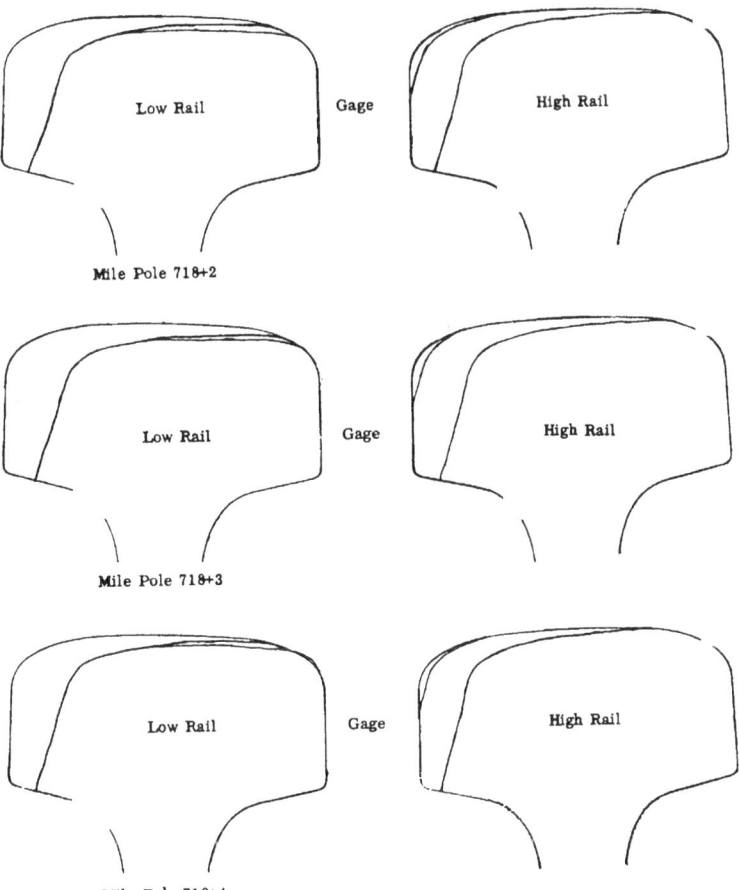

Low Rail   Gage   High Rail

Mile Pole 718+2

Low Rail   Gage   High Rail

Mile Pole 718+3

Low Rail   Gage   High Rail

Mile Pole 718+4

Fig. 2 (Cont'd)—Rail profiles, Curve C–717–36 to 718–12, no lubrication.

8 30' Curvature, 4 Inch Elevation - High Rail Laid January 20, 1959
First Profile March 5, 1959
Seventh Profile April 4, 1961
Gross Traffic January 21, 1959 to April 4, 1961 - 51,726,000 tons

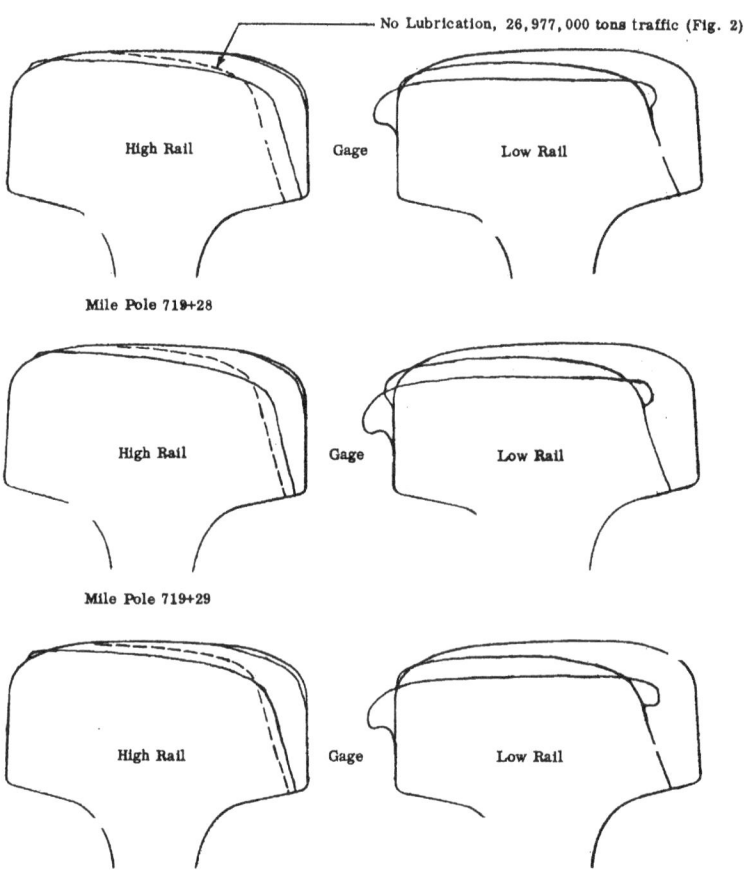

Fig. 3—Rail profiles, Curve C-719-26 to 719-35, track lubricator.

8° 30' Curvature, 4 Inch Elevation - High Rail Laid January 20, 1959
First Profile March 5, 1959
Seventh Profile April 4, 1961
Gross Traffic January 21, 1959 to April 4, 1961 - 51,726,000 tons

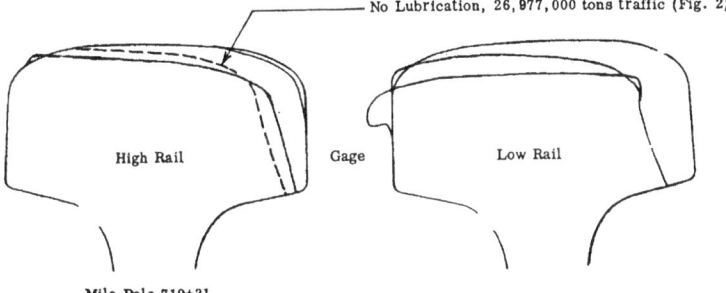

No Lubrication, 26,977,000 tons traffic (Fig. 2)

High Rail   Gage   Low Rail

Mile Pole 719+31

Fig. 3 (Cont'd)—Rail profiles, Curve C–719–26 to 719–35, track lubricator

8° 12' Curvature, 3 Inch Elevation - High Rail Laid March 8, 1960
First Profile March 17, 1960
Third Profile April 4, 1961
Gross Traffic March 9, 1960 to April 4, 1961 - 24,749,000 tons

No Lubrication, 26,977,000 tons traffic (Fig. 2)

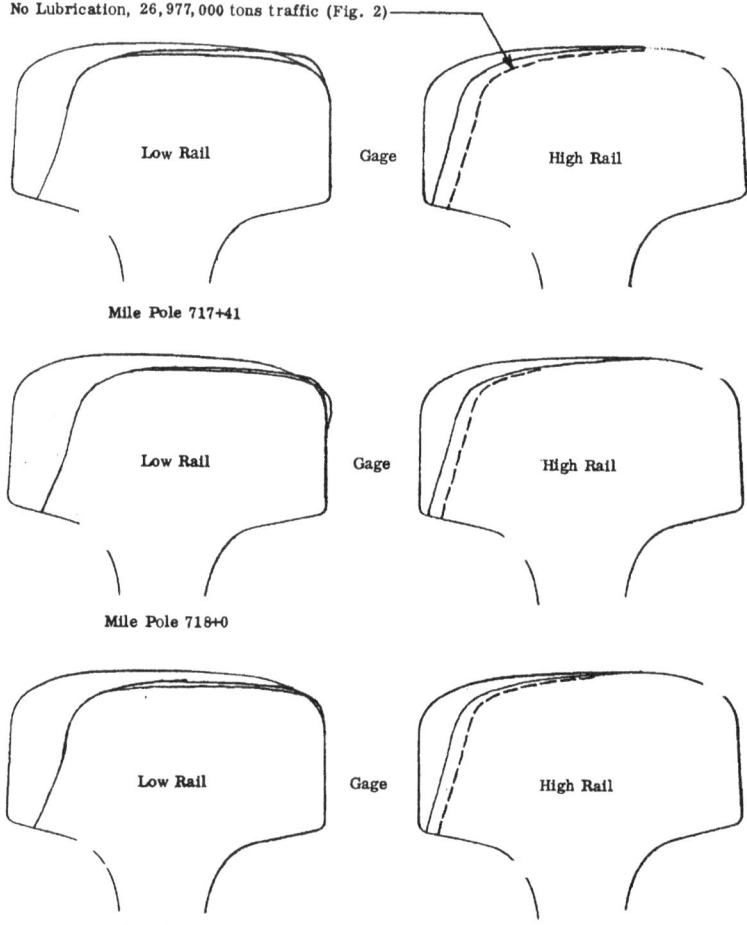

Fig. 4—Rail profiles, Curve C–717–36 to 718–12 insufficient
moly-disulphide lubrication.

8° 12' Curvature, 3 Inch Elevation - High Rail Laid March 8, 1960
First Profile March 17, 1960
Third Profile April 4, 1961
Gross Traffic March 9, 1960 to April 4, 1961 - 24,749,000 tons

No Lubrication, 26,977,000 tons traffic (Fig. 2)

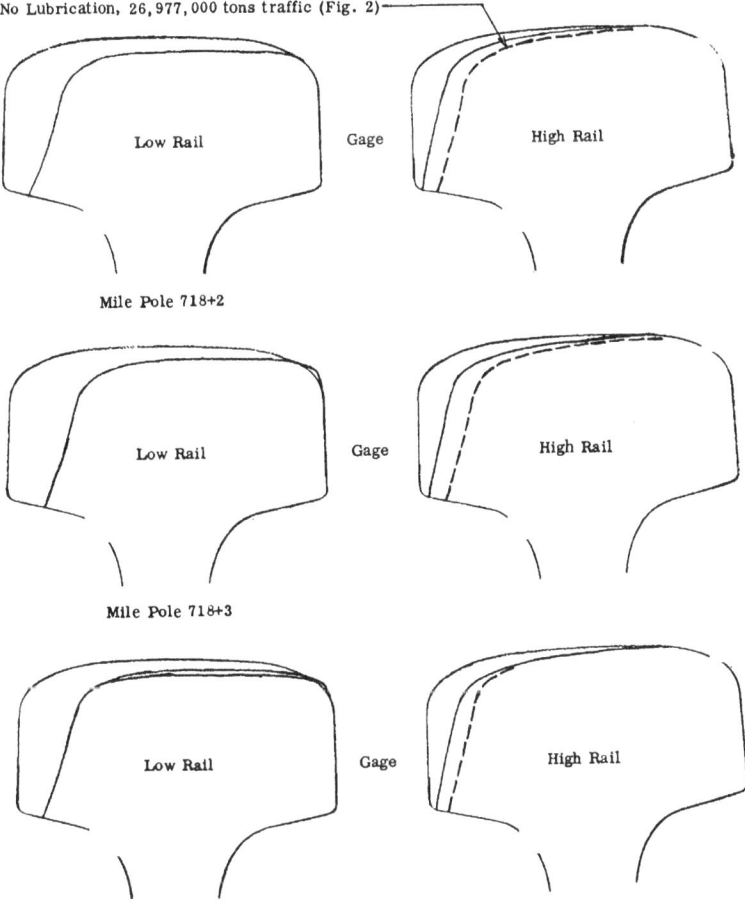

Fig. 4 (Cont'd)—Rail profiles, Curve C-717-36 to 718-12, insufficient moly-disulphide lubrication.

Advance Report of Committee 1—Roadway and Bàllast

Report on Assignment 6 (c)

Performance of Filter Materials in Subdrains

A. E. Lewis (chairman, subcommittee), E. W. Bauman, K. W. Bradley, B. H. Crosland,
J. B. Farris, H. O. Ireland, W. G. Murphy, J. E. Newby, A. J. Pacelli, G. W. Payne.

Second Progress Report on Performance of Filter Materials

By John C. Guillou and Richard F. Lanyon*

INTRODUCTION

This is the second progress report dealing with the use of concrete sand as filter material around perforated corrugated metal subdrains. The study is being conducted on a cooperative basis by the Association of American Railroads and the University of Illinois under one of the assignments of AREA Committee 1—Roadway and Ballast. The purpose of the study is to evaluate the requirements of filter materials and, ultimately, to develop definite design information for field conditions.

The first report considered the hydraulic capacity of coated and uncoated pipe, the general operating characteristics of concrete sand filters, and presented the results of a questionnaire regarding field practice (1).** The purpose of the second report is to present the results of a series of compaction tests in a concrete sand filter, and to develop some concept of probable maximum rates of inflow to a subdrain under a hypothetical design condition.

It should be understood that the reports are only progress reports; they should not be interpreted as indications of proposed changes in design standards. It is a purpose of the research program to aid in the development of new, more economical, and more efficient standards, but the progress reports are intended only to record activity during the particular period of study.

The experimental work just completed considered tests with the perforations in the drain pipe upward and symmetrical about the vertical. This orientation is contrary to generally accepted practice. Studies with the holes up were undertaken for two specific reasons:

1. It is difficult to compact filter material beneath the spring line of the pipe, and this is in the portion of the filter subjected to maximum velocity when the holes are down.

2. When the perforations are up the longitudinal flow through the drain pipe does not affect the filter stability as it may when the holes are down.

Tests to be completed during the next period of study will include the "holes down" situation. This will enable direct comparison of the results of the two orientations.

Tentative results of the first phase of the cooperative program indicated that the stability of the filter adjacent to the pipe perforation was not a function of material

---

* Respectively, research associate professor and research assistant in hydraulic engineering, University of Illinois.
** Numbers in parenthesis indicate references in the bibliography at the end of this report.

sorting and consequent bridging of large grain sizes.† Further study of this tentative finding was undertaken by means of a series of flow tests in which the filter material was compacted to six different specific weights.

The major conclusions developed during the present phase of the investigation are as follows:

1. There is an optimum compaction of concrete sand when used as a filter material. The optimum compaction is a function of velocity of flow through the filter. The optimum compaction applies to both filter stability and flow capacity.

2. Compaction of the filter in excess of the optimum results in filter destruction, while compaction less than the optimum leads to readjustment of the filter and eventual stability.

3. Readjustment of the filter material is accomplished by the migration of fine particles toward the pipe perforations. This migration continues until the number of fine passing through the coarser material adjacent to the perforations causes physical binding. After binding the filter is stable for all velocities equal to or less than the velocity at the time of binding.‡

4. Determination of the required size of drain pipe for a hypothetical case indicates that many subdrains are being installed with pipe diameters at least 75 percent oversize.

EXPERIMENTAL APPARATUS

The experimental activity in this program has deliberately been directed toward evaluation of general operating characteristics of concrete sand filters. This work necessarily precedes detailed studies of permeabilities, movement in pore spaces and theoretical studies of the phases of filter flow which are now becoming apparent.

The test apparatus, which was modified from the initial test series, is essentially a plywood tank 52 in wide, 40 in deep and 48 in long. A filter section, 24 in wide and 36 in deep, was located inside the tank. The filter section extended the full length of the test tank, and the tops of both sections were at the same elevation. This arrangement causes the filter area to be surrounded by a U-shaped section which is utilized for the introduction of water to the filter material. The walls and floor of the filter section are composed of open-type subway grating overlaid on the filter side with $\frac{1}{16}$-in hardware cloth to prevent loss of filter material to the water supply chamber.

An 8-in-diameter, perforated, corrugated metal drain pipe was located midway between the walls of the filter section so the minimum thickness of filter material on the sides was 8 in. The drain pipe was installed on a zero longitudinal slope, and the concrete floor of the filter section was built up to the spring line of the pipe. Thus, water entered the filter material only from the sides. In this series of tests the pipe was installed with the perforations upward and symmetrical about the vertical. This was done to eliminate the effect of flow in the drain pipe on saturated filter material outside the pipe.

† Since this report was first issued by the AAR Research Department (Report No. ER-12), some confusion has apparently developed in regard to this statement, which was initially presented in the first progress report. This pertained to the fact that the size gradation of material which passed into the drain pipe was essentially the same as that of the filter material itself, and thus it was concluded that a volume of coarse material did not form around the holes to cause filter stability.

‡ Questions about this statement have also been raised since the report was first issued. Based upon the experimental work just completed it appears that in certain cases filter stability is caused by the migration of relatively fine particles to the vicinity of the pipe perforation. Upon reaching this area these particles obstruct the intergranular passages and become physically bound. After this binding has occurred the filter is stable, and it would be expected to remain stable even with higher or lower intergranular velocities unless the physical arrangement of particles is altered. It is anticipated that this phase of the study will be investigated further during the next period of experimental activity.

Fig. 1—Outflow measuring apparatus. Note brass cloth basket in first compartment of weir tank, hook gage, and water level recorder. The two flexible hoses near top of photo connect water supply chamber with adjustable-height overflow weir.

An adjustable-height overflow weir was connected to the water supply chamber to insure that tests were conducted with conditions of constant head. A water stage recorder was used to provide a continuous record of depth of water outside the filter.

Flow intercepted from the filter by the drain pipe passed through the pipe and into a 10-deg V-notch weir tank for measurement. The weir tank was also equipped with a water stage recorder to provide a continuous discharge record and with a standard hook gage for checking purposes. After the water passed over the weir it was pumped to a drain. A basket of 80-mesh brass cloth was located at the entrance to the weir tank to intercept filter material that was washed out of the pipe. Arrangement of the equipment in the laboratory is shown by Fig. 1.

A special dead-weight compactor was constructed to provide adequate control and uniformity of density in the filter volume. This device consisted of 50 truncated cones, each 3.5 in high and with top and bottom diameters of 2.9 and 2.2 in, respectively, attached to a steel frame work. The concrete cones were cast around a threaded stud which, in turn, was screwed into a channel-iron frame work such that the cones were approximately 0.4 ft center to center in both directions. The entire assembly was attached to a movable, overhead chain hoist by means of a suspension link. During compaction work the assembly was lowered to 0.5 ft from the filter surface, and then the link was tapped open and the compactor allowed to fall freely to the sand surface. The weight of the compactor was 216 lb, and the force exerted by each of the cones was approximately 50 psi. The force varied with penetration. Different degrees of compaction were obtained by repeating the number of blows from the compactor. When sufficient blows were required to cause the channel sections to bear on the sand, the sand was raked to a smooth surface and then the final compaction was completed. The compactor, with link closed, is shown in Fig. 2.

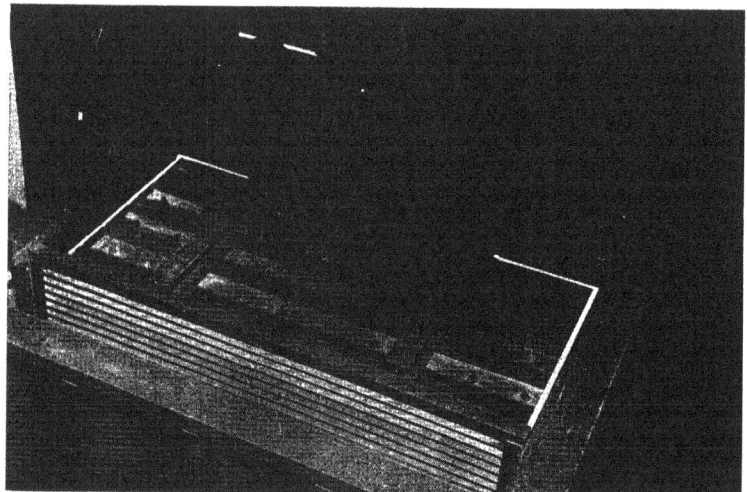

Fig. 2—View of compactor ready to drop. Compactor, which is out-lined with white tape, is carefully centered over filter bed. Tapping clevis at left side of link permits free fall to filter surface.

COMPACTION STUDIES

Concrete sand, with uniform moisture content of 5 percent, was placed in the filter box in charges of approximately 160 lb. The material was raked into a uniform layer about 6 in thick and was then compacted. The specific weight of the filter bed was determined from the known weight of filter material and the final volume. Uniformity of moisture content was assured by mixing each charge in a small concrete mixer, deter-mining the moisture, and then adding sufficient water to obtain the desired final moisture content. The grain size distribution curve for the test sand is shown in Table 2.

Flow tests were conducted under a mean head of 0.5 ft as measured between the free surface in the water chamber and the center line of the lower hole at the mid-length of the corrugated metal pipe.

Six different degrees of filter compaction, with head equal to 0.52 ft and moisture content of 5 percent, were studied. Three of the runs were repeated to evaluate reproduci-bility of results. Table 1 presents pertinent data for the flow tests and comments regarding the filter condition.

Compaction figures in the table are relative to the zero blow, or loose fill, test condition.

All flow tests, execpt those with complete failure, were continued for 72 hr to permit evaluation of filter stability. This period of time was found to be sufficient during the work previously reported (1).

In all cases the initial flow through the drain pipe was very turbid and contained some filter material. Tests in which the filter did not fail were characterized by cessa-tion of flow of filter material prior to absence of turbidity. In cases where the filter did fail, turbidity cleared up before the movement of filter sand ended.

TABLE 1—SUMMARY OF FLOW TEST, 0.5-FT HEAD

No. of Blows	Specific Weight PCF	Compaction Percent	Rate of Flow		Flow Volume in 72 Hr CFS-Hr	Remarks
			Peak CFS	72 Hr CFS		
0	84.1	0.0	0.0071	0.0047	0.432	No surface failure
2	91.6	9.1	0.0092	0.0044	0.510	No surface failure
5	94.5	12.5	0.0103	0.0070	0.593	Slight surface failure
8	96.5	14.9	0.0098	0.0013	0.334	Some surface failure
10	98.5	17.3	0.0083	0.0035	0.422	No surface failure
25	100.8	20.0	0.0050	---------	---------	Complete surface failure

Representative photographs showing the surface configuration of the filter bed after several of the tests are shown as Fig. 3. In all tests the first external sign of deformation in the filter was development of a crack in the filter surface along the drain pipe. For compactive values less than 12.5 percent the cracking was accompanied by general settling of the filter material. Reference to the photos of Fig. 3, and in particular to the trace of the original surface on the Lucite end wall, shows that the settlement was inverse to the initial compaction.

Quantitative study of the amount of material washed out of the filter, and observation of conditions within the filter, indicate that flow through the sand causes an increase in bed density when initial compaction is less than 12.5 percent. This adjustment of the bed is caused by transverse movement of some particles and probably by reorientation of other particles. Study of the filter during the 20 percent compaction test showed the same type of settlement failure, but in this case the filter did not stabilize, and sand continued to be carried into the drain pipe. In all cases the finding of the initial program, that the grain-size distribution for material washed out of the filter was the same as that of the original filter material, was substantiated.

Flow data of Table 1 are presented graphically as Fig. 4 to illustrate the two distinct types of action within the filter. It will be noted that in the case of peak rate of discharge, and 72-hr volume of discharge, the 12.5 percent compaction was the optimum condition. For compactive amounts less than optimum the rate of change of discharge relative to compaction was only about half that of compactive amounts greater than optimum.

Determination of total amounts of sand washed out of the filter during the 72-hr tests showed that about 30 lb of material was removed during the 0, 9.1 and 17.3 percent compaction runs, and about 107 lb was removed during each of two 12.5 percent compaction tests. In all five tests the loss of filter material terminated after about 30 min. In the 20 percent compaction test the filter continued to lose material and the test had to be terminated.

Conditions within the drain pipe are illustrated by Fig. 5 which was made during the first 30 min of the 20 percent compaction run. It will be noted that a steep sand surface had formed inside the pipe, and near the outlet material was being carried away as quickly as it was supplied. In the upstream end of the pipe the rate of material supply was greater than the flow could remove, and the pipe was almost completely full of sand.

FILTER READJUSTMENT

Hydrograph curves of all tests that have been conducted include two principal features—the initial discharge is substantially less than the peak discharge, and peak

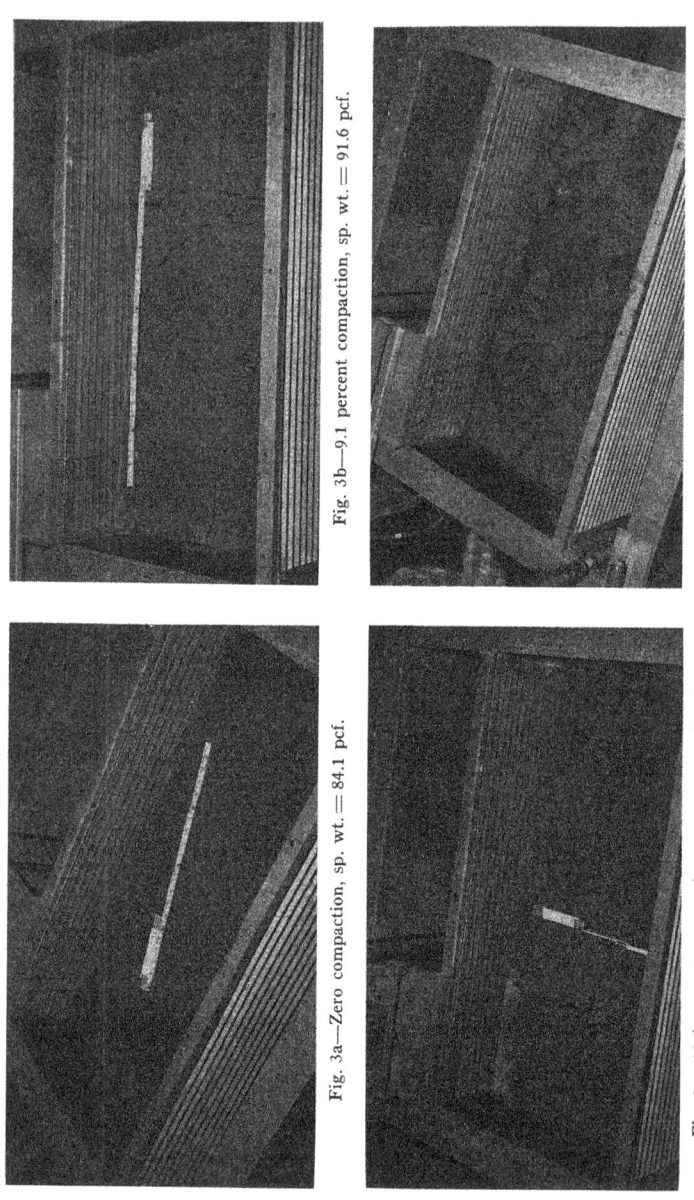

Fig. 3b—9.1 percent compaction, sp. wt. = 91.6 pcf.

Fig. 3d—20 percent compaction, sp. wt. = 100.8 pcf.

Fig. 3a—Zero compaction, sp. wt. = 84.1 pcf.

Fig. 3c—12.5 percent compaction, sp. wt.= 94.5 pcf.

Fig. 3—Bed configurations after flow tests.

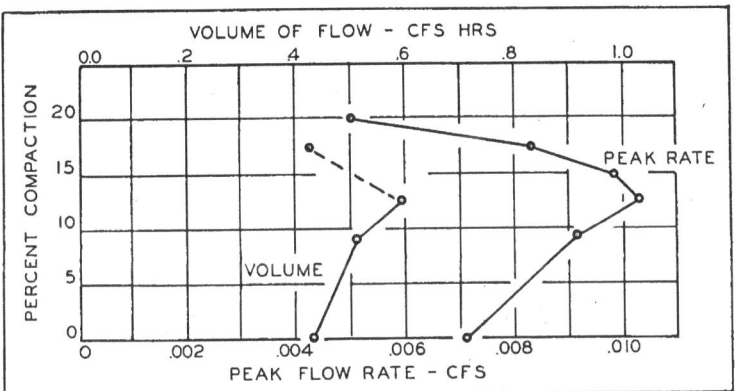

FIG. 4   OPTIMUM FLOW CURVES

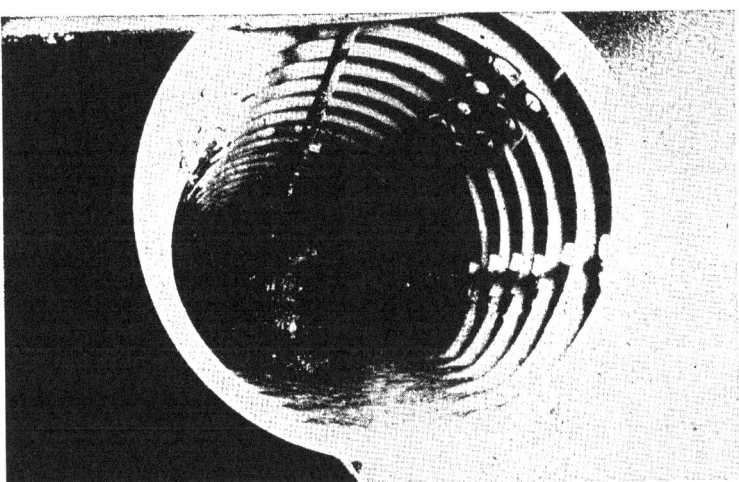

Fig. 5—Looking into drain pipe during 20 percent compaction test. The rate of sand movement is quite high. The filter failed completely.

discharge continues for 4–8 hr and then decreases. This action is illustrated by the representative hydrographs of Fig. 6.

It appears that the explanation of this action is in the reorganization of particles within the filter. It is certain that wash silts are being removed from the filter during the initial operating period. This is evidenced by the very turbid discharge from the drain pipe. As these fines move out of the filter the permeability increases, and because of the constant head, the discharge increases. Increase in discharge increases the com‾petence to move particles into the drain pipe, and if the discharge continued to increase there is no doubt that the filter would fail completely. It is believed that this is precisely the occurrence when the filter is compacted substantially over the optimum. In this case original inter-granular pressure is high, and any material movement causes a rela‾tively large pressure change. The resulting freedom of movement after transport of the fines begins, when coupled with the steep hydraulic gradient that accompanies high initial compaction, leads to complete failure of the filter in the vicinity of the pipe perforation. Absence of this material leads to settlement of the filter surface where it is not supported by the drain pipe itself (see Fig. 3d).

For compactive rates equal to or less than the optimum the filter is able to readjust to the volumetric change caused by removal of the very fine particles. Here, the inter-granular pressure is not so high, the permeability is greater, and therefore the hydraulic gradient is not so steep, although the rate of flow is considerably greater than would be true under conditions of high compaction. The increased discharge and the lower inter-granular pressure cause movement of larger particles toward the pipe perforations. These particles move from all saturated parts of the filter toward the drain pipe. As these particles concentrate in the vicinity of the pipe perforations, the inter-granular channels become congested and finally a physical binding occurs. Thus, a sand structure begins to form which is in equilibrium with the rate and velocity of flow, and with the size and number of particles which cause the obstruction. The obstruction raises the hydraulic gradient from the water chamber and thus the velocity and rate of flow, and therefore particle movement, approaching the obstructed area diminish. This process, which would occur over a considerable period of time, leads to a completely stable filter.

Based on the preceding it appears that the increase in filter capacity, up to the peak rate of flow, is caused by a net increase in permeability throughout the filter. Con-versely, the decrease in discharge after the peak is caused by decrease in permeability in the near vicinity of the pipe perforations.

Sieve analysis of material removed from the filter verifies the formation of a zone of low permeability adjacent to the pipe perforations. The sieve data are presented as Table 2.

Both samples were taken from the entire length of the test chamber on one side of the pipe. The sample "near holes" was from a triangular volume all of which was within 2 in of the pipe. The "near wall" sample was taken from a strip about 4 in further from the pipe and about 2 in from the edge of the water chamber.

TABLE 2—SIEVE ANALYSIS OF FILTER ZONES

Sieve No.	Opening mm	Percent of Sample Retained			Change in % Retained*
		Original	Near Holes	Near Wall	
200	0.074	----------	1.03	0.89	+0.14
120	.125	----------	0.66	0.54	+0.12
100	.149	----------	1.35	1.23	+0.12
80	.177	----------	7.3	6.6	+0.7
50	.297	13.0	14.0	13.4	+0.60
40	.420	20.4	21.8	21.4	+0.40
30	.590	36.1	35.4	36.1	—0.70
16	1.190	19.3	16.3	17.7	—1.40
8	2.380	1.8	1.5	1.4	+0.10

*Difference between Cols. 4 and 5. Positive sign indicates increase in percent of material near holes.

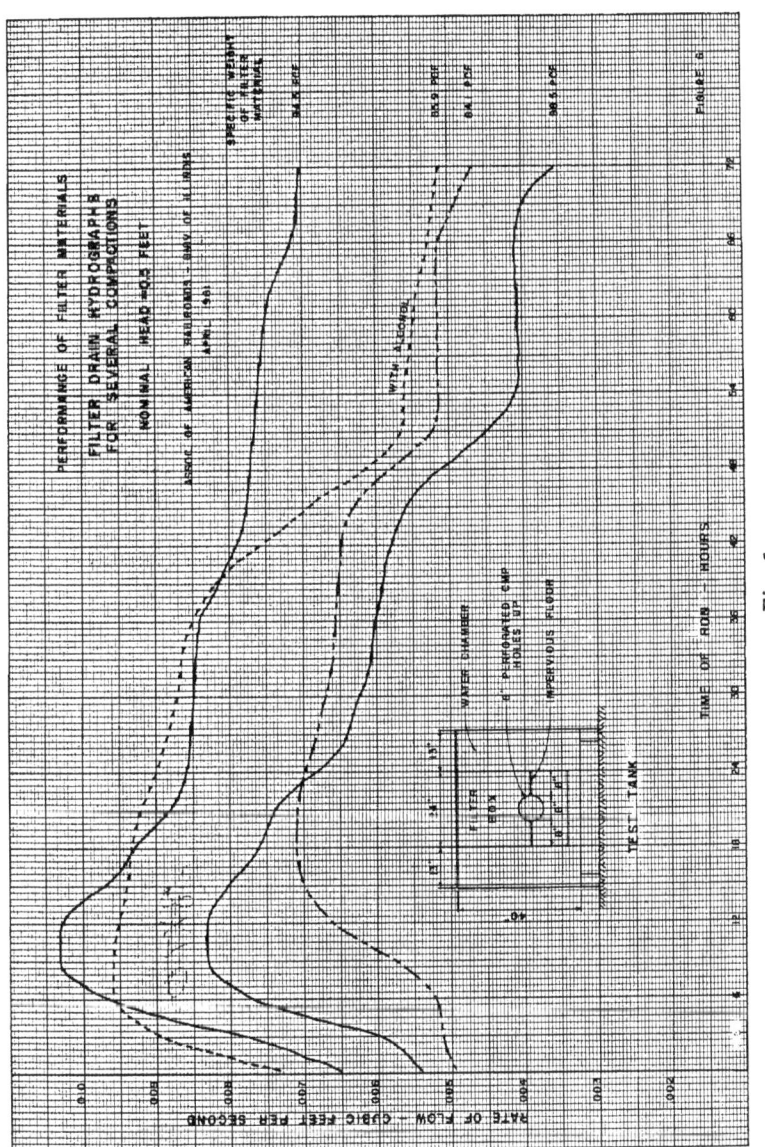

Fig. 6.

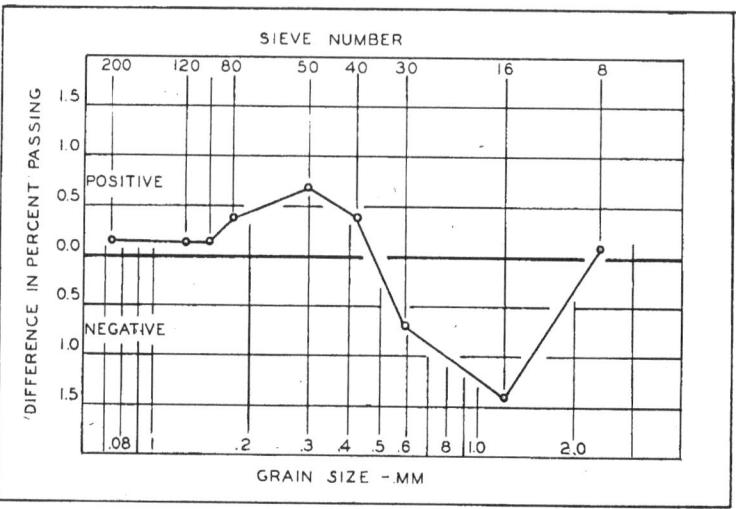

FIG. 7 PARTICLE MIGRATION CURVE

The variance of change in percent retained as a function of particle size is presented as **Fig. 7**. The curve shows that there was a measurable increase in percent of sample with particle size between 0.15 and 0.45 mm at the location near the pipe perforations. This is accompanied by a corresponding decrease in the percent of the sample composed of particles with size between 0.45 and 2.0 mm. Thus, the sieve analysis substantiates the hypothesis of filter readjustment.

A further investigation of filter stability was performed by mixing a small amount of alcohol with the filter material before placement. The alcohol was added to reduce the surface tension of the water and thus reduce the contact pressure within the filter. Alcohol tests were completed with sand compacted to specific weights of 85.9 and 92.6 lb per cu ft. In both cases the filter failed in the early few minutes of the tests and filter material continued to be carried out of filter for over an hour. The flow hydrograph for the 85.9 lb per cu ft run is shown on **Fig. 6**. It will be noted that except for the shorter time of rise to peak flow, the hydrograph is similar in shape to the no-alcohol tests. This is evidence that air-binding is not of significant importance in the test work. Had air-binding been important, the hydrograph for the alcohol run would have been relatively flat because of the reduced surface tension. It is concluded that the contact pressure is of very significant proportions in the test work. This belief is in harmony with the general concept of filter readjustment.

HYDROLOGIC CONSIDERATIONS

The following material is presented as an aid in establishing the proper perspective relative to required capacity of the drain pipe, and realistic flow conditions within the filter material.

As an example, assume that subsurface drainage is being designed for a classification yard. Hooghoudt (2) derived the following equation for drain spacing which has been widely used in many parts of the world. It has recently been evaluated by Talsma and Haskew (3) for Australian practice.

$$s^2 = 8(k/q) \, d(m-h) + 4(k/q) \, (m^2 - h^2)$$

where s is the drain spacing in feet; k is the hydraulic conductivity, feet per day; q is the rate of discharge per unit of surface area, cubic feet per square feet per day; d is a function of depth to impermeable layer, the radius of the tile, and the drain spacing; m is the height of the ground water above the tile lines midway between parallel drains, feet; and h is the height of the water surface immediately above the drain, feet.

This equation may be transposed to provide a solution for discharge per unit of surface area, as follows:

$$q = 8(k/s^2) \, d(m - h) + (4k/s^2)(m^2 - h^2)$$

and if the following values for the several variables are assumed: $k = 5$ ft/day, $d = 5$ ft, $h = 1$ ft, $s = 100$ ft, and $m = 6$ ft, the discharge is calculated to be 0.17 cu ft per sq ft of area served per day.

Since each drain serves an area 100 ft wide, the discharge into the pipe, in cu ft per sec, under conditions of equilibrium is

$$q^1 = \frac{100 \, q}{24 \times 3600} = 0.000197 \text{ cfs per ft of length,}$$

and if the drain pipe is 1000 ft long the total discharge at exit will be, neglecting storage, 0.197 cfs. Furthermore, if the design velocity in the drain is 3.0 ft per sec to avoid silting, the required diameter of pipe at outlet is found to be only 3.5 in. For comparison it is noted that based upon the questionnaire of the previous report (1) only 7 percent of the drains installed by the railroads is 4 in. in diameter or less. It is also interesting to note that agricultural tile with internal diameter less than 2 in is used very extensively in the Netherlands (4).

For conditions of the example an 8-in-diameter drain should be capable of serving an area 100 ft wide and 5300 ft long. According to the questionnaire almost three-fourths of the underdrains that are being installed by the railroads are 8 in or more in diameter.

The rate of surface infiltration that is required to satisfy the discharge rate of 0.17 cu ft per sq ft of area per day is equal to 12(0.17) or 2.04 in per day. If it is assumed that only one-half of the precipitation event which causes the infiltration event is surface runoff, then the depth of precipitation required is 4.08 in per day. In central Illinois such a rainfall has a return period of over 10-years if the rain falls during 12 hr, and over 100 years if it falls during a 2-hr period.

The preceding example does not consider an increase in soil moisture during the precipitation event. The Hooghoudt equation applies to steady flow, and the characteristic values assumed for the example are for maximum condition of saturation with the drain located about 7 ft below the ground surface. For the conditions of the example it may be shown that the allowable rate of precipitation could be doubled if storage of moisture in soil prism is considered.

The example is not patterned after a particular prototype installation. It is only intended to illustrate the very conservative designs which result when relatively large pipe is used for subsurface drainage. It appears that a major use of the pipe area is for storage of filter material.

ACKNOWLEDGMENTS

This study is being conducted on a cooperative basis by the Association of American Railroads and the University of Illinois. The contract officer for the AAR is G. M. Magee, director of engineering research, and for the University, Professor Ross J. Martin, director of the Engineering Experiment Station.

Technical and operating supervision of the study is provided by Rockwell Smith, research engineer—roadway, AAR, and by Professor J. C. Guillou, associate professor of hydraulic engineering. Special appreciation is due Mr. Smith for his continued interest and cooperation in the prosecution of the program.

The project is being conducted in the Hydraulic Engineering Laboratory. Richard Lanyon has been in direct charge of both the testing program and the development of special apparatus.

BIBLIOGRAPHY

1. Guillou, John C., *First Progress Report on Performance of Filter Materials*, AREA Proceedings, Vol. 61, 1960, pp. 677–692.

2. Hooghoudt, S. B., Bijdragen tot de kennis van eenige natuurkundige grootheden van den grond, Verslag. Landbouwk. Onderzoek., 46, pp. 515–707, 1940.

3. Talsma, T. and Haskew, Henry, *Investigation of Water-Table Response in Comparison with Theory*, Journ. Geophysical Res., Vol. 64, No. 11, pp. 1933–1944, Nov. 1959.

4. Kirkham, D. and deZeeuw, J. W., *Field Measurements for Tests of Soil Drainage Theory*, Soil Science Soc. of Am., Proc. 16, pp. 286–293, 1952.

Advance Report of Committee 16—Economics of Railway

Location and Operation

Report on Assignment 4

Potential Applications of Electronic Computers to Railway Engineering and Maintenance Problems in Research, Design, Inventory, Etc.

Collaborating with Committees 11 and 30, and Informally with
the Railway Systems and Procedures Association

F. Wascoe (chairman, subcommittee), L. P. Diamond, W. J. Dixon, S. B. Gill, C. A. James, T. J. Lamphier, R. J. Lane, A. S. Lang, M. B. Miller, V. J. Roggeveen, George Rugge, G. S. Sowers, J. J. Stark, Jr., C. L. Towle, T. D. Wofford, Jr.

The Digital Terrain Model Approach to Railroad Route Location

By Paul O. Roberts

Assistant Professor of Civil Engineering, Massachusetts Institute of Technology

THE ROUTE LOCATION PROBLEM

Early Practice

Route location engineering was first performed by American railroad engineers in the 1800's. Many of these early engineers practiced their profession from horseback. Their task was to build a railroad in their mind's eye as they rode, then transform their vision to stakes in the ground. They were hampered by what they could not see. There was always the possibility that there was a better route which they had overlooked.

Route location engineering has changed markedly since those days. The engineer of today does almost all of his work in an office. Advanced map-making techniques, particularly photogrammetry, permit him to see, in minutes, routes that would have taken his grandfather months to find. Advances in construction techniques have made it possible, moreover, to build railroads on many more locations than could ever before have been considered. The engineer's problem now is not to find a possible alinement, but to find the best alinement out of the almost infinite number of possibilities that lie before him. It is unfortunate that in making this choice, trial lines must be partially "designed" before their faults are revealed or before a choice can be made on any but a qualitative basis. The sheer magnitude of this task for each of many alternative route possibilities usually discourages the evaluation of more than one or two trial lines.

New Tools

Over the past few years, many new tools have become available which the location engineer can use to solve his problems. Photogrammetry, as mentioned above, gives him a vantage point from which to view the terrain easily for the selection of alternate alinements. It can furnish him large amounts of topographic information at low cost. Air photo analysis can give him qualitative information such as soil type, pipelines, transmission towers, stream development, and community facilities.

39

The electronic computer is a third development of importance to the railroad location engineer. Whereas photogrammetry and air photo analysis assist the engineer in selecting trial lines and in gathering data for their evaluation, the computer provides a powerful tool for the analysis of these data and the choice between the trial lines. It is the specific purpose of this paper to describe one way in which computers can be used for such an analysis.

The Location Problem

To date, engineers have most frequently used the electronic computer simply for the computation of cut and fill quantities during the final design phase of the location process. While it is true that this is the point in the engineering process where the most time and drudgery can be saved, it is not in the design phase, but in the location phase, that a computer can return the maximum economic benefit. An improvement in the location capability can save not only engineering cost, but construction costs and railway operating costs as well. Construction and operating costs can run 10 to 60 times the cost of the engineering alone. As the location of a line becomes fixed, moreover, the opportunity for making major savings in any of these costs diminishes.

Ideally, the location engineer will compare alternative locations on the basis of the total cost associated with each. Since one of the most important elements of total cost is the cost for construction, it must be evaluated for every line being considered. Furthermore, it is not enough merely to look at a line and mentally determine all the costs qualitatively, because the results generally would not provide a sufficient basis for a location decision.

The Digital Terrain Model System

The Digital Terrain Model (D.T.M.) System of integrated electronic computer programs was conceived with these points in mind. First and foremost, the system was designed to assist the location engineer during each step of route selection and evaluation, furnishing automatic computation, plots of useful data, and a basis for comparison between alternatives. Carefully detailed input forms facilitate this process. The program runs quickly and with a minimum of machine passes so as to increase the number of trial lines that can be practicably examined. The output has been set up to furnish only pertinent information in the most convenient form. Output, for instance, appears as plots rather than as long numerical listings.

The Digital Terrain Model itself is simply a discrete numerical description of the terrain in terms of Cartesian coordinates. It involves the recording of sample points with respect to a base line rather than a route center line, a feature which gives the system its most significant capabilities. The engineer need take terrain data only once. When he has recorded these data on punched cards, he can then use them for the rapid evaluation of many trial lines.

The first computer programs using this concept were developed for highway location analysis. Programs have subsequently been developed for many other applications, ranging from borrow pits to earth dams. In theory, any problem requiring the quantitative evaluations of alternative geometric solutions can make use of the concepts embodied in the original D.T.M.

Cost Factors

A breakdown of railway construction costs would include the following:

Right-of-way

Clearing and grubbing

Earthwork
Track (including ballast)
Structures
Drainage
Signaling
Miscellaneous (signs, fencing, etc.)

The magnitude of these costs for any given route location depends primarily upon the "states of nature" which exist prior to construction. These states of nature include such items as:

Topography
Soil type and bearing capacity
Rock location
Present system of roads, railroads, etc.
Drainage systems
Urban development
Land use
Material availability

To determine construction costs, then, the engineer must first establish the spatial relationships between the terrain or environment (those items in the second list) and the proposed alinement. Of course, some cost factors are more important than others. A relatively constant cost, such as track, will affect the choice between alinements only through length and, therefore, is a less important consideration in location than widely variant costs such as earthwork, structures, or right-of-way. The line for which these variant costs are a minimum is the one for which the engineer is actually looking.

The primary function of the D.T.M. System is to assist the engineers in this process. Given a simple numerical description of a trial alinement, the computer can use the D.T.M. programs to determine all of the geometric relationships between the ground and the alinement. On this basis it outputs the parameters from which the major cost elements can be estimated.

USING THE D.T.M. SYSTEM

Manner of Operation

Location practice using the D.T.M. Systém is broken down into the following phases:

1. Terrain Data Procurement
2. Horizontal Alinement
3. Vertical Alinement
4. Roadbed Design

Each phase is dependent on work performed in previous phases and is implemented with one or more computer programs depending on the type and capability of the machine being used. See Fig. 1.

Terrain Data Procurement

After the overall feasibility of a route has been established, a band of interest is selected by the engineer. The width of this band varies depending upon the location phase and the latitude the location affords. The coordinates for terrain points in this band of interest are then digitized for use by the computer.

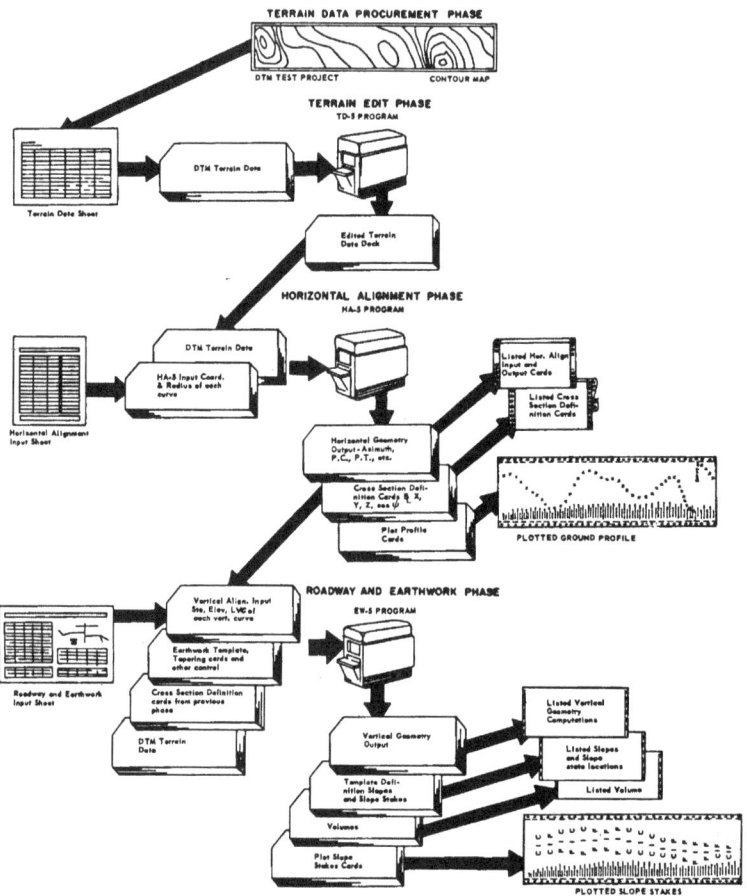

Fig. 1.

Terrain data for D.T.M. can be taken from standard contour maps or directly from photogrammetric stereomodels. Automatic take-off equipment may be used if available. Because terrain data are taken only once and can be used again and again for the evaluation of many lines the lack of special take-off equipment does not make the use of the system impractical.

Terrain Cross Section Lines

As a first step in the digitization process, a convenient base line forming the X-axis of the data coordinate system is located within the band of interest. (See Fig. 2) When the band of interest changes direction, a P.I. or a curve is introduced into the base line to cause it to follow the band of interest.

Next, the surface of the ground is represented by a series of points. (See Fig. 3) Each point is specified by its elevation and its offset distance along a cross section or scan line located at right angles to the base line. A terrain cross-section line is defined by its base line X-coordinate. The spacing of these lines and the density of the terrain points will depend on the location stage, the maps available, and the nature of the terrain. Terrain data taken along these cross sections are punched into cards, edited, and placed on file ready for use by the location engineer.

Selection of Alinements

Based on a study of the maps, photographs, soil conditions, and other location data, the engineer selects horizontal alinements within the digitized band. Each trial alinement is defined by specifying the state plane coordinates of each P.I. and one curve parameter on each. The base line is defined in the same manner. (See Fig. 2 again)

Horizontal Alinement Phase

Input forms containing the data for each trial alinement are punched into cards and fed into the computer along with the terrain data. The horizontal alinement program then computes the center line geometry and its spatial relation to the base line and the terrain at each cross section. From this a machine-made plot showing the ground profile beneath the center line is automatically prepared.

Trial Grade Lines

Using the plot of the ground profile, the engineer can establish the vertical relationships of his proposed line to the ground. He can develop the vertical design directly on the machine-prepared plot and scale the station and elevation of each V.P.I. from it for entry into a second data input form. At the same time, he decides on the shape of the typical template and the specifications governing such variables as side slope and compaction factors. These are also entered into the form. (See Fig. 4)

Roadway and Earthwork

Input for the vertical alinement and the specifications governing the shape of the typical section are run, along with output from the previous phases and the terrain data. At each terrain section, the template is automatically located, first horizontally and then vertically, to establish the correct spatial relationship between the digital model and the alinement. After the relationships are established the computer has a basis for selecting the proper typical section. It then establishes cuts or fill slopes and computes slope intercepts with the terrain. Quantity of cut and fill are also evaluated at each station, and the incremental and accumulated volumes are computed and output. Cards

(Text continues on page 47)

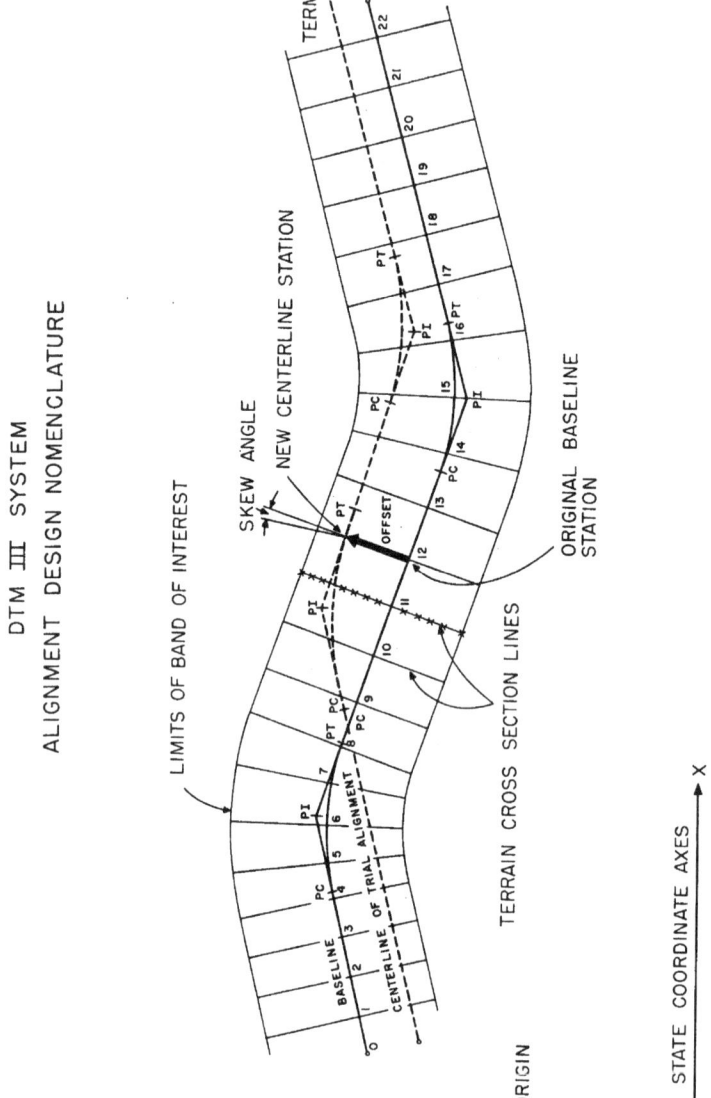

DTM III SYSTEM
ALIGNMENT DESIGN NOMENCLATURE

Fig. 2.

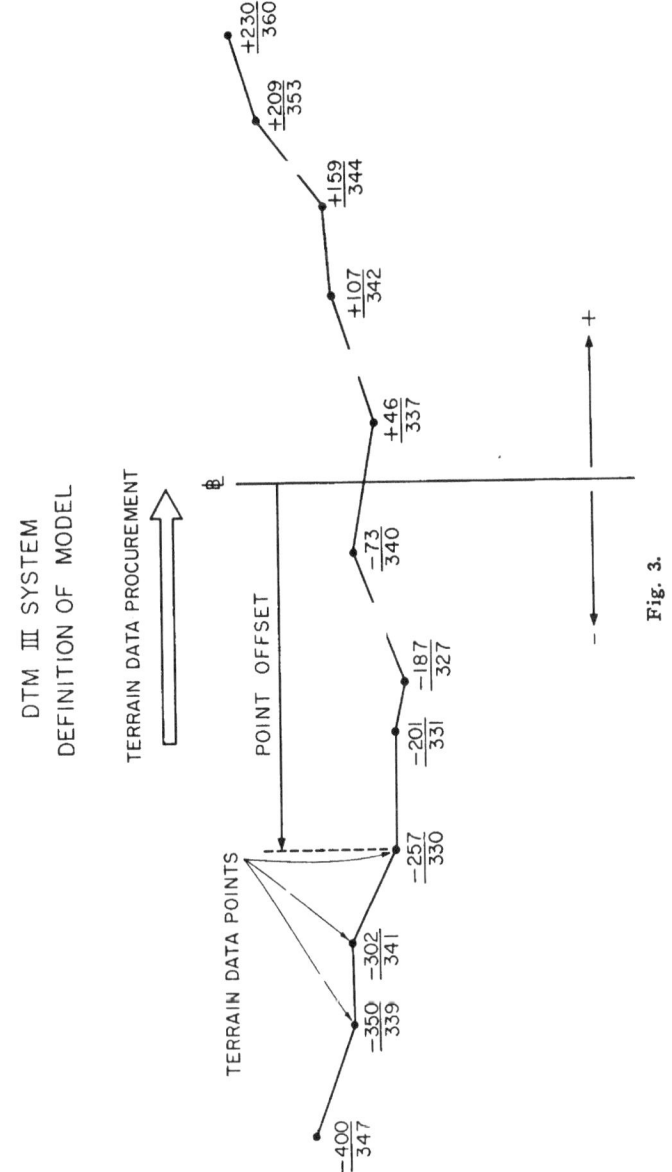

DTM III SYSTEM
DEFINITION OF MODEL

Fig. 3.

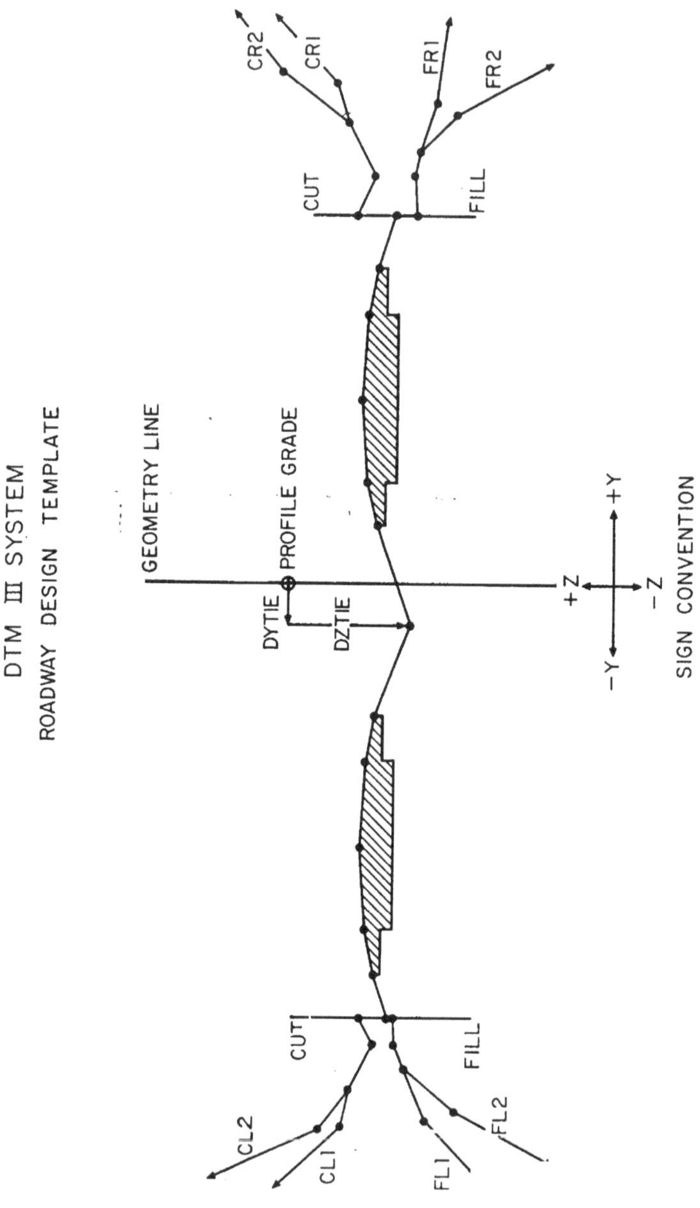

DTM III SYSTEM
ROADWAY DESIGN TEMPLATE

GEOMETRY LINE

PROFILE GRADE

DYTIE

DZTIE

CR2
CR1
CUT
FR1
FILL
FR2

CL2
CL1
CUT
FL1
FILL
FL2

SIGN CONVENTION

+Y

−Y

+Z

−Z

Fig. 4.

containing station number and slope intercepts are then used by another program which plots them.

Earthwork

The programs output earthwork volumes and mass haul ordinates directly. If cut and fill compaction factors are specified, the volumes produced will incorporate these corrections. The cost of embankment can be computed directly from the volumes. If additional trial lines are contemplated, a plot of the mass diagram can be produced by machine for use by the engineer in his analysis of the effects that changes will have on his original grade line.

Slope Plots

The plot of slope limits can be used by the engineer to establish the relationship of his line to the planimetric features shown on the map. The effect which the alinement is going to have on adjacent structures, highways, adjoining properties, stream crossings, and existing railroad facilities will be immediately obvious. (See Fig. 5)

Structures

Using the slope plots in conjunction with the topographic maps and the profile plots the location engineer can readily determine the length of each structure. Using a cost per linear foot, the costs for structures can be quickly estimated. Clearance not previously prescribed can be checked. The bridge designer can also use the system output if a more complete design of the structure is necessary.

Drainage

The slope plots are particularly useful where drainage costs have a significant effect on the location process. The length of each culvert can be scaled directly from them. The flow of water across the roadbed can be predetermined and counter measures such as down spouts, ditch paving, underdrain, and the location of catch basins and cross drains can be accurately determined. Costs can be estimated on major items or on a cost-per-mile basis depending upon the accuracy required.

Land Costs

Land costs are becoming increasingly important to the location decision. The high cost of land in urban situations particularly is forcing the designer to consider land with poor foundations or extreme slopes. Slope limits plots can be used to establish additional slope easements beyond what is normally taken. From them the spatial relationships between the alinement and contiguous parcels can also be seen.

Programs for computing the costs of land have also been written and used in conjunction with the D.T.M. System. Data acquisition still remains an obstacle, however, to fully automatic land cost evaluation.

Making the Location Decision

In this fashion the location engineer can prepare construction cost summaries for each trial alinement and use them in making the location decision. Each alinement is thus supported by quantitative, factual information rather than by an engineering "guesstimate." During the early phases of preliminary location, less detailed, faster-running programs can be used. Programs utilizing a simple template allow the engineer to try many alinements and narrow the selection process down to perhaps two or three lines. He can then tailor these carefully by small line and grade shifts to produce the

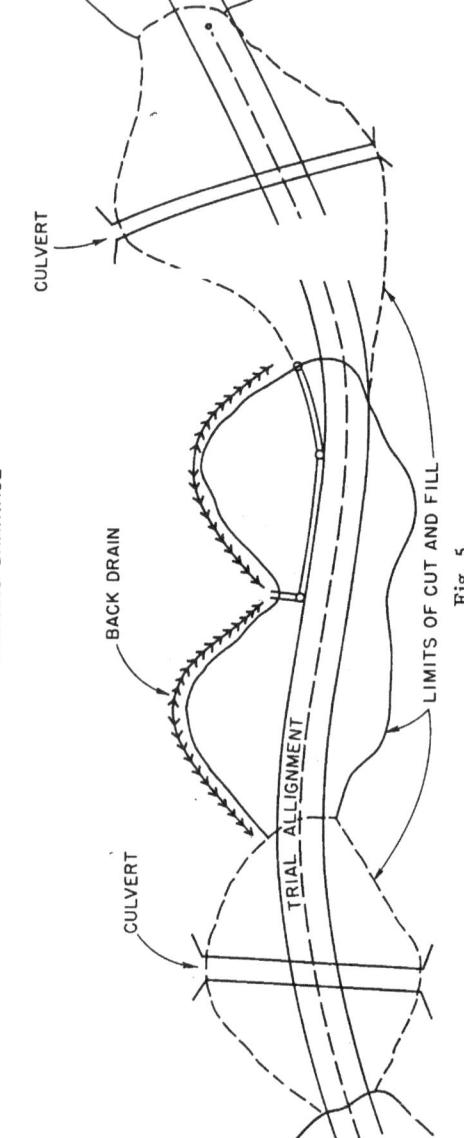

DTM III SYSTEM
SLOPE LIMITS PLOT
SHOWING DRAINAGE

CULVERT

BACK DRAIN

CULVERT

TRIAL ALLIGNMENT

LIMITS OF CUT AND FILL

Fig. 5.

best alinement. Notice again that he may analyze many trial alinements without the necessity for gathering new terrain data.

Present Development of D.T.M.

The present set of computer programs in the D.T.M. System is the result of continuous revision and reprogramming of the original series which was programmed for the I.B.M. 650. These programs were supplemented by the plot programs, the rapid reconnaissance program, and a zone cost program in 1959 and 1960. A new series written for the I.B.M. 1620 and the I.B.M. 709 has now been undertaken. It is based on the original system, but incorporates the results of all the research carried out since then. It should be noted that the system could easily be programmed for any large or medium scale digital computer.

COST IMPLICATIONS OF THE D.T.M. APPROACH

Engineering Cost with the D.T.M. System

Costs for using the system vary widely depending on the amount of data required, the cross section spacing, the output required, and the machine being used. A general idea of the costs can be obtained by considering the following cost schedule for a recently completed 27.9-mile highway location problem analyzed with the I.B.M. 650. The band of interest was over one-half mile wide. Terrain data were taken along scan lines spaced approximately 200 ft apart. Part of the terrain data were taken by semi-automatic take-off devices and part by hand. The engineering cost figures shown are for four trial lines.

TERRAIN DATA PROCUREMENT COST (200-FT SECTIONS, 2000-FT WIDE)

Manual Data Procurement (2-man teams—150 pts./hr)	
140 man hr at $2.50/hr ....$ 350.00	
Terrain Data Recorder	
21 hr at $10.00/hr 210.00	
Key Punching and Verifying	
53.5 hr at $4.00/hr 214.00	
Total for Terrain Data Procurement ....$ 774.00	
Terrain Data Procurement	
$774.00/27.9 miles	$26.40/route mile

ENGINEERING AND COMPUTING COSTS

Computer and Peripheral Equipment	
24 hr at $50.00/hr $1200.00	
Engineering (Base line selection, preparation of input	
forms and cards, machine operation, analysis, super-	
vision and overhead) $1000.00	
Total for Engineering and Computing $2200.00	
Running Costs	
$2200.00/111.6 miles (4 lines)	$19.71/trial line mile
Total Costs	
$2974.00/111.6 miles (4 trial lines)	$26.64/trial line mile

The same amount of earthwork computation performed by conventional electronic computer techniques (as opposed to D.T.M. System technique) would have cost between $25/mile and $32/mile or perhaps even more, depending upon the efficiency of the programs being used. The output would not have included horizontal geometry, profile plots, or slope-stake plots. With the best of conventional computer techniques the trial of an additional line would cost as much as the first (i.e., $25 to $32 per mile). With D.T.M. the trial of another line would cost less than $20/mile. The difference in cost is a reflection of the necessity for retaking terrain data in the non D.T.M. case.

It would be impracticable to perform calculations for the same number of trial lines by hand. Profiles would have to be stripped from a contour map. After the vertical alinement had been computed, cross sections would have to be stripped, drawn, and planimetered. Volumes would have to be computed and accumulated. Slope stake points would have to be plotted, working from the cross-section sheets. Each new line tried would require a similar effort. Altogether the task would be a formidable one, requiring (at one mile per man day) perhaps 22 man-weeks of work. To avoid expending this kind of time and effort, most engineering organizations not using a computer would estimate earthwork volumes on the basis of a profile take-off, assuming that the ground is flat, and using a chart or graph of quantities versus depth of cut or fill. Depending on the terrain, these estimates could be up to 20 percent in error. Depending on the percent which earthwork is of the total cost, this could produce bad location decisions.

The Significance of the D.T.M. System

All route location is essentially a trial-and-error process in which the engineer is looking for that line which will minimize total construction, engineering, and operating costs. With each successive trial the chances of finding a line with lower construction and operating costs are better; but engineering costs, of course, must go up as more trials are made. The engineer's problem is to know where the point of diminishing returns on his effort sets in.

With conventional location engineering techniques, whose cost is high, there is a tendency to stop looking after a small number of trials, perhaps only one or two. Where the engineering cost (in both time and money) is low, it becomes feasible to think in terms of many more trials and the possibility of finding a better and better line. The D.T.M. highway location projects with which the author has been associated have involved no fewer than 7 and as many as 35 distinctly different trial lines. In each case the small cost of additional trials was repaid several fold by savings in earthwork cost alone.

The significant thing about a production approach to location engineering, such as the D.T.M. System, is not that its cost is lower than other engineering techniques (which it generally is), but that it provides the engineer with a practical means of finding a better final location. The importance of this capability is no less today than it was 80 years ago when Wellington wrote that:

> . . . "one may fairly say that the locating engineer has but the one end before him to justify his existence as such—to get the most value for a dollar which nature permits; and but one failure to fear—that he will not do so."

Like aerial photogrammetry, the Digital Terrain System is a major step forward in helping the engineer to meet that injunction.

ACKNOWLEDGEMENTS

The author gratefully acknowledges the sponsorship of the Massachusetts Department of Public Works and the U. S. Bureau of Public Roads. Their support has made possible the development of the Digital Terrain Model System in the M.I.T. Civil Engineering Systems Laboratory, C. L. Miller, director.

This paper was sponsored by Subcommittee 4 (Potential applications of electronic computers to railway engineering and maintenance problems) of Committee 16— Economics of Railway Location and Operation.

REFERENCES

(1) "Using New Methods In Highway Location", Paul Roberts, Photogrammetric Engineering, June 1957.
(2) Digital Terrain Model Approach To Highway Earthwork Analysis, C. L. Miller, MDPW–BPR Research Project Report, Massachusetts Institute of Technology, August 1957.
(3) The Skew System For Highway Earthwork Analysis, C. L. Miller, Massachusetts Institute of Technology, MDPW–BPR Research Project Report, September 1957.
(4) Earthwork Data Procurement By Photogrammetric Methods, C. L. Miler and T. H. Kallstad, Highway Research Board Bulletin 199.
(5) "The Digital Terrain Model—Theory And Application", C. L. Miller and R. A. Laflamme, Photogrammetric Engineering, June 1958.
(6) Digital Terrain Model System Manual, C. L. Miller and R. A. Laflamme, MDPW–BPR Research Project Report, Massachusetts Institute of Technology, December 1958.
(7) The Digital Terrain Model System of Providing Highway Location And Design Information, C. L. Miller, Proceedings Of The American Association Of State Highway Officials, Committee Meeting, San Francisco, December 1958.
(8) Preliminary Report on The Digital Terrain Data Recorder, C. L. Miller And E. P. Gladding, MDPW–BPR Research Project Report, May 1959.
(9) Zone Cost Evaluation Program EA–2, C. L. Miller and D. E. Weisberg, MDPW–BPR Research Project Report, Massachusetts Institute of Technology, October 1960.
(10) Digital Terrain Model System DTM 11 Manual, P. O. Roberts, C. L. Miller and R. A. Laflamme, MDPW–BPR Research Project Report, Massachusetts Institute of Technology, November 1960.
(11) A New Technique For The Prediction Of Vehicle Operating Costs In Connection With Highway Design, A. S. Lang, And D. H. Robbins, MDPW–BPR Research Project Report, Massachusetts Institute of Technology, November 1960.
(12) Use Of Digital Computers In Land Cost Evaluation, D. E. Weisberg, Unpublished S. M. Thesis, Massachusetts Institute Of Technology, June 1961.
(13) Highway Location Evaluation, P. O. Roberts, Proceedings Of The Electronics Committee Of The American Association Of State Highway Officials, Boston, August 1961.

R. B. RADKEY, *Chairman*

Termite Control Investigation—Inspection of Specimens After 40 Months of Exposure

DIGEST

This report describes and analyzes data secured during the inspection of oak, fir and pine untreated control specimens and specimens treated with nine different preservatives after about 40 months of exposure to severe decay and termite attack in the Austin Cary Memorial Forest of the University of Florida near Gainesville, Fla. The preservatives under investigation are:

1. Coal tar creosote
2. Chromated zinc chloride
3. Tanalith
4. Pentachlorophenol
5. Copper napthenate
6. Acid copper chromate (Celcure)
7. Ammoniacal copper aresenite (Chemonite)
8. Chromated zinc arsenate (Boliden salt)
9. Chromated copper arsenate (Greensalt)

Three different retentions were used for each preservative as follows: (1) a retention 50 percent of that recommended by the AREA or AWPA for the particular preservative, (2) 100 percent of the recommended retention, and (3) 200 percent of the recommended retention.

About 24 months after the installation of the specimens treated with the nine preservatives, a supplemental installation was made of specimens treated with different mixtures of creosote with coal tar and creosote with petroleum. The three retentions used were the same as in the original installation. The four preservative combinations are:

10. 60 percent creosote with 40 percent coal tar
11. 80 percent creosote with 20 percent coal tar
12. 50 percent creosote with 50 percent petroleum
13. 25 percent creosote with 75 percent petroleum

Each specimen was removed from the ground and inspected. A system of grading for both decay and termite attack, as recommended by the American Wood Preservers Association, was used. After the grading of a specimen was completed, the specimen was replaced in the ground in its original position.

A brief summary of the analysis of the data as found after 40 months of exposure is as follows:

1. The amount of decay and termite attack in the untreated controls for the three species has increased with exposure time. Table 3 lists the rating or re_ maining quality of the wood after 14 months, 25 months and 40 months ex_ posure on the basis of 100 for the specimens when installed. The fir specimens show the greatest resistance to both termite attack and decay with the pine specimens showing the least resistance.

2. The amount of decay and termite attack in the treated specimens is shown in Table 8. The data show that only the specimens treated with coal tar creosote have resisted all decay and termite attack in all three species.

3. The rate of decay and termite attack in the untreated control specimens and those specimens treated with the lower retention of chromated zinc chloride and tanalith are shown in Figs. 1 and 2. The preservative treatment of only 50 percent of the recommended retentions has reduced the rate of both decay and termite attack, except for the fir specimen treated with chromated zinc chloride.

INTRODUCTION

This report describes the method of inspecting the specimens, including the system of grading, to determine the rate of decay and termite attack, and presents the results of an inspection of most of the specimens after about 40 months of exposure.

The initial installation of treated specimens and untreated control specimens placed in the Austin Cary Memorial Forest is described in the 1959 Proceedings, Vol. 60, page 131. This detailed report covers the treatment of the specimens, chemical analysis of the preservatives and installation of the 1296 treated specimens and 30 untreated controls. A total of nine different preservatives were used to treat the oak, pine and fir specimens, using three different retentions for each preservative.

A supplemental installation involving the use of four additional preservatives was made in November 1959. A detailed report covering this installation of 576 treated specimens and 15 untreated controls is presented in the 1961 Proceedings, Vol. 62, page 95.

A brief report covering the inspections of the specimens after 14 months and 25 months of exposure is published in the 1961 Proceedings, Vol. 62, page 89, with the general conclusion that the treated specimens had not received sufficient exposure to show any indication of the relative merits of the different preservatives.

This investigation is being conducted under the general direction of G. M. Magee, director of engineering research, Engineering Research Division, AAR. The conduct of the investigation and preparation of the report were in the charge of E. J .Ruble, research engineer structures, research staff, AAR. Invaluable assistance was extended by J. O. Blew of the Forest Products Laboratory and Dr. J. B. Huffman of the University of Florida in planning and conducting the investigation. Funds for the investigation are being provided by the AAR.

GRADING

The system of specimen grading for both decay and termite attack used during inspections corresponds to that recommended by the American Wood Preservers Association. The code numbers and letters of the system are listed below, with the corresponding ratings developed by such grades:

Code No.	Decay Grades	Rating-Percent
1	Sound, no evidence of decay	100
2	Localized superficial decay	75
3	Slight but positive decay	50
4	Deep or severe decay	25
5	Failure, almost complete loss of strength	0

Code Letter	Termite Attack Grades	Rating-Percent
A	No attack	100
B	Slight termite attack	75
C	Moderate termite attack	50
D	Severe termite attack	25
E	Failure caused by termites	0

FIELD INSPECTIONS

The field inspections of the specimens are usually conducted by a member of the AAR research staff assisted by Dr. J. B. Huffman of the School of Forestry of the University of Florida and a technician or student from the University. The inspection consists of the removal of each 2- by 4- by 18-in specimen, which is buried in the ground for half its length, cleaning off the dirt and sand with a spatula, and then determining its grade for both decay and termite attack. The specimen is then replaced in its original position.

The decay and termite attack grades are recorded on "Field Inspection Data Sheets", an example of which is shown on Table 1. The example shows the results of all field inspections of the specimens in Rows M, N and O of Plot 2. For example, in position 4 of row M, there is shown specimen O12D/21. As explained in the report covering the installation, the letter "O" designates oak species, "12" the stick number, and "D" the fourth specimen cut from stick 12. The figure "2" designates the preservative number, which in this case is chromated zinc chloride. The figure "1" represents the retention used in the treatment, and in this case is the lowest retention. The inspection of February 1961 indicates this specimen had a grade of 3C, indicating a decay grade of 3, which means slight but positive decay, and a termite attack grade of C, which means moderate termite attack.

The first inspection of the specimens was conducted in January 1959, and since the specimens had only been exposed about 14 months, it was decided to look only at those treated with retention 1, the lowest retention, and the untreated controls. The second inspection was conducted in November 1959 or about 25 months after their installation. At this time all the specimens having No. 1 and 2 retentions and the untreated controls were inspected.

An inspection of the test installation was conducted in February 1961, after 40 months of exposure. This inspection consisted of a rating of all untreated controls, all specimens treated with No. 1 and 2 retention of all nine of the original preservatives and those treated with preservative 2 (chromated zinc chloride) and preservative 3 (tanalith) having retention No. 3. A few of the specimens treated with preservative 13 (25 percent creosote and 75 percent petroleum with No. 1 retention) were inspected. These specimens were installed in November 1959 and had only been exposed about 15 months. The inspection of these specimens did not reveal any decay or termite attack so no further comments about their condition are included in this report.

ANALYSIS OF FIELD INSPECTION DATA

The data on the decay and termite attack grades as recorded on the Field Inspection Data Sheets were summarized and recorded on "Classified Data Sheets", an example of which is shown on Table 2. The example shows the results of the inspection of all specimens treated with preservative 2 (chromated zinc chloride having a No. 1 retention). It can be seen in Table 2 that previously mentioned specimen O12D/21 is included under the oak species, the second from the top, with a grade of 3C assigned to it during the inspection of February 1961. The other nine specimens of like species and preservative retention are also shown, accompanied by their respective grade. A summary of the grades for this preservative retention and species is as follows:

  1 specimens having a grade of 1A
  2 specimens having a grade of 1C
  1 specimen having a grade of 2B
  4 specimens having a grade of 2C
  1 specimen having a grade of 3A
  1 specimen having a grade of 3C

The ratings for decay and termite attack are obtained in the following manner:

Decay	*Termite Attack*
three (1) at 100 = 300	two (A) at 100 = 200
five (2) at 75 = 375	one (B) at 75 = 75
two (3) at 50 = 100	seven (C) at 50 = 350
775 ÷ 10 = 77.5 percent	625 ÷ 10 = 62.5 percent

The average index rating is obtained by averaging the decay rating with the termite attack rating, resulting, in this particular case, in 70.0 percent.

RESULTS OF INSPECTIONS

Untreated Controls

The decay and termite attack ratings with their average index ratings for the untreated controls installed in November 1957 are shown in Table 3 for the 14 months, 25 months and 40 months exposure. The values shown in this table are the average of 10 untreated specimens of each species, and it can be seen that after 40 months, four of the oak, three of the fir and nine of the pine specimens had completely failed. The four failures in the oak species consisted of two specimens with a rating of 5D and two with a rating of 5E. It can be seen that both decay and termite attack contributed about equally to the failure. The three failures in the fir species consisted of one specimen with a rating of 3E, one with 5E and one with 5C. The nine failures in the pine species consisted of two specimens with a rating of 5C, two with 5D, two with 3 E, one with 5E, one with 3E and one with 4E. Again, it can be seen that decay and termite attack contributed about equally to the failure.

Additional untreated controls were installed in January 1959 and November 1959, and the ratings for these specimens are shown in Tables 4 and 5.

Treated Specimens

The decay and termite attack ratings with their average index ratings for the treated specimens having the No. 1 or lowest retentions after 14 months exposure are shown in Table 6. Retention No. 1 is only 50 percent of the recommended retention for the particular preservative. The ratings for the same specimens after 25 months of exposure as well as the ratings for the specimens having No. 2 retention are shown in Table 7. Retention No. 2 is the recommended retention for the particular preservative. It is obvious from the values shown in Table 7 that there is no positive indication of the relative merits of the different preservatives except coal tar creosote, which had prevented all decay and termite attack in all three species for the 25-month exposure period.

The decay and termite attack ratings for the treated specimens after 40 months of exposure are shown in Table 8. Complete Index Ratings data are shown for retention

No. 1 and No. 2 for all preservatives and also for the specimens having No. 3 retention of preservative 2 (chromated zinc chloride) and preservative 3 (Tanalith). No inspection was made of the specimen preserved with the No. 3 retention of the other seven preservatives. Specimens having No. 3 retention have twice the amount of preservative recommended by the AREA. The data shown in Table 8 for the specimens having the No. 1 retention indicate that only coal tar creosote has resisted all decay and termite attack in all three species for the 40-month period. The specimens treated with the No. 2 retention of copper napthenate, acid copper chromate (Celcure) and ammoniacal copper arsenite (Chemonite) show negligible decay or termite attack for the 40 months of exposure.

The data shown in Table 8 for specimens treated with No. 1 retention of preservative 2 (chromated zinc chloride) and in Table 3 for the untreated controls are shown graphically in Fig. 1 for all three species with the ratings, in percent, shown as the ordinate and the exposure time, in months, as the abscissa. The data for preservative 3 (Tanalith) and the untreated controls are shown in Fig. 2. The values shown by the open triangles indicate the rate of decay of the untreated specimens and by the closed triangles, the rate of termite attack on the untreated specimens. The values shown by the open circles indicate the rate of decay of the treated specimens and by the closed circles, the rate of termite attack on the treated specimens. It is obvious that the preservative treatment of only 50 percent of the recommended retention has reduced the rate of both decay and termite attack, except possibly for the fir specimens treated with preservative 2 (chromated zinc chloride) where the rate of termite attack is slightly greater than the rate of decay. However, it appears that the decay rate is increasing while the rate of termite attack is decreasing.

CONCLUSIONS

From the data secured during the inspection of treated specimens of oak, fir and pine species after 40 months of exposure, it seems logical to conclude that coal tar creosote is the only preservative affording complete resistance to both decay and termite attack.

TABLE 1

TERMITE CONTROL INVESTIGATION

AAR RESEARCH CENTER
FIELD INSPECTION DATA SHEET

Plot   2

Location		Code	Rating of Condition of Specimens on Indicated Date									
Row	Position	Number	1-8-59	11-13-69	2-21-61							
M	1	P94F/32		1A	1C							
	2	P40G/92		1A	1A							
	3	F103A/32		1A	1A							
	4	O12D/21	1B	3C	3C							
	5	O107E/81		2A	3A							
	6	F24H/23			1A							
	7	F56C/93			-							
	8	P70I/63			-							
	9	P58I/72		1A	1A							
N	1	O103E/42		1A	1A							
	2	O61B/83			-							
	3	O53G/53			-							
	4	O63E/33			3A							
	5	F104E/31		1A	1B							
	6	P48B/12		1A	1A							
	7	Q7I/63			-							
	8	P42F/81		1A	3A							
	9	O73B/31	1A	2B	3B							
O	1	F76D/31	1A	1A	1A							
	2	P101C/11	1A	1A	1A							
	3	F76C/32		1A	1A							
	4	F10A/43			-							
	5	F51C/83			-							
	6	O13G/32		2A	3A							
	7	O15G/91		1A	1A							
	8	O66A/13			-							
	9	P42G/41		1A	2A							
	1											
	2											
	3											
	4											
	5											
	6											
	7											
	8											
	9											

AWPA Rating Code:
  Numbers 1 to 5 indicate degree of decay with Number 1 being sound - no evidence of decay
  Letters A to E indicate degree of termite attack with A having no attack.

TABLE 2
TERMITE CONTROL INVESTIGATION

AAR RESEARCH CENTER
CLASSIFIED DATA SHEETS

Preservative No. 2 - Chromated Zinc Chloride
Retention No. 1

Species	Code Number	Plot Location	Retention lb. cu. ft.	Rating of Condition of Specimens on Indicated Date						
				1-8-59	11-13-59	2-22-61				
Oak	O 7H	3J9	0.52	--	3C	2C				
	12D	2M4	0.48	1B	3C	3C				
	15D	1A5	0.51	1A	1A	1C				
	20D	8N9	0.48	1A	1A	1C				
	21C	6J1	0.50	1A	1B	2C				
	23G	4C7	0.52	1A	1A	2B				
	40G	1F5	0.53	1A	1A	1A				
	41D	5L5	0.52	1A	1A	2C				
	67F	7C8	0.51	1A	2A	3A				
	84B	9F6	0.48	1B	2B	2C				
	Months of Service			14	25	40				
	Av. Termite Rating			94.4	85	62.5				
	Av. Decay Rating			100	85	77.5				
	Av. Index Rating			97.2	85	70.0				
	No. of Failures			0	0	0				
Fir	F 31F	5I1	0.42	1A	1A	1B				
	33G	1K6	0.55	1A	1C	1C				
	44G	9G1	0.59	1B	1C	2D				
	45G	9H6	0.48	1A	1B	1C				
	53G	7B1	0.44	1B	1C	1C				
	55E	10J7	0.49	1A	1A	1B				
	56G	4C5	0.55	1A	1A	1A				
	61E	8O5	0.55	1A	1C	1C				
	70C	4O7	0.54	--	1C	1C				
	74G	10L5	0.52	1A	2C	1C				
	Months of Service			14	25	40				
	Av. Termite Rating			94.4	67.5	57.5				
	Av. Decay Rating			100	97.5	97.5				
	Av. Index Rating			97.2	82.5	77.5				
	No. of Failures			0	0	0				
Pine	P 27I	2D3	0.50	1B	2C	1C				
	28H	8L7	0.51	1A	1A	1B				
	31I	9I1	0.50	1C	3E	* 5E				
	33H	10F2	0.51	1B	2C	1C				
	34I	4I7	0.51	1A	1A	1C				
	37I	5C9	0.49	1C	2C	1C				
	38I	7D1	0.50	1A	1C	2C				
	39J	3F8	0.51	1A	1C	1C				
	47H	10D3	0.50	1A	1C	2C				
	48H	3J8	0.49	--	1C	2C				
	Months of Service			14	25	40				
	Av. Termite Rating			83.3	55	47.5				
	Av. Decay Rating			100	87.5	82.5				
	Av. Index Rating			91.65	71.25	65.0				
	No. of Failures			0	1	1				

AWPA Rating Code:
Numbers 1 to 5 indicate degree of decay with No. 1 being sound - no evidence of decay.
Letters A to E indicate degree of termite attack with A having no attack.

* Failed & Removed

Table 3
SUMMARY OF INDEX RATINGS
UNTREATED SPECIMENS
INSTALLED: NOV. 1957

T=Termite or D=Decay	14 Months Exposure			25 Months Exposure			40 Months Exposure		
	Oak	Fir	Pine	Oak	Fir	Pine	Oak	Fir	Pine
T	77.5	67.5	60.0	37.5	40.0	30.0	27.5	40.0	10.0
D	72.5	95.0	70.0	40.0	72.5	25.0	22.5	55.0	12.5
AV.	75.0	81.25	65.0	38.75	56.25	27.5	25.0	48.75	11.25
No. of Failures	1	1	0	2	1	6	4	3	9

Table 4
SUMMARY OF INDEX RATINGS
UNTREATED SPECIMENS
INSTALLED: JAN 1959

T=Termite or D=Decay	11 Months Exposure			26 Months Exposure		
	Oak	Fir	Pine	Oak	Fir	Pine
T	62.5	60.0	62.5	47.5	37.5	37.5
D	67.5	67.5	65.0	42.5	57.5	17.5
AV.	65.0	63.75	63.75	45.0	47.5	27.5
No. of Failures	0	0	0	2	2	5

Table 5
SUMMARY OF INDEX RATINGS
UNTREATED SPECIMENS
INSTALLED: NOV: 1959

T=Termite or D=Decay	15 Months Exposure		
	Oak	Fir	Pine
T	70.0	67.5	55.0
D	75.0	77.5	70.0
AV.	72.5	72.5	62.5
No. of Failures	0	0	0

Table 6
SUMMARY OF QUALITY INDEX-RATINGS
TREATED SPECIMENS
14 MONTHS EXPOSURE

Preservative, Number	T=Termite or D=Decay	No. 1 Retention		
		Oak	Fir	Pine
Coal Tar Creosote	T	100.0	100.0	100.0
No. 1	D	100.0	100.0	100.0
	AV.	100.0	100.0	100.0
Chromated Zinc,	T	94.4	94.4	83.3
Chloride,	D	100.0	100.0	100.0
No. 2	AV.	97.2	97.2	91.65
Tanalith,	T	100.0	100.0	100.0
No. 3	D	97.2	97.5	97.5
	AV.	98.6	98.75	98.75
Pentachlorophenol,	T	100.0	100.0	100.0
No. 4	D	100.0	100.0	100.0
	AV.	100.0	100.0	100.0
Copper Napthenate,	T	100.0	100.0	100.0
No. 5	D	100.0	100.0	100.0
	AV.	100.0	100.0	100.0
Acid Copper	T	97.2	95.0	94.4
Chromate (celcure),	D	97.2	100.0	91.7
No. 6	AV.	97.2	97.5	93.05
Ammoniacal Copper	T	100.0	100.0	100.0
Arsenite (chemonite),	D	100.0	100.0	100.0
No. 7	AV.	100.0	100.0	100.0
Chromated Zinc	T	100.0	100.0	100.0
Arsenate (Boliden Salt),	D	88.9	100.0	96.9
No. 8	AV.	94.45	100.0	98.45
Chromated Copper	T	100.0	100.0	100.0
Arsenate (Green Salt),	D	92.9	100.0	100.0
No. 9	AV.	96.45	100.0	100.0

Installation Date: Nov. 1957
Inspection Date: Jan. 1959

Table 7
SUMMARY OF QUALITY INDEX-RATINGS
TREATED SPECIMENS
25 MONTHS EXPOSURE

Preservative Number	T=Termite or D=Decay	No. 1 Retention			No. 2 Retention		
		Oak	Fir	Pine	Oak	Fir	Pine
1	T	100.0	100.0	100.0	100.0	100.0	100.0
	D	100.0	100.0	100.0	100.0	100.0	100.0
	AV.	100.0	100.0	100.0	100.0	100.0	100.0
2	T	85.0	67.5	55.0	75.0	70.0	87.5
	D	85.0	97.5	87.5	92.5	100.0	97.5
	AV.	85.0	82.5	71.25	83.75	85.0	92.5
3	T	67.5	95.0	62.5	97.5	97.5	95.0
	D	72.5	97.5	87.5	77.5	97.5	90.0
	AV.	70.0	96.25	75.0	87.5	97.5	92.5
4	T	100.0	100.0	92.5	97.5	100.0	97.5
	D	90.0	97.5	92.5	97.5	100.0	92.5
	AV.	95.0	98.75	92.5	97.5	100.0	95.0
5	T	97.5	97.5	100.0	95.0	100.0	97.5
	D	87.5	100.0	100.0	100.0	100.0	100.0
	AV.	92.5	98.75	100.0	97.5	100.0	98.75
6	T	82.5	77.5	87.5	97.5	97.5	92.5
	D	87.5	92.5	87.5	95.0	100.0	100.0
	AV.	85.0	85.0	87.5	96.25	98.75	96.25
7	T	100.0	100.0	97.5	100.0	100.0	100.0
	D	100.0	100.0	100.0	97.5	100.0	100.0
	AV.	100.0	100.0	98.75	98.75	100.0	100.0
8	T	97.5	100.0	92.5	97.5	100.0	97.5
	D	70.0	92.5	90.0	80.0	92.5	100.0
	AV.	83.75	96.25	91.25	88.75	98.75	98.75
9	T	85.0	95.0	85.0	100.0	97.5	95.0
	D	80.0	97.5	100.0	95.0	85.0	100.0
	AV.	82.5	96.25	92.5	97.5	91.25	97.5

Installation Date: Nov. 1957
Inspection Date: Nov. 1959

Table 8
SUMMARY OF QUALITY INDEX-RATINGS
TREATED SPECIMENS
40 MONTHS EXPOSURE

Pres.	T=Termite or D=Decay	No. 1 Retention			No. 2 Retention			No. 3 Retention		
		Oak	Fir	Pine	Oak	Fir	Pine	Oak	Fir	Pine
1	T	100.0	100.0	100.0	100.0	100.0	100.0			
	D	100.0	100.0	100.0	100.0	100.0	100.0			
	AV.	100.0	100.0	100.0	100.0	100.0	100.0			
2	T	62.5	57.5	47.5	62.5	62.5	70.0	75.0	80.0	82.5
	D	77.5	97.5	82.5	95.0	100.0	92.5	95.0	100.0	100.0
	AV.	70.0	77.50	65.0	78.75	81.25	81.25	85.0	90.0	91.25
3	T	65.0	80.0	60.0	85.0	90.0	67.5	100.0	100.0	92.5
	D	60.0	85.0	70.0	55.0	90.0	87.5	62.5	95.0	97.5
	AV.	62.50	82.50	65.0	70.0	90.0	77.5	81.25	97.5	95.0
4	T	95.0	87.5	82.5	100.0	97.5	80.0			
	D	87.5	90.0	85.0	97.5	97.5	95.0			
	AV.	91.25	88.75	83.75	98.75	97.5	87.5			
5	T	87.5	87.5	97.5	97.5	100.0	100.0			
	D	80.0	97.5	97.5	100.0	100.0	100.0			
	AV.	83.75	92.50	97.5	98.75	100.0	100.0			
6	T	72.5	62.5	67.5	97.5	100.0	90.0			
	D	87.5	95.0	87.5	100.0	100.0	100.0			
	AV.	80.0	78.75	77.50	98.75	100.0	95.0			
7	T	97.5	100.0	92.5	100.0	100.0	100.0			
	D	87.5	100.0	95.0	90.0	100.0	100.0			
	AV.	92.5	100.0	93.75	95.0	100.0	100.0			
8	T	80.0	95.0	82.5	97.5	97.5	85.0			
	D	67.5	80.0	80.0	70.0	92.5	95.0			
	AV.	73.75	87.50	81.25	83.75	95.0	90.0			
9	T	75.0	80.0	82.5	85.0	92.5	90.0			
	D	75.0	95.0	97.5	95.0	92.5	100.0			
	AV.	75.0	87.50	90.0	90.0	92.5	95.0			

Installation Date: Nov. 1957
Inspection Date: · Feb. 1961

FIG. 1
QUALITY INDEX RATING
PRESERVATIVE: NO. 2 (CHROMATED ZINC CHLORIDE)
RETENTION: NO. 1

SYMBOLS

    TREATED SPECIMENS   • TERMITE ATTACK
                          o DECAY

  UNTREATED SPECIMENS   ▲ TERMITE ATTACK
                           △ DECAY

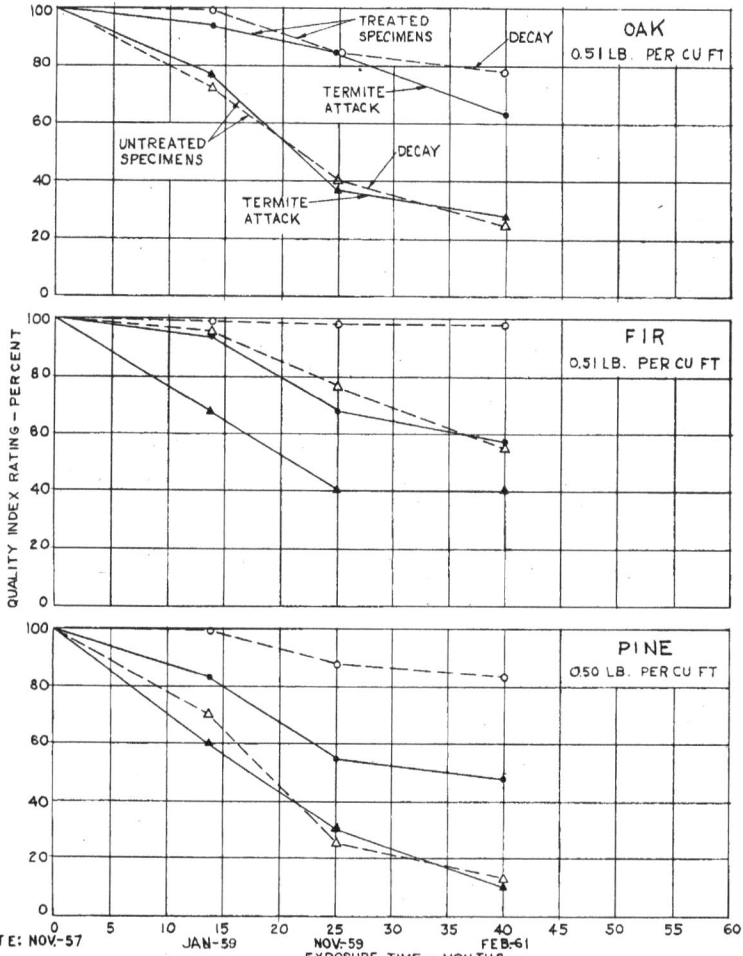

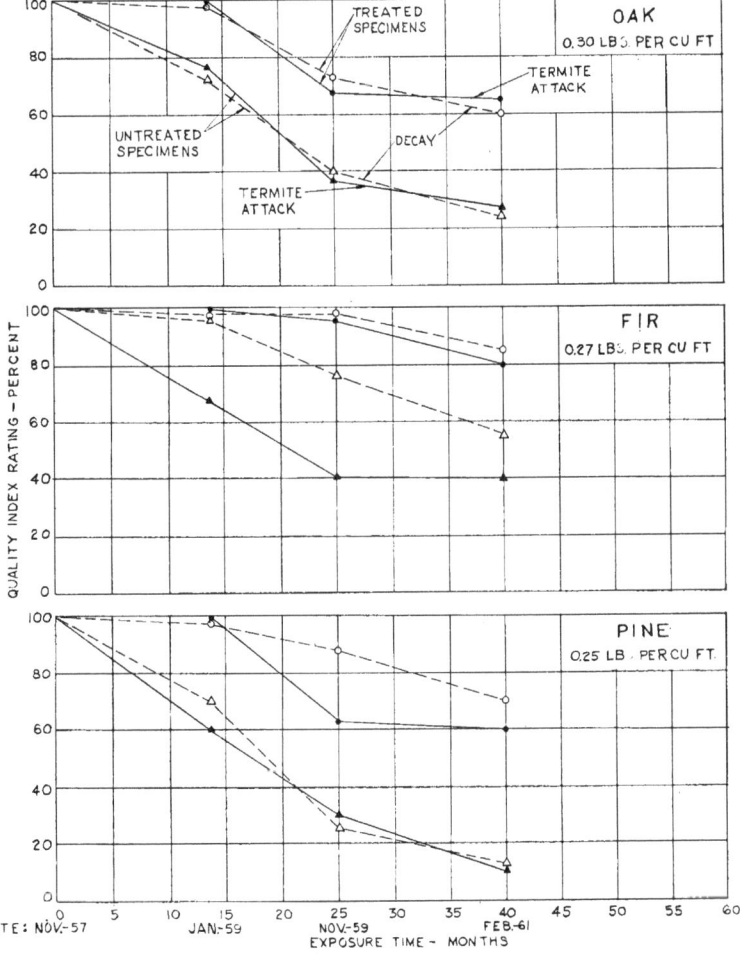

FIG 2
QUALITY INDEX RATING
PRESERVATIVE· NO.3   TANALITH
RETENTION: NO.1

SYMBOLS
    TREATED SPECIMENS   • TERMITE ATTACK
                          o DECAY
    UNTREATED SPECIMENS   ▲ TERMITE ATTACK
                          △ DECAY

Critical Review of the Subject of Speed on Curves as Affected by Present-Day Equipment

Collaborating with the AAR Joint Committee on Relation Between Track and Equipment

L. H. Jentoft (chairman, subcommittee), J. E. Armstrong, Jr., W. R. Bjorklund, P. H. Croft, A. D. DeMoss, R. G. Garland, L. W. Green, R. J. Hollingsworth, H. W. Jensen, R. E. Kuston, J. E. Martin, S. H. Poore, R. E. Simpson, V. M. Schwing, J. R. Talbott, Jr., J. B. Wilson.

Speeds of Trains Through Turnouts

Purpose of Test

In connection with the assignment of the AREA Track committee to recommend a limited number of standard turnouts for adoption by the Association, the AAR Research Center was requested to develop and recommend to the committee a formula to give the maximum comfortable train speed through turnouts, particularly with respect to the switch-point angle and length of point. The determination of the maximum comfortable riding speed through turnouts does not lend itself well to analysis, but must be made by the "seat-of-the-pants" method or by actual measurement of the lateral accelerations developed. Since there was little available experimental data on this subject, arrangements were made with the Santa Fe to conduct a test on this road's double-track main line between Chicago and Kansas City.

Test Procedure

This test was made using two three-way ride recorders manufactured by the Impact Register Company of Champaign, Ill. These recorders make a continuous record of the fore-and-aft, vertical, and lateral accelerations and have a satisfactory chart speed, sensitivity, and frequency response for this purpose. One recorder was placed in a roomette opposite the center pin at each end of a regular Pullman car on the rear of the Santa Fe Chief. The Pullman car was of modern-type construction, having four-wheel trucks with long-travel springs, snubbers and cross stabilizers. The two axles of each truck were on 8-ft 9-in centers.

On the regularly scheduled run of the Chief between Chicago and Kansas City on December 12, 1960, arrangements were made to operate alternately on the westbound and eastbound main tracks in order to pass through several crossovers. Some of the crossovers had No. 24 frogs with 39-ft curved switch points, and the train was run through them at a speed of approximately 50 mph. Others had No. 20 frogs with 30-ft straight points and were operated through at a speed of approximately 40 mph.

Test Results

Fig. 1 is a typical record obtained from the ride recorder in one end of the car passing through a No. 20 crossover at M.P. 14.2 at a speed of 40 mph. It will be observed that there are four traces on the record. The top trace shows the vertical accelerations and the bottom trace, which almost parallels the sixth horizontal line from the bottom, shows the fore-and-aft accelerations. The second and third traces are the ones of particular interest for this study. The second trace shows the lateral accelerations towards the left from the center line of the track. Each horizontal line on the record for the calibration of the lateral accelerometer is equivalent to 0.035g acceleration. The chart traveled at a speed of 6 in per min for the original record which has been enlarged in the reproduction so that the time interval between each vertical dotted line in Fig. 1 is 1/6 min or 10 sec. The chart travel was in the direction of decreasing numbers as shown at the bottom of the figure. For this particular record the train was traveling on the eastbound main. When it entered the No. 20 crossover a large lateral jerk to the right occurred, as indicated on the third trace, followed by two jerks of decreasing intensity. As the car passed through the crossover and entered the lead curve into the westbound main, two jerks of moderate intensity to the left occurred as will be noted in the second trace, followed by a severe jerk decreasing to a line with very small variations when the car got back on the tangent track on the westbound main. By measuring the distance on the chart between the beginning of the large jerk to the right indicated on the third trace and the beginning of the large jerk to the left indicated on the second trace, it will be found that this distance is equivalent to 6½ seconds of chart travel, which at a speed of 40 mph approximates closely the length of the crossover from actual point of switch on the eastbound to the actual point of switch on the westbound. This indicates, therefore, that the large jerk at the end of the car where the recorder was placed came first just as the truck was deflected from the tangent rail to the switch point angle and the second large jerk came as the truck was deflected from the leaving switch point angle onto the westbound tangent track. It is thus evident from this record that the switch point angle has an imporant influence or effect on the riding condition in the car.

Fig. 2 shows a corresponding record for the train passing through a No. 24 crossover at Argyle at a speed of 50 mph. Here again the train was crossed over from the eastbound main to the westbound main, and it will be observed that the pattern of lateral acceleration or jerk is similar to that shown in Fig. 1 on the No. 20 crossover. The first large jerk as shown in the third trace occurs to the right when the truck is deflected from the eastbound tangent onto the switch point angle and the second large jerk as shown in the second trace occurs to the left when the truck is deflected from the switch point angle back to the tangent on the westbound main. It will be noted particularly that the maximum lateral acceleration on the lead curvature is only about half as much as the lateral jerk experienced right at the switch point.

In making the test run the train ran through 6 No. 20 crossovers at a speed of approximately 40 mph and 5 No. 24 crossovers at a speed of approximately 50 mph. Table A shows the results obtained in going through these 11 crossovers. The maximum lateral acceleration and average lateral acceleration are given for the Pullman car with the two ride recorders passing through the turnout on the westward main and on the eastward main. It will be observed that there is considerable variation in the amount of lateral acceleration among the different turnouts. This is to be expected because of the play between the wheel gage and the track gage. As a result of this play the flange of the leading wheel of the truck might be close to the gage of the rail when it strikes the turnout point, which would give a higher acceleration, or it might be near the opposite rail, which would give a lower acceleration. Also, the truck might be "hunting", and if it were hunting in the direction of the turnout at the time the leading wheel flange contacted the switch point the acceleration would be lower than if it were hunting in the opposite direction at this instant. The figures shown in the table are the number of lines on the chart. As previously stated each line is equal to 0.035g. It will be further noted that for both the No. 20 and No. 24 turnouts the maximum lateral acceleration measured varied from 3 lines to 8 lines. The 10 maximum values for number of lines registered average 6.2 lines for the No. 20 turnouts at 40 mph. and 7.1 lines for the No. 24 turnouts at 50 mph. Converting lines to rate of acceleration, 6.2 lines times 0.035 times 32.2 ft per sec per sec equals an average maximum lateral acceleration of 7 ft per sec per sec for the No. 20 turnouts and 7.1 lines times 0.035 times 32.2 ft per sec per sec equals 8 ft per sec per sec for the No. 24 turnouts.

Discussion of Test Results

The No. 20 turnouts had 30-ft straight switch points with an angle at the point of 0° 57′ 18″.. The No. 24 turnouts had 39 ft curved switch points with an angle at the point of 0° 32′ 45″ and a curvature of 1° 07′ 22″. A method which seems logical for computing the lateral accelerations produced as the truck passes from the straight rail to the switch point is as follows: For the 30-ft straight switch point, compute the radius for a curve having a central angle equal to the switch point angle and a long chord equal to the truck wheel base. Then compute the acceleration by the formula for centrifugal force for a curve of this radius at the speed of operation. It will be observed from Table B that the computed lateral acceleration by this method for the 30-ft straight point at a speed of 40 mph is 6.55 ft per sec per sec. The value determined from the ride recorder of 7 ft per sec per sec for the 30-ft straight point is therefore only 6 percent in excess of the computed value.

It might be well to explain why the foregoing method of calculation seems logical. When the truck is on tangent just at the point of entering the switch point, its lateral acceleration is zero. When it is just fully on the switch point angle of the straight point, its lateral acceleration is also zero, and on the curved point its lateral acceleration is that for the degree of curvature and the speed. At the time the leading wheel flange strikes the switch point angle the lateral acceleration theoretically becomes infinitely high, but we know that it does not actually do this because of yielding in the track structure and in the various components of the car. Because of the small angle and displacement involved, it may be expected that the center pin would tend to follow the path of a curve while the truck passes the 8 ft 9 in from just entering to being just fully on the switch point.

The same procedure is followed for the 39-ft curved point except that after the radius and degree of curvature are determined for the switch point angle, the degree of curvature of the switch point is added, and the acceleration is then calculated for the combined curvature. The acceleration so obtained as shown in Table B is 6.92 ft per sec per sec and the measured acceleration of 8 ft per sec per sec is 15 percent in excess of the computed value. Although this is not as close agreement as for the 30-ft straight point, nevertheless it is good agreement considering the very small angles involved.

In adding the curvature of the switch point it is recognized that only the leading axle of the truck is actually on the curvature during this period. However, in the normal tracking pattern of a truck going around a curve the outer leading wheel hugs the outer rail and the trailing axle tends to maintain a radial position with the outer wheel generally away from the outer rail. It is presumed, therefore, that the centrifugal force would be about the same as it would be if the entire truck were on the switch point curvature.

It would appear, therefore, that this is a reasonably accurate method of determining the rate of acceleration which a passenger riding in a car of this type and particularly having this type of truck would be subjected to in passing through a turnout or cross-over with any given switch point angle and curvature at any selected speed.

Permissible Lateral Acceleration

In order to obtain information on what lateral acceleration would be acceptable, some records were taken traveling at high speed on tangent track where the ride was very comfortable. A duplication of one such record is shown in Fig. 3. It will be observed that the peaks of lateral acceleration are moderate, having a maximum value of about three lines or 0.10g. Records were also taken traversing curves at high speed. A reproduction of one such record is shown in Fig. 4. It will be observed here that in passing this curve a maximum lateral acceleration of approximately 8 lines was measured equivalent to 0.28g, although generally on the curve the maximum acceleration was 5 lines or 0.175g. A 3-in unbalanced speed combined with the tilting of the car body on the springs is equal to about 0.10g lateral acceleration. Although Fig. 3 shows an unusually good riding condition at high speed on tangent track, other records were obtained on tangent track where riding conditions were still acceptable and maximum lateral accelerations of 7 lines or 0.25g were measured.

In Table C information is given of measured maximum lateral accelerations and average lateral accelerations as obtained with the ride recorders with both the A and B instruments, one in each end of the car on several curves for which records were obtained. It will be observed in general that where the actual speed was near the maximum recommended comfortable speed on the curve the maximum lateral acceleration approximates the maximum values measured in the runs through the crossovers. Accordingly, it may be concluded that at the 40 mph operating speed through the No. 20 turnouts with 30-ft straight switch points and 50-mph operating speeds through the No. 24 turnouts with the 39-ft curved points, the maximum lateral accelerations are on the order of maximum lateral accelerations that would be occasionally encountered in high-speed running on tangent track and curves.

Conclusions

Although more extensive test data, including a wider range of conditions, would be desirable before drawing definite conclusions, the results of these tests indicate that the recommended comfortable speed through turnouts may be established on the basis of the following criteria, whichever gives the lower speed:

1. The calculated lateral acceleration for a curve having a long chord or length equal to the truck wheel base and a central angle equal to the switch point angle, plus the degree of curvature of the switch point, if curved, not to exceed 0.22g or 7 ft per sec per sec.
2. The calculated unbalance of the lead curve not to exceed 3 in.

Acknowledgement

It is desired to express appreciation to the Santa Fe Railway for making these test data available for this report. E. C. Honath, engineer of track design, made all arrangements for the test run, operated the two ride recorders with the assistance of F. H. Smith, assistant engineer, and furnished the tabulated readings from the records. R. J. Yost, superintendent, arranged for the special operation of the train to pass through as many crossovers as possible. H. L. Lewis and H. W. Gibson, trainmasters, J. G. Hynes, assistant supervisor air brakes, G. S. Shaffer and E. B. Reynolds, road foremen of engines, rode the train on their respective territories to supervise the train operation and speed through the crossovers.

G. M. Magee, director of engineering research, and Ralph Schinke, stress analyst, from the AAR Research Center staff were present on the test run, analyzed the data, and prepared this report as a part of the research activities of the AAR Research Department, W. M.. Keller, vice president.

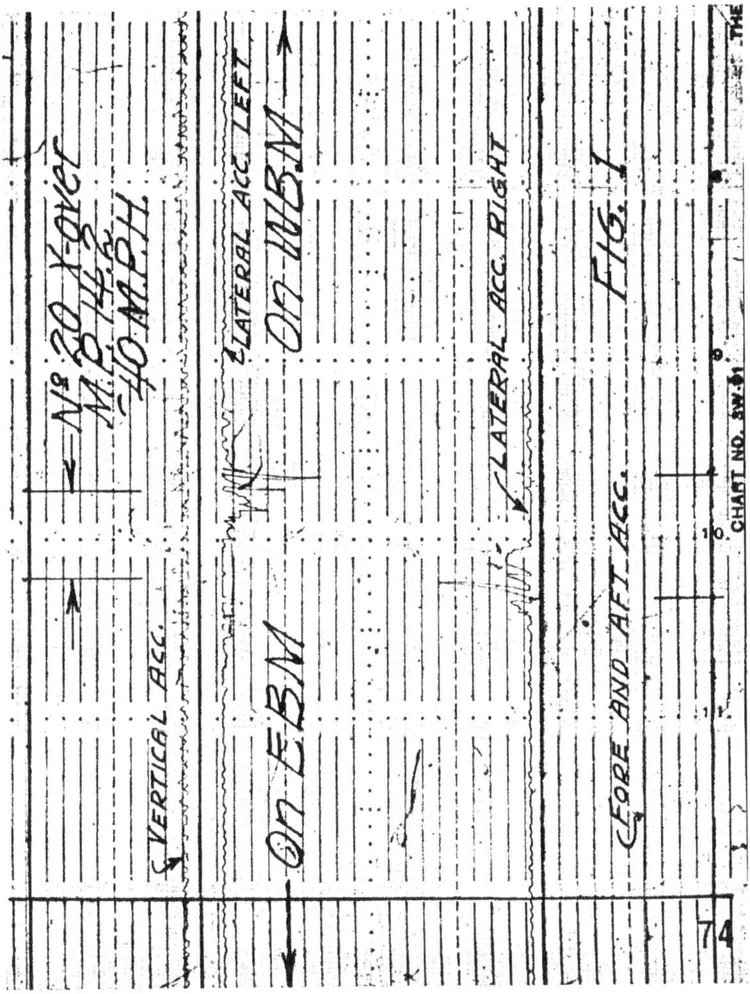

No. 20 X-over
M.P. 4.3
40 M.P.H.

VERTICAL ACC.

On E.B.M.

LATERAL ACC. LEFT

On W.B.M.

LATERAL ACC. RIGHT

FORE AND AFT ACC.

FIG. 1

CHART NO. SW.

74

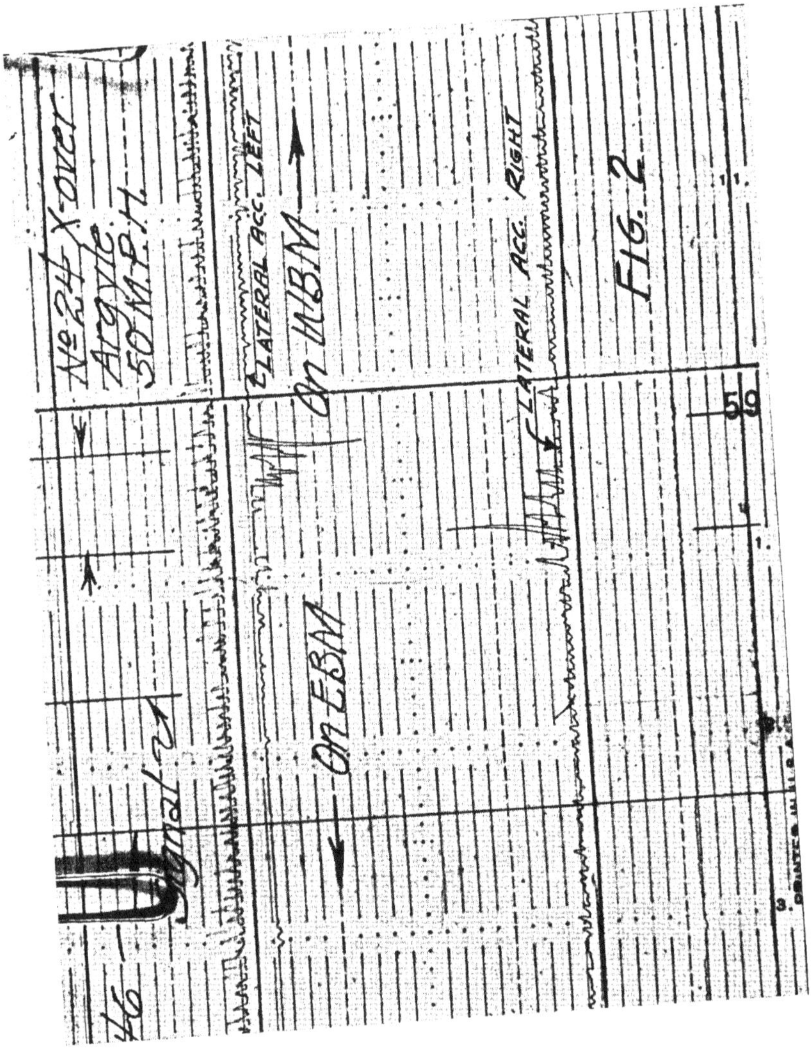

Nº 24 X-over
Argylc
50 M.P.H.

← LATERAL ACC. FEET

ON HBM →

← LATERAL ACC. RIGHT

← ON EBM

← Signal →

FIG. 2.

59

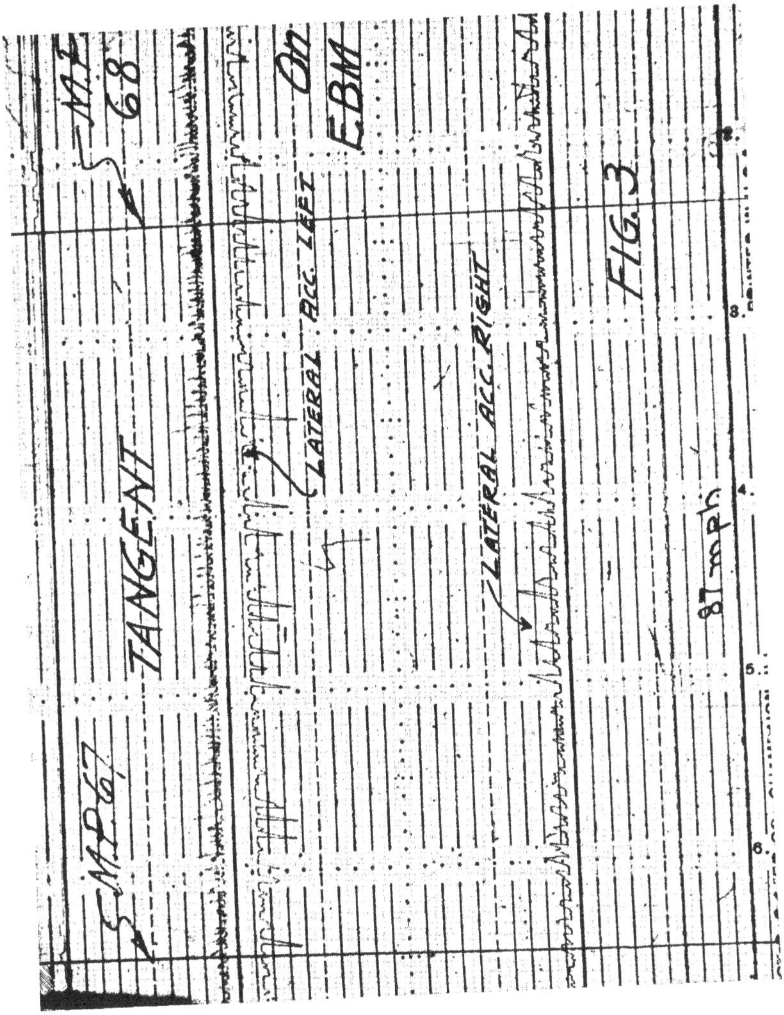

FIG. 3

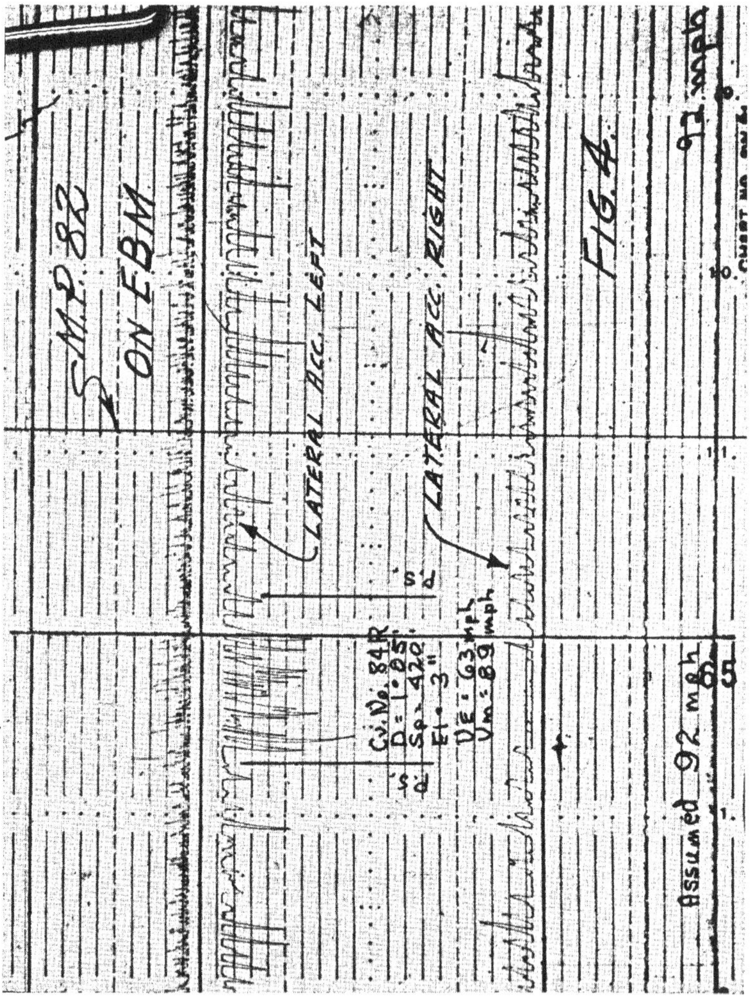

FIG. 4.

TABLE A

STATEMENT SHOWING LATERAL ACCELERATION RECORDED ON VARIOUS
CROSSOVERS ON AT&SF RY BETWEEN CHICAGO AND KANSAS CITY-DEC. 12, 1960

Location	Instrument	Kind of Turnouts	Speed MPH	On Westward ML		On Eastward ML	
				Max. Lat. Accel.	Average Lat. Accel.	Max. Lat. Accel.	Average Lat. Accel.
MP 6	A	14	30	6	3/4	4	1/2
MP 14	A	20	40	6	1/2	5	1/2
MP 14	B	20	40	8	3/4	4	1/2
MP 17	A	20	40	7	1/2	3	1/2
MP 17	B	20	40	6	1/2	3	1/4
MP 29	A	20	40	5	1/2	5	3/4
MP 29	B	20	40	6	1	3	1/4
MP 57	A	20	40	4	3/4	7	1/4
MP 57	B	20	40	6	3/4	5	1/2
MP 84	A	20	38	3	1/2	3	1/2
MP 84	B	20	38	5	1/4	3	1/4
MP 236	A	20	40	6	1/2	4	1/2
MP 236	B	20	40	5	1/2	5	1/2
MP 246	A	24	50	7	1/2	7	1/2
MP 246	B	24	50	6	1/2	8	1/2
MP 262	B	24	45	6	1/2	4	1/2
MP 262	A	24	45	7	1/2	4	1/2
MP 265	A	24	51	5	3/4	7	3/4
MP 265	B	24	51	4	3/4	7	3/4
MP 290	A	24	50	8	1-3/4	3	3/4
MP 292	A	24	50	8	1-1/2	6	1-1/4

TABLE B

CALCULATION OF LATERAL FORCES ON TURNOUTS

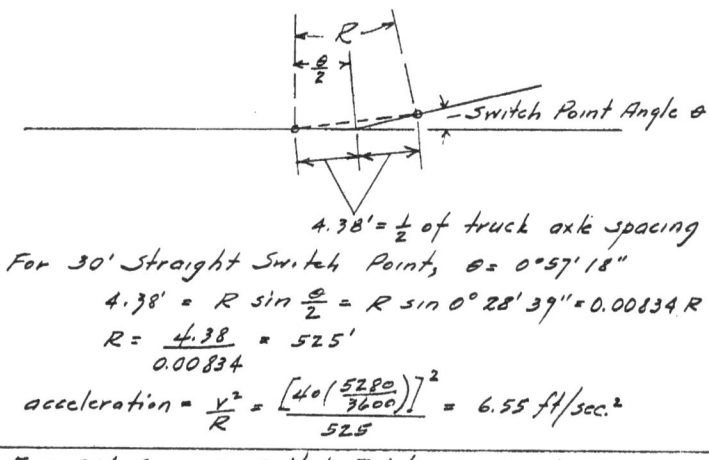

$4.38' = \frac{1}{2}$ of truck axle spacing

For 30' Straight Switch Point, $\theta = 0°57'18"$

$4.38 = R \sin\frac{\theta}{2} = R \sin 0°28'39" = 0.00834 R$

$R = \dfrac{4.38}{0.00834} = 525'$

acceleration $= \dfrac{V^2}{R} = \dfrac{\left[40\left(\frac{5280}{3600}\right)\right]^2}{525} = 6.55\ ft/sec.^2$

For 39' Curved Switch Point, $\theta = 0°32'45"$

$4.38' = R \sin\frac{\theta}{2} = R \sin 0°16'22\frac{1}{2}" = 0.00476$

$R = \dfrac{4.38}{0.00476} = 920' = 6°14'$ Curve

$6°14' + 1°07'$ (Curvature of Switch Point) $= 7°21'$

$7°21'$ Curve $= 778'$ Radius

acceleration $= \dfrac{V^2}{R} = \dfrac{\left[50\left(\frac{5280}{3600}\right)\right]^2}{778} = 6.92\ ft/sec.^2$

TABLE C
STATEMENT SHOWING LATERAL ACCELERATION RECORDED ON VARIOUS
CURVES ON AT&SF RY BETWEEN CHICAGO AND KANSAS CITY- DEC. 12, 1960

Location and Track	Instrument	Curve No.	Degree of Curve	Spiral Length ft.	Super Elevation Inches	Equilibrium Speed-MPH (Cal.)	Maximum Speed-MPH (Cal.)	Av. Speed per Mile mph	Max. Lat. Accel. In	Max. Lat. Accel. Out	Average Lat. Accel. In	Average Lat. Accel. Out	Remarks
MP 10 EW	A	28L	3°12'	360	3	37	52	43	1		1/2		Speed less than equil. speed
MP 11 EW	A	29L	2°04'	410	5-1/2	62	77	57	3		3/4		Speed less than equil. speed
MP 12 EW	A	30L	2°02'	210E, 230W	3	46	65	63		7		3/4	
MP 16 WW	B	31R	1°00'	270	2-1/2	80	89	62	3		1/2		Speed less than equil. speed
MP 19 EW	B	33R	1°30'	680	5-1/2	72	90	63	5		1-1/4		
MP 20 EW	B	34L	1°00'	270	2-1/2	60	89	63		4		1/2	
MP 21 EW	B	35L	1°00'	270	2-1/2	60	89	72		6		3/4	Also lat. accel. in.
MP 21 EW	B	36R	0°29'55"	170	1-1/2	66	113	77		4		1/4	
MP 24 EW	B	37L	1°23'57" to 2°00'	100 to 460	2-1/2 to 5-1/2	46 to 63		41	5		1-1/4		
MP 27 EW	B	42L	3°06'	430	5-1/2	50	63		7			1-3/4	Curve limit not located
MP 27 EW	A	42L	3°06'	430	5-1/2	51	63	61		6		1-1/4	Curve not located, sp. greater than equil. sp.
MP 28 EW	B	43L	3°04'	440	5-1/2	51	63	50		4		1/2	Curve not located, sp. greater than equil. sp.
MP 28 EW	B	43L	3°04'	440	5-1/2	51	63	51		5	1/2	1-1/4	Curve not located
MP 29 EW	B	44L	1°50'	290	3-1/2	52	71	40			3/4		Curve not located
MP 29 EW	A	44L	1°50'	290	3-1/2	52	71	40	1-1/2				Curve not located
MP 32 WW	B	47R	2°02'24"	540	5-1/2	62	77	77		5		1-1/4	
MP 33 WW	B	48R	1°00'	270	2-1/2	60	89	75		4		3/4	
MP 34 WW	B	49L	0°47'55"	220	2	60	95	75-82	3	9		1-1/4	
MP 35 WW	B	50R	0°44'	270	2-1/2	70	104	51	4		1/2		Curve not located
MP 35 WW	B	51R	1°30'	340	3-1/2	58	79	51			3/4		Curve not located
MP 58 EW	B	74L	1°01'	180	1	46	79	57		2		1/4	
MP 58 EW	B	75R	2°00'	140	2	38	60	57		6	1	1-1/2	Curve not located
MP 81 EW	B	84L	1°05'	420	3	63	89	92		6	1	1	Curve not located
MP 81 EW	A	84L	1°05'	420	3	63	89	92		8	1-1/2	1-1/4	
MP 242 EW	B	169R	1°29'11"	630	5-1/2	72	89	77	6	4	1/2	1/2	Equil. speed
MP 242 EW	B	170L	1°25'	530	5-1/2	74	92	77	5		3/4		Indicates less than equil. speed
MP 243 EW	B	171L	1°17'	380	4	67	89	70	2		1/4		Indicates less than equil. speed
MP 244 EW	B	172L	1°01'	300	2	53	84	61-70					
MP 244 EW	B	122L	1°01'	300	2	53	84	61		4	3/4	1/4	
MP 245 EW	A	173R	0°59'40"	570	3	66	93	51	4		3/4		
MP 245 EW	B	173R	0°59'40"	570	3	66	93	51	4	2	3/4	1/4	

TABLE C (Continued)

Location and Track	Instrument	Curve No.	Degree of Curve	Spiral Length ft.	Super Elevation Inches	Equilibrium Speed-MPH (Cal.)	Maximum Speed-MPH (Cal.)	Av. Speed per Mile mph	Max. Lat. Accel. In	Max. Lat. Accel. Out	Average Lat. Accel. In	Average Lat. Accel. Out	Remarks
MP 247 WW	B	174L	0°59.8'	500	3	66	93	50-50	2			3/4	Speed less than equil. speed
MP 248 WW	B	175R	0°49'40"	250	2	53	83	67		4	3/4	3/4	
MP 249 WW	B	176L	1°06'	350	2	51	81	51	2		1/2		
MP 250 WW	B	177L	1°02'	270	2	53	83	48	3		1/2		
MP 251 WW	B	180R	3°04'	400	5-1/2	51	63	45	4		3/4		
MP 251 WW	B	181L	4°03'	250	5-1/2	44	55	45	2	5		1/2	Speed nearer max. speed
MP 252 WW	B	182R	1°55'	340	4	55	72	53		1	1/4	1/4	Equil. speed
MP 253 WW	B	183L	2°57-1/2' to 3°08-1/2'	430	5-1/2	50	62	53-58		5		3/4	
MP 253 WW	B	184L	3°08'	350	5-1/2	50	62	58		5		1-1/4	
MP 254 WW	B	185R	3°06'	360	5-1/2	50	63	55		4		3/4	
MP 254 WW	B	186R	4°03'	360	5-1/2	44	55	55		7		1-1/4	
MP 255 WW	B	187L	2°02'	230	3	46	65	55-50		2		1/2	
MP 255 WW	B	188L	4°07'	360	4-1/2	40	51	50		6		1-1/2	
MP 256 WW	B	189R	4°07'	360	4-1/2	40	51	50		5		1	
MP 256 WW	B	190R	0°40'	320	2	65	103	61	2	1	1/4	1/4	Equil. speed
MP 256 WW	B	191L	1°00'	570	3	66	93	76		6		1/2	
MP 260 WW	A	192R	1°00'	570	3	66	93	87		6		1	
MP 260 WW	A	192R	1°00'	570	3	66	93	90		5		1	
MP 260 WW	B	193L	1°02'	570	3	65	91	83-86		6		3/4	
MP 261 WW	A	194R	1°00'	570	3	66	93	84-89		6		3/4	
MP 261 WW	A	194R	1°00'	570	3	66	93	63-64	2		1/2		Speed greater than equil. speed
MP 261 WW	A	195L	1°00'	570	3	66	93	62-83	1		1/4		Speed greater than equil. speed
MP 266 WW	B	195L	1°00'	570	3	66	93	62				1/2	
MP 267 WW	A	197R	1°00'	570	3	66	93	63		3		3/4	
MP 269 WW	A	198L	1°00'	570	3	66	93	78		5		3/4	
MP 270 WW	A	199R	1°30'	680	5-1/2	72	90	78		6		3/4	
MP 282 WW	A	200L	1°30'	680	5-1/2	72	90	84		5		1/2	
MP 283 WW	A	209R	1°29'30"	680	5-1/2	73	90	84		4		3/4	
MP 284 WW	A	210L	1°30'	680	5-1/2	72	90	80		3		1/2	
MP 284 WW	A	211R	1°30'	680	5-1/2	72	90	86		6		1-1/4	
MP 285 WW	A	212L	1°30'	680	5-1/2	72	90	86-91		5		1-1/4	
MP 286 WW	A	213R	1°29'	680	5	70	88	91		6		1-1/2	
MP 287 WW	A	214L	1°30'	680	5	69	87	90		5		1-1/4	
MP 288 WW	A	215R	0°59'30"	570	3	66	93	87		3		3/4	
MP 289 WW	A	216L	1°18'	570	4	67	89	78		2		1/2	
MP 289 WW	A	217R	6°30'	250	1-1/2	66	113	78	2		1/4		Curve limit not located

View of 71-ft 6-in prestressed concrete girder spans in Santa Fe
Bridge 672.1 near Colorado Springs, Colo.

Advance Report of Committee 30—Impact and Bridge Stresses

D. W. Musser, *Chairman*

Field Investigation of Santa Fe Railway Prestressed Concrete Girders

A. DIGEST

This report contains a description and analysis of a span consisting of four 71-ft
6-in post-tensioned, prestressed concrete girders tested under controlled operating
conditions that included a range of train speeds between 5 and 55 mph.

Strains were measured on the top of curbs, bottom of the deck slab and the bottom of the four girders at the center line of span, and horizontal shearing strains were
measured on both sides of one girder. Vertical accelerations were measured at the front
and rear of the first unit of the test locomotive .All measurements were recorded under
a test train consisting of four locomotive units.

From an analysis of the data obtained from this test under test locomotives having
a rating of E 38.7, the following features are significant:

1. Using a modulus of elasticity of 4,630,000 psi, the ratio of recorder static flexural strain to calculated strains ranged from a minimum of 0.59, which
   occurred in girder 2, to a maximum of 0.76, which occurred in girder 4.

2. Using a modulus of rigidity of 1,930,000 psi, the ratio of recorded static horizontal shearing strains to calculated horizontal shearing strains ranged from a
   minimum of 0.85 to a maximum of 1.12 for the east side of the web and a
   minimum of 0.56 to a maximum of 0.70 for the west side of the web, both on
   girder 2, 12 ft 0½ in from the center line of the north bearing.

79

3. Using a modulus of elasticity of 4,630,000 psi, the maximum recorded reduction of compressive prestress on the bottom of the girders at center line of span was 280 psi on girder 4, under the test train operating at maximum recorded speed of 55 mph.

4. Load distribution to the girders, as measured by strains on the bottom of the girders at midspan, was approximately equal.

5. The maximum total impact, as calculated from recorded strains on the bottom of the girder at midspan, was 13.5 percent and occurred on girders 2 and 3 at speeds of 41 and 55 mph, respectively. AREA impact for this span is 34.2 percent using the nominal axle loads on the test locomotives.

6. The maximum recorded horizontal shearing strain was 0.0000536 in per in and occurred at a speed of 37 mph on the east side of girder 2, 12 ft 0½ in from the center line of the north bearing. Using a modulus of rigidity of 1,930,000 psi, the maximum shearing strain was equivalent to 100 psi.

7. The maximum recorded impact, as determined from recorded horizontal shearing strain, was 52.7 percent on the east side of girder 2 at a speed of 37 mph.

8. The maximum recorded static deflection, measured on the bottom of girder 2 at the center line of span, was 0.094 in under a speed of 7 mph. This is 1/8910 of the span.

9. The vertical strain distribution indicated composite action between the curb and slab and the slab and girders.

10. The maximum recorded tie play was 5/16 in. The maximum acceleration of the first unit of the test train was $0.53g$, (g being the gravitational acceleration of 32.2 ft per sec^2), and occurred in the vicinity of a low spot behind the backwall of the north abutment.

B. FOREWORD

The purpose of this investigation was to determine the static and dynamic effects and the vertical accelerations of the test locomotive on the 71-ft 6-in prestressed concrete girders of Santa Fe Railway Bridge 672.1 near Colorado Springs, Colo. This bridge was selected because the girders are the longest in this country carrying railway loading.

The Association of American Railroads under the sponsorship of Committee 30—Impact and Bridge Stresses, conducted this investigation in May 1958 as part of the committee's assignment on Concrete Structures, collaborating with Committee 8—Masonry. This investigation was under the general direction of G. M. Magee, director of engineering research, Engineering Division, AAR, with funds provided by the AAR.

The conduct of the investigation, analysis of data and preparation of the report were under the direction of E. J. Ruble, research engineer structures, assisted in the field by F. P. Drew, assistant research engineer structures, and M. F. Smucker, assistant electrical engineer. The office work was under the direction of F. P. Drew, assisted by D. W. Wilki, chief draftsman and J. S. Nardi, assistant structural engineer. This report was prepared by Mr. Nardi.

C. INSTRUMENTS

The instruments used in this investigation to determine the strains consisted essentially of two 12-channel oscillographs which recorded the strains on photographic paper in response to SR–4 electrical resistance strain gages. A detailed description of the oscillographs and their auxiliary units is given in AREA Proceedings, Vol. 46, 1945, page 201, and a description of the SR–4 strain gage with necessary equipment is in the Proceedings, Vol. 52, 1951, page 152.

Deflection readings were taken under slowly moving locomotives with an Ames dial mounted on a telescoping steel rod at the center line of span at the bottom of girder 2.

Rail profile readings were taken with a surveyor's level and rod.

Vertical accelerations of the test locomotive were recorded by means of Statham unbonded strain gage accelerometers and associated amplifiers which produced a continuous record of accelerations on a 2-channel Brush pen writing oscillograph.

D. TEST SPAN AND LOCATION OF GAGES

The present structure, completed in 1958, is within the boundaries of the Air Force Academy near Colorado Springs. The bridge consists of two 71-ft 6-in ballasted-deck simple spans resting on a center pier and U type abutments 59 ft 7 in and 45 ft 7 in at the north end and south end of the bridge, respectively (see **Fig. 1.**)

The deck slab, curb and sidewalk were cast in place over the four post-tensioned girders spaced 4 ft 0 in on center. The top of each girder is 2 ft 10 in wide, tapered to 3 ft 0 in at 4 in from the top of the girder and to 8 in at 1 ft 0 in from the top of the girder. Each girder is 2 ft 2 in wide at the bottom. The girder webs are thickened near the ends. Diaphragms, the full depth of the girders, are 10 in wide (except 1 ft 0 in at the ends) and are spaced at 8 ft 6 in centers and staggered, as the bridge is on a 32 deg 15 min skew with the roadway below. There were 11 cables used to post tension each girder.

The curbs are 1 ft 0 in wide at the bottom and 9 in wide at the top. The sidewalk slabs projecting out beyond the girders on each side of the bridge are 10 in thick. The deck slab is 6 in thick over the girders and 9 in thick between the girders. Composite action between the slabs and girders was obtained through the use of shear keys and hold-down stirrups and by projecting the top flange of the girder 3 in into the cast-in-place deck slab. Through the use of stirrups composite action was also obtained between the slab and curbs.

The center pier is skewed and supported on six 5 ft diameter cast-in-place caissons. The abutments are also skewed and each supported on ten 2 ft diameter cast-in-place concrete caissons.

A more detailed discussion of the design, description and construction of the bridge was presented by the Portland Cement Association in its report No. 57, "Concrete for Railways", pages 8–14.

The design criteria for the girders were:

    (a) Cooper loading .... E 72
    (b) AREA live load impact factor 50.8%
    (c) Ultimate tensile strength of prestressing steel ....245,000 psi
    (d) Steel tensile stress due to prestressing at transfer ....161,070 psi
    (e) Steel tensile stress due to effective prestress force after
        deduction of all losses (15%) ....136,500 psi
    (f) Cylinder compressive strength at 28 days f'_c .... 5,500 psi?
    (g) Cylinder compressive strength at the time of prestressing .... 4,500 psi
    (h) Permissible compressive stress at the time of prestressing
        (0.45 × 5500 psi) .... 2,475 psi

(i) Final permissible compressive stress (after losses) .... 2,200 psi
(j) Working compressive stress at bottom flange .... 214 psi

The average compressive strength of the concrete used in all the girders as determined from 6- by 12-in cylinders was 6,480 psi at 28 days.

Due to diaphragms and the skew of the bridge, SR–4 strain gages could not be
located at the exact center span, but were placed as near to the center as conditions
permitted (see Fig. 2.) SR–4 strain gages were located on the top of the curbs, bottom
of the slab and at the bottom of each girder.

Two 45-deg SR–4 strain rosettes were placed on girder 2 (see Fig. 2) 4 ft 7¾ in
from the bottom of the girder and 12 ft 0½ in from the center line of the north
bearing. The usual practice is to locate the rosettes at a distance from the center line
of bearing equal to the depth of the beam. However, due to the diaphragms and thick-
ening of the girders at the ends it was necessary to locate the rosettes as shown.

Accelerometers were located on the steel frame of the first unit of the test locomotive at the front and rear, 5 ft 1 in above the top of rail and 2 ft 0 in from the longitudinal center line. The front and rear accelerometers were 1 ft 6 in and 2 ft 0 in
from the center line of the respective couplers.

E. TEST TRAINS

A test train operating under controlled conditions and consisting of four diesel,
4-axle 1500-hp locomotive units was used to secure data for this investigation. The gross
weight of one unit was 244,000 lb and for each of the other three units the gross weight
was 250,000 lb.

The distance between wheels was 9 ft 0 in on each truck and from center to center
of trucks, 30 ft 0 in. From center to center of couplers the distance was 50 ft 0 in.
The weights and spacings were provided by the Santa Fe Railway Company.

The rating of the test train for moment at the center line of the test span was
E 38.7.

F. STATIC AND DYNAMIC EFFECTS

The data recorded on the oscillograms were analyzed to obtain simultaneous maximum recorded tensile strains and impacts per rail and simultaneous maximum recorded
horizontal shearing strains and impacts. The Brush oscillograms were analyzed to obtain
maximum accelerations of the test locomotive.

In the following discussion the flexural stresses reported were calculated by multiplying the recorded flexural strains by an assumed modulus of elasticity of concrete
of 4,630,000 psi. This was the value recommended by the designers of the structure
as a result of their deflection recordings obtained under static loading of a similar highway bridge girder. This compares with 5,040,000 psi based on the AREA equation of
$1,800,000 + 500 f'_c$, using $f'_c = 6,480$ psi.

Horizontal shearing stresses were obtained by multiplying recorded horizontal shearing strains by the modulus of rigidity of concrete. The modulus of rigidity was calculated
from the equation:

$$G = \frac{E}{2(1+u)} = \frac{4,630,000}{2(1+0.20)} = 1,930,000 \text{ psi}$$

From information available Poisson's ratio, u, for concrete can vary from 0.15 to
0.25, therefore an average value of 0.20 was used in calculating the modulus of rigidity
of concrete.

1. Comparison of Calculated and Simultaneous Maximum Recorded Static Flexural and Shearing Strains

A comparison of simultaneous maximum recorded static and calculated flexural tensile strains is shown in Table 1.

The maximum calculated flexural tensile strain of the test locomotives on the test span was 0.0000769 in per in, assuming that the live load was distributed equally to the four girders and using the moment of inertia based on one girder and the corresponding portion of the slab. The ratio of recorded to calculated strain ranged from 0.59 for girder 2 to 0.76 for girder 4.

Part of the difference in ratios between the girders can be attributed to the skew of the bridge combined with a redistribution of axle loads associated with the low spot in the track discussed in Art. 7, below.

The other maximum calculated flexural tensile strains of the test locomotive on the test span was 0.0000785 in per in based on the assumption that the girders supported equal loads and the moment of inertia based on two girders, slab, curb and sidewalk. Using this calculated value had little effect on the recorded over calculated ratios. The ratios were reduced 0.01 or 0.02 and followed the same pattern as mentioned above. The ratio of recorded to calculated strains ranged from 0.58 for girder 2 to 0.74 for girder 4. The magnitude of the ratios for the individual girders indicated that the diaphragms between girders were effective in equalizing the load distribution to each girder.

Table 2 shows the comparison of two sets of calculated (VQ/Ib) horizontal shearing strains to the recorded maximum simultaneous horizontal shearing strains determined from the two rosettes located on the east and west sides of girder 2, 12 ft 0½ in from the center line of the north bearing.[1] The two sets of calculated values were based on the assumption that the girders supported equal loads, and full composite action was attained between the girder and deck slab.

One set of calculated values was based on the positions of the test locomotives on the test span that produced maximum simultaneous horizontal shearing strains in girder 2 for each run. The ratio of recorded to calculated strains ranged from 0.85 to 1.12 for the east rosette and 0.56 to 0.70 for the west rosette.

The other calculated value was based on the position of the test train on the test span that produced the maximum shear at the location of the rosettes. The ratio of recorded to calculated strains ranged from 0.73 to 0.95 for the east rosette and 0.50 to 0.58 for the west rosette.

2. Average Simultaneous Maximum Recorded Strains in Girders

Fig. 3 shows the average simultaneous maximum strains recorded in the bottom of the girders under the test locomotives at a speed range of 5 to 55 mph. It can be seen from Fig. 3 that the load distribution to the two girders under each rail was approximately equal.

The average simultaneous maximum recorded stresses in girders 1, 2, 3 and 4 were 263, 252, 277 and 281 psi, which occurred at speeds of 43, 41, 55 and 55 mph, respectively. Although the gages at the bottom of the girders in the field recorded tension under load, the recorded strains were a measurement under live load of the relief of compressive strains due to the prestressing force in the bottom of the girders. To know

---

[1] Horizontal shearing strains were determined as described in the Engineering Research Division's Report ER-1, InVestigation of 60-Ft Glued Laminated Beams on the Weyerhaeuser Timber Company Railroad, January 1961, page 7.

the actual state of strain under live load at the bottom of the girders, it would have been necessary to obtain strain readings as the prestressing force was being applied, and to keep a record of all losses.

3. Total Impacts Per Girder and Per Rail Calculated from Average Simultaneous Maximum Recorded Tensile Strains

The maximum total impacts for the bottom of each girder are shown in Fig. 4. The total impacts plotted include roll and other vertical effects, and no attempt was made to separate these various effects. The total impact percentage in each test run for a particular speed is the increase in strain in the girder over the average of the strains recorded for the test locomotives at speeds less than 10 mph.

The maximum total impacts for girders 1, 2, 3 and 4 were 12.7, 13.5, 13.5 and 10.0 percent, which were recorded at speeds of 43, 41, 55 and 55 mph, respectively. The maximum speed obtained during this investigation was 55 mph.

The maximum total impacts per rail, shown on Fig. 5, follow the same trend as the maximum impacts for each girder. The maximum total impacts for the east and west rail were 10.5 and 11.1 percent occurring at 43 and 45 mph, respectively.

Using the present AREA impact equation, $L/(L + D)$, and the nominal axle loads of the test locomotives, the calculated impact is 34.2 percent.

4. Simultaneous Maximum Recorded Horizontal Shearing Strains and Impacts

Fig. 6 shows that the horizontal shearing strain on the east side of girder 2, rosette B, was higher than the west side of girder 2, rosette A. This difference was due, in all probability, either to the skew of the bridge or unequal bearing.

The simultaneous maximum recorded horizontal shearing stresses for rosettes A and B were 53 and 103 psi, which occurred at speeds of 30 and 37 mph, respectively.

The total impacts were calculated in the same manner as explained in Art. 2. The impacts in rosette A increased when the speed increased from 14 to 30 mph and then decreased from 30 to 55 mph. Except for two impact plots in rosette B, there was a slight increase in impact with an increase in speed. The maximum recorded impacts for rosettes A and B were 31.9 and 52.7 percent occurring at speeds of 30 and 37 mph, respectively.

5. Deflection

The deflections were recorded as mentioned in Sec. C. The maximum recorded deflections of static runs, speeds less than 10 mph, for the center line of the test span at the bottom of girder 2 are shown in the following table:

DEFLECTIONS

Speed MPH	Deflection Inches	Speed MPH	Deflection Inches
5R	0.088	7R	0.094
5F	0.087	7F	0.087
5R	0.086	8F	0.090
6R	0.089	8R	0.089

The letters F and R after the speed designate forward in the northbound direction and reverse in the southbound direction, respectively.

As can be seen in the preceding table, the maximum deflection was 0.094 in at 7 mph. This recorded deflection is 1/8910 of the span.

6. Vertical Strain Distribution

SR–4 strain gages were placed at the bottom of the girder, bottom of the slab and top of the curb as shown on Fig. 2, section AA, and with the average recorded flexural strains obtained from these gages vertical strain distributions were plotted for the east and west sides of the bridge during static and dynamic test runs (see Fig. 7.)

The static and dynamic test runs were at 8 and 54 mph. The calculated strains and center of gravity were based on the moment of inertia of half the bridge, which includes two girders, slab, sidewalk and curb, and the assumption that the modulus of elasticity was 4,630,000 psi.

The significant results obtained from the vertical strain distribution diagram were that there was composite action between the slab and girders and between the curbs and slab.

7. Recorded Vertical Tie Displacement

During the field investigation, it was noticed that the ties at the north end of the bridge under the east rail were pumping. The amount of tie play under the east rail was measured with a surveyor's level and rod. The play between the tie and rail was secured by forcing the tie down with a lever under the rail and recording its position. The tie was then forced up against the rail with a lever under the end of the tie and its position again recorded. The difference between the positions is tie play. The results are shown on the top diagram of Fig. 8.

It can be seen that the tie play was in the vicinity of 30 to 75 ft south of the north end of the bridge which includes part of the north test span and north abutment. The maximum recorded tie play was 0.025 ft, approximately 5/16 in, and occurred behind the backwall of the north abutment.

8. Maximum Acceleration of the Test Locomotive

The lower diagrams of Fig. 8 show the maximum accelerations of the front and rear of the first unit of the test locomotive for northbound and southbound (reverse) runs. The plotting of the accelerations of the front of the locomotive are relative to the center line of the front truck with respect to the bridge, and the plotting of the rear accelerations are relative to the center line of the rear truck with respect to the bridge. Individual maximum accelerations were read for each run with the test train off the south end of the bridge, on the south abutment, on the north abutment and on either the north test span or south span, depending on where the maximum occurred. Accelerations for southbound (reverse) runs were not reported for the north test span or south span because of the influence of the low spot on the north abutment.

The accelerations for the test train off the south end of the south abutment and on the south abutment appeared to be within the same range. The maximum accelerations for the front axles of the test train on the south span or the north test span were lower than the maximum accelerations of the test train off the south end of the south abutment or on the south abutment. Maximum accelerations for all accelerations that were plotted occurred when the test locomotive was in the vicinity of the low spot mentioned in Art. 7, which would place the test train on the north abutment and the north test span. The plotting indicates that the low spot had a bearing on the degree of acceleration of the test locomotive.

Fig. 9 plots accelerations against speed for the front and rear of the first unit of the test locomotive during northbound and southbound (reverse) runs off the bridge and for northbound runs on the bridge. On the bridge accelerations occurred when the test train was on the south span or north test span. Off the bridge accelerations occurred when the test train was on the north or south abutment or off the south end of the south abutment. Recording accelerometers were calibrated in terms of gravitational acceleration, g, equivalent to 32.2 ft per sec².

Both diagrams show that there was a noticeable increase in accelerations with an increase in speed, and that the recorded accelerations of the test locomotive off the bridge were higher than when on the bridge. The diagrams also show that the majority of the accelerations of the front of the locomotive were higher than the rear of the locomotive.

The following table summarizes maximum accelerations and where they occurred:

MAXIMUM ACCELERATIONS IN G

		Maximum Acceleration	Frequency CPS	Speed MPH	Position for Maximum
Northbound	Front Axles	0.533	2.0	54	North Abutment
	Rear Axles	0.327	2.0	40	North Abutment and Off the Bridge
Southbound	Front Axles	0.289	2.3	47	Off the Bridge
	Rear Axles	0.197	2.2	36	Off the Bridge

The preceding table shows that all the maximum accelerations occurred when the first unit of the test locomotive was on the north abutment or off the bridge. The maximum acceleration for the test locomotive occurred during a northbound run when the test locomotive was in the vicinity of the low spot mentioned in Art. 7.

G. CONCLUSIONS

The static and dynamic effects and the vertical accelerations of diesel locomotive loadings on a post-tensioned prestressed concrete girder bridge were analyzed, and from the test data it can be concluded that:

1. Recorded static flexural and horizontal shearing strains were less than the calculated values.
2. There was little increase in recorded strains with increase in speed even though the trains operated over a speed range of 5 to 55 mph.
3. The maximum recorded impact in flexure was 40 percent of that indicated by current AREA specifications. All recorded impacts in horizontal shear were also less than specified except for one value which was 56 percent greater.
4. Composite action was attained between the curb and slab and between the slab and girders.
5. The live load was distributed approximately equal to the girders.
6. Accelerations of the first unit of the test locomotive increased with speed, and higher accelerations were recorded off the bridge than on it.

TABLE 1 - COMPARISON OF CALCULATED AND AVERAGE MAXIMUM RECORDED SIMULTANEOUS
STATIC TENSILE FLEXURAL STRAINS AT CENTER LINE OF SPAN
(All Strains in Inches/Inch)

Recorded Static Flexural Strains Girders				Calculated Maximum Strain (a)	Recorded / Calculated Girders (a)				Calculated Maximum Strain (b)	Recorded / Calculated Girders (b)			
1	2	3	4		1	2	3	4		1	2	3	4
.0000518	.0000473	.0000519	.0000548		.67	.62	.67	.71		.66	.60	.66	.70
.0000526	.0000495	.0000540	.0000569		.68	.64	.70	.74		.67	.63	.69	.72
.0000506	.0000460	.0000530	.0000539		.66	.62	.69	.70		.64	.61	.68	.69
.0000499	.0000484	.0000533	.0000557		.65	.63	.69	.72		.64	.62	.68	.71
.0000486	.0000470	.0000511	.0000562	.0000769	.63	.61	.66	.73	.0000785	.62	.60	.65	.72
.0000505	.0000470	.0000525	.0000545		.66	.61	.68	.71		.64	.60	.67	.72
.0000498	.0000488	.0000513	.0000535		.65	.63	.67	.70		.63	.62	.65	.68
.0000532	.0000505	.0000559	.0000582		.69	.66	.73	.76		.68	.64	.71	.74
.0000515	.0000492	.0000543	.0000561		.67	.64	.71	.73		.66	.63	.69	.71
.0000477	.0000454	.0000517	.0000519		.62	.59	.67	.67		.61	.58	.66	.66
.0000496	.0000468	.0000518	.0000542		.64	.61	.67	.70		.63	.60	.66	.69
			Average		.66	.62	.69	.72	Average	.64	.61	.67	.70

(a) Based on one girder and corresponding portion of slab acting as a composite unit.

(b) Based on two girders, slab, curb and sidewalk acting as a composite unit.

TABLE 2 – COMPARISON OF CALCULATED AND SIMULTANEOUS MAXIMUM RECORDED HORIZONTAL SHEARING STRAINS ON GIRDER 2
(All Strains in Inches/Inch)

Recorded Horizontal Shearing Strains		Calculated Max. Hor. Shearing Strain (a)	Recorded / Calculated Rosettes (a)			Calculated Max. Hor. Shearing Strain (b)	Recorded / Calculated Rosettes (b)		
Rosettes									
East	West		East	West	Average		East	West	Average
.000035	.000023	.000033	1.06	.70	.88		.88	.58	.73
.000035	.000021	.000033	1.06	.64	.85		.88	.53	.71
.000038	.000022	.000034	1.12	.65	.89		.95	.55	.75
.000034	.000020	.000034	1.00	.59	.80	.000040	.85	.50	.68
.000029	.000020	.000034	.85	.59	.72		.73	.50	.62
.000035	.000020	.000035	1.00	.57	.79		.88	.50	.69
.000033	.000023	.000034	.97	.68	.83		.83	.58	.71
.000038	.000020	.000036	1.06	.56	.81		.95	.50	.73
.000038	.000020	.000034	1.12	.59	.86		.95	.50	.73
.000035	.000020	.000033	1.06	.61	.84		.88	.50	.69

(a) Based on position of test train on the test span that produced simultaneous maximum recorded horizontal shearing strains.

(b) Based on position of test train on the test span that produced the maximum calculated horizontal shearing strain at the rosettes.

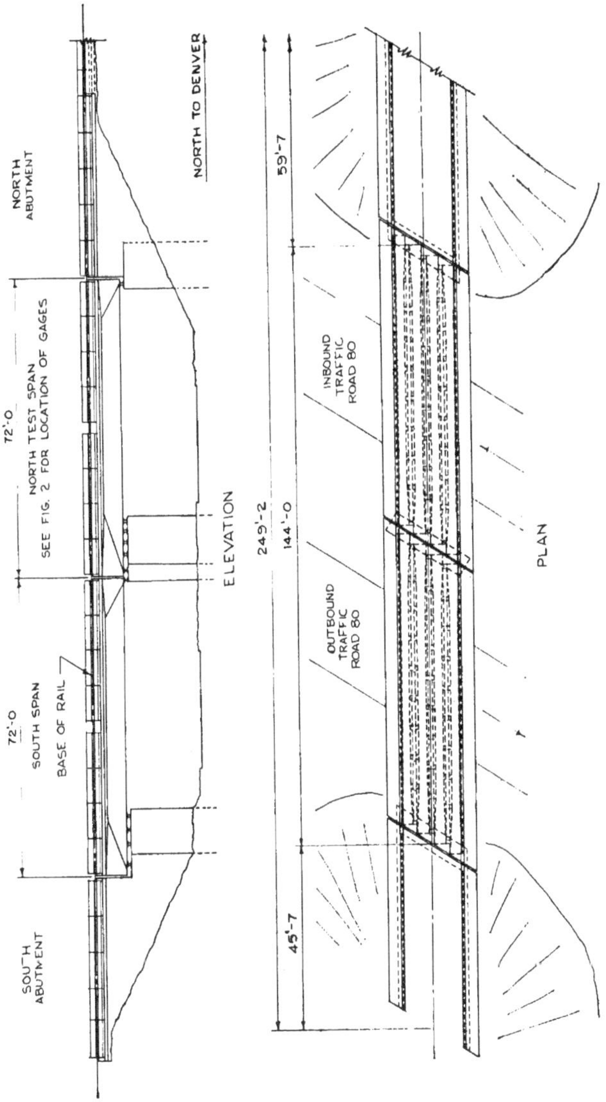

FIG. 1
A.T. & S.F. RY. BRIDGE TEST
LOCATION OF TEST SPAN
BRIDGE 672.1, COLORADO SPRINGS, COLORADO

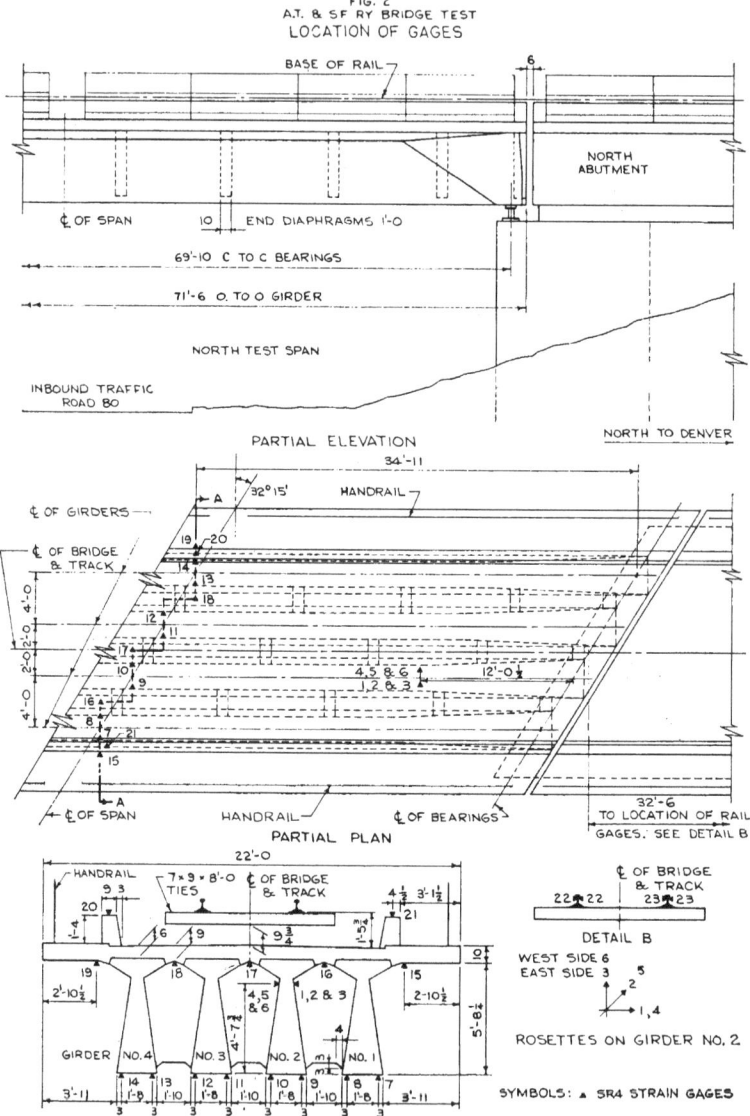

FIG. 2
A.T. & SF RY BRIDGE TEST
LOCATION OF GAGES

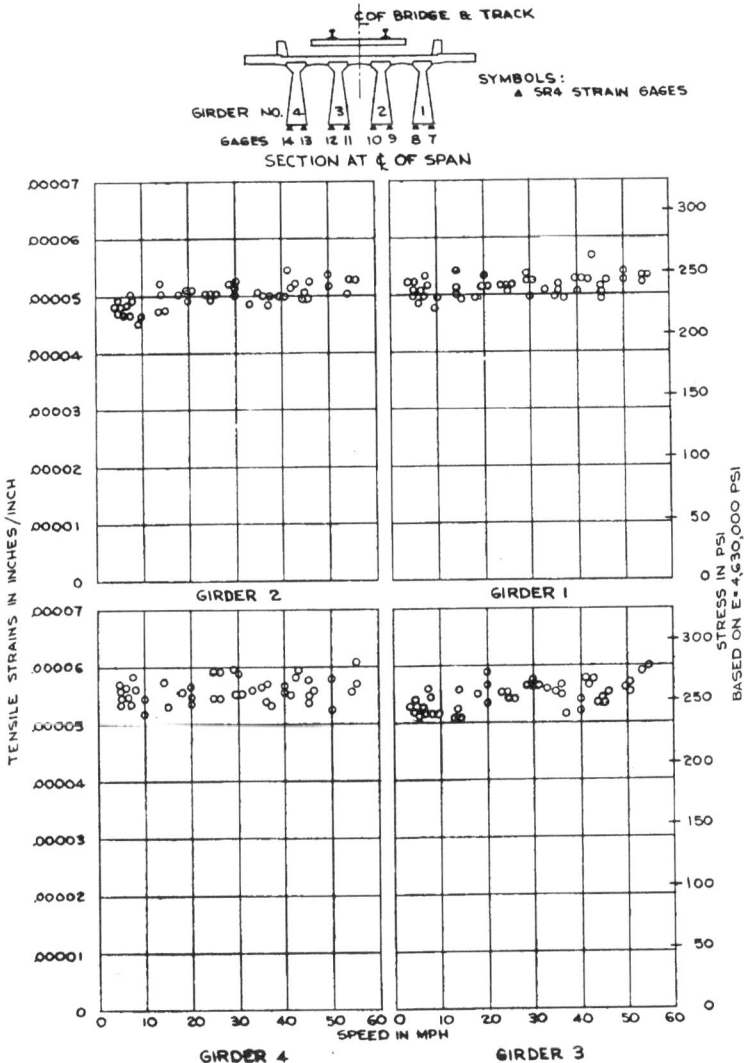

FIG. 3
A.T. & S.F. RY. BRIDGE TEST
SIMULTANEOUS MAXIMUM STRAINS IN GIRDERS

FIG. 4
A.T. & S.F. RY. BRIDGE TEST
TOTAL IMPACTS RECORDED IN CONCRETE PER GIRDER

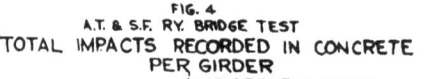

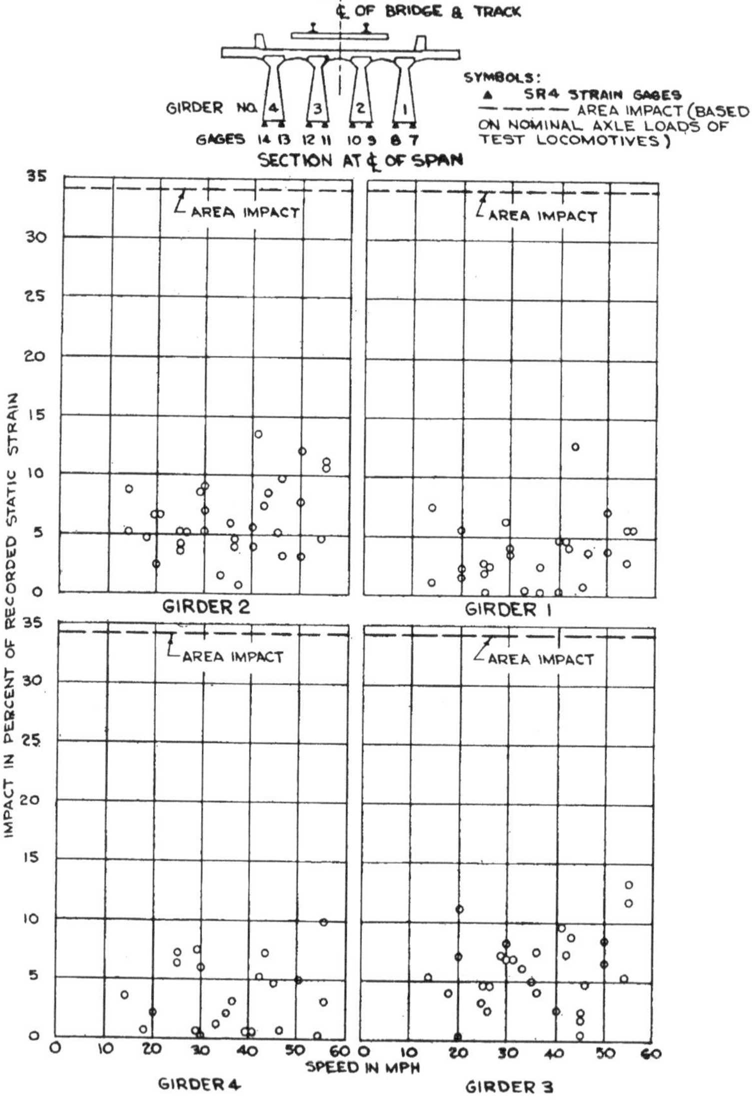

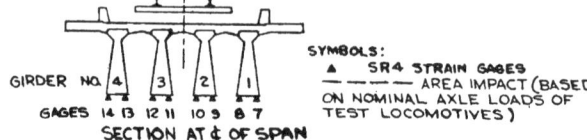

℄ OF BRIDGE & TRACK
GIRDER NO. 4 3 2 1
GAGES 14 13 12 11 10 9 8 7
SECTION AT ℄ OF SPAN
SYMBOLS:
▲ SR4 STRAIN GAGES
— — — — AREA IMPACT (BASED ON NOMINAL AXLE LOADS OF TEST LOCOMOTIVES)
AREA IMPACT
GIRDER 2
GIRDER 1
GIRDER 4
GIRDER 3
IMPACT IN PERCENT OF RECORDED STATIC STRAIN
SPEED IN MPH
0
5
10
15
20
25
30
35
40
50
60

FIG. 5
A.T. & S.F. RY BRIDGE TEST
RECORDED TOTAL IMPACTS PER RAIL

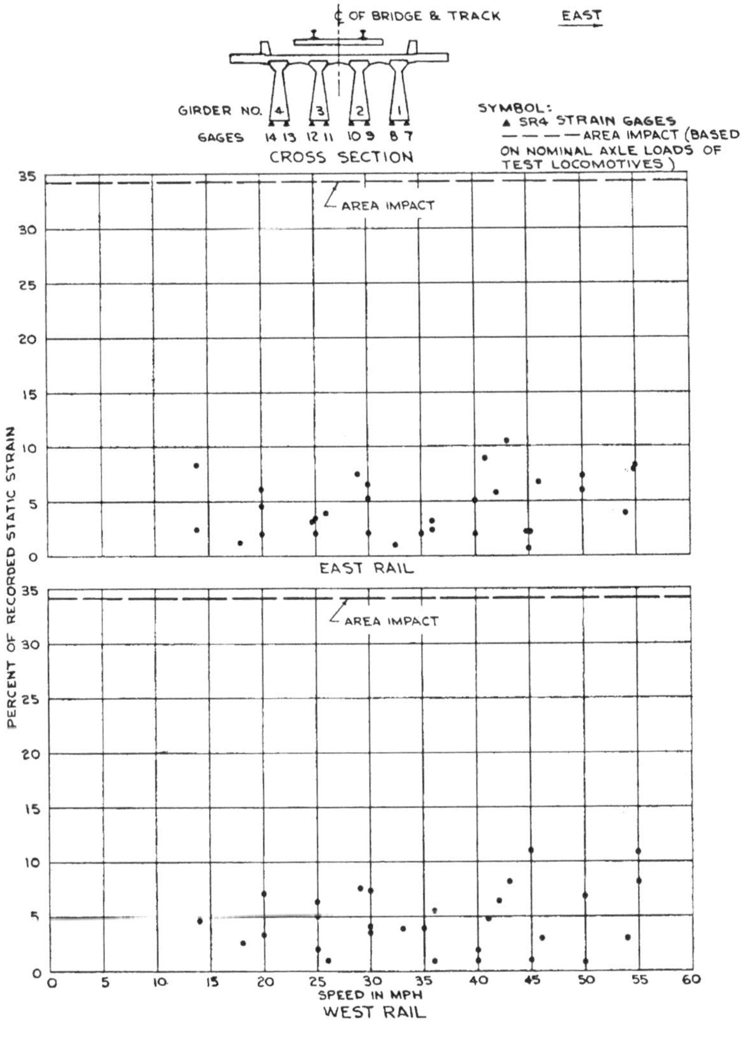

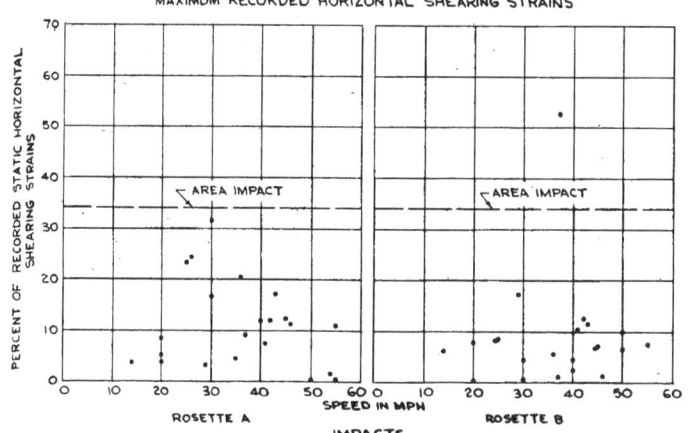

FIG. 6
A.T. & S.F. RY BRIDGE TEST
MAXIMUM RECORDED HORIZONTAL SHEARING STRAINS
AND IMPACTS

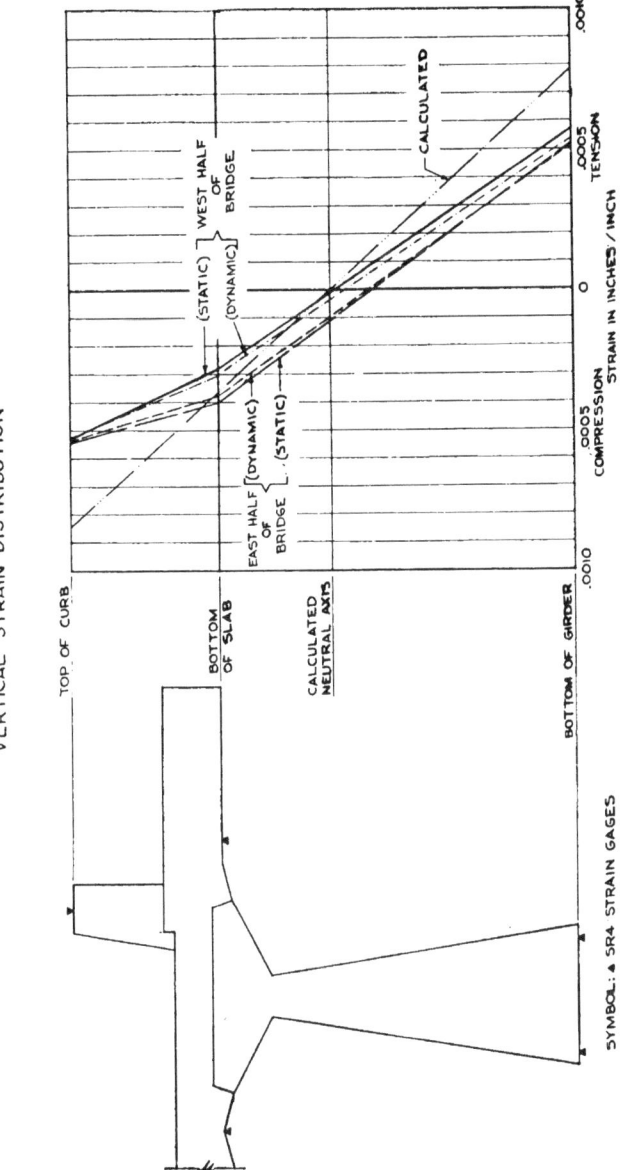

FIG. 7
A.T. & S.F. RY. BRIDGE TEST
VERTICAL STRAIN DISTRIBUTION

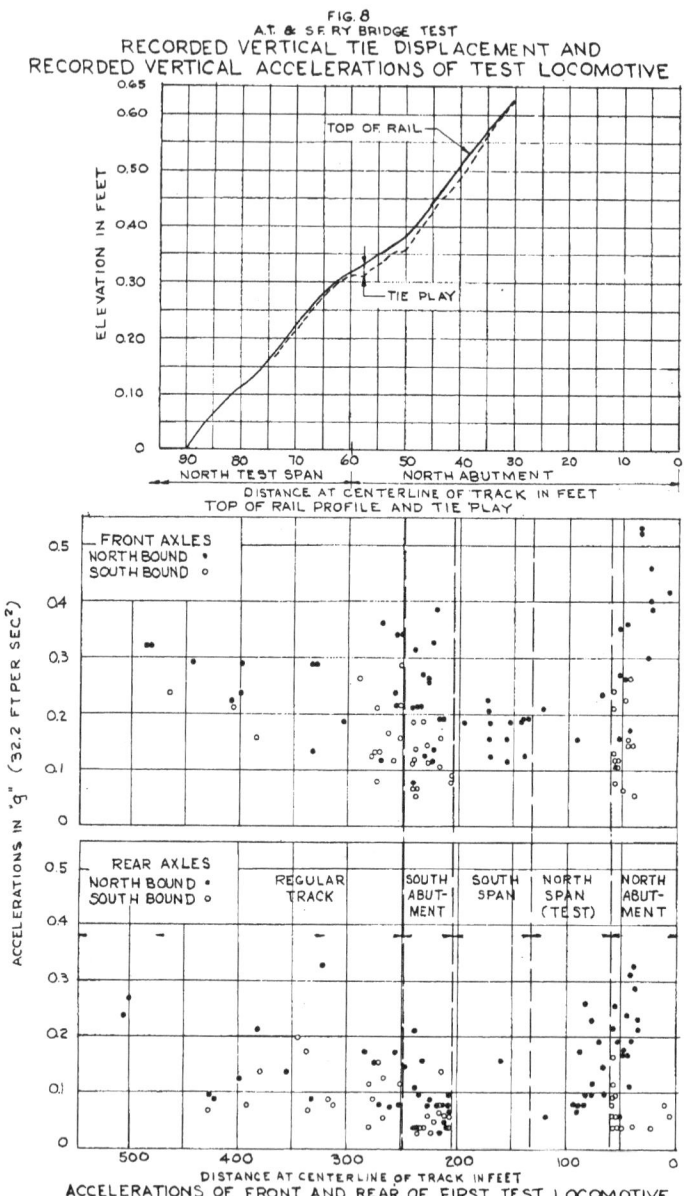

FIG. 8
A.T. & S.F. RY BRIDGE TEST
RECORDED VERTICAL TIE DISPLACEMENT AND
RECORDED VERTICAL ACCELERATIONS OF TEST LOCOMOTIVE

ACCELERATIONS OF FRONT AND REAR OF FIRST TEST LOCOMOTIVE

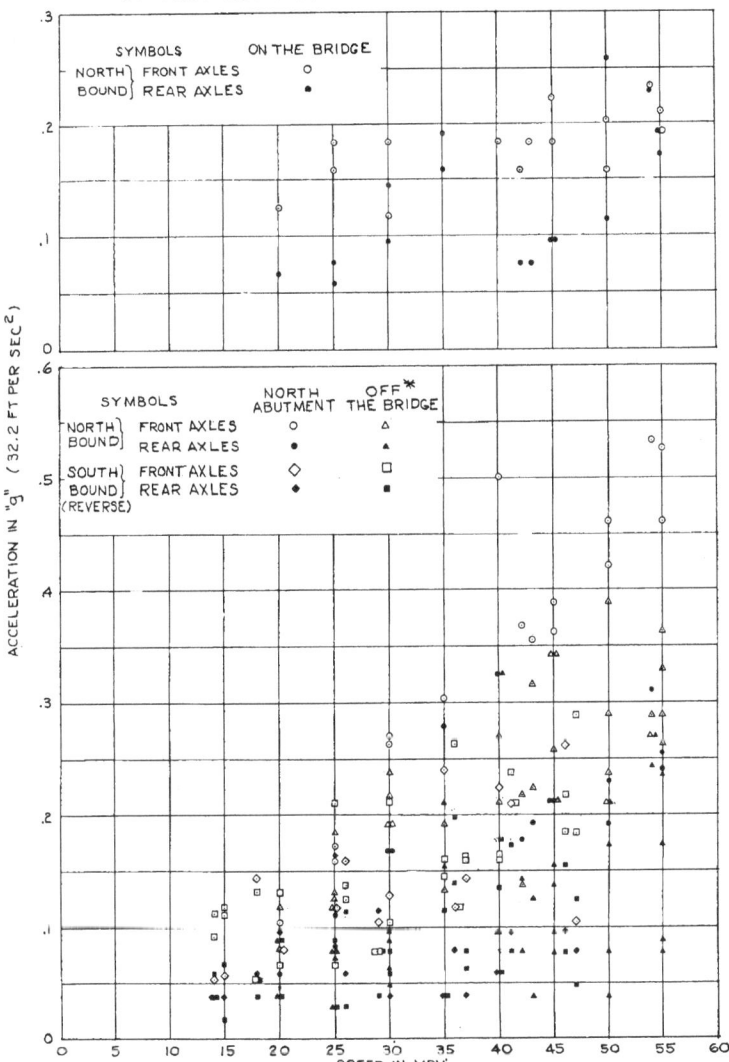

FIG. 9
AT & S.F. RY. BRIDGE TEST
MAXIMUM RECORDED VERTICAL ACCELERATIONS
OF TEST LOCOMOTIVE ON AND OFF THE BRIDGE

* REGULAR TRACK SOUTH OF SOUTH ABUTMENT AND SOUTH ABUTMENT
ACCELERATIONS.

Model N U Tie Cutter

HERE IS THE WINNING TEAM

The Woolery NU Tie Cutter and the Woolery Tie-end Remover preserve the line and surface of the track and at the same time reduce the cost of tie renewals. Ties can be removed without trenching, jacking up track or adzing tops of rail-cut ties. With this team you simply cut both ends of tie, pry out center piece, insert in its place the tie-end remover and out go the tie ends pushed by the double acting, double ended hydraulic cylinder of the Tie-end remover.

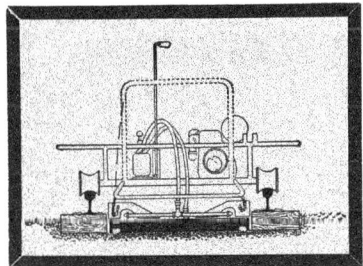 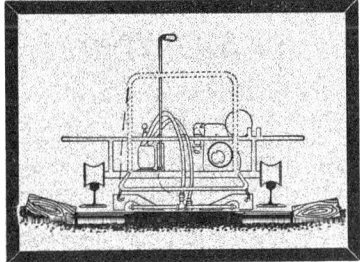

FOR HIGHEST EFFICIENCY USE TWO TIE CUTTERS WITH ONE TIE-END REMOVER

WOOLERY MACHINE COMPANY
MINNEAPOLIS, MINN.

MASSIVE STRENGTH

CHANNELOC rail anchors pack a lot of metal—*working* metal.

This massive, one-piece anchor is forged from a solid bar of tempered steel. Its U-channel provides rugged double-flange support for greatest holding power. It has thorough and even heat treating. Plenty of reserve gripping power for reapplication. And you can install the CHANNELOC with sledge, maul or machine.

True Temper will be glad to help you see that anchors are properly applied. Contact True Temper, Railway Appliance Division, 1623 Euclid Avenue, Cleveland 15, Ohio.

TRUE TEMPER®
CHANNELOC RAIL ANCHOR

PROGRESS REPORT

Here are the up-to-date facts on the SPENO Ballast Cleaning and the SPENO Rail Grinding Services.

BALLAST CLEANING

SPENO Engineering and Research has developed a superior screening arrangement so that we are now using an improved Ballast Cleaner with greater efficiency.

RAIL GRINDING

Our Rail Grinding Service has been so well received we are now building a *THIRD* Rail Grinding Train to take care of the increased demand

SPENO is constantly developing means for better service to make sure that the Railroads receive everything they pay for — and more

Just Ask the Railroads That have used us!

Machine knocks off, ejects ties, lines track behind plow while above subgrade.

Mannix AUTO-TRACK
Eliminates 15 to 20 Men

PLOWING, TIE REPLACEMENT, TRACK ALINEMENT are accomplished faster and with fewer men using the new Mannix AUTO-TRACK unit.

Hydraulic hammer on each side knocks tie down. Conveyor belt ejects tie to either side. ▶

Lining head controlled by levers near rear end where operator observes alinement. One man replaces lining crew.

◀

NOW, lease or purchase MANNIX Auto-Track Equipment to operate with railroad crew for greatest convenience and maximum savings. Write for details. Arrange showing of operating films with no obligation.

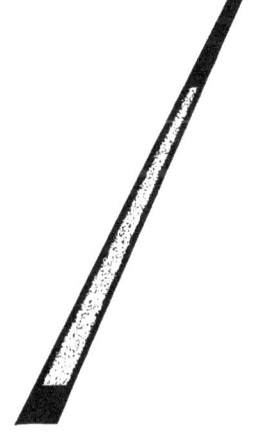

VEGETATION CONTROL
WITH
CHEMICALS

READE MANUFACTURING COMPANY, INC.

Jersey City—Chicago—Minneapolis—Kansas
City—Birmingham—Stockton

SERVING RAILROADS OF AMERICA FOR
MORE THAN FORTY YEARS

W
E
E
D

A
N
D

B
R
U
S
H

C
O
N
T
R
O
L

for effective
weed control...

- Concentrated BORASCU®
- POLYBOR-CHLORATE®
- UREABOR®
- MONOBOR-CHLORATE®

These borate weed killers are proving best for roads in every way...*efficiency, safety, economy, convenience, easy application.*

Today's use of borates for maximum control of vegetation began years ago with our pioneer work in the field. Continued research has developed the group of herbicides, listed above, which most roads now favor for every phase of weed control. These four weed killers are nonselective. They are widely used for year-round maintenance of weed-free conditions about trestles, tie piles, yards, signals, switches, and rights of way. Find out how you, too, can do a better job on weeds...write today.

AGRICULTURAL SALES DEPARTMENT

U.S.BORAX ®

630 SHATTO PLACE · LOS ANGELES 5, CALIFORNIA

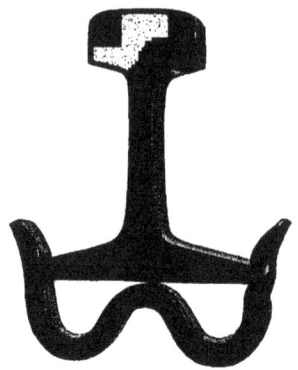

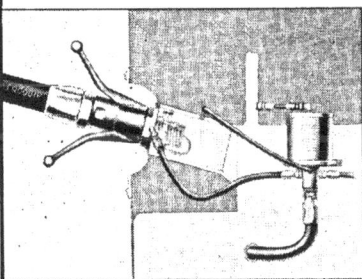

AREA Publications—Price List

The following include some of the Association publications available from the secretary's office on order. Prices shown are for Members only:

	Member Price
Manual of Recommended Practice, complete in 2 volumes, including binders (first copy)	$15.00
Extra binders, each	4.50
Annual Supplements, each	1.00

Separate Chapters

1—Roadway and Ballast°	$ 1.50
3—Ties°	0.25
4—Rail°	0.75
5—Track°	0.75
6—Buildings	1.50
7—Wood Bridges and Trestles	1.00
8—Masonry	1.00
9—Highways	0.50
11—Engineering and Valuation Records	1.25
13—Water, Oil and Sanitation Services	1.00
14—Yards and Terminals	1.00
15—Iron and Steel Structures	1.25
16—Economics of Railway Location and Operation	0.75
17—Wood Preservation°	0.50
20—Contract Forms	1.25
22—Economics of Railway Labor	0.50
25—Waterways and Harbors	0.25
27—Maintenance of Way Work Equipment	0.50
28—Clearances	0.25
29—Waterproofing	0.25
Flexible-cover, loose-leaf binder for separate chapters, each	0.40
Portfolio of Trackwork Plans—119 plans, 8 sheets of specifications, 5 sheets definitions of terms, complete with leatherette cover	$12.50
Track Scale Pamphlet—109 pages, flexible cover	0.80
Federal Valuation of Railroads—87 pages, flexible cover	1.00
Instructions for Mixing and Placing Concrete—24 pages, flexible cover	0.40
Notes on Railroad Location and Construction Procedures from the School of Experience—43 pages, flexible cover	0.50
Handbook of Instructions for the Care and Operation of Maintenance of Way Equipment—149 pages, hard cover	0.85
Instructions for Care and Safe Operation of Welding and Grinding Equipment—23 pages, flexible cover	0.30
Specifications for Steel Railway Bridges (fixed spans)—70 pages, flexible cover	0.75
Specifications for Movable Railway Bridges—73 pages, punched sheets	1.00

° Limited Number of Copies Available.

the JACKSON TRACK MAINTAINER

offers this unbeatable combination

unmatched versatility: The JACKSON MAINTAINER is the ONLY machine that will give you perfect, long lasting tamping in all production work regardless of height of track lift in any kind and size of ballast. It is, also, by far the best tamper for every kind of smoothing and spotting job where top quality is demanded. **top speed:** True maximum footage rates of highest quality tamping is the day in and day out performance of the MAINTAINER. **true economy:** The exceptionally high quality of tamping achieved with the MAINTAINER is longer-lasting . . . and hence, considerably more economical in the long run. Machine maintenance and repair are the lowest by comparison. **outstanding reliability:** More than any machine in its class, the MAINTAINER is ready to go and stay on the job with only ordinary routine care and minor adjustment to job conditions. **unequalled service:** JACKSON backs the MAINTAINER with field engineers who have the experience and know-how to capably and wholeheartedly assist you with all phases of use, care, operation, application and methods for getting everything possible from your JACKSON TRACK MAINTAINER.

JACKSON
VIBRATORS, INC.
LUDINGTON, MICHIGAN

Write, wire or phone for complete information and details of our attractive acquirement plan.

RI-6A

THE TRASCO
AUTONOMIC CAR RETARDER

CLAMPS IN PLACE
ANYWHERE IN TRACK

SIMPLE — EFFECTIVE — INEXPENSIVE

TRACK SPECIALTIES CO.
GENERAL MOTORS BLDG.
NEW YORK 19, N. Y.

REPORTS OF COMMITTEES

The reports in this issue of the Bulletin will be presented to the 1962 convention of the Association at the Conrad Hilton Hotel, Chicago, March 9–10. Comments and discussion with respect to any of the reports are solicited, and should be addressed to the chairman of the committee involved, in writing in advance of the convention, or from the floor of the convention.

Published by the American Railway Engineering Association, Monthly, January, February, March, November and December; Bi-Monthly, June–July, and September–October, at 2211 Fordem Avenue, Madison, Wis.; Editorial and Executive Offices, 59 Van Buren Street, Chicago 5, Ill. Second class postage paid at Madison, Wis. Accepted for mailing at special rate of postage for in Section 1103, Act of October 3, 1917, authorized on June 29, 1918. Subscription $10 per annum.

Report of Committee 13—Water, Oil and Sanitation Services

D. C. TEAL, *Chairman*
E. C. HARRIS,
Vice Chairman
T. L. HENDRIX, *Secretary*

J. J. DWYER
C. F. MUELDER
W. F. ARKSEY
R. A. BARDWELL
T. A. TENNYSON, JR.
V. C. BARTH
R. S. GLYNN
R. O. BARDWELL
R. C. ARCHAMBEAULT
R. C. BARDWELL (E)
J. M. BATES
M. V. BIAR
W. E. BILLINGSLEY
M. R. BOST
I. C. BROWN
T. W. BROWN
P. J. CALZA
C. E. DEGEER
D. E. DRAKE
A. E. DULIK
C. E. FISHER
H. E. GRAHAM

T. I. GRAY
W. C. HARSH
G. E. HARTSOE
H. M. HOFFMEISTER
A. W. JOHNSON
F. O. KLEMSTINE
H. L. MCMULLIN (E)
G. F. METZDORF
E. T. MYERS
C. W. OWENS
A. B. PIERCE (E)
R. D. POWRIE
J. C. ROBERTS
J. P. RODGER
E. R. SCHLAF
H. E. SILCOX (E)
R. M. STIMMEL
A. G. TOMPKINS
H. W. VAN HOVENBERG (E)
C. B. VOITELLE
E. M. WALTERS
J. E. WIGGINS, JR.
Committee

(E) Member Emeritus.
Members listed in bold face are the official representatives of the Engineering Division, AAR.

To the American Railway Engineering Association:

Your committee reports on the following subjects:

1. Revision of Manual.

   Progress report, with recommendations submitted for adoption .... page 100

2. Prevention of corrosion in hot and cold water systems.

   Report on corrosion prevention in potable hot water systems submitted as information with the understanding that the material will be reviewed and condensed into appropriate form for submission as Manual material next year .... page 115

3. Design, construction and operation of coach servicing facilities to comply with regulation of U.S. Public Health Service.

   Progress report, presented as information .... page 119

4. Cathodic protection of pipe lines and steel storage tanks, collaborating with Committee 18.

   Progress has been made in study of the second phase of this assignment— Cathodic Protection for Steel Storage Tanks—but not enough information has been developed to justify a formal report.

5. Economic justification of and methods for conditioning water for use in steam generators and engine cooling systems of diesel locomotives.

   Final report, submitted as information .... page 122

99

6. Railway waste disposal.

Progress report, submitted as information .... page 132

8. Methods of controlling spillage of fuel oil at diesel fueling and unloading stations.

A survey is being made of general practices and types of equipment used for unloading diesel fuel oil from tank cars. Progress in study, no report.

9. Disinfectants, deodorants, fumigants and cleaning materials.

This subject was continued from last year for the purpose of condensing previously published informative reports into Manual material. Disagreement as to the nature of the material that should go into the Manual has delayed progress. No report.

10. Railroad aspects of radioactive substances.

Lack of personnel on the committee with a working knowledge of this subject has precluded progress. No report.

THE COMMITTEE ON WATER, OIL AND SANITATION SERVICES,

D. C. TEAL, *Chairman.*

AREA Bulletin 567, November 1961.

Report on Assignment 1

Revision of Manual

E. C. Harris (chairman, subcommittee), R. A. Bardwell, C. E. DeGeer, J. J. Dwyer, H. E. Graham, T. L. Hendrix, A. W. Johnson, F. O. Klemstine, T. A. Tennyson, Jr., C. B. Voitelle.

Your committee has reviewed Chapter 13 of the Manual, with special attention given to Parts 4, 5, 6, and 7, and recommends that the following action be taken with respect to each document.

Pages 13–1–1 and 13–1–2

SUPPLY

Reapprove without change.

Pages 13–2–1 to 13–2–15, incl.

PUMPING PLANTS

Reapprove with the following revisions:

On page 13–2–1 under Sec. A, Water Station Facilities, delete the third paragraph beginning with "In southern climates . . ."

On page 13–2–2 under Sec. C, Oil Engine Equipment, change the material now shown to read:

"The oil engine is the most economical and practical type of power where electricity is not available. There are two main types: (a) the gasoline engine and (b) the diesel engine.

"Oil engine houses should be ventilated at the floor line, and whenever possible the floor should be above the ground."

On page 13-2-2 under Sec. D, Electric Motor Equipment, Art. 1. General, delete the second sentence reading "Where reliable . . . automatically controlled."

On page 13-2-6 delete Sec. I. Hydraulic Rams.

On page 13-2-6 change Sec. J. Pipe Lines to Sec I. Pipe Line.

On page 13-2-7, second line from top, delete the words "under high or low water conditions."

On page 13-2-7, second paragraph, second sentence, delete the words "bell and spigot."

On page 13-2-7, under Art. 2. Suction Lines, revise the fourth paragraph to read as follows:

"Steel or wrought iron pipe with welded or flanged joints is preferred for suction lines. Care should be taken to make all joints tight, and the suction line should be laid with a slight down grade toward the source of supply."

On page 13-2-7, under Art. 3. Discharge Lines, delete second sentence, reading "When possible, two 45-deg. bends should be used in place of 90-deg." Delete from the last sentence on page 13-2-7 the words "and should not be neglected."

On page 13-2-8, change Art. 4. Gravity Lines, to read:

"Care must be exercised in locating and laying gravity pipe line so that, if possible, no portion of it lies above the hydraulic gradient. In case this cannot be done, special provisions must be made."

On page 13-2-8, under Art. 5. Service Lines, change last sentence under Par. (c) to read "However, galvanized pipe should not be used for water treated with caustic soda as the high pH will dissolve the zinc coating."

Change Par. (e) to read:

"(e) For service lines 2 in. or less in size, the use of copper water tubing is suggested. Underground lines should be Type K, soft temper, with either flared or soldered connections."

On pages 13-2-9 through 13-2-15, delete entire Secs. K—Friction Losses, L—Water Column Delivery Charts, and M—Water Meters, Methods for Testing, Reading and Checking Consumption.

Page 13-3-1

SPECIFICATIONS FOR CAST IRON PIPE AND SPECIAL CASTINGS

Reapprove without change.

SPECIFICATIONS FOR HYDRANTS AND VALVES

Reapprove without change.

Pages 13-3-2 to 13-3-5, incl.

SPECIFICATIONS FOR LAYING CAST IRON PIPE

Reapprove without change.

Pages 13-3-6 and 13-3-7

SPECIFICATIONS FOR WOOD WATER TANK

Delete.

Pages 13–3–8 to 13–3–10, incl.

SPECIFICATIONS FOR TANK HOOPS

Delete.

Pages 13–3–11 to 13–3–13, incl.

SPECIFICATIONS FOR STEEL SUBSTRUCTURES FOR WOOD WATER TANK

Delete.

Pages 13–3–14 to 13–3–16, incl.

SPECIFICATIONS FOR TIMBER SUBSTRUCTURES FOR WATER TANKS

Delete.

Pages 13–3–17 to 13–3–27, incl.

SPECIFICATIONS FOR WELDED STEEL TANKS FOR WATER OR OIL STORAGE

Reapprove without change.

Delete the material on Water Treatment—General, pages 13–4–1 to 13–4–6, incl., substituting therefor the following rewritten version:

GENERAL

A. DESIGN

In most uses of water some form of chemical treatment is necessary and its selection must be based on a study of the requirements. If the use of softening plants is indicated, such plants should have adequate capacity for present and future needs. Any mechanical apparatus should be simple in construction and operation, stable in adjustment and should accurately and uniformly introduce the chemicals into the water being conditioned.

Apparatus for chemical treatment of water is available commercially or can be designed by qualified railroad personnel.

B. OPERATION, MAINTENANCE AND SUPERVISION

Adequate and capable supervision of the water treating plant by a chemist or an engineer experienced in water treatment is essential for the best results.

Frequent analyses of both the untreated and the treated water should be made by a competent chemist. Such analyses will serve as a check on the chemical treatment, keeping it properly adjusted with any changes in the quality of the natural water being supplied for conditioning.

C. METHODS OF WATER TREATMENT

There are various methods of removing or neutralizing the scale-forming and cor-rosive substances from water to render it suitable for use in steam generators. The three most common systems used on railroads are (1) the complete, or lime-soda ash process (cold), (2) the wayside, or internal system, and (3) the ion exchange system.

1. Lime-Soda Ash Method

The lime-soda ash method of water treatment is a process by which the hard scal-ing impurities of water (calcium and magnesium salts) are removed from the water prior to its use in steam generating, or other systems. This is accomplished by intro-ducing into the water being treated a mixture of hydrated lime (calcium hydroxide) and soda ash (sodium carbonate) in amounts sufficient to react with all the calcium and magnesium impurities present. The water is then retained in a treating basin for a suitable period of time, not less than 2 hr for the cold process, to allow for the com-pletion of the chemical reactions and the precipitation of the sludge containing the calcium and magnesium impurities.

The two general types of lime-soda ash water treating plants are the continuous and the intermittent. In the continuous system the treating chemicals are continuously added to the water as it continues to flow through the water treating and settling basin.

In the intermittent system the water and chemicals are mixed as a batch into a treating tank and then allowed to stand for a suitable period of time to allow for com-pletion of the chemical reactions and the precipitation of the sludge. The intermittent plant is outmoded and is being replaced by the continuous system.

In the continuous softeners, the capacity of the tank reserved for addition of the treatment and sedimentation of the sludge should be not less than three or four times the hourly capacity of the plant, depending on the temperature of the water, low tem-perature requiring the larger capacity. With the use of sludge-blanket-type equipment the capacity of the tank reserved for treatment may be reduced to one and one-half times the hourly demand of treated water from the plant.

At water temperature above 70 deg F an efficiently operating lime-soda ash plant will produce water practically free of its calcium and magnesium content and thus ap-proach zero hardness. The only minerals left in the treated water will be the non-scaling salts of sodium, including both the sodium salts natural to the water, and those formed by the chemical reactions between the alkali carbonate (generally soda ash) and the acid impurities present in the water prior to treatment.

2. Wayside or Internal Method

In the wayside, or internal, system of water conditioning an alkali carbonate treat-ing reagent, generally soda ash (sodium carbonate), is uniformly added to the water as it passes into either a storage tank for immediate use, or directly into the steam-generating system itself. The alkali carbonate neutralizes the so-called permanent hard-ness of the water, viz., sulfates, nitrates, chlorides of calcium and magnesium. Comple-tion of the chemical reactions and precipitation of the sludge is carried out inside the steam-generating boiler. Supplementary boiler-water treatments such as lignin, tannin, sulfite, and others are also introduced into the water being conditioned by means of the chemical proportioning equipment employed in the wayside or internal treating process.

Reagents Used in Water Softening

The quantity of reagents required per unit of corroding or scaling substances held in solution by the water is as tabulated below:

Incrusting or Corrosive Substances In Solution	Pounds of Reagent Required	Increase in Alkali Salts
Sulfuric acid	0.76 lime plus 1.08 soda ash	1.45 lb
Free carbon dioxide gas	1.68 lime	None
Calcium carbonate	0.74 lime	None
Calcium sulfate	0.78 soda ash	1.04 lb
Calcium chloride	0.96 soda ash	1.05 lb
Calcium nitrate	0.65 soda ash	1.04 lb
Magnesium carbonate	0.88 lime	None
Magnesium sulfate	0.61 lime plus 0.88 soda ash	1.18 lb
Magnesium chloride	0.78 lime plus 1.11 soda ash	1.23 lb
Magnesium nitrate	0.50 lime plus 0.71 soda ash	, 1.15 lb

Note—Since lime and soda ash are commercial chemicals and are not 100 percent pure, the above factors for these chemicals must be divided by the percentage of purity of the material used.

Given the analysis of a water, the pounds of incrusting or corrosive substances held in solution per 1000 gallons can be obtained by dividing the grains per gallon of each substance by 7, or the parts per million by 120.

3. Ion Exchange Method

Insoluble hydrated silicates of sodium and aluminum, known as sodium zeolite, have the unique property of removing the hard scaling calcium and magnesium content from water in exchange for the non-scaling sodium constituent contained by the zeolite mineral. This transfer of ions occurs when the water is passed through a closed tank containing a bed of granulated zeolite mineral at a flow regulated to the size of the mineral bed.

After the zeolite bed has become exhausted, i.e., has released all its sodium ions in exchange for calcium and magnesium ions from the water being treated, the plant is regenerated by passing a concentrated solution of common salt (sodium chloride) through the exhausted mineral bed. The process of ion exchange is now reversed, the magnesium and calcium ions being displaced from the zeolite mineral and replaced with sodium ions furnished by the saturated solution of sodium chloride. The regenerated zeolite mineral bed is then rinsed free of the excess brine solution and returned to service.

Some of the advantages of the treatment of water by the sodium zeolite water treating process are listed below:

(a) The treated water will not cause post precipitation of scale in feed water lines and appurtenances of steam generators as all the hardness of calcium and magnesium are removed by this treatment.

(b) The waste effluent from the plant containing soluble calcium and magnesium chlorides is harmless and may be discharged to the sewer.

(c) The salt required for regeneration of such plants is dust-free, easy to handle and store, thus allowing the location of such plants in buildings used for commercial purposes.

(d) The treating equipment, being compact and simple in design, can be located in existing buildings with minimum of changes and at minimum expense of space.

(e) The operation of a zeolite plant is simple, consequently there is small likelihood of irregular results.

(f) Waters of relatively high hardness can be more economically conditioned by the zeolite process than by other methods.

Disadvantages of the sodium zeolite treatment of water are:

(a) Water to be softened by the process must be practically clear.

(b) Waters containing iron and manganese in excess of 4 ppm are not suitable for zeolite softening without prior treatment.

(c) Where the natural water is high in soluble salts, particularly in bicarbonates, zeolite treatment may be impracticable due to the tendency of the sodium salts remaining in the conditioned water to concentrate in the boiler and release corrosive carbon dioxide gas.

Carbonaceous Zeolites or Organic Cation Exchangers

An organic cation exchanger consisting of a sulfonated synthetic resin has the property for removing the calcium and magnesium content from natural waters in exchange for either sodium ions or hydrogen ions, depending upon the reagent used for regenerating the mineral bed, whether sodium chloride or acid.

Anion Exchangers

A synthetic amine-formaldehyde resin has the property of removing the acid radicals, or anions, from water containing acids in exchange for the hydroxial (OH) ion. Regeneration of this unit for replacement of its OH ions is made with sodium carbonate or sodium hydroxide.

Demineralization of Water

Water approaching the quality of distilled water can be produced by the cation-anion exchange method, known as the demineralization process. This is accomplished by first passing the water through a cation exchanger unit operating on the hydrogen cycle where the calcium, magnesium, sodium and other cations (silica not included) present in the water are moved from solution in exchange for hydrogen ions. The partially conditioned water is then passed through an anion exchanger for removal of the acid radicals, or anions, in exchange for OH ions furnished by the mineral bed. The demineralized water containing only the silica which was originally present in the natural water and carbon dioxide is then discharged through a decarbonater, or degasifier, for the removal of the carbon dioxide gas and then treated with a sufficient amount of alkali carbonate, or hydroxide, to bring its alkalinity to accepted levels for use in boilers.

Demineralized water treated with an alkali to inhibit corrosion is the highest quality water that can be produced by chemical means. Because of its exceedingly high cost its use on railroads, when available, has been mostly limited to steam generators of the flash boiler type.

With the advent of the diesel locomotive, ion-exchange water treatment has acquired an important position among the methods of water softening and should receive concerted study by qualified railroad personnel before a definite decision is reached regarding the method of water conditioning that may be planned for adoption at any location.

Pages 13–4–7 to 13–4–16, incl.

TREATMENT OF WATER FOR COOLING PURPOSES

Reapprove with the following revisions:

On page 13–4–7 change the title of this recommended practice to "Treatment of Water for Cooling Purposes Other Than Engines."

On pages 13–4–14 and 13–4–15 delete Sec. F—Slime Formation, in its entirety.

On page 13–4–15, Under Sec. G, delete item 1. *Chlorine,* and item 3. *Copper Sulfate.*

On page 13–4–16, add at the end of present Sec. G the following paragraphs:

Biological destruction of wood cooling towers can occur under certain conditions and is caused by a micro-organism of fungi that utilizes the wood for food. Most wood fungi attack the cellulose of the wood, leaving a dark residue known as "brown rot". It is usually found at points where the wood is in contact with iron fastenings and in sections of a tower kept moist but not flooded by the recirculating water.

Another type of wood fungi attack is that known as "white rot" which destroys the lignin, reducing the wood to a light-colored stringy mass. "White rot" is uncommon but is more severe than "brown rot" as it will not remain confined to localized areas, but will spread to all parts of the cooling tower.

Wood fungi attack on towers may be prevented by treatment of the cooling water with chlorophenate materials as outlined above.

On page 13–4–16, under Sec. H, delete all of item 1. *Occurrence,* including the paragraph beginning "For new installations . . ."

On page 13–4–16, Delete Sec. I—Biological Attack and Its Control in Cooling Towers, in its entirety.

Pages 13–5–1 to 13–5–11, incl.

Delete the following specifications for chemicals to be used in water treatment:

Specifications for Soda Ash to Be Used in Water Treatment, pages 13–5–1 and 13–5–2.

Specifications for Hydrated Lime to Be Used in Water Treatment, pages 13–5–3 and 13–5–4.

Specifications for Quicklime to Be Used in Water Treatment, pages 13–5–5 and 13–5–6.

Specifications for Aluminum Sulfate to Be Used in Water Treatment, pages 13–5–7 and 13–5–8.

Specifications for Ferrous Sulfate to Be Used in Water Treatment, pages 13–5–8 and 13–5–9.

Specifications for Salt to Be Used in Regeneration of Zeolite Water Softening Plants, pages 13–5–10 and 13–5–11.

Substitute for the foregoing the following new specifications:

SPECIFICATIONS FOR CHEMICALS TO BE USED IN WATER TREATMENT

Specifications covering all chemicals used in the treatment of water for steam generating, engine cooling and other purposes may be found in the Journals of the American Water Works Association as designated below:

Standard for Sodium Chloride, AWWA Specification B200-53, published in AWWA Journal of March 1960.

Standard for Soda Ash, AWWA Specification B201-53, published in AWWA Journal of February 1952.

Standard for Quicklime and Hydrated Lime, AWWA Specification B202-54, published in AWWA Journal of January 1953.

Standard for Ferrous Sulfate, AWWA Specification B402-53, published in AWWA Journal of October 1950.

Standard for Aluminum Sulfate, AWWA Specification B403-58, AWWA Journal of November 1952.

Standard for Liquid Sodium Silicate, AWWA Specification B404-58, published in AWWA Journal of October 1955.

Standard for Caustic Soda, AWWA Specification B501-53, published in AWWA Journal of December 1951.

Delete the material on Water Analysis and Interpretation of Results, pages 13-6-1 to 13-6-10, substituting therefor the following rewritten version:

WATER ANALYSIS AND INTERPRETATION OF RESULTS

A. GENERAL

For years it has been the general practice for the American railroads to report water analyses in terms of grains per U. S. gallon. This system of units is equivalent to parts per 58,300, being the product of the number of grains in one pound (7,000) and the number of pounds in one gallon of water (8.33), which allows for the direct conversion of the substances found present in the water into pounds per 1000 gallons by using a 58.3 cc sample of the water for the analysis. Such a system of expressing results tended to convey a clear impression of the conditions of the water when used in steam locomotive boilers and was developed at a time when water analysis standards were confined to specialized applications.

The railroads have also adhered to the practice of reporting results of the chemical analyses in the hypothetical combinations form. In the hypothetical form of reporting, each of the positive and negative ions or radicals are shown to combine as probable chemical compounds. In contrast, most laboratories of other industries use the ionic form of reporting results of an analysis wherein each element or radical is shown as a separate item with no attempt being made to combine them into chemical compounds.

In recent years the use of chemically treated water has undergone changes on the railroads. Not only is it used in the conventional type of steam generator boilers, but special types of chemically conditioned water are required for diesel engine cooling systems, flash-type steam generators, air conditioning systems, drinking purposes, and others. In view of these varied uses of treated water it now becomes expedient for the railroads to adopt industry-wide standards of reporting results of analyses and also to adopt the procedures for performing the analyses, as these are the basis of technical development, research and control and will enable the railroads to translate their problems into the accepted practices.

In order to conform to accepted industrial practices, it is recommended that railroad laboratories report all concise laboratory analyses of water in the ionic form and in terms of parts per million (ppm), and follow the chemical procedures as outlined in the latest publications of the American Society for Testing and Materials or the American Public Health Association. The only exception to these systems of reporting results and conducting the analysis will be when using the "rapid field analyses", a system developed by the railroads for approximating results of tests on mineral content of water made in the field. When using these methods of testing water the original system of reporting results will be continued as outlined under the rapid field procedures in Sec. C.

B. FULL AND COMPLETE LABORATORY ANALYSES

Methods of performing the laboratory water analyses for each of the elements or groupings listed below, including the methods of preparing the standard solutions, can be found in the following books of standards on the pages given in the tabulation:

Standards Methods for the Examination of Water, Sewage, and Waste Water, Eleventh Edition (1960), published by the American Public Health Association.

ASTM Book of Standards 1958, Part 10, or lastest edition, published by the American Society for Testing and Materials.

CHEMICAL LABORATORY PROCEDURES

Item	Reference
Acidity	APHA, pp. 42–43 ASTM, D-1067, non-referee method B
Alkalinity	APHA, pp. 44–47 ASTM, D-1067, non-referee method B
Aluminum	APHA, pp. 48–51 ASTM, D-857
Boron	APHA, pp. 56–57, method A
Ammonia	APHA, ASTM, D-1426, non-referee method
Calcium	APHA, pp. 64–69 ASTM, D-511
Carbon Dioxide	APHA, pp. 69–77 ASTM, D-513, non-referee method
Chloride	APHA, p. 78, method A ASTM, D-512, method B
Color	APHA, pp. 111–113
Dissolved Solids	APHA, pp. 214–215, B-Filterable residue ASTM, D-1069, method B
Hardness	APHA, pp. 133–137, method B (EDTA) ASTM, D-1126, either method
Iron	APHA, pp. 140–143, method A ASTM, D-1068, non-referee method
Magnesium	APHA, pp. 153–155, method B ASTM, D-511
Manganese	APHA, pp. 155–157, method A ASTM, D-858
Nitrate	APHA, pp. 178–179, method B ASTM, D-992
Nitrite	APHA, pp. 180–182 ASTM, D-1254
Nitrogen (organic)	APHA, 182–184

Item	Reference
Oxygen (dissolved)	APHA, pp. 309–311 ASTM, D-888, non-referee method B
pH Value	APHA, pp. 193–196 ASTM, D-1293
Phosphate	APHA, pp. 198–206 ASTM, D-515, non-referee methods A or B
Potassium	APHA, pp. 230–232, method A ASTM, D-1127
Silica	APHA, pp. 222–230 ASTM, D-858, non-referee method B
Sodium	APHA, pp. 230–232, method A ASTM, D-1127
Sulfate	APHA, pp. 240–243, method A or B ASTM, D-516
Sulfide	APHA, pp. 243–244 ASTM, D-1255
Sulfite	APHA, pp. 244–245 ASTM, D-1339
Tannin and Lignin	ALPHA, pp. 252–253
Turbidity	APHA, pp. 261–265

1. Reagents

Methods of preparing the reagents used in the analysis of water may be found in *Standard Methods for the Examination of Water and Waste Water,* 11th Edition, 1960, published by the American Public Health Association and in the *ASTM Book of Standards,* 1958, Part 10, published by the American Society for Testing and Materials. Methods of preparing the more common reagents may be found in the 11th Edition of the APHA on the pages as indicated in the list that follows:

Standard Solutions	APHA 11th Edition
Barium Chloride	p. 239
Calcium Chloride (for Soap. Sol.)	p. 136
EDTA, buffer	p. 134
EDTA, 0.01 M (for Calcium)	p. 136
Hydrochloric Acid .02N	p. 266
Indicator Solutions	
Methyl Orange	p. 46
Methyl Red	p. 64
Phenolphthalein	p. 46
Potassium Chromate	p. 78
Silver Nitrate 0.014N	p. 78
Sodium Carbonate 0.02N	p. 45
Sodium Chloride 0.014N	p. 78
Sodium Hydroxide 0.02N	p. 42
Sodium Thiosulfate 0.025N (for dissolved oxygen)	p. 310
Sulfuric Acid 0.02N	p. 153

C. RAPID FIELD TESTS

The railroads have long used a system known as rapid field tests for analyzing both natural and treated waters. As the name implies, this scheme of analysis was designed for use at the location in the field where the water samples were collected to enable the analyst immediately to adjust the chemical treatment of the water at the time such formula adjustments were found to be necessary.

The rapid field tests are confined to a means of approximating the total hardness, which is basically the total amount of scale-forming matter, and the division of this figure into alkalinity (carbonate hardness) and the non-carbonate hardness, as well as an approximation of the ratio of the calcium and magnesium salts. All results obtained by these methods of analyses are expressed in terms of grains per U. S. gallon as calcium carbonate in the hypothetical combination form. In using the hypothetical combination form of expressing results, each positive ion is shown to combine with one or more of the negative ions or radicals in the following order of sequence:

Positive Radicals (Cations)	Negative Radicals (Anions)
Calcium	Carbonate
Magnesium	Hydroxide
Ammonium	Sulfate
Sodium	Chloride
Potassium	Nitrate

With experience in manipulation on the part of the analyst, the rapid field procedure as outlined below will produce very close and satisfactory results:

Procedure

1. Raw Water Tests

(a) *Total Hardness.*—Measure 58.3 ml (or fractional sample if hardness is greater than 15) of the water to be examined into an 8 oz bottle. If desired add 1 ml of a saturated sodium tetraborate ($Na_2B_4O_7$) solution. If a fractional sample is used make up to 58.3 ml with distilled water. Add the standard soap solution in 0.5 ml amounts shaking vigorously after each addition until a stable lather is secured which will stand up for 5 min. As the end-point is approached reduce the additions of soap solution to 0.1 ml depending upon the accuracy desired and experience of the manipulator. Note and record any false end-point (Ghost Point) which is the dividing line between the calcium and the magnesium salts. Deduct lather factor from maximum false end-point reading to obtain the calcium hardness in grains per gallon as $CaCO_3$. The final burette reading after deducting the lather factor gives the total hardness in grains per gallon as $CaCO_3$. The difference between the total hardness and the false end-point gives the magnesium hardness expressed as calcium carbonate. If a fractional sample is used deduct the lather factor and then multiply the reading by the proper factor.

To avoid mistaking the false end-point for the true one when adding the soap solution to waters containing magnesium salts read the burette after the titration is apparently finished and add about 0.5 ml more of soap solution. If the end-point was due to calcium the lather will disappear. Soap solution must then be added until the true end-point is reached. Usually the false lather persists for less than 5 min.

Note 1—Acid waters should be neutralized with N/50 sodium carbonate using methyl orange indicator before the addition of sodium tetraborate and proceeding with test.

Note 2—Winkler's method for determination of calcium. If the false end-point is indefinite and careful determination of calcium is necessary take a 58.3 ml sample and add 2 ml Winkler's solution. Run test as usual and final end-point gives calcium hardness. The lather formed at the end-point in this case is less stable than in the regular soap test. If the calcium hardness is greater than 10 a fractional sample should be used, adding the Winkler's solution before making up to 58.3 ml with distilled water.

(b) *Alternate Schwarsenbach Test for Total Hardness.*—Measure 58.3 ml of the water under investigation into a white porcelain casserole. Add 0.5 ml of the hardness buffer solution and stir. Add 4 to 6 drops of the hardness indicator solution and stir. If hardness is present the sample will turn red.

Add hardness titrating solution slowly from burette with continued stirring. The end-point is reached when color changes from red to blue. (Color change at end-point from red to blue is sharp and rapid and should be approached carefully).

The number of milliliters of hardness titrating solution used gives total hardness direct as grain per gallon in terms of calcium carbonate.

(c) *Alternate Schwarzenbach Test for Calcium.*—Measure 58.3 ml of water under investigation into a white porcelain casserole. Add 1 ml of 2.0 N NaOH solution and stir. Add 0.20 g of calcium indicator from calibrated dipper, and stir. If calcium is present the sample will turn salmon-pink.

Add hardness titrating solution slowly with continued stirring. The end-point is a final change to orchid-purple.

The number of milliliters of hardness titrating solution used gives calcium direct as grains per gallon in terms of calcium carbonate.

(d) *Alkalinity.*—Titrate 58.3 ml of the water under investigation with N/50 sulfuric acid solution, using methyl orange indicator. The number of milliliters of N/50 sulfuric acid used gives the alkalinity (A) directly as grains per gallon in terms of calcium carbonate.

In case of alkaline waters, if the total hardness is greater than the alkalinity, the difference represents the noncarbonate hardness. If the total hardness is less than the alkalinity, the difference is sodium carbonate, all of the hardness being then in the form of carbonate hardness. In case of acid waters, all of the hardness will be noncarbonate hardness.

(e) *Acidity.* In case the water under investigation is acid to methyl orange, titrate 58.3 ml sample with N/50 sodium carbonate solution, using methyl orange indicator. The acidity is obtained directly as grains per gallon in terms of calcium carbonate.

2. Treated Water Tests

(a) *Hardness.*—Determination is made in same manner as indicated in (a) for raw water omitting sodium tetraborate.

(b) *Alkalinity.*—Titrate 58.3 ml of the water with N/50 sulfuric acid solution, using phenolphthalein indicator. Multiply the reading by 2.0 and record as causticity modulus (C).

Add methyl orange indicator to the colorless solution and continue the titration with N/50 sulfuric acid solution, recording the final reading (that is, the sum of the phenolphthalein and methyl orange requirements) as alkalinity (A).

If the alkalinity (A) is greater than the total hardness (H) then the difference between them is due to the presence of alkali carbonate. If the hardness (H) is greater than the alkalinity (A) then the difference between them shows the absence of alkali carbonate, this figure being due to an acidic ion of calcium or magnesium, or both, most generally being considered as calcium or magnesium sulfate. Likewise, excess of the causticity modulus (C) over the alkalinity (A) indicates the presence of the OH originating from either calcium hydroxide (hydrated lime) or sodium hydroxide (caustic soda), while an excess of alkalinity (A) over causticity modulus (C) shows an absence of OH ions. In other words, this difference shows the amount of the excess or deficiency of caustic alkalinity. In any water analysis when phenolphthalein (P) and methyl orange (M) indicators are used as in this case, one of the following conditions must exist:

1. When $P = O$, then $M =$ total bicarbonate.
2. When P is less than $\frac{1}{2}$ M, then 2P gives carbonate and M minus 2P gives bicarbonate alkalinity.
3. When $P = \frac{1}{2}$ M, then $2P =$ carbonate alkalinity.
4. When P is greater than $\frac{1}{2}$ M, then 2P minus M gives caustic and M minus caustic gives carbonate alkalinity.
5. When $P = M$, caustic alkalinity only is present.

3. Boiler Samples

(a) *Hardness.*—All suspended matter should be removed by filtration. Then proceed as under (a) for treated water.

(b) *Alkalinity.*—Test same as Art. 2 (b) for treated water tests, unless watε٫ is colored and Barium Chloride Method is indicated.

Barium Chloride Method. When a boiler water is colored, indicating the presence of organic material such as tannin, lignin, or decayed ve٫ etable matter, false readings with methyl orange indicator are obtained. To offset the effect of this organic matter in the titration, the following method, although not absolutely accurate, should be used: 58.3 ml of water is titrated in the usual way with N/50 sulfuric acid, using phe-. nolphthalein indicator. Call this reading P. A second sample of 58.3 ml is then taken and 10 ml of a 10 percent solution of neutral barium chloride ($BaCl_2$) is added. This precipitates the carbonate, leaving the caustic alkalinity. It is not necessary to filter. Titrate with N/50 sulfuric acid, using phenolphthalein as an indicator. The end-point is reached when the pink color disappears for a few seconds. The reappearance of the pink color should be disregarded. The reading should be recorded as P ($BaCl_2$).

$P(BaCl_2) =$ Caustic alkalinity in grains per gallonas $CaCO_3$.
$2[P — P(BaCl_2)] =$ Carbonic alkalinity in grains per gallon as $CaCO_3$.
$2\ P — P(B_aCl_2) =$ Total or methyl orange alkalinity in grains per gallon as $CaCO_3$.

(c) *Dissolved Solids.*—For boiler water analyses modern conductivity instruments and hydrometers are available from several manufacturers and with these it is possible to make rapid determination of dissolved solids with reasonable accuracy. When using these instruments manufacturer's instructions should be carefully followed and the instruments shall be calibrated at frequent intervals against gravimetric determinations. Conductivity instruments do not measure organic matter.

4. Reporting of Results

The following methods can be used in calculating hypothetical combinations in average waters:

(a) If the noncarbonate hardness exceeds the magnesium, then the magnesium is present as magnesium sulfate. Calculate the remainder of the noncarbonate hardness to calcium sulfate. The alkalinity is then due entirely to calcium bicarbonate and expressed as grains per gallon in terms of $CaCO_3$.

(b) If the noncarbonate hardness is less than the magnesium, calculate the noncarbonate hardness to magnesium sulfate and the remainder of the magnesium to magnesium carbonate. The magnesium carbonate is subtracted from the total alkalinity. The difference between the total alkalinity and the magnesium carbonate is then reported as calcium carbonate in grains per gallon expressed as $CaCO_3$.

(c) If the hardness is less than the total alkalinity, then the magnesium is present as magnesium carbonate, the calcium as calcium carbonate and the difference between the hardness and the total alkalinity is sodium carbonate.

(d) If the total alkalinity is zero, as in the case of acid waters, then the total hardness is calcium and/or magnesium sulfate and/or chlorides.

(e) The difference between the total dissolved solids and the sum of magnesium sulfate, magnesium carbonate, calcium sulfate, calcium carbonate, sodium chloride, sodium carbonate, iron and aluminum oxides, and silica, gives the amount of non-incrusting sulfates, nitrates and organic matter. If the difference is negative, the result indicates calcium and magnesium chlorides or nitrates in which case it will be necessary to make the sulfate determination to secure the most probable hypothetical combination.

5. Special Reagents for Rapid Field Analysis

Sodium Chloride Solution—Dissolve 1 g fused sodium chloride in 1 liter of distilled water. One ml is equivalent to 1 mg NaCl.

Silver Nitrate Solution—Weigh about 2.90 g of silver nitrate and dissolve in 1 liter of distilled water. Adjust so that 1 ml of this solution is equivalent to 1 ml of standard sodium chloride solution.

Winkler's Solution—Dissolve 200 g sodium potassium tartrate and 12 g sodium hydroxide in distilled water making up to 1 liter.

Delete the material on Analysis of Chemicals Used in Water Treatment, pages 13–6–11 to 13–6–14, incl., substituting therefor the following rewritten version:

ANALYSIS OF CHEMICALS USED IN WATER TREATMENT

Methods of sampling and analyzing water softening chemicals may be found in the ASTM Books of Standards, 1958 Edition, or latest publication thereof, published by the American Society For Testing and Materials, as outlined below:

Analysis of Soda Ash ....ASTM Method D 501
Analysis of Caustic Soda ....ASTM Method D 501
Analysis of Trisodium Phosphate ....ASTM Method D 501
Analysis of Hydrated Lime ....ASTM Method C 25

Page 13–7–1

FOAMING AND PRIMING

Delete.

Page 13–7–2

WASHOUTS, WATER CHANGES, AND BLOWDOWN OF LOCOMOTIVE BOILERS AS INFLUENCED BY WATER CONDITIONS

Delete.

Page 13–7–2

RECOMMENDED MEANS OF CONTROL OF INTERCRYSTALLINE CORROSION

Delete.

Page 13–8–1

WATER FOR DRINKING PURPOSES

Reapprove without change.

Pages 13–8–2 to 13–8–21, incl.

RAILWAY SEWAGE DISPOSAL FACILITIES

Reapprove without change.

Pages 13–8–21 and
13–8–22

STERILIZATION OF NEW AND REPAIRED WATER WELLS PIPE LINES AND OTHER EQUIPMENT USED IN HANDLING DRINKING WATER

Reapprove without change.

Pages 13–8–23 and 13–8–24

WASTE DISPOSAL

Reapprove with the following revisions:

On page 13–8–24, under Sec. G, substitute the word "should" for the word "must" in the first and fourth lines of the first paragraph.

Pages 13–9–1 to 13–9–14, incl.

DIESEL LOCOMOTIVE FUEL AND WATER SERVICES

Reapprove with the following revision:

On page 13–9–3, substitute the word "should" for the word "must" in the first line of Sec. B, Art. 1.

On page 13–9–9 delete the last sentence of Sec. G, Art. 1 (last sentence on the page).

Pages 13–M–1 to 13–M–6, incl.

WATER SERVICE RECORDS

Delete.

Pages 13–M–7 and 13–M–9

WATER SERVICE ORGANIZATION

Delete.

Report on Assignment 2

Prevention of Corrosion in Hot and Cold Water Systems

J. J. Dwyer (chairman, subcommittee), R. C. Archambeault, M. V. Biar, W. E. Billingsley, T. W. Brown, P. J. Calza, C. E. DeGeer, T. I. Gray, A. W. Johnson, F. O. Klemstine, J. P. Rodger, R. M. Stimmel, A. G. Tompkins.

Corrosion Prevention in Potable Hot Water Systems

There has been a tendency in the past to ignore corrosion damage in potable water systems, and simply to replace failed piping and tanks. This attitude might have been created by the fact that we were dealing with *potable* water, and that *treatment*, to railroad water supply personnel, meant the use of such chemicals as lime, soda ash, caustic soda, and the like, which have little appeal as additives to water for human consumption. Labor and material costs have now risen to such a level, however, that this attitude is no longer tenable.

There has been a trend to switch from galvanized iron to copper piping in the potable water field. Copper costs more than iron, but it should last much longer; it is quicker and easier to install, so labor costs are reduced; and it does not plug with deposits nearly as fast as iron pipe.

There are certain conditions under which copper pipe may fail faster than iron. Corrosion of copper water piping is accelerated rapidly by (1) increased dissolved oxygen, (2) increased carbon dioxide, (3) increased temperature, (4) increased velocity, and (5) soft water. These facts have been verified in studies by Obrecht and Quill at Michigan State University. In one example, part of an old copper pipe potable water system was replaced with new copper pipe, and water softeners were installed. In the new Type L copper water tube 40 leaks developed in 30 months. These leaks appeared in the cold water lines, but appeared faster and in greater numbers in the hot water piping. The old copper piping, now carrying soft water also, did not fail. Investigation disclosed that the old tubing was coated on the inside with a protective film of adherent deposit, while the new tubing had no coating on its inner surface.

The failures were always at the top inner surface of the tube in horizontal position. They occurred from accumulations of carbon dioxide and oxygen. There was also a direct correlation between the frequency of failures and the demand for water. Where leaks were numerous, high flow rates were necessary to meet demand. All the failed copper tube samples in the study indicated a velocity effect in addition to chemical attack. Velocities should be kept below 5 ft per sec if possible. In failures of heat exchangers utilizing potable hot water, severe localized pitting occurred facing the path of hot water flow through the exchangers. The causes were (1) agressive water, (2) high temperatures, and (3) high velocities (exceeding 5 ft per sec).

Another cause of failures in potable hot water systems by corrosion is the galvanic couple. Whenever two dissimilar metals, such as copper or brass piping connected to a galvanized water tank, are placed in electrical contact with each other, an electric current flows. The water flowing in or out of the tank is the electrolyte. The current produced tends to cause one of the metals to be corroded away. The rate at which corrosion proceeds is dependent upon a number of factors, such as (1) the particular metals present, (2) the temperature, (3) the substances dissolved in the water, and (4) its conductivity. One solution to this problem would be to use copper piping with a copper or monel tank, and galvanized piping with a galvanized tank.

In spite of the disadvantages mentioned, it sometimes becomes necessary to use components of two or more different metals or alloys in a potable hot water system. Corrosion problems so introduced can be diminished by the use of rubber or plastic couplings, so-called dielectric unions, bushings, nipples, or gaskets placed between the two different metals so as to separate them electrically. This will tend to reduce greatly the rate of corrosion at critical points at which dissimilar metals are close together in the system. It should be noted, however, that a dielectric union or other insu'ator will not stop or greatly mitigate corrosion if the water has a substantial content of copper dissolved in it. The copper ions will plate out as metallic copper on galvanized surfaces, forming miniature galvanic cells, which produce a corrosive action. Water containing ammonia or carbon dioxide is very prone to cause any copper in the system to go into solution and form copper ions.

A method of reducing corrosion in potable hot water tanks is to place a piece of magnesium in the tank. Such magnesium anodes are normally effective for use only in water supplies with mineral content from 7 to 15 grains per gallon. Very pure soft water has such low conductivity that there is no great need to provide protection against galvanic corrosion. And high dissolved solids content gives such a high conductivity rate that the magnesium anode will be dissolved too quickly. The anode method cannot be recommended as a means of protecting ordinary galvanized steel tanks because of the large area which must be protected. It is a valuable accessory in a vitreous-enameled or "glass-lined" steel tank because of the small amount of actual bare metal exposed in such a tank.

It has been mentioned that the primary causes of corrosion in potable hot water systems are dissolved gases of corrosive nature. A means of diminishing the dissolved gas content of water is the mechanical degasifier or deaerator. Deaeration is accomplished by dividing the water into small particles and thus facilitate gas removal. On the deaerator a vacuum should be maintained which corresponds to the boiling pressure for the water temperature involved. If necessary to reach extremely low gas concentrations, multi-stage units can be employed. Deaeration can lower incoming concentration of oxygen from the range of 6–12 ppm to 1–2 ppm, and can lower carbon dioxide from around 40 ppm to 5.0–6.5 ppm, or reductions of up to 80 percent for these two gases. In further MSU test studies of soft water at high temperatures (170 and 200 F) and high velocities (8.1 and 13.1 ft per sec), corrosion losses in nondeaerated water were up to 50 percent higher than in deaerated water. Deaerators for potable hot water supply systems are designed mainly for use in large buildings such as hospitals and hotels.

A mechanical means of reducing the corrosive gas content of systems where deaerators might not be practical is the application of simple air-release valves to hot water heaters and other high points in the hot water system where gases collect.

The influence of pressure often is overlooked. For example, water at 70 F and atmospheric pressure could hold 8 ppm oxygen in solution. At 140 F and the same pressure it could hold only 4 ppm oxygen. At 140 F and 60 psig, however, it could hold 26 ppm oxygen. Under this condition, since it had only 8 ppm oxygen originally, it is now undersaturated, and won't release any oxygen when heated to 140 F. Now suppose a pneumatic tank had been installed and air pressure was used to raise the water to 60 psig. At 70 F and 60 psig the water could dissolve 46 ppm oxygen. When heated to 140 F about 20 ppm will be driven off, but 26 ppm remains in solution to accelerate cor_rosion. Thus design has added three times as much oxygen in solution, and has left tremendous amounts of oxygen to be vented, or to cause circulation problems, air binding, etc.

Temperature control in hot water systems is very important. Galvanizing is not recommended for hot water lines at temperatures above 140 F. Above this temperature zinc reverses its potential and begins to accelerate the corrosion of iron at any "holidays" which exist. At lower temperatures, zinc is sacrificial and provides cathodic protection for the iron or steel pipe. Copper pipe is excellent for temperatures not exceeding 140 F. For higher temperature water, red brass (85 percent copper, 15 percent zinc) or a better alloy is required. For example, in a dishwashing system, copper tubing could be used for the normal hot water supply (140 F), but for the short run from booster heater to dishwasher (180–190 F) red brass should be used. Temperatures above 140 F present the following disadvantages: (1) corrosiveness increases rapidly; (2) reversal of potential occurs in galvanized pipe, and accelerated attack will take place at locations where galvanizing has been lost; (3) rapid dezincification of brass pipe occurs; (4) pinhole pitting in copper pipe may be expected; (5) expansion and contraction strains are magnified, and contribute to leaks at screwed fittings. Where temperatures exceeding 140 F are required for a specific use, such as 180–190 F for dishwashing, a booster heater close to the equipment should be provided, and the piping from the booster heater to the point of use must be red brass or better to withstand such high temperatures. *In considering temperatures, a most important point to remember is that rates of corrosion in hot water systems are doubled with each 17 F increase or rise.*

One of the chemical methods of protecting potable hot water systems suggested is the carbonate balance (Saturation Index) system of treatment, which aims at maintaining a protective layer of calcium carbonate on the inside of the pipe or container. The Saturation Index is equal to the difference between the actual measured pH of a water and the value of pH calculated by the Langlier equation. It is an indication of the tendency for a calcium carbonate scale to be deposited. Two possible disadvantages to this method are (1) that it requires constant attention and analyses to control the amount of calcium carbonate and the pH, and (2) what is satisfactory for cold water will deposit calcium carbonate scale in hot water, and if the treatment is set up to be satisfactory for hot water, it will not be very satisfactory for cold water. Obviously, this treatment cannot be used on waters softened to zero hardness.

Two straight chemical treatments are used without injuring the water for domestic use. One of these is sodium silicate, which has been used since 1920 to protect iron, lead, and brass water pipe. The dosage for the initial month is generally 12 to 16 ppm as silica, after which it may be reduced to 8 ppm, or even lower. This simple treatment has reduced the corrosion rate of iron pipe by 70 percent, and has practically stopped dezincification of brass pipe. The same treatment may be used to retard solution of copper by regulating the silicate to give the water a pH of about 8. The common method of application is the bypass feeder, but proportioning pumps can be used as well.

In general, three types of silicates are used for the inhibition of corrosion in potable waters. They have Na_2O/SiO_2 percentage compositions of 8.9/28.7, 18.0/36.0, and 14.7/29.4 percent. The first is used for waters with a pH above 6, while the latter two are for waters with pH of 6 or below. These two are basically the same silicate, differing in concentration and in viscosity. For hot water only, sodium silicate glass, a very slowly dissolving solid, may be used.

The protection afforded by silicate treatment is due to a film which forms on the inside surface of the pipe. It is not a scale and is invisible when the pipe is wet. The thickness of this film is about that of a relatively small colloid. If the silicate feed is discontinued, the film gradually disappears and the corrosion will begin again—but there is a considerable lag, just as there was in building up the film initially. Sodium silicate treatment may provide 90 percent protection or better.

Sodium metaphosphate glass has proved to be highly useful in preventing the formation of scale deposits due to excess calcium and magnesium salts in water, utilizing a dosage of only 2 to 5 ppm. Higher dosages of 8 to 10 ppm have materially retarded corrosion when flow rate in pipes has been 1 ft per sec or higher. It should be noted that the water must be circulated to get benefits from polyphosphates; they *do not* protect in stagnant systems. Another requirement is that calcium must be present. Thus, calcium is built into and is a part of some of the slowly dissolving metaphosphate glasses. A mixture of sodium metaphosphate glass and sodium silicate fed together is reported to give better protection to potable water systems, both hot and cold, than either separately.

SUMMARY

Causes of Hot Water System Corrosion

(1) Corrosive dissolved gases.

    (a) Carbon dioxide.
    (b) Oxygen.
    (c) Hydrogen sulfide.*
    (d) Ammonia.*

(2) Dissolved copper.
(3) High temperature.
(4) High velocity and turbulence.
(5) Galvanic couples.
(6) Soft water.

Remedies for Potable Hot Water System Corrosion

(1) Use deaerators where practicable.
(2) Use air release valves at appropriate locations.
(3) Do not use pneumatic tanks to maintain pressure in hot water systems.
(4) Use Type K copper pipe and fittings for temperatures up to 140 F.
(5) Confine temperatures in hot water system to 130–140 F.
(6) If necessary to use higher temperatures, use red brass pipe or better.
(7) Confine velocities to a maximum of 5 ft per sec.
(8) Insulate galvanic couples.
(9) Use magnesium anodes in hot water tanks where practicable.
(10) Use appropriate chemical treatment.

    (a) Carbonate balance system.
    (b) Sodium silicate.
    (c) Sodium metaphosphate.
    (d) Mixture of (b) and (c).
    (e) pH adjustment.*
    (f) Dealkalization.*

REFERENCES

(1) *Heating, Piping and Air Conditioning,* Jan. 1960, p. 165; April 1960, p. 131; May 1960, p. 105; July 1960, p. 115.

* See AREA Bulletin 560, Vol. 62, November 1960, page 237.

(2) *Jour. AWWA,* Aug. 1960, p. 1033.

(3) H. H. Uhlig, *The Corrosion Handbook* (1948).

(4) F. N. Speller, *Corrosion—Causes and Prevention* (1951).

(5) Ulick R. Evans, *The Corrosion and Oxidation of Metals* (1960).

(6) *Corrosion (NACE),* Sept. 1960, p. 453t.

(7) *Consumer Bulletin,* Sept. 1960, p. 22.

(8) Correspondence from J. F. Wilkes.

(9) Correspondence from Dr. William Stericker.

This report of progress is submitted as information. Your committee plans to reassemble this information into a format suitable for subsequent submission for inclusion in the Manual as recommended practice.

Report on Assignment 3

Design, Construction and Operation of Coach Servicing Facilities to Comply with Regulations of U. S. Public Health Service

C. F. Muelder (chairman, subcommittee), W. F. Arksey, J. M. Bates, M. R. Bost, I. C. Brown, P. J. Calza, H. E. Graham, A. W. Johnson, G. F. Metzdorf, R. D. Powrie, J. C. Roberts, J. E. Wiggins, Jr.

No new developments concerning design or construction of coach serving facilities have come to the attention of the committee. However, a summary of the research project on coach watering hydrants being carried out at the AAR Research Center under the sponsorship of Committee 13 is presented as Part 2 of this report.

Another function of this committee is to advise AREA members of any changes in or new regulations of the U. S. Public Health Service, and in Part 1 of this report it submits as information certain revisions made in the drinking water standards of the Service.

Part 1—Revision of 1946 Drinking Water Standards of the U. S. Public Health Service

1. Revisions of the 1946 Drinking Water Standards have been published and were made available September 1, 1961. The principal changes are as follows:

(a) *No. 72.202 Source and Protection*—Rewritten and expanded to be more specific.

(b) *No. 72.203 Bacteriological Quality*—Minimum number of samples per month required from distribution system serving a population of less than 2,000 is 2.

(c) *No. 72.204—Physical Characteristics*—Expanded to include a recommended sampling frequency and lower recommended limits for turbidity and color.

(d) *No. 72.205 Chemical Characteristics*—Recommended limits for alkyl benzene sulfonate, arsenic, carbon chloroform extract, cyanide, fluoride, manganese and nitrate, have been included. Recommended limits for copper and zinc have been lowered. Mandatory limits for barium, cadmium, cyanide and silver have been included. The mandatory limits for lead and selenium have been lowered. Limits for fluoride concentrations are based on temperature ranges. A table comparing previous and revised limits is presented in Appendix A.

(e) *No. 72.206 Radioactivity*—A section on radioactivity has been introduced for the first time, including a recommended sampling frequency and establishing limits for specific radionuclides and gross beta activity.

2. A new policy has been placed into effect by the U. S. Public Health Service for grading dining car sanitation. A summary of this grading program for railroad dining cars is presented in Appendix B.

APPENDIX A

REVISION OF CHEMICAL LIMITS

	Recommended Maximum Limits* (Milligrams per Liter)		Concentrations Which Constitute Grounds for Rejection of Supply (Milligrams per Liter)	
	1946	*1961 Revision*	*1946*	*1961 Revision*
Alkyl benzene sulfonate (detergent)		0.5		
Arsenic		0.01	0.05	0.05
Barium				1.0
Cadmium				0.01
Carbon chloroform extract (exotic organic chemicals)		0.2		
Chloride	250.	250.		
Chromium			0.05	0.05
Copper	3.0	1.0		
Cyanide		0.01		0.2
Fluoride		1.7**	1.5	2.2**
Iron+manganese	0.3			
Iron		0.3		
Lead				0.05
Manganese		0.05		
Nitrate		45.		
Phenola	0.001	0.001		
Selenium			0.05	0.01
Silver				0.05
Sulfate	250.	250.		
Total dissolved solids	500.	500.		
Zinc	15.	5.		

*Concentrations in water should not be in excess of these limits when more suitable supplies can be made available.

**Fluoride temperature concentration relationships are discussed in detail in the text.

APPENDIX B—SUMMARY OF PUBLIC HEALTH SERVICE GRADING PROGRAM FOR RAILROAD DINING CARS

After July 1, 1961, no dining car will receive a numerical score on inspection by the Public Health Service. The grade of a dining car shall be given as A, B, or C, based upon compliance with the recommendations given in the "Handbook on Sanitation of Railroad Passenger Car Construction" and the "Handbook on Sanitation of Dining Cars in Operation." Since these recommendations and the items listed on the inspection report are considered to be minimum requirements, *Grade* A will be given only to those dining cars on which there are no defects. *Grade B* dining cars are those which fail to comply with item 1, 2, 4, 5 or 17 on the inspection report, but which conform with all other items of sanitation required for Grade A dining cars. *Grade C* dining cars are those which fail to comply with either the Grade A or the Grade B requirements.

A car bearing a Grade A placard on July 1, 1961, will retain this rating pending the results of subsequent inspections. A car that does not have a grade shall be given a

Grade A rating and a placard on the first inspection, if no defects are found. The appropriate grade rating shall be given on the second inspection, if any defect noted on the first inspection is repeated.

A dining car will not be degraded upon the first violation of any item. A notification of defects noted shall be given in writing to the responsible railroad official by means of an inspection report. A specific period of time will be given in which corrections may be made, but not to exceed 30 days before an appropriate grade is established.

The period between inspection and reinspection of a car may be considered as a "provisional" period. This interval should provide adequate time for correction of any but the most serious structural defects. Reinspection will have to be conducted as working schedules of the Public Health Service will allow. However, a statement in writing from a company official that the reported defect or defects have been corrected will be accepted as evidence of correction and will remove the "provisional" status of the car rating. A form will be attached to each inspection report which may be used in reporting the corrections of defects. A copy of the statement or form used in reporting the corrections to the Public Health Service should be forwarded immediately to the car and attached to the salmon copy of the inspection report that was left on the car at the time of inspection.

If written evidence of correction has not been received by the end of the "provisional" period, the Public Health Service will reinspect the car as soon as practicable. On this reinspection, a violation of any part of an item that was "circled" on the previous report will result in the degrading of the car. In the event that violations are found in both the "B" and "C" categories on the reinspection, the lower grade shall prevail in rating the car. A dining car may be graded or regraded on submission of a written statement by an official of the railroad company to the Public Health Service that the defects have been corrected. In such incidences, the appropriate grade placard will be forwarded for posting in the car.

The grade placard must be prominently posted in the dining area of each dining car and in the customer service area of each bar car, or at each passenger entrance. Frames shall be of the type which will allow easy removal of the placard without the use of tools and shall be of such construction as to protect the placard from soiling.

Grade placards have no expiration date and remain current until removed by a Public Health Service representative. The salmon carbon copy of the latest inspection report *shall not be removed* from the car. When the salmon copy of the latest inspection report is not aboard, the dining car grade earned at the time of the new inspection will be posted immediately.

Part 2—Summary of Research Project on Coach Watering Hydrants

This past year some basic research was carried out in the AAR Research Laboratory, experimenting with several ideas to provide a new approach to a frost-proof, sanitary water hydrant which will meet the requirements of the railroads and the U. S. Public Health Service. These experiments were helpful in approaching the basic problems, but were not extensive enough to permit designing a new hydrant without further study and practical evaluation.

With the increasingly urgent need for a hydrant that would be immediately available to the railroad industry, it was decided that efforts should be directed in such a manner that a certain post-type hydrant formerly available commercially would again be manufactured and placed on the market. This was done. Your committee in conjunction with the AAR Research Center staff and the information developed during their investigation, was able to interest an established hydrant manufacturer. As a result of these negotiations, this manufacturer has acquired the patent rights, and purchased the manufacturing drawings and patterns.

This frost-proof sanitary water hydrant is now available as an improved model for various depths of bury. Repairs for previous models, as well as the new, improved models, are also immediately available.

Additional information concerning this project may be obtained from the director of engineering research, Association of American Railroads, 3140 S. Federal St., Chicago 16.

Report on Assignment 5

Economic Justification of and Methods for Conditioning Water for Use in Engine Cooling Systems and Steam Generators of Diesel Locomotives

R. A. Bardwell (chairman, subcommittee), M. V. Biar, T. W. Brown, C. E. DeGeer, D. E. Drake, A. E. Dulik, J. J. Dwyer, R. S. Glynn, W. C. Harsh, F. O. Klemstine, C. W. Owens, R. M. Stimmel, T. A. Tennyson, E. M. Walters.

The extent of the savings realized from the use of various methods of water treatment for diesel locomotives has been gathered from various railroads. The figures show that treatment is economically justified in all cases, with the best methods giving the largest savings.

Methods to calculate individual savings are offered in this report. The report indicates that an approximate return of $13 can be realized for each $1 judiciously spent for cooling water treatment and at least $2.60 for each $1 spent for steam generator water treatment. Net annual material savings can average almost $500 for each diesel cooling system and over $1000 for each steam generator under the conditions reported. These figures, exclusive of labor, are felt to be conservative. To assure continuity of operations, the best water treatment methods available are recommended to effect these savings. This report is offered as information.

Engine Cooling Systems

A. WATER PROBLEMS ENCOUNTERED

1. Corrosion

The potential due to dissimilar metals in contact in the electrolyte of cooling water system can lead to localized or general pitting, the resulting current flow tending to force the metals with the highest potential into solution. These potentials can vary from the highest—zinc and aluminum coupled to copper—down to the concentration cells such as at the junction of liner and engine block or the bare iron under seals with adjacent iron areas made passive by water treatment or corrosion products.

There is a form of physical loss of metal, sometimes termed cavitation erosion, which does not respond readily to any form of water treatment, although increased doses of certain types of inhibitors have been known to alleviate the situation. This metal loss is thought to be due to high-frequency vibration, which is diminished by a better fit of liner and jacket.

Although any amount, especially over 10 ppm, of sodium chloride will hasten corrosive action of waters, a limit is usually sought of not over 65 ppm. The smaller amounts aggravate aluminum corrosion whereas solubilization of iron hydroxide to eliminate protective film is caused by higher concentrations.

Metal corrosion from oxygen and other gases dissolved in water can occur, requiring neutralizing treatment. Acidic gases include, in addition to carbon dioxide, sulfur dioxide which may enter from minute cracks into the combustion chamber. Such gas entrance has been reported to materially lower the pH of cooling water, which condition will also indicate the presence of cracks.

2. Scale Formation

Deposition of scale or sludge is most apt to occur at points of low circulation and on high heat transfer surfaces. This can lead to overheating of castings and metal failure. Any non-adhering sludge or scale is still liable to plug the small radiator or cooler tubes, leading to poor cooling.

3. Foaming

Some inhibitor formulations and some waters have been known to cause foaming. In either case immediate correction is necessary to retain coolant and cooling water circulation.

B. METHODS OF TREATMENT AND CONTROL

1. Types of Chemicals

a. Alkaline Chromates. These materials, potassium or sodium salts, buffered with sodium carbonate and sometimes with phosphates added, is the best and most commonly used corrosion passivator. A 1:1 mixture by weight of sodium dichromate and soda ash will react in water to form 55 percent of its weight of sodium chromate plus the necessary excess soda ash. The mixture is available in powder or pellet form or in a 40 percent liquid solution. Ready solubility of pellets or powder makes direct application to cooling systems feasible. The solution form in either quart or gallon sizes is even more easily applied. The solution must be kept in warm storage during cold weather to prevent crystalization. Optimum dosage recommended is 1800 ppm as sodium chromate, or 1½ lb/100 gal, at a pH 8.5–9.3. A smaller dosage is sometimes used effectively in conjunction with phosphates, although to assure protection an excess over this is usually initially applied.

b. Sodium Nitrite. This compound is seldom used alone but is buffered to pH 9 with borax and other special inhibitors are incorporated in the mixture. A copper passifier, such as mercaptobenzothiazole, is necessary to protect copper alloys such as used in oil cooler, radiators, and pump impellers. The mixtures may also contain complexing agents to neutralize the galvanic effect of traces of metals in solution, phosphates, and silica, along with anticaking and antifoam chemicals plus a color indicator. There is some difficulty in dissolving the solid form of this treatment due to the low solubility of borax, and various methods of application have been devised. Direct application of the solid to the diesel vent pipe is not recommended due to the possibility of plugging not only this pipe but interior passages. Methods of application include a slurry form; a compact applicator which has a steam entrance into the container of concentrated water mixture and a pump to lift the dosage into the water fill pipe; heating of the concentrated water mixed dose in a bucket with steam; or large-batch treatment of make-up water. A minimum dosage of this buffered solution should produce at least 0.020 percent nitrite in water for iron passivation. This treatment is compatible with anti-freeze glycols, as are organic corrosion inhibitors, but not the chromates.

c. Organic. Soluble-oil-type inhibitors, which plate out on metal surfaces to give metal protection, have been used since the early days of diesel engines. These emulsifiable inhibitors, if used in recommended dosages in properly pre-conditioned water, provide good protection to metals. Due to deterioration of rubber hose and seals, use of this type was almost completely abandoned. However, new developments now being marketed evidently contain components to abate this attack and are now being used. Being in liquid form, this type of treatment is easily applied.

2. Application of Chemicals

a. Direct feed into cooling systems. The liquid forms of alkaline chromate, nitrite-borate, and organic inhibitors are easily applied through the air vent directly into the cooling systems of diesel engines.

b. Small feeders on each diesel unit have proved to be a handy method of application and more accessible than vent pipe application.

c. Predissolving individual dosages using hot water and agitation is usually done with nitrite-borate inhibitors. A portable feed device allowing the dosage to be pumped into the water fill pipe from ground-floor level will eliminate the labor of pouring concentrate in at roof level..

d. Complete proportioning of the required dosage into the make-up water is found to be economically justified at some larger facilities and is usually preceded by zeolite or demineralization treatment.

3. Methods of Concentration Control

a. Electrical conductivity of the cooling water measured with suitable instruments gives a simple, effective means of estimating dosage treatment.

b. Spot comparison of a chemically impregnated blotter which gives a color reaction in proportion to the water treatment salt concentration provides another means whereby personnel can estimate the amount of inhibitor present and the dose needed.

c. Titration of alkalinity accompanying the inhibitor mixture will give a close check on the total amount of inhibitor present.

d. More complete chemical analysis is sometimes used even though compounded inhibitors should contain the necessary proportions and buffering. Analyses may include titration for chromate; colorometric analyses for chromate, nitrite, phosphates, and silica; pH determinations; and titration for chloride concentration. Incrusting compounds are determined by hardness analyses. Investigations may include analyses for concentration of corrosion products.

C. COSTS OF CHEMICAL TREATMENT

1. Cost of Chemicals

Table 1 gives the cost range of available treatments and cost range per unit per year. This latter figure is based on an actual use figure where consumption of chemicals gave an indicated average of 10 changes or approximately 2400 gal of treated water per unit per year.

TABLE 1

	Alkaline Chromate	Nitrite-borate	Organic type
a. Cost range—per pound	$0.12*–0.26	$0.12*–0.29	
—per gallon	$0.89—1.12 (40%)		$1.76–2.14
b. Dose range—Average	0.5 oz/gal	1.0 oz/gal	0.6 fluid oz/gal
—Maximum	1.0	1.5	0.6
—Minimum	0.25	0.5	0.13
Dose per 100 gal—Average	3 lb.	6 lb.	½ gal
c. Cost per unit per year—range	$7–$13	$13–$32	$15–$18

*Non-proprietary.

2. Cost for Further Conditioning of Water

a. Many roads have existing lime-soda ash treating plants, and their effluent, free of incrusting hardness, was and is used for diesel cooling water make-up. The average cost of such treatment is estimated to be 5 cents/1000 gal, which would increase the yearly cost per unit by only 12 cents.

b. Zeolite softening to remove incrustments is especially desirable for high-hardness waters where difficulty from scaling or sludge in cooling systems can be serious. The range in cost varies with the hardness of the water from about 20 cents up to 60 cents per 1000 gal. Based on 50 cents/1000 gal, such conditioning would raise the annual cooling water cost of each unit about $1.40.

c. Demineralizing to remove all ions from the water has the advantage of removing salts which might aggravate corrosion as well as incrustants. In one case, where an original 30-grains-per-gallon water is used, the cost amounts to $1.16/100 gal. This would cost $27.85 per diesel unit based on 2400 gal each year.

3. Cost for Chemical Application

Any method of individual application should not take over 10 min, which amounts to 50 cents per application or $5 per year per unit based on 10 fillings. Investment costs for mechanical proportioning into make-up water would probably not be much under this.

4. Cost for Control Tests

The time consumed in taking water samples when diesel lube oil is also sampled added to the time to read the conductivity of the sample should be not over 1 min, or 5 cents per test. Present practice tending to bring diesel units in only once per month for maintenance would give a minimum of 12 such tests per unit per year, or 60 cents/unit/year. Spot comparisons would cost about the same as conductivity tests. Periodical laboratory tests for more complete analysis are recommended. Such control is estimated to cost a railroad a minimum of $1 per unit per year.

5. Summation of Water Costs

Table 2 gives the annual total cost range per unit per year using the three types of treatment with varying degrees of conditioning.

TABLE 2

	Alkaline Chromate	Nitrite-Borate	Organic Type
a. No pre-conditioning—Min	$13.	$19.	$21.
Max	$20.50	$38.50	$24.50
b. Zeolite treatment —Max	$22.	$40.	$26.
c. Demineralizing —Max	$47.50	$66.50	$52.50

D. BENEFITS TO DIESEL COMPONENTS

1. Liners

Two widely used makes of diesel locomotive engines differ in sensitivity of certain parts to attack by corrosive cooling water. In what follows these engine makes will be referred to as Locomotive A and Locomotive B.

Locomotive A, switch and road unit, liners are most prone to cavitation attack. By proper treatment, liner life may be extended from two to at least eight years. On Locomotive B, road units, although not as critical, if liners are to be chrome plated the water side must be kept free from corrosion or scale. Following are examples of the savings to be expected:

TABLE 3

	Locomotive A Switcher 6 liners	Locomotive A Road 12 liners	Locomotive B Road 16 liners
Value of new liner	$ 155	$ 112	$ 100
Cost to chrome liner	$ 125	$ 75	$ 75
Cost per 8-year cycle per unit:			
a. No treatment	$3,720*	$5,376*	$3,200**
b. Treatment and chrome	$1,680***	$2,244***	$2,800***
c. Savings/unit/8 years	$2,040	$3,132	$ 400
Savings/unit/year	$ 255	$ 391	$ 50

*Locomotive A liners replaced every two years.
**Two replacements on Locomotive B liners.
***Cost of one new set plus one chrome plated set of liners.

2. Heads

Although head cracking around the valve seat is probably more of a mechanical problem, it will be worsened by overheating if cooling is poor due to scale or sludge adhering in the intricate passageways of this casting. The labor to clean the scale off both the engine block seat and head itself in the seal ring area will be a sizeable item if scaling is not controlled. With the original cost of a Locomotive B head being $105 or $1,680 per engine set, even allowing for welding cracked heads at one-half the new price, the savings will be $84/unit/year based on a 10-year head life with proper treatment and only 5 years with scaling conditions.

3. Oil Coolers

Due to the high temperature and velocity of the water, the water side of coolers will deteriorate rapidly with improperly treated water by removal of solder, impingement attack at entrance to tubes, dezincification by alkaline waters, and penetration of tubes by corrosion. Oil coolers used on locomotive B cost $350 to $496. Extension of life from 3 to 12 years is entirely feasible with proper water conditioning, giving an average savings of $100/unit/year. It is stressed that oil cooler leaks allow water to go on the oil side directly to the engine's main bearings which will cause water wiping, overheating, and possible crankshaft scoring. Such failures cost from $10,000 to $20,000 each.

4. Radiators

The fin-type radiators used to cool water in Locomotive B units are valued at $148 to $199, and there are either 8 or 10 per road unit. Although results of attack or resulting leaks are not as drastic as with oil coolers, radiators will deteriorate and leak with improperly conditioned water. Also, any loose scale or adherent sludge can easily plug the slotted openings, necessitating expensive cleaning. Estimating an extension of life of from 6 to 12 years with proper water conditioning, this gives an estimated savings of from $100 to $170/unit/year.

5. Pumps, Connections and Hoses

Impellers of water pumps may be eroded where there is improper water conditioning. This can impair efficiency of cooling water circulation, allowing engine parts to overheat and fail. Properly treated water will reduce this attack of bronze parts. Value of Locomotive B impellers is $20.25, with two per unit. By proper water treatment, extension of life from 3 to 12 years is ordinary, giving annual savings of $10 per unit. Hose connections have quickly deteriorated when soluble oil treatments were used, and even inorganic treatments cause some hardening, swelling, or cracking and disintegration. Loosened pieces of rubber are liable to plug radiators. Leaks may result in water entering the crankcase or combustion space and at the very least will result in frequent low water. Bronze accordion-type connections are subject to impingement attack which is aggravated by improper water treatment. The latest flanged, gasketed types now being installed should eliminate some of the water problems which have previously been encountered.

6. Engine Blocks

There are some thin plates in the lower top decks of Locomotive B engines and these have been known to pit through allowing water to run into the air box. These pitted areas are very difficult to repair and a unit exchange engine, complete, is usually recommended. Assuming this exchange costs from $15,000 to $20,000 and basing the decrease in the life of the crankcase assembly from 20 to 15 years, this gives an annual cost of from $187.50 to $250 per unit for corrosion damage from this source.

7. Headers

Some style engines had aluminum headers which were prone to attack by corrosive waters, especially those containing appreciable amounts of chloride. Most of these have been replaced with a cast iron type. These still require inhibited water for maximum life.

8. Thermostats

Thermostats inserted in cooling water lines will malfunction and need early replacement under adverse water conditions,

E. SUMMARY OF ECONOMIC JUSTIFICATION FOR TREATMENT OF COOLING
WATER FOR DIESEL LOCOMOTIVES

1. Summary of Estimated Annual Costs Without Treatment

Estimated annual costs resulting from use of non-treated water for a 16-cylinder
Locomotive B unit, labor not included, taken from previous estimations is shown in
Table 4.

TABLE 4

	Additional Annual Cost per Unit Without Treatment	Total Value of Parts on Locomotive B
Cylinder liners	$ 50.00	$1,600
Cylinder heads	$ 84.00	$1,680
Oil coolers	$100.00	$ 400
Radiators	$100.00	$1,500
Water pump impellers	$ 10.00	$ 40
Crankcase assembly	$187.50	
Total	$531.50	

A minimum figure is taken for oil cooler, radiators, and engine block. It is estimated that annual labor to replace these parts would be an additional $340, which
would mean a minimum additional cost of about $870/unit/year if average corrosive
and scaling water were used. This figure is rated as conservative. The maximum with
more aggressive waters would be considerably greater.

2. Water Costs

The annual total cost of water conditioning is seen to range from a minimum of
$13 per unit up to a maximum of $66.50, depending on the type of pretreatment and
inhibitor used. A mean of $40/unit/year may be used for estimating purposes.

3. Conclusions and Justifications

The above figures show that an average of over 1000 percent return on expenditures, not including labor, is a conservative estimate of the direct savings available from
proper conditioning and control of water for engine cooling systems of Locomotive B
road diesel locomotives. On Locomotive A the savings from liner deterioration alone
will give a minimum of 650 percent return on annual expenditures. If all the labor costs
involved are added to parts costs, plus value of time out of service for these expensive
engines, the savings which result from good water treatment will be impressively higher.

The facts outlined above indicate that all makes of diesel locomotive engines will
show comparable savings as the result of good cooling water treatment. The savings are
large enough to justify the best possible treating facilities and enough qualified personnel to prescribe, control, and supervise the treatment.

Steam Generators

A. WATER PROBLEMS ENCOUNTERED

1. Scale Formation

Due to high heat transfer rates and the low ratio of water to steam in the inner
coils, conditions are set up for easy deposit of scale. Even soluble salts are liable to de-

posit if the steam/water ratio is very high, such as sodium sulfate which becomes less soluble in hot water. For this reason, non-adhering sludge and highly soluble reaction compounds are the aim in any interior water treatments. Apart from coils, scale formation in heat exchangers, by-pass regulator and servo controls, blowdown tank and valve, and lines to and from the water tank will cause malfunctioning.

2. Tube Corrosion and Erosion

By measuring the weight loss of tubes during the rebuild period, a definite trend as to metal loss can be determined. This loss is due to the combination of acid cleaning and erosion by solids carried through the tubes during operation, plus the metal loss from the exterior of tubes from corrosion by the combustion gases. Many tubes on rebuild are not reapplied due to their thin shell.

3. Storage Tank Corrosion and/or Sludging

Corrosion of tanks has been reported mainly in cases where demineralized or zeolite water was used without further pH adjustment. Periodical rinsing of tanks is necessary, especially when internal treatment is used and sludging occurs due to return of solids carried over which are not blown-down.

B. METHODS OF CHEMICAL TREATMENT

1. Zeolite or Demineralization

Zeolite or demineralization is primarily recommended to remove incrusting salts. Both methods require slight post-treatment for corrosion inhibition. With some waters, deionizing results in some benefit through decreasing salt concentration, necessitating less blowdown.

2. Internal Treatment

Potassium carbonate is the primary softening agent used due to its own complete solubility and the solubility of its other salts, especially the silicate. These solubilities increase with temperature rise. Oxygen fixatives and organic sludge conditioners may be included in a complete treating formula. A large number of railroads simply slug-feed a dose of this treatment into the steam generator tank along with the steam generator water at the supply points. Diaphragm pumps on diesel units working off generator supply pumps, or other forms of metering pumps on diesels, can apply treatment from a small storage tank on the diesel unit. In some instances wayside metering of treatment into the water supply has been used, although in most cases this has been where water was pre-conditioned by external treatments.

3. Use of Existing Lime-Soda Ash Plants

Use of existing lime-soda ash plants for water supply gives a non-incrusting water source where such plants are available. Instances of condensate water use have also been reported.

C. COSTS OF CHEMICAL TREATMENT AND CONTROL

1. Cost of Chemicals per 1000 Gal

For zeolite waters, costs will run between 20–60 cents/1000 gal and will be much greater for deionized water. The after-treatment for these waters amounts to about 4 cents/1000 gal. For internal treatments, costs vary from 10–18 cents/1000 gal for slug feed but may run considerably higher, depending upon excess alkalinity carried. With injector pumps set to take care of the worst water conditions, this has amounted to 30 cents/1000 gal in one case.

2. Investment Costs for Plants or Proportioners

Zeolite plants and demineralizers vary in costs according to size capacity. These also require a storage tank and booster pump to water diesels, as flow through exchangers is not great enough to supply generator tanks directly unless filling is done at lay-over points. Average cost for such plants is $12,000 to $15,000 each, and it is estimated that each such plant supplies an average of 11 generators daily. Small proportioners and tanks range from approximately $100 per diesel unit up to $600 for installations at diesel facilities.

3. Cost for Application of Chemicals

Make-up to treating tanks at terminals is usually daily, costing approximately $1.50 each, while slug feed proportioning costs are negligible.

4. Control Tests

Control tests should be made periodically, especially where complete treatment is in effect. Such analyses should include tests for hardness, alkalinity, total solids and perhaps phosphate. Cost would depend upon the volume of samples run.

D. BENEFITS TO STEAM GENERATORS AND APPURTENANCES

1. Extended Coil Life

There are 3–4 sets of coils in a steam generator system, with a total value from a minimum of $625 on small generators to a maximum of $1,083 for the larger sizes. The inner coil, with a minimum value of $180 to a maximum of $298, which runs hottest is the one giving most trouble. With poor water conditioning, only about one year's service is secured from these inner coils. With mediocre conditioning, two years are secured. With complete zeolite or demineralizing and after-treatment, five year's service or more is easily reached.

2. Extension of Period of Acid Wash

All internal methods of treatment will require some period of acid wash to remove the scale plugging the tubes. The time period between washes will depend on the amount of hardness and the volume passing through the coils. For example, a generator evaporating 5000 gal/day of 7-grain-per-gallon hardness water might require 15-day washouts, while if only 1000 gal/day are evaporated the period might be stretched to 75 days, using internal treatment in both cases. Zeolite treatment, if properly controlled, can extend the period of acid washout indefinitely if all hardness is removed. Regulations require locomotive generators to have at least clear water pumped through their coils every 30 days. The acid for washing a generator plus the extra labor to perform this operation is estimated to cost an additional $10–$15 per wash over a clear-water rinse.

3. Elimination of Sludging and Corrosion of Storage Tanks

Periodic draining of tanks and the necessity to install plugs and remove baffles to facilitate flushing of sludge are requirements with internal treatment. Time out of service for cleaning, plus costs, liability of plugging generator feed lines and appurtenances, are definite disadvantages for internal treatment which are eliminated by the use of zeolite or demineralized waters. Corrosion of storage tanks has caused very expensive replacements, which action can be nullified by proper adjustment of feedwater treatment.

4. Other Generator Parts

Heat exchangers have to be cleaned or renewed periodically if scaling or adherent sludge is present. By-pass regulator and servo control, unless washed with inhibited phosphoric or sulfamic acids, are liable to become plugged with the use of internal treatment and will cause generator malfunctioning.

E. SUMMARY OF ECONOMIC JUSTIFICATION FOR TREATMENT OF STEAM GENERATOR WATER

Due to the larger volume of water, even at lower unit treatment cost, than is used for diesel cooling water, it is difficult to economically justify complete treatment on cost of parts replacement alone. However, with no water treatment generators just could not be operated with continuity, and steam must be furnished diesel-powered passenger trains without failure. The best water treatment to this end should be used.

There is a definite fuel oil savings in operating a clean-tubed generator over one with any amount of scale. A variation depending upon the amount of time which generator tubes are becoming plugged compared with a 30 percent increase in fuel consumption for severely scaled tubes is the basis for the savings shown in Table 5. Since it is not considered reasonable to expect coils to operate under severely scaled conditions or even be in that condition at the time of washouts, a figure of 50 percent of the severely scaled condition is considered at the time of the washout, which indicates a 15 percent increase in fuel consumption at that time. An average of the build-up from clean to this partially scaled condition is considered as the mean. The other assumption made is that one generator evaporates about 1,000,000 gal of water per year and operates about 25 percent of the time under full load. Specific savings for particular conditions can be estimated from these averages.

TABLE 5

The following examples are based on one generator evaporating 1,000,000 gal of water/year with properly adjusted fuel/water ratio:

	Example A	Example B	Example C	Example D
Type of water treatment	Zeolite	Individual unit proportioners	Slug feed	None
Cost of chemical treatment (1,000,000 gal/year)	$450	$270	$100–$180	0
Investment in plants/generator	$1,100	$100	nil	
Cost @ 10% of capital	$110	$10		
Total cost of treatment	$560	$280	$160 avg	
Inner coil life, years	5 min	3½	2	1 max
Inner coil savings/year	$144–$238	$128–$220	$90–$149	
Number of acid washes/year	0	4	12	24
Savings in cost of acid wash per year @ $15 each	$360	$300	$180	0
Total coil and acid wash savings/year	$504–$598	$428–$520	$270–$329	
Time running partially scaled 50% of severely scaled condition	0%	12½%	50%	75%
Mean increased fuel consumption	0%	1⅞%	7½%	11¼%
Total fuel: 95 gal/1000 when clean, 123¼ gal/1000 when severely scaled	95,000	96,781	102,125	105,687
Fuel savings, gallons	10,687	8,906	3,562	
Cost @ 9c/gal	$961	$801	$320	
Total savings/generator/year	$1,465–$1,559	$1,229–$1,321	$590–$669	
Gross savings/1000 gal water	$1.46–$1.56	$1.23–$1.32	$0.59–$0.67	
Cost of water/1000 gal	$0.56	$0.28	$0.16	

Based on these figures, the minimum net return for treatment of steam generator feed water will amount to at least 160 percent with about 65 percent of savings in fuel. Interpretation that investment for zeolite treatment will yield only a slight return is misleading to the degree that more time out of service is required even with the best internal treatment and that only inner coils are considered in the material savings.

Report on Assignment 6

Railway Waste Disposal

T. A. Tennyson (chairman, subcommittee), R. C. Archambeault, R. O. Bardwell, J. M. Bates, I. C. Brown, D. E. Drake, R. S. Glynn, W. C. Harsh, G. E. Hartsoe, G. F. Metzdorf, J. C. Roberts, E. R. Schlaf, A. G. Tompkins.

Since one important phase of handling waste disposal problems is to establish a proper understanding of the applicable requirements, your committee feels that the following up-to-date list of control agencies will be of value. This report is submitted as information.

State	Agency and Address	Title of Official
Alabama	Water Improvement Commission State Office Building Montgomery 4, Ala.	Technical Secretary
Alaska	Division of Health Department of Health & Welfare Alaska Office Building Juneau, Alaska	Director
Arizona	State Department of Health State Office Building Phoenix, Ariz.	Commissioner of Public Health
Arkansas	State Water Pollution Control Commission 921 West Markham Little Rock, Ark.	Technical Secretary
California	State Water Pollution Control Board Room 316, 1227 "O" St. Sacramento 14, Calif.	Executive Officer
Colorado	State Department of Public Health 4210 East 11th Ave. Denver 20, Colo.	Director of Public Health
Connecticut	Water Resources Commission State Office Building 165 Capitol Ave. Hartford 15, Conn.	Director
Delaware	State Water Pollution Commission State House Annex Governor's Ave. & Division St. Dover, Del.	Executive Officer
District of Columbia	District of Columbia Department of Public Health Washington 1, D. C.	Chief, Food and Public Health Engineering
Florida	State Board of Health 1217 Pearl St. Jacksonville 1, Fla.	State Health Officer
Georgia	Georgia Dept. of Public Health State Office Building Atlanta 3, Ga.	Director
Hawaii	Hawaii Department of Health Kapuaiwa Building P. O. Box 3378 Honolulu 1, Hawaii	Director of Health

State	Agency and Address	Title of Official
Idaho	Idaho Department of Health Statehouse Boise, Idaho	Administrator of Health
Illinois	State Sanitary Water Board State Office Building 400 South Spring St. Springfield, Ill.	Technical Secretary
Indiana	Stream Pollution Control Board State Board of Health 1330 West Michigan St. Indianapolis 7, Ind.	Technical Secretary
Iowa	State Department of Health State Office Building Des Moines 19, Iowa	State Commissioner of Health
Kansas	State Board of Health State Office Building Topeka Ave. at 10th Topeka, Kans.	Executive Secretary
Kentucky	Water Pollution Control Commission State Department of Health Frankfort, Ky.	Executive Director
Louisiana	Stream Control Commission Box 9055, University Station Baton Rouge 3, La.	Executive Secretary
Maine	Water Improvement Commission c/o Dept. of Health & Welfare State House Augusta, Me.	Secretary
Maryland	State Department of Health State Office Building 301 W. Preston St. Baltimore 1, Md.	Chief, Bureau of Environmental Hygiene
Massachusetts	Massachusetts Dept. of Public Health 546 State House Boston 33, Mass.	Commissioner of Public Health
Michigan	Water Resources Commission Station B Reniger Building 200 Mill St. Lansing 13, Mich.	Executive Secretary
Minnesota	Water Pollution Control Commission State Department of Health Building Campus, University of Minnesota Minneapolis 14, Minn.	Secretary
Mississippi	State Board of Health Felix J. Underwood State Board of Health Building P. O. Box 1700 Jackson 5, Miss.	Secretary and Executive Officer
Missouri	State Water Pollution Board 112 West High St. Jefferson City, Mo.	Executive Secretary
Montana	State Board of Health Laboratory Building Helena, Mont.	Executive Officer and Secretary
Nebraska	Department of Health State Capitol Building Lincoln 9, Nebr.	Director of Health
Nevada	State Department of Health Carson City, Nev.	State Health Officer
New Hampshire	Water Pollution Commission 61 So. Spring St. Concord, N. H.	Technical Secretary

State	Agency and Address	Title of Official
New Jersey	State Department of Health 129 E. Hanover St. Trenton 25, N. J.	State Commissioner of Health
New Mexico	New Mexico Department of Public Health Santa Fe, N. Mex.	Director of Public Health
New York	State Department of Health 84 Holland Ave. Albany 8, N. Y.	Commissioner of Health
North Carolina	State Stream Sanitation Committee State Department of Water Resources North McDowell St. Raleigh, N. C.	Director and Executive Secretary
North Dakota	State Department of Health Capitol Building Bismarck, N. Dak.	Executive Officer
Ohio	Water Pollution Control Board 306 Ohio Departments Building Columbus 15, Ohio	Chairman
Oklahoma	State Department of Health 3400 Block of North Eastern Oklahoma City 5, Okla.	Commissioner of Health
Oregon	State Board of Health 1400 S. W. 5th Ave. Portland 1, Ore.	State Health Officer
Pennsylvania	Sanitary Water Board Pennsylvania Department of Health State Capitol Health and Welfare Building Harrisburg, Pa.	Chairman
Puerto Rico	Department of Health Ponce de Leon Ave. San Juan 18, Puerto Rico	Secretary of Health
Rhode Island	Department of Health State Office Building Providence 3, R. I.	Director of Health
South Carolina	State Water Pollution Control Authority Room 417, Wade Hampton Building Columbia 1, S. C.	Executive Director
South Dakota	Committee on Water Pollution State Capitol Pierre, S. Dak	Secretary and Executive Officer
Tennessee	Stream Pollution Control Board Cordell Hull Building Sixth Ave., North Nashville 3, Tenn.	Director
Texas	State Department of Health 1100 W. 49th St. Austin 5, Tex.	Commissioner of Health
Utah	Water Pollution Control Board 45 S. Fort Douglas Blvd. Salt Lake City 13, Utah	Executive Secretary
Vermont	State Water Conservation Board State Office Building Montpelier, Vt.	Commissioner
Virginia	State Water Control Board 415 W. Franklin St. Richmond 20, Va.	Executive Secretary
Washington	State Pollution Control Commission 224 Old Capitol Building Olympia, Wash.	Director

State	Agency and Address	Title of Official
West Virginia............	State Water Resources Commission 1709-A Washington St., East Charleston 1, W. Va.	Director
Wisconsin................	State Committee on Water Pollution State Office Building Madison 2, Wis.	Director
Wyoming.................	State Department of Health State Office Building Cheyenne, Wyo.	Director of Public Health

The committee was able to locate the address of only one agency in Canada, which is:

> Ontario Water Resources Commission
> 801 Bay St.
> Toronto 5, Ont.

In addition to the state control agencies, some railroads operate in territories also under interstate control agencies. These are:

Cooperating States	Interstate Water Pollution Control Agency	Title of Official
Illinois Missouri	Bi-State Development Agency 915 Olive St. St. Louis 1, Mo.	Chairman
Delaware New Jersey New York Pennsylvania	Interstate Commission on the Delaware River Basin Suburban Station Building Philadelphia 3, Pa.	Executive Secretary
District of Columbia Maryland Pennsylvania Virginia West Virginia	Interstate Commission on the Potomac River Basin Transportation Building 815-17th St., N. W. Washington 6, D. C.	Director
Connecticut New Jersey New York	Interstate Sanitation Commission 10 Columbus Circle New York 19, N. Y.	Director and Chief Engineer
California Oregon	Klamath River Compact Commission P. O. Box 388 Sacramento 5, Calif.	Executive Secretary
Connecticut Maine Massachusetts New Hampshire New York Rhode Island Vermont	New England Interstate Water Pollution Control Commission 73 Tremont St. Boston 8, Mass.	Secretary
Illinois Indiana Kentucky Ohio Pennsylvania New York Virginia West Virginia	Ohio River Valley Water Sanitation Commission 414 Walnut St. Cincinnati 2, Ohio	Executive Director and Chief Engineer

Report of Committee 16—Economics of Railway Location and Operation

C. L. Towle, *Chairman*

T. J. Lamphier,
Vice Chairman

C. W. Sooby, *Secretary*

J. W. Bolstad

F. Wascoe

G. Rugge

T. D. Wofford
J. W. Barriger
A. L. Sams
L. E. Ward
Q. K. Baker
G. A. Bennewitz, Jr.
C. H. Blackman (E)
I. C. Brewer
D. E. Brunn
H. S. Bull
B. Chappell
J. L. Charles
H. B. Christianson
P. J. Claffey
L. P. Diamond
W. J. Dixon
A. J. Gellman
S. B. Gill
R. A. Gleason
R. L. Gray
F. E. Gunning
L. W. Haydon
H. C. Hutson
J. E. Inman
C. A. James
T. D. Kern
R. J. Lane

A. S. Lang
R. F. Lark
H. A. Lind
A. E. MacMillan
J. C. Martin
Raymond McCann
D. K. McNear
M. B. Miller
R. L. Milner
T. C. Nordquist
F. N. Nye
G. C. Payne
W. E. Quinn
J. P. Ray
F. L. Rees
F. J. Richter
V. J. Roggeveen
H. F. Schryver (E)
L. K. Sillcox
G. S. Sowers
J. J. Stark, Jr.
D. S. Sundel
J. E. Teal (E)
D. L. Vercelote
K. A. Werden
H. L. Woldridge
Committee

(E) Member Emeritus.
Members listed in bold face are the official representatives of the Engineering Division, AAR.

To The American Railway Engineering Association

Your Committee reports on the following subjects:

137

5. Methods of reducing time of freight cars between loading and unloading points, collaborating with Car Service Division, AAR, Communication and Signal Section, AAR, Operating-Transportation Division, AAR, and American Association of Railroad Superintendents.
   Committee has completed a plan of approach covering various phases of problem and is assembling information for analysis and study.

6. Features of economic and engineering interest in the study, design, construction and operation of new railway line projects, or major line relocations, proposed, in progress, or recently completed.
   Progress report, submitted as information .... page 148

7. Engineering, maintenance and operating benefits to be derived from increased joint use of railway facilities, collaborating with Committees 11, 14 and 20.
   Final report, submitted as information .... page 160

8. Innovations in railway operations
   No report at this time. Committee is assembling data for a report covering various aspects of containerization.

11. Review of developments in new methods and modes of transport
    No report at this time. Committee is gathering data and studying material for possible future reports on pipeline operations and air-flow vehicles.

THE COMMITTEE ON ECONOMICS OF RAILWAY LOCATION AND OPERATION;
C. L. TOWLE, *Chairman.*

AREA Bulletin 567, November 1961.

Report on Assignment 1

Revision of Manual

A. S. Lang (chairman, subcommittee), J. W. Bolstad, H. S. Bull, R. A. Gleason, R. L. Gray, C. A. James, T. J. Lamphier, R. F. Lark, R. L. Milner, T. C. Nordquist, J. P. Ray, V. J. Roggeveen, Geo. Rugge, L. E. Ward.

Your committee submits the following material to replace that now in Chapter 16 of the Manual, as follows:

Part 1, Secs. A and B of ECONOMICS (pages 16–1–1 to 16–1–4), and Sec. C, Art. 5, Par. d4 (pages 16–1–8 and 16–1–9); GRADES AND ALINEMENT THROUGH TUNNELS (page 16–1–16); FORMULA FOR SIDING LENGTH FOR FREIGHT TRAINS (page 16–1–17); and the material on train resistance now appearing in Part 3, pages 16–3–27 and 16–3–28.

It is recommended that the new material be adopted and published in the Proceedings but not incorporated in the Manual until the complete revision and rearrangement of Chapter 16 permits its insertion in proper order. It is also recommended, however, that a note giving the Proceedings reference to the new material, if adopted, be added to the contents sheet of Chapter 16 in the 1962 Manual Supplement.

RAILWAY LOCATION

A. BASIC CONSIDERATIONS

1. Location

A line (or alinement) may be a projected, reconnaissance, preliminary, final location, or constructed line.

A line is said to be located when its position is fixed horizontally and vertically at all points along its length.

2. Traffic Estimates

Locating a railway in effect constitutes designing a physical plant which can produce as economically as possible railroad transportation whose quality of service is commensurate with the nature of the traffic at hand.

Almost all decisions regarding the design and location of a railway depend in some way upon the quantity and class of traffic it is to handle. A railway plant that is satisfactory for a given quantity and class of traffic may not be satisfactory for a different quantity or class of traffic. It is therefore of the utmost importance that estimates of present and future traffic be as complete and accurate as possible. Where substantial uncertainties exist, separate economic analyses for different levels and classes of traffic may be necessary to assess the benefits or losses which would result from major changes in traffic conditions.

3. Capital-Return Criterion

Rate of return on invested capital is generally the most satisfactory criterion for judging the suitability of a line location or its detailed design features. An approximate formula for rate of return is:

$$p = \frac{R-E}{C}$$

where $p =$ approximate rate of return on investment.
$R =$ annual revenue from operation.
$E =$ annual expense of operation (including taxes and depreciation).
$C =$ capital investment.

A more precise value for rate of return is that interest rate which, together with an appropriate value for the economic life of the project, yields a capital recovery factor (CRF) that satisfies the same ratio as in the equation above. That is, where

$$CRF = \frac{R-E}{C}$$

This true rate of return is preferable as an economic criterion. Where capital recovery factors are large and economic life long, however, the approximate rate of return gives an almost identical answer.[1]

4. Fundamental Economic Principles

The following four rules enunciated by Wellington are generally applicable in the design of any location:[1]

---

[1] For a more detailed discussion of capital return computations, see E. L. Grant and W. G. Ireson, *Principles of Engineering Economy* (New York, 1960), Chapter 8.
[2] See A. M. Wellington, *The Economic Theory of the Location of Railways*, (6th ed., corr., New York, 1910), pages 15–19.

a. Regardless of how profitable a line promises to be, no expenditure over the absolute minimum is justifiable which is not of itself a profitable investment.

b. Conversely, any additional expenditure which *is* profitable of itself should always be made.

c. The exception to the rule immediately above is that no additional $_{ex}$Penditure is wise which endangers completion of a project with the funds available.

d. Unless the traffic volume can be predicted quite exactly, it is often best to postpone any expenditure where that can be done without great loss.

B. SPECIAL CONSIDERATIONS

1. Temporary Construction

When the traffic on a line is expected to be small in the years immediately following construction but is expected to increase significantly in the future, it may be advantageous to introduce extra distance, sharp curvature, or steeper gradients wherever these are so situated as to be capable of reduction at reasonable expense when justified by the growth in traffic.

The desirability of such temporary construction depends upon whether the interest saved by deferring construction of the final line will offset the cost of temporary work ultimately to be abandoned and the higher train operating costs encountered in the interim.

In general, stations should not be located on such temporary lines. A receiving and delivery point for local traffic once established is generally moved only against public opposition.

2. Operating Districts

a. Length

The length of locomotive districts depends upon the type of motive power used and upon changes in traffic and ruling gradient. A principal restriction is imposed by the requirements for servicing of locomotives: long servicing stops may make locomotive change desirable. A break in ruling gradient (or in some cases speed of operation) may require a different locomotive for the same train. A change in traffic volume may require a change in the type, or size, or number of locomotives. Within the limits of these general restrictions, locomotive districts should be as long as possible so as to maximize the utilization of motive power units.

Crew districts should be of such length that freight trains can cover the maximum distance possible within the normal hours of service of the crew. Allowance should be made for traffic delays and unfavorable weather conditions.

b. Terminal Points

An adequate water supply for both locomotives and domestic use is an important requisite for a terminal point. It is also desirable to locate terminal points on minor summits, where possible.

Facilities for full servicing of engines should be provided at all locomotive change points. Any heavy repair facilities for locomotives and cars should always be located at a terminal point and preferably at a locomotive change point.

In acquiring terminal property, particular attention should be given to possible future needs, not only because of the probable long-term increase of neighboring land values, but also to provide for potential industrial development and to avoid encroachments which could hamper operating efficiency.

3. Ruling Gradients

a. Definition

The ruling gradient is that section of adverse gradient which limits the tonnage a locomotive can haul over an operating district. Thus, the ruling gradient determines the minimum number of trains that can handle any given volume of traffic between two terminals.

It should be carefully noted that the maximum gradient is not necessarily the ruling gradient. Momentum gradients (see below) may be steeper than the ruling gradient. Similarly, the short-time ratings of locomotives with electric drive (see Sec. B, Art. 2, and Sec. 3, Art. 2, Part 3, this Chapter) may permit them to negotiate short sections of gradient steeper than the ruling gradient on the district.

b. Importance of Ruling Gradient

Ruling gradient is the most important single consideration in line location. Because it is the major determinant of freight train size, ruling gradient largely determines the amount of motive power and the number of crews required to operate a line. It is less likely to determine the size of passenger, local, or fast freight trains, but it will affect the speed of all trains.

Ruling gradients should be made uniform over as many adjacent operating districts as possible. Preferably, they should be uniform over the entire distance between major traffic generation or classification points or over entire locomotive districts.

Wherever possible, sections of ruling gradient should be so located as to facilitate their future reduction if traffic volumes warrant.

4. Helper Districts

Ideally, locomotives should be working at or near capacity over as much of their district as possible. Where a reasonable balance cannot be struck between a proposed ruling gradient and the rest of the line, it may be best to concentrate adverse gradients into one relatively short, steep section. This section can then be operated as a helper district with one or more helper locomotives on each train.

Helper gradients should be such as to permit utilization of the full capacity of the helper and train locomotives on a full tonnage train as determined by the ruling gradient on the remainder of the district.

Helper service is less desirable for modern, high-speed freight operation, because of the delays (and thus costs) associated with cutting the additional engines in and out. This is especially true where train lengths or physical limitations of the line require that helper engines be cut into the middle of trains.

The advantages of helper districts must also be weighed against the costs of maintaining locomotive servicing facilities and special engine crew assignments, and against the possible reduction in line capacity due to downbound light engine movements. If it is absolutely necessary to resort to helper service, consideration should be given to running helper engines through between terminals, preferably as additional units operated in multiple with the regular road locomotives.

5. Balanced Profiles

Characteristically, the total tonnage moving over a line in one direction will not be the same as that moving in the opposite direction. If the ruling gradient is the same in both directions, locomotives (and thus crews) will not be working to full capacity in the light-tonnage direction.

Where the extent of this traffic unbalance can be predicted in advance, it may be advantageous to employ a steeper ruling gradient in the light-tonnage direction and a flatter ruling gradient in the heavy-tonnage direction. These gradients should be so balanced that the average tractive effort required in each direction is the same. Notice that in computing train resistance (and thus required tractive effort) it is necessary to specify a desired train speed.

6. Momentum Gradients

Momentum gradients are sections of gradients which are steeper than the ruling gradient but short enough so that trains can overcome them with momentum.

Momentum gradients may often be used to effect economy in construction costs without reducing train size or the overall operating efficiency of a line. Such gradients should not be located at points where train stops or reduced speeds (below that required to operate the gradient) are likely to be necessary. The number and character of these gradients should in any case be such as to minimize the number and severity of train slack run-ins and run-outs.

Momentum gradients should not exceed that gradient over which a locomotive loaded for the ruling gradient could handle its train in two parts if stalled in the sag.

The length of momentum gradients should be such that the maximum speed of tonnage freight trains at the bottom of the sag need not exceed the speed limit for such trains and such that the minimum speed at the top of the gradient is within the limits permitted by short-time ratings of electric-drive locomotives in consideration of the power requirements on adjacent sections of the line.

7. Compensation for Curvature

Ruling gradients should be compensated for curvature so that the total train resistance on curves will not exceed that on tangent track. The general rule is that ruling gradients should be reduced throughout the length of each curve by 0.04 percent per degree of curvature.

Where a gradient is less than the ruling gradient, the gradient should, where necessary, be reduced sufficiently on each curve to keep the total train resistance equal to or less than that on the ruling gradient.

8. Gradient and Alinement Through Tunnels

The gradient through long tunnels should be kept as low as is economically practicable, so as to minimize the ventilation problem caused by locomotive exhaust gases and increase the overall reliability of train operation. It is recommended that in general the gradient through tunnels more than 1000 ft in length should not exceed 75 percent of the ruling gradient on the district. This reduced gradient should preferably be carried some distance above and below the tunnel as well.

A minimum gradient sufficient to ensure proper drainage in tunnels should also be established. It is recommended that this be no less than 0.3 percent. A somewhat steeper gradient may be necessary in wet tunnels where extremely cold weather can cause freezing of the drains.

Summits and sags should be avoided in tunnels which have no provisions for forced air ventilation. The seriousness of the ventilation problem will, in any case, depend upon the type of locomotives used, the number of locomotives on each train, the frequency of train movements, and the natural air currents in the vicinity.

Curvature should be avoided in tunnels wherever practicable. If curvature is required, the gradient should be compensated for it.

9. Passing Sidings

a. General

Passing sidings should be located on level track or minor gradients wherever this is consistent with the requirements for siding spacing. If passing sidings must be located on ruling gradients, the gradient should be so compensated throughout the entire length of the siding as to permit tonnage trains to be started from a full stop. Where trains must be stopped to operate hand-thrown switches at the entrance and exit to the siding, such compensation should also be carried a full train length beyond each end of the siding. Consideration should further be given to the compensation required on the turn-out curves at each end of the siding. An alternative to such heavy compensation is to so reduce siding spacing that downbound trains only can take the siding without causing an unacceptable rise in the average train delay.

Where possible, turnouts to auxiliary tracks which require switching by road crews should be located between switches on passing sidings. Such tracks should ideally connect with the passing siding rather than with the main track.

Road crossings should be avoided between the clearance points of passing sidings.

In determining the length of individual passing sidings for freight trains, consideration should be given to the following: the total length of trains including locomotives, cars, and caboose; a margin of stopping distance in siding; the length of turnouts to clearance points; the required signal clearance; and an allowance where standing trains must be cut to clear road crossings.

b. Siding Spacing

In general, the spacing between passing sidings (and thus the number of passing sidings on a district) should be determined on the basis of the traffic volume (number of trains in both directions per unit of time) and the resulting line capacity required, the average acceptable delay per train due to meets and passes, and the capital and maintenance costs of the sidings and their associated signaling.

In the interest of line capacity the time-spacing between each pair of adjacent passing sidings on single-track lines should be approximately equal over an entire district between traffic generation, traffic classification, or train make-up points. Time-spacing in this context is the time an average freight train in one direction requires to run between two sidings *plus* the time an average freight train in the other direction requires to run between the same two sidings plus the time the inferior train loses in making a meet.[3]

TRAIN PERFORMANCE

A. TRAIN RESISTANCE

1. Level Tangent Resistance

a. Characteristics

A railway vehicle moving on level, tangent track at constant speed and in still air encounters a total specific resistance (in pounds per ton of vehicle weight) that is the sum of the following components:

---

[3] For a discussion of the subject of line capacity see Part 4, Sec. B, of this chapter, as well as the various references cited therein.

(1) Rolling friction between the wheels and the rail. This can be considered constant for a given rail weight and track condition.

(2) Journal bearing friction. This varies with the weight on each axle and at low speeds with the type of bearing. Rolling resistance and journal bearing resistance are sometimes known collectively simply as journal resistance.

(3) Flange resistance. This is associated with the lateral motion of the vehicle and the friction between the wheel flanges and the gage side of the rail. It varies approximately as the first power of train speed.

(4) Air resistance. This includes head-end resistance for the locomotive and skin friction for both locomotives and cars. It is generally assumed to vary as the second power of train speed.

b. Davis Formula for Train Resistance[1]

In 1926 W. J. Davis, Jr., proposed a set of formulas for computing level tangent train resistance as a function of car locomotive weight, axle loading, vehicle cross-section, and train speed. The more important of these are:

Locomotives ....

$$R = 1.3 + \frac{29}{w} + 0.03V + \frac{0.0024AV^2}{wn}$$

Passenger cars ....
(trailing)

$$R = 1.3 + \frac{29}{w} + 0.03V + \frac{0.00034AV^2}{wn}$$

Freight cars ....
(trailing)

$$R = 1.3 + \frac{29}{w} + 0.045V + \frac{0.0005AV^2}{wn}$$

where R = resistance in pounds per ton.

w = average weight per axle in tons.

n = number of axles per vehicle.

A = vehicle cross-section in square feet (usually taken as 120 sq ft for locomotives and 90 sq ft for cars).

V = train speed in miles per hour.

In general it is felt that the Davis formulas give satisfactory results for both roller- and friction-bearing equipment for speeds ranging from 5 to 40 or 50 mph.

c. Level Tangent Resistance at High Speeds

At speeds greater than 40 to 50 mph the Davis formulas give values which are considered somewhat unreliable. At higher speeds the Davis values for freight equipment should probably be reduced somewhat to account for better track construction and maintenance.[2] For passenger equipment it is possible to use the more recent results published by A. I. Totten.[3]

2. Starting Resistance

The journal resistance of friction bearings is much higher at starting than when in motion. Depending upon the weight per axle and the temperature of the bearings (which is in turn a function of both the ambient temperature and the length of time the equipment has been stopped), starting resistance may be as high as 30 lb or more per ton. An average for light and heavy cars of 20 lb per ton at starting is a conservative assumption for above-freezing ambient temperatures.

Unlike friction bearings, the journal resistance of roller bearings at starting is essentially the same as when they are in motion. In general, the resistance given by the Davis formulas for 0 mph should be satisfactory for roller-bearing equipment when starting at above freezing ambient temperatures.

[1] See W. J. Davis, Jr., "Tractive Resistance of Electric Locomotives and Cars," *General Electric Review*, October 1928, pp. 685–708.

[2] See W. W. Hay, *Railroad Engineering* (Wiley, 1953), pages 72–75.

[3] A. I. Totten, "Resistance of Lightweight Passenger Trains," American Society of Mechanical Engineers, *Transactions*, Vol. 60, No. 2, Sec. 1 (February 1938), pages 206–8.

3. Curve Resistance

In general it can be assumed that the additional train resistance due to curvature amounts to 0.8 lb per ton per degree of curvature.

It should be noted that this resistance may be reduced somewhat by the use of curve oilers.

4. Grade Resistance

The additional resistance encountered on ascending gradients is equal to 20 lb per ton per percent of grade.

5. Wind Resistance

Though on most lines trains do not move in a constant direction with respect to winds, the possible effect of winds on train resistance should not be ignored.

The additional resistance due to head winds can be accounted for by adding the wind velocity to the train speed in computing air resistance.

Though full-scale train tests have not been conducted, wind-tunnel tests show that side winds can increase train resistance significantly. Winds quartering from ahead offer the most resistance. Winds of 30 to 40 mph with a 60-deg quartering angle may increase total air resistance by 40 to 60 per cent.

6. Other Resistance Factors

A discussion of other factors which may affect train resistance can be found in the following references:

	AREA Proceedings	
	Vol.	Page
Air conditioning (axle-driven)	42	83
Air conditioning	43	53
Atmospheric temperature	42	76
Atmospheric temperature	43	53
Equipment weight	42	77
Locomotive contour	42	78
Passenger train contour	42	80
Rail support	43	52
Rail weight and condition	42	74
Rail weight and condition	43	52
Speed on curves	42	75
Track conditions	42	76

Your committee submits for adoption the following additional recommendations with respect to Chapter 16 of the Manual, these recommendations, however, if adopted, to be incorporated in the 1962 Manual Supplement.

Pages 16–1–11 to 16–1–16, incl.

OPERATING DATA REQUIRED FOR A STUDY OF THE ECONOMIC JUSTIFICATION OF LINE AND GRADE REVISIONS

On page 16–1–11 change the first paragraph to read as follows:

"The fundamental requirement to justify either line or grade revisions is that when completed, the resulting operation will show a profit sufficient to pay a return on the total net cost of the improvement. In lieu of a profit there may be special cases in which other conditions will justify, for safety or competition will force an improvement irrespective of its cost, but for whatever reason a revision is made, estimates of the total cost of conducting traffic before and after changes should be made for the information of all concerned."

On page 16–1–11, second paragraph, change the first line to read "Ways in which operation may be bettered."

Change Item (c) to read "More effective supervision." Also, change the word "intensive" in the last line of the third paragraph to "effective."

On page 16–1–12, in the first paragraph after Item (26), delete the last sentence, reading "Also, heavier trains generally mean more road time which may increase overtime."

Page 16–2–12

COMPARATIVE FREIGHT TRAIN PERFORMANCE CHARTS

Change Par. B1 to read "More effective supervision, including scientific study and thoughtful effort, may increase the capacity of a railway."

In Par. B3, fourth line, substitute the words "more effective" for the word "increased."

Pages 16–4–1 and 16–4–2

TRAFFIC CLASSIFICATION OF RAILWAY MAIN TRACKS
Delete.

Page 16–4–2

SCHEDULE OF CLASSES OF COMPLETE ROADWAY AND TRACK STRUCTURE
Delete.

Report on Assignment 4

Potential Applications of Electronic Computers to Railway Engineering and Maintenance Problems in Research, Design, Inventory, Etc.

Collaborating with Committees 11 and 30, and informally with the Railway Systems and Procedures Association

F. Wascoe (chairman, subcommittee), L. P. Diamond, W. J. Dixon, S. B. Gill, C. A. James, T. J. Lamphier, R. J. Lane, A. S. Lang, M. B. Miller, V. J. Roggeveen, George Rugge, G. S. Sowers, Jos. J. Stark, Jr., C. L. Towle, T. D. Wofford, Jr.

Your committee calls attention to its advance report "The Digital Terrain Model Approach to Railroad Route Location," published in Bulletin 566, September–October 1961.

Your committee also submits the following report of progress in the compilation of information relative to the use of electronic computers by railways for engineering and maintenance problems.

On December 11, 1959, a questionnaire was submitted to 70 railroads, including one railroad which was subsequently merged into another company. Sixty-seven replies were received up to June 19, 1961, including a reply from the railroad subsequently merged. The following information was tabulated from these replies.

(1) ANALYSIS OF REPLIES:

a. Question 2—"Are you presently utilizing computers for solution of engineering and maintenance problems?"   Yes 7
      No 60

(Note: One "yes" respondent worked through consultant firm.)

b. Question 3—"Do you anticipate use of computers for these problems?"
      No 47

(Notes: Three "no" respondents had possible uses under investigation. One anticipated use was through the facilities of a consultant.)

(2) ANALYSIS OF "YES" REPLIES TO QUESTION 2 (1-A ABOVE):

Types of Problems	Number Using
1. Deed description	1
2. Tonnage rating	1
3. Curve data	1
4. Transition spirals	1
5. Cooper E ratings	2
6. Cut and fill calculations	1
7. Time study, maintenance gangs	1
8. Analysis of rail failures	1
9. Laying out 150 KC transposition points	1
10. Track capacity; siding length and spacing	1
11. Train performance	1
12. Details not furnished	1

(3) ANALYSIS OF "YES" REPLIES TO QUESTION 3 (1-B ABOVE):

Types of Problems	Number Using
1. Traverse closures	1
2. Triangulation grids	1
3. Bridge ratings	3
4. Bridge design	9
5. Building design	3
6. Economics of line changes	1
7. Grading quantities	7
8. Simulated train operation	1
9. Not yet determined	4
10. Line changes and alinement problems	3
11. Complicated calculations	2
12. Time and motion studies	1
13. Processing data on engineering records	1
14. Tonnage rating	1
15. Train performance	1
16. CTC studies	1
17. Clearance calculations	1

(4) ANALYSIS OF REPLIES ABOUT TYPES OF EQUIPMENT:

a. Now in use:

IBM 650	4
IBM 405	1
Sperry Univac	1
Consultant's machine	1

b. To be used:

IBM 650 8
IBM 705 1
IBM 1401 1
Sperry Rand Univac 120 1
Undecided 6
Through consultant 1
Through central bureau 1

(5) RESPONSE TO QUESTION 6:

'Would your company be willing to contribute duplicate programs to a central library for general use?"

Yes 13
Would consider 1
Plan to use central bureau 1
Use consultants 1
Plan to use existing programs from library 1

Total 17

Report on Assignment 6

Features of Economic and Engineering Interest in the Study, Design, Construction and Operation of New Railway Line Projects, or Major Line Relocations, Proposed, in Progress, or Recently Completed

T. D. Wofford, Jr. (chairman, subcommittee), J. L. Charles, A. J. Gellman, R. A. Gleason, F. E. Gunning, L. W. Haydon, H. C. Hutson, J. E. Inman, H. A. Lind, A. E. MacMillan, J. C. Martin, F. N. Nye, G. C. Payne, F. L. Rees, G. S. Sowers, D. L. Vercelote.

Vancouver to Squamish and the Peace River Extension on the Pacific Great Eastern Railway

History

The Pacific Great Eastern Railway was conceived in the railway boom days prior to 1912 as the Vancouver outlet of the Grand Trunk Pacific Railway and to develop the interior of British Columbia. Railways are still required to pioneer and to transport the products of agriculture, forests, mineral and other bulky commodities.

In 1909 the Grand Trunk Pacific Branch Lines Company was seeking a rail connection to the port of Vancouver. This Company preceded the Canadian Northern Pacific Railway in a survey down the Thompson River to Kamloops. The latter, under Provincial bond guarantee, constructed the present Canadian National Railway line to Port Mann and gained access over the Great Northern track to Vancouver.

In 1910, the GTP surveyed a route to Harrison Lake which was found to be impracticable, especially along the cut banks of the Fraser. Furthermore, it was unable to obtain financial support from the Province of British Columbia.

The PGE was incorporated under provincial charter in 1912 for the purpose of constructing and operating a railway from Vancouver to Prince George by way of Squamish, Lillooet and the Cariboo plateau, a distance of 468 miles, with bonds guaranteed by the Provincial Government at $35,000 per mile.

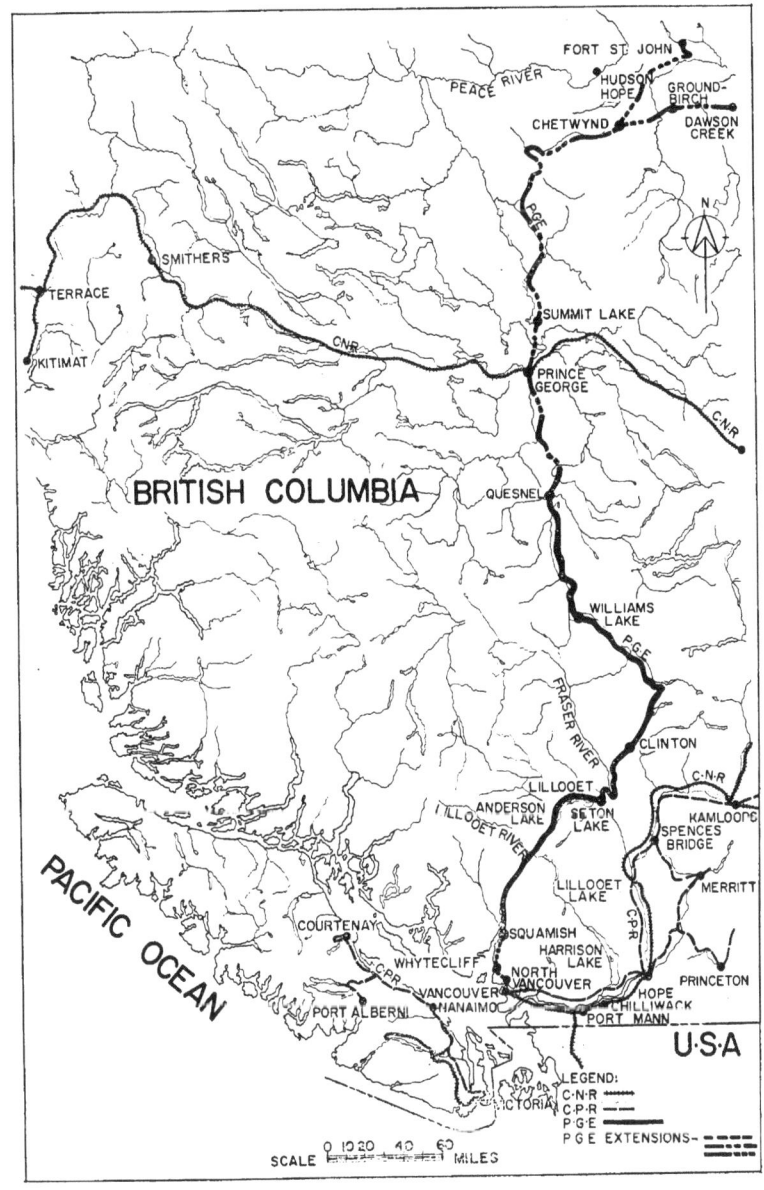

At Prince George terminal, a traffic agreement was made whereby all freight destined for Vancouver originating on the GTP was to be routed over the PGE; reciprocally, all freight originating on the PGE destined for eastern points was to be routed over the GTP lines.

Due to construction and financial difficulties in 1914, the bond guarantee was increased to $42,000 per mile as far as Prince George. A ncrthern extension into the Peace River Country was authorized, with a guarantee of $35,000 per mile.

When war effort curtailments and heavy construction costs exhausted available funds, the railway company was able to build only 13 miles from Vancouver to Whytecliff, and 165 miles from Squamish to Clinton. In 1917, the Provincial Government took over the company and appointed a board of directors.

The line was completed for operation into Quesnel in 1921.

This 347 miles of railway continued to operate in conjunction with its own freight barge service and a passenger service on the Union Steamship between Vancouver and Squamish, and has continued to operate, now that the Vancouver to Squamish extension, Quesnel to Prince George extension, and Prince George to Fort St. John and Dawson Creek extension are complete.

The PGE serves communities and industries which it has develcped. It also contributes to the economic welfare and indirect revenue of the Province of British Columbia apart from freight and passenger service.

This is the Pacific Great Eastern Railway.

Proposed Extensions

In 1945 a committee on resources and railways was asked to submit a report to the Government of the Province of British Columbia, in which they were to review reports of the resources of those areas in the north which would contribute traffic to the PGE and its extension, and to indicate by which route such extensions might be projected to form part of the transportation system for northern development.

The summary of this report stated that:

1. With the present amount of traffic (1945) in sight, extension will not improve the financial balance sheet of the PGE. A million tons a year of available railway freight would alter this conclusion.
2. The rate of settlement of the country and the rate of development of the natural resources should be increased.
3. When justified, build to Hudson Hope, a future junction of lines beyorid north and south Peace River.
4. The southern connection with Vancouver should be altered as business warrants.
5. The railway location surveys should be completed and revised before construction.

It was on these conclusions that each stage of the extensions, that is, the Quesnel to Prince George, the Squamish to North Vancouver, and the Peace River extensions north from Prince George, was undertaken.

A further report on the PGE extensions was made in 1954, and it was noted at that time that traffic potential was constantly improving due to at least two factors:

1. Accelerated development of the northern mineral wealth—which wou'd provide additional basic tonnage.
2. Increased oil and gas exploration and discovery which would at least provide additional high revenue "merchandise" traffic.

It could be seen that with the assumptions made, the Peace River project could be considered more or less economic, but only if low interest capital or preferred stock type capital was available. On the other hand, even at full current rates of interest, it would be difficult to refute the economic and practical advantages of a southern rail link over the present method of tug and barge. Therefore, it did appear that the most opportune time had come to get the southern extension under way, since capital would have to be expended on one type of link or the other.

Economic Considerations

Items considered were:

1. National defense function.
2. Theory that transportation facilities should precede resource development, not follow it.

It was further considered that if normal development would accrue anyway, then the extension was premature.

Further, the non-economic benefits are primarily those associated with the opening of a new frontier area, not those associated with "diversification" because the principal products of the Peace River are already produced in quantity in Western Canada. Recent mineral discoveries may add benefits of "diversification."

The precedent created by government subsidy of the Canadian Pacific Railway Company seems to bear out the foregoing argument. The CPR was subsidized in the hope that the long run development which it would encourage would finally make the railroad economic. However, the use of the CPR as a precedent is complicated by the great non-economic factor—the unifying of a country. It was suggested at the time the extension was contemplated that the non-economic benefits need not be on the scale of those achieved by the CPR to justify aid of the Peace River extension. The mere shortening of the route to the coast by 500 miles and the development of larger centers of population in the "hubs" of Prince George, Dawson Creek and Fort St. John, which could provide more and cheaper amenities of life for the people in our frontiers, was justification enough.

The total investment for the North Vancouver to Squamish and the Prince George to Fort St. John and Dawson Creek was $91,247,000. This is divided by the southern extension, costing $17,498,000, and the northern extension, costing the remainder of $73,749,000. This includes the Road General, the Track General, the Track Structures General, Wharves and Docks, Communications, Miscellaneous, Equipment, and Interest.

The total added revenue from traffic on the whole line was estimated at $11,640,000, while the out-of-pocket costs and fully distributed costs were estimated at $6,165,700, and $9,466,400, respectively. It was estimated that adjustment for Crow's Nest grain rates would decrease the operating revenue by $950,000. But this would be only if the line were to come under Federal jurisdiction.

A further estimate of divisions of revenue the northern extension might expect was prepared in 1954, and this estimated revenue totalled approximately $5,149,000. This was made up of such commodities as grain, forest products, pulpwood, coal, livestock, asbestos, merchandise and express. It was further estimated that the total operating expense less interest on capital invested would amount to $4,333,000. This made it apparent that the extension could pay all costs including depreciation, and would also be able to pay $816,000 on interest charges when tonnages equalled the estimated traffic.

The following authoritative statement was made by the 1945 Committee on Resources and Railways:

"From the studies that have been made, it would appear that the ship and barge service will be able to take care of the traffic until it reaches about four times the present quantity."

By 1953, the traffic had increased by six times the 1944 level.

Location

The preliminary and actual field location is a very interesting subject, and an endeavour will be made here to give the highlights surrounding the actual work.

Keeping design features in mind, the GTP in 1906 made an original survey of the Pine Pass route. This route, with some variations, was adopted by the PGE as the shorter and more direct route to the north, and was chosen over the Peace River Canyon route and the Monkman Pass route.

In 1946 the Department of Railways and the PGE made a preliminary location of the Pine Pass route using as much of the 1906 location information as possible. Forty years had elapsed since the ground control had been established and the field location had to be rerun.

The actual field location commenced in 1954 until construction started in the spring of 1955 and continued until the end of major construction in 1958. In conjunction with the field location, a preliminary location was spearheading the survey by the use of topographical maps, aerial photographs and a complete reconnaissance, so as to determine the most feasible and economical line between the terminal points; to locate the controlling points, both natural and artificial; to determine the maximum grade and the maximum rate of curvature; to ascertain the kind of material likely to be encountered in the construction of the road, and to determine the effect of the material on the cost of maintenance; to note the resources of the country and its capabilities for future development, and to calculate the probable effect of the building of the road on this development; to obtain a general idea of the approximate cost per mile and the total cost of the completed road.

The field parties for the preliminary and field location operated from base camps and fly camps using power wagons, river boats, horses and snowmobiles for transportation. Aerial photographs were used during preliminary location of the line between the East Pine River and Groundbirch, this being a semi-prairie cross-country route.

On the southern extension into Vancouver, since the route follows the shores of Howe Sound, water taxis were used for transporting crews during winter and summer. This water access had great advantages since the terrain was rugged, and for 25 miles no other access was available.

The location of the line for construction purposes was done mostly during winter months, all transportation again was by water taxis, and it was not until the grade was partly completed and tote roads spearheading construction were made, that power wagons could be used.

Design Features

1. Ruling grades 2.2 percent.
2. Curvature maximum 12 deg.
3. Spirals—Sullivans spirals (similar to Holbrook, Weatherby and other basic spirals):

   No. of chords = degree of curve.
   Total length = 200 ft (approx.)

Deflection $= \dfrac{N^2 L}{10}$

Where $N =$ chord number.
  $L =$ chord length.

4. Vertical curves:
     Summits 0.20
     Sags   0.10
5. Superelevation maximum 2½ in.
6. Grade compensation for curvature
     0.04 percent per degree of curvature.

The total line miles on the southern extension are 41 miles, and it has a total of 7800 deg in curvature, while 42 percent of the line is tangent. The total line miles on the northern extension are 325 miles, total curvature 17,121 deg, and approximately 75 percent of the line is tangent.

Construction

The problems of construction on the southern extension were many and varied. Access was difficult, beginning with clearing and grubbing and ending with distribution of materials for track laying.

Most of the equipment was brought in by barges, and wharves had to be constructed for unloading of same. The transportation of men and material was accomplished by using water taxis which had their headquarters at the different construction camps.

The construction on Howe Sound was divided into five residences, and was completed by four major contractors.

The quantities are as follows:

Clearing .... 510 acres
Grubbing .... 280 acres
Grading solid rock .... 2,531,000 cu yd
         other material .... 1,355,000 cu yd
Foundation excavation .... 29,500 cu yd
Tunnels, 5, total length .... 3,136 lin ft
Culverts .... 16,400 lin ft
Trestles 32
Major steel bridges 4

To make the project more difficult and interesting, a highway was being built simultaneously by the same general contractors. This highway is located above the railway and the problem of disposing of and placing material was ever present. Much of the solid rock was decomposed and continually threatened the construction crews with slides; scaling had to be carried on continually.

The laying of the track was done by company forces; 85-lb rail was used, and No. 8 and No. 10 turnouts were used in yards and sidings.

Access on the northern extension was made possible by the Hart Highway. Other access was accomplished by building access laterals from the highway to the railway right-of-way. Where the railway was too far away from the highway, access roads were constructed along the right-of-way. Weather and muskeg were the conditions most difficult to overcome during roadbed construction, while ice and foundation problems made bridge construction more difficult.

The northern extension was divided into twelve residences. Six major contractors completed the subgrade, with the following quantities:

Clearing	6,340	acres
Grubbing	2,360	acres
Grading solid rock	1,259,000	cu yd
other material	22,722,000	cu yd
overhaul	8,400,000	cu yd
Foundation excavation	102,000	cu yd
Tunnels, 1, length	950	lin ft
Trestles, 43, length	9,000	lin ft
Culverts	83,000	lin ft
Bridges, major steel	5	
timber	1	
overpasses	4	

The laying of track was done by company forces, while some of the surfacing was done by contract.

To overcome the muskeg problem, many drainage ditches were built, some of them paralleling the right-of-way, while fill of good granular material was placed after the muskeg had been excavated inside the roadbed fill slope stakes. Wet cuts were overcome by thorough drainage. In some cases where the foundation would not carry the fill load superimposed on it, trestles were built to overcome the problem. River erosion was prevented by rip-rapping, and the problem of soft roadbed was overcome by proper drainage.

The construction of the links between Vancouver and Fort St. John and Dawson Creek is now complete. The railroad is already making its impact on the economy of the North. The future looks bright for the Pacific Great Eastern Railway since it will grow with the North.

Santa Fe Relocation of Main Line in Arizona

The Santa Fe in August 1959 began construction of a 44-mile double-track line change in its Chicago–California route between Williams and Crookton, Ariz. This segment of the old line had been one of the most critical spots in this route because of heavy grades, sharp curvature, limited tunnel clearance, and expensive maintenance due to its location in steep Johnson Canyon. Construction of the new line was completed to such a point that operation was started on December 19, 1960, a matter of 68 weeks.

For the benefit of those not familiar with the Southwest, a little of the history of this line is worth mentioning. The original track was placed in operation in 1883 on a 2.6 percent grade ascending eastward and a 1.58 percent grade ascending westward. There were numerous sharp curves up to 10 deg 34 min. This location was established, even though the grades were heavy and curvature was sharp, because it was near to Prescott, a center of extensive mining activities. It provided ready access to water for locomotives, and required the least amount of grading through the volcanic materials. This route, despite the tunnel and the winding canyon, was a comparatively easy way to get down from elevation 6748 at Williams to elevation 5128 at Ash Fork, in a distance of about 23 miles.

By 1911 annual traffic had grown to the point that greater track capacity was needed and a second track was built. This track was constructed in such a manner that by use of non-parallel segments, lefthand operation and split grades, the maximum grade eastward was reduced to 1.85 percent from the previous 2.6 percent, the westward maximum grade remaining 1.58 percent.

Today the Santa Fe is in an era of competitive transportation which places great emphasis on fast and reliable freight service. During 1959 both eastward and westward

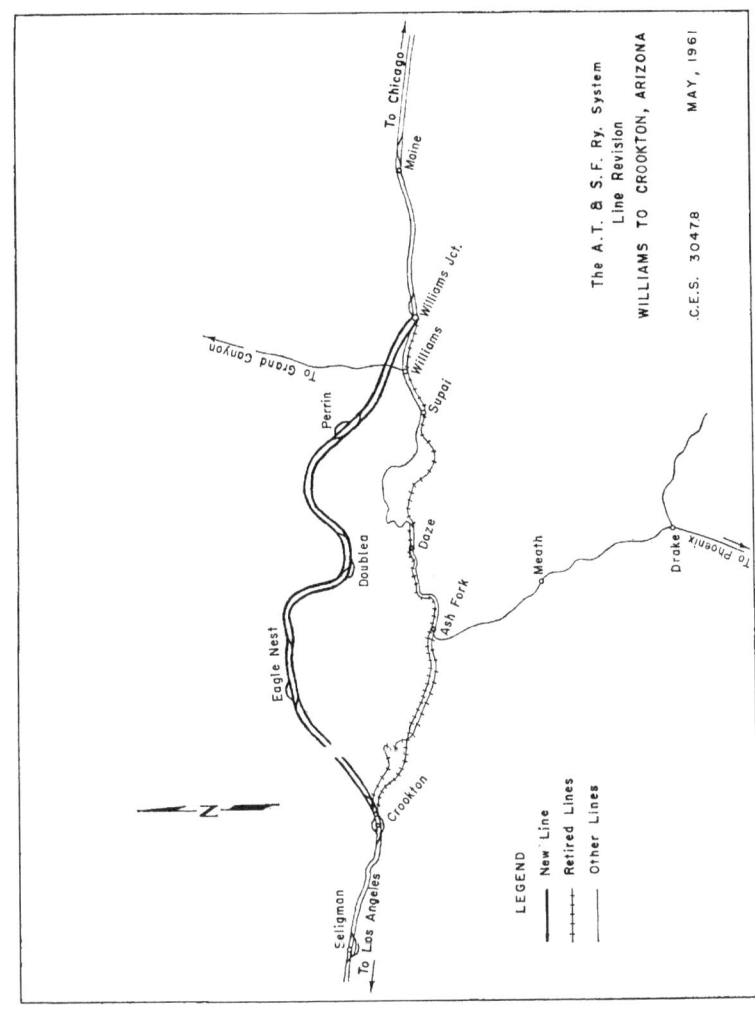

The A.T. & S.F. Ry. System
Line Revision
WILLIAMS TO CROOKTON, ARIZONA

.C.E.S.   30478   MAY, 1961

LEGEND
———— New Line
+++++ Retired Lines
———— Other Lines

freight schedules were stepped up to give one day earlier delivery on freight between California and the East. To make these time reductions it was necessary to adopt operating changes such as caboose pooling, engine terminal changes, maintracking of freight trains, higher speed limits, increased horsepower per train ton, shortened trains, and improved terminal operations.

A new national system interstate highway is being constructed almost parallel to the Santa Fe lines across New Mexico, Arizona and California, which presents an extremely competitive rival for the business. This necessitated a very close examination of all phases of operation from track to equipment. Fast, more powerful diesel locomotives had been put into service; new, longer, and special types of rolling stock had been added to the car fleet; and new, mechanized maintenance equipment had been developed. The clearance restrictions, heavy grades and sharp curvature of the mountain territory route, which limited the ability to utilize this new equipment to its greatest efficiency, could no longer be tolerated.

There were some places in the Chicago-California route in which curvature and grades could be improved to great advantage, although the necessary work would be expensive. The following three conditions had to be examined to determine which spot should be the first for consideration:

1. Could schedules be reduced and maintained over the line.
2. Which location had the greatest traffic involvement.
3. Which location had the most difficult operating and maintenance conditions.

The stretch of track between Williams and Crookton, Ariz., with its restrictive tunnel clearances, canyon wall location, helper grades, excessive curvature, severe speed limitations and considerable rise and fall, met all of these conditions and was chosen to be the first for improvement. This section for the past five years carried approximately 50 percent of the gross tons of revenue freight which moved over the railway system.

As previously stated the Williams–Crookton segment of the main line had many restrictions which hampered operation. Some of them were:

1. Westward freight trains had 19 miles with maximum allowable speed of 15 to 20 mph.
2. Westward track was located in Johnson Canyon wall for 10 miles.
3. There was an eastward 1.8 percent helper grade from Ash Fork to Supai which was the steepest ruling grade east of Barstow on the Santa Fe freight line.
4. The westward track had the Johnson Canyon Tunnel on a 10-deg 6-min curve which was one of the most restrictive clearances on the main line, making it necessary to move many trains against traffic on the eastward track.
5. Welded rail was not practical because of the sharp curves.
6. Maintenance-of-way forces for the most part could not be motorized.

In the area in which the relocation took place, mining had lost its significance. The change from steam to diesel power had eliminated the vital need for water, and modern grading equipment and the advance in engineering knowledge had surmounted the volcanic material obstacle. Therefore, the lines could be moved out of the canyon to improve the grade and curvature condition.

There were some controlling factors which had to be considered as a new route was being selected: the line had to stay within a reasonable distance of Ash Fork so a connection could be built to protect the Phoenix business or leave in a portion of present line to handle it; get shortest route around bad sections with minimum new construction; keep grades and curvature to a minimum. Consideration of simple curve and grade reductions was out of the question as about 7 miles additional distance would be necessary between Williams and Ash Fork for a continuous 1 percent grade line. It was mandatory to find a higher elevation for the low point on the line. To do this a reconnaissance was made by foot and automobile; and with what poor maps could be obtained, three feasible routes were determined.

Aerial surveys were made of the three routes, and photogrammetric maps on a scale of 1 in to 400 ft, with contour interval of 10 ft, were produced. About 15 lines were projected with grades varying up to a maximum of 1.42 percent and curves up to 2 deg. Construction costs of the most promising line over each route were analyzed for single-track TCS and double-track TCS. Exhibit "A" shows a brief cost comparison of the single- and double-track lines which were most economical from the construction standpoint. When these estimate figures were determined, management decided the double-track TCS line with high-speed traffic reversal was best for only a 12 percent increase in A&B cost over the single-track TCS line.

An economic study was necessary to show how such a large capital expenditure could be justfied. Exhibits "B", "C" and "D" show the items which were used in this analysis. Other items, such as savings per car day as related to per diem, cost of train stops and delays, cost of routing trains against current of traffic, dollar value of time saved on basis of tonnage added, and brake shoe wear, were not considered even though they have considerable value.

A time study and redispatch of trains on selected average traffic density days, were made, and time saving was calculated as follows:

Westward freight ....Average time saving 1 hr 11 min per train
Eastward freight ....   "   "   "   1 " 6   "   "   "
Westward passenger ....   "   "   "   " 49   "   "   "
Eastward passenger ....   ,"   "   "   " 25   "   "   "

The operating cost savings and the time savings above mentioned convinced management that the relocation was justifiable, and construction of a double-track TCS line with automatic train stop and high-speed crossovers for traffic reversal was authorized. These were to be no passing tracks as such, but crossovers were to be so spaced that main tracks could serve as passing tracks. Short setout sidings were provided. The traffic reversal signalling was extended both east and west beyond the limits of new construction to Maine and Seligman, giving approximately 66 miles of traffic reversal. One track between Williams and Ash Fork would remain in service for the Phoenix connection.

The engineering design used in this construction called for 58-ft base. including cut ditches for cuts in cinder or common material with slopes of 1 to 1. In limestone, sandstone or malpais (basalt) a 44-ft base including ditches, was used with slopes of ½ to ¼ to 1. Embankment width is 36 ft with slopes of 1½ to 1. Berms and benches were placed when deemed necessary by the construction engineer because of conditions encountered. Only three steel bridges were necessary, one to carry the track over U. S. Highway 66, one to carry the railroad over the road to Williams airport, and one to carry gas pipelines over the tracks in a deep cut. All other structures are corrugated pipe from 36 to 120 in. in diameter, or concrete boxes up to a triple 16- by 14-ft opening, many of which serve as drainage structures and cattle passes. Maximum cut is 115 ft and maximum fill 120 ft. All curves are compensated 0.04 percent per degree. Rate of change for all vertical curves is 0.04 except at the lowest point on line where 0.0182 rate was used to change from a minus 0.94 percent to a plus 0.88 percent grade,

The high-speed crossovers are No. 24 lateral turnouts with 39-ft curved switch points and rail bound manganese frogs.

Some difficulty was encountered in many spots in the cuts. Drilling indicated the types of material which underlaid the surface, but in the sandstone and limestone the structure was so heterogeneous that cuts had to be opened before enough information was available for design. In many places the dip of the strata varied in three directions up to 15 deg from horizontal, and the material was so badly shattered that it could be picked out by hand. One cut which is approximately 3 miles long, containing over 1¼ million yards had to be reworked, placing 25 ft wide benches every 20 ft on one side and 22½ ft wide benches every 30 ft on the other side. This added considerable yardage but makes for safety against crippling slides in the future.

At one section in this same cut the upper crust of solid sandstone about 25 ft thick was under such heavy pressure that after blasting the entire block had moved towards the center of the cut about 8 to 10 in on each side.

In many of the cuts the malpais was underlain with volcanic cinders which would not stand under the weight of the malpais, and the upper strata had to be cut back to relieve the pressure.

Select material for roadbed in rock cuts was not readily available at all locations, and pits were located and opened which gave the necessary material with haul sometimes up to 2 miles.

Rail laying started on July 6, 1960. This was the first time the Santa Fe had laid long welded rail on new construction and some interesting innovations were developed. Ties were plated at the company's Albuquerque treating plant and bundled in groups of 16. The bundles were trucked out on the grade and placed on a sled which was towed behind an unloading crane. The ties were spaced at a rate of 24 ties per 39 ft. The ties were lined to precise position before rail was taken off the car.

To get the rail in proper position over the tie plates, a roller device was constructed and placed about every 35 ties. A pipe-laying machine pulling two 1440-ft rails at a time, dragged the rails from the cars, guided them into these rollers and kept them in relative position by a spreader attachment on the puller boom. The rail was lifted from the rollers by a crane, the rollers were removed and rail lowered into final position on the tie plate. Every fifth tie was spiked so as to permit the train to move to next position for rail unloading.

Each day's rail laying was followed up by ballasting with a minimum of 4-in lift of volcanic cinders from the company's own ballast pit. Rail laying and first lift of ballast progressed at a rate of about three-quarters of a mile per day, including installation of turnouts and setout sidings. Second lift of ballasting went at a little slower rate.

The accompanying Exhibit "E" and map will give some idea of why this line relocation was given priority in the selection of possible improvements.

Exhibit "A"

Comparative Construction Cost Estimate

	Double Track TCS	Single Track TCS
Grading ....$ 8,497,719		$ 8,341,647
Bridging 3,700,000		3,275,000

	Double Track TCS	Single Track TCS
Track in place, ballasted	5,854,543	4,022,619
Signals and communications	2,184,197	2,145,763
All other and contingencies	3,779,042	3,168,534
Cost of New Work	$24,015,501	$20,953,563
Incidental operating expense	490,370	275,620
Gross cost	$24,505,871	$21,229,183
Less retirements and I.O.E.	5,157,690	4,025,620
Net A&B	$19,348,181	$17,203,563

EXHIBIT "B"

COST OF WORK NECESSARY TO KEEP PRESENT RAILROAD IN CONDITION FOR NEXT 5 YEARS

Rail, including ties and resurfacing $3,868,447
Bridges, maintenance only 11,800
Signals, TCS, EW, Welch to Supai, and modernize balance 1,067,000
Daylight Johnson Canyon Tunnel 870,000
Maintenance of retaining walls in Johnson Canyon 180,000

Grand total $5,997,247

EXHIBIT "C"

COST OF INSTALLING RAIL BRACES AND GAGE RODS ON 14.5 MILES OF CURVES

Material $199,738
Labor 22,620

Total $222,358

EXHIBIT "D"

ANNUAL SAVINGS DOUBLE TRACK—TCS LINE

Fuel
Road freight ....$ 257,922
Road passenger 63,164
Helper 28,932
Reduction in helpers, hostler and crews at Ash Fork 113,102
Reduction in mechanical forces at Ash Fork 51,839
Reduction in station forces at Ash Fork 63,542
Due to reduction in road locomotive hours 146,214

Maintenance of Way
Reduced forces 96,000
Rail transposition 36,000
Use of continuous welded rail 43,500
Bridge maintenance 5,100
Annual tie renewals 12,900
Reduction in car wheel wear 36,000
Reduction in perishable claims 45,844

Grand Total $1,000,119

EXHIBIT "E"

PHYSICAL CHARACTERISTICS

	Present Lines		Single Trk.	Double Trk.
	WW	EW	TCS	TCS
Length of line (miles)	43.0	43.9	44.0	44.0
Maximum curve	10°34′	10°01′	1°00′	1°00′
Total central angle	4614°	3490°	1391°	1510°
Maximum grade	1.46%	1.80%	1.0%	1.0%
Maximum elevation	6949		6949	6949
Minimum elevation	5035		5457	5457

Report on Assignment 7

Engineering, Maintenance and Operating Benefits to be Derived from Increased Joint Use of Railway Facilities

Collaborating with Committees 11, 14 and 20

J. W. Barriger (chairman, subcommittee), I. C. Brewer, D. E. Brunn, B. Chappell, S. B. Gill, J. E. Inman, T. D. Kern, R. J. Lane, R. F. Lark, R. McCann, F. J. Richter, K. A. Werden, H. L. Woldridge.

This is a final report, presented as information.

Previous reports have dealt with the overall scope of economic benefits that may be anticipated from joint facility arrangements and thereby permitting discontinuance of duplicate and often obsolete facilities, that represent an additional burden on railway operating expenses; also, specific examples of coordination, outlining the physical and economic benefits, as well as the general terms of such agreements, have been cited.

There are many notable coordination projects which have been consummated within the past few years, especially those between the Erie and Lackawanna Railroads for joint use of Hoboken Station as well as main trackage and appurtenant facilities, all prior to complete consolidation of the two properties in 1960.

Of course, greater benefits are derived through consolidation of railway properties, with three major mergers recently accomplished and many others being considered throughout the country; nevertheless, so many coordination opportunities remain that the possibilities are worthy of continuous exploration by railway officers.

For the reason that there are so many criteria available for use as guides in determining the principles involved and which serve as a basis for negotiating joint facility possibilities, and because the citing of other examples would merely be repetitive, we are closing this assignment with the plea for continued attention to this important phase of railway management as a means for producing economies and also improving service.

Report of Committee 30—Impact and Bridge Stresses

J. A. Erskine
P. L. Montgomery
C. V. Lund
E. R. Andrlik
A. R. Harris
N. E. Ekrem
D. S. Bechly
E. R. Bretscher
J. S. Carter
K. L. DeBlois
W. E. Dowling
C. E. Ekberg, Jr.
D. J. Engle
R. J. Fisher
A. T. Granger
J. F. Hoss, Jr.
K. H. Lenzen
J. F. Marsh
James Michalos
W. H. Munse
C. H. Newlin

N. M. Newmark
L. P. Nicholson
A. L. Piepmeier
M. J. Plumb
E. W. Prentiss
E. D. Ripple
C. A. Roberts
M. B. Scott
A. P. Smith
C. B. Smith
J. E. South
L. F. Spaine
C. A. Still
F. W. Thompson
G. S. Vincent
J. R. Williams
E. N. Wilson
L. W. Wood
J. D. Woodward
L. T. Wyly

D. W. Musser, *Chairman*
J. W. Davidson,
Vice Chairman
E. S. Birkenwald

Committee

Members listed in bold face type are the official representatives of the Engineering Division, AAR.

To the American Railway Engineering Association:

Your committee reports on the following subjects:

2. Steel truss spans.

At the expense of the Pittsburgh & Lake Erie Railroad, a service test was made during 1961 of the 1409 ft long cantilever truss span of its Ohio River Bridge at Beaver, Pa. The investigation was conducted principally to determine the magnitude of stress reversal in certain members of the anchor arm and to determine if heavier ore cars could be permitted on the structure.

3. Viaduct columns, collaborating with Committee 15.

No progress was made on this assignment this year. Generally this assignment has been progressed in conjunction with tests on other subjects. However, the type of structures which were tested and analyzed this year precluded the possibility of any progress on this assignment.

4. Longitudinal forces in bridge structures, collaborating with Committees 7, 8 and 15.

Due to the reduction in research appropriations, no testing for longitudinal forces in bridge structures was done during the past year. Also, prior tests which were reviewed by the committee during the year contained no data on this subject.

5. Distribution of live load in bridge floors:
   (a) Floors consisting of transverse beams,
   (b) Floors consisting of longitudinal beams.

Progress report, presented as information .... page 163

6. Concrete structures, collaborating with Committee 8.

   Progress report, presented as information .... page 163

7. Timber structures, collaborating with Committee 7.

   Progress report, presented as information .... page 164

8. Vibrational characteristics of bridges affecting deflections and depth ratios.

   An analytical study was started in the Department of Civil Engineering, New York University, in January 1959, to determine the effects of various parameters on the vibrational characteristics of railroad bridges. A method of analysis has been developed and programmed for solution by use of a digital computer and a few preliminary studies completed. However, this study has been suspended for the present, because of lack of funds.

9. Use of electronic computers for railroad bridge problems.

   Seventeen additional moment and shear tables for special heavy railroad cars were calculated with the moving load program and distributed to the chief engineers of Member Railroads. No new computer programs could be developed because of the reduced research budget. Depending on funds available in 1962, more tables for special loads that are in demand may be published, and perhaps a truss rating program can be developed.

10. Steel continuous structures, collaborating with Committee 15.

    Work during the year has consisted of assembling ideas regarding the type of research on this subject that will be of greatest value and interest to railroad engineers. Actual testing and preparation of reports has been deferred because of lack of funds.

11. Composite design of steel structures having concrete decks, collaborating with Committees 8 and 15.

    Research carried out by others down through the years on composite steel-concrete beams and progress reports of the Joint ASCE–ACI Committee on Composite Construction have been reviewed and reported on in committee meetings. Because of lack of funds no actual testing was carried out or recommended by the subcommittee.

    THE COMMITTEE ON IMPACT AND BRIDGE STRESSES,
    D. W. MUSSER, *Chairman.*

AREA Bulletin 567, November 1961.

Report on Assignment 5

Distribution of Live Load in Bridge Floors

(a) Floors Consisting of Transverse Beams
(b) Floors Consisting of Longitudinal Beams

N. M. Newmark (chairman, subcommitee), D. S. Bechly, E. S. Birkenwald, J. W. Davidson, K. L. DeBlois, R. J. Fisher, J. F. Hoss, Jr., C. H. Newlin, E. W. Prentiss, J. E. South, C. A. Still, E. N. Wilson, L. W. Wood.

The work on this assignment during the last year has dealt mainly with the evaluation and revision of the previously developed design procedures for the floor systems of steel railway bridges.

In November 1960 the final report of the University of Illinois on its investigation of the load distribution in steel railway bridges was presented to the committee. Although the design procedures presented in the report showed an excellent correlation with solutions obtained from the analytical methods and with the results of the AAR bridge tests, it was felt that these procedures were too complicated for use in design offices.

Since that time the suggested specifications and design procedures have been simplified considerably. However, in this simplification it has been necessary to make a number of assumptions concerning the range of each variable in the design equations. With the aid of typical and unusual bridge designs submitted by various railroads, the validity of the assumptions is being checked.

The revised specifications are now being prepared and will be ready shortly for review by the entire committee.

Report on Assignment 6

Concrete Structures

Collaborating with Committee 8

P. L. Montgomery (chairman, subcommittee), J. W. Davidson, W. E. Dowling, C. E. Ekberg, Jr., N. E. Ekrem, J. A. Erskine, J. F. Hoss, Jr., K. H. Lenzen, C. V. Lund, J. Michalos, N. M. Newmark, L. P. Nicholson, A. L. Piepmeier, E. D. Ripple, A. P. Smith, C. A. Still, F. W. Thompson, J. R. Williams, J. D. Woodward.

The field investigation by the AAR research staff on one of two identical 144 ft long prestressed concrete railroad bridges constructed in conjunction with the Air Force Academy near Colorado Springs, Colo., was published in October 1961, under Report No. ER–18. Each bridge consists of two 71 ft 6 in long spans, and each span has four modified "T" girders. These bridges, which carry the Santa Fe track over entrance roads to the Academy, have the longest spans in America today.

Test trains operated over the bridge at speeds up to 55 mph, and it was observed that there was little increase in strain with increase in speed. Maximum reported impact was only about 40 percent of that specified by the current AREA specification. The recorded static strains were less than the calculated values, and deflection values were found to be well within design limits.

The static and dynamic effects of trains operating over two prestressed concrete beam spans of a Florida East Coast bridge near Pampano Beach, Fla., has also been investigated by the research staff and a report prepared for review by the committee.

Each test span is composed of six 30 ft 6 in long beams and are identical except that the beams in one span have longitudinal shear keys, while those in the other span do not. An analysis was made on the spans before and after transverse post tensioning. From a preliminary review, it can be concluded that either shear keys or transverse post tensioning is effective in distributing the load across the deck, with some improvement if both are present. It is expected that a final report will be published next year.

Report on Assignment 7

Timber Structures

Collaborating with Committee 7

C. V. Lund (chairman, subcommittee), E. R. Bretsher, K. L. DeBlois, D. J. Engle, C. H. Newlin, F. W. Thomson, J. R. Williams, L. W. Wood.

Under this assignment the AAR Research Department prepared and released Report ER-1 titled "Investigation of 60-Ft Glued Laminated Beams on the Weyerhaeuser Tim- ber Company Railroad".* This investigation is of particular significance in that the beams are the longest glued laminated beams in the United States carrying regular diesel locomotives, and are of dimensions that could conceivably reflect future design practice. The bridge is shown in photograph 1. Fig. 1 shows the layout of the bridge and gives its significant dimensions. The diesel units in use at the time of test were of the 1200-hp 125-ton class.

The report compares actual with calculated stresses in flexure and in horizontal shear. Since strength in horizontal shear may be critical in the design of bridges utilizing long timber beams, the report presents some new information with respect to the posi- tion of wheel loads for maximum shear which is of special interest.

Photograph 1.

* Copies of the report may be obtained from the director of engineering research, Association of American Railroads, 3140 South Federal Street, Chicago 16, Illinois.

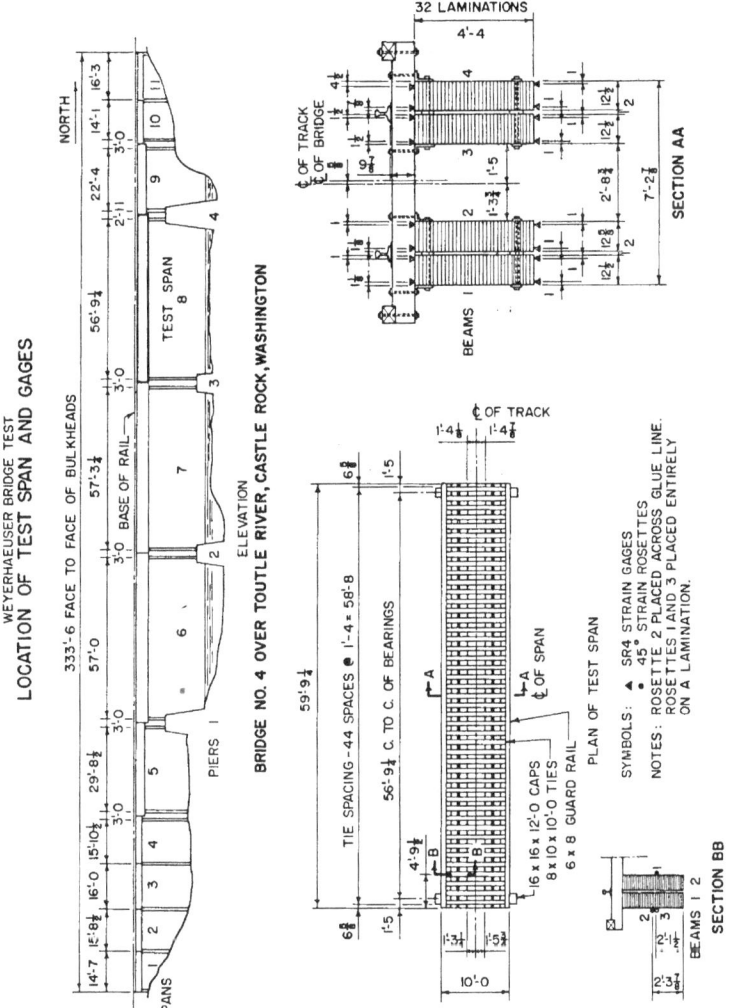

FIG. I
WEYERHAEUSER BRIDGE TEST
LOCATION OF TEST SPAN AND GAGES

ELEVATION

BRIDGE NO. 4 OVER TOUTLE RIVER, CASTLE ROCK, WASHINGTON

PLAN OF TEST SPAN

SYMBOLS: ▲ SR4 STRAIN GAGES
         ⬦ 45° STRAIN ROSETTES
NOTES: ROSETTE 2 PLACED ACROSS GLUE LINE.
       ROSETTES I AND 3 PLACED ENTIRELY
       ON A LAMINATION.

SECTION AA

SECTION BB

The following conclusions appear in the report.

(1) The simultaneous recorded maximum static flexural strains in the four glulam beams average approximately 13 percent less than calculated. The maximum live-load tensile stress was 1320 psi and the maximum live-load compressive stress was 1350 psi.

(2) The simultaneous recorded maximum static horizontal shear stresses in the two beams for which measurements were made varied considerably, being approximately 12.5 percent less and 47.0 percent greater than calculated according to accepted design equation. The maximum live-load horizontal shear stress was 118 psi.

(3) Measurements of the shear stress indicated an effective transfer of stress across a glue line.

(4) The study indicated that horizontal shear becomes a maximum when the first wheel of a truck was 1.7 to 2.0 times the depth of the beam from the bearing as the locomotive moved across the span.

(5) Impact tended to increase with speed, with the maximum recorded impact in flexure 12½ percent at a speed of 34 mph.

(6) Distribution of maximum simultaneous flexural stress to the four beams was reasonably consistent with the eccentricity of the track. Distribution of maximum recorded shear stress to the two beams tested was unequal due to a difference in reaction to each beam.

Report of Committee 20—Contract Forms

D. F. LYONS, *Chairman*
J. L. PERRIER,
    Vice Chairman

A. F. HUGHES, *Secretary*
W. D. KIRKPATRICK
J. F. HALPIN
E. M. HASTINGS, JR.
E. W. SMITH
MANUEL GARCIA
K. J. SILVEY
J. J. BAFFA
V. J. BONNER, SR.
E. E. BRADY
H. F. BROCKETT
R. F. CORRELL
A. B. COSTIC
C. R. DuBOSE
C. L. GATTON
C. E. GIPE
E. A. GRAHAM
R. C. HECKEL
F. M. JONES
J. S. LILLIE (E)

L. W. LINDBERG
F. B. MALLAS
W. J. MALONE
F. J. McMAHON
J. C. MILLER
H. K. MODERY
P. A. MOFFITT
W. G. NUSZ (E)
G. W. PATTERSON (E)
C. M. SHERMAN
W. B. SMALL
C. W. SMITH
W. R. SWATOSH (E)
D. S. TAYLOR
J. D. TAYLOR
W. B. TITTSWORTH, JR.
J. W. WALLENIUS
CLARENCE YOUNG
H. L. ZOUCK
    Committee

(E) Member Emeritus.
Members listed in bold face type are the official representatives of the Engineering Division, AAR.

To the American Railway Engineering Association:

Your committee reports on the following subjects:

1. Revision of Manual.
   Progress report, with recommendations submitted for adoption .... page 168

2. Form of agreement covering purchase and application of weed-control chemicals on railway property, collaborating with Committee 1.
   Deferred due to review of Manual material.

3. Form of lease covering right to strip-mine on railway miscellaneous physical property.
   Final report, submitted for adoption .... page 173

4 Form of agreement to cover disposal of surplus railway property.
   Progress report, submitted as information .... page 173

5. Form of lease for railway property used for unloading and storing liquified petroleum gases, anhydrous ammonia, and other flammable or dangerous materials.
   Deferred due to review of Manual material.

7. Bibliography on subjects pertaining to contract forms.
   Brief status statement .... page 175

THE COMMITTEE ON CONTRACT FORMS,
DONALD F. LYONS, *Chairman.*

AREA Bulletin 567, November 1961.

Report on Assignment 1
Revision of Manual

In keeping with the policy of the association that Manual material be reviewed periodically and kept up to date, Committee 20 last year divided its membership into seven subcommittees to study Chapter 20. Several of these subcommittees made partial or complete reports last year. The following report completes the review.

Report on Assignment 1 (a)
Revision of Manual (Construction Agreements)

Clarence Young (chairman, subcommittee), V. J. Bonner, Sr., E. E. Brady, R. F. Correll, C. R. DuBose, J. F. Halpin, E. M. Hastings, Jr., A. F. Hughes, L. W. Lindberg, F. J. McMahon, H.K.Modery, C. M. Sherman, C. W. Smith, J. D. Taylor, H. L. Zouck.

Your subcommittee submits the following recommendations with respect to Part 1, Chapter 20 of the Manual.

Page 20–1–12
FORM OF BOND
Reapprove without change.

Pages 20–1–13 to 20–1–22 incl.

FORM OF COST-PLUS PERCENTAGE CONSTRUCTION CONTRACT
Reapprove without change.

Pages 20–1–23 to 20–1–28 incl.

FORM OF CONSTRUCTION CONTRACT FOR MINOR PROJECTS
Reapprove without change.

Report on Assignment 1 (b)
Revision of Manual (Agreements Covering Passenger and Freight Facilities)

W. D. Kirkpatrick (chairman, subcommittee), J. J. Baffa, Manuel Garcia, C. E. Gipe, E. M. Hastings, Jr., R. C. Heckel, A. F. Hughes, F. B. Mallas, J. C. Miller, P. A. Moffitt, W. B. Small, E. W. Smith, D. S. Taylor, J. W. Wallenius.

Your committee submits the following recommendations with respect to Part 2, Chapter 20 of the Manual.

Pages 20–2–1 to 20–2–27 incl.

FORM OF AGREEMENT FOR THE ORGANIZATION AND OPERATION OF A JOINT PASSENGER TERMINAL PROJECT
Reapprove without change.

Pages 20-2-27 to 20-2-30 incl.

FORM OF AGREEMENT FOR CAB STAND AND BAGGAGE TRANSFER PRIVILEGES

Reapprove without change.

Pages 20-2-31 to 20-2-35 incl.

FORM OF AGREEMENT FOR JOINT USE OF PASSENGER STATION FACILITIES

Reapprove without change.

Pages 20-2-36 to 20-2-42 incl.

FORM OF AGREEMENT FOR JOINT USE OF FREIGHT TERMINAL FACILITIES

Reapprove without change.

Report on Assignment 1 (c)

Revision of Manual (Electrical Agreements)

E. M. Hastings, Jr. (chairman, subcommittee), D. F. Lyons, J. L. Perrier.

Your committee submits the following recommendation with respect to Part 3, Chapter 20 of the Manual.

Pages 20-3-25 to 20-3-30 incl.

FORM OF AGREEMENT FOR WIRE OR CABLE LINE CROSSING

Reapprove with the following revision:

Page 20-3-27, Art. 7. Insurance—Delete "$50,000" in line two; "$100,000" in line three; "$25,000" in line four; and "$50,000" in line five, and substitute therefor "$. . . . ." in each instance.

Report on Assignment 1 (d)

Revision of Manual (Agreements Covering Track)

J. L. Perrier (chairman, subcommittee), H. F. Brockett, A. B. Costic, C. L. Gatton, E. A. Graham, E. M. Hastings, Jr., A. F. Hughes, L. W. Lindberg, H. K. Modery, P. A. Moffitt, K. J. Silvey, W. B. Small, W. B. Tittsworth, Jr.

Your committee submits the following recommendations with respect to Part 4, Chapter 20 of the Manual.

Pages 20-4-1 to 20-4-6.1 incl.

FORM OF AGREEMENT FOR TRACKAGE RIGHTS

Reapprove with the following revisions:

Page 20-4-2, Art. 2. Description—Change the second paragraph to read:

"The tracks in said joint section are shown in red lines on the map dated . . . .

·········, ·········· designed as Exhibit ·········· attached hereto and made a part hereof."

Page 20–4–4, Art. 15. Removal of Employees—Change last line to read: " "A" Company to the "B" Company shall be prohibited from working upon or over said joint section."

Page 20–4–6.1—Change Art. 26. Execution, to read as follows:

IN WITNESS WHEREOF, the parties hereto have executed this agreement in ········ ······················, as of the day and year first above written.

·· Company

Witness ····························· By ·································

·· Company

Witness ····························· By ·································

Pages 20–4–7 to 20–4–9 incl.

FORM OF AGREEMENT FOR INDUSTRY TRACK

Reapprove with the following revisions:

Page 20–4–7—Change first two lines of WHEREAS clause to read: "WHEREAS, the Industry desires track facilities, hereinafter called side track, at or near . . .", and last line of this clause to read: "marked Exhibit "A" attached hereto and made a part hereof."

Page 20–4–8. Art. 7. Clearances—Add the following to this article:

"The minimum clearances herein specified may be changed by the Railway Company to meet changes in operating requirements and conditions or legal requirements, and the industry shall, upon written notice by the Railway company, make such changes in its tracks and facilities at the expense of the Industry, as may be necessary."

Page 20–4–9—Add article heading before the "Witness" clause and change the first two lines of this clause to read as follows:

11. Execution

"IN WITNESS WHEREOF, the parties hereto have executed this agreement in ········ ····················, as of the day and year above written."

Pages 20–4–10 to 20–4–14, incl.

FORM OF AGREEMENT FOR CROSSING OF RAILWAYS AT GRADE

Page 20–4–10—Change last four lines of the second "WHEREAS" clause to read: ". . . shown upon the plan marked Exhibit "A", dated ·············· attached hereto and made a part hereof."

Delete the third "WHEREAS" clause.

In Art. 1. Definition, change semicolons to commas in: third line, after word "installed", fifth line, after word "crossings", sixth line, after word "towers", and seventh line, after word "appliances".

Page 20–4–14, Art. 15. Duration and Succession, change second paragraph to read:

"All the covenants and agreements herein contained shall be binding upon and inure to the benefit of the successors and assigns of the respective parties hereto, provided however, that the "A" Company shall not assign or transfer the rights hereby granted to it, without the written consent of he "B" Company."

Change the "Witness" clause to read as follows:

16. Execution

In Witness Whereof, the parties hereto have executed this agreement in ......
...................., as of the day and year first above written.

.... Company

Witness .... By ....

.... Company

Witness .... By ....

Pages 20–4–15 and 20–4–16

FORM OF LICENSE FOR PRIVATE ROAD CROSSING

Reapprove with the following revisions:

Page 20–4–16, Art. 6. Liability—In the sixth line place a period after the word "crossing" and delete balance of sentence.

Change "Witness" clause to read:

8. Execution

In Witness Whereof, the parties hereto have executed this agreement in ....
................., as of the day and year first written.

.... Company

Witness .... By ....

.... Company

Witness .... By ....

Pages 20–4–17 to 20–4–20, incl.

FORM OF AGREEMENT FOR OPERATION OF COMMISSARY AND BOARDING OUTFITS

Reapprove with the following revisions:

Page 20–4–18, Art. 12. Liability and Risk—In Par. (b), fourth line, place a period after the word "damage" and delete balance of sentence.

Page 20–420—Change "Witness" clause to read:

7. Execution

In Witness Whereof, the parties hereto have caused this agreement to be executed in ...., as of the day and year first above written.

.... Company

Witness .... By ....

.... Company

.... Contractor

Witness .... By ....

Pages 20–4-*21* and 20-4-*22*

FORM OF AGREEMENT FOR FURNISHING WATER FROM RAILWAY WATER SYSTEMS TO EMPLOYEES AND OTHERS

Reapprove with the following revisions:

Page 20–4–22—Change "WITNESS" clause to read:

11. Execution

IN WITNESS WHEREOF, the parties hereto have executed this agreement in ......
....................., as of the day and year first above written.

                       .... Company

Witness ....   By ....

                       .... Consumer

Witness ....   By ....

Page 20–4–23

FORM OF AGREEMENT FOR PURCHASE OF WATER

Reapprove with the following revisions:

Change "WITNESS" clause to read:

5. Execution

IN WITNESS WHEREOF, the parties hereto have executed this agreement in ....
...................., as of the day and year first above written.

                       .... Company

Witness ....   By ....

                       .... Company

Witness ....   By ....

Pages 20–4–24 and 20–4–25

FORM OF AGREEMENT FOR PLACING SNOW OR SAND FENCES OFF THE RAILWAY COMPANY'S PROPERTY

Reapprove with the following revisions:

Page 20–4–25—Change "WITNESS' clause to read:

4. Execution

IN WITNESS WHEREOF, the parties hereto have executed this agreement in ....
...................., as of the day and year first above written.

                       .... Company

Witness ....   By ....

                       .... Licensor

Witness ....   By ....

Pages 20–4–25 to 20–4–28, incl.

FORM OF LEASE COVERING THE USE OF RAILWAY TRACKS FOR STORAGE OF TANK CARS CONTAINING LIQUIFIED PETROLEUM GASES, ANHYDROUS AMMONIA AND OTHER FLAMMABLE OR DANGEROUS MATERIALS

Reapprove without change.

It is recommended that the following agreements be transferred from Part 4 to Part 7, Miscellaneous Agreements.

Form of Agreement for Operation of Commissary Outfits
Form of Agreement for Furnishing Water
Form of Agreement for Purchase of Water

Report on Assignment 3

Form of Lease Covering Right to Strip-Mine on Railway Miscellaneous Physical Property

E. M. Hastings, Jr. (chairman, subcommittee), J. J. Baffa, R. F. Correll, C. L. Gatton, A. F. Hughes, F. M. Jones, F. B. Mallas, J. C. Miller, P. A. Moffitt, C. W. Smith, W. B. Tittsworth, Jr., H. L. Zouck.

Last year your committee submitted a draft of agreement covering this subject, which was published in Bulletin 560, pages 351–356, incl., and requested comment and criticism thereon.

Having received no adverse comment, the committee recommends that this agreement, as published in the above-mentioned Bulletin, be adopted and published in the Manual at the end of Part 5, Chapter 20.

Report on Assignment 4

Form of Agreement to Cover Disposal of Surplus Railway Property

E. W. Smith (chairman, subcommittee), E. E. Brady, H. F. Brockett, A. B. Costic, C. E. Gipe, E. M. Hastings, A. F. Hughes, L. W. Lindberg, W. J. Malone, F. J. McMahon, W. B. Small, Chas. W. Smith, D. S. Taylor, W. B. Tittsworth, J. W. Wallenius, Clarence Young.

Your committee submits (below) a form of agreement to cover disposal of surplus railway property, with the understanding that such property would consist of minor items such as small buildings and miscellaneous facilities of small monetary value where no unusual hazard would be involved in their removal. It is recommended that the agreement be published in the Bulletin this year for information only, in order to give members of the Association an opportunity to review it critically, looking to submitting it next year for adoption and publication in the Manual.

Next year the committee will also attempt to work out a form of agreement to cover removal of larger items of surplus railway property—miscellaneous facilities of considerable monetary value and involving considerable hazard in their removal, which would warrant the inclusion of a suitable railroad protective liability insurance clause.

FORM OF AGREEMENT TO COVER DISPOSAL OF SURPLUS RAILWAY PROPERTY

In consideration of the payment of $ ...., receipt of which is hereby acknowledged, ... Railway Company (hereinafter called Seller) does hereby sell to .... (hereinafter called Purchaser) the following item(s) located at ....

....

....

....

....

In consideration of such sale, Purchaser agrees, at Purchaser's sole cost and expense, to remove said item(s) from Seller's property not later than .... 19 ...., and leave said property of Seller in a condition satisfactory to Seller, and when necessary cap any and all water, sewer and gas mains affected by such removal, and not to permit any person or persons, other than Purchaser, Purchaser's agents, servants or employees, to be upon or about said property of Seller.

Purchaser further agrees to comply with all local, state and federal laws applicable to the dismantling and removal of said item(s) from Seller's property.

As an additional consideration for such sale, Purchaser hereby agrees to protect, indemnify and save harmless Seller from and against any and all claims, demands, causes of action, damages and costs and expenses of every kind for or on account of injury to or death of persons whomsoever, or loss or destruction of or damage to property whatsoever or to whomsoever belonging, arising from, growing out of or incidental to the removal and dismantling of said item(s) or the violation by Purchaser of any of the provisions hereof.

In Witness Whereof, the parties hereto have executed this agreement in ...., as of the day and year first above written.

.... Company

Witness ....   By ....

.... Company

Witness ....   By ....

Note: The above form of agreement is to cover disposal of minor items of surplus railway property, small buildings, miscellaneous facilities of small monetary value, and where no unusual hazard is involved. .

Instructions: If Blanket Sale Order, show number and date and list items. If a Specific Sale Order, attach to contract without listing items, and insert in blank space the words "as described in Sale Order No. .... dated .... hereto attached and made a part hereof."

Report on Assignment 7

Bibliography on Subjects Pertaining to Contract Forms

K. J. Silvey (chairman, subcommittee), G. K. Davis, Manuel Garcia, J. F. Halpin, A. F. Hughes, W. D. Kirkpatrick, J. S. Lillie, D. F. Lyons, O. K. Morgan, W. G. Nusz, G. W. Patterson, J. L. Perrier, W. R. Swatosh, Clarence Young.

Your committee submits the following report of progress on Assignment 7.

All bibliographies that have been printed in the Proceedings were reviewed in order to determine the type of bibliography most suitable for Committee 20. It was decided that this bibliography should be in the form of reference material and latest legal literature pertaining to contracts, and the committee will start to assemble such information for a further report on this subject.

Report of Committee 9—Highways

E. R. Englert
R. E. Nottingham
J. A. Jorlett
F. C. Cunnninham
Raymond Dejaiffe
W. S. Autrey
F. N. Barker
H. E. Bartlett
G. B. Blatt
Bernard Blum (E)
W. A. Buckmaster
R. B. Carrington, Jr.
A. C. Cayou
C. A. Christensen
H. B. Clarkson
M. H. Corbyn
J. T. Fitzpatrick
C. R. Fears
T. L. Gibson
C. I. Hartsell
Wm. J. Hedley
J. T. Hoelzer

J. A. Holmes
W. H. Huffman
Maro Johnson (E)
N. M. Kelly
J. E. K. Krylow
J. R. Latimer
H. L. Michael
E. S. Miller
H. G. Morgan (E)
R. J. Pierce
W. C. Pinschmidt (E)
P. H. Slack
K. E. Smith
J. E. Spangler
R. F. Spars
C. W. Traister
T. M. Vanderstempel
H. W. Walbright
V. R. Walling (E)
H. J. Wilkins
K. E. Wyckoff

J. M. Trissal, *Chairman*
R. W. Mauer,
   Vice Chairman
R. E. Skinner, *Secretary*

Commitee

(E) Member Emeritus.
Members listed in bold face type are the official representatives of the Engineering Division, AAR.

To the American Railway Engineering Association:

Your committee reports on the following subjects:

1. Revision of Manual.

   Progress report, including recommended revisions .... page 178

2. Merits and economics of prefabricated types of highway–railway grade crossings.

   Installation and maintenance costs of the various types of prefabricated crossings are being accumulated and studied. Analysis of these costs will provide the basis for future committee recommendations.

3. Merits of various types of highway–railway grade crossing protection, collaborating with Communication and Signal Section AAR.

   Progress report, submitted as information .... page 185

5. Recommended method of developing annual maintenance cost of the various types of highway–railway grade crossing protection (collaborating with Communication and Signal Section AAR.

   Progress report, submitted as information .... page 187

6. Methods of providing additional advance warning to highway traffic approaching a highway–railway grade crossing.

   Progress report, submitted as information .... page 187

THE COMMITTEE ON HIGHWAYS,
J. M. Trissal, *Chairman.*

AREA Bulletin 567, November 1961.

MEMOIR

Raymond Westcott

Raymond Westcott, senior engineer, Reading Company, passed away at his home, 441 Oley Street, Reading, Pa., January 21, 1961.

Mr. Westcott was born at Dennisville, N. J., April 18, 1898, and received his higher education at Drexel Institute, Philadelphia, Pa.

Mr. Westcott entered the service of the Reading Company on May 21, 1917. After working in various positions in the engineering department as rodman, levelman and transitman, he was transferred to the operating department as supervisor of track and division engineer. On December 1, 1952, he returned to the engineering department as construction engineer. On Septembr 1, 1957, he was made crossing engineer and on August 1, 1959, senior engineer, the position. he held at the time of his death.

Mr. Westcott joined the AREA in 1945 and had served on Committee 9 since 1955.

The members of Committee 9 sincerely regret Mr. Westcott's passing and will miss his service and friendly association.

G. B. BLATT, *Chairman,*
J. M. TRISSAL,
R. E. SKINNER,
Committee on Memoir.

Report on Assignment 1

Revision of Manual

E. R. Englert (chairman, subcommittee), F. N. Barker, A. C. Cayou, C. A. Christensen, H. B. Clarkson, F. C. Cunningham, C. R. Fears, Wm. J. Hedley, J. T. Hoelzer, J. A. Jorlett, J. E. K. Krylow, J. R. Latimer, R. W. Mauer, E. S. Miller, R. E. Nottingham, W. C. Pinschmidt, J. E. Spangler, T. M. Vanderstempel, H. J. Wilkins, K. E. Wyckoff.

Your committee has completed a review of its entire Chapter of the Manual and submits the following recommendations:

Page 9-1-2

GENERAL SPECIFICATIONS FOR HIGHWAY GRADE CROSSING OVER RAILROAD TRACKS

Reapprove with the following revision:

Art. 1—Width of Crossing. Change to read "In order to reduce the hazard of vehicles running off the crossing, it is recommended that the crossing be wider than the adjacent approach pavement or roadway surface."

Page 9-2-2

Fig. 1—Highway Crossing Sign, Painted, 6-Ft, 50-Deg Type

Reapprove with the following revisions:

Change note reading "ground level" to "roadway level." Change the dimension of 10 ft 6 from the "roadway level" to the center line of the crossbuck to read 9 ft min.

Page 9–2–3

Fig. 2—Highway Crossing Sign, Reflector, 6-Ft 50-Deg Type

Reapprove with the following revisions:

Change note reading "ground level" to "roadway level." Change the dimension of 10 ft 6 from the "roadway level" to the center line of the crossbuck to read 9 ft min.

Page 9–2–4

Fig. 3—Highway Crossing Sign, Painted, 4-Ft 90-Deg Type

Reapprove with the following revisions:

Change note reading "ground level" to "roadway level." Change the dimension of 10 ft 6 from the "roadway level" to the center line of the crossbuck to read 9 ft min.

Page 9–2–5

Fig. 4—Highway Crossing Sign, Reflector, 4-Ft 90-Deg Type

Reapprove with the following revisions:

Strike out the third note "Paint post white or aluminum." Strike out the second sentence of the next to last note and add to the first sentence the words "but other types of posts may be used." Strike out the side view sketch of the crossing sign. Change note reading "ground level" to "roadway level." Change the dimension of 10 ft 6 from the "roadway level" to the center line of the crossbuck to read 9 ft min.

Page 9–2–6

Fig. 5—Reflector Watchman Off Duty Sign

Reapprove without change.

Page 9–2–7

Fig. 6—Reflector Gates Not Working Sign

Reapprove without change.

Page 9–2–8

Fig. 7—Mounting of Reflector Signs

Reapprove without change.

Page 9–2–9

Fig. 8—Cover Plates for Signs

Reapprove without change.

Page 9–3–1

RECOMMENDED USE OF HIGHWAY-RAILWAY GRADE CROSSING SIGNALS

Reapprove without change.

Pages 9–3–2 and 9–3–3

REQUISITES FOR HIGHWAY GRADE CROSSING SIGNALS

Reapprove with the following revisions:

Change title from "Requisites" to "Specifications."

Art. 1—Purpose. In the first line change the word "requisites" to "specifications." In the second line change the word "requirements" to "provisions."

Page 9-3-4

Fig. 1—Typical Location Plan for Automatic Crossing Signals With or
    Without Gates for Two-Way Highway Traffic–Right Angle

Reapprove with the following revisions:

Change drawing to show crossing material extending slightly past edges of pavement. Strike the two words "shoulder" on each side of pavement and remove the two
outside lines representing the outer edges of the shoulders. Change the 12 ft dimensions
from the center line of tracks to center line of the gate mechanisms to read "12 ft min."

Page 9-3-5

Fig. 2—Typical Location Plan for Automatic Crossing Signals With or
    Without Gates for Two-Way Highway Traffic–Acute Angle

Reapprove with the following revisions:

Change drawing to show crossing material extending slightly past edges of pavement. Strike the two words "shoulder" on each side of pavement and remove the two
outside lines representing the outer edges of the shoulders. Change the 12 ft dimensions
from the center line of tracks to center line of the gate mechanisms to read "12 ft min."

Page 9-3-6

Fig. 3—Typical Location Plan for Automatic Crossingn Signals With or
    Without Gates for Two-Way Highway Traffic–Obtuse Angle

Reapprove with the following revisions:

Change drawing to show crossing material extending slightly past edges of pavement. Strike the two words "shoulder" on each side of pavement and remove the two
outside lines representing the outer edges of the shoulders.

Page 9-3-6.1

Fig. 3A—Typical Location Plans for Automatic Flashing-Light Signals With
    or Without Gates for One-Way Highway Traffic—Right Angle

Reapprove with the following revisions:

Change drawing to show crossing material extending slightly past edges of pavement. Strike the two words "shoulder" on each side of pavement and remove the two
outside lines representing the outer edges of the shoulders. Change the 12 ft dimensions
from the center line of tracks to center line of the gate mechanisms to read "12 ft
min." In the title of the drawing replace the words "Flashing-Light" with "Crossing."

Page 9-3-6.2

Fig. 3B—Typical Location Plans for Automatic Flashing-Light Signals
    With or Without Gates for One-Way Highway Traffic—Right Angle

Reapprove with the following revisions:

Change drawing to show crossing material extending slightly past edges of pavement. Strike the two words "shoulder" on each side of pavement and remove the two
outside lines representing the outer edges of the shoulders. Change the 12 ft dimensions
from the center line of tracks to center line of the gate mechanisms to read "12 ft
min." In the title of the drawing replace the words "Flashing-Light" with "Crossing."

Page 9–3–6.3

Fig. 3C—Typical Location Plans for Automatic Flashing-Light Signals With or Without Gates for One-Way Highway Traffic—Right Angle

Reapprove with the following revisions:

Change drawing to show crossing material extending slightly past edges of pavement. Strike the two words "shoulder" on each side of pavement and remove the two outside lines representing the outer edges of the shoulders. Change the 12 ft dimensions from the center line of tracks to center line of the gate mechanisms to read "12 ft min." In the title of the drawing replace the word "Flashing-Light" with "Crossing."

Page 9–3–6.4

Fig. 3D—Typical Location Plans for Automatic Flashing-Light Signals With or Without Gates for One-Way Highway Traffic Divided Highway—Right Angle

Reapprove with the following revisions:

Change drawing to show crossing material extending slightly past edges of pavement. Strike the two words "shoulder" on each side of pavement and remove the two outside lines representing the outer edges of the shoulders. Change the 12 ft dimensions from the center line of tracks to center line of the gate mechanisms to read "12 ft min." In the title of the drawing replace the word "Flashing-Light" with "Crossing."

Page 9–3–6.5

Fig. 3E—Typical Location Plans for Automatic Flashing-Light Signals With or Without Gates for One-Way Highway Traffic Divided Highway—Right Angle

Reapprove with the following revisions:

Change drawing to show crossing material extending slightly past edges of pavement. Strike the two words "shoulder" on each side of pavement and remove the two outside lines representing the outer edges of the shoulders. Change the 12 ft dimensions from the center line of tracks to center line of the gate mechanisms to read "12 ft min." In the title of the drawing replace the words "Flashing-Light" with "Crossing."

Page 9–3–6.6

Fig. 3F—Typical Location Plans for Automatic Flashing-Light Signals With or Without Gates for One-Way Highway Traffic Divided Highway—Right Angle

Reapprove with the following revisions:

Change drawing to show crossing material extending slightly past edges of pavement. Strike the two words "shoulder" on each side of pavement and remove the two outside lines representing the outer edges of the shoulders. Change the 12 ft dimensions from the center line of tracks to center line of the gate mechanisms to read "12 ft min." In the title of the drawing replace the word "Flashing-Light" with "Crossing."

Page 9–3–6.7

Fig. 3G—Typical Location Plans for Automatic Flashing-Light Signals With or Without Gates for One-Way Highway Traffic Divided Highway—Right Angle

Reapprove with the following revisions:

Change drawing to show crossing material extending slightly past edges of pavement. Strike the two words "shoulder" on each side of pavement and remove the two outside lines representing the outer edges of the shoulders. Change the 12 ft dimensions from the center line of tracks to center line of the gate mechanisms to read "12 ft min." In the title of the drawing replace the words "Flashing-Light" with "Crossing."

Page 9–3–6.8

Fig. 3H—Typical Location Plan (Composite) for Automatic Flashing-Light Signals and Gates for One-Way Highway Traffic Divided Highway—Acute Angle

Reapprove with the following revisions:

Change drawing to show crossing material extending slightly past edges of pavement. Strike the two words "shoulder" on each side of pavement and remove the two outside lines representing the outer edges of the shoulders. Change the 12 ft dimensions from the center line of tracks to center line of the gate mechanisms to read "12 ft min." In the title of the drawing replace the words "Flashing-Light" with "Crossing."

Page 9–3–6.9

Fig. 3J—Typical Location Plan (Composite) for Automatic Flashing-Light Signals and Gates for One-Way Highway Traffic Divided Highway—Obtuse Angle

Reapprove with the following revisions:

Change drawing to show crossing material extending slightly past edges of pavement. Strike the two words "shoulder" on each side of pavement and remove the two outside lines representing the outer edges of the shoulders. Change the 12 ft dimensions from the center line of tracks to center line of the gate mechanisms to read "12 ft min." In the title of the drawing replace the word "Flashing-Light" with "Crossing."

Page 9–3–7

Fig. 4—Typical Curb and Gutter Location Plan for Automatic Crossing Signals With or Without Gates

Reapprove without change.

Page 9–3–8

Fig. 5—Highway Crossing Sign—Flashing-Light Type With Stop on Red Signal Sign

Reapprove without change.

Page 9–3–9

Fig. 6—Highway Crossing Signal—Flashing-Light Type With Stop Sign
Reapprove without change.

Page 9–3–10

Fig. 7—Highway Crossing Signal—Wig-Wag Type With Stop When Swinging Sign

Reapprove without change.

Page 9–3–11

Fig. 8—Highway Crossing Signal—Wig-Wag Type with Stop Sign
Reapprove without change.

Page 9–3–12
Fig. 9—Highway Crossing Signal—Flashing-Light Type, 6-Ft and 8-Ft Cantilever Span

Reapprove with the following revision:
Change the 14 ft 0 clearance to read 14 ft min.

Page 9–3–13
Fig. 10—Highway Crossing Signal—Flashing-Light Type, 10-Ft and 12-Ft Cantilever Span

Reapprove with the following revision:
Change the 14 ft 0 clearance to read 14 ft min.

Page 9–3–14
REQUISITES FOR "NO RIGHT TURN" OR "NO LEFT TURN" SIGNALS

Reapprove with the following revisions:
Change title from "Requisites" to "Specifications."
Art. 1—Purpose. In the first line change the word "requirements" to "specifications."
In the second line change the word "requirements" to "provisions."
Art. 5—Painting. Strike out the letter (a) at the beginning of the article and the last two words "when installed."

Page 9–3–15
Fig. 11—"No Right Turn" or "No Left Turn" Signal

Reapprove without change.

Pages 9–4–1 and 9–4–2
REQUISITES FOR AUTOMATIC CROSSING GATES

Reapprove with the following revision:
Change title from "Requisites" to "Specifications."

Page 9–4–3
LAMPS ON MANUALLY OPERATED CROSSING GATES

Reapprove with the following revision:
Change title to read "Specifications for Lamps on Manually Operated Gates."

Page 9–4–4
Fig. 1—Highway Crossing Signal—Flashing-Light Type with Suspended Lights and Mast-Mounted Gate

Reapprove without change.

Page 9–4–5
Fig. 2—Highway Crossing Signal—Flashing-Light Type with Extended Lights and Pedestal-Mounted Gate

Reapprove without change.

Page 9–5–1
RECOMMENDED USE OF FLOODLIGHTING

Reapprove without change.

Page 9–5–2

REQUISITES FOR FLOODLIGHTING OF HIGHWAY—RAILWAY GRADE CROSSINGS

Reapprove with the following revision:
Change title from "Requisites" to "Specifications."

Page 9–M–1

PROTECTING HIGHWAY—RAILWAY GRADE CROSSINGS AND FLANGEWAYS

Reapprove without change.

Page 9–M–2

USE OF CENTER COLUMNS FOR HIGHWAY GRADE SEPARATION

Delete all the material on this page.

Page 9–M–3 to 9–M–6 incl.

TYPES OF BARRIERS FOR DEAD-END STREETS

Reapprove with the following revisions:

Page 9–M–3. Change first sentence to read "These plans show two types of barriers and the locations recommended for their use." Strike out "3· Concrete wall barrier—recommended . . ." and "4· Sign for barrier." In Fig. 1 remove the "Stop" sign and the 3'6" dimension to its center line. In the "Specifications" at bottom of page, replace the words "shall" with "should" in the first and second lines of the first paragraph and in the first, second and third lines of the second paragraph.

Page 9–M–4. Remove the "Stop" sign and the 3'6" dimension to its center line. Remove note reading "8" square wood post." Change Specifications to read as follows: "Posts shall be selected treated wood cross ties. Railings and bracing shall be of seasoned, planed spruce or yellow pine. Timber shall be sound, clear and cut from live trees. The joints of the rails, bracing and posts shall be given one coat of approved white paint before assembling. After assembling, the surfaces of the rails and bracing shall be given three coats of approved white paint and the stripes shall be formed with one coat of black paint. No paint shall be applied in wet or freezing weather."

Page 9–M–5. Delete all the material on this page.

Page 9–M–6. Delete all the material on this page.

Pages 9–M–7 and 9–M–8

HIGHWAY CROSSING ACCIDENT REPORT

Reapprove without change.

Pages 9–M–9 to 9–M–12 incl.

HIGHWAY GRADE CROSSING RECORD

Reapprove without change.

Pages 9–M–15 to 9–M–17 incl.

LOCATION OF HIGHWAYS PARALLEL WITH RAILWAYS

Reapprove with the following revision:
Page 9–M–17, Art. 5—Crossing Protection. Delete last sentence.

Page 9-M-18
PROBLEMS RELATED TO LOCATION AND CONSTRUCTION OF LIMITED ACCESS HIGHWAYS IN VICINITY OF OR CROSSING RAILWAYS

Reapprove without change.

Pages 9-M-19 and 9-M-20
LICENSE OR EASEMENT APPLICATIONS

Reapprove without change.

Report on Assignment 3

Merits of Various Types of Highway—Railway Grade Crossing Protection

Collaborating with Communication and Signal Section, AAR

J. A. Jorlett (chairman, subcommittee), W. S. Autrey, G. B. Blatt, A. C. Cayou, M. H. Corbyn, Raymond Dejaiffe, E. R. Englert, Wm. J. Hedley, J. A. Holmes, J. E. K. Krylow, R. W. Mauer, H. L. Michael, R. F. Spars, K. E. Wyckoff.

The Armour Research Foundation of the Illinois Institute of Technology, with funds provided through the Research Department of the Association of American Railroads, has produced the final report of an "Analysis of Railroad Crossings and Accident Data for the State of Ohio During the 10-Year Period, 1949 through 1958," ARF Project E 672, dated March 10, 1961, revised September 1, 1961. The Ohio Department of Highways and 12 railroad companies operating in the state of Ohio assisted invaluably by providing information concerning the physical characteristics, the volume of vehicle traffic, the accident record, number and speed of trains for 7416 crossings.

A progress report of this study which appeared on page 364 of Bulletin 560, November 1960, gave a tabulation of the accident records and a summary of the types of protection for 7758 crossings.

The principal objectives of the report were:

1. To determine relationships between accident rates and characteristics of the selected highway grade crossings, and
2. To obtain risk factors for these crossings.

The program was carried out in three well defined phases: (1) the data were examined and summary tables constructed, (2) correlations between accidents and crossing characteristics were investigated, (3) a regression surface was fitted to observational data to provide risk factors. A further study had been planned to determine the effect of changed protection on accident rates, but because of financial limitations it was not possible to accomplish this study. The calculations required were performed on Armour Research Foundation's Univac 1105 digital computer and auxiliary equipment.

The risk factor for a given crossing is defined as the expected accident rate over a period of 10 years. The prediction equations shown in the report were derived for each of the following three crossing categories:

1. Crossings with painted crossbucks.
2. Crossings with automatic flashing lights.
3. Crossings with automatic flashing lights and gates.

The risk factors were computed from a knowledge of six crossing characteristics, namely:

1. Average visibility from vehicle driver's view.
2. Highway grade.
3. Rail traffic volume.
4. Rail traffic speed.
5. Highway traffic volume.
6. Number of tracks and spurs.

In addition to the prediction equations, six risk-factor graphs appear in the report. They were derived from the relevant equations and enable the reader to select risk factors for the following situations:

Fig. No.	Page No.	Type of Protection	No. of Tracks	Average Visibility Feet	Highway Grade %	Speed Rail Traffic MPH
3	82	Painted Crossbucks_____	1	735	4.75	40
4	83	Painted Crossbucks_____	2	735	3.75	45
5	84	Flashing Lights_____	1	425	3.75	45
6	85	Flashing Lights_____	2	425	3.75	45
7	86	Automatic Gates_____	2	425	3.75	45
8	87	Automatic Gates_____	5	100	3.00	40

Two additional graphs are presented in Fig. 9, page 88, for constant risk factors of 1.0 and 2.0 based on number of vehicles over a single-track crossing protected by flashing lights and with train speeds between 10 and 80 mph.

Conclusions deduced from the study are:

1. In general, the accident rates decrease with improved visibility, with increasing percentage of highway grade, and with mounting speed of rail traffic.
2. Accident rate increases with increasing volumes of rail and highway traffic.
3. Volume of rail traffic and volume of highway traffic are the only two known factors which are statistically significant predictors of accidents at all three crossing categories.
4. Multiple correlation coefficients for each of the three crossing categories imply that unknown and unaccounted factors, such as the behavior of the motorists, contribute considerably to accidents at grade crossings.
5. Conclusion No. 4 is particularly true for crossings with more elaborate types of protection.

The report has been distributed to AAR Member Roads, Highway Department of the State of Ohio, and members of Committee 9. Because the statistical analysis is unique, some of the procedures have stirred up controversies. It is the feeling of Committee 9 that this report represents the best information on this subject now available and it should be distributed to railroad and public authorities and their comments solicited. There are, of course, weaknesses in the report which call for careful evaluation of all factors before this method of determining risk factors is used. For example, information was not available concerning the speed of vehicular traffic, the weather, the time of day or year the accidents occurred. These factors could be very significant for a particular crossing. Your committee recommends that the statistical work be continued when funds become available, especially the investigations of the accident experiences for those crossings where the type of protection has been improved in the study period.

This is a progress report submitted as information. Your committee recommends 'hat the assignment be continued.

Report on Assignment 5

Recommended Method of Developing Annual Maintenance Cost of the Various Types of Highway–Railway Grade Crossing Protection

Collaborating with Communication and Signal Section, AAR

F. C. Cunningham (chairman, subcommittee), W. S. Autrey, G. B. Blatt, W. A. Buck-master, A. C. Cayou, M. H. Corbyn, C. R. Fears, T. L. Gibson, C. I. Hartsell, J. A. Holmes, J. A. Jorlett, J. E. K. Krylow, E. S. Miller, R. E. Nottingham, P. H. Slack, K. E. Smith, J. E. Spangler, R. F. Spars, C. W. Traister, H. W. Walbright.

Your committee has been assembling information from various railroads on their experience in regard to the maintenance cost of the various types of highway–railway grade crossing protection devices. The Communication and Signal Section, AAR, is progressing a concurrent study of this subject and has initiated an actual cost record study on a number of railroads, which will last for a period of at least one year. Committee 9 is holding its report in abeyance pending the outcome of this cost study. An evaluation of the results of this study, along with the information the committee now has, will form the basis of future recommendations.

This is a progress report, submitted as information. Your committee recommends that the subject be continued.

Report on Assignment 6

Methods of Providing Additional Advance Warning to Highway Traffic Approaching a Highway–Railway Grade Crossing

Raymond Dejaiffe (chairman, subcommittee), W. S. Autrey, H. E. Bartlett, G. B. Blatt, W. A. Buckmaster, H. B. Clarkson, F. C. Cunningham, C. R. Fears, T. L. Gibson, C. I. Hartsell, J. A. Holmes, W. H. Huffman, H. L. Michael, E. S. Miller, R. J. Pierce, W. C. Pinschmidt, R. E. Skinner, K. E. Smith, J. E. Spangler, R. F. Spars, T. M. Vanderstempel, K. E. Wyckoff.

Last year your committee reported on the requirements of the Manual on Uniform Traffic Control Devices of the U. S. Bureau of Public Roads for advance warnings at highway–railway grade crossings. It also reported on the use of special signs with continuously flashing yellow lights in the State of Ohio and distinctive street lights used at special locations. This year reports are given on other special advance warning devices which have been used at selected locations.

At Tracy, Calif., where U. S. Highway 50 crosses the main track of the Southern Pacific Company, flasher lights and automatic gates were installed in March 1955. During the remainder of that year there were 39 accidents involving collisions with the gates. During 1956 there were 49 such accidents. In May 1957 special advance warning devices were placed in service. These devices consisted of a flashing yellow light with a 24-in by 48-in neon sign displaying the symbol "R X R" over the word "GATE". These signs are located about 800 ft in advance of the gates and are tied in with the track circuits. During the next three years there were an average of 16 accidents annually. It is believed that the improved accident record was due principally to the installation. The entire installation was paid for and is maintained by the state highway department.

At Los Angeles, Calif., where Roscoe Boulevard crosses the Southern Pacific Company track, flasher-light signals and automatic gates were installed in September 1959. In accordance with the Public Utility Commission order a special advance warning device was provided approximately 400 ft on each side of the crossing. These devices consist of three yellow lights in a horizontal position similar to a three-unit traffic signal laid on its side. The center light flashes continuously except when a train is approaching. When a train reaches the approach circuit the center light stops and the other two lights flash alternately. Damage to the gates at this location has not been abnormally high and there is no way of judging the effectiveness of these advance warnings. The entire cost of this installation was paid for and is maintained by the city.

At Flint, Mich., where Coldwater Road crosses the Chesapeake & Ohio near a large automotive plant, a crossing having a very bad accident record was given special study and treatment. Vehicular traffic on the three-lane road exceeds 10,000 per day and is concentrated at times of shift changes in the automotive plant. Rail traffic, including a large amount of switching, is about 700 movements per day over 4 tracks. The crossing is protected by reflector-button, 4-ft blade, 90-deg crossbuck signs. Because of the special traffic conditions, automatic protection was not considered practical. Standard 30-in octagonal red "STOP" signs have been placed on both sides of the tracks. Above and below the word "STOP" on the signs are two automatic flashing red lights which flash alternately at all times. Approximately 350 ft from the nearest track on each side is a standard round "RR" approach sign. Approximately 500 ft from the nearest track is an advance warning sign which reads "STOP AHEAD." The roadway is well lighted by modern street lights on both sides of the crossing. There have been no accidents at this crossing since the installation was made in March 1958. One of the most important features of this protection is the regular patrol of state police who issue tickets to drivers who do not observe the stop signs. All special signs were installed and are maintained at county expense.

At Flint, Mich., where Pierson Road crosses the Chesapeake & Ohio near a large automotive plant, the crossing was treated similarly with good results. The traffic pattern was similar to the above crossing except that the highway had four lanes. The main difference is that standard 30-in octagonal, red-background "STOP" signs are placed to the right of the highway on both sides of the crossing to stop all traffic on the outside lanes. Cantilever posts located at the same point suspend 30-in electrically lighted "STOP" signs over the inside lanes of the highway. This protection has been in service for over eight years and no fatalities have occurred since the extra protection was placed. The signs were installed and are maintained at city expense. Again the effectiveness of the protection is largely attributable to the city police, who effectively enforce observance of the stop signs.

Cook County, Ill., since 1955 has been experimenting with "rumbler strips" in approaches to "STOP" signs at the more hazardous county highway intersections. The central idea of the rumbler strip is an irregular surface on the road at a stop sign approach that will produce an audible rumble and thus will, at the least, prompt the driver to wonder what goes on. The rumbler strip is 300 ft long and is placed only on the lane of the highway approaching the stop sign. The surface of the highway is primed, and a special asphalt mixture having about $\frac{3}{4}$ in aggregate is applied and rolled. The mixture is such that the voids in the aggregate are not filled and therefore gives the surface a rough appearance. Actual before-and-after surveys of observance of stop signs have shown the rumbler strips very effective. The county is now considering a variation of this treatment of highway surfaces approaching highway-railway grade crossings. This application would be in cross-wise strips alternating with clear pavement. Both the strips and the intervals would be about 25 ft long. Further development of this idea may prove that the rumbler strip may make our highway crossing protection more effective than additional signs.

Your committee hopes to canvass the different states during the coming year concerning their requirements for advance warning to highway-railway grade crossings and recommends the continuance of this assignment.

This is a progress report, submitted as information.

Report of Committee 6—Buildings

K. E. HORNUNG,
Chairman

J. W. HAYES,
Vice Chairman

D. J. MURRAY, *Secretary*

W. G. HARDING
H. T. SEAL
G. A. MORISON
J. W. GWYN

J. H. ADAMS, JR.
J. L. AGEE
C. M. ANGEL
W. F. ARMSTRONG
F. R. BARTLETT
D. A. BESSEY
S. M. BIELSKI
G. J. BLEUL
J. R. BOWMAN
W. L. BURGESS
R. R. CAHAL
H. M. CHURCH (E)
D. W. CONVERSE
J. S. COOPER
F. D. DAY
A. G. DORLAND (E)
L. A. DURHAM, JR.
V. E. ELSHOFF
T. J. ENGLE
H. E. FERTIG
R. L. FLETCHER
I. G. FORBES
C. S. GRAVES
G. W. GUINN, JR.
A. T. HAWK (E)
H. R. HELKER
J. F. HENDRICKSON
W. C. HUMPHREYS
W. R. HYMA

E. J. HYNES
B. J. JOHNSON, JR.
S. E. KVENBERG
H. J. LIESER
R. E. LILLISTON
G. H. MCMILLAN
I. A. MOORE
J. D. MOORE, JR.
C. W. MORRISON
L. S. NEWMAN
L. J. NICHOLS
W. H. PAHL, JR.
W. C. PANARESE
C. L. ROBINSON
J. T. ROWAN
J. B. SCHAUB (E)
J. T. SCHOENER
T. H. SEEP
H. A. SHANNON, JR.
LOREN SHEDD
E. R. SHULTZ
R. C. SMITH
M. L. THORNBROUGH, JR.
R C. TURNBELL
S. G. URBAN
J. W. WESTWOOD
O. G. WILBUR (E)
T. S. WILLIAMS
Committee

(E) Member Emeritus.
Members listed in bold face type are the official representatives of the Engineering Division, AAR.

To the American Railway Engineering Association:

Your committee reports on the following subjects:

1. Revision of Manual.

   Progress report, with recommendations submitted for adoption .... page ...

2. Specifications for railway buildings.

   No report because of concentration of effort on bringing Manual chapter up to date.

4. Wind loading for railway building structures.

   No report because of concentration of effort on bringing Manual chapter up to date.

8. Infra-red ray heating, collaborating with Committee 18.

   No report because of concentration of effort on bringing Manual chapter up to date.

THE COMMITTEE ON BUILDINGS,
K. E. HORNUNG, *Chairman.*

AREA Bulletin 567, November 1961.

Report on Assignment 1

Revision of Manual

W. G. Harding (chairman, subcommittee), J. H. Adams, Jr., C. M. Angel, F. R. Bart-lett, S. M. Bielski, W. L. Burgess, R. R. Cahal, D. W. Converse, J. S. Cooper, R. L. Fletcher, I. G. Forbes, C. S. Graves, G. W. Guinn, Jr., J. F. Hendrickson, W. C. Humphreys, S. E. Kvenberg, R. J. Lieser, J. D. Moore, Jr., W. H. Pahl, Jr., C. L. Robinson, J. T. Schoener, Loren Shedd, R. C. Smith, R. C. Turnbell, J, W. Westwood.

During the past year your committee continued its review of Chapter 6 of the Manual and now submits for adoption the following recommendations in connection therewith.

Delete General Conditions, pages 6–1–1 to 6–1–4, incl., substituting therefor the following:

SPECIFICATIONS FOR GENERAL CONDITIONS

1. General

These general conditions are intended to be used in connection with the Form of Construction Contract, Part 1, Chapter 20, of the AREA Manual, and shall apply to all sections of these specifications with equal force.

2. Company, Engineer and Contractor Defined

Refer to Form of Construction Contract, Part 1, Chapter 20, Art. 4.

3. Plans and Specifications

The plans furnished by the company shall be considered as part of and illustrating the specifications. The specifications are intended to supplement the plans. The plans and specifications are complementary each to the other and what is called for by one shall be as binding as if called for by both.

The plans and specifications show the location, extent, and general character of the work. Additional or working drawings may be furnished from time to time as required in exemplification of the drawings and specifications hereto attached. All such additional or working drawings and all the information shown thereon shall be considered a part of this contract and as illustrating the work covered thereby. The contractor shall not deviate from the drawings or specifications except upon the written consent of the chief engineer.

Parts not detailed shall be constructed in the customary manner for that class of work so as to maintain the strength and complete the structure.

Where parts only of the work are shown on the plans, the balance shall be considered as a repetition, and where any detail is started on a plan it shall, in the construction, be carried the full length of the part and similar parts that it details.

Figures on plans shall take precedence over measurements -by scale, detail plans over small-scale plans, and full-size details over all other plans. The decision of the engineer shall be final as to the interpretation of plans and specifications.

The Company will furnish without cost to the contractor a reasonable number of sets of plans and specifications as determined by the chief engineer. Additional plans and specifications desired by the contractor will be furnished at the expense of the contractor. One or more copies of all plans and specifications shall be kept on file by the contractor at the site of the work for ready reference.

4. Errors or Discrepancies

Refer to Form of Construction Contract, Part 1, Chapter 20, Art. 28.

5. Shop Plans and Working Drawings

The contractor shall prepare all necessary shop plans and working drawings required for the work of the various trades. He shall submit prints of these drawings, in triplicate, including prints of drawings covering all prefabricated items such as doors, windows, etc., to the engineer for approval. Such drawings shall include list of all parts of equipment with pattern members or other necessary designation in order that repair parts may be readily ordered in the future.

Drawings shall be .... inches in size, including margins of .... on all sides. The title shall be placed in the lower right-hand corner. Plans covering prefabricated items such as doors, windows, etc., may be on the manufacturer's standard-size sheets.

All shop plans and working drawings must be approved by the engineer before the work involved is started. The approval of said drawings by the engineer shall not imply any change in the specifications or relieve the contractor from the responsibility of any errors thereon. No change shall be made on any approved drawings without the written consent of the engineer. The contractor shall supply additional copies of drawings on request.

Upon completion of the work, the contractor shall furnish a complete set of corrected shop plans and working drawings, showing a true record of the work as constructed, to the engineer. Whenever the contractor has made original tracings, such tracings shall be the corrected drawings which are furnished to the engineer.

6. Laying Out Work

Necessary lines, corners and elevations will be established on the site of the building by the engineer. The contractor shall erect permanent batter boards and protect the points so established until the work is completed and accepted. Using the points established by the engineer, the contractor shall lay out his own work and be responsible for its accuracy.

7. Prosecution of the Work

(a) *Schedule of Progress:* As soon as practicable after the contract is awarded, the contractor and engineer shall agree upon a schedule of procedure which shall comprehend:

(1) The dates when each major division of the work will be started and completed.

(2) The dates when the shop plans for each division of the work will be submitted for approval.

The contractor shall thereafter so prosecute the work as to conform to the agreed schedule, unless changed with the written consent of the engineer.

(b) *Cooperation:* When the work of the contractor engages with the work of another contractor or of the company, cooperation and extraordinary care will be required to prevent injury to work or material. The contractor shall not endanger or alter the work of any other contractor without the written consent of the engineer.

(c) *Cutting, Fitting and Patching:* The contractor shall do all cutting, fitting, and patching of his work that may be required to make its several parts come together properly and to receive or be received by work of other contractors shown upon, or reasonably implied by, the plans and specifications for the complete structure. Any cost caused by defective or ill-timed work shall be borne by the party responsible therefor.

8. Materials

All materials shall be new and of the grade specified, and shall be the best of their respective kinds for the uses intended.

(a) *Priced Materials:* Where the quality or kind of material cannot be definitely specified, and the amount of money the contractor is to pay for it is given in these specifications, the sum so given is intended to cover the purchase price of the material and freight charges; but this sum shall not include any cost of hauling, cartage, supervision, preparatory work, profit, or the cost of erection, it being intended that the contractor shall include such foregoing items in his contract price. The engineer will select such materials and notify the contractor of his selection and the price agreed upon, but the contractor shall purchase such material and supervise its delivery and erection as fully as other parts of the work.

If the required payment for such priced material is more than the sum specified, the difference will be paid by the company, and if it is less the difference shall be deducted from the sum to be paid the contractor under the contract.

(b) *Approved Materials:* The term "Approved Material" if used in this contract, signifies that the engineer must be consulted as to the source from which the material is to be purchased as well as its general quality and construction, but such approval does not mean the acceptance of the material actually furnished if it should be defective.

(c) *Special Materials:* Special brands of materials and devices mentioned in the specifications or shown on the plans are named for the purpose of establishing a standard or criterion of quality and character desired and a uniform basis for bidding. The base price in the contract is for construction with the special brands of materials and devices named. Other material of equal quality and adaptability to the purpose for which it is intended will be considered and may be substituted only with the written approval of the chief engineer.

Where a specific make or kind of apparatus is called for and furnished by the contractor, the furnishing of such apparatus does not relieve the contractor of liability until he shall make such apparatus or appliance operative so that it will successfully perform the function for which it is intended.

(d) *Samples:* Where required, the contractor shall furnish samples of materials for approval. The materials used shall be in accordance with approved samples.

9. Equipment

The contractor shall provide all equipment required for the execution and completion of the work, including all staging, scaffolding, apparatus, tools, etc., which are necessary. All equipment must meet with the approval of the company, and the engineer may require the removal of any portion of equipment which is defective or unsuitable for the proper prosecution of the work, and the contractor will be required to substitute therefor satisfactory equipment without delay.

10. Permits, Laws and Ordinances

All work shall conform to the federal, state or municipal laws, ordinances or regulations governing such work. The contractor shall give all requisite notices in connection with his work to the proper authorities, and shall procure at his own expense all

permits, licenses, etc., of every description, necessary for the construction and completion of the work, and shall deliver to the company all certificates of inspection for
plumbing, electric wiring, or any other branch of the work for which such certificates
may be required in connection with this contract.

Whenever these specifications, or any document which they supplement, conflict
with the building code of the city or town in which the work is executed, the contractor shall submit the question as to which to follow to the engineer and abide by his
decision.

11. Temporary Facilities

(a) *Construction:* The contractor shall furnish at his own cost, risk and expense,
all pumping, bailing, sheet piling, temporary bridging or other temporary work of any
nature whatever, required for the prosecution of the work.

(b) *Protection:* The contractor shall continually maintain adequate protection of
all his work from injury due to weather, frost, accident or other cause.

(c) *Pumping:* The contractor shall take care of any water occurring within the limits
of the work and shall pump same to nearest sewer or other outlet.

(d) *Temporary Heat:* The contractor shall at his sole cost and expense furnish
all necessary temporary heat. Such heat shall include heat required to dry plaster or
paint or to prevent damage by freezing of materials or equipment or for any other
purpose.

All temporary heating appliances used to furnish temporary heat shall be of approved make and the contractor shall, when installing them, comply with the National
Board of Fire Underwriters regulations stipulating safe clearances for the installation
of all temporary heating appliances. The use of temporary heating appliances shall
meet with the approval of the engineer.

In the event the use of temporary heating appliances results in damage by fire,
then the contractor shall at his sole cost and expense, and to the satisfaction of the
engineer, do the necessary restoration work.

(e) *Temporary Light and Power:* The contractor shall at his sole cost and expense
furnish all necessary temporary light and power for whatever purpose required.

(f) *Water:* The contractor shall provide and pay for an abundant supply of water
for building purpose during the entire progress of the work.

(g) *Temporary Toilet Facilities:* The contractor shall establish and maintain in a
good sanitary condition, free from filth and rubbish, temporary toilet facilities for the
accommodation of his employees, meeting the requirements of the local health officers
or other public authorities having jurisdiction, unless existing company's facilities are
available and their use approved by the engineer.

12. Cleaning Up

Refer to Form of Construction Contract, Part 1, Chapter 20, Art. 49.

13. Force Account Work

Whenever any work is done or material furnished on a force-account basis, that is,
for a price based upon the actual cost and an added percentage, the percentage shall
include profit, overhead, general supervision, all taxes or profits, such as income taxes,
also any contractor's license fees and the use of all tools and equipment, except as
noted in the definition of cost. The actual cost shall include all material, labor, small
tools actually used up in the work; also compensation, public liability and contingent

insurance, unemployment and old age insurance and sales or other taxes on items entering into the work; also the agreed rental of special power equipment. Where work is
done on this basis, the time of all employees shall be entered by the contractor on
forms supplied for that purpose and checked and signed in duplicate daily by the contractor and the engineer and only time so entered and checked will be allowed.

14. Accounting Requirements

At the completion of the work, the contractor shall furnish a complete list of all
quantities in accordance with the company's classification for all work underground
for each item or structure, and shall furnish in lump sum form, the cost of the superstructure for each item or building, this cost to include the proportionate part of the
contractor's overhead and profit.

Where the work is of such nature that existing facilities are removed or remodeled
by the contractor, he shall furnish the company with a statement showing in detail
the cost of such work, the materials removed and the disposition of the materials.
The above information shall be furnished in order to comply with Interstate Commerce
Commission accounting requirements.

Delete Excavation, Filling and Backfilling, pages 6–2–1 to 6–2–3, incl., substituting
therefor the following:

SPECIFICATIONS FOR EXCAVATION, FILLING AND
BACKFILLING

1. General

The contractor shall furnish all labor, materials, tools and equipment, except as
otherwise noted, necessary entirely to complete all excavation for foundation walls,
piers, footings, pits, ducts, tunnels, basements and any other excavation which may
be implied or shown on the drawings to receive the subsequent work.

Any excavation paid for or deducted on a unit price basis shall be for the actual
measured yardage.

No allowance shall be made on account of slope to the sides of excavation, but
measurements for quantities of excavation shall be taken to outside of sheeting.

The unit price paid or deducted shall include the whole value of the sheeting,
bracing or any other material actually used in connection with the work, either as a
form for concrete foundations, as a protection against caving during the process of
excavating, or as a cofferdam, and shall also include any pumping or bailing which
may be necessary.

2. Classification

All material excavated shall be classified as rock excavation, wet excavation and
common excavation.

3. Rock Excavation

Rock excavation shall include all rock in solid beds or in compact, stratified masses
which, in the judgment of the engineer, should be removed by continuous blasting.
Boulders or detached rock measuring 1 cu yd or more, shall be classified as solid rock.

4. Wet Excavation

Wet excavation shall comprise that material, not included under rock excavation,
which requires pumping or sheet piling to overcome seepage and overflow.

5. Common Excavation

Common excavation shall include all materials that do not come under the classifications of rock or wet excavation.

6. Soil Test

Soil tests shall be made before any foundation work is placed. Foundation soils, exploration and tests shall conform to current specifications as outlined in Physical Properties of Earth Materials, Part 1, Chapter 1, of the AREA Manual.

7. Beds for Footings

The beds for footings shall be leveled and free of all loose material before any foundations are put in place. No footings shall rest on filled ground except where absolutely necessary, and all filling under such footings shall be sand or other approved filling, puddled and tamped in place. No such footings shall be put in place by the contractor without first obtaining permission from the engineer.

8. Quicksand Pockets

If any quicksand pockets or other soft spots are encountered beneath foundation walls, piers, or footings, the same shall be excavated and filled with concrete, the extra work being paid for on the basis of unit prices provided in contract.

9. Pumping and Bailing

The contractor shall perform all pumping and bailing necessary to keep all excavation entirely free from water during the progress of the work under all circumstances and contingencies which may arise, using such means as may be best adapted to conditions. The cost of pumping and bailing shall be included in the contractor's bid for excavation.

10. Blasting

The contractor shall do all blasting necessary in connection with the excavation as shown on the drawings. All drilling, placing of charges and shooting, together with the covering of blasts, shall be done in an approved manner. All work in connection with blasting shall be done in strict accordance with any laws or ordinance in effect where the work is located.

11. Shoring, etc.

The contractor shall do all shoring, bracing, etc., which is necessary to support adjoining soil, streets, walks, buildings, tracks, etc., and shall remove such shoring, bracing, etc.

12. Guard Rails and Lights

Guard rails and lights and other protection deemed necessary by the engineer shall be furnished, erected, and maintained by the contractor until their removal shall be approved by the engineer.

13. Disposal of Excavated Material

Excavated material shall be used for backfilling around all underground work. After forms for such work have been removed and the work has been inspected by the engineer, the contractor shall fill up to the finished grade as shown on the drawings.

Only material suitable for backfilling shall be so used. Large frozen lumps, boulders, etc., shall not be used. Backfilling must be placed in layers not to exceed 6 in, each layer being thoroughly compacted.

The contractor, when so required, shall haul and place surplus excavated material within a distance not to exceed 300 ft from the building as directed by the engineer.

Any surplus excavated material which cannot be disposed of within 300 ft of the building shall be disposed of by the contractor, unless otherwise released by the company.

14. Filling

Sand, crushed stone, crushed slag or other granular filling, where called for on the drawings, shall be thoroughly tamped, rolled and compacted in place by the contractor. Where floors are on fill, the fill shall be placed in layers and thoroughly puddled, tamped and rolled or flooded. Wherever such fill occurs it shall be included in the lump sum price for the structure in which it occurs. Sand fill shall be clean sand free from sticks or other foreign matter.

No filling or backfilling shall be done at a time when there is danger of frost entering the material, except at the discretion of the engineer.

Backfilling around foundation basement walls shall not be done until first floorbeams or floor slabs have been set in place, unless such prior backfilling is approved by the engineer.

15. Grading and Final Cleaning

All grading that may be necessary around the buildings as shown by the drawings shall be done by the contractor. Crushed stone, crushed slag, sand, clean dirt or top soil shall be used for the work as called for by the drawings.

At the completion of the work the contractor shall thoroughly clean up and remove any rubbish, dirt or excavated material from site as called for under disposal of excavated material, and leave the site clean and graded to finish grades as shown by the drawings.

16. Underground and Overhead Structures

All pipes, sewers or conduits, shall be supported in place by the contractor and all expense attending their renewal shall be borne by him. All telegraph, electric light or telephone wires, signals, etc., which in the judgment of the engineer interfere with the progress of the work shall be removed without expense to the contractor. During construction the contractor shall maintain in safety, permanent poles, wires, sewers, pipes or conduits affecting his work or with which it may interfere. If damaged through his negligence, all expense attending repairs thereto shall be borne by him.

17. General Conditions

All materials entering into the work and all methods used by the contractor shall be subject to the approval of the engineer, and no part of the work shall be considered as finally accepted until all the work is completed and accepted.

The General Conditions as given in Part 1, this Chapter, shall be considered to apply with equal force to this specification.

Delete Specifications for Pile Foundations for Railway Buildings, pages 6-3-1 and 6-3-2, substituting therefor the following:

SPECIFICATIONS FOR PILE FOUNDATIONS FOR RAILWAY BUILDINGS

For wood pilings, see Specifications for Wood Piles, Part 1, Chapter 7, and Specifications for Driving Wood Piles, Part 3, Chapter 7.

For concrete and various types of metal piles, see Specifications for Pile Foundations, Part 4, Chapter 8.

Delete Architectural Terra Cotta, pages 6–4–1 to 6–4–3, incl., substituting therefor the following:

SPECIFICATIONS FOR ARCHITECTURAL TERRA COTTA

1. General

The contractor shall furnish all labor, materials, tools, scaffolding and equipment, except as otherwise noted, necessary entirely to complete any or all classes of architectural terra cotta work herein specified, according to the class of building, and as shown or implied on the drawings.

2. Quality of Material

Terra cotta shall be made from suitable selected clays grog, and fusible minerals carefully proportioned and mixed, and properly burned to produce a strong homogeneous body which will give a sharp, metallic, bell-like ring when struck.

3. Defective Work

All work shall be carefully modeled by skilled workmen in strict accordance with detail drawings. All pieces shall be perfect when set in place, and any work damaged after installation shall be replaced before final acceptance.

4. Drawings

The architectural terra cotta contractor shall prepare and submit to the engineer, for his approval, complete detail and setting drawings (in triplicate) for all terra cotta work covered by this contract. Such drawings shall show in detail, jointing, bonding, anchoring and other construction features. All blocks shall be numbered serially.

5. Models

If desired by the engineer, full-size plaster models prepared by experts shall be submitted for his approval. The price of such models shall be agreed upon in advance.

6. Molding and Fitting

Templates for molded work shall be made according to details and models. Carving and molding work must be sharp, straight, true and well undercut. Blocks must be straight, true and out of wind. A reasonable number of additional blocks must be provided to prevent delay from defective materials or injury. So far as possible all grinding of joints and fitting of material shall be done at the factory. Washes and drips shall be provided for all projecting courses. Wherever flashing occurs raglets shall be provided. Proper provision shall be made for anchors, tie rods, etc.

7. Cement, Sand, Lime, Mortar, Water

(a) *Cement*—The cement shall meet the requirements of current AREA specifications for portland cement, Part 1, Chapter 8. Cement that has hardened or partially set shall not be used.

(b) *Sand*—All sand used for mortar shall be clean, washed, hard and well graded and shall contain not more than 3 percent by weight of such organic impurities as loam, clay, mica, etc., determined by decantation, and shall be tested in accordance with the current ASTM Methods of Tests, designations C 40 and C 117.

Mortar sand shall be free from salt, alkalies and other deleterious substances.

The sand shall have a fineness modulus ranging between 2.00 and 2.50. In general, a sieve analysis shall show the sand to come within the following limits:

Passing a No.   8 sieve ....100 percent
Passing a No.   50 sieve .... 30 percent
Passing a No. 100 sieve .... 10 percent

(c) *Lime*—Lime shall conform to current ASTM Specifications, designation C 5. If the use of hydrated lime is authorized by the engineer, it shall conform to current ASTM Specifications, designation C 6.

(d) *Mortar*—Mortar for architectural terra cotta shall be composed of 1 part portland cement, 3½ parts sand, and ½ part lime putty by volume.

(e) *Water*—Water for mortar shall be clean and free from injurious amounts of oil, acid, alkali, organic matter or other deleterious substances, and wherever possible shall be taken from the mains of the municipality.

8. Setting and Anchoring

All blocks must be cleaned and wetted before setting except in freezing weather.

Mortar shall be kept ½ in from the face of the terra cotta to allow for pointing. Splashing exposed faces of the terra cotta with mortar shall be avoided.

All beds and vertical joints shall be of maximum width of ⅜ in unless otherwise indicated. The terra cotta shall be set accurately, true to line and level. Face blocks shall be set on thoroughly wetted wooden wedges, which are not to be removed until the building is cleaned and pointed.

All terra cotta shall be thoroughly bonded to masonry backing. Cornices, column caps and blocks with greater projection than bed shall be thoroughly anchored.

Anchors and dowels, rods and hooks shall be of the proper size and shape and thoroughly galvanized or coated with asphaltum paint.

This contractor shall do all cutting and fitting of terra cotta to accommodate other trades.

9. Protection

Wherever necessary, all projecting courses or individual blocks shall be protected against injury during the setting process by wooden covering, which shall be maintained in good and substantial condition until removed for the purpose of cleaning down the work.

10. Cleaning and Pointing

The face of the terra cotta work under this contract shall be thoroughly cleaned upon completion, such cleaning to be done with soap powder boiled in clean water and applied vigorously with stiff fiber brushes. If necessary, clean sharp, fine white sand may be added to the soap and water mixture. The use of wire brushes or acids will not be permitted for cleaning terra cotta work.

All face joints shall be brushed out ½ in. in depth and pointed flush with mortar consisting of 1 part stainless cement, 2 parts clean white sand and sufficient cold lime putty to make a mixture as stiff as can be worked. All joints shall be wetted before pointing.

As an alternate use non-shrinking cement mixed with sand in accordance with manufacturer's instructions for pointing mortar.

11. General Conditions

All material entering into the work and all methods used by this contractor shall be subject to the approval of the engineer, and no part of the work will be considered as finally accepted until all the work is completed and accepted.

The General Conditions, as given in Part 1, this Chapter, shall be considered to apply with equal force to this specification.

Delete Brickwork, pages 6–4–4 to 6–4–10, incl., substituting therefor the following:

SPECIFICATIONS FOR BRICKWORK

1. General

The contractor shall furnish all labor, materials, tools, scaffolding and equipment, except as otherwise noted, necessary entirely to complete any or all classes of brickwork herein specified, according to the class of building and as shown or implied on the accompanying drawings, including all backing, covering of iron and steel, all piers, walls, chimneys and other special work shown, specified, or otherwise implied.

2. Brick

All bricks used, if of clay or shale, shall preferable be side-cut. All bricks used shall at least conform to the requirements shown in the following table:

REQUIRED STRENGTH OF BRICK

Part of Structure	Compressive Strength Pounds per Square Inch Brick Tested Flatwise		Modulus of Rupture	
	Individual Minimum	Average 5 Specimens	Individual Minimum	Average 5 Tests
Walls	5000	Not less than 6000	450 lb	600 lb or over
Foundation	3500	Not less than 4000	400 lb	500 lb or over

3. Classification of Brickwork

Brickwork shall be classified as either common brickwork or face brickwork. The class of brickwork to be used shall be determined by the class of the building or by notations on the accompanying drawings. Unless otherwise specified common brickwork shall be used on all buildings of mechanical terminals, shops, storehouses, isolated freight houses and similar buildings. In general, face brick should be used for passenger stations and auxiliary buildings, combination passenger and freight stations and auxiliary buildings, combination passenger and freight stations and freight houses built in conjunction with passenger stations.

4. Common Brickwork

All common brickwork shall be laid even and true to line, plumb, level and with all joints accurately kept. All brickwork shall be laid with joints not more than ⅜ in thick and bonded together with full headers every sixth course. All brick shall be good, hard, well burned brick free from cracks and uniform in size, shape and quality and shall not absorb more than 10 percent of their weight in water. They shall be laid in a full bed of mortar, with all joints on exposed walls tooled to give a concave finish. The bricks used on the face of the wall shall be selected whole bricks of a uniform size and with true, rectangular face.

Porous or salmon brick shall be thoroughly wetter either by immersion or by sprinkling before being laid, except in freezing weather.

5. Face Brickwork

The exterior face brickwork shall be laid up with a selected and approved No. 1 pressed, hardburned or rough face brick as specified by the engineer. The contractor, as a basis for his proposal, shall figure on a face brick to cost $.......... per thousand, F.O.B. building site or company's lines as provided in the letter of invitation, and any variation from this price more or less will be adjusted according to the actual cost of the brick. Face brickwork shall. be laid with all stretchers, unless otherwise shown, and shall be bonded either by blind headers or an approved metal wall tie every sixth course.

All face brick shall be laid true to line, plumb, level and with all joints accurately kept. All work shall be laid so that 4 courses shall not exceed 11 in or 3 courses 8 in. in height for modular construction and joints tooled to give a concave finish unless otherwise shown on the drawings or ordered by the engineer.

The company reserves the right to deviate frcm the type of joint specified above so as to conform with the type of brick selected. All brick courses shall be so proportioned that they will work out evenly with height of windows and doors. No split or fractional courses will be permitted. All backing up of face brick shall be as specified under common brickwork.

Porous or salmon brick shall be thoroughly wetted either by immersion or by sprinkling before being laid, except in freezing weather.

6. Detail of Brickwork

All brickwork details such as lintels, belt courses and other trim shall be laid up according to details shown on accompanying drawings and as specified under either common brickwork or face brickwork.

7. Samples

The contractor will furnish samples of all brick to be used, together with prices for the various kinds of face brick submitted for approval of the engineer. The engineer also shall have the option of obtaining samples and prices for face brick. The samples selected and approved will be filed with the engineer and taken as a standard of material to be furnished, and all material in the work must be equal in all respects to the approved samples.

8. Cement, Sand, Lime, Mortar, Mortar Color, Water

(a) *Cement*—The cement shall meet the requirements of current AREA specifications for portland cement, Part 1, Chapter 8. Cement that has hardened or partially set shall not be used.

(b) *Sand*—Sand used for mortar shall be clean, washed, hard and well graded and shall not contain more than 3 percent by weight of such organic impurities as loam, clay, mica, etc., determined by decantation, and shall be tested for such impurities in accordance with current ASTM Methods of Tests, designations C 40 and C 117.

Mortar sand shall be free from salt, alkalies and other deleterious substance.

The sand used shall have a fineness modulus ranging between 2.00 and 2.50. In general a sieve analysis shall show the sand to come within the following limits:

Passing a No.   8 sieve ....100 percent
Passing a No.   50 sieve .... 30 percent
Passing a No. 100 sieve not over .... 10 percent

Where so required for face brickwork, sand shall be white, but shall otherwise conform to this specification.

(c) *Lime*—Lime shall conform to current ASTM Specifications, designation C 5. If the use of hydrated lime is authorized by the engineer, it shall conform to current ASTM Specifications, designation C 6.

(d) *Mortar*—Mortar shall be mixed in the ratio of 1 part portland cement, 1 part lime putty, and 6 parts sand, measured by volume. When colored mortar is to be used, the lime putty content shall be reduced to $\frac{1}{2}$ part.

An approved mortar cement may be substituted for cement and lime and mixed according to the manufacturer's directions.

(e) *Mortar Color*—A mortar color of an approved brand shall be used to color mortar for face brickwork; color and mixture shall be as approved by the engineer. The contractor shall, upon request of the engineer, lay up samples of face brickwork with different shades of mortar in order that the engineer may decide by comparison the proper shade of mortar to use. These samples shall be of a size not to exceed 6 sq ft in area, and the contractor shall build if requested, not to exceed 6 such samples. In general, unless otherwise specified, or ordered by the engineer, the mortar shall be colored slightly darker than the face brick used.

(f) *Water*—Mixing water shall be clean and free from injurious amounts of oil, acid, alkali, organic matter or other deleterious substances, and wherever possible shall be taken from the mains of the municipality. The contractor shall arrange for his own water supply at his own expense.

9. Integral Waterproofing

No waterproofing materials shall be added to the mortar except by permission of the engineer.

10. Wood Centerings

The contractor shall provide wood centers for all openings wherever necessary. Centers shall be strongly constructed, made to fit accurately to the work, be well supported and rigidly braced so as to carry all loads until the brickwork has set. At the completion of the work all centering shall be removed from the premises.

11. Scaffolding, Protection, Etc.

The contractor shall provide all scaffolding, staging, ladders, etc., necessary for the work. All walls or other parts shall be securely braced and protected against damage by wind and storm during construction.

12. Anchors, Steel, Etc.

The contractor shall provide chases for all pipes, set bearing plates for beams, etc., as indicated on drawings and will be responsible for accurate location of same.

13. Backing

Where so shown, iron, steel and other material shall be backed up with brickwork in a manner indicated on details.

14. Proportioning and Mixing Mortar

(a) All mortar used shall be mixed by volume in the proportion of 1 part of portland cement, 1 part of slaked quick lime putty or of soaked hydrated lime putty, and 6 parts of sand. No lime putty shall be used which has not been slaked or soaked at least 12 hr before being mixed into the mortar. All mortar shall be mixed with the minimum amount of water consistent with maximum density and workable plasticity.

An approved mortar cement may be substituted for cement and lime and mixed according to the manufacturer's directions.

(b) The method of measuring mortar materials shall be such that the specified proportions thereof can be controlled and accurately maintained at all times.

(c) All mixing of mortar shall be done in a mechanically operated batch mixer of the drum type for a period of at least 3 min after all materials for a batch are in the drum. The drum must be completely emptied before the succeeding batch of materials is placed therein. Continuous mortar mixers and hand mixing will not be allowed.

(d) The use of retempered mortar will not be permitted.

15. Brick Laying

(a) *Wetting Bricks*—All bricks immediately before being laid shall be sprinkled in the stock pile, or elsewhere as may be suitable, except in freezing weather, for not less than 5 min, or for such additional time or wetted in such other manner as the engineer may decide is necessary to supply the bricks with sufficient moisture to effect a proper bond between the bricks and mortar.

(b) *Mortar Beds and Other Joints*—All bricks shall be laid on a thickly spread bed of mortar with furrow shallow and not deep, such that there will be enough excess mortar in bed joint to completely fill furrow when bricks are bedded to the lime. All head and side joints shall be completely filled by applying sufficient mortar to brick already in place on brick to be placed.

(c) *Condition of Equipment*—All equipment used for mixing or transporting mortar and bricks shall be clear and free from set mortar, dirt or other injurious foreign substances.

(d) *Laying Brick Masonry in Foundations*—Before laying bricks in a foundation, a layer of not less than 1 in of mortar shall be spread over the surface of the soil. Immediately thereafter the first course of bricks shall be laid.

(e) *Joining Work*—When fresh masonry is to join with masonry that is partially or entirely set, the exposed joining surface of the set masonry shall be cleaned, roughened and wetted so as to effect the best possible bond with the new work. All loose bricks and mortar shall be removed.

(f) *Disturbance of Completed Work*—When any portion of the brickwork has been completed, such work shall remain undisturbed until thoroughly set, except in the case where work left off at the end of a day is recommenced on the following morning, or as soon thereafter as practicable.

(g) *Finishing of Work*—All brickwork shall be finished in a workmanlike manner with a thickness of joints and manner of striking or tooling indicated on the drawings or as described in the specifications.

(h) *Cleaning and Tuck Pointing*—All exterior brick masonry shall be thoroughly cleaned and tuck pointed. If so specified, a 5 percent solution of muriatic acid shall be used for cleaning down, but this must be followed by copious baths of clean water.

16. Laying Bricks in Freezing Weather

(a) *Protection of Bricks*—All bricks delivered for use in freezing weather shall be fully protected immediately upon delivery by a weather-tight covering such as will prevent the accumulation of water, snow or ice on the bricks. Loose board covering will not be permitted.

(b) *Heating of Sand*—All sand shall be heated in such a manner as will remove all frost, ice or excess moisture. The methods and equipment used shall be of such character as will prevent the burning or scorching of the sand.

(c) *Heating of Bricks*—All frosted bricks shall be defrosted by heating to a temperature of approximately 180 deg F.

(d) *Heating of Water*—During freezing weather, or when so directed by the engineer, all water used shall be heated to a temperature of approximately 180 deg F.

(e) *Slaking or Soaking of Lime*—All slaking of quick lime or soaking of hydrated lime shall be done at a temperature of at least 60 deg F, and this temperature shall be maintained until the lime is incorporated into the mortar.

(f) *Protection of Mortar Against Freezing*—After the mortar is mixed, it shall be maintained at such temperature as will prevent its freezing. Mortar on the boards shall be kept from freezing at all times, and if necessary the contractor shall use metal mortar boards equipped with banjo-type oil or gas torches. No anti-freeze liquid, salt or other substance shall be used in mortar except by permission of the engineer.

(g) *Enclosures and Artificial Heat*—All work under construction shall be protected against freezing for a period of 48 hr by means of enclosures, artificial heat, or by such other protective methods as will meet the approval of the engineer. In general, the methods now commonly accepted and used for the protection of reinforced concrete construction in freezing weather shall be used.

17. Bricklaying in Hot Weather

All finished or partly completed work shall be covered or wetted in such manner as will prevent too rapid drying of the masonry.

18. Compression Tests

(a) *Brick Masonry*—At least 3 compression test specimens, each nominally 8 in square and 16 in high, shall be made and tested before actual construction is commenced. These test specimens shall be built up of unselected bricks from the stock pile and laid in the same mortar mixture and in the same manner proposed to be used on the job. The specimens shall be moist cured for 27 days, exposed to the atmosphere of the laboratory for 1 day, and then tested in a vertical position. The average compression strength of such test specimens shall not fall below the requirements of Art. 2 according to the allowable unit stress to be used.

In preparing compression test specimen, care shall be taken that the top and bottom bearing areas are exactly parallel and that the mortar joints do not exceed $\frac{3}{4}$ in. The method of capping and testing shall be that presented in current ASTM Specifications, designation C 67.

(b) *Mortar Cubes*—Mortar test cubes shall be 2 in by 2 in by 2 in and shall be tested in accordance with current ASTM Specifications, designation C 150. Such cubes shall develop a compressive strength of at least 900 psi at 7 days and 2000 psi at 28 days. At least 3 cubes shall be made and tested for each lot of 50,000 bricks.

19. Vitrified Tile Wall Coping

Vitrified tile wall coping shall be provided where indicated on the accompanying drawings. It shall be best hard-burned, salt-glazed tile, laid in full bed of mortar of 1 part cement to 3 parts sand, omitting all lime.

20. Cast Concrete Coping

All walls where so indicated on the drawings shall be coped with cast concrete coping. This to be of the section as detailed and made in lengths as shown on the drawings or as directed by the engineer.

21. Cast Concrete Sills, Lintels, Etc.

Where so indicated on drawings, window and door sills, lintels, chimney caps, etc., shall be of cast concrete according to details shown for them.

22. Requirements for Cast Concrete

Concrete for cast copings, lintels, sills, caps, etc., shall be composed of 1 part portland cement, 2 parts sand and 3 parts crushed stone or gravel of a size to pass a $1\frac{1}{4}$ in ring. Exposed surfaces shall be troweled smooth and edges shall be smooth and unbroken. Cast concrete copings, sills, lintels, caps, etc., shall be set true, level and plumb and carefully pointed out. No cast concrete member shall be set until the concrete is sufficiently hard to prevent damage. Copings, sills and caps shall be provided with drips.

All cast concrete members shall be reinforced with steel rods according to details shown on the drawings.

23. New Masonry Joining to Old

The contractor shall use special precaution where new masonry work joints up with old masonry work, to see that the old work is sufficiently roughed up, anchors provided and work keyed so that an absolutely tight and neat bond is assured between old and new work.

The contractor shall do all work in connection with cutting out old brickwork, stone work or concrete where required. Care shall be exercised to see that only such portion of the masonry is disturbed as is necessary.

24. Protection and Pointing Up

The contractor must keep his work covered and protected from the action of the weather or frost. He shall also protect by boxing all dressed or ornamental work liable to damage. At the completion of the work or at any time when so ordered he shall do all patching in a satisfactory manner, clean down and point up all brickwork, etc., removing all surplus mortar and stains. All window and door frames shall be carefully caulked with oakum and pointed up after they have been inspected and before the staff beam is applied.

25. General Conditions

All materials entering into the work and all methods used by the contractor shall be subject to the approval of the engineer and no part of the work will be considered as finally accepted until all of the work is completed, and accepted.

The General Conditions as given in Part 1, this Chapter, shall be considered to apply with equal force to this specification.

Page 6-4-11

CONCRETE

Reapprove with the following revisions:

Delete the paragraph under Art. 2. Design and Installation, substituting therefor the following two paragraphs:

Concrete shall be designed and installed in accordance with current AREA specifications for concrete and reinforced concrete, Part 1, Chapter 8.

All buildings constructed of prestressed concrete shall be designed in accordance with the latest report of ACI–ASCE Joint Committee 323 "Tentative Recommendations for Prestressed Concrete", unless otherwise specified by city ordinance or state code.

Pages 6-4-17 to 6-4-19, incl.

STONE MASONRY AND CUT STONE WORK

Reapprove without change.

———————————

Delete Structural Steel, pages 6-5-1 to 6-5-5, incl., substituting therefor the following:

SPECIFICATIONS FOR STRUCTURAL STEEL

All contracts entered into and work performed under this specification shall conform to the Specification for the Design, Fabrication, and Erection of Structural Steel for Buildings of the American Institute of Steel Construction, current edition, and/or the Specification for the Design of Light-Gage Cold-Formed Steel Structural Members of the American Iron and Steel Institute, current edition.

Pages 6-5-6 to 6-5-12, incl.

ORNAMENTAL AND MISCELLANEOUS METAL WORK

Reapprove with the following revisions:

Page 6-5-7, Art. 5. Wrought Iron—Change to read:

"Wrought iron shall comply with current ASTM Specifications, designation A 207 for rolled shapes and bars."

Page 6-5-7, Art. 9. Pipe Railings—Change to read:

"Where shown on drawings pipe railing shall be made of seamless standard-weight steel pipe with either screwed connections or welded connections ground smooth. Where set on masonry or concrete, the railing shall be set in pipe sleeves not less than 4 in deep and leaded in, or where designated to be set with standard based flanges, the flanges shall be securely anchored with ½-in by 4-in expansion bolts. Where attached to steel or iron work they shall be securely bolted and lock nuts used. Where attached to wood, they shall be securely fastened with appropriate wood or lag screws or bolts. Railing shall be designed to withstand a horizontal pressure of not less than 50 lb per lin ft.

Page 6-5-10, Art. 22. Metal Covered Sash—Delete this article on the grounds that little, if any, of this type of sash is now being used in railroad building construction. Substitute therefor the following new Art. 22:

22. Metal Screens

Window and door screens where called for on the drawings shall be of stock design as recommended and furnished by the manufacturer for installation with the specified window or door units.

Page 6–5–10, Art. 23. Metal Sash Operators—Change first paragraph to read:

"Sash operators shall be furnished and installed as called for on the drawings and shall be in strict accordance with the manufacturer's specifications.

Pages 6–5–13 and 6–5–14

METAL SCREENS

Delete.

Pages 6–6–1 to 6–6–5, incl.

YARD LUMBER SPECIFICATIONS FOR RAILWAY BUILDINGS

Reapprove with the following revisions:

Delete Tables I and II, pages 6–6–3 to 6–6–5, incl., substituting therefor revised Tables I and II which have been corrected to conform to the latest specifications and grading rules of the various lumber associations. (Revised Tables I and II are not presented herewith but will be included in the 1962 Supplement to the AREA Manual.)

Delete Carpentry and Millwork, pages 6–7–1 to 6–7–6, incl., substituting therefor the following:

SPECIFICATIONS FOR CARPENTRY AND MILLWORK

1. General

Carpentry and millwork shall include all framing and woodwork which form part of the completed building. Unless otherwise noted, sized called for in the specifications or on the drawings shall be nominal dressed sizes conforming to American Lumber Standards (AREA Manual, Chapter 7). The sizes of all timber and lumber shall conform to the sizes shown on the drawings or specified herein, and where sizes are not so indicated the contractor shall request the engineer to furnish this information before beginning the work affected. All lumber throughout the work shall be graded and classified:

    a. Structural (stress-grade) timber, in accordance with grading rules and classi-fication of timber and lumber appearing in Part 1, Chapter 7 of the AREA Manual.

    b. All other lumber shall conform to current recommendations for yard lumber as outlined in Yard Lumber for Railway Buildings, Part 6, this Chapter.

2. Inspection

All lumber shall be subject to inspection on delivery at the site. All lumber not bearing the grade stamp of a recognized grading agency for the grade and species specified and other rejected lumber shall be promptly removed from the site by the contractor.

3. Species and Grades

The lumber used in the various parts of the building shall be of the species and shall conform to the grades as set forth in the specifications. Species and grades shall be included in specifications for the various items of the work or a list of the items with species and grades of lumber for each item shall be made a part of or an addendum to the specifications.

4. Seasoning

All framing, lumber sheathing and timbers shall be thoroughly air seasoned before being used, and all finishing lumber, flooring, ceiling, siding, and millwork shall be kiln dried to a moisture content consistent with the conditions of service. After delivery at the site, all kiln dried lumber shall be protected from the weather and other damage until the final completion and acceptance of the building.

5. Treated Lumber

Lumber impregnated with a wood preservative shall be called treated lumber. Unless otherwise provided in the specification, the railway company shall furnish all treated lumber pre-cut and pre-framed prior to treatment, and the contractor shall provide for unloading and erecting such lumber in his proposal. Boring, cutting or otherwise fabricating pieces of treated lumber after treatment shall be done only with the express permission of the engineer who shall prescribe a method for treating such field fabrication.

6. Protection Against Decay and Termites

All wood members in contact with the ground, foundations, basement walls, piers and slabs on grade, shall be treated with a preservative as specified and shall be furnished by the railroad company unless otherwise provided. Where shown on the drawings, termite shields shall be of at least 26-gage galvanized steel or other corrosion resisting metals and shall be applied continuously with joints soldered or locked together across the entire joint length. The cross section and projection shown on the drawing shall be maintained for their entire length. When required, soil poisoning for the control of subterranean termites shall be done in accordance with recommendations of U. S. Department of Agriculture Bulletin 1911. Service life guarantee, or bond, for the performance of the chemicals shall be furnished to the engineer upon completion of the soil-treating work.

7. Framing

All framing throughout shall be of the dimensions shown on the drawings and shall be placed as indicated. The framing shall be done in a neat, workmanlike manner to provide closed joints and shall be securely nailed or fastened. Studs shall be doubled at all openings and opposite each cross partition, and all corners and angles shall be made solid and well braced with proper fasteners designed for this purpose. In areas of high winds, studs shall be fastened to sole plates and top plates anchored to studs with special metal fasteners which are designed to provide for the additional strength of nails in lateral resistance. Details for anchors and fasteners shall be furnished to the contractor. All studs shall be in one piece from sill to plate and shall be horizontally bridged at intervals not to exceed 4 ft. The contractor shall provide and set all hangers, straps, anchors, shoes, timber connectors and bolts as required and shown on the drawings. Horizontal joist supports shall be carefully notched or fastened to studs, and all wall plates on top of studs shall be doubled with joints broken over studs.

8. Joists

Joists shall be of the grade, species and size specified and shall be placed as shown on the drawings. Header and trimmers shall be as shown and shall be securely spiked or fastened together. Joists framing into beams, or headers, shall bear on ledgers se-curely fastened to the beams or headers or shall be fastened by metal fasteners made for this purpose. Joists spanning across a center bearing shall be lapped and securely nailed, or they shall be cut to length and fastened by a scab of the same size material nailed to each joist and to the bearing. Wood bridging not less than 1- by 3-in nominal size or compression type metal bridging not less than 18-gage galvanized steel shall be installed between floor joists where the clear span exceeds 8 ft. Ceiling joists shall be securely fastened to rafters and shall be joined to each other across the span as shown. After installation, joists shall not be notched, drilled or cut without express permission from the engineer or architect. Openings in floor shall be framed with headers and trimmers of the size shown.

9. Partitions

All partitions shall be framed and located as shown on the drawings and shall be straight, plumb and well braced. Top plates of all bearing partitions shall be doubled while non-bearing partitions may be constructed with single top plates. Openings for doors shall be framed with double studs and headers.

Where partitions run parallel to the direction of the joists, they shall be located on doubled joists and securely fastened thereto. Sufficient blocking for nailing and fastening the covering material specified shall be provided.

10. Roof Framing

Roofs shall be framed and built in accordance with the detail drawings, accurately fitted and securely fastened. Rafters shall be in one piece unless otherwise shown and shall bear on wall plates and ridge piece and be securely attached thereto with adequate nails or fasteners as shown on the drawings. Valleys, hips and ridges shall be straight and true intersections of roof planes. Hip and valley rafters shall be anchored to plates as shown, and rafters longer than available lumber shall be doubled with pieces lapped at least 4 ft and well spiked. Trussed rafters or trusses shall be accurately fabricated and assembled and shall be erected, anchored and braced for withstanding the loads encountered. No truss or trussed rafter shall be assembled or erected until the architect or engineer shall have approved shop drawings for same. All trusses shall be adequately braced and purlins or joists shall be fastened to trusses with fasteners as shown. Roof joists shall be bridged between trusses with 1- by 3-in wood pieces or compression-type metal buildings, not less than 18 gage galvanized steel.

11. Sheathing and Siding

Sheathing may be wood, plywood, or treated fiberboard as specified. Fiberboard sheathing shall be a minimum of ¾ in. in thickness, and sheets shall be installed with joints on the studs. The boards shall be treated on both sides with asphalt or other waterproofing compound. Wood sheathing, where shown, shall be placed diagonally or horizontally and double nailed at each bearing. Horizontally sheathed and fiberboard-covered walls shall be braced at the corners by means of let-in braces which are not necessary where diagonal wood sheathing is used. Diagonal sheathing shall be applied at approximately 45 deg, and the direction of the sheathing shall be changed on each side adjoining at a corner. Plywood shall be applied with joints at studs and shall be well nailed around the perimeter as indicated on the drawing. Plywood sheathing shall be of a minimum thickness of ⅜ in and, where the possibility of excess moisture is encountered, plywood should be of "exterior" type.

Drop or beveled siding, shiplap or other special wood sheathing shall be placed truly horizontal with tight, square, butt joints, the ends of which have been covered with white lead, closely and accurately fitted against all casing, sills, water table and exterior trim. All siding shall be drawn tight and fastened with blind nails or special fasteners designed for this purpose. Siding shall be installed to provide a tight surface against wind and rain. Where exposed nails are used in fastening siding, they shall be either galvanized steel, aluminum, or other corrosion-resistant metal.

12. Flooring

Sub-flooring may be of wood or plywood as shown on the drawing. Wood subfloors shall be a minimum of 1 in nominal thickness and 4 in nominal width, square edge or tongue-and-grooved and shall be securely nailed at each joist. Butt joints shall be staggered over supports with end cuts parallel to joists. Plywood sub-flooring shall be grade C–D when finished floor is to be wood and grade B–D when resilient floor covering is used. Minimum thickness of plywood subflooring shall be ½ in when wood finished floor is shown and ⅝ in thick for resilient type finished floors. Plywood subflooring shall be securely nailed 6 in on centers around edges and 10 in centers at intermediate bearings, and edges of sheets shall be securly blocked between joists. Top of all subflooring shall form a true even plane with tight joints and solid bearing. Fnished wood strip flooring shall be of the size and grade specified. It shall be kiln dried and matched evenly laid and blind nailed with wire or cut nails, hand or power driven. Joints shall be alternated so that there will be at least two boards between them. An expansion space of 1/16 in per ft of width shall be left between flooring and walls or partitions. All wood finished flooring shall be machine finished to an even smooth surface.

13. Building and Sheathing Papers, Etc.

Where called for on the drawings, storm sheathing and subflooring shall be covered with one layer of waterproof building paper, weighing not less than 5 lb per 100 sq ft. Paper shall be lapped at least 2 in all joints, and carried underneath all corner boards, casing, etc., making a wind tight finish throughout.

14. Furring and Grounds

Unless otherwise specified or shown on the drawings, interior surfaces of stone, brick or concrete walls which are to be plastered shall be furred with 1-in by 2-in furring strips placed 16 in on centers and securely nailed. Furring on masonry walls shall provide a plumb surface for lathing, and shall be nailed to wood bricks or inserts built into the walls by the mason. Grounds ¾ in thick shall be provided among all openings and along base, and shall be in true planes.

15. Window and Door Frames

Window and door frames shall conform to details and shall be substantially built, using kiln-dried lumber securely framed into sill and heads. Frames shall be dip treated with a wood preservative in accordance with the provisions of the National Millwork Manufacturers Association and shall be prime painted before delivery to the site. Corners shall be braced and material protected for shipment. All frames damaged in shipment shall be promptly removed and replaced.

Frames shall be set plumb and true, heads level, and shall be securely attached to rough framing. Where wall construction is masonry, rough bucks shall be set in the wall with metal anchors attached with screws. Frames and transoms and mullions shall be made in one frame, with transom bar and mullion mortized in. All frames shall be of proper size to receive sash and doors, and shall be weathertight. Frames for double-hung windows shall have sash pulleys or spring balances built in as specified under Part 12, this Chapter. Where called for, frames shall be built to receive fixed thermal glass units. The frame shall be designed for setting of the glass units in accordance with the manufacturer's specifications. Where called for on the drawings, window frames shall be built to receive storm sash and door frames to receive storm doors. Frames shall be built to receive screens where specified. Plank frames for masonry walls shall have a break strip built into the wall and nailed to frame around head and jambs.

16. Wood Screens

Window, transom and door screens shall be made of kiln-dried dressed lumber as specified and built according to details. Lumber shall be dip treated in accordance with National Millwork Manufacturers Association standards. Frame stiles and rails to be of thickness and such widths to match windows, transoms and door frames. Corners and joints to be rigid and substantial. Suitable screen molding shall be applied after screen is attached. Wire cloth shall be 16-mesh copper bronze unless otherwise specified. A special sun-ray shade screen may be used on those exposures where it is desirable to reduce thermal build-up in rooms because of sun light. Copper-bronze wire cloth shall be given one coat of clear lacquer. All removable screens shall be provided with a duplicate set of non-corrodible metal number tags, one applied to each screen and one applied to the corresponding opening.

17. Stairs

Stairs shall be strongly and rigidly built in location shown, and as detailed. Rough work for all stairs shall be self-supporting without the aid of angle posts. Treads shall have molded nosings, be plowed into risers, and risers into the underside of treads, and both housed into the wall stringer and tightly wedged and glued. Unless otherwise specified or shown on the drawings, treads shall be 1¼ in thick and risers 1 in thick, both of hardwood, and shall be in one piece. All newels, balusters and handrails shall be as detailed. Landings and platforms shall be finished to match treads, and all finish on stairways shall match general finish throughout the building. Cellar and porch stairs on minor buildings may be open without risers where directed by the engineer. Outside steps shall be framed with proper waterfall.

18. Outside Finish and Trim

Outside finish and trim shall be neatly and accurately fitted. All necessary baseboards, watertable, corner trim, casings, fascias, frieze boards, cornice and moldings, and everything necessary to make a complete finished piece of work, shall be furnished and erected.

19. Interior Finish

Interior trim, wainscoting, chair rail base, picture moldings, etc., shall be kiln dried and conform to the details, be neatly and accurately fitted, joints mitered and secret nailed with fine finishing nails. If face nailed, all nails shall be set for puttying. Interior finish shall be free from hammer marks and shall be hand dressed and sandpapered where required. No splicing of the window or door trim will be permitted and joints of bases, chair rail and moldings must be carefully matched.

20. Cabinets, Counters, Etc.

All cabinets, counters, drawers, lockers, shelving, etc., called for on the drawing shall be provided in place and fitted up with all hardware as specified. These facilities shall be preassembled units or job-built as specified. Preassembled units shall be securely attached to the floors or framing and shall be fitted to other finish work in a workmanlike manner. Job-built cabinet work shall be made in a workmanlike manner and securely and rigidly built in place, supported by necessary brackets and cleats. All lumber for this work shall be kiln dried and of the species and quality specified.

21. Toilet Partitions

- Where wood water closet partitions are called for on the drawings, they shall be provided with approved metal fittings and hardware, and door in accordance with the details.

22. Sash

Wood sash, including storm sash, shall be accurately made to fit openings, dressed and sanded to a smooth finish, pinned and through-tenoned with muntins, etc., as detailed. All sash shall be dip treated in accordance with the National Woodwork Manufacturers Association standards. They shall be rabbeted for glass and molded, and shall be properly hung, hinged, or pivoted as required. Sash shall be so made and installed as to provide for a watertight fit. Double-hung windows shall have the sash counterbalanced, using spring balance suspension, or by counterweights of lead or cast iron hung on approved sash cord or sash chains of proper strength. Sash shall be fitted to operate easily, but shall not be so loose as to rattle. Casement windows shall be made watertight by grooving the bottom rails and providing rabbets at jambs, head and meeting stiles. Glass sizes, thicknesses, widths of rails and stiles will be as shown on the drawings. Where glass sizes only are given, widths of rails, stiles and muntins shall be in accordance with standard mill practice. Storm sash shall be provided with a non-corrodible number tag to correspond with the numbered opening.

23. Doors

Doors shall be of the sizes and types shown on the drawings, properly and neatly hung so as to fill openings, freed from warp, and fully equipped with all hardware necessary for their operation. All exterior wood doors shall be dip treated in accordance with National Woodwork Manufacturers Association standards. Sliding doors in warehouse and baggage rooms shall have suitable protection built to protect the doors when in an open position, shall have all necessary stops, shall be so hung that the doors cannot be lifted off the track from the outside and shall be hung and fitted so that no lateral motion will exist. Heavy and special doors shall be built to details with frames mortized together, backing rigidly fastened, and fitted with sash where shown.

A special schedule of hinged doors, showing thicknesses, sizes, design, panelling, glazings, etc., will be furnished to supplement this specification where needed. Unless otherwise specified or called for on the drawings, all panelled doors shall be 1¾ in thick, except interior doors in minor buildings, which may be 1⅜ in thick. Stiles and rails shall be mortized and tenoned, and pinned or doweled. In either case, joints, shall be solidly glued up. Doors shall be hung with the proper size and number of butts to prevent sagging. Double-acting doors and gates shall swing clear and fill openings. Hardwood carpet strips or thresholds shall be provided for all doors unless otherwise shown on the drawings

24. Miscellaneous Carpentry

The carpenter shall provide, in place, all miscellaneous woodwork not above specified, such as wood foundation blocks and posts, fencing, atticing, coal bins, walkways in attics, wood gutters, signs, notice boards, etc., and do all necessary cutting, fitting and patching, and special framing necessary for the proper installation of work of other trades. Upon completion of the work, the carpenter shall remove all temporary work, scrap lumber and debris, draw all projecting and temporary nails, and leave the work in a complete, finished and orderly condition.

25. Detailed Shop Drawings

Detailed shop drawings for all millwork shall be submitted for approval before the millwork is performed.

26. General Conditions

All materials entering into the work and all methods used by the contractor shall be subject to the approval of the engineer, and no part of the work will be considered as finally accepted until all of the work is completed and accepted. The General Conditions as given in Part 1, this Chapter, shall be considered to apply with equal force to this specification.

Delete Lathing and Plastering, pages 6–8–1 to 6–8–5, incl., substituting therefor the following:

SPECIFICATIONS FOR LATHING, PLASTERING AND STUCCOING

1. General

Under this heading shall be included all furring, wood, metal and gypsum lathing, all plain and ornamental plastering and stucco work.

This contractor shall provide, maintain, erect and move all necessary scaffolding, staging, tools, labor and materials, as required for the proper execution of plastering and stucco work shown or implied on the drawings and as called for in these specifications.

2. Furring

This contractor shall furnish all furring, forms, anchors, ties, hangers for all suspended ceilings, cornices, coves, moldings, etc., shown on the drawings. Where ceilings are suspended below floor or roof system, they shall be formed on metal lath secured to furring and carrying channels of sizes indicated and spaced as shown, and securely fastened by means of hangers to the floor or roof members.

3. Hangers

Hangers shall be 1- by 3/16-in flat iron securely bolted to connecting members, or more commonly used 7/32 in round mild steel rods secured to carrying channels by means of a saddle tie or three twists of hanger around channel.

Hangers shall be spaced on not more than 4 ft centers.

4. Carrying and Furring Channels

Carrying channels shall be not less than $1\frac{1}{2}$ in cold-rolled channels and shall be spaced on not more than 4 ft centers. Channels shall be placed true and level and properly positioned for the required ceiling height.

Furring channels shall be ¾-in cold-rolled channels and shall be spaced on 12 in centers. Channels shall be securely saddle-tied to the carrying channels with two strands of No. 16 gage, galvanized, soft annealed wire ties, at each intersection.

5. Metal Lathing

Metal lath shall be expanded metal, diamond mesh weighing not less than 3.4 lb per sq yd, or ⅜-in hy-rib lath weighing not less than 4.0 lb per sq yd. Metal lath shall be drawn tight over surface to which it is to be applied, lapped not less than 3 in at all sides where joints occur; laced together and to the furring at intervals not to exceed 6 in with No. 18 gage galvanized soft wire. Lathing shall be left in perfect condition to receive plaster—level and rigid.

6. Solid Plaster Partitions

Solid metal lath and plaster construction is recommended for partition work when fire-resisting material is required and the saving of floor space is desirable. This type of partition is 2 in thick with a finish coat of plaster both sides and is permissible for heights up to 18 ft.

The framing shall consist of ¾-in cold-rolled channel studs spaced on 12 in centers with ¾-in channel horizontal bracing on the flat between studs and spaced not over 4 ft 6 in apart. Studs to be secured to ¾-in channel runners bolted to floor or the more commonly used method of driving the studs crimped end into ¾-in round holes drilled into floor 1 in deep. Top of studs to terminate 3 in from ceiling. Splice top of stud 10 in to 16 in long channel having 3 in long right-angle bend at ceiling. Use No. 18 gage annealed galvanized wire for fastening at ceiling and splice. Stud and splicing channel to be of same size.

Expanded metal lath, 3.4 lb, to be placed on one side of studs and secured with No. 18 gage galvanized wire.

Begin plastering the scratch coat on metal lath side of studs and follow up the same day with plastering on opposite side of lath to fill in the full thickness of stud. The remainder of the full 2 in of plaster shall be placed on both sides of studs the following day.

7. Painting

All metal furring, channels, ties, hangers and clips shall be painted on all sides before erection with one coat of approved lead and oil paint.

8. Gypsum Lath

Gypsum lath shall be ⅜ or ½ in thick and shall conform to current ASTM Specifications, designation C 37.

Application can be made with nails or more preferably, recommended clips and tie wire. Nails shall be 1⅛ in by 13 gage, blue lath nails set ⅛ in below surface of lath. Clips and tie wire shall be galvanized and of type recommended by the lath manufacturer.

Gypsum lath shall be applied with the long dimension across the studs with end joints staggered in successive courses and be closely fitted together at all angles. When used over metal furring, joists or studs, gypsum lath shall be attached with manufac_ turer's recommended metal clips.

Provide metal fabric over all joints and in all angles fastened with staples.

9. Corner Beads

Expanded metal corner beads 2⅝ in wide and galvanized shall be installed on all external plaster angles, both vertical and horizontal, including all door and window openings with plaster return at jambs and head.

The corner beads shall be extended full height from floor to ceiling or soffit of openings. The flanges of all beads must be kept clear of plaster finish lines and anchored wherever necessary to insure rigidity.

10. Materials

All materials shall be the best of their respective kinds, mixed and applied strictly according to manufacturer's direction.

Plaster shall be delivered at the site in the original unbroken packages and stored in a dry place until used.

White-finish gaging plaster shall consist of lime properly prepared to a smooth putty and mixed with gypsum gaging plaster; one part of gypsum gaging plaster to two parts of lime. Careful application and sufficient troweling during and after set are essential to produce a smooth glossy surface. With the addition of sand, it is adaptable for sand float finishes.

Base gypsum plaster shall meet the requirements of ASTM specifications, designation C 28–58 fibered (hair, not wood) and unfibered.

Sand shall be clean, sharp, free from clay, or impurities that will stain the plaster. It shall be screened, and any material retained on a 10 mesh screen or passing through a 30 mesh screen shall be rejected.

Water shall be clear, and free from oil, acid, alkali, organic matter or other deleterious substances.

When light weight aggregates are used in basecoats in lieu of sand, they shall meet the requirements of ASTM specifications, designation C 35–57 T.

The proportions of mix are as follows:

First or scratch coat on all types of lath—2 cu ft of lightweight aggregate to 100 lb of gypsum plaster.

First coat on masonry surfaces and second or brown coat in all three-coat work—3 cu ft of lightweight aggregate to 100 lb of gypsum plaster.

Keene's cement shall be used in the finish coat where it is subject to moisture. The proportions of mix are 25 lb of dry hydrated lime per 100 lb of Keene's cement.

11. Interior Plastering

In general, all lastering on lathed surfaces shall be three-coat work consisting of a scratch coat, a brown coat, and the finishing coat. On masonry and gypsum block it shall be two-coat work consisting of a base coat and a finishing coat. Each coat shall be allowed to dry thoroughly before the next coat is applied. Before beginning his work the plasterer shall test and prove the lathing and grounds so that the finish plaster will be plumb true, level and waveless. Plaster shall extend behind all sill aprons, wainscoting, bases, etc.

Masonry walls to be plastered shall be thoroughly drenched with water before applying the first coat of plaster.

Necessary precaution should be taken to prevent the plaster from freezing or drying too rapidly until the plaster has hardened.

The scratch coat shall be well rubbed in and troweled against masonry and into lathed surfaces so as to form a perfect bond and shall be scored and scratched in both directions to form a key for the brown coat.

The brown coat shall be applied to the scratch coat and brought to a true level plane with rod and darby flush with the grounds, and broomed or otherwise roughened to receive the finishing cost. Angles and corners shall be left straight, true and plumb.

On gypsum lath or masonry surfaces, the base (first) coat shall be applied with sufficient material and pressure to form a good bond and to cover well. Then, before this scratch coat has set, double back to bring the plaster out to grounds, straightened to a true surface with rod and darby, and leave rough, ready to receive the finish (second) coat.

The finishing coat shall be applied to the brown coat and may be a sand float or white trowel finish as specifically designated. If a white trowel finish is called for, it shall be made of Keene's cement and lime putty troweled to a smooth hard finish free from trowel or brush marks. The finishing coat shall be applied in two coats over a hard dry gypsum plaster base. Scratch in a thin coat, then go over a second time, filling all imperfections and catfaces. Trowel to a smooth finish, applying water with a damp brush. The trowelling should be completed before the finish has set. Avoid joinings by working top and bottom of the wall at the same time. The general thickness shall be $\frac{1}{16}$ to $\frac{1}{8}$ in unless the finish coat is marked off or jointed, in which case the thickness may be increased as required by the depth of the marking or jointing

Sand-floated finishing coat shall be mixed in the proportions of 100 lb of Keene's cement to 50 lb of dry lime and 400 lb of sand.

The plasterer shall run all plaster molds, cornices, coves, etc., in accordance with models or full-sized profiles; all angles to be carefully and accurately mitred. Run work shall be carefully and accurately formed from templates to form continuous, unbroken level lines. Ornamental enrichments shall be firmly secured in place with plaster of paris white lead and galvanized wire nails.

12. Exterior Stucco Work

Material for the stuccoing of exterior wall surfaces shall be portland cement for the under coats and white portland cement for the finish coat. Aggregates for the under coats shall be clean sand, graded from fine to coarse grains, with the coarse grains predominating, and shall be free from loam, salt, vegetable and deleterious matter. Aggregate for the finish coat shall be clean yellow gravel grit, marble or granite screenings, as directed by the engineer. Hydrated lime and coloring compounds shall be first quality. Hair shall be first quality long cattle or goat hair.

Mortar for the first and second coats shall be composed of one part portland cement, three parts sand and one-tenth part of hydrated lime by volume, with sufficient hair added to bond the mortar to the lath.

Mortar for the finishing coat shall be composed of one part white portland cement, three parts aggregate and one-tenth part of volume of hydrated lime This coat shall be brought to the selected tone by the addition of dry coloring compound not exceeding 10 percent of the weight of the cement.

The different constituents shall be thoroughly mixed dry to a uniform color, water then added to obtain the proper consistency, and the whole turned over until it becomes uniform in color and consistency. No more mortar shall be mixed than can be used within 30 min. The dry color in the finish coat shall be weighed or measured and thoroughly mixed with the sand. The cement and lime shall then be added and the entire mass thoroughly mixed by shoveling from one side of platform to the other through a ¼-in mesh screen, and when the batch is of uniform color, the water shall be added.

The stucco shall be applied in three coats, each coat not less than ¼ in thickness, the full thickness being ⅞ in beyond masonry line or 1 in thick over furring strips. The plastering shall be carried on continuously in one general direction without allowing the mortar to dry at the edge. Where this is impossible the joints shall be made at an opening or other natural division of the surface. Stucco shall not be applied when the temperature is below freezing. Masonry surfaces shall be clean and wet before first coat is applied, and brick walls shall have the joints raked out ½ in. After the first coat has been applied under pressure but before it has set, the second coat should be applied and floated to a true plane. The under coats shall be cross scratched before the initial set has taken place and thoroughly wetted before the succeeding coats are applied. After the second coat has set, but before it has dried, the finishing coat shall be applied and kept damp for four days, either by sprinkling or by hanging wet burlap over the surface.

Various methods of finishing can be accomplished as hereinafter specified.

Exposed Aggregate (Integral Method)—Within 24 hr after the finish coat has been troweled to an even surface, it shall be scrubbed with a stiff brush until the aggregate has been uniformly exposed. Should the cement be too hard to be readily removed by water, a solution of one part muriatic acid to five parts of water may be used; but as soon as the aggregate has been exposed, particular care should be taken to remove all trace of acid by spraying thoroughly with clean water.

Stippled—Finishing coat shall be smoothed with a metal trowel, with as little rubbing as possible, and then shall be lightly patted with a brush or broom straw to give an even stippled surface.

Sand-Floated Finishing Coat after being brought to a smooth even surface, shall be rubbed in a circular motion with a wood float. The floating shall be done when mortar has partially set.

Rough Cast or Spatter Dash—After the finishing coat has been brought to an even surface and before attaining its final set, it shall be uniformly coated with a mixture of one part white cement to two parts white sand, thrown forcibly against the wall in such a manner as will produce a rough surface of uniform texture.

Pebble Dash—After the finishing coat has been brought to an even surface and before attaining its initial set, clean pebbles shall be forcibly thrown against the mortar and embedded therein. Pebbles shall vary in size from ¼ in to ⅜ in, shall be well wetted before being cast, and uniformly distributed over the surface. They may be pressed into the mortar with a wooden paddle without disturbing the surface.

Note: Samples of the surface finish shall be laid up well in advance of the work, and the approved sample shall be carefully preserved during the performance of the work and used as a standard.

13. General Conditions

The General Conditions as given in Part 1, this Chapter, are a part of this specification and the Contractor shall consult them in detail for instructions pertaining to his work.

14. Completion

At the completion of the finish plaster work, clean all plaster from beads, screeds, metal base and metal trim, leaving work ready for decoration by others. Remove all plaster rubbish from the building, leaving floors broom clean. Remove excess material, scaffolding, tools and other equipment from the job site.

Delete Floors for Railway buildings, pages 6-9-1 to 6-9-3, incl., substituting therefor the following:

FLOORS FOR RAILWAY BUILDINGS

1. Transfer Platforms

Wood plank platforms should be laid preferably with the planks parallel to the line of trucking traffic. Metal plates may be used for a runway to produce easier trucking and to reduce wear on the plank. Concrete floors are used in some cases, and for extremely heavy traffic a concrete base with creosoted wood or asphalt block or asphalt mastic wearing surface is used.

2. Freight Storage Houses

For freight storage houses, which are usually of fireproof construction, concrete floors are generally approved.

3. Freight Piers

Floors of freight piers must, of necessity, largely conform to the style of construction used in the pier. They should be fire-resistant and have enough flexibility to take up the vibration caused by boats being moved along the pier.

4. Machine Shops

In small buildings a wood plank floor, of thickness suited to the severity of service, is common practice. For buildings of a higher grade, wood blocks (preferably treated) or mastic give excellent results. Concrete floors may be used where local conditions justify this construction as economical, although their lack of resiliency may result in discomfort to employees and their hard surface may damage tools dropped upon them.

5. Paint Shops

In passenger car paint shops a concrete floor meets all requirements and it is doubtful if a more expensive type of floor is justified. In freight car shops, where paint is sprayed on, earth floors may be used in shops where the use of concrete is not justified.

6. Freight Car Repair Shops

Concrete floors are satisfactory and are in common use; however, where wheels are handled the use of wood block flooring should be considered.

7. Storehouses

Concrete floors are satisfactory and are in common use. In small storehouses, at outlying points, the ordinary wood plank floor is commonly used.

8. Oil Houses

Because of the necessity for fireproof construction, concrete is recommended for oil houses.

9. Carpenter Shops

In carpenter shops, where considerable bench work is done, wood plank floors are desirable because of the comfort they afford to workmen. Concrete floors are more easily kept clean and are sometimes used.

10. Office Buildings

Office buildings may have floors of a resilient type of material applied over either a concrete or wood floor system.

Corridors or areas of heavy traffic may be surfaced with materials more resistant to wear.

Floors in toilet room areas should be of an impervious material extending up on the walls to form a cove base.

11. Passenger Stations

Floor areas of passenger stations subject to heavy traffic should be surfaced with materials resistant to wear.

Office areas should have a resilient type of flooring.

Floors in toilet rooms should have an impervious material extending up on the walls to form a cove base.

12. Signal and Communication Buildings

Floors in signal towers may be of concrete, composition or wood, depending upon the type of construction of the building. When concrete is used in connection with electrical machinery, precaution should be taken to secure a non-dusting surface.

13. Freight Houses

In small houses, which are usually of frame construction, a plank floor laid on wooden joists is satisfactory and economical. In larger and more important houses, where much trucking is done, a floor of greater first cost is justified, and appreciable economies in operation can be obtained by the selection of a suitable trucking surface. Concrete floors are fairly permanent, sanitary and easy to keep clean, but have as disadvantages a possible failure of the wearing surface, especially at expansion joints, and an unyielding surface which occasionally produces complaints from truckers. Expansion joints should be located outside of the heavily used area wherever practical. If a concrete surface is not considered suitable, some different type of wearing surface such as square-edge maple, wood, or asphalt mastic may be laid on the concrete. Asphaltic concrete may be applied over a sound wood base with good results.

14. Enginehouses

Heavy concrete floors or floors of creosoted wood blocks on a concrete base should be used.

Pages 6–9–4 to 6–9–6, incl.

HOT ASPHALT MASTIC FLOORS

Reapprove without change.

Pages 6–9–7 to 6–9–18, incl.

BRICK PAVEMENTS AND FLOORS

Delete.

Pages 6–9–31 to 6–9–33, incl.

MARBLE AND TILE WORK

Reapprove without change.

Pages 6–9–44 to 6–9–51, incl.

CONCRETE PAVEMENTS

Reapprove without change.

Pages 6–9–51 to 6–9–54, incl.

PLATFORM SURFACES

Reapprove without change.

Pages 6–11–1 to 6–11–5, incl.

SHEET METAL WORK

Reapprove without change.

Delete Hardware, pages 6–12–1 and 6–12–2, substituting therefor the following:

SPECIFICATIONS FOR HARDWARE

1. General

The contractor shall furnish and set all rough and finish hardware necessary for the operation of all doors, windows, blinds, screens, screen doors, toilet partition doors, partitions, partition doors, cabinets, drawers, gates, and ticket windows completely to equip the building. Hardware shall be neatly and accurately fixed in place with screws or bolts, which shall match the hardware and be in perfect working order, free from rust and scratches and other defects. The container shall provide such miscellaneous articles of hardware as screws, bolts, nails and other fastening devices, although not specifically mentioned or shown, but necessary for the ordinary operation of the building.

2. Finish Hardware

The contractor shall furnish and set all finish hardware in accordance with the schedule on detailed plans.

As a basis for bids the contractor shall include in his proposal the sum of $........ to cover the purchase of all finish hardware. Any difference between actual cost and this sum will be added or subtracted from the lump sum amount of the contract as the case may require. The cost of placing the finish hardware shall not be covered by the above amount but shall be included by the contractor in his proposal. All hardware in connection with prefabricated partitions shall be furnished as shown on the detailed plans and as specified.

3. Rough Hardware

The contractor shall furnish all rough hardware of every description and shall include the cost to furnish and set such hardware in his proposal. Rough hardware shall include nails, spikes, screws, bolts and washers, sash pulleys, sash weights, sash cord or chain, sliding door hardware, fire door hardware, special operating devices for rolling doors, horizontal cross folding doors, sliding doors and all miscellaneous rough hardware such as track, hangers, bumpers, stops, stay rollers, chafe and binder strips, door pulls and locks.

Hardware for fire doors shall be of an automatic type approved by the National Board of Fire Underwriters.

Sash balances for wood windows shall be of cast iron or lead and of proper weight to counterbalance the sash, and sash pulleys shall be of an anti-friction type of proper size and with approved face. Sash weights and pulleys shall be fitted to the sash and frames in the factory.

Hardware for pressure-sealed-type windows shall be installed in the factory, including operating mechanism and sash locks for ready application.

Hardware for special doors, such as locomotive entrance doors, are shown in details on the plans.

4. General Conditions

All materials entering into the work and all methods used by the contractor shall be subject to the approval of the engineer. The work will be considered as finally approved when all of it is completed and accepted.

The General Conditions as given in Part 1, this Chapter, shall be considered to apply with equal force to this specification.

Pages 6–13–4 to 6–13–6, incl.

DIFFERENT TYPES OF PAINT AND THEIR ECONOMICAL SELECTION

Reapprove with the following revision:

Page 6–13–4, Art. 2. Definitions—Add the following definition:

k. Alligatoring is a term used to describe breaks in the paint film that have developed to the point where deep and wide fissures are prominent in a paint film, seperating the film surface into amorphous sections. This condition is best illustrated by a mud field that has been flooded with water and which, in drying out, cracks in all directions.

Pages 6–13–7 to 6–13–12, incl.

PAINTING AND GLAZING

Reapprove with the following revisions:

Page 6–13–7, Art. 1. General—Delete the second paragraph, starting with "Under the heading of 'glazing' shall be included . . .", as this, in general, is repeated in Art. 24, page 6–13–12.

Page 6–13–8, Art. 7. Priming Coats—Add paragraph as follows:

"The priming coat shall be mixed with sufficient lamp black to contrast with the finish coat in two-coat work. In three-coat work the first two coats of paint shall contrast with each other as well as the finish coat."

Page 6–13–11, Art. 20. Painting Interior of Enginehouse—Delete all references to cold-water paint.

Page 6–13–11, Art. 21. Staining Shingles—Add paragraph as follows:

"The first coat shingle stain shall be mixed with sufficient pigment to produce a contrast with the final coat of stain."

Pages 6–16–12 to 6–16–18, incl.

VENTILATION OF RAILWAY SHOP BUILDINGS

Reapprove with the following revisions:

Page 6–16–15, Art. 6. Tuning and Light Repair Section—Eliminate the third paragraph. In fourth paragraph, first line, substitute "exhaust gases" for "fumes." In second line, eliminate "the required number of." Reword the second and third sentences to read as follows: "It is recommended that one fan ventilator be provided for every 20 ft, with a minimum capacity of 10,000 cfm. Provision of seven to eight fan ventilators for a 2-unit 4000-hp locomotive having a total length of about 140 ft should provide adequate ventilation."

In fifth paragraph, first line, substitute "gases" for "fumes."

Page 6–16–16, Art. 13. Paint Shops—In second line substitute "desirable" for "necessary."

Delete Art. 14. Roundhouses, on pages 6–16–16 and 6–16–18, but retain the drawings on page 6–16–17 under a new Art. 14. Ventilators, correcting these drawings, however, as follows: In notes in third and fourth diagrams on left side of page, change "must" to "should." Also, in third diagram, left side, change words "close East side" to read "open East side."

————

Delete Oil Burning Equipment, page 6–16–19, substituting therefor the following:

SPECIFICATIONS FOR OIL BURNING EQUIPMENT

The contractor shall furnish and install automatic fuel oil burning equipment, including all necessary electrical connections for automatic operation. The oil burner shall be as manufactured by the .... Company or approved equal. It shall have an ample output capacity to handle the boiler requirements. It shall be designed to burn fuel or diesel oil, commercial standard No. ...... of .... to .... degrees specific gravity, with a flash point of .... deg F, and a viscosity of .... deg at .... deg F. Integral electric fuel oil preheaters to be furnished as required for proper combustion.

Burner may be of the pressure atomizing or rotary type with mechanical automatic forced draft and be designed to permit the use of refractory material in the boiler combustion chamber. The design of the burner shall be such that the fuel oil is completely burned in suspension and a constant and invariable forced flow mixture of oil and air be supplied for proper combustion under all firing conditions.

Electric spark ⎫
Gas Electric ⎬ combination electric spark ignition shall be used as the method
No.—fuel oil ⎭ of igniting the main fuel oil.

The fuel oil burner shall be equipped with a device to prevent the burner from starting up after it has been shut off momentarily and to prevent the flow of oil into the combustion chamber under any conditions of incomplete combustion or explosion. Operation of the oil burner shall be automatically controlled by an approved device operated by either steam pressure or water temperature installed in location approved by the engineer.

For storage of fuel oil, the contractor shall furnish and install a fuel oil tank of .... capacity. This tank shall be buried underground in location shown on drawing, and equipped with the necessary openings in tank to permit filling, measuring, connections to oil burner pump, etc. The burner and storage tank hereinbefore described shall conform in every respect to the rules and regulations of the National Board of Fire Underwriters, and to the laws and ordinances of the state and city in which the equipment is installed.

The electrical equipment on the burner shall be designed for and connected to the electric current available, and all electric wiring and equipment shall conform to the National Electrical Code. The electric equipment used on the oil burner must have the approval of the engineer before installation. Contractor shall state in his proposal the name of manufacturer, and size and capacity of the oil burner on which his proposal is based. After a burner is accepted, no substitution shall be made.

Pages 6–17–1 to 6–17–3, incl.

ELECTRIC LIGHT WIRING

Reapprove without change.

Pages 6–18–1 to 6–18–9, incl.

HYDRAULIC ELEVATORS—BAGGAGE OR FREIGHT

Delete.

Pages 6–18–10 to 6–18–16, incl.

ELECTRICALLY OPERATED FREIGHT OR BAGGAGE ELEVATORS

Delete, substituting therefor the following:

SPECIFICATIONS FOR ELEVATORS

1. General

The contractor shall furnish all labor, materials, tools and equipment entirely to complete the elevator system as hereinafter specified and as indicated on drawings.

The contractor shall furnish and install, under his lump sum price, cars, doors, guides, hoisting equipment, foundations and supports, car safeties, signal systems, and all other machinery, equipment, materials and supports necessary to make a complete and finished system in running order, as herein described and specified.

2. Checking Drawings

Contractor shall check all drawings of his own work and of work with which his work engages. Dimensions on drawings shall be verified on the site of the work by the contractor, and he shall assume all responsibility for their accuracy. He shall report all discrepancies before the work has been started. The contractor shall submit a layout of the system and complete details of all parts showing equipment to be installed. Drawings shall conform to rules and regulations of the local and state authorities, and if necessary, submitted to such authorities for approval.

3. System

The system shall consist of .... (Traction-Type Elevator; Direct Plunger Hydraulic Elevator); or specify type ...., and shall be provided with the necessary equipment to meet the requirements hereinafter specified:

Each elevator shall be installed in hatchway shown on accompanying drawing. Each elevator shall lift a load of .... lb, exclusive of the weight of the car, at a speed of .... ft per min. The capacity shall be shown in large letters in a prominent position on the car. The travel of the car platform shall be .... ft, which distance is between .... and .... which are the lower and upper terminals.

4. Hatchways, Shaft Enclosure and Equipment Supports

The size of the hatchway and shaft enclosure shall be as shown on the drawing, but some slight modifications may be made if the contractor should consider it necessary. Contractor shall state in his proposal whether or not such changes will be necessary. The hatchway and shaft enclosure will be constructed by another contractor, but the elevator contractor shall submit drawings of the apparatus he proposes to furnish within 1 week after he has received notification to commence work, showing details of openings or other items which he will require. He shall also show the space required for elevator doors, guides, counterweights, hoisting machinery, control board and all other apparatus requiring additional space, including the space required for over-run, and pit dimensions required in locality where elevator is to be installed. The contractor will be required to do all cutting of any material to permit the installation of the apparatus in the building, and he will also be required to make good against any damage resulting from same. The contractor will be required to install all foundations and supports for elevator equipment.

5. Car

Car frame shall be wrought iron or steel and be capable of withstanding without undue distortion any stress that may be induced, either by eccentric loading or by the action of the buffers. The design of the car shall be for a load of .... lb per sq ft and consist of material suitable for the purpose for which it will be used. Adequate lighting shall be provided for the car.

The specifications for car shall be included with proposal submitted by contractor.

6. Elevator Doors

Elevator doors shall be designed to close the entire opening in the shaft enclosure and shall be as wide as the distance between car guards. Door guides shall be of steel and well anchored to the shaft enclosure. Doors must be provided with full interlocks arranged so that all doors must be closed and locked before elevator can operate, opening of any door shall stop elevator.

Contractor shall furnish and install all necessary equipment for doors and shall submit with proposal specifications for doors showing construction and method of operation.

7. Hoisting Equipment

The hoisting equipment shall be designed in accordance with good engineering practice, to include all safety features and necessary appurtances and accessories to make a complete and finished system in running order. The specifications and method of operation for same shall be included with the proposal submitted by contractor.

8. Electrical Equipment

The electrical equipment shall be designed to suit the current which will be delivered to the service box in the building and the contractor shall make the necessary connections to this point. The current available at the location shall be ascertained by the contractor. All wiring shall conform to the requirements of the National Board of Fire Underwriters. Contractor shall include in his proposal the manufacturers' names and catalog number of all electrical material he proposes to use in this installation. All material furnished shall be subjected to the approval of the engineer before final acceptance.

9. Signal System

A suitable method of signaling shall be furnished and installed by the contractor in a manner satisfactory to the engineer.

10. Prosecution of Work

The work as outlined in these specifications shall be prosecuted in such a manner that it will cause the least possible interference with the business of the railroad or the work being done by other contractors

11. Painting

The contractor shall, after installation, paint all machinery and other iron work, in connection with this elevator, with two coats of lead and oil, the final coat to be tinted as directed. The second coat shall be put on after the work included in these specifications has been installed, and before turning same over to owners.

12. Instructing Operator and Railroad

After the installation of the elevators, the contractor shall instruct the railroad company's operator in every detail of the handling and controlling of same. The contractor shall give the railroad company all instructions as to management of the elevators and equipment, showing in detail all points that require attention. The contractor shall leave a six-month supply of all lubricants for use of the railroad.

13. List of Parts

The contractor shall furnish three complete lists of all parts, including pattern numbers or other necessary designation, in order that any repair parts required may be readily ordered. One list shall be framed, glazed, and hung in a convenient location to be designated by the railroad company. The list shall be arranged as follows:

Description of Parts	Pattern Number	Remarks
....	....	....
....	....	....
....	....	....

14. Guarantee

The contractor hereby guarantees all material, workmanship and the successful economical and safe operation of the elevators for a period of 12 months after the completion of same, and agrees to repair or replace at his own expense any part of this apparatus which may show defects during that time, provided such defects are, in the opinion of the engineer, due to imperfection in material or workmanship, as specified, and not caused by carelessness or improper operation. He guarantees also that each of the elevators will be capable of lifting the live load specified herein, exclusive of weight of car, at the specified speed. The checking and approval of shop drawings by the railroad company, or any omission in these specifications or the drawings accompanying same, do not relieve the contractor of his obligation to install the elevators complete in every respect, and to fulfill his guarantee.

15. Inspection

The contractor shall furnish at his own expense necessary certificates of inspection for the entire elevator plant put in under these specifications good for 12 months from date of final certificate from the engineer.

16. Final Inspection and Test

At the completion of the work the contractor shall test the elevator and shall arrange for final inspection. The test shall be made during the final inspection and shall consist of subjecting the elevator plant to 60 min running as in service conditions with the loads and speeds hereinbefore set forth, and shall consist of such other tests as may be deemed necessary by the engineer to show compliance with these specifications. These tests shall be made by and at the expense of the contractor, and under the direction and supervision of the engineer. The contractor shall furnish all the labor, instruments and weights necessary to make these tests.

17. Laws and Ordinances

All designs, clearances, construction workmanship and material shall be in accordance with the requirements of the American Standard Safety Code for Elevators A–17.1 (latest edition) and any federal, state, city or municipal laws or ordinances. Contractor shall be solely responsible for any violations and shall save the railroad company free and harmless from any penalty attaching to the violation of any legal regulations affecting his work. If there are any conditions shown on the drawings or mentioned in the specifications that conflict with laws or ordinances, the contractor shall so state in a letter accompanying his proposal.

18. Cleaning

At the completion of the work, the contractor shall remove all construction equipment, scaffolding, staging, erection platforms and all surplus material from the premises, leaving the premises in a clean and acceptable condition. If any equipment or debris is not removed promptly, such material may be removed at the expense of the contractor.

19. General Conditions

All materials entering into the work and all methods used by the contractor shall be subject to the approval of the engineer, and no part of the work will be considered as finally accepted until all of the work is completed and accepted.

The General Conditions in Part 1, this chapter, shall be considered to apply with equal force in this specification

Pages 6–19–1 to 6–19–9, incl.

CHIMNEYS, FLUES AND VENTS

Reapprove with the following revisions:

Page 6–19–3, Sec. B, Art. 1. General—Change first sentence of first paragraph to read: "*Materials.* Chimneys shall be constructed of brick of a quality at least equal to that required by current ASTM Specifications, designation C 62, Grade MW, for clay or shale brick, or of perforated radial brick of special design adapted to chimney construction where the average amount of perforations does not exceed 30 percent."

Change last sentence of this paragraph to read: "Masonry shall be laid in portland cement, cement-lime, or masonry cement mortar."

Page 6–19–4, Sec. B, Art. 1. General—Add at the beginning of the last paragraph of the article: "All brickwork shall comply with Specifications for Brickwork, Part 4, this Chapter.

Page 6–19–14, Sec. B, Art. 4. Isolated Chimneys—Change the table of unit stresses and the paragraph following it at the top of page 6–19–5 to read as follows:

UNIT STRESSES

                                                                                    PSI
Brick masonry in tension 0
Brick masonry in compression (mortar strength = 2500 psi or more) 250
Brick masonry in compression (mortar strength = 1800 to 2500 psi) 225
Brick masonry in compression (mortar strength = 750 to 1800 psi) 200

The working stresses given above are for brickwork in which brick having a compressive strength of 4500 to 8000 psi and mortar having the above indicated 28-day average compressive strength are used. For brick and mortar having a greater or lesser compressive strength, these unit stresses should be changed.

Page 6-19-5—Add to the paragraph on Foundations, between the second and third sentences, the following sentence: "The allowable unit bearing values for soil or piles supporting chimney foundations shall be consistent with sound engineering practice based upon adequate subsoil investigations. Particular care should be observed to avoid unequal settlement, causing tilting of the chimney."

Change the last sentence of the paragraph on Foundations to Read: "Where piles are used, they should conform to the Specifications for Pile Foundations, Part 3, this Chapter."

Change the paragraph on concrete to read as follows: "Concrete for foundations shall comply with the specifications for concrete, Parts 1 and 2, Chapter 8."

Change the paragraph on Lightning Protection to read as follows: "A lightning protection system that will meet the requirements of the Code for Protection Against Lightning, designation C 5, as approved by the American Standards Association, shall be installed on each isolated masonry chimney."

Change the paragraph on Lettering to read as follows: "When specified, lettering shall be applied to the chimney shaft by using built-in permanently colored kiln-burnt brick, by painting, or by other acceptable methods."

Page 6-19-6, Sec. C. Reinforced Concrete Chimneys—Delete Arts. 1 to 4, incl., substituting therefor the following:

1. Design and Construction

Reinforced concrete chimneys shall be designed and constructed in accordance with the latest American Concrete Institute Standard Specification for the Design and Construction of Reinforced Concrete Chimneys (ACI 505).

For clearances, excavation, lettering, draft gage, pyrometer, lights and markings, see Sec. B. Masonry Chimneys.

Delete Freight Houses, pages 6-20-6 to 6-20-10, incl., substituting therefor the following:

FREIGHT HOUSES

1. General

The decision as to the provision of separate houses for inbound and outbound freight is dependent upon the volume of freight handled, the size and configuration of the land available, and the layout of the tracks serving the facility.

The best arrangement utilizing separate houses may be one in which the inbound and outbound houses are adjacent and parallel, separated by house tracks.

The other extreme is a combination of inbound and outbound platforms laid end to end, with the freight office near the center on the driveway side. This arrangement may be dictated where the space available is of limited width.

The width and length of platforms are no longer of prime concern since the use of power trucks and conveyors have become so prevalent.

Where inbound and outbound platforms are separated by tracks, they should be connected by a platform at one end and one or more connecting retractible bridges spanning the tracks. One variation of this arrangement includes a transfer platform located between the inbound and outbound platforms and connected to each by a bridge which retracts under the platform when cars are being spotted.

It may be desirable to consider leaving the long sides of the outbound freight house open when it is separate from the inbound house.

For other factors affecting the design, such as location, overall layout, capacity, supporting tracks and other general considerations, refer to Part 3, Chapter 14.

2. Materials

In general, freighthouses should be built of fire-resistant materials throughout.

Floors should be of concrete with suitable application of floor-hardener to provide a relatively dust-free and maintenance-free wearing surface.

Superstructure materials may vary with location, economics and architectural requirements. The use of so called "standard" metal buildings is increasing and may be initially most economical. New building systems, including the use of precast or prestressed concrete framing, roof systems, and side walls either cast in place or erected through use of the tilt-up method, are being utilized to an increasing extent.

3. Fire Walls and Fire Protection Devices

Where fire walls are required they should conform to the requirements of the local building codes and the National Board of Fire Underwriters.

Openings in fire walls should be as limited in number as is consistent with operating conditions. Each opening shall be equipped with an approved type automatic fire door. Size of these doors should be sufficient to provide clearances not less than those provided by freight doors on the platforms.

Standpipes and hose racks and chemical extinguishers should be provided and located in accordance with Underwriters' and local code recommendations. Locations should be suitably marked and so laid out that ready access may be had at all times. Where possible, consideration should be given to the use of sprinkler and public fire-fighting services.

4. Platforms, Doors and Roof

The best arrangement is one in which all tracks are inside the building, especially in colder climates. Where this layout is not used, a continuous platform protected by an overhanging roof should be provided. The least desirable arrangement is that which only provides door openings spaced one car-length apart, in which case each individual door should be protected by a canopy.

The height of the platform on the driveway side should be such that it will match the tailgate level of the trucks generally in use at the terminal. It may be advisable to provide special ramps adjustable to various truck-bed heights at some of the doors.

Doors should be of the overhead type, at least 10-ft high, and should be continuous on the truck side so that all of the house can be opened except for the space occupied by columns.

Door jambs in masonry walls should be protected by metal guards for their full height.

On the driveway side, platforms and walls should be protected from damage by trucks through the use of suitable longitudinal fenders. These fenders may be of timber or steel and rubber, and should be continuous along the platform regardless of door location. They should be so attached that they are easily removed for maintenance purposes.

Outside platforms should slope away from the building to prevent accumulation of water.

Doors on the driveway side should be protected by a roof overhang or canopy at least 12-ft in width and with a minimum height of 14-ft above the driveway. Downspouts should not be located inside the building. Placing them outside requires that they be properly protected from damage by trucks. The alternate is to use heavier material which will absorb considerable impact throughout the lower 15 ft of the downspout.

It is desirable to have the platforms as free of columns as possible. Increasing use of rigid-frame construction and other new building systems now makes it possible to provide clear spans of over 100 ft. The extent to which the designer must limit the clear floor space is governed by economic considerations and the overall layout of the individual freight house.

5. Lighting

(a) *Artificial Lighting*—Artificial lighting is needed for operation at night and during dark winter days. The circuits should be carefully planned, so as to allow maximum flexibility and economy in the use of lights. Wiring should meet the minimum local code and Underwriters' requirements.

Mercury-vapor-type fixtures are coming into increasing use for general lighting. Exterior platforms should be well lighted. Consideration should be given to the installation of spot lights oriented so as to shine into trucks and cars. Receptacles should be provided at doorways for the accomodations of extension cord lighting.

(b) *Natural Light*—Since the introduction of plastic panels with conformation fitting the various types of roofing sheets, previous objections to the use of skylights because of maintenance problems have largely been removed. These translucent panels are easily installed, nesting with the roofing sheets into an unbroken roof line. They require no more maintenance than the roofing itself, and in most cases replace daytime artificial light on all but the darkest days.

6. Offices

Where inbound and outbound platforms are parallel and separated, the office should be located at one end on a platform joining them. Offices are usually of masonry construction, and should include space for agent, clerks, foremen, and for the communication system.

Where the inbound and outbound platforms are end-to-end, the office should be located near the center of the street side of the freight house.

In many instances, division offices may be combined with the freight office. Toilet and locker facilities for office personnel and freight handlers should be provided adjacent to the office.

7. Communication Facilities

No new freight house should be built without a thorough study of the advantages that may be gained through the use of a centralized checking system. Likewise, expenditures for the installation of such facilities could probably be justified for many existing freight houses. It has been demonstrated that savings realized from recent installations have more than offset the total cost during the first year of operaiton. At the same time, better and faster service to shippers can be accomplished.

Under this system, all checkers are in a central office, where better working conditions increase their efficiency. Each checker has a small console connected with each loading or unloading spot on the platforms by telephone or loudspeaker lines with outlet jacks. As the freight handler must turn up a few packages at a time to read the consignees' names and addresses, the checker works only a few minutes at a time with one freight handler. Thus the checker can work with several freight handlers in rotation, and three to five cars can be checked in the time ordinarily required to check one car under the old system. In addition, the checker does not have to walk between the office and platforms picking up and returning waybills, as all paper work is kept at his console.

Either paging and talk-back speakers or telephones can be used. Telephone handsets have the advantage in noisy areas, in that they do not pick up as much ambient sound as loudspeakers. On the other hand, loudspeakers enable the freight handler to have his hands free while working the car. In some cases, both systems may be justified.

Its flexibility may make the use of radio justifiable. Its initial cost is also claimed to be less than that of the conventional system. With portable radios, a two-way conversation can be carried on between any two points in the freight-house, including the inside of cars.

A pneumatic tube system for handling interchange routing, bills of lading and miscellaneous papers may also be desirable.

8. Mechanized Freight-Handling Facilities

The possibility of installing an underfloor conveying system which tows platform trucks throughout the freight house should not be overlooked.

Where inbound and outbound platforms are separated, there should be a connecting loop which carries trucks from one platform to the other. Where the conveyors serving each platform come together and/or parallel each other, one system should run clockwise and the other counterclockwise, to facilitate transfer in safety.

Control switches should be located at convenient intervals to allow stopping of the conveyor in an emergency or when operations require.

Where the platform trucks cross the tracks, a special crossing may be made by ramping down to the level of the tracks, or a tunnel may be utilized. Proper signalling should be provided for the former arrangement.

In the design of a new freight house where the expense of a conveyor system cannot be immediately justified, provision may be made in the floor for its inclusion later. Channels can be provided in the floor, filled in and topped with a steel plate and a thin layer of concrete, which can be removed for installation of the conveyor when justified.

9. Scales

Scales are a necessary part of freight house equipment. They should be located so that freight can be weighed as received and trucked to cars without rehandling.

With modern handling methods, it is no longer required that scales be located at such close intervals as they have been in the past, when hand-truck methods were used. Scales spaced 300 ft apart have recently been installed. The arrangement must be determined after a study of the requirements of the particular operation.

Scales should have a minimum capacity of 5 tons. Higher capacity scales may cost little more and be economical from a maintenance standpoint, inasmuch as they will stand up better under the abuse to which they are usually subjected. Scale platforms should be as small as practicable to accommodate the trucks used.

10. Tracks

Track clearances for platforms and buildings, as well as heights of platforms and floors above top of rail, should conform with the recommendations of Chapter 28— Clearances, or with state or other legal restrictions governing such clearances.

11. Appurtenant Facilities

Where unusually heavy or bulky shipments or wheeled equipment are to be handled, an outside platform at car-floor level, allowing both side loading and end loading, should be provided. Such platforms should be combined with a ramp to street level. Provision of a crane to handle heavy lifts should also be considered.

Where tracks enter the building, a power-operated, lightweight rocker or similar type door should be considered. This push-button operated door opens or closes in a matter of seconds to allow cars to be switched into or out of the building. Suitable safety devices should be included.

In addition to the offices and locker rooms, consideration should be given to the provision of an equipment maintenance room, a cooperage room, and salvage room. Provision of a special room for perishable freight and one for bonded freight may also be indicated. All these facilities should be located near the offices.

Further reference should be made to Part 3, Chapter 14 for recommendations concerning the provision of other appurtenant facilities.

Pages 6–21–1 to 6–21–4, incl.

ENGINEHOUSES—GENERAL

Delete.

Pages 6–21–4 to 6–21–8, incl.

DROP PITS, JACKS AND TABLES

Delete

Pages 6–23–1 to 6–23–4, incl.

LOCOMOTIVE SANDING FACILITIES

Reapprove with the following revisions:

Page 6–23–1, Art. 1. General—Change second paragraph to read:

"If dried sand is to be delivered to servicing stations, consideration should be given to the drying of the sand at some centrally located railroad plant from which distribution can be made to several stations. The purchase of commercial dried sand has in some instances proven to be the most economical means of procurement. The use of quarry waste of the fineness of sand, if available, on a railroad, should be considered."

Page 6–23–2, Art. 3. Storage Tank and Service Tanks—Change second paragraph to read:

"The service tank at track side is generally of steel construction mounted on a steel column or as an integral part of the sand storage silo, at a suitable elevation to permit loading sand into locomotive sand boxes by gravity through pipe and hose connections. Capacities of 5 or 10 tons are satisfactory for these latter tanks, the size being determined by the quantity of sand handled."   .

Page 6–23–2, Art. 4, Unloading—Change third paragraph to read:

Probably the most economical method of unloading dry bulk sand is from totally closed hopper-bottom cars or from box or rebuilt tank cars with hopper bottoms. These cars can be spotted over a hopper built into a pit, into which the sand can be discharged, or the sand can be drawn from the car by the use of air-operated vacuum devices and discharged into the storage hopper or directly to the sand elevating tank. The sand can be elevated from the hopper into the dry storage tank either by conveyor or by the use of air.

Page 6–23–4, Art. 13. Air Loading—Add the following sentence: "While this method has been used successfully it is not generally accepted as the most desirable system due to the close control of air required."

Page 6–23–4, Art. 17. Location—Change to read:

"If space is available, the sanding and fueling spots should preferably be on the same track, but not closer than 50 ft. In this way the operations may be performed quickly and in sequence without fouling one another."

Pages 6–25–1 to 6–25–2

PASSENGER CAR SHOPS

Reapprove without change.

Pages 6–25–2 and 6–25–3

FREIGHT CAR REPAIR SHOPS

Reapprove with the following revision:
Change the title to read "Freight Car Heavy Repair Shops."

Pages 6–25–4 and 6–25–5

CAR PAINT SHOP

Reapprove without change.

Pages 6–25–6 to 6–25–19, incl.

SHOP FACILITIES FOR DIESEL LOCOMOTIVES

Reapprove with the following revisions:

Page 6–25–10—In Art. 13, first sentence, substitute "suitable" for "adequate" and eliminate "to meet requirements."

Page 6–25–12. In Art. 6. Painting, fifth line, eliminate remainder of sentence after "eye arresting."

Page 6–25–14—Replace the second, third and fourth paragraphs beginning with "As the ventilation system......" and ending "......supply air," with the following "To ventilate a shop where a three-unit 6000 hp diesel has all of its engines idling, consider a maximum acceptable concentration for oxides of nitrogen at 25 parts per million parts of air. It would require 5.8 times the volume of a mixture of fresh air and exhaust gas to bring a concentration of 145 parts per million down to 25. Therefore, the dilution volume to be supplied will be 4.8×5604, or about 27,000 cfm. In addition, the air for combustion is: $4.27 \times 0.344 \times 6000 = 8808$ cfm.

"The combustion air plus the ventilating air will combine to equal 35,800 cfm to be furnished by the fans. Where it is desired to reduce the gas concentration still further, additional air can be supplied."

On same page, fifth line from bottom, change "must" to "should."

Page 6–25–15—Under 1. Lighting and Electrical Outlets, first sentence of first paragraph, change "it is a requirement of utmost importance" to "should be provided."

Under G. Overhead Cranes, Art. 1. General, second line, change the words "an essential" to "a desirable."

Page 6–27–17—In first line of first paragraph, Sec. H. Fire Protection, change "shall" to "should."

It is further recommended that all specifications in Chapter 6 of the Manual not presently specifically designated as "specifications" in their headings be given such designation by the addition of the words "Specifications for."

Report of Committee 14—Yards and Terminals

A. S. KREFTING, *Chairman*
D. C. HASTINGS,
   Vice Chairman
A. E. BIERMANN, *Secretary*
F. E. AUSTERMAN
F. R. SMITH
HUBERT PHYPERS
F. S. KING
B. E. BUTERBAUGH
H. J. McNALLY
J. J. TIBBITS
C. J. MORRIS

M. H. ALDRICH
R. O. BALSTERS
R. F. BECK
H. R. BECKMANN
W. O. BOESSNECK
H. M. BOOTH
E. G. BRISBIN
W. P. BUCHANAN
G. H. CHABOT
J. F. CHANDLER
H. P. CLAPP
E. H. COOK
V. R. COPP
B. E. CRUMPLER
J. L. DAHLROT
A. V. DASBURG
OSCAR FISCHER
C. M. FRAZIER
W. H. GILES (E)
W. H. GOOLD
C. W. HAMILTON
G. F. HAND·(E)
WM. J. HEDLEY
H. W. HEM
F. A. HESS
J. E. HOVING
C. C. JACOBSON*
V. C. KENNEDY
B. LAUBENFELS
F. C. LARSEN, JR.
GLEN LICHTENWALNER
J. L. LOIDA

E. T. LUCEY
L. L. LYFORD (E)
S. N. MAC ISAAC
G. W. MAHN, JR.
T. F. MALONEY
J. C. MILLER
R. F. MORIS
C. H. MOTTIER (E)
B. G. PACKARD
R. H. PEAK, JR.
G. A. PEIRCE
L. F. POHL
W. H. POLLARD
L. J. RIEKENBERG
L. W. ROBINSON
R. E. ROBINSON
H. T. ROEBUCK
H. H. RUSSELL
R. A. SKOOGLUN
C. E. STOECKER
R. F. STRAW
T. D. STYLES
J. G. SUTHERLAND
JACK SUTTON
J. B. SUTTON
L. G. TIEMAN
J. N. TODD (E)
J. W. TUCKER
P. P. WAGNER, JR.
W. E. WEBSTER, JR.
W. A. WOOD
C. E. ZEMAN
Committee

(E) Member Emeritus.
* Deceased.
Members listed in bold face are the official representatives of the Engineering Division, AAR.

To the American Railway Engineering Association:

Your committee reports on the following subjects:

1. Revision of Manual.

   Progress report, including recommendations submitted for adoption (See
   report on Assignment 3 for additional Manual recommendations) .... page 236

2. Classification yards, collaborating with Committee 16.

   Under this assignment a subcommittee has been trying to secure information
   on rollability of freight cars. The work has been hampered by the fact that
   funds requested by the committee for research and testing have not been
   provided. Some progress has been made in analyzing data previously secured
   from tests made in 1960. It is the hope of the committee that funds for
   the necessary test work to secure pertinent information on the subject will
   soon be made available.

6. Facilities for loading and unloading rail-truck freight equipment, collaborating with Committee 6.
Progress has been made this year in developing typical layout arrangements for rail-van terminals, but the report is being held up until next year to develop more complete information.

7. Waterfront terminals.

8. Present trends in yard maintenance.
Two progress reports have been previously submitted on this subject. While progress has been made in developing a third and final report, it requires further study and will not be submitted until next year.

THE COMMITTEE ON YARDS AND TERMINALS,
A. S. KREFTING, *Chairman.*

AREA Bulletin 567, November 1961.

Report on Assignment 1

Revision of Manual

F. E. Austerman (chairman, subcommittee), M. H. Aldrich, A. E. Biermann, G. H. Chabot, W. H. Goold, C. W. Hamilton, D. C. Hastings, A. S. Krefting, C. J. Morris, L. F. Pohl, R. E. Robinson, J. J. Tibbits, W. E. Webster, Jr., C. E. Zeman.

Your committee submits for adoption the following recommendations with respect Chapter 14 of the Manual, and the Glossary.

Pages 14–2–1 to 14–2–19, incl.

PASSENGER TERMINALS

Reapprove with the following revision:
On page 14–2–18 change the word "required" in the seventh line to read "needed", making the line read "Other facilities, some or all of which may be needed, include:".

Pages 14–4–1 to 14–4–8, incl.

LOCOMOTIVE TERMINALS

Reapprove with the following revisions:
Page 14–4–6. Under Art. 1. Enginehouse, revise second paragraph to read:

"The track layout should be designed so that locomotives which do not require turning may be serviced without crossing the turntable."

Page 14-4-7. Under Art. 8. Cinder-Handling Facilities, add a new item (d) reading:

"(d) A depressed track beside or between the incoming tracks, deep enough to accommodate cars into which cinders are chuted directly from the locomotives."

Reletter present item (d) to (e).

Page 14-4-8. Revise the paragraph under Art. 3. Communication Facilities, to read:

(a) Telephone
(b) Radio
(c) Paging—Talk-back speakers
(d) Pneumatic tube (if necessary)

Add a new Art. 6 reading as follows:

6. Boiler Washing Facilities

(a) In diesel, diesel-electric, and electric locomotive terminals, boiler-washing facilities should be provided for heat generating boilers on locomotives used in passenger, mail and express service.

(b) In steam-locomotive terminals, boiler-washing facilities should be provided for all locomotives.

Add a new Art. 7 reading as follows:

7. Wreck Equipment Track

Locomotive terminals should be provided, where required, with a double-ended track for the storage of a wreck train.

Renumber present Art. 6. General, to 8.

Delete the present material on pages 14-4-14 to 14-4-16, incl., substituting therefor the following rewritten version:

STORES FACILITIES, INCLUDING RECLAMATION, SCRAP AND MATERIAL YARDS

The stores department is responsible for the ordering, care, control and economic distribution, and in some instances for the accounting of materials and supplies needed for, or reclaimed from, the construction, maintenance and operation of the railroad. The size and extent of its facilities will vary in accordance with the requirements of the road. It is important to consult the chief stores officer and receive his approval concerning any plans for the construction, alteration or elimination of stores facilities

1. Stores

There are three types of stores, namely, general, district, and local.

a. The general stores, also known as a system or regional store, is the largest store unit of the stores department. It should be located on available railroad property and usually at a convenient point where large quantities of materials and supplies can be efficiently received, handled, stored and shipped. The location of this store will also be greatly influenced by the traffic problem created in the handling of these shipments and by the freight charges involved on off-line items received. The general store will also operate reclamation and scrap yards where needed.

b. The district and local stores have the same characteristics and functions as the general stores, except that they are much smaller. These stores are generally established on larger railroads at various points to expedite the handling of materials and supplies. Such stores operate under the jurisdiction of the general store.

Stationary, office supplies and maintenance of way materials are normally handled by the general store. Maintenance-of-way materials, however, are generally handled in separate facilities. Dining car service supplies, including foodstuffs, may be handled by the general store, but in many instances such items are handled separately at major terminals in a local store, called a commissary.

2. Buildings and Structures

Storehouse buildings for the handling of all materials requiring inside storage should be constructed so as to create the most efficient and expeditious material storage and handling methods. Office space to house the necessary personnel to handle the records and accounting for the store's operation may be part of a storehouse building if suitable. Platform, docks, ramps, racks and shelters are erected according to the needs. All storage buildings and related facilities should be served with tracks and hard-surface driveways for the efficient handling of materials by rail or by truck. It is often possible to pave the track area so that one platform at car-floor level can serve both means of handling.

3. Material Yards

There are numerous items used in maintenance of way and of equipment that can be stored out of doors; these items are handled in material yards. Whenever possible, such yards should be located adjacent to the storehouse area so that trackage can be kept to a minimum. Material is stored on permanent racks and platforms, and the areas between should be paved to facilitate the operation of rubber-tired handling equipment such as trucks, loaders, cranes, etc.

The storage of heavy items in a material yard is usually at a separate location served by at least two tracks and an overhead crane or other types of cranes of suitable capacity. The material is stored in the area between the tracks, one track being used for receiving, the other for shipping.

The ideal scrap yard has a receiving and a shipping track with the sorting area in between and served by an overhead crane of suitable capacity. The sorting area should be hard surfaced and the driveways serving it paved to support the heavy wheel loadings of truck cranes and trailers used to handle scrap within the yard. All cranes should be equipped with magnets.

4. Lumber and Timber Yards

Lumber products are not generally kept in large quantities at the general storehouse, but are frequently shipped direct from the dealer to the point of application. However a certain quantity of lumber, cross and switch ties, bridge timber and poles must be stored. These products require outside storage; unseasoned materials should be stored on permanent racks in covered storage so they can season properly; treated timbers should be stored in the manner approved by the stores department to prevent loss by fire. The areas between the racks should be paved and the piles so arranged that fork-lift tractors or truck cranes can handle these materials into and out or onto and off freight equipment on a track serving the storage yard. This track is usually in the center of the yard unless the area is too large, then two or more tracks serving storage areas on both sides of each track are required, and the tracks, if possible, are connected at both ends.

5. Reserve Oil Storage

The stores department may be called upon to provide large storage reserves for fuel oils. When the size and location of the facilities have been determined, the tanks should be installed in accordance with the requirements set forth by the governing ordinances, building and fire codes.

6. Reclamation Plant

The reclamation plant is usually located at the same point and adjacent to the scrap yard to minimize handling of materials. The reclamation shop building should be situated between a receiving and a shipping track, the latter depressed to facilitate the loading of materials coming out of the plant for forwarding to points of application or storage. The size of the shop will vary with the amount and type of reclamation to be done. Paved roads parallel to the tracks are needed for the operation of truck cranes; a large area adjacent to the tracks and the shop building should be paved so that materials can be transported in and out of the building with motorized equipment.

7. Rail Reclamation Plants

Rail requires special handling in general reclamation, and the plant to handle it should be separate from other plants. The plant layout should be designed for the rapid turnover of rail and would consist of receiving and shipping tracks served by overhead or other types of cranes, with the area between the tracks used for the straightening presses, the cropping operation, drilling rack, hardening apparatus, welding and classifying prior to loading.

GLOSSARY

Add the following definitions:

RETARDER, CAR.—A braking device, usually power operated, built into a railway track to reduce the speed of cars by means of brake shoes which, when set in position, press against the sides of the lower portions of the wheels. *14*

YARDS, RETARDER.—A hump yard provided with retarders to control the speed of the cars during their descent to the classification tracks. *14*

RETARDER, INERT.—A braking device, without external power, built into a railway track to reduce the speed of cars by means of brake shoes against the sides of the lower portions of the wheels and sometimes provided with means for opening it to nullify its braking effect. *14*

VELOCITY HEAD.—The vertical distance through which a body would fall to obtain a given velocity. A term used for the rating measure of the braking power of a car retarder. *14*

Report on Assignment 3

Scales Used in Railway Service

Collaborating with Committee 18

H. Phypers (chairman, subcommittee), W. P. Buchanan, J. H. Dahlrot, D. C. Hastings, H. W. Hem, C. C. Jacobson, V. C. Kennedy, A. S. Krefting, F. C. Larsen, G. A. Peirce, H. H. Russell, R. F. Straw, J. W. Tucker.

Your committee submits its report on Assignment 3 in two parts. In Part 1 the committee offers for adoption recommendations with respect to Part 5, Chapter 14, of the Manual. Part 2 is a progress report on weighing freight cars by the two-draft method, submitted as information.

Part 1—Manual Recommendations

Your committee submits for adoption the following recommendations with respect to the scale specifications in Part 5, Chapter 14, of the Manual:

Pages 14–5–1 to 14–5–15, incl.

RULES FOR THE LOCATION, MAINTENANCE, OPERATION AND TESTING OF RAILWAY TRACK SCALES

Page 14–5–4, Sec. B, Art. 1—In the first sentence delete the last five words, reading "at least every 3 months."

Pages 14–5–19 to 14–5–37, incl.

SPECIFICATIONS FOR THE MANUFACTURE AND INSTALLATION OF FOUR-SECTION KNIFE-EDGE RAILWAY TRACK SCALES

Reapprove with the following revisions:

Page 14–5–19, Sec. A. Introduction—In the third line of the paragraph, second sentence, after the word "specifications" replace the comma with a period and delete remainder of sentence.

Page 14–5–21, Sec. BB. Art. 3—Change paragraphs (a) and (b) to read as follows:

(a) For scales of 60 tons sectional capacity or less, the load per inch of knife edge shall not exceed 6000 lb for high-carbon steel (SAE 1095) or 7000 lb for alloy steel (SAE 52100) or for steels which will give equivalent performance.

(b) For scales of greater than 60 tons sectional capacity the load per inch of knife edge shall not exceed 5000 lb for high carbon steel (SAE 1095) or 6000 lb for alloy steel (SAE 52100) or for steels which will give equivalent performance.

Page 14–5–22, Sec. BB. Table 2—Delete reference to SAE 6195 steel.

Page 14–5–24, Sec. D, Art. 1—Change paragraphs (a) and (b) to read as follows:

(a) Alloy steel (SAE 52100) or a steel which will give equivalent performance, hardened to Rockwell C scale not less than 58 nor more than 62, or

(b) High-carbon steel (SAE 1095) or a steel which will give equivalent performance, hardened to Rockwell C scale not less than 60 nor more than 62.

Page 14–5–31, Sec. HH, Art. 11—Change second sentence to read: "Full-length rails or welded sections shall be used."

Page 14-5-36, Sec. LL, Art. 15—In the first sentence, first line, delete the word "weighbeam" and substitute the word "scale". In the first sentence, second line, change the comma after the word "wall" to a period and delete the words "preferably the latter."

Page 14-5-36—Change the title of Sec. MM from "Weighbeam House" to "Scale House." Change the paragraph under Art. 1 to read:

"Except where the indicating elements are mounted in a separate building, a scale house large enough to install, observe and service the indicating elements shall be provided. It shall have windows of sufficient size and so located as to give the weigher an unobstructed view of the scale deck and approaching cars."

Pages 14-5-38 to 14-5-55, incl.

SPECIFICATIONS FOR THE MANUFACTURE AND INSTALLATION OF TWO-SECTION, KNIFE EDGE RAILWAY TRACK SCALES

Reapprove with the following revisions:

Page 14-5-39, Sec. BB, Table 1—Delete reference to SAE 6195 steel.

Page 14-5-42, Sec. D, Art. 1—Change to read:

"The material to be used for pivots and bearings shall be alloy steel (SAE 52100) or a steel which will give equivalent performance, hardened to Rockwell C scale not less than 58 nor more than 62."

Page 14-5-50, Sec. HH, Art. 8—Change second sentence to read "New rails or welded sections shall be used."

Page 14-5-50, Sec. I, Art. 1—Add new second sentence reading as follows: "For 20 ft two-draft scales of 100-ton sectional capacity the section modulus shall not be less than 144.

Page 14-5-51, Sec. KK, Art. 1—Substitute the words "indicating elements" for the word "weighbeam" wherever it appears in this article.

Page 14-5-53, Sec. LL, Art. 14—In first sentence, first line, delete the word "weighbeam", substitute the word "scale." In second line change the comma after the word "wall" to a period and delete the words "preferably the latter."

Page 14-5-54—Change the title of Sec. MM from "Weighbeam House" to "Scale House." Change the paragraph under Art. 1 to read as follows:

"Except where the indicating elements are mounted in a separate building, a scale house large enough to install, observe and service the indicating elements shall be provided. It shall have windows of sufficient size and so located as to give the weigher an unobstructed view of the scale deck and approaching cars."

Pages 14-5-56 to 14-5-78, incl.

SPECIFICATIONS FOR THE MANUFACTURE AND INSTALLATION OF TWO-SECTION MOTOR TRUCK SCALES, AND BUILT-IN, SELF-CONTAINED AND PORTABLE SCALES FOR RAILWAY SERVICE

Page 14-5-56, Sec. AA—Change the first two questions to read:

"If a motor truck scale, what type of indication is required and what minimum graduation should be furnished?"

"If a built-in scale, what type of indication is required and what minimum graduation should be furnished?"

Page 14–5–59, Sec. CC, Table 3—Delete reference to SAE 6195 steel.

Page 14–5–60, Sec. CC, Art. 3—Change to read:

"The load per inch of knife-edge shall not exceed 5000 lb for high-carbon steel (SAE 1095) or 6000 lb for alloy steel (SAE 52100) or for steels which will give equivalent performance."

Page 14–5–65, Sec. E, Art. 1—Change paragraphs (a) and (b) to read as follows:

"(a) Alloy steel (SAE 52100) or a steel which will give equivalent performance, hardened to Rockwell C scale not less than 58, nor more than 62."

"(b) High-carbon steel (SAE 1095) or a steel which will give equivalent performance, hardened to Rockwell C scale not less than 60, nor more than 62."

Page 14–5–72, Sec. I—Substitute the word "indicating element" for the word "weighbeam."

Page 14–5–73—Change the title of Sec. L from "Weighbeam House or Box", to "Scale House or Box." Also, change the title of Art. 1 to "Scale House or box." In the first line of Art. 1, substitute the words "indicating elements" for the word "weighbeam."

Page 14–5–74, Sec. L, Art. 2—Change to read:

"A scale house large enough to install, observe and service the indicating elements shall be provided. It shall have windows of sufficient size and so located as to give the weigher an unobstructed view of the scale platform. If the indicating elements are required to be boxed, the box shall be of such size as to suitably enclose the indicating element. It shall be provided with a hinged door or doors of such size and in such location as to give the weigher a clear and unobstructed access to the indicating elements."

Page 14–5–74, Sec. L, Art. 13—In the third line, after the word "or", delete the word "weighbeam", substitute the word, "scale."

Page 14–5–78, Sec. MM, Art. 1—Substitute the words "indicating elements" for the words "scale weighbeam."

Pages 14–5–7 to 14–5–92, incl.

SPECIFICATIONS FOR THE MANUFACTURE AND INSTALLATION OF HAND-OPERATED GRAIN HOPPER SCALES

Page 14–5–80, Sec. BB, Table 1—Delete reference to SAE 6195 steel.

Page 14–5–81, Sec. BB, Art. 3—Change to read:

"The load per inch of knife-edge shall not exceed 5000 lb for high-carbon steel (SAE 1095) or 6000 lb for alloy steel (SAE 52100) or for steels which will give equivalent performance."

Page 14–5–82, Sec. CC, Art. 1—Change paragraphs (a) and (b) to read as follows:

(a) Alloy steel (SAE 52100) or a steel which will give equivalent performance, hardened to Rockwell C scale not less than 58, nor more than 62, or

(b) High carbon steel (SAE 1095) or a steel which will give equivalent performance, hardened to Rockwell C scale, not less than 60, nor more than 62.

Pages 14–5–93 to 14–5–109, incl.

SPECIFICATIONS FOR THE MANUFACTURE AND INSTALLATION OF FOUR-SECTION MOTOR TRUCK SCALES

Page 14–5–93, Sec. AA—Change items 2 and 5 to read as follows:

"2. Type of indicating element required."

"5. Any special requirements for clearance between the platform and indicating elements."

Page 14–5–93, Sec. B—In the third paragraph, substitute the words "indicating elements" for the word "weighbeam."

Page 14–5–95, Sec. C, Table 1—Delete reference to SAE 6195 steel.

Page 14–5–95, Sec. C, Art. 3—Change to read:

"The load per inch of knife-edge shall not exceed 5000 lb for high-carbon steel (SAE 1095) or 6000 lb for alloy steel (SAE 52100) or for steels which will give equivalent performance."

Page 14–5–100, Sec. DD, Art. 1—Change paragraphs (a) and (b) to read as follows:

(a) Alloy steel (SAE 52100) or a steel which will give equivalent performance, hardened to Rockwell C scale, not less than 58 nor more than 60, or

(b) High-carbon steel (SAE 1095) or a steel which will give equivalent performance, hardened to Rockwell C scale, not less than 60 nor more than 62.

Page 14–5–101, Sec. EE, Art. 3—Substitute the word "indicating elements" for the word "weighbeam" in second line and third lines.

Page 14–5–106—Change the title of Sec. KK from "Weighbeam House or Box" to "Scale House or Box". Also, change the title of Art. 1 to read "Scale House or Box". In Art. 1 substitute the words "indicating elements" for the word "weighbeam."

Page 14–5–106, Sec. KK, Art. 2—Change to read:

"A scale house large enough to install, observe and service the indicating elements shall be provided. It shall have windows of sufficient size and so located as to give the weigher an unobstructed view of the scale platform. If the indicating elements are required to be boxed, the box shall be of such size as to suitably enclose the indicating elements. It shall be provided with a hinged door or doors of such size and in such location as to give the weigher a clear and unobstructed access to the indicating elements."

Page 14–5–108, Sec. M, Art. 1—Substitute the words, "indicating elements" for the word "weighbeam."

Part 2—Weighing Freight Cars by the Two-Draft Method

Your committee presents as information the following progress report on its study of the subject of weighing freight cars by the two-draft method.

Two-draft weighing should be clearly defined, as there are two distinct methods involved:

1. Two-draft gravity motion weighing uncoupled.

2. Two-draft coupled-in-motion weighing.

Both of these methods are for weighing standard two-axle-truck freight cars of 29-ft inside length or longer; they are not suitable for weighing shorter wheel-base cars, such as ore cars. Shorter wheel-base cars can be weighed in a similar manner on specially designed scales, however.

This report covers the first method only. Some of the advantages of this method are as follows:

1. Cars may be weighed at a rate of approximately three cars per minute.
2. Human errors in weighing, such as weight determination and proper spotting of car on the scale, are reduced.
3. In addition to being able to determine if a complete car is overloaded, information concerning the overload by individual trucks is provided at time of weighing.
4. Savings in cost of certain scale components, superstructure, foundation and installation can be realized.

A small hump must be constructed on the scale track, with apex 50 ft ahead of the end of the scale weighrail, the grade from the apex falling at the rate of 0.5 percent over the scale and beyond for 60 ft. From this point the grade can be determined to suit local conditions in running cars well down the track.

Track scales conforming with AREA specifications for two-section knife-edge railway track scales, equipped with regular weighbeam and an automatic printer recorder adapted to motion weighing must be used. The weighbeam is used for testing, checking the recorder, and static weighing, but is cut out when using the recorder for weighing in motion.

Each truck weight of a car is automatically printed on a tape or ticket, as desired, in 100-lb increments. The weights of each truck are printed directly below each other, with a space between each pair of truck weights to write in the total, the car number, the tare and the load limit. A flange-actuated tripper is located at each end of the beamside scale weighrail, which together control the cycling of the recorder. The upper tripper counts the wheels coming onto the scale and actuates the timer. The lower tripper actuates the printer, which is kept locked in the event that the speed is excessive so that inaccurate weights are not recorded.

It must be clearly understood that a high standard of maintenance is necessary on all types of motion weighing installations to provide continually good weighing. The following tables show test information covering two-draft gravity motion weighing on track scales built to meet AREA specifications. This information covers lever-type mechanical track scales equipped with mechanical recording elements.

New fully electronic installations have been made. Such scales should be required to perform in a manner comparable to the mechanical-type scale of similar sectional capacities, weighbridges and approaches. Specifications for electronic installations will be compiled when necessary information is available.

Report on two-draft gravity motion weighing for AREA Committee 14, Chicago Great Western Railway, Fort Dodge, Iowa, September 20, 1960, 20-ft weighrail track scale. Gradient of weigh rail 0.5 percent. Scale installed March 29, 1958, last tested August 1960.

Track scale to AREA Specifications except that length of approach wall at entering end is deficient by 10 ft. Scale is of two-section design equipped with type-registering weigh beam and automatic printer-recorder. Main girders, 30 in @ 230 lb.

Car Number	Static-Weights Beam A	Recorder B	Motion Weights taken by Recorder Run 1	Run 2	Run 3	Run 4	Run 5	Run 6	Run 7	Run 8	Run 9	Run 10	Run 11
C-G-W. 425 Test Car FROM A	L 42620 R 42630 M 42625	42600 42575 42590 -35	42650 +25	42625 0	42600 -25	42650 +25	Excessive Speed	42600 -25	42600 -25	42650 +25	42625 0	42600 -25	42625 0
I.C. 72925 Hopper FROM A	21710 23010 44720	21675 23000 44675 -15	21700 23025 44725 +5	21650 22950 44600 -120	21675 22950 44625 -95	21700 23000 44700 -20	21650 22950 44600 -120	21700 23000 44700 -20	21700 23000 44700 -20	21700 23050 44750 +30	21700 22975 44675 -45	21700 22950 44650 -70	21700 23025 44725 +5
SHPX. 21122 Tank FROM A	20560 20790 41750	20800 20740 41650 -100	20850 20825 41675 -75	20825 20800 41625 -125	20875 20800 41675 -75	20825 20825 41650 -100	20675 20800 41675 -75	20850 20775 41625 -125	20850 20775 41625 -125	20850 20800 41650 -100	20850 20800 41650 -100	20850 20800 41650 -100	20875 20850 41725 -25
C&NW. 68200 Box FROM A	43210 42950 86160	43100 43025 86175 -5	43150 43100 86250 +70	43200 43025 86225 +45	43200 43050 86250 +70	43175 43050 86225 +45	43150 43050 86200 +20	43125 43125 86325 +145	43200 43050 86250 +70	43250 43025 86275 +95	43200 43100 86300 +120	43200 43125 86325 +145	43300 43100 86400 +220
C&GW. 52052 Box FROM A	25270 24760 50030	25200 24775 49975 -55	25200 24850 50050 +20	25150 24850 50000 -30	25200 24800 50000 -30	25100 24800 49900 -130	25175 24800 49975 -55	25175 24850 50025 -5	25200 24800 50000 -30	25200 24800 50000 -30	25200 24825 50025 -5	25200 24825 50025 -5	25200 24850 50050 +20
M&StL. * 70331 Hopper FROM A	27040 25360 52400	26850 25175 52025 -375	26775 25325 52100 -300	26700 25325 52025 -375	26700 25300 52000 -400	26675 25225 51900 -500	26700 25300 52000 -400	26700 25200 51900 -500	26650 25225 51875 -525	26725 25375 52100 -300	26850 25275 52125 -275	Excessive Speed	26825 25300 52125 -275

Sheet 1 of 2 ...

C&O. 293056 Box FROM A.														
27600	27625	27575	27625	27675	27725	27800	27800	27800	27800	27800	27800	27825	27850	27900
26070	26025	26125	26025	25975	25925	25875	25850	25850	25850	25850	25900	25850	25800	25775
53670	53650	53700	53650	53650	53650	53675	53650	53650	53650	53700	53675	53650	53675	
	=20	+30	=20	−20	−20	+5	−20	−20	−20	+30	+5	−20	+5	
C&GW. 93443 Box FROM A.														
24410	24475	24300	24525	24500	24525	24700	24800	24800	24800	24800	24875	24900	25000	
24110	24075	24200	23850	23900	23850	23750	23700	23600	23675	23625	23550	23500	23450	
48520	48550	48500	48375	48440	48375	48450	48500	48400	48475	48425	48425	48400	48450	
	+30	−20	−145	−120	−145	−70	−20	−120	−45	−95	−95	−120	−70	
I.C. 93695 Hopper FROM A.														
76920	76475	76850	76300	76350	76350	76200	76200	76125	75875	75575	75475	75400	75375	
28030	28250	28100	28375	28300	28300	28350	28300	28500	28800	29200	29225	29290	79350	
154950	154725	154950	154675	154650	154650	154550	154500	154625	154675	154775	154700	154700	154725	
	−225	0	−275	−300	−300	−400	−450	−325	−275	−175	−250	−250	−225	
I.C. 66068 Hopper FROM A.														
74950	75175	74675	75225	75300	75300	75500	75650	76100	76350	76700	76775	76775	76825	
80800	80525	81050	80450	80350	80350	80175	79925	79625	79300	79100	78975	28990	78900	
155750	155700	155725	155675	155650	155650	155625	155625	155725	155650	155800	155750	155725	155725	
	−50	−25	−75	−100	−100	−75	−125	−25	−100	+50	−0	−25	−25	
N.Y.C. 87980 Hopper FROM A.														
80810	80775	80700	80900	80900	80900	81125	81125	81775	82700	No weight	83675	83850	83975	
79350	29400	29350	29225	29200	29200	28975	28900	28425	27425		26625	26400	26200	
160160	160175	160050	160125	160100	160100	160100	160175	160200	160175		160200	160250	160175	
	+15	−110	−35	−60	−60	−60	+15	+40	+15		+40	+90	+15	

* N&STL. 70131 – 1st. wheels on front truck shelled, causing excessive vibration of recorder.

57.6%	of Motion Weights fall within	0.1%	or less of Static	(A)
75.4%	" " " "	0.2%	" " "	(A)
91.5%	" " " "	0.3%	" " "	(A)
* 8.5%	" " in excess of	0.3%	of Static	(A)

AREA Special Committee –   C.C. Jacobson
                               H.H. Russell
                               W.P. Buchanan

Present and Assisting
Committee –
H.K. Franklin – Streeter Amet Co.
A.F. Ayer   –   "   "   "
R. Doty   –   C. & G.W. RR

CHICAGO, GREAT WESTERN RAILROAD:
Three new installations, completed in 1958, designed to A R E A Specifications
for two-draft gravity motion weighing. Initial test witnessed by various officials.

MASON CITY, IOWA, MARCH 20, 1958

Car Number	Weight by beam Spotted	Weight by Recorder Spotted	Running Weights taken on Recorder:				
			1st.	2nd.	3rd.	4th	5th
AT&SF. 34039	78,400	78,400	78,300	78,150	77,850		
	74,500	74,500	74,700	74,800	75,050		
	152,900	152,900	153,000	152,950	153,000		
			+100	+50	+100		
NYC. 162975	77,060	77,050	77,050	76,950	76,800		
	74,300	74,300	74,300	74,400	74,450		
	151,360	151,350	151,350	151,350	151,250		
			-10	-10	-110		
R.I. 145546	55,400	55,400	55,400	55,350	55,350		
	53,440	53,450	53,450	53,450	53,400		
	108,840	108,850	108,850	108,800	108,750		
			-10	-40	-90		
C.G.W. 8778	66,340	66,350	66,300	66,600	66,900		
	63,200	63,200	63,200	62,950	62,500		
	129,540	129,550	129,500	129,550	129,400		
			-40	+10	-140		

COUNCIL BLUFFS, IOWA, APRIL 3rd, 1958:

Car Number	Weight by beam Spotted	Weight by Recorder Spotted	1st.	2nd.	3rd.	4th	5th
C.G.W. 1055	27,060	27,000	27,050	27,050	27,050	27,000	27,000
	27,620	27,600	27,600	27,600	27,600	27,600	27,550
	54,680	54,600	54,650	54,650	54,650	54,600	54,550
			-30	-30	-30	-30	-130
C.G.W. 719	25,300	25,400	25,250	25,300	25,300	25,400	25,300
	23,800	23,700	23,950	23,900	23,750	23,700	23,800
	49,100	49,100	49,200	49,200	49,050	49,100	49,100
			+100	+100	+50	-0	-0
C.G.W. 715	23,700	23,750	23,650	23,700	23,650	23,700	23,700
	25,160	25,100	25,200	25,200	25,200	25,150	25,250
	48,860	48,850	48,850	48,900	48,850	48,850	48,950
			-10	+40	-10	-10	+90
C.G.W. 70097	57,100	56,900	56,350	56,150			
	55,380	55,500	56,000	56,200			
	112,480	112,400	112,350	112,350			
			-130	-130			

FORT DODGE, IOWA, APRIL 16th, 1958:

Car Number	Weight by beam Spotted	Weight by Recorder Spotted	1st.	2nd.	3rd.	4th	5th
C.G.W. 6012	97,320	97,250	97,200	96,950	96,750	96,600	
	95,180	95,350	95,400	95,700	95,900	96,000	
	192,500	192,600	192,600	192,650	192,650	192,600	
			+100	+150	+150	+100	
C.G.W. 9005	104,760	104,800	104,700	104,800	104,800	104,900	
	98,060	98,000	98,050	98,050	97,800	97,850	
	202,820	202,800	202,750	202,850	202,600	202,750	
			-70	-30	-220	-70	
C.G.W. 1255	112,820	112,800	112,900	113,000	113,000	113,500	
	106,460	106,400	106,300	106,200	Missed	105,800	
	219,280	219,200	219,200	219,200		219,300	
			-80	-80		+20	
SOU. 287456	82,980	83,000	83,000	83,000	83,200	83,300	
	77,480	77,500	77,400	77,400	77,250	77,000	
	160,460	160,500	160,400	160,400	160,450	160,300	
			-60	-60	-10	-160	

CHICAGO, MILWAUKEE, ST. PAUL & PACIFIC R.R. CO.

Record of test weighing on new track scale at Dell Rapids, S.D. for
two-draft gravity motion weighing, grade 0.5 per cent.
Two-section design, 20-ft weighrail, 100-ton section capacity,
equipped with regular weighbeam and an automatic mechanical printer
recorder. Recorder graduated in 100-lb increments which can be read
to 50-lb min.
Regular run of cars used, weighed loaded by two-draft method, then
hauled twenty miles and reweighed at Sioux Falls, S.D. single draft
static to nearest 20-lbs. Witnessed test, attended by S.D. Public
Utilities Comm. Chairman, six inspectors, also some twelve other interested
railroad officials - June, 1959.

Car No.	Dell Rapids 2 draft beam weights static	Recorder Weights Running		Soo Falls, S.D. Weights static single draft on weighbeam	Weights compared with recorder	
		1st. Run	2nd. Run		Over	Under
C&O.32986	150800	150750	150800	150780	30	
MILW.95215	153780	153750	153800	153700	10	
" 73577x	145080	145150	145050	144760		390x
" 96394	152460	152400	152400	152500		100
" 36691	135600	135600	135600	135640		40
" 70018	189840	189700	189650	189650		80
" 73446	151940	151900	152000	151750		140
" 46464	150940	150850	151000	150950	110	
" 96149		148600		148730	180	
" 83147		149900		149950	60	
" 71035		198600		198440		160
MP. 12080		166300		166320	20	
SPS. 12361		145300		145320	20	
MILW.70792		175000		174980		20
" 69190		145700		145500		0
" 70002		197700		197740	40	
" 69102		153500		153380		120
" 89118		140100		140020		80
" 90200		143300		143360	60	
" 85229		144150		144120		30
" 89155		148300 (Holes in car flr.)		148080		220
" 68505		145850 (" " " ")		145420		430
" 73478		150550 (" " " ")		150260		290
					550	2200

Note "x": MILW. 73577 leaking water at both Dell Rapids and Soo. Falls.

Average of 24 cars, including leakers, Soo.Falls Wts. are 68-lb per car under.

"   " 20 " excluding leakers, "   "   "   " 30-lb "   "   "

Above test made after checking scale, both beam and recorder with RR.Co. 80,000 lb
scale test car.

CHICAGO, MILWAUKEE, ST. PAUL & PACIFIC RR. CO.
Record of test, new track scale at Dell Rapids, S.D. 20-ft weighrail, 2-draft
gravity, then reweighed single-draft static at Sioux Falls, S.D.   August, 1959

Car No.	Beam Weight Dell Rapids	Running Wght. Dell Rapids	Static Wght. Sioux Falls	Weights compared with Recorder	
				Over	Under
MILW.					
73259	...	145,500	145,380	-	120
69093	...	142,900	143,060	160	-
69081	...	146,400	146,280	-	120
29265	...	141,700	141,620	-	80
68011	...	150,100	150,200	100	-
69893	...	143,000	142,900	-	100
70198	...	195,300	195,340	40	-
69360	...	139,700	139,600	-	100
69196	...	139,200	139,000		200
83360	156,580	156,400	156,400	-	-
69071	...	146,800	146,800	-	-
90706	155,440	155,200	155,340	140	
89166	156,180	156,000	156,080	80	
83239	160,500	160,200	160,480	280	

It should be noted that Beam Weights at Dell Rapids are two-spot weights with cars
uncoupled. It is impossible to make spot on scale for weighing the "Rear" set of
trucks, without shifting the rock. For this reason we expect the Dell Rapids Beam
Weight to be a little more than actual. In any event, all static two-draft weights
have been greater than one-draft Sioux Falls weights.
. . . .

CANADIAN NATIONAL RAILWAYS:
Check weighing 10 cars various loads at Danforth Yard, Toronto, Ont. 2-draft gravity
motion weighing scale, 20-ft weighrail, cars weighed spotted two-draft uncoupled on
the weighbeam, then running two-draft gravity weights taken on Recorder:

	CN.210104	CN.211444	CN.510624	CN.524283	CN.520740	CN.148150
	78740	79040	39880	30060	55840	94660
Static on	80120	79760	40320	32560	55700	94820
Weighbeam	158860	158800	80200	62620	111540	189480
Running on	78500	78750	39800	30000	55700	94450
Recorder	80250	79800	40300	32600	55650	94850
	158750	158550	80100	62600	111350	189300
	-110	-250	-100	-20	-90	-180
	CN.142243	CN.142257	CN.143313	ONT.90734		
	95360	87980	88920	46980		
Static on	92820	95800	97220	51480		
Weighbeam	188180	183780	186140	98460		
Running on	95275	87950	88950	46825		
Recorder	92900	95550	96900	51500		
	188175	183500	185850	98325		
	-5	-280	-290	-135		

. . . .

CANADIAN NATIONAL RAILWAYS:
Check weighing six empty refer. cars for repeatability of two-draft motion weighing,
uncoupled on 20-ft weighrail track scale with Recorder - three runs each car :

CN.210566	CN.211864	CN.210853	CN.211296	CN.211243	CN.211186
33600	33800	34200	32800	32700	32500
34000	33300	33900	32800	32400	32700
67600	67100	68100	65600	65100	65200
33500	33800	34200	32800	32700	32400
34000	33300	33800	32800	32400	32800
67500	67100	68000	65600	65100	65200
33500	33900	34200	32800	32700	32500
34000	33300	33800	32700	32500	32800
67500	67200	68000	65500	65200	65300

Maximum variation 100-lb   Recorder graduations 100-lb

Report on Assignment 4

One-Spot Repair Track Facilities

Collaborting with Committee 6

F. S. King (chairman, subcommittee), A. E. Biermann, M. H. Aldrich, R. O. Balsters, H. R. Beckmann, W. O. Boessneck, E. G. Brisbin, B. E. Buterbaugh, E. H. Cook, Vern Copp, C. M. Frazier, C. W. Hamilton, Wm. J. Hedley, A. S. Krefting, Glen Lichtenwalner, E. T. Lucey, S. N. MacIsaac, H. J. McNally, R. H. Peak, Jr., L. W. Robinson, R. E. Robinson, R. A. Skooglun, J. G. Sutherland, James B. Sutton, Jack Sutton, Leo G. Tieman, C. E. Zeman.

Your committee submits the following report as information, with the recommendation that the subject be discontinued.

Many railroads, in an effort to improve their operation, have installed one-spot repair track facilities. The basic principle of this system of car repairs is the moving of cars to the repair facility or facilities, rather than men carrying tools and materials to cars requiring repairs. Generally, this system is used for running car repairs. The rate of repairing cars requiring the light type of repairs ranges from 10 to 20 cars per spot per 8-hr trick.

Description

One-spot repair track facilities vary in size and complexity from single-track, single-spot, single-trick, manual outdoor installations, to multiple-track, multiple-spot, continuously operated, semi-automatic covered installations. They consist of: (1) an inbound track or tracks; (2) a shop area equipped with tools, supplies and repair parts; (3) an outbound track or tracks of about the same capacity as the inbound track or tracks; (4) a system for moving the bad order cars from the inbound track or tracks to the shop area and then moving the repaired cars to the outbound track or tracks; and (5) safeguards on the inbound and outbound tracks to eliminate the possibility of an unexpected movement of a locomotive or cars into the shop area while workmen are engaged in repair work to a car on the spot.

The inbound tracks have a car capacity commensurate with the production rate of repaired cars and the frequency of yard engine deliveries of bad-order cars to the shop. Where a mechanical system of moving cars from these tracks to the shop area is utilized, the inbound tracks are laid on an ascending grade of from 0.05 percent to 0.25 percent to the shop area. This ascending grade is desirable to prevent cars from drifting away from the pushing device. At smaller installations the inbound track is laid on a descending grade to the shop area in order to move the cars by gravity.

The tracks through the shop area are laid on a level grade. For a single-spot operation the length of level track is from 80 to 100 ft. For multiple spots on one track the length of level track is increased accordingly.

There are several methods commonly employed mechanically to move cars from the inbound tracks to the shop area and beyond to the outbound tracks. One method utilizes a mobile self-contained unit equipped to operate both on and off tracks. It is readily moved to or from tracks, and is capable, depending upon its size and capacity, of pushing or pulling up to 15 loaded cars at a time.

Another method makes use of a rabbit-type mechanical puller. A rabbit is essentially an axle-engaging device, running between the running rails and pulled by an electric winch. Each track is equipped with two or more rabbits. The inbound or large rabbits move cuts of cars from the storage portion of the inbound tracks up to the shop as they are needed. A small rabbit moves the cars into and through the shop to the outbound tracks. These rabbits are controlled with precision from a push-button panel centrally located in the spot shop area.

Outbound tracks generally have about the same capacity as the inbound tracks. Where a mechanical method of moving cars is utilized, the grade is ascending for a short distance beyond the shop, and then descending into a "saucer." The repaired car is then moved by the mobile unit, rabbit or other device to the top of the grade, whereupon gravity provides the energy to move it down the outbound track. The elevation at the end of the outbound track or tracks, is equal to, or greater, than the elevation at the shop or car stoppers, if used. This eliminates any possibility of repaired cars fouling ladders or leads. In some smaller spot shops where no mechanical method of moving cars is employed to place cars on the spot, the inbound and outbound tracks are laid on a descending grade to and from the spot shop area. In these cases the cars move into the shop area by gravity, controlled by the manual car brake. After repairs are completed, they are moved from the level track section to the descending grade of the outbound track by means of a simple cable winch or manual car movers.

The larger spot shop areas are fully equipped and protected from the weather by shed-type buildings. Generally, the ends of these buildings are open. However, in some instances the ends are closed in and provided with rolling doors of such height and width to permit entrance of the highest and widest carloads to be handled and also meet legal restrictive clearance regulations. Translucent plastic panels in both the sides and roof provide better daylight illumination. These buildings vary in length from 80 to 175 ft, and in width from 40 ft for a single track to that necessary for multiple-track arrangements.

Floors are constructed of concrete, reinforced along the full length of the track in the shop area to provide for jacking pads. Provisions are made adequately to drain these floors, both inside and outside the track, of water, solvent and oil. Some installations are constructed with the base of rail at floor level. Other roads prefer the top of the rail at floor level, in which case pits or recessions are built into the floor to provide shop employees with easier access under cars.

Artificial light is usually furnished with mercury vapor lights and/or fluorescent lamps. The most common method of heating is by means of radiant floor heat and/or overhead gas or electric infrared heaters.

The following equipment is located in the shop building or at the shop site: jib cranes that swing across the track; electric or hydraulic 30 to 75 ton capacity transversing or fixed jacks recessed into the floor located both inside and outside the rails at about the mid-point of the shop area; portable air or hydraulic jacks; all necessary fluids and gases including air, oxygen, acetylene, lubricating oil and solvent which is piped to overhead retractable hose reels at convenient locations; adequate compressed air outlets provided along the rail; a sufficient number of electric and welding outlets; electric rivet heater; tanks for journal lubricating pads; bins or trays for small material; tool racks, scrap bins and toilet facilities. Other facilities or equipment which are in use at some spot shops include (1) wheel storage tracks for both new or reconditioned and defective wheels; (2) a track for handling material and scrap; (3) fork lift truck or mobile crane to handle material, wheels between the wheel tracks and the shop, and scrap from bins into scrap car; (4) auxiliary building for office, locker and wash rooms, air brake shop, journal pad reconditioner and other supporting operations; (5) load straighteners; and (6) equipment for accurate, expeditious handling of AAR billing on foreign road cars including audio recording systems.

At the larger mechanically operated spot shops derails, signals and/or smash boards are interlocked with the shop operation to eliminate the possibility of locomotives or cars making any unexpected movement into the shop area. Some companies protect both ends of the inbound and outbound tracks with interlocked derails and signals to exclude any unauthorized movements into these tracks, as well as into the shop area itself. An illuminated panel at the spot shop controls the operation and indicates the position of the derails and signals which are sometimes interlocked with the operation of swinging jib crane, the rabbit, and the stationary jacks. Protection for employees and equipment in the small manual-type shops is usually afforded by hand-thrown derails on both ends of the shop area.

Operation

With the protective devices in the proper position, a locomotive places bad order cars on the inbound track or tracks. When the locomotive has cleared this track, and the protective devices are properly reset to protect the shop operation, the cars are moved to the shop area. Where rabbits are used, entire cuts of cars are moved to a point just outside the shop limits by the large rabbits. The small rabbit then moves, in most cases, one car at a time onto the spot. Generally, where gravity is utilized, cars are moved directly from their original location on the inbound track to the "spot". Procedures vary where other car-moving methods are employed depending on local requirements and conditions.

Necessary repair work is then performed on the car. When truck work is needed, the outside jacks lift the car body, permitting the truck to be rolled out from under the car. The jib crane is utilized in renewing of wheels and the handling of other heavy truck components. At the same time other shop employees make necessary running repairs to the body of the car. Journals are serviced if and as required, and the car is ready to move to the outbound track.

The force usually employed at a one-spot, one-track operation consists of one to five men per trick. As the number of spots increase, the force increases, although not proportionately. In other words, a larger number of spots at one location will decrease the average number of employees per spot.

After a car is repaired it is moved to the outbound track, when at a prescribed time, with the protective devices in the proper position, it is picked up in cuts by a locomotive. These cuts of repaired cars are then returned to the yard for classification and dispatchment.

Benefits and Conclusions

Railroads contacted have reported actual annual savings in car repair costs on a single-spot, single-track operation from $20,000 to $50,000 compared to the old-style rip-track operation. The variation in savings, to a great extent, results from production requirements. Obviously, at a point where only a minimum output of repaired cars is necessary, a small, simple, outdoor, single spot can be justified, which shop will show less savings than a more complex installation with the higher production. Savings on multiple-spot operations increase more than proportionally to the same number of comparably equipped single-spot operations.

Other benefits result from less delay to loaded cars, thereby improving service, a reduction in per diem payments, reduced switching costs and better working conditions for shop employees.

There are apparently some savings to be made and other benefits to be gained by using one-spot repair track facilities, or some modification thereof, where as few as 10 cars per day require running repairs.

Fig. 1 is a sketch of the New York Central's one-spot repair track facility at Big Four Yard, Indianapolis, Ind.

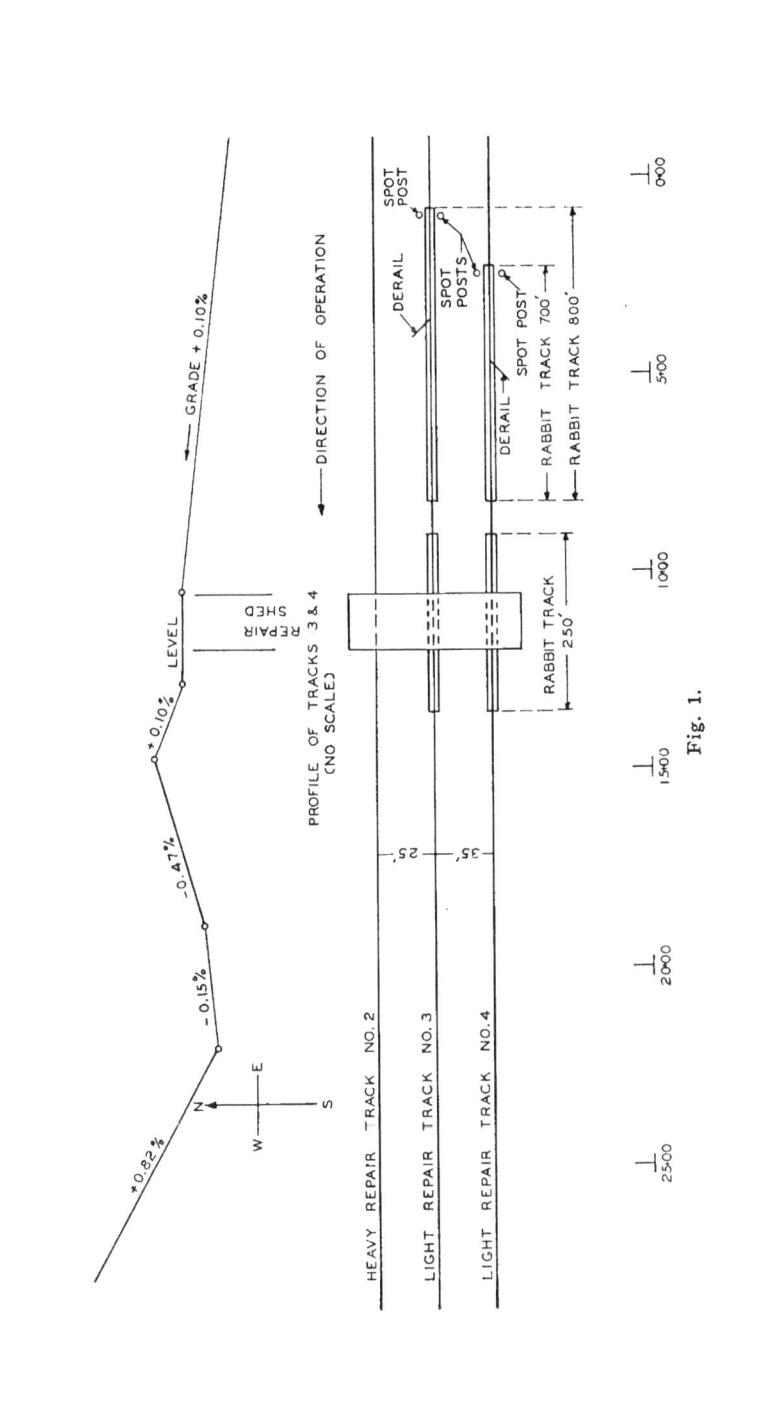

Fig. 1.

Report on Assignment 5

Application of Car Retarders at Locations in Yards Other Than on a Hump

B. E. Buterbaugh (chairman, subcommittee), M. H. Aldrich, A. E. Biermann, R. O. Balsters, R. F. Beck, W. O. Boessneck, E. G. Brisbin, G. H. Chabot, J. F. Chandler, E. H. Cook, V. R. Copp, B. E. Crumpler, A. V. Dasburg, C. M. Frazier, W. H. Goold, C. W. Hamilton, Wm. J. Hedley, F. A. Hess, A. S. Krefting, B. Laubenfels, G. Lichtenwalner, S. N. MacIsaac, G. W. Mahn, Jr., R. F. Moris, B. G. Packard, W. H. Pollard, L. J. Riekenberg, H. T. Roebuck, R. A. Skooglun, C. E. Stoecker, J. G. Sutherland, T. D. Styles, Jack Sutton, J. J. Tibbits, W. E. Webster, Jr. W. A. Wood.

Your committee submits the following report as information on the applications of car retarders at locations in yards other than on a hump, with the recommendation that the subject be discontinued.

Three types of mechanical retarders which produce retardation forces are available.

One type is a device consisting of spring-loaded braking rails which apply retarding force simultaneously to both sides of each wheel. The retarder is usually made in 39-ft units but is available in 19-ft 6-in lengths.

The second type is of spring-loaded rail design applying retarding force to the flanges of each pair of wheels. This retarder is available in 13-ft long sections.

The third type being manufactured is of the weight-responsive variety, having braking rails applying retarding force to one wheel of each pair of car wheels. Slewing of trucks is eliminated by a guard rail on the opposite rail of the location of the braking rails. This retarder is available in multiples of 6-ft 6-in sections.

Mechanical retarders of the types described above are being used by several railroads at pull-down ends of classification tracks. At locations where skates and skatemen may be eliminated by their use, appreciable savings may be realized. Some railroads report additional car capacity may also be obtained.

One railroad has installed a mechanical retarder on one track in a receiving–departure yard. This installation, on a 0.40 percent grade, permits head-end switching without setting of hand brakes, eliminating delay to switch engines.

Use of retarders at car-repair tracks and car-cleaning tracks eliminate the possibility of rollout from these tracks fouling switching leads.

Consideration is being given to the installation of retarders in advance of bumping posts to reduce damage to lading, and at approches to end-loading and unloading ramps to give enginemen a sense of feel when shoving cars to the ramps, to reduce damage to ramp, lading and equipment.

Savings are to be realized by the use of retarders in the reduction of the number of derailments caused by skates being caught in frogs, and rehandling cars fouling leads.

Several railroads are testing these retarders, and preliminary reports indicate satisfaction with their use under proper conditions.

Report on Assignment 7

Water Front Terminals

J. J. Tibbits (chairman, subcommittee), F. E. Austerman, J. F. Chandler, B. E. Crumpler, J. E. Hoving, A. S. Krefting, T. F. Maloney, Jr., J. C. Miller, B. G. Packard, W. H. Pollard, H. T. Roebuck, W. E. Webster, Jr.

Your committee submits as information the following report on car float facilities as a phase of its assignment on water front terminals:

Car float or car ferry facilities are utilized by railroads to transport cars and other rolling stock across bodies of water which are impossible or impracticable to cross by any other means.

When the body of water is a sheltered harbor or a river with not too strong a current, the cars are usually transported on car floats propelled by tugs, the car floats being uncovered steel or timber barges with rails fastened to their decks. Some railroads use self-propelled car floats with open decks except for a navigating bridge. The installation to serve such marine equipment is termed a car float facility.

When it is necessary to transport rolling stock across a lake or larger body of water, the craft employed are large, enclosed, self-propelled vessels designed to withstand rough water and severe weather conditions. Such traffic is handled by a car ferry facility.

The components of the car float or car ferry facility consist of a device by means of which cars are transferred between land and vessel, a fender system, ferry rack, mooring platforms or some other means of guiding and holding the craft to the transfer device, a track layout on the approach to the transfer device so designed as to ensure flexible operation, and a supporting yard with a capacity sufficient to handle the traffic expected to use the facility.

The transfer device may be one of several types. The factors governing the selection of the type to be used are the range in the variation of the water level at the particular location and the density of traffic to be handled.

At locations where the fluctuations in water level vary within a range of 14 or 15 ft, a transfer device known variously as a float bridge, transfer bridge, transfer table, transfer apron or transfer platform may be used.

In its simplest form, this type of transfer device is a single-span structure, usually about 100 ft long and wide enough to accommodate two tracks with switches at the outshore end to handle car floats with three or more tracks. The inshore end is supported on rocker or hinge bearings anchored to a suitable foundation, the design of which is governed by the subsurface conditions underlying the site. The outer end is supported on a timber or steel pontoon designed with a fixed buoyancy so as to maintain a height above the level of the water such that the end of the bridge will be slightly higher than the freeboard of the lightest vessel likely to use the facility.

In order to adjust the level of the fixed flotation type of pontoon bridge to that of a car float, several weighted cars are backed onto the structure by a switch engine until the rails on the bridge and the rails on the float are in alinement, the rails on the bridge usually being provided with some means of horizontal adjustment to suit variable track centers on the car floats. Using this method of vertical adjustment, each side of the bridge is depressed separately. Bridge and float are maintained in alinement by means of steel latch or toggle bars thrust into sockets on the deck of the car float and are finally fastened together with ropes or steel cables. The pontoon type of transfer bridge is often designed so that its outboard end may be raised or lowered to suit any variation in vessel freeboard by pumping sufficient water into the pontoon to lower it and by dewatering the pontoon to raise it.

A second type of single span transfer device is similar to the first, differing from it only in the manner of supporting the outshore end of the bridge. In this case, the pontoon is dispensed with and the outer end is supported by means of suspender bars attached to the ends of pivoted, counterweighted balance beams, the raising and lowering of the end of the bridge being accomplished by a hand- or power-operated winch and cables.

A more advanced type of end suspension employs a structural steel tower on each side of the transfer bridge, the towers being connected by a light, overhead framework. On each tower is mounted a large sheave over which are passed several cables. One end of the cables is attached to the transfer bridge, the other end to a system of counterweights within the towers. Smaller sheaves on each tower carry several lighter cables, one end of which is fastened to a set of the counterweights, the other end being wound on the drum of an electric winch mounted at the base of one of the towers.

The most efficient type of single-span transfer device utilizes a tower on each side of the outer end of the bridge connected by a heavy overhead structural framework to support and house the operating machinery. In this type, the greater part of the weight of the outboard end of the bridge is supported by steel suspender bars, the remainder of the weight being taken by a system of cables, sheaves and counterweights. The upper ends of the suspender bars are threaded, the threaded ends being engaged by large nuts which are revolved by power driven gears, causing the suspender bars, together with the end of the bridge to be raised and lowered. This type of transfer device is often designed with a separate bridge for each track, permitting fast and efficient adjustment of bridge to car float.

Sometimes, an auxiliary apron, 25 to 30 ft long, hinged to the outer end of the transfer bridge and independently supported by counterweighted cables from a separate, overhead structure, is used as a buffer to spare the main transfer bridge the severe strains due to the lateral and vertical movement of the float. The apron also serves as a transition grade between the transfer bridge and the deck of the car float at extremes of high and low water.

A radically different type of transfer device, in use where there are extremely large variations in elevation betwen the land and water, is the traveling cradle and incline, developed for use along the Mississippi River. This device consists of a timber trestle incline with a single track, constructed diagonally down the river bank from the crest to an elevation 8 or 9 ft below the low-water elevation of the river. The incline is usually built on a 3 or 4 percent grade, and since the difference in elevation between the crest and the river end may be 60 to 70 ft, its length will range between 1500 and 2000 ft. Mounted on wheels and running on the rails of the incline trestle is a timber cradle, 200 to 300 ft long, built on an ascending grade toward the river, provided with short transition grades at its inshore and outshore ends and having an adjustable apron to engage the deck of the car ferry.

The cradle also carries a single track, the inshore rails of which terminate in track fittings known as "feather rails" which are planed to $\frac{1}{2}$ in thickness and are free to slide along the rails of the incline. The cradle is moved up or down the incline by power-operated cables to suit the variable water level of the river.

A final type of transfer device is that used by certain steamship companies for transferring cars to and from ocean-going ships. In this type of installation, the car is moved onto a cradle, the rails of which are in alinement with the rails on the pier, the car being prevented from rolling by checking the wheels. The cradle and car are then lifted and transported laterally by a traveling bridge crane to a cantilever structure overhanging the ship and lowered down the ship's hatch to its assigned deck level. The car is freed of the chocks and hauled off the cradle to its position on the deck by cables operated by power winches.

Car float or car ferry facilities, regardless of the various methods of transfer employed, must be provided with some form of structure to guide the float or vessel to the transfer device and to restrain movement of the craft after it is moored. This structure may be a fender system or ferry rack, a group of mooring platforms or a series of stone-filled timber cribs.

The conventional type of fender system consists of a timber structure comprised of two or three lines of wood piles separated by laminated timber wales with vertical lagging fastened to wales on the slip side. The vertical timbers act as a structural web to distribute the shock of the vessel's impact upon the fender. The timbers also provide a smooth surface to ease the float toward the transfer bridge and, after it is moored, act as a buffer to prevent abrasion of the piling and damage to the vessel.

For a single car float facility, two lines of fenders are usually employed to form a slip with the ends of the fenders near the transfer device formed to the shape of the float so as to hold its movement to a minimum. In a group of float bridges, two long lines of fendering are used to form one large slip with short fenders between each bridge.

A fender and mooring system, now in the planning stage by a port authority for the use of an eastern railroad, will have short lengths of conventional timber fenders close to the transfer bridge. Along each side of the slip, on 100 ft centers, will be constructed three reinforced concrete mooring platforms, 21 ft 6 in long, 15 ft wide, 3 ft deep and 10 ft above mean low water. Each platform is to be founded on a group of 10 prestressed concrete piles, battered in 4 directions, each pile to have a bearing value of 70 tons. The platforms will be provided with timber protection consisting of pile, wale and chock fenders on three sides, the slip side to be faced with vertical timber lagging, with rubber bumpers between the top wale and the concrete platform. Extensive use of greenheart piling will be made throughout the installation.

At the outshore ends of long fenders of conventional construction, pile clusters or dolphins are usually installed to protect the fenders from the impact of car floats approaching or leaving the slip. Piles in the dolphins are generally driven with the small ends down. However, good results have been reported by driving dolphin piles with the butt end down. The advantages claimed for this method are that it provides a greater diameter of pile at the mud line with a higher resistance to failure in bending and with greater resilience in the upper portion.

Fenders for car ferry facilities utilizing transfer bridges are similar to those for car float facilities but must be built more substantially in order to withstand the greater impact of heavier, self-propelled vessels. The slip face of the fender is sometimes reinforced with steel rails placed between the vertical timber lagging.

One system of fenders used for car ferries consists of a long pile and timber ferry rack extending nearly the full length of the vessel with a much shorter rack on the opposite side near the transfer bridge. Another car ferry fender system utilizes two short lines of pile and timber racks on one side of the slip and, near the transfer bridge, a structure termed a "knuckle" consisting of 6 or 7 lines of wood piles separated by several tiers of 12- by 12-in wales, faced with 6-in vertical timber lagging on the slip side and the outshore end. At locations where a rock bottom precludes the driving of piles, stone filled timber cribs, faced with hardwood lagging above the water line, are used in place of conventional pile and timber fenders.

Since car ferry facilities serve large, self-propelled vessels, they must be provided with much heavier protection of the longer fenders at their outer ends. In addition to large pile clusters made up of 40 to 60 wood piles, a turning pier is usually provided. This turning pier may consist of a cluster of 200 or more piles, or· it may be a cylindrical, reinforced concrete structure about 35 ft in diameter, faced with vertical lagging and provided with mooring bollards or bitts.

The fender system used for the cradle and incline type of car ferry facility generally consists of a single line of pile clusters installed on the river bank side, parallel to and a uniform distance from the center line produced of the incline trestle. The tops of the pile clusters are 20 to 30 ft above the top of rail of the incline and serve to guide the vessel to the cradle and prevent it from going around regardless of the level of the water.

The track layout on the approach to the transfer device should be so designed that it will produce the most flexible operation possible. With a double-track bridge, where space permits, a double crossover, also known as a "diamond" or "scissors" crossover, is sometimes employed. This layout enables the locomotives serving the facility to move cars most expeditiously between the car float or ferry and the receiving or storage tracks. It is especially useful on the approach to a bridge supported by a pontoon of fixed buoyancy where the height above water level of each side of the outboard end of the bridge must be adjusted separately to the deck level of the vessel by backing several weighted cars onto it. On the approach to a group of transfer bridges, the track layout may become quite complicated, sometimes to the extent of utilizing single and double slip switches.

Yard layouts serving car float and car ferry facilities vary widely. In most locations on the water front, due to lack of space, a large or even a moderate size yard is impracticable. Only sufficient storage tracks are usually provided to handle the traffic using the facility, the cars being made up into short trains and hauled to a remote classification yard where they are made up into road trains.

At some water front freight stations, the float bridge facility is the sole means of handling cars into and out of the freight house yard. The track layouts serving such stations are necessarily very cramped, requiring the use of No. 3 and No. 4 turnouts and crossovers.

Where there is sufficient space, the ideal yard for serving car float or car ferry facilities is one having receiving, departure and storage tracks of such capacity as to permit a full-size freight train to be yarded, the road engine cut off, the train broken up and the cars, in suitable cuts, backed onto the car float or ferry. In the reverse operation, the cars as they are hauled off the float or ferry, may be made up into road trains and be dispatched directly from the water front yard.

The car float or car ferry facility, while it is a useful and indispensable operating feature, is also recognized as expensive to construct, operate and maintain. In designing a track layout or yard to serve such a facility, every effort should be made to have it as flexible as possible, keeping in mind that the density of the traffic using it generally occurs in peaks, and any malfunction such as a derailment or failure in the mechanical equipment will result in costly delays. The proper design of the approach track layout and the yard to serve a car float or car ferry facility is, therefore, an important factor in the economical operation of the installation.

How GRS Syncroscan®
Paid Off Between
Buffalo and Cleveland

In 1956, the New York Central initiated its modern cTc program by placing a 163-mile stretch in service between Buffalo and Cleveland. The installation demanded a traffic control system with enormous capacity and speed. GRS Syncroscan was the answer. Syncroscan continuously checks—with electronic speed—the location of every train, the position of each switch and signal. Fast-acting relays control switches and signals with ample speed to keep up with operator's decisions.

By using Syncroscan—like the Central —you can handle many miles of busy multiple-track main line, including major interlockings, from a single office. And you can do it better, faster, and more economically than with your present method.

FOUR TRACKS CUT TO TWO The $6 million project produced $3 million in salvage.

FREIGHT TIME SAVED Train speed increased from 20-30 mph to 60 mph—freight time reduced from 7 hours to 3½.

GREATER FLEXIBILITY Two tracks give better operation—fast trains can now be run around slower trains.

EXCELLENT RETURN Estimated saving of 87% on the investment.

3100

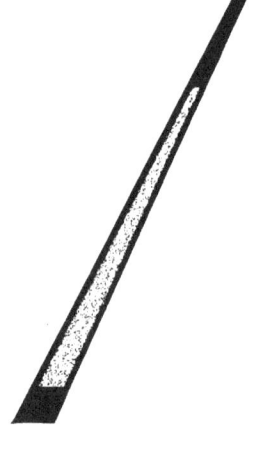

VEGETATION CONTROL
WITH
CHEMICALS

READE MANUFACTURING COMPANY, INC.

Jersey City—Chicago—Minneapolis—Kansas
City—Birmingham—Stockton

SERVING RAILROADS OF AMERICA FOR
MORE THAN FORTY YEARS

W
E
E
D

A
N
D

B
R
U
S
H

C
O
N
T
R
O
L

PROGRESS REPORT

Here are the up-to-date facts on the SPENO Ballast Cleaning and the SPENO Rail Grinding Services.

BALLAST CLEANING

SPENO Engineering and Research has developed a superior screening arrangement so that we are now using an improved Ballast Cleaner with greater efficiency.

RAIL GRINDING

Our Rail Grinding Service has been so well received we are now building a *THIRD* Rail Grinding Train to take care of the increased demand.

SPENO is constantly developing means for better service to make sure that the Railroads receive everything they pay for — and more

Just Ask the Railroads That have used us!

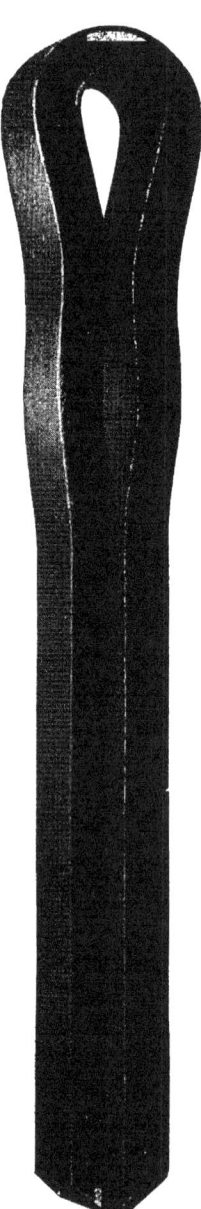

Model N U Tie Cutter

HERE IS THE WINNING TEAM

The Woolery NU Tie Cutter and the Woolery Tie-end Remover preserve the line and surface of the track and at the same time reduce the cost of tie renewals. Ties can be removed without trenching, jacking up track or adzing tops of rail-cut ties. With this team you simply cut both ends of tie, pry out center piece, insert in its place the tie-end remover and out go the tie ends pushed by the double acting, double ended hydraulic cylinder of the Tie-end remover.

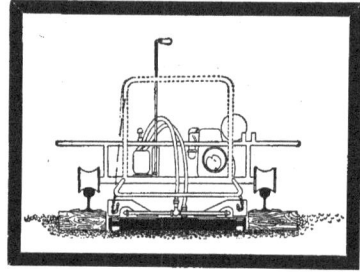 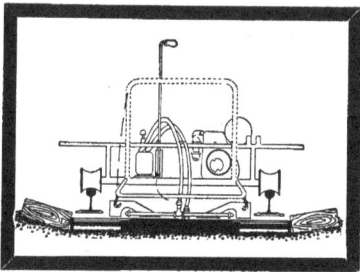

FOR HIGHEST EFFICIENCY USE TWO TIE CUTTERS WITH ONE TIE-END REMOVER

WOOLERY MACHINE COMPANY
MINNEAPOLIS, MINN.

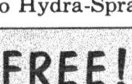

the JACKSON TRACK MAINTAINER 1962

- MORE POWERFUL TAMPERS . . . FASTER PENETRATION

- FASTER WORKHEAD ACTION . . . FASTER INDEXING

- FULL ELECTRIC OPERATION WITH PUSH-BUTTON CONTROL

- IMPROVED SUSPENSION OF TAMPERS provides direct application of greatly increased vertical and horizontal vibratory forces plus workhead weight.

- POSITIVE MOTOR AND GENERATOR PROTECTION against short circuits, single-phasing and low voltage.

- FASTER, MORE POSITIVE BRAKING

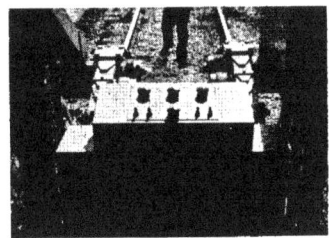

Complete push-button Control

With these highly important improvements, the Jackson Track Maintainer continues to be unequalled in versatility and its ability to put up and maintain the finest track at the lowest cost per mile . . . in any kind of ballast and lift of track. Let us give you the complete facts.

JACKSON VIBRATORS, INC.
LUDINGTON, MICHIGAN

As always, direct sales, leases and service to all U.S. railroads.

R1-8A

Kill more weeds *per mile...per dollar*
with *Liquid* **UROX**®!

Liquid Urox Weed Killer is the *first* liquid — substituted urea-type herbicide ever developed for railroads. It's fast-acting . . . withers annual and perennial grasses as well as broadleaved weeds within 12 hours after application, regardless of weather. It's long-lasting . . . just *one* application wipes out weeds and brush for 8 to 18 months. What's more, control can be continued economically each year with small "booster" doses.

Liquid Urox is ideal for railroad spray trains . . . doesn't need continuous agitation . . . won't clog spray nozzles . . . won't settle out . . . can be mixed with fuel oil, diesel oil or ordinary weed oils. Write today for the complete story on railroad-proved liquid Urox Weed Killer.

GENERAL CHEMICAL DIVISION
40 Rector Street, New York 6, N. Y.

for effective
weed control...

- **Concentrated BORASCU®**
- **POLYBOR-CHLORATE®**
- **UREABOR®**
- **MONOBOR-CHLORATE®**

These <u>borate weed killers</u> are proving best for roads in every way... *efficiency, safety, economy, convenience, easy application.*

Today's use of borates for maximum control of vegetation began years ago with our pioneer work in the field. Continued research has developed the group of herbicides, listed above, which most roads now favor for every phase of weed control. These four weed killers are nonselective. They are widely used for year-round maintenance of weed-free conditions about trestles, tie piles, yards, signals, switches, and rights of way. Find out how you, too, can do a better job on weeds...write today.

AGRICULTURAL SALES DEPARTMENT

U.S.BORAX®

630 SHATTO PLACE · LOS ANGELES 5, CALIFORNIA

the BorTunCo GROUP

BorTunCo has an established record with engineers and contractors for capable sub-contract job performance*

ROAD BORING AND TUNNELING COMPANY, INC. Road Boring for Municipal and Industrial Utilities in Texas, Oklahoma and Arkansas.	**TEXAS ROAD BORING COMPANY OF LA.-MISS.** Road Boring and Tunneling for Municipal and Industrial Utilities in Louisiana and Mississippi.

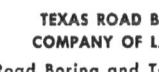

BORING AND TUNNELING COMPANY OF AMERICA Road Boring and Tunneling for Municipal and Industrial Utilities throughout the remainder of the U. S.	**TEXAS TUNNELING COMPANY** Tunneling — Pipejacking — Multiplate Installations; also furnishes equipment and supervision for other divisions throughout the U. S.	**HORIZONTAL HOLES, INC.** Specializes in complete turnkey road-crossings throughout the U. S. for Cross-Country Gas, Oil and Products Pipelines.

BORING AND TUNNELING CO. OF AMERICA
2902 Ricks Road   ●   P. O. Box 14214   ●   Houston, Texas   ●   JA 6-2755

3920 Monroe Road Charlotte 5, North Carolina	P. O. Box 4755 Audubon Station Baton Rouge 8, La.	5515 Redfield Dallas 35, Texas

*Negotiations and inquiries strictly confidential.

THE TRASCO
AUTONOMIC CAR RETARDER

CLAMPS IN PLACE
ANYWHERE IN TRACK

SIMPLE — EFFECTIVE — INEXPENSIVE

TRACK SPECIALTIES CO.
GENERAL MOTORS BLDG.
NEW YORK 19, N. Y.

REPORTS OF COMMITTEES

The reports in this issue of the Bulletin will be presented to the 1962 convention of the Association at the Conrad Hilton Hotel, Chicago, March 9–10. Comments and discussion with respect to any of the reports are solicited, and should be addressed to the chairman of the committee involved, in writing in advance of the convention, or from the floor of the convention.

Published by the American Railway Engineering Association, Monthly, January, February, March, November and December; Bi-Monthly, June–July, and September–October, at 2211 Fordem Avenue, Madison, Wis.; Editorial and Executive Offices, 59 Van Buren Street, Chicago 5, Ill.
Second class postage paid at Madison, Wis.
Accepted for mailing at special rate of postage for in Section 1103, Act of October 3, 1917, authorized on June 29, 1918.
Subscription $10 per annum.

Report of Committee 22—Economics of Railway Labor

J. E. EISEMANN, *Chairman*
J. S. SNYDER,
   Vice Chairman
L. C. GILBERT, *Secretary*
E. J. SIERLEJA
H. J. FAST
H. W. KELLOGG
W. E. CHAPMAN
T. L. KANAN
W. J. JONES
LEM ADAMS (E)
A. D. ALDERSON

M. B. ALLEN
J. A. BARNES
J. F. BEAVER
O. C. BENSON
E. J. BROWN
R. F. BUSH
J. L. CANN
R. H. CARPENTER
J. A. CAYWOOD
J. L. CHAFIN
R. E. CLANCY
S. A. COOPER
P. A. COSGROVE
C. G. CRAWFORD
C. G. DAVIS
M. H. DICK
L. E. DONOVAN
W. M. S DUNN
J. L. FERGUS
R. L. FOX
R. R. GUNDERSON
K. H. HANGER (E)
V. C. HANNA
GENE L. HARRIS
W. W. HAY
K. E. HENDERSON
CLAUDE JOHNSTON
R. H. JORDAN
H. E. KIRBY
N. W. KOPP

L. A. LOGGINS
J. M. LOWRY
T. D. MASON
R. L. MAYS
F. H. McGUIGAN III
J. R. MILLER
H. C. MINTEER
G. M. O'ROURKE (E)
R. W. PEMBER
C. T. POPMA
R. W. PREISENDEFER
GRIFFITH RAY
M. S. REID
D. E. RUDISILL
H. W. SEELEY
R. G. SIMMONS
N. E. SMITH
JOHN STANG
A. H. STIMSON*
G. M. STRAWHUN
J. T. SULLIVAN
W. B. THROCKMORTON
JOHN T. WARD
G. E. WARFEL
H. J. WECCHEIDER
N. H. WILLIAMS
H. E. WILSON
F. R. WOOLFORD
C. R. WRIGHT (E)
D. H. YAZELL
Committee

---

* Died October 12, 1961.
(E) Member Emeritus.
Members listed in bold face are the official representatives of the Engineering Division, AAR.

To the American Railway Engineering Association:

Your committee reports on the following subjects:

1. Revision of Manual.

2. Analysis of operations of railways that have substantially reduced the cost of labor required in maintenance of way work.

3. Labor economies to be derived from work measurement standards for comparison of work performance among various gangs or divisions

4. Labor economies to be realized through the use of a combined surfacing-timbering gang vs. separate gangs for surfacing and timbering.

5. Labor economies to be effected through the use of power tools and mechanized equipment by bridge and building gangs.

7. Labor economies in track maintenance to be derived through the use of combination on-off-track equipment vs. on-track equipment only.
Progress report, presented as information .... page 272

8. Labor economies to be derived from the welding, distributing, laying, and maintenance of continuous welded rail, collaborating with the Special Committee on Continuous Welded Rail.
Questionnaires were sent to numerous railroads, and some replies have been received. Collaboration with the Special Committee on Continuous Welded Rail is being carried out. When fully representative replies have been received from a sufficient number of roads, a progress report will be prepared.

THE COMMITTEE ON ECONOMICS OF RAILWAY LABOR,
J. E. EISEMANN, Chairman.

AREA Bulletin 568, December 1961

Report on Assignment 1

Revision of Manual

W. W. Hay (chairman, subcommittee), M. B. Allen, J. F. Beaver, E. J. Brown, J. A. Caywood, S. A. Cooper, C. G. Davis, L. E. Donovan, W. M. S. Dunn, J. L. Fergus, L. C. Gilbert, V. C. Hanna, W. J. Jones, N. M. Kelly, N. W. Kopp, L. A. Loggins, J. R. Miller, H. C. Minteer, C. T. Popma, Griffith Ray, D. E. Rudisill, H. W. Seeley, John Stang, G. M. Strawhun, W. B. Throckmorton, G. E. Warfel, H. J. Wechheider, H. E. Wilson.

Your committee submits for adoption the following recommendations with respect to Chapter 22 of the Manual.

Delete the material on Recruiting, Training and Welfare, pages 22–1–1, 22–1–2, and the top of page 22–1–3, and replace with the following.

RECRUITING AND TRAINING

The development of highly organized mechanized gangs, the rapid evolution of new methods and new machinery for doing maintenance and construction work, and the necessity for increasing efficiency to reduce expenditures in the face of rising labor and material costs have, in recent years, placed greater demands upon railroad personnel. These demands will continue to grow. Only by an organized system, similar in its character to those employed in the industrial field, for carefully recruiting and selecting candidates for employment and for definite training both new and experienced employees for secondary supervisory positions and for specially skilled jobs can the railroads meet the demands of the future with the full assurance that they will be able to operate efficiently and economically.

Capable and efficient foremen and other secondary supervisor personnel are essential to supervise and organize the mechanized maintenance and construction gangs of the present day and to justify the high cost of operation. Skilled operators and mechanics are needed to operate and maintain the increased amount of machinery. In order to get these results, it is desirable that capable young men be attracted into the service. Employing officers should recruit such men from graduates of high schools and technical institutes, especially in the smaller communities.

After capable young men are recruited and employed, there should be a thorough and systematic program of training and educating them for the job they are on as well as advancement to higher positions. The ambitious, capable, and qualified employees who are already in the service and who are willing to make the effort to succeed should not be overlooked in the training program and should certainly be considered as prospects for promotion to higher positions. In promotion, fitness and ability should govern, but under our system, seniority must be considered, and employees having necessary qualifications should be given every legitimate opportunity and encouragement to obtain necessary training and experience.

Gradual on-the-job training, with rotation in service where possible, is effective. Technical training combined with that of organizing, directing, and supervising work, along with education in certain rules and practices of operating and accounting departments, is necessary. Give promising employees the opportunity to obtain more technical education by attending schools and training sessons offered by many of the manufacturers and suppliers.

To accomplish the best results, install methods to promote individual effort and interest. Personal contact and interest by supervisory people go far to bring this about. Encourage employees to seek further education from outside sources on general principles of railway maintenance and operation through correspondence and night schools, railway periodicals, manufacturers' manuals, etc.

Establish definite qualifications for the individual positions. Also make standard methods of performing the work as well as standard units of measurement for production and quality available to the employee. Evaluate the results of the training program periodically to determine the progress being made by the individual.

Page 22-1-3

ANNUAL INSPECTION AND PRIZE AWARDS

Reapprove with the following revisions:

Change title to read "Annual Inspection for Checking Progress and Planning Future Programs."

Delete the first two paragraphs and replace with the following:

"A well-devised plan for an annual inspection provides a desirable means for checking progress and making future programs. This can be combined with a properly directed competition to increase the interest and activities of maintenance of way forces."

Pages 22-1-4 and 22-1-5

OUTFIT CARS FOR HOUSING

Reapprove with the following revisions:

Under Sec. A. Design, item 4, delete the words: "both as a matter of convenience and safety."

Page 22-1-5. Delete item 4 of Sec. B and replace with:

"4· Inspection should be made at sufficiently frequent intervals to ensure the proper physical and sanitary condition of the equipment."

Pages 22-2-1 and 22-2-2

PROGRAMMING WORK AND DIVERSION OF TRAFFIC

Reapprove with the following change:
Page 22-2-2. Add new section, as follows:

C. USE OF RADIO*

The use of radio in maintenance of way work can materially reduce lost man-hours and idle time of expensive equipment, facilitate coordination between various track machines, thus increasing gang efficiency, and reduce delay to trains. Radio is most effective when machines are spread over one-quarter to one-half mile or more of territory.

---

* Reference, Vol. 57, 1956, pp. 626–629.

Page 22-2-3

WEIGHT OF RAIL

Reapprove with the following revision:
Delete Par. (e) and replace with:
"(e) For lines of high traffic density, the saving in track labor following the installation of heavier rail sections may reach 40 percent of the total expenditure for this item."

Page 22-2-3

VEGETATION CONTROL

Reapprove with the following revision:
Delete the words "fire hazard and" from the third line of the second paragraph.

Page 22-2-5

MECHANICAL EQUIPMENT

Reapprove with the following revisions:
Delete second paragraph and replace it with the following:
"Special gangs organized to take advantage of mechanical tools and work equipment in rail laying, ballast renewal work, tie renewals, and other maintenance tasks increase production and promote uniformity of work.*"
Add new fifth paragraph, as follows:
"The use of trucks in lieu of motor cars can effect substantial savings in various phases of maintenance of way work where geographical location and traffic of the railroad warrants their use.**"

---

* Reference, Vol. 59, 1958, pp. 593–597.
* Reference, Vol. 57, 1956, pp. 424–428.

Pages 22-3-1 to 22-3-4, incl.

PROGRAMMING WORK

Reapprove with the following revision:
Page 22-3-1. In the second paragraph, third from the last line of that paragraph, delete the phrase "it is necessary" and substitute the phrase "it is suggested,"

Page 22–3–5

MECHANICAL EQUIPMENT

Reapprove with the following revisions:

Delete the second and third paragraphs and replace with the following:

"Experience has shown that some types of power tools can well be used by nearly all types of bridge and building crews. Special gangs organized to take advantage of mechanical tools and work equipment in bridge work, concreting, painting, water service work, and other bridge and building maintenance increase production and promote uniformity of work."

Delete the word "mentioned" in the first line of the fourth paragraph.

Add a new paragraph reading as follows:

"The use of on-track-off-track types of bridge and mechanical equipment makes for greater maneuverability and saving."

Report on Assignment 2

Analysis of Operations of Railways That Have Substantially Reduced the Cost of Labor Required in Maintenance of Way Work

E. J. Sierleja (chairman, subcommittee), A. D. Alderson, O. C. Benson, E. J. Brown, R. F. Bush, J. A. Caywood, J. L. Chafin, R. E. Clancy, S. A. Cooper, C. G. Craw-ford, L. E. Donovan, J. L. Fergus, V. C. Hanna, G. L. Harris, W. W. Hay, K. E. Henderson, Claude Johnston, R. H. Jordan, H. W. Kellogg, N. M. Kelly, H. E. Kirby, L. M. Lowry, H. C. Minteer, C. T. Popma, R. W. Preisendefer, Griffith Ray, R. G. Simmons, N. E. Smith, J. S. Snyder, W. B. Throckmorton, H. J. Wechheider, N. H. Williams, H. E. Wilson, F. R. Woolford, D. H. Yazell.

This report, submitted as information, is the 20th report of a series on this subject, which has been reassigned annually since 1935.

This analysis is of the track rehabilitation process currently being used by the Baltimore & Ohio Railroad. The committee inspected the organization performing this work on the main line of the Buffalo Division at a point approximately four miles east of Bradford, Pa.

Purpose

The purpose of the rehabilitation is to prepare one track of two for carrying greater annual tonnage and greater train frequency after the removal of the second track. The track now carries approximately 10.9 million tons of traffic annually; after removal of the second track it will carry approximately 16.7 million tons. The rehabilitation consists of cleaning bed, shoulder and crib ballast, renewing ties, raising, and a later date, renewing rail. It also includes increasing the ballast section to meet AREA specifications.

Track Specifications

The track before rehabilitation had 112-lb and 100-lb rail, 4-hole and 6-hole joint bars, 2 rail-holding spikes per tie plate, hardwood treated ties, 22 per 39-ft rail, and 1½-in stone ballast of less than AREA standard section. The shoulder ballast was last cleaned in 1957. The crib ballast was fouled with normal dirt accumulation without a large amount of mud in the observed area.

Labor and Equipment Organization

The work of rehabilitation is performed by a large mechanized gang. This organization is provided with many types of specialized machinery and works in an integrated in-line type of set-up. The details of its operations, equipment and labor are shown in Table I. Unless otherwise specified, all equipment is of the on-track type.

Material distribution is performed by work train at time intervals as required. The labor source for the work train service is the tie renewal unit, which has excess capacity as organized in the line.

Scheduling and Line Balance

The job scheduling of the rehabilitation is performed by the division engineer. A notable feature is the arrangements for use of the track. The track being worked is removed from service between adjacent block stations for the duration of the job in that block, or blocks when the unit occupies two blocks. This feature maximizes the productive time of the organization to about 7 hr during the regular tour of duty of 8 hr from 8 am to 5 pm. This tour begins and ends at the job site (normally the grade crossing closest to the point of work), and not at any camp or fixed headquarters. As noted in Table I the ballast cleaner operates a second trick from 5 pm to 2 am. This second trick is necessary to balance the line.

The tie renewal unit has been renewing approximately 343 ties per mile, and at this rate it has more lineal dimension capacity than the rest of the line. For this reason this unit is used to install switch timber, crib out turnouts, repair crossings, etc.

Communications   ·

The organization is equipped with two-way radio communication facilities. There are five stations so equipped—the general foreman's truck, the ballast cleaner, each ballast remover, and the preceding block station, and they may communicate with any station. All trains on the adjacent track receive train orders to stop at the point of work and proceed on receipt of a hand signal. This maximizes the productive time of the ballast removers which foul the adjacent track. The radio facilities also provide fast communication when mechanical failure occurs and parts and/or mechanics are required for the operation in trouble.

Quality

The track is restored to service at normal scheduled speed after the unit clears the complete block.

It has been necessary to schedule a second tamping after traffic has used the track to correct for irregular settlement of the track. This work is done by a unit independent of the above described organization and was not observed by the committee.

Costs

The cost history of the rehabilitation was developed for a period of 39 work days from April 17 to June 9, 1961. During this period the cost was unfavorably influenced by 16 partial or whole days of rain or snow. The costs as recorded are shown on page 268 for the distance of 12.64 miles.

Table I

Rehabilitation Organization

No	Operation Description	Equipment Quantity	Equipment Description	Labor Fore-man	Labor Oper-ator	Labor La-borer
A	Supervise operations 1 - 11, tie renewal unit.			1		
1	Plow off ballast shoulder both sides of track to depth of tie bottom. Operation performed on alternate days with operation 22.	1	Ballast regulator.		1	
2	Remove spikes from ties to be renewed. Mark rail for subsequent tie spacing.	1	Hydraulic spike puller.			2
3	Raise track approximately 2½ in and tamp jack points. Quality profile not maintained.	1	Hydraulic jack tamper.		1	
4	Remove worn out ties. Ties left perpendicular to track. Remove old tie plates and place in clear for reuse.	1	Hydraulic piston tie remover.			2
5	Eject ballast from exposed tie bed to shoulder of track.	1	Hydraulic scarifier with lateral stroke tools.			1
6	Loosen crib ballast between remaining ties in track.	1	Same as operation 5.			1
7	Pile old ties for subsequent burning. Pre-position new ties at track shoulder.	1	Tie handler.			1
8	Remove fouled ties, load scrap, final cleaning old tie bed.		Hand tools.			2
9	Place new ties in final position.	1	Tie crane.			2
10	Position tie plates on new ties. Spike joint ties.		Hand tools.			2
11	Spike new ties and tighten spikes as required in old ties.	1	Spiker with material feeding and positioning fixtures.			2
	Labor sub-total for tie renewal unit.			1	2	15

Table I (Continued)

No	Operation		Equipment		Labor		
	Description	Quan-tity	Description		Fore-man	Oper-ator	La-borer
B	Supervise operations 12 - 17, cleaning and spacing unit, including second trick of operation 17.				2		
12	Remove ballast from cribs and beneath ties to a depth of six in. Ballast deposited in windrow on track shoulder. Remove rail anchors and place in clear for reuse. Work alternate 1200 ft lengths of track with operation 14.	1	Ballast remover with continuous chain tools.			1	1
13	Space all ties to provide 22 ties per 39 ft rail. Work alternate 1200 ft lengths of track with operation 15.	1	Tie spacer.				1
14	Same as operation 12 working alternate 1200 ft lengths of track.	1	Same as operation 12.			1	1
15	Same as operation 13 working alternate 1200 ft lengths of track.	1	Same as operation 13.				1
16	Apply rail anchors previously removed. Redrive bent spikes.		Hand tools.				2
17	Clean ballast previously windrowed and distribute in track. This operation is performed on two tricks. Labor shown is for both tricks.	1	Ballast cleaner with vibrating screens.			4	4
	Labor sub-total for cleaning and spacing unit.				2	6	10
C	Supervise operations 18 - 22, raising and lining unit.				1		
18	Raise track four in and tamp jack point ties.	1	Hydraulic tamping jack with telescopic indicator to foresight unit.			1	1
19	Throw in ballast for operation 18 and 20.		Hand tools.		From operation D		
20	Tamp all ties.	1	Tamper with air operated tamping tools.			1	1

Table I (Continued)

No	Operation — Description	Quantity	Equipment — Description	Foreman	Operator	Laborer
21	Line track.	1	Liner with wire type indicator.			1
22	Regulate, and shape ballast section. Performed on alternate days with operation 1.		From operation one.	From operation one		
	Labor sub-total for raising and lining unit.			1	2	2
D	Haul and distribute miscellaneous material and supplies. Prepare grade crossings for work and resurface same.	1 1	Track car. Highway truck.	1	2	6
E	Inspect, service, and repair machinery.	1	Highway truck.	Gen'l Engr. Fore, Ass't, Mech.		3
F	General supervision and engineering.	1	Highway truck.	1	1	

Direct Labor
Operator 10
Laborer 27
Total 37

Labor Summary

Indirect Labor
General Foreman 1
Engineering Ass't. 1
Mechanic 3
Foreman 5
Laborer 6
Total 10

	Cost per mile
Labor	
Rehabilitation organization	$2,528
Unload ballast	115
Temporary block station operators	428
Miscellaneous	59
Total labor	$3,130
Material	
Cross ties (343 per mile)	$1,412
Miscellaneous track material	83
Stone ballast (11 cars per mile)	1,015
Fuel and supplies	161
Total material	$2,671
Equipment	
Fees for rented equipment (depreciation for owned equipment not included)	$1,041
Replacement parts	227
Total equipment	1,268
Work train services	327
Total Costs	$7,396

It should be pointed out that approximately 55 percent of the consumed ballast was used to increase the ballast section to standard AREA section.

CONCLUSIONS

The work and organization inspected by the committee reflect the cost benefits in long-range planning by the B&O management. This is especially true in the features noted, which result in almost 90 percent of the tour of duty being used in productive effort.

Report on Assignment 3

Labor Economies to Be Derived from Work Measurement Standards for Comparison of Work Performance Among Various Gangs or Divisions

H. J. Fast (chairman, subcommittee), M. B. Allen, O. C. Benson, J. L. Cann, R. H. Carpenter, J. A. Caywood, W. E. Chapman, R. E. Clancy, S. A. Cooper, M. H. Dick, R. L. Fox, W. W. Hay, C. Johnston, R. H. Jordan, H. W. Kellogg, N. M. Kelly, H. E. Kirby, N. W. Kopp, T. D. Mason, J. R. Miller, C. T. Popma, R. W. Preisendefer, M. S. Reid, H. W. Seeley, E. J. Sierleja, N. E. Smith, J. Stang, J. S. Snyder, G. M. Strawhun, G. E. Warfel, N. H. Williams, D. H. Yazell.

Your committee submits the following progress report. Senior engineering officers on 52 railways in the United States and Canada were contacted for a report on measured work and production control practices. Replies were received from 33. The questionnaire was intended to make a survey of the "state of the art."

From the replies it can be concluded that all railways are scheduling track maintenance programs to a pre-determined target. The production targets are usually based on previous performance.

A number of roads are also developing production standards in structures mainte-nance. However, improvements in this activity are associated with mechanization rather than production against engineered standards.

It is recommended that this study be continued by a close examination of specific operations for information and a recommendation for refinement in measured work.

Report on Assignment 4

Labor Economies to Be Realized Through the Use of a Combined Surfacing-Timbering Gang vs. Separate Gangs for Surfacing and Timbering

H. W. Kellogg (chairman, subcommittee), A. D. Alderson, O. C. Benson, E. J. Brown, J. L. Cann, R. H. Carpenter, W. E. Chapman, S. A. Cooper, P. A. Cosgrove, C. G. Crawford, L. E. Donovan, W. M. S. Dunn, L. C. Gilbert, R. R. Gunderson, V. C. Hanna, K. E. Henderson, W. J. Jones, R. H. Jordon, T. L. Kanan, H. E. Kirby, J. M. Lowry, R. L. Mays, J. R. Miller, H. C. Minteer, R. W. Pember, Griffith Ray, M. S. Reid, H. W. Seeley, R. G. Simmons, N. E. Smith, J. S. Snyder, A. H. Stim-son, G. M. Strawhun, J. T. Ward, D. H. Yazell.

It was learned as a result of a survey that opinion apparently is about evenly divided as to whether ties should be installed with the track-surfacing operation or separately. Various reasons were given, but for the most part they were based on local conditions affecting the economy of the operation, including density or rate of renewal, and traffic conditions, as would be expected.

Regardless of the practice followed by the various railroads, there seems to be a pronounced similarity as to equipment, number of men, production rate and man-hours per tie. All roads used mechanized equipment, with an average of about ¾ man-hours per tie.

Although in some instances precise details were not given, the data received enable generalization as follows:

(1) *Separate forces usually are more economical—*

    (a) When replacements are in such numbers that the surfacing gang will be delayed or the tie gang will get ahead, thus resulting in unproductive waits by one or the other.

    (b) When tie renewals are on a separate program not coordinated with the surfacing.

Random comment from the "separatists" include:

"Difficult to synchronize production of two gangs—ties and surfacing."

"Track surfaced where required by surface condition, not by tie replacements."

"Tie gang usually can take time for special jobs—crossings, switches, etc."

(2) *Ties are usually renewed more economically in connection with surfacing—*

    (a) When the average density of tie replacements is such that renewals can be made during a given day over a distance about the same as the length of track surfaced that day.

    (b) When the work may be performed by having only one detour if on mul-tiple track, or only one slow order if working under traffic.

Typical comment by those railroads using a combined gang included:

"The time lapse between the tie and surfacing operations is conducive to rougher riding track and to reducing the smoothing and surfacing cycle."

"Reduces cost of supervision, flag protection, etc."

"On single-track roads, usually the most economical job with least interference to traffic is performed by one gang for both operations."

It is considered to be impractical to attempt to devise a set rule or formula for general application because of the great number of variables inherent in the problem, both as among railroads and the districts of a single railroad. One of the variable factors is the cost of owning the machinery, which should be taken into consideration together with direct labor costs, although mechanization as fully as possible is recommended.

The importance of machinery and urgency of its use under the general conditions of modern track maintenance increase the importance of attempting to anticipate machine developments. While some machines appear to be sound and almost fundamental, and might be expected to be used indefinitely with comparatively slight change in design, the field now is highly competitive. Innovations are constantly sought. This requires an objective and realistic view in which consideration will be given to multipurpose equipment, and the questions of owning or leasing, particularly in areas of rapid obsolescence.

We must conclude from analysis of comment from all sources that overall economy is the prime factor governing the choice. Thorough analysis of conditions affecting the work should be made prior to making the decision. This will include weighing the effect of the factors of size and extent of programs, density of tie renewals, traffic, track availability, mechanical aids and their cost of ownership, and—in some geographical areas—the season of the year.

This report is submitted as information; it is recommended that the subject be discontinued.

Report on Assignment 5

Labor Economies to be Effected Through Use of Power Tools and Mechanized Equipment by Bridge and Building Gangs

W. E. Chapman (chairman, subcommittee), J. F. Beaver, E. J. Brown, J. L. Cann, J. A. Caywood, J. L. Chafin, R. E. Clancy, C. G. Davis, M. H. Dick, L. E. Donovan, H. J. Fast, R. L. Fox, V. C. Hanna, K. E. Henderson, L. A. Loggins, T. D. Mason, R. L. Mays, F. H. McGuigan, III, J. R. Miller, R. W. Pember, C. T. Popma, Griffith Ray, M. S. Reid, D. E. Rudisill, H. W. Seeley, John Stang, G. M. Stwahun, W. B. Trockmorton, G. E. Warfel, H. E. Wilson, and D. H. Yazell.

Your committee submits the following report of progress in determining the labor economies to be effected through the use of new or improved power tools and mechanized equipment for assignment to bridge and building gangs.

Cleaning Unit with 5 to 1 Ratio Pump

This piece of equipment, using warm water and detergent, can be assigned to washing any permanent installations, such as windows, and to clean motors in B&B trucks and other equipment, both interior and exterior, and for various similar and miscellaneous jobs. As an example, this pump can be used to operate washing equipment mounted on an aerial hoist to clean high windows in diesel engine houses. In one specific case, through use of the described equipment, highly located windows were cleaned with only 72 man-hours of labor. In 1957 these same windows were cleaned by a contractor using acid, wire brushes, scrub brushes and water at a cost to the railway 50 times greater.

Permanent Stay-in-Place Corrugated Forms

These forms are considered dispensable, and no effort is made to recover or strip them from the completed concrete job. Consequently, there is no need to erect scaffolding or perform other work to accommodate form removal. It is stated that, under certain circumstances, slow order requirements can also be minimized. Use of this material represents a possible savings of approximately 15 percent in the overall cost of concrete structures.

Grouting Equipment

Newly devised equipment can be used to obtain greater economies in grouting beneath piers and abutments of bridges and in performing grouting work required in the piers or abutments themselves. The total assignment of new and conventional equipment and personnel for such work is as follows:

1. A spud driving and pulling machine, activated by air, mounted on a two-wheel rubber-tired cart, which pulls itself up or is let down the sides of fills by the air-operated winch with which the cart is provided. Any reasonable manner of tying down the loose end of the cable can be followed, giving full consideration to safety of operation. The cart has an aluminum superstructure designed to cradle the air-operated driving hammer and spuds in order that they can be placed in position for driving. This aluminum tower supports the driving hammer during the period that the spuds are being driven and is also used as an improved means of pulling the spuds out of the ground.

2. A conventional mud jack operated in the usual manner with a preference towards mounting the mud jack on rubber-tired wheels.

3. A conventional self-propelled off-track air compressor, also mounted on rubber-tired wheels.

4. Conventional water storage vats and water handling equipment.

5. A foreman and four men are assigned to this class of work as opposed to a heavier assignment for normal roadbed grouting. It is estimated that the newly designed specialized equipment aids in increasing the speed of grouting by approximately 20 percent.

Continuing use of the hydraulic tools and other equipment described in the 1959 report has provided further data on time-savings. With added experience, it has been found that driving piling under main-line traffic conditions has been improved from an average 20 piles daily to an average of 33.

For earlier and more complete information on this study, it is recommended that reference be made to reports published in Vol. 60, page 407; Vol. 61, page 449, and Vol. 62, page 494, of the AREA Proceedings.

CONCLUSIONS

This report is submitted as information with the recommendation that the subject be dropped at this time, with full consideration being given toward continuing the study at an appropriate later date.

Report on Assignment 7

Labor Economies in Track Maintenance to be Derived Through the Use of Combination On-Off-Track Equipment vs. On-Track Equipment Only

T. L. Kanan (chairman, subcommittee), L. C. Gilbert, A. D. Alderson, R. H. Carpenter, P. A. Cosgrove, W. M. S. Dunn, R. L. Fox, R. R. Gunderson, V. C. Hanna, G. L. Harris, W. W. Hay, K. E. Henderson, Claude Johnston, R. H. Jordan, N. W. Kopp, J. M. Lowry, T. D. Mason, F. H. McGuigan III, R. W. Pember, C. T. Popma, H. W. Seeley, R. G. Simmons, J. S. Snyder, A. H. Stimson, J. T. Sullivan, F. R. Woolford.

Your committee submits the following progress report as information. Initial appearances in this study indicate that experience in the use of on-off-track equipment is relatively limited, with a minimum of railroads using this type of equipment to any great extent.

.On-off-track equipment, in most instances, can be used without pilots and train crews and eliminates the need for flagmen. Most roads indicate that rules at the present time, when on-track equipment is used, require one pilot or pilot and engineer and, in some instances, engineer, conductor and brakeman, depending on assignment of work, whether track or B&B construction or maintenance.

Rapid changes in design and engineering of mobile cranes being placed on the market, available with compensating steel flanged guide wheels, and track retirements in which better all-weather roadways are automatically made available, along with improved highways adjacent to railroad rights-of-way, are accentuating use of this type of equipment. Guide wheels provide off-track equipment with highway mobility on railroad tracks where right-of-way conditions will not readily permit off-track operation.

It is stated that at least 33 percent more working hours can be obtained using this class equipment, with daily savings of one to three man-hours of work for each individual at the job location.

Maneuverability of on-off-track cranes on a job site gives greatly increased productivity, and the cranes can be left at the work location to avoid traveling excessive distances with slower moving machines. Construction of temporary set-out trackage for storage of equipment can thus be eliminated in many instances, as well as the need of loading, unloading and train hauling machines from job to job. Availability of equipment would, in that manner, improve. The number of work equipment flat cars, and the need to maintain them, would be reduced appreciably. Switching and road haul needs would be minimized resulting in worthwhile economies. Many newer types of on-off-track equipment do eliminate requirements for specialization of operators. In the case of mobile cranes, they have more diversified ability and are not restricted to specific operations.

Tangible figures on savings have not been developed this early in this study. It is anticipated that information about this will be forthcoming during the ensuing year. Not enough advice is as yet at hand from various railroads to determine the status of various items, including train crew requirements, new equipment assignments (to replace strictly on-track machines) plus numerous other pertinent factors. Information is still required about the availability and suitability of necessary access and adjoining vehicular routes for the ready and rapid movement of rubber-tired specialized equipment.

The abandonment of second main and other trackage on numerous railroads has made available adequate roadways on the right-of-way while, on other railroads, this feature is not quite so easily solved.

Although work programs on many railroads are not comprehensive enough to justify ownership of certain items of equipment, the savings possible may justify making lease arrangements that permit obtaining on-off-track equipment in lieu of on-track whenever possible. In this respect, constant improvement of models and types of equipment is a strong argument for exploring lease arrangements in lieu of ownership.

Reduction of train delays, on-job productivity, the diversification of on-off-track equipment, compared with machines having special applications, minimizing travel time to work site and realistic reductions in work train costs, either direct or indirect, all comprise factors for further study under this assignment.

Report of Committee 8—Masonry

D. H. Dowe, *Chairman*

J. R. Willams,
Vice Chairman

E. P. Wright
F. A. Kempe, Jr.
E. A. McLeod
J. M. Gilmore
J. W. Dolson
R. J. Brueske
W. P. Hendrix
A. L. Becker
L. B. Boyd
D. E. Bray
J. W. Briscoe

H. C. Brown
M. W. Bruns
A. W. Carlson
A. E. Cawood
G. W. Clarvoe
Maurice Coburn (E)
G. W. Cooke
L. F. Currier
E. J. Daily
H. M. Dalziel
G. H. Dayett, Jr.
B. M. Dornblatt
W. J. Eney
J. A. Erskine
J. U. Estes
W. J. Galloway
N. O. Geuder, Sr.
R. W. Gilmore
S. E. Griffin
G. P. Hayes, Jr.
S. B. Holt
H. W. Hopkins
J. R. Iwinski
A. C. Johnson
T. R. Kealey
T. N. Khoury
R. J. Klueh
A. P. Kouba
A. N. Laird (E)
E. F. Manley
P. R. Matthews
L. M. Morris

L. H. Needham
L. P. Nicholson
R. F. Noll
M. S. Norris
R. E. Pearson
R. B. Peck
J. E. Peterson
Milton Pikarsky
E. D. Ripple
W. H. Robertson
R. I. Rollings
F. A. Russ, Jr.
J. H. Sawyer, Jr.
B. J. Shadrake
M. Schifalacqua
D. H. Shoemaker
C. H. Splitstone (E)
S. A. Stutes
Anton Tedesko
R. A. Ullery
G. R. Vanderpool
Neil Van Eenam
K. J. Wagoner
J. W. Weber
D. V. Wigal
William Wilbur
J. M. Williams
W. R. Wilson
S. G. Wintoniak
G. A. Wolf
K. B. Woods
Committee

(E) Member Emeritus.
Members listed in bold face are the official representatives of the Engineering Division, AAR.

To the American Railway Engineering Association:

Your committee reports on the following subjects:

1. Revision of Manual.
   Progress report, including recommendations submitted for adoption. (Other Manual recommendations submitted under Assignments 2, 7 and 8) .... page 276

2. Design of masonry structures, collaborating with Committees 1, 5, 6, 7, 15, 28 and 30.
   Progress report recommending reapproval with revisions of Manual material covering plain and reinforced concrete members, reinforced concrete arches, reinforced concrete pipe and concrete transmission poles .... page 278

3. Foundations and earth pressures, collaborating with Committees 1, 6, 7, 15 and 30.
   Pile foundation specifications are being reviewed and some revisions will be proposed. However, the work is not completed and recommendation cannot be made at this time.

4. Deterioration and repair of masonry structures.
   Progress report, presented as information .... page 280

6. Prestressed concrete for railway structures, collaborating with Committee 6.
   Progress report, presented as information .... page 281

7. Quality of concrete and mortars, collaborating with Committee 6.
   Progress report, recommending reapproval with revisions of Manual material covering concrete and reinforced concrete railroad bridges and other
   structures; also revision of ASTM specification references .... page 281

8. Waterproofing for railway structures, collaborating with Committees 6, 7
   and 15.
   Progress report, including recommended revisions of Manual material covering specifications for membrane waterproofing and specifications for
   waterproofing coatings for exposed concrete surfaces, and deletion of subject matter under specifications for bituminous emulsions for dampproofing .... page 313

10. Methods of construction with precast-concrete structural members col-
    laborating with Committee 6.
    Progress is being made on the preparation of a report on this assignment
    which will be in the form of a discussion of present practice, but no formal
    report can be made at this time.

THE COMMITTEE ON MASONRY,

D. H. DOWE, *Chairman*.

AREA Bulletin 568, December 1961.

Report on Assignment 1

Revision of Manual

E. P. Wright (chairman, subcommittee), R. J. Brueske, J. W. Dolson, J. M. Gilmore,
W. P. Hendrix, F. A. Kempe, Jr., E. A. McLeod, J. R. Williams.

Your committee recommends the adoption of the revisions to the Manual as set
forth in the reports on Assignments 2, 7 and 8.

Your committee also submits, as the result of a special review of Chapters 8 and
29, the following report recommending revisions to certain material now in these
chapters.

Pages 8–4–1 to 8–4–12, incl.

PILE FOUNDATIONS

Change title to read "Specifications for Pile Foundations."

Pages 8–7–1 to 8–7–7, incl.

REINFORCED CONCRETE ARCHES

Page 8-7-1. Change title to read "Specifications for Reinforced Concrete Arches."
Page 8-7-6, Sec. C, Art. 2 (a). Change fifth sentence to read: "When it is desired

to use a long span for centering to obtain specific clearances under the structure during construction, allowance shall be made when setting the forms for the deflection of such span due to its own dead load and to the dead weight of the wet concrete to be supported, and suitable means shall be provided for adjusting the forms to their proper contour."

Pages 8–8–1 to 8–8–6, incl.

SPECIFICATIONS FOR REINFORCED CONCRETE RIGID-FRAME BRIDGES OF ONE SPAN

Page 8–8–5, Sec. E, Art. 2. Change fifth sentence to read: "When it is desired to use a long span for centering to obtain specific clearances under the structure during construction, allowance shall be made when setting the forms for the deflection of such span due to its own dead load and to the dead weight of the wet concrete to be supported, and suitable means shall be provided for adjusting the forms to their proper contour."

Pages 8–11–1 to 8–11–7, incl.

SPECIFICATIONS FOR LINING RAILWAY TUNNELS WITH CONCRETE

Page 8–11–1, Sec. A. Change to read: "These specifications, where conditions permit, cover the lining of new tunnels and those portions of old tunnels which are completely relined through ordinary rock formations which involve no extraordinary side pressure or special features."

Pages 8–11–8 to 8–11–12, incl.

SPECIFICATIONS FOR LINING RAILWAY TUNNELS WITH BRICK

Page 8–11–8, Sec. A, Art. 1—Change to read: "These specifications, where conditions permit, cover the lining of new tunnels and those portions of old tunnels which are completely relined through ordinary formations which involve no special features."

Pages 8–14–1 to 8–14–8, incl.

REPAIRING AND SOLIDIFYING MASONRY STRUCTURES

Page 8–14–1. Change title to read "Specifications for Repairing and Solidifying Masonry Structures."

Pages 29–1–1 to 29–1–4, incl.

PRINCIPLES GOVERNING THE WATERPROOFING OR DAMPPROOFING OF RAILWAY STRUCTURES

Page 29–1–1, Sec. A, Par. 2. Change the word "must" to "should."

Pages 29–1–1 to 29–1–4, incl. Change the word "shall" to "should" wherever it appears on these pages.

Report on Assignment 2

Design of Masonry Structures

Collaborating with Committees 1, 5, 6, 7, 15, 28 and 30

F. A. Kempe, Jr. (chairman, subcommittee), H. C. Brown, A. W. Carlson, E. J. Daily, H. M. Dalziel, J. U. Estes, N. O. Geuder, Sr., A. N. Laird, R. E. Pearson, F. A. Russ, Jr., B. J. Shadrake, Anton Tedesko, D. V. Wigal.

Your committee submits the following report recommending the adoption of revisions, deletions, and additions to certain material now in the Manual.

Pages 8–2–1 to 8–2–27, incl.

SPECIFICATIONS FOR DESIGN OF PLAIN AND REINFORCED CONCRETE MEMBERS

Reapprove with following revisions:

Page 8–2–1, Sec. A, Art. 3. Buildings. Delete the last six words: "and the unit stresses hereinafter specificed."

Page 8–2–2, Sec. B, Art. 3. Change heading "Volume Change" to "Shrinkage", and at end of paragraph add another sentence reading "Under normal conditions 0.0002 may be considered an acceptable value."

Page 8–2–3, Sec. C, Art. 3, Live Load, third paragraph. Change to read: "Live load from a single track acting on the top surface of a structure with ballasted deck or under fills shall be assumed to have uniform lateral distribution over a width equal to the length of track tie plus the depth of ballast and fill below the bottom of tie unless limited by the extent of the structure." Fourth paragraph, change to read: "The lateral distribution of live load from multiple tracks shall be as specified for single tracks and further limited so as not to exceed the distance between centers of adjacent tracks."

Page 8–2–4, Sec. C, Art. 4, Impact Load. Add paragraph just before last paragraph, reading: "Maximum steam engine impact shall not exceed 80 percent. Maximum diesel engine impact shall not exceed 60 percent."

Pages 8–7–1 to 8–7–7, incl.

REINFORCED CONCRETE ARCHES

Reapprove with the following revisions:

Page 8–7–5, Sec. B, Art. 6, Drainage. Add paragraph at end reading: "Where longitudinal deck slope less than ⅛ in per ft is necessary, provide adequate transverse slope to half-round deck drains at the curb lines and provide a 1/16 in per ft minimum slope of the half-round deck drains. Additional lines of deck drains may be placed between tracks of multi-track structures."

Page 8–7–6, Sec. B, Art. 7, Waterproofing. Delete words "sheet metal" in lines seven and nine.

Pages 8–10–1 to 8–10–5, incl.

SPECIFICATIONS FOR THE PLACEMENT OF CONCRETE CULVERT PIPE

Reapprove with the following revisions:

Page 8–10–1, Art. 3 (b). Change last sentence to read: "No section of pipe shall be placed higher than the adjoining upstream section."

Page 8–10–2, Figs. 1 to 3, incl. Delete and replace with revised Figs. 1 to 3, incl., presented herewith.

Page 8–10–5, Art. 7, Joints. Change entire article to read:

"(a) When watertight joints are required, mortar, grout, rubber gaskets, oakum or other material shall be used as specified by the engineer.

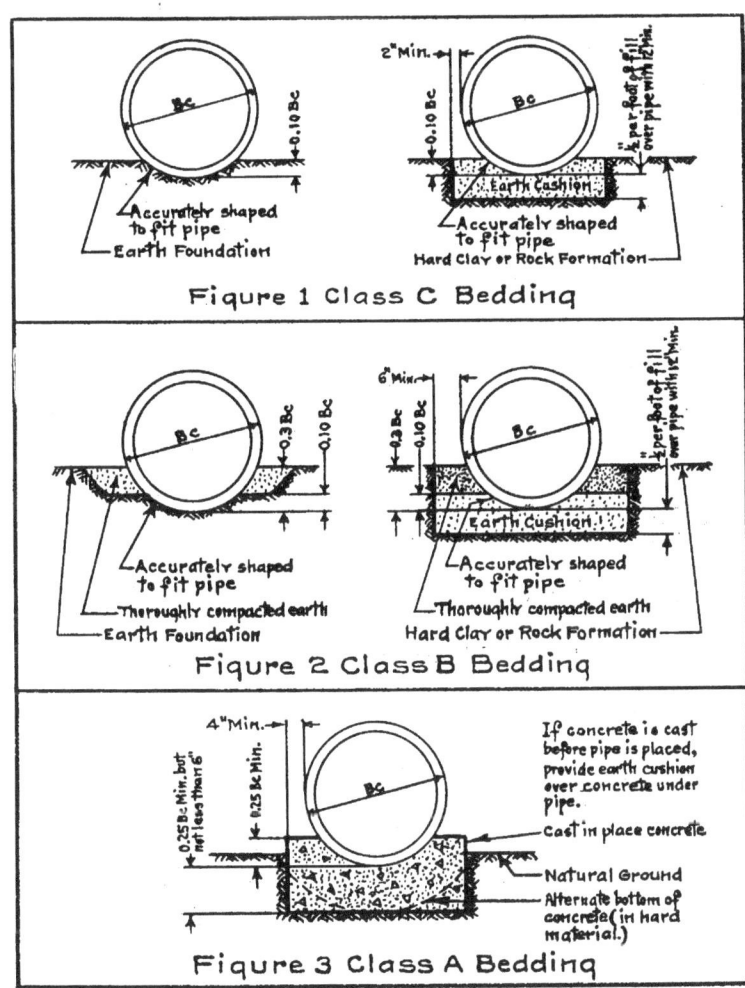

Figure 1 Class C Bedding

Figure 2 Class B Bedding

Figure 3 Class A Bedding

"(b) In areas where a tendency exists for pipe sections to separate, suitable ties shall be installed.

"(c) Where mortar joints are used, bell and spigot pipe shall be bedded with the bell end upstream. The interior surface of the bell shall be thoroughly cleaned and wetted, and the lower portion filled with a stiff mortar of sufficient thickness to make the inner surfaces of the abutting sections flush. The spigot end of the next section shall also be cleaned and wetted and fitted into the bell so that the sections are closely matched. The annular space in the bell shall then be filled with mortar and the inner surface of the pipe brushed smooth at the joint.

"(d) Where mortar joints are used, tongue-and-groove pipe shall be laid with the groove upstream. A shallow excavation shall be made underneath the pipe at the joint and filled with a stiff mortar into which the next section of pipe is laid. The groove end of the first pipe shall be cleaned and wetted and a layer of mortar applied to the lower half of the groove. The tongue end of the next section shall also be cleaned and wetted, and a layer of mortar applied to the upper half of the tongue. The tongue shall then be fitted into the groove and shoved in place so that part of the mortar is squeezed out and the sections closely matched. The inner surface shall then be brushed smooth at the joint.

"(e) Joints in pipe that is jacked in place shall not be sealed with mortar until the jacking of the culvert is completed."

Pages 8–10–6 to 8–10–14, incl.

SPECIFICATIONS FOR REINFORCED CONCRETE CULVERT PIPE

Reapprove with the following revisions:

Pages 8–10–6 to 8–10–14, incl. Delete all material on these pages and substitute the following paragraph:

"Pipe shall conform to ASTM Standard Specifications for Reinforced Concrete Culvert Pipe (Designation C 76). The engineer shall specify the required class, and if pertinent, the wall type. When a stronger pipe is required, the engineer shall so specify in accordance with Par. 10 of the ASTM specifications on alternate and special design."

Pages 8–12–1 to 8–12–6, incl.

SPECIFICATIONS FOR DESIGN OF CONCRETE
TRANSMISSION POLES

Delete the entire subject matter under this heading and add: "Temporarily with-drawn. Subject matter to be studied and rewritten."

Report on Assignment 4

Deterioration and Repair of Masonry Structures

J. M. Gilmore (chairman, subcommitee), L. B. Boyd, G. H. Dayett, Jr., H. W. Hopkins, R. J. Kluch, P. R. Matthews, L. H. Needham, J. E. Peterson, J. H. Sawyer, Jr., Mariano Schifalacqua, D. H. Shoemaker.

Your committee submits the following report of progress in gathering data pertaining to the repair of deteriorated concrete using epoxy resins.

As bonding agents, epoxy resins have proven to be superior to all other bonding agents used heretofore. Even though quite costly by the gallon, their cost as bonding agents becomes insignificant when taken into the overall cost of labor and material, and considering the overall cost of the structure in which bond plays a critical part. However, when used in lieu of cement and water in the making of a concrete, their cost becomes more significant, and the properties may not be those which are desired.

As a product for repairing cracks and fissures where expansion and contraction will not recur, epoxy resins have considerable merit in making a structure whole.

The committee will continue to gather information on repair projects using epoxies, and pertinent data and results will be covered in future reports.

Report on Assignment 6

Prestressed Concrete for Railway Structures

Collaborating with Committee 6

J. R. Williams (chairman, subcommittee), L. F. Currier, W. J. Eney, R. W. Gilmore, T. N. Khoury, E. D. Ripple, S. A. Stutes, G. R. Vanderpool, W. R. Wilson, G. A. Wolf.

Your committee is currently preparing the design of a standard prestressed concrete trestle. The supporting bents will consist of 24-in square prestressed concrete piles with a center void and precast caps. Two widths of box girders are being incorporated into the design. These widths will be 3 ft and 4 ft. The design will provide for deck widths of 12 ft, 14 ft, 15 ft and 16 ft by utilizing various combinations of the two girder widths.

Report on Assignment 7

Quality of Concrete and Mortars

Collaborating with Committee 6

J. W. Dolson (chairman, subcommittee), M. W. Bruns, W. J. Galloway, S. B. Holt, A. C. Johnson, L. M. Morris, L. P. Nicholson, M. S. Norris, R. I. Rollings, R. A. Ullery.

Your committee submits the following report recommending the adoption of revisions and additions to certain material now in the Manual.

Delete the Specifications for Concrete and Reinforced Concrete Railroad Bridges and Other Structures, pages 8-1-1 to 8-1-26, incl., substituting therefor the following rewritten version:

SPECIFICATIONS FOR CONCRETE AND REINFORCED CONCRETE RAILROAD BRIDGES AND OTHER STRUCTURES

A GENERAL CONDITIONS

1. Purpose

These specifications are intended to be used in connection with the American Railway Engineering Association Contract Form, Part 1, Chapter 20, or when called for on plans or specifications or when directed by the engineer, for work carried out by railroad or railway companies or by contractors for the railroad or railway companies. These specifications shall apply to all sections of the specifications with equal force.

2. Scope

These specifications describe the selection, sampling and testing of materials to be used; the composition of concrete; and the mixing, transporting, placing, finishing and curing of concrete.

3. Definitions

a. *Concrete*—Concrete is an intimate mixture of cement, aggregates and water of the qualities herein specified. It shall be proportioned, mixed, transported, placed and cured by the methods herein specified.

b. *Company*—Company shall be understood to mean the railroad or railway company.

c. *Engineer*—Engineer shall be understood to mean the chief engineer of the company or his duly authorized representative.

d. *Contractor*—Contractor shall be understood to mean the person, firm or corporation agreeing to perform the work covered by the specifications.

e. *Approved or Approval*—Approved or approval shall be understood to mean written approval.

f. *Workmanship*—Workmanship shall be understood to mean the best class obtainable with modern equipment and skilled labor, and shall conform in every respect to the requirements of these specifications.

4. Acceptability

All materials used on the work shall be subject to the approval of the engineer who shall be the sole judge of their quality and suitability. The engineer shall be notified in advance whenever the preparation for manufacture of any material for the work is to be commenced at any place.

5. American Society for Testing and Materials—Specification References

Wherever, in these specifications, reference is made to American Society for Testing and Materials, the number indicating the year of the approved issue is intentionally omitted from the ASTM designation. For the year of the approved issue of the ASTM designations see Part 18 of this Chapter.

6. Selection of Materials

The concrete materials shall be selected and combined in such a manner as to produce uniformity of color and texture in the surface of any structure or group of structures in which they are to be used. No change shall be made in the brand or type of cement, character and source of aggregate, water, class of concrete and method of transporting, placing, finishing or curing concrete without approval of the engineer.

7. Defective Materials

All porous or defective concrete of any kind, occurring prior to final acceptance of the work, shall be remedied by the contractor at his own expense and to the satisfaction of the engineer. All damaged or rejected materials shall be immediately removed from the site of the work by the contractor.

8. Equipment

The contractor shall provide all equipment required for the execution and completion of the work, including all staging, scaffolding, apparatus, tools, etc., which are necessary. All equipment must meet with the approval of the company, and the engineer may require the removal of any portion of equipment which is defective or unsuitable for the proper prosecution of the work, and the contractor will be required to substitute therefore satisfactory equipment without delay. Upon request, the contractor shall furnish for approval a statement of the method and equipment he expects to use for handling, storing, and proportioning materials and for mixing, transporting, placing, protecting, and curing the concrete, but whether the engineer exercises this authority or not the contractor shall not be relieved from his sole responsibility for the safe, proper, adequate and lawful construction, maintenance and use of such a method and equipment.

B. CEMENT

1. General

Cement shall be furnished by the contractor or the railroad as provided for in the contract. Only one brand or type of cement shall be used in any part of the structure, and cements of the same brand from different mills shall not be mixed or used in any part of the structure except by written permission of the engineer.

2. Specifications

Cement shall be portland cement meeting the requirements of ASTM Specification C 150, air-entraining portland cement meeting the requirements of ASTM Specification C 175, portland blast-furnace slag cement meeting the requirements of ASTM Specification C 205, portland-pozzolan cement meeting requirements of ASTM Specification C 340, or masonry cement meeting the requirements of ASTM Specification C 91, as specified on the plans or by the engineer.

The entrainment of air in concrete shall be obtained by the use of air-entraining cements as referred to in the ASTM specification for cement in the above paragraph, or by the use of an air-entraining admixture meeting the requirements of ASTM Specification C 260.

3. Quality, Sampling and Testing

The quality of cement and the methods of sampling and testing it shall be as required by ASTM Specification C 150 for portland cement, C 175 for air-entraining portland cement, C 205 for portland blast-furnace slag cement, C 340 for portland-pozzolan cement or C 91 for masonry cement.

C. AGGREGATES

1. Sampling and Testing

a. Representative samples shall be selected and sent to the testing laboratory at frequent intervals as directed by the engineer. Aggregates may not be used until the samples have been approved by the designated laboratory as being in conformity with the following requirements, and their use authorized by the engineer.

b. Sampling and testing shall be in accordance with the following standard methods of the American Society for Testing and Materials:

	Designation
(1) Unit Weight of Aggregate	C 29
(2) Concrete Aggregates	C 33
(3) Organic Impurities in Sands for Concrete	C 40
(4) Surface Moisture in Fine Aggregate	C 70

c. The required tests shall be made on test samples that comply with requirements of the designated test methods and are representative of the grading that will be used in the concrete. The same test sample may be used for sieve analysis and for determination of material finer than the No. 200 sieve. Separated sizes from the sieve analysis may be used in preparation of samples for soundness or abrasion tests. For determination of all other tests and for evaluation of potential alkali reactivity where required, independent test samples shall be used.

d. The fineness modulus of an aggregate is the sum of the percentages of a sample retained on each of a specified series of sieves divided by 100, using the following standard sieve sizes: No. 100, No. 50, No. 30, No. 16, No. 8, No. 4, ⅜ in, ¾ in, 1½ in and larger, increasing in the ratio of 2 to 1. Sieving shall be done in accordance with ASTM Method C 136.

2. Soundness

a. Except as provided in Pars. b and c, aggregates subjected to five cycles of the soundness test, shall show a loss, weighted in accordance with the grading of a sample complying with the limitations set forth herein, not greater than the following percentages:

Aggregate	Sodium Sulphate	Magnesium Sulphate
Fine	10	15
Coarse	12	18

b. Aggregate failing to meet the requirements of Par. a may be accepted, provided that concrete of comparable properties, made from similar aggregate from the same source, has given satisfactory service when exposed to weathering similar to that to be encountered.

c. Aggregate not having a demonstrable service record and failing to meet the requirements of Par. a may be accepted, provided it gives satisfactory results in concrete subjected to freezing and thawing tests

Fine Aggregate

3. General

Fine aggregate shall consist of natural sand or, subject to the approval of the engineer, other inert materials with similar characteristics.

4. Grading

a. Sieve Analysis—Fine aggregate, except as provided in Pars. b and d shall be graded within the following limits:

Sieve Size	Total Passing Percentage by Weight
⅜-in	100
No. 4 (4760 micron)	95 to 100
No. 8 (2380 micron)	80 to 100
No. 16 (1190 micron)	50 to 85
No. 30 (590 micron)	25 to 60
No. 50 (297 micron)	10 to 30
No. 100 (149 micron)	2 to 10

b. The minimum percentages shown above for material passing the No. 50 and No. 100 sieves may be reduced to 5 and 0, respectively, if the aggregate is to be used in air-entrained concrete containing more than 4½ bags of cement per cubic yard, or in non-air-entrained concrete containing more than 5½ bags of cement per cubic yard, or if an approved mineral admixture is used to supply the deficiency in percentages passing these sieves. Air-entrained concrete is here considered to be concrete containing air-entraining cement or an air-entraining agent and having an air content of more than 3 percent.

c. The fine aggregate shall have not more than 45 percent retained between any two consecutive sieves of those shown in Par. a, and its fineness modulus shall be not less than 2.3 nor more than 3.1.

d. For walls and other locations where smooth surfaces are desired, the fine aggregate shall be graded within the limits shown in the table of Par. a except that not less than 15 percent shall pass the No. 50 sieve and not less than 3 percent shall pass the No. 100 sieve.

e. To provide uniform grading of fine aggregate, a preliminary sample representative of the material to be furnished, shall be submitted at least 10 days prior to actual deliveries. Any shipment made during progress of the work which varies more than 0.20 from the fineness modulus of the preliminary sample shall be rejected or, at the option of the engineer, may be accepted provided suitable adjustments are made in concrete proportions to compensate for the difference in grading.

5. Deleterious Substances

a. The amount of deleterious substances in fine aggregate shall not exceed the following limits:

Item	Maximum Limit Percentage by Weight
Clay lumps	1.0
Coal and lignite	0.5*
Material finer than No. 200 sieve:	
Concrete subject to abrasion	3.0**
All other classes of concrete	5.0**

* Does not apply to manufactured sand produced from blast furnace slag.
** For manufactured sand, if the material finer than the No. 200 sieve consists of the dust of fracture, essentially free from clay or shale, these limits may be increased to 5.0 and 7.0 percent, respectively.

b. *Organic Impurities*

(1) Fine aggregate shall be free of injurious amounts of organic impurities. Except as herein provided, aggregates subjected to the test for organic impurities and producing a color darker than the standard shall be rejected.

(2) A fine aggregate failing in the test may be used, provided that the discoloration is due principally to the presence of small quantities of coal, lignite, or similar discrete particles.

(3) A fine aggregate failing in the test may be used, provided that, when tested for mortar-making properties, the mortar develops a compressive strength at 7 and 28 days of not less than 95 percent of that developed by a similar mortar made from another portion of the same sample which has been washed in a 3 percent solution of sodium hydroxide followed by thorough rinsing in water. The treatment shall be sufficient to produce a color lighter than standard with the washed material.

(c) Fine aggregate for use in concrete that will be subject to wetting, extended exposure to humid atmosphere, or contact with moist ground shall not contain any materials that are deleteriously reactive with the alkalies in the cement in an amount sufficient to cause excessive expansion of mortar or concrete, except that if such materials are present in injurious amounts, the fine aggregate may be used with a cement containing less than 0.6 percent alkalies as measured by percentage of sodium oxide plus 0.658 times percentage of potassium oxide, or with the addition of a material that has been shown to prevent harmful expansion due to the alkali-aggregate reaction.

6. Mortar Strength

Fine aggregate shall be of such quality that when made into a mortar and subjected to the mortar strength test prescribed in ASTM Method of Test, C 87, the mortar shall develop a compressive strength not less than that developed by a mortar prepared in the same manner with the same cement and graded standard sand having a fineness modulus of 2.4 0 ± 0.10. The graded sand shall consist of approximately equal parts by weight of standard Ottawa sand and graded Ottawa sand as defined in ASTM Methods of Test, C 190 and C 109, respectively.

Coarse Aggregate

7. General

a. Coarse aggregate shall consist of crushed stone, gravel, crushed slag, or a combination thereof, or subject to the approval of the engineer, other inert materials with similar characteristics, having hard, strong durable pieces, free from adherent coatings, and shall conform to the requirements of this specification.

b. Crushed slag shall be rough cubical fragments of air-cooled iron-blast-furnace slag which, when conforming to the grading to be used in the concrete, shall have a compact weight of not less than 70 lb per cu ft. It shall be obtained only from sources approved by the engineer.

8. Grading

a. Coarse aggregate shall be graded between the limits specified and shall conform to the requirements prescribed in the table on page 288 for the designated sizes.

b. The maximum size of aggregate shall be not larger than one-fifth of the narrowest dimension between forms of the member for which the concrete is used, nor larger than one-half of the minimum clear space between reinforcing bars.

9. Deleterious Substances

a. The amount of deleterious substances in coarse aggregate shall not exceed the following limits:

Item	Maximum Limit Percent by Wt.
Clay lumps	0.25
Soft particles	5.0
Chert that will readily disintegrate (soundness test, five cycles)	1.0
Material finer than No. 200 sieve	1.0*
Coal and lignite	1.0**

* In the case of crushed aggregates, if the material finer than the No. 200 sieve consists of the dust of fracture, essentially free from clay or shale, this percentage may be increased to 1.5.
** This requirement does not apply to blast-furnace slag coarse aggregate.

b. Coarse aggregate for use in concrete that will be subject to wetting, extended exposure to humid atmosphere, or contact with moist ground shall not contain any materials that are deleteriously reactive with the alkalies in the cement in an amount sufficient to cause excessive expansion of mortar or concrete, except that if such materials are present in injurious amounts, the coarse aggregate may be used with a cement containing less than 0.6 percent alkalies as measured by percentage of sodium oxide plus 0.658 times percentage of potassium oxide, or with the addition of a material that has been shown to prevent harmful expansion due to the alkali-aggregate reaction.

10. Abrasion Loss

a. Coarse aggregate to be used in concrete when subjected to test for resistance to abrasion (ASTM Method C 131) shall show a loss of weight not more than the following:

(1) For concrete subject to severe abrasion such as concrete in water, precast concrete piles, paving for sidewalks, platforms or roadways, floor wearing surfaces, the loss in weight shall not exceed 40 percent.

(2) For concrete subject to medium abrasion such as concrete exposed to the weather, the loss of weight shall not exceed 50 percent.

(3) For concrete not subject to abrasion, the loss in weight shall not exceed 60 percent,

Size No.	Nominal Size (Sieves with Square Openings)	Amounts Finer than Each Laboratory Sieve (Square Openings), Percent by Weight												
		4 in	3½ in	3 in	2½ in	2 in	1½ in	1 in	¾ in	½ in	⅜ in	No. 4 (4760 micron)	No. 8 (2380 micron)	No. 16 (1190 micron)
1	3½ to 1½ in.	100	90 to 100		25 to 60		0 to 15		0 to 5					
2	2½ to 1½ in.			100	90 to 100	35 to 70	0 to 15		0 to 5					
357	2 in to No. 4				100	95 to 100		35 to 70		10 to 30		0 to 5		
467	1½ in to No. 4					100	95 to 100		35 to 70		10 to 30	0 to 5		
57	1 in to No. 4						100	95 to 100		25 to 60		0 to 10	0 to 5	
67	¾ in to No. 4							100	90 to 100		20 to 55	0 to 10	0 to 5	
7	½ in to No. 4								100	90 to 100	40 to 70	0 to 15	0 to 5	
8	⅜ in to No. 8									100	85 to 100	10 to 30	0 to 10	0 to 5
3	2 to 1 in.				100	90 to 100	35 to 70	0 to 15	0 to 5					
4	1½ to ¾ in.					100	90 to 100	20 to 55	0 to 15	0 to 5				

11. Rubble Aggregate

Rubble aggregate shall consist of clean, hard, durable stone or gravel, retained on a 6-in square opening and with individual pieces weighing not more than 100 lb.

12. Cyclopean Aggregate

Cyclopean aggregate shall consist of clean, hard, durable stone or gravel with individual pieces weighing more than 100 lb.

D. MIXING WATER

1. General

Preferably, water used in mixing concrete shall be potable and free from pronounced taste and odor. Non-potable water may be used if mortar strength tests satisfy the requirement of the following paragraph.

2. Mortar Strength

When subjected to the mortar strength test, ASTM Method of Test C 87, the 7- and 28-day strengths of mortar specimens made with the water in question shall be not less than 90 percent of the strength of similar specimens made with the same cement and potable water.

E. METAL REINFORCEMENT—MATERIAL

1. General

Metal reinforcement shall include plain and deformed steel bars and cold-drawn steel wire or fabricated forms of these materials.

2. Quality

All steel reinforcing bars shall be of billet, rail or axle steel as specified on the drawings.

The material shall conform in quality to the following ASTM Specifications:

Bars: A 15, A 16, A 160, A 408, A 431, A 432
Deformations: A 305, A 408
Wire: A 82
Fabricated Materials: A 184, A 185

When cast iron or structural steel sections are used, as in composite or combination columns, these materials shall conform in quality to the standard ASTM Specifications of the following applicable serial designations: A 7, A 377.

3. Deformed Bars

Deformed bars shall be manufactured in accordance with ASTM Specifications.

4. Size of Bars

Reinforcing bars shall have minimum section areas not less than shown in the following table, and the unit weights shown in this table shall be used in computing weights:

SIZES AND AREAS OF REINFORCING BARS
Dimensions are for Round Sections

Bar Number		Nominal Diameter Inches	Cross-Sectional Area Square Inches	Perimeter Inches	Weight Pounds Per Foot
(a)					
(b)	2	0.250	0.05	0.786	0.167
	3	0.375	0.11	1.178	0.376
	4	0.500	0.20	1.571	0.668
	5	0.625	0.31	1.963	1.043
	6	0.750	0.44	2.356	1.502
	7	0.875	0.60	2.749	2.044
	8	1.000	0.79	3.142	2.670
(c)	9	1.128	1.00	3.544	3.400
(c)	10	1.270	1.27	3.990	4.303
(c)	11	1.410	1.56	4.430	5.313
(c)	14s	1.693	2.25	5.32	7.65
(c)	18s	2.257	4.00	7.09	13.60

Notes: (a) Bar numbers denote nominal diameters of round bars in eighths-of-an-inch. The nominal diameter of a deformed bar is equivalent to the diameter of a plain bar having the same weight per foot as the deformed bar.
(b) No. 2 Bar in plain round only.
(c) Bars of designation 9, 10, 11, 14s and 18s correspond to the former 1-in, 1⅛-in, 1¼-in, 1½-in and 2-in square sizes and are equivalent to those former standard bar sizes in weight and nominal cross-sectional areas.

F. STORAGE OF MATERIALS

1. Cement

Cement delivered in cars must be promptly stored in houses and the cars released. Sacked cement must be stored in weathertight buildings having floors above ground to protect against dampness. Sacks should be stacked close together to reduce circulation of air and in such a manner as to permit easy access for proper inspection. They should not be stacked against outside walls. If in bulk it shall be stored in weathertight bins, and for the proportioning of batches it shall be weighed rather than measured by volume. All cement shall be subject at any time to retest. If under retest it fails to meet any of the requirements of the specifications, it will be rejected and shall be promptly removed from the site of the work by the contractor.

Where the railroad company furnishes the cement and the injury is due to negligence on the part of the contractor properly to care for it, the cost of such cement shall be charged to the contractor.

2. Aggregates

Fine and coarse aggregates shall be stored separately and in such manner as to avoid the inclusion of dirt and other foreign material in the concrete. Aggregates shall be unloaded and piled in such manner as to maintain the uniform grading of the sizes. Stock piles of coarse aggregates shall be built in horizontal layers to avoid segregation. Crushed slag shall be wet down when necessary to insure not less than 3 percent moisture content.

3. Steel Reinforcement

Steel reinforcement shall be stored in racks in such a manner as to avoid contact with the ground. If steel is to remain in storage on the site for more than 1 month it shall be covered to protect it from weather. If steel accumulates rust and scale during storage, it shall be cleaned before being used. Severe deterioration of this kind may be a basis for rejection.

G. FORMS

1. General

Forms shall be constructed of wood, steel, or other approved material, and of a type, size, shape, quality, and strength which will produce true, smooth lines, and surfaces, conforming to the line and dimensions of the concrete as called for on the plans. They shall be substantial and designed to resist the pressures to which they are subjected. Lumber in forms for exposed surfaces shall be dressed to uniform thickness. Undressed lumber may be used in forms for unexposed surfaces. Forms may be omitted for foundation concrete if, in the opinion of the engineer, the sides of the excavation are sufficiently firm so that the concrete may be thoroughly rammed without causing the adjacent earth to yield—in which case the actual dimensions of the excavation shall be slightly greater than the plan dimensions of the foundation so as to insure design requirements.

2. Construction

Forms shall be constructed mortar-tight, and shall be made sufficiently rigid by the use of ties and bracing to prevent displacement or sagging between supports and to withstand the pressure, ramming and vibration without deflection from the prescribed lines during and after placement of the concrete.

Joints in forms shall be horizontal or vertical, and suitable devices shall be used to hold adjacent edges together in accurate alinement.

All forms shall be so constructed that they can be removed without hammering or prying against the concrete.

Bolts and rods preferably shall be used for internal ties. They shall be so arranged that, when the forms are removed, no metal shall be within 1 in of any surface.

Any material once used in forms shall be thoroughly cleaned and re-oiled before erection in a new location. All rough surfaces shall be smoothed and repairs made to put the sections in first-class condition; and forms which have been used repeatedly and are not acceptable to the engineer for further use shall be discarded.

In the case of long spans where no intermediate supports are possible, the probable deflection in the forms due to the weight of the fresh concrete shall be compensated for, so that the finished members shall conform accurately to the desired line and grade. If adequate foundation for shores cannot be secured, truss supports shall be provided.

Shores supporting successive stories shall be placed over those below, or so designed and placed that the load will be transmitted directly to them.

3. Moldings

Unless otherwise specified or directed by the engineer, suitable moldings or bevels shall be placed in the angles of forms to round or bevel the edges of the concrete, including abutting edges of expansion joints.

4. Oiling

The inside of forms shall be coated with non-staining mineral oil or other approved material which shall be applied before the reinforcement is placed.

5. Temporary Openings

Temporary openings shall be provided at the base of column and wall forms, and at other points where necessary, to facilitate cleaning and inspection immediately before depositing concrete. Forms for walls or other thin sections of considerable height shall be provided with openings or other devices which will permit the concrete to be placed in a manner to avoid accumulation of hardened concrete on the forms or metal reinforcement.

6. Removal

Forms shall be removed in such a manner as to insure the complete safety of the structure. Care shall be taken to preserve form sections and not to injure the corners or surfaces of the concrete. Prying between forms and concrete shall not be permitted.

Form shall not be disturbed until the concrete has adequately hardened and has acquired sufficient strength to support its weight and any construction load upon it.

The removal of forms shall depend on the character of the concrete, the location of the form, and the temperature and moisture conditions which affect the strength of the concrete.

If not otherwise specified by the plans, specifications or engineer, the following table shall be used for determining the minimum safe concrete strength required before the forms can be removed.

STRENGTH OF CONCRETE FOR SAFE REMOVAL OF FORMS

Classification	Required Minimum Strength PSI
a. Concrete not subject to appreciable bending or direct stress, not reliant on forms for vertical support, not liable to injury from form-removal operations	500
b. Concrete subject to appreciable bending and for direct stress, and partially reliant on forms for vertical support:	
(1) Subject to dead load only	750
(2) Subject to dead and live loads	1,500
c. Concrete subject to high bending-stress, and wholly or almost wholly reliant on forms for vertical support	2,000

The age-strength relation shall be determined from tests of representative samples of concrete used in the structure and cured under job conditions.

H. METAL REINFORCEMENT PLACING

1. Cleaning

Metal reinforcement, before being positioned, shall be thoroughly cleaned of mill and rust scale, dirt, grease or coating that will tend to destroy or reduce the bond, and after being placed, shall be maintained in a clean condition until completely embedded in concrete. Reinforcement appreciably reduced in section shall be rejected.

2. Bending

Reinforcement shall be accurately formed to the dimensions indicated on the plans. Stirrups and tie bars shall be bent around a pin having a diameter not less than 2 times the minimum thickness of the bar. Bends for other bars shall be made around a pin having a diameter not less than 6 times the minimum thickness except for bars larger than 1 in. in which case the bends shall be made around a pin of 8 bar diameter. All bars shall be bent cold.

Field bending of rail or axle steel shall not be permitted except as approved by the engineer.

3. Straightening

Metal reinforcement shall not be bent or straightened in a manner that will injure the material. Bars with kinks or bends not shown on plan shall be rejected. Hot bending of reinforcement will be permitted only when the entire operation is approved by the engineer.

4. Placing

Metal reinforcement shall be accurately positioned and secured against displace-- ment by using annealed iron wire ties or suitable clips, or spot welding at intersections, and shall be supported by concrete or metal supports, spacers or metal hangers. Metal clips or supports shall not be placed in contact with forms for exposed surfaces. Where reinforcing is supported by concrete blocks they shall be of a quality equal to the concrete into which they are to be incorporated and shall be tapered to insure unyielding incorporation in the finished structure.

The minimum center to center distance between parallel bars shall be $2\frac{1}{2}$ times the diameter of round or 3 times the side dimensions for square bars, but in no case shall the clear spacing between the bars be less than twice the maximum size of the coarse aggregate, nor shall the clear distance from the face of the finished surface be less than 1 in.

5. Moisture Protection

At those surfaces of footings and other principal structural members in which the concrete is deposited directly against the ground, metal reinforcement shall have a minimum covering of 3 in of concrete. At other surfaces of concrete exposed to the ground or to severe weathering conditions, metal reinforcement shall be protected by not less than 2 in of concrete for bars over $\frac{5}{8}$ in. in diameter, and $1\frac{1}{2}$ in for bars $\frac{5}{8}$ in. in diameter or less.

In concrete exposed to sea water the metal reinforcement shall be placed not less than 3 in from any plane or curved surface, and at corners it shall be not less than 4 in from adjacent surfaces.

6. Splicing

When it is necessary to splice reinforcement at points other than shown on the plans, the character of the splice shall be determined by the engineer. Splices shall not be made at points of maximum stress, and splices in adjacent bars shall be staggered.

7. Future Bonding

Exposed reinforcement intended for bonding with future extensions shall be pro- tected from corrosion in an approved manner

I. JOINTS

1. Location of Joints

When the structures or portions of the structures are designed to be monolithic, they shall be cast integrally, except as hereinafter modified. When necessary to provide construction joints not indicated or specified, such joints shall be located as directed by the engineer and formed so as not to impair the strength and least to impair the appearance of the structure.

Joints in columns shall be made at the underside of floor members and at floor levels. Haunches and column capitals shall be considered as part of and continuous with the floor or roof. At least 2 hr shall elapse after depositing concrete in columns or walls before depositing the concrete in the floor system.

2. Procedure in Forming Construction Joints[1]

The procedure specified in Sec. L, Art. 9, for bonding new concrete to old shall be followed in the formation of all construction joints. The reinforcement shall continue through the joint. For concrete without reinforcement, shearing strength shall be provided by means of a concrete key or dowel bars as the engineer may direct.

3. Extra Steel at Construction Joints

Where a construction joint is required in a section of a building more than 100 ft long or more than 100 ft between expansion joints, special reinforcement shall be placed at right angles to the joint and extending in both directions from the joint 40 diameters in the case of deformed bars and 50 diameters in the case of plain bars. This reinforcement shall be placed near the face of the member opposite from the main tensile reinforcement. The cross-sectional area of such reinforcement shall not be less than 0.5 percent of the section of the members cut by the joint.

4. Watertight Construction Joints

Construction joints shall not be made in watertight construction unless shown on the plans or authorized by the engineer.

Where a horizontal construction joint is required to resist water pressure, special care shall be taken in finishing the surface to which the succeeding concrete is to be bonded. The consistency of the concrete shall be carefully controlled so that it can be placed with a minimum of puddling with no free water showing. The surface shall be protected from loss of moisture and from mechanical injury. In applying the new concrete the procedure specified in Sec. L, Art. 9, shall be followed.

Where construction joints are required to be watertight, a continuous keyway shall be constructed in the face of the first section of concrete placed, and continuous sheet of non-corrosive metal not less than 12 in wide placed so as to extend the full length of the joint and be embedded equally in the concrete on each side thereof. Metal shall be deformed or perforated to insure bond on each side of the joint. Water stops of rubber or other material approved by the engineer may be used instead of metal. Before continuing with the placement of concrete, the joint shall be thoroughly cleaned of laitance or other foreign material and saturated with water. The concrete shall then be placed in such a manner as will insure an excess of mortar over the entire surface of the joint.

5. Procedure in Forming Control Joints[2]

The procedure specified in Chapter 6 for forming control joints in unit masonry buildings shall be followed.

Control joints for slabs-on-ground shall be made by one of the methods shown in Fig. 1 or as shown on the plans.

---

[1] Construction joints allow for no differential movement across joint. They are provided at locations where casting was temporarily suspended.

[2] Control joints allow for differential movement across the joint only in one direction, usually in the plane of the finished surface. They are provided to allow for drying shrinkage of the concrete.

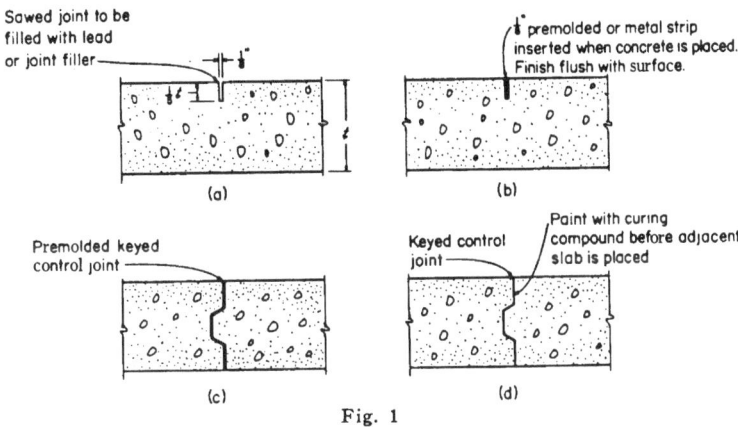

Fig. 1

Sawing of control joints shall be done as soon as the concrete hardens sufficiently to prevent raveling of the concrete edges. Sawing shall not be done when the concrete temperature is falling. Sawed control joints shall be cleaned and filled as shown on the plans or as specified by the engineer.

6. Steel at Control Joints

If welded wire fabric is used in the slab, alternate wires shall be cut where they cross the control joints.

If reinforcing bars are used in the slab, they shall be reduced by one-half at the predetermined control joint. Reinforcing bars shall not be lapped at control joints.

7. Procedure in Forming Isolation Joints[3]

Isolation joints will be installed around columns, walls, footings, etc., in the manner shown on the plans or as specified by the engineer.

They shall be formed with a premolded joint filler or similar material as shown on the plans or as specified by the engineer. They shall be made straight, and there shall be no connection across the joint by reinforcement, keyways or bond.

J. PROPORTIONING

1. General

The ingredients shall be thoroughly mixed and brought to a proper consistency. The proportions in which these materials are to be used for different parts of the work shall be determined by the engineer from time to time during the progress of the work as analyses and tests are made of samples of the aggregate and the resulting concrete. In general the proportions shall be designed to produce a concrete of maximum practical economy to the railroad company.

2. Measurement of Materials

In the measurement of cement, 94 lb, 1 bag, or ¼ bbl shall be assumed as 1 cu ft.

---

[3] Isolation joints allow for differential movement in all directions of the material on either side of the joint. They are sometimes called expansion joints.

Materials shall be measured by weighing, except as otherwise specified or where other methods are authorized specifically by the engineer. The apparatus provided for weighing the aggregates and cement shall be suitably designed and constructed for this purpose. The coarse aggregate, fine aggregate and cement shall be weighed separately. The accuracy of all weighing devices shall be such that successive quantities can be measured to within 1 percent of the desired amount. Cement in standard packages (bags) need not be weighed, but bulk cement and fractional packages shall be weighed. The mixing water shall be measured by volume or by weight. The water-measuring device shall be susceptible to control accurate to plus or minus ½ percent. All measuring devices shall be subject to approval by the engineer.

Where volumetric measurements are authorized by the engineer, the weight proportions shall be converted to equivalent volumetric proportions. In making this conversion, suitable allowance shall be made for variations in the moisture condition of the aggregates, including the bulking effect in the fine aggregate.

3. Water Cement Ratio

The proportioning of materials shall be based on the requirements for a plastic and workable mix suited to the conditions of placement containing not more than the specified amount of water, including the free water contained in the aggregate, per 94-lb bag of cement. The maximum specified amount of water shall not exceed the quantities shown in Table 1 for the type of structure and the condition of exposure to which it will be subjected. Moisture in the aggregate shall be measured by methods satisfactory to the engineer.

Free water content of the aggregates is included in the quantities specified and must be deducted from the amounts given in the table to determine the amount to be added at the mixer. Allowance may be made for absorption when aggregates are not saturated.

4. Air Content of Air-Entrained Concrete

The volume of entrained air in concrete shall be within the limits shown in the following table.

Maximum Size Coarse Aggregate Inches	Air Content % by Volume
1½, 2, or 2½	5 ± 1
¾, 1	6 ± 1
⅜, ½	7½ ± 1
¼, or less	9½ ± 1

The air content shall be determined by one of the following methods:
  a. The gravimetric method, ASTM C 138
  b. The volumetric method, ASTM C 173
  c. The pressure method, ASTM C 231

5. Strength of Concrete Mixtures

When preliminary tests of the materials to be used are not available, the required water-cement ratio shall be determined in accordance with Method 1. When strengths in excess of 4250 psi are required, or where lightweight aggregates or admixtures (other than those exclusively used for the purpose of entraining air) are to be used, the required water-cement ratio shall be determined in accordance with Method 2.

TABLE 1—MAXIMUM PERMISSIBLE WATER-CEMENT RATIOS (GAL PER SACK) FOR DIFFERENT TYPES OF STRUCTURES AND DEGREES OF EXPOSURE

Type of structure	Exposure Conditions*					
	Severe wider range in temperature or frequent alternations of freezing and thawing (air-entrained concrete only)			Mild temperature rarely below freezing, or rainy, or arid		
	In air	At the water line or within the range of fluctuating water level or spray		In air	At the water line or within the range of fluctuating water level or spray	
		In fresh water	In sea water or in contact with sulfates**		In fresh water	In sea water or in contact with sulfates**
Thin sections, such as railings, curbs, sills, ledges, ornamental or architectural concrete, reinforced piles, and pipe	5.5	5.0	4.5†	6	5.5	4 5†
Moderate sections, such as retaining walls, abutments, piers, girders, beams	6.0	5.5	5.0†	††	6.0	5.0†
Exterior portions of heavy (mass) sections	6.5	5.5	5.0†	††	6.0	5.0†
Concrete deposited by tremie under . . . water		5.0	5.0		5.0	5.0
Concrete slabs laid on the ground . . .	6.0			††		
Concrete protected from the weather, interiors of buildings, concrete below ground	††			††		
Concrete which will later be protected by enclosure of backfill but which may be exposed to freezing and thawing for several years before such protection is offered	6.0			††		

*Air-entrained concrete should be used under all conditions involving severe exposure and may be used under mild exposure conditions to improve workability of the mixture.
**Soil or ground water containing sulfate concentrations of more than 0.2 percent.
†When sulfate resisting cement is used, maximum water-cement ratio may be increased by 0.5 gal per bag.
††Water-cement ratio should be selected on basis of strength requirements.

Method 1—Without Preliminary Tests

Concrete made with Type I or Type II portland cement with the quantities of water as specified in Sec. J, Art. 3, may be assumed to have the following minimum compressive strength at the age of 28 days:

Total Water U.S. Gal per 94-Lb Bag of Cement	Assumed Minimum Compressive Strength at the Age of 28 Days PSI
7½	2500
7	2750
6½	3000
6	3300
5½	3700
5	4250

The above values are based on the use of cement and aggregates meeting the requirements of these specifications and the concrete being sufficiently protected from loss of moisture and from low temperatures to insure that proper hardening will develop. When Type III portland cement is used in lieu of Type I or Type II portland cement, it may be assumed that the above values for compressive strength will be obtained at the age of 7 days. The above strengths may be somewhat lower when air-entrained concrete is used.

The strength of cylinders made with Types I, IA, II or IIA portland cement and tested at the age of 7 days should not fall below 65 percent of the assumed compressive strength at the age of 28 days. The strength of cylinders made with Types III or IIIA portland cement and tested at the age of 3 days should not fall below 65 percent of the assumed minimum compressive strength at the age of 28 days shown for Types I, IA, II and IIA portland cement.

Method 2—With Preliminiary Tests

The strength of the concrete shall be established by test made with representative samples of the materials to be used in the work. The results of the tests shall be submitted to the engineer in advance of the beginning of operations. These tests shall be made using the consistencies suitable for the work and in accordance with ASTM Method of Making and Curing Concrete Compression and Flexure Test Specimens in the Laboratory, C 192, and with Method of Test for Compressive Strength of Molded Concrete Cylinders, C 39. A curve representing the relation between the water content and the average 28-day compressive strength or earlier strength at which the concrete is to receive its full working load shall be established for a range of values including all the compressive strengths called for by the plans or specifications. The curve shall be established by at least three points, each point representing average values from at least four test specimens. The maximum permissible water-cement ratio for the concrete to be used shall be that shown by the curve to produce a strength 15 percent greater than called for by the plans or specifications. If any changes are to be made in the materials, new curves shall be established by tests as described above.

6. Workability

The concrete shall be of such consistency and composition that it can be worked readily into the corners and angles of the forms and around the reinforcement without the segregation of materials or the collection of free water on the surface. Subject to the limiting requirements of Sec. J, Art. 3, the contractor shall, if the engineer requires, adjust the proportions of cement and aggregates so as to produce a mixture which will be easily placeable at all times, due consideration being given to the methods of placing and compacting used on the work.

7. Slump

The slump test may be used as a control measure to maintain the consistency suitable to the work. When mechanical vibrators are used to compact the concrete, the consistency suitable to that method shall be used. The slump test shall be made in accordance with the ASTM Method of Test, C 143. In lieu of slump test, the ball penetration Method of Test, ASTM C 360, may be used.

8. Compression Tests

Specimens for compression tests shall be made and stored in accordance with ASTM Specification C 31. These specimens shall be tested in accordance with ASTM Method of Test C 39.

9. Field Tests

During the progress of construction the engineer will have tests made to determine whether the concrete which is being produced compares to the quality specified in Art. 5 or as specified by the plans or specifications. The contractor shall cooperate in the making of such tests to the extent of allowing free access to the work for the selection of samples and storage of specimens and in affording protection to the specimens against injury or loss through his operations.

Four cylinders will generally be made for each class of concrete used in any one day's operation. In special cases this normal number of control specimens may be exceeded when in the opinion of the engineer such additional tests are necessary. The contractor, however, shall not be required to furnish for such additional tests more than 2 cu ft of concrete for each 100 cu yd of concrete being placed.

Samples of concrete for test specimens shall be taken at the mixer, or in the case of ready-mixed concrete, from the transportation vehicle during discharge. When, in the opinion of the engineer, it is desirable to take samples elsewhere, they shall be taken as directed by him. Such specimens shall be molded immediately after the sample is taken, placed in a protected spot and kept under moist curing conditions at approximately 70 deg F for 24 hr, whereupon they shall be removed to the testing laboratory.

The air content of freshly mixed air-entrained concrete shall be checked at least twice daily for each class of concrete. Changes in air content above or below the amount specified shall be corrected by adjustments in the mix design or quantities of air-entraining material being used.

Should the strengths shown by the test specimens fall below the values given in Art. 5 or as specified by the plans or specifications, the engineer shall have the right to require changes in proportions to apply on the remainder of the work.

K. MIXING

1. Equipment

The mixing equipment shall be capable of combining the aggregates, cement, and water within the specified time into a thoroughly mixed and uniform mass, and of discharging the mixture without sugregation.

2. Machine Mixing (at Site or at Central Mixing Plant)

Unless otherwise authorized by the engineer, the concrete shall be mixed in a batch mixer of approved type and size which will insure a uniform distribution of the material throughout the mass. The equipment at the mixing plant shall be so constructed that all materials (including the water) entering the drum can be accurately measured and weighed and be under control. The entire batch shall be discharged from the mixer before recharging. The volume of the mixed material per batch shall not exceed the manufacturer's rated capacity of the mixer. Except as qualified in Sec. K, Art. 4, mixing of each batch shall continue for the periods indicated below*, during which time the drum shall rotate at a peripheral speed of about 200 ft per min. The mixing periods shall be measured from the time when all of the solid materials are in the mixer drum, provided that all of the mixing water shall have been introduced before one-fourth of the mixing time has elapsed.

Minimum mixing time shall be as follows:

a. For mixers of a capacity of 1 cu yd or less—$1\frac{1}{2}$ min.

b. For mixers of capacities greater than 1 cu yd, the time of mixing shall be increased 15 sec, for each cubic yard capacity or fraction thereof.

3. Truck Mixing

Truck mixers shall be watertight and so constructed that the concrete can be mixed to insure a uniform distribution of materials throughout the mass and, unless otherwise authorized by the engineer, shall be of the revolving drum type. All solid materials for the concrete shall be accurately measured in accordance with Sec. J, Art. 2, and charged into the drum at the proportioning plant. The truck mixer shall be equipped with a tank for carrying mixing water. The tank shall have a device by which quantity of water added to the concrete can be positively controlled and readily verified. No water shall be admitted to the drum until the mixer is at the site of the work unless authorized by the engineer. If, with the approval of the engineer, the water is added and the batch is mixed in transit, the mixer shall be equipped with a device by which the engineer can determine the amount of mixing that has been done on arrival of the truck at the site of the work. The maximum size of batch shall be in accordance with the specified rating of the mixer. The mixing shall be continued for not less than 50 revolutions after all ingredients, including water, are in the drum. The speed shall be not less than 4 rpm, nor more than the speed resulting in a peripheral velocity of the drum of 225 ft per min. Not more than 150 revolutions of mixing shall be at a speed in excess of 6 rpm. Mixing shall begin within 30 min after the cement has been added either to the water or the aggregate. In case the required amount of mixing has been completed before the time of discharging the batch, mixing shall be resumed briefly to agitate the batch just before discharging. Mixers shall be dispatched from the proportioning plant so as to provide a uniform rate of delivery. Each batch of concrete delivered at the job shall be accompanied by a time slip issued at the batching plant, bearing the time of departure therefrom and the signature of the inspector.

4. Partial Mixing at the Central Plant

When a truck mixer, or an agitator provided with adequate mixing blades, is used for transportation, the mixing time at the stationary machine mixer may be reduced to 30 sec and the mixing completed in a truck mixer or agitator. The mixing time in the

---

* When, in the judgment of the engineer, longer mixing times are necessary he should so specify.

truck mixer or agitator equipped with adequate mixing blades shall be as specified for truck mixing in Sec. K, Art. 3. The maximum volume of mixed concrete transported in an agitator shall be in accordance with the specified rating.

5. Time of Hauling Ready-Mixed Concrete

Concrete transported in truck mixers, agitators or other transportation devices shall be delivered to the site of the work in a plastic and workable condition, satisfactory for the placement in the work without the addition of water. Delivery and discharge shall be made within $1\frac{1}{2}$ hr after the cement has been added to the water or the aggregates. In hot weather or under conditions contributing to quick stiffening of the concrete, a lesser time may be specified by the engineer.

L. DEPOSITING CONCRETE IN AIR

1. General

Before beginning placement of concrete, hardened concrete and foreign materials shall be removed from the inner surfaces of the mixing and conveying equipment. Before depositing any concrete all debris shall be removed from the space to be occupied by the concrete, and mortar splashed upon the reinforcement and surfaces of forms shall be removed. Reinforcement shall be checked for position and fastening and approval of the engineer obtained. Where concrete is to be placed on a rock foundation, all loose rock, clay, mud, etc., shall be removed from the surface of the rock. Any unusual conditions or excess fissures shall be treated as directed by the engineer. Water shall be removed from the space to be occupied by the concrete before concrete is deposited, unless otherwise directed by the engineer. Any flow of water into an excavation shall be diverted through proper side drains to a sump, or be removed by other approved methods which will avoid washing the freshly deposited concrete. If directed by the engineer water ventpipes and drains shall be filled by grouting or otherwise after the concrete has thoroughly hardened. All temporary runways for delivery of concrete must be supported free from all reinforcing steel.

2. Handling and Placing

Concrete shall be handled from the mixer, or in case of ready-mixed concrete, from the transporting vehicle, to the place of final deposit as rapidly as practicable by methods which will prevent the separation or loss of the ingredients. Special care shall be taken to fill each part of the forms by depositing concrete as near final position as possible, to work the coarser aggregates back from the face and to force the concrete under and around the reinforcement without displacing it. Concrete shall not have a free fall of more than 4 ft unless permitted by the engineer. Depositing a large quantity at any point and working it to final position, will not be permitted.

Concrete shall be placed in horizontal layers and each layer shall be placed and compacted before the preceding layer has taken initial set so as to prevent formation of a joint. It shall be so deposited as to maintain, until the completion of the unit, a plastic surface approximately horizontal, except in arch rings. Temporary struts or braces within the form shall be removed when concrete has reached an elevation rendering their further service unnecessary. These temporary members shall be entirely removed from the forms and not buried in the concrete. After the concrete has taken its initial set, care shall be exercised to avoid jarring the forms or placing any strain on the ends of the projecting reinforcement. Under no circumstances shall concrete that has partially hardened be deposited in the work.

In placing concrete for an arch ring, the work shall be carried on symmetrically with respect to the center line, and the working faces of the completed courses shall be on approximately radial planes. This requirement applies whether or not the arch is placed in voussoir sections with allowance for key sections for final placement.

In order to allow for shrinkage or settlement, at least 2 hr shall elapse after placing concrete in walls, columns or stems of deep T-beams before depositing concrete in girders, beams or slabs supported thereon, unless otherwise specified or shown on the plans. If the columns are structural steel encased in concrete, the lapse of time to allow for shrinkage or settlement need not be observed.

Concrete in girders, slabs and shallow T-beam construction shall be placed in one continuous operation for each span, unless otherwise provided. Concrete shall be deposited uinformly for the full length of the span and brought up evenly in horizontal layers.

Concrete in columns shall be placed in one continuous operation and allowed to set at least 2 hr before the caps are placed.

No concrete shall be placed in the superstructure until the column forms have been stripped sufficiently to determine the character of the concrete in the columns, and the load of the superstructure shall not be allowed to come upon abutments, piers and column bents until they have been in place at least 7 days, unless otherwise permitted by the engineer.

3. Chuting

When concrete is conveyed by chuting, the plan shall be of such size and design as to insure a practically continuous flow in the chute. The chutes shall be of metal or metal lined. The angle of the chute with the horizontal and the shape of the chute shall be such as to allow the concrete to slide without separation of the ingredients. The delivery end of the chute shall be as close as possible to the point of deposit. When the operation is intermittent, the chute shall discharge into a hopper. The chute shall be thoroughly flushed with water before and after each run; the water used for this purpose shall be discharged outside the forms. Chutes must be properly baffled or hooded at the discharging end to prevent separation of the aggregates.

4. Pneumatic Placing

Pneumatic placing of concrete shall be done with a concrete placing machine which forces the concrete through a pipeline by means of compressed air. The equipment shall be suitable in kind and adequate in capacity for the work. There shall be available a working pressure of 120 psi. The air pressure and the consistency of the concrete shall be so regulated, without varying the water-cement ratio, that the coarse aggregate shall not be blown away from the mortar matrix when the concrete issues from the end of the delivery pipe. The machine shall be located as close as practicable to the place of deposit. The size and design of the pipeline and the cleaning of it when operation ceases shall be as specified in Sec. L, Art. 5. Flexible pipe branches shall be provided when necessary to deliver the concrete at approximately its final position in the forms. The position of the nozzle shall be such that the concrete is directed to the point of deposit without impinging directly on the reinforcement and without excessive rebound of material. The nozzle of the delivery pipe shall be as close to the mass of concrete as conditions will permit, and in no case shall it be more than 5 ft away. On important work duplicate pneumatic equipment and additional pipe shall be provided to prevent delay due to breakdown of equipment.

5. Pumping Concrete

The pump and all appurtenances shall be so designed and arranged that the specified concrete can be transported and placed in the forms without segregation. The pump shall be capable of developing a working pressure of at least 300 psi, and the pipeline and fittings shall be designed to withstand twice the working pressure.

Where it is necessary to lay the pipe on a down grade, a reducer shall be placed at the discharge end of the pipe to provide a choke and thus produce a continuous flow of concrete. When the type of pump is such that it discharges the concrete in small batches, or by "belching", a baffle box shall be provided into which the concrete shall be discharged. This box should preferably be of metal, about 2 ft square, with open sides so as to permit the concrete to flow into the forms at right angles to line of discharge. The pipe shall be not less than 6 in nor more than 8 in outside diameter, and the line shall be laid with as few bends as possible. When changes in direction are necessary they should be made with bends of 45 deg or less, unless greater bends are specifically permitted. If greater bends are permitted in special cases, they shall be long-radius bends. The maximum distance of delivery of concrete by pumping shall be 1000 ft horizontally and 100 ft vertically, unless otherwise specifically permitted by the engineer. (A 90-deg bend is figured as equivalent to 40 ft of horizontal piping. A 45-deg bend is equivalent to 20 ft. A $22\frac{1}{2}$-deg bend is equivalent to 10 ft.) When pumping is completed, the concrete remaining in the pipe line, if it is to be used, shall be ejected in such a manner that there will be no contamination of the concrete or separation of the ingredients. The pipeline and equipment must then be thoroughly cleaned. The pipeline can be cleaned by either water or air. If water is used, a pump shall be provided with a capacity of at least 80 gpm and capable of developing a pressure of 400 psi. Cleaning of the pipe can also be accomplished by the use of a "go-devil" which is propelled through the line by water or air pressure. (The "go-devil" is a dumb-bell shaped piece with a rubber cup on each end. The cups are turned toward the liquid, or air, and the seal is the same as in a simple plunger pump.) If water is used, it must be discharged outside of the forms. On important work duplicate pumping equipment and additional pipe shall be provided to prevent delay due to breakdown of equipment.

6. Compacting

Concrete shall be thoroughly compacted during and immediately after depositing by vibrating the concrete internally by means of mechanical vibrating equipment, unless otherwise directed by the engineer.

Internal mechanical vibrators shall be of a type approved by the engineer. They shall be of sturdy construction, adequately powered, capable of transmitting vibration to the concrete in frequencies of not less than 3500 impulses per minute and shall produce a vibration of sufficient intensity to consolidate the concrete into place without a separation of the ingredients.

The vibratory element shall be inserted into the concrete at the point of deposit and in the areas of freshly placed concrete. The time of vibration shall be of sufficient duration to accomplish thorough consolidation, complete embedment of the reinforcement, the production of smooth surfaces free from honeycomb and air bubbles, and to work the concrete into all angles and corners of the forms. However, over-vibration shall be avoided, and vibration shall continue in a spot only until the concrete has become uniformly plastic and shall not continue to the extent that pools of grout are formed. The length of time of vibration depends upon the frequency of the vibration (impulses per minute), size of vibrators and the slump of the concrete. This length of time must be determined in the field.

The internal vibrators shall be applied at points uniformly spaced, not farther apart than the radius over which the vibration is visibly effective, and shall be applied close enough to the forms effectively to vibrate the surface concrete. The vibration shall not be dissipated in lateral motion but shall be concentrated in vertical settlement in consolidation of the concrete.

The vibrator shall not be used to push or distribute the concrete laterally. The vibrating element shall be inserted in the concrete mass a sufficient depth to vibrate the bottom of each layer effectively, in as nearly a vertical position as practicable. It shall be withdrawn completely from the concrete before being advanced to the next point of application.

To secure even and dense surfaces, free from aggregate pockets or honeycomb, vibration shall be supplemented by working or spading by hand in the corners and angles of forms and along form surfaces while the concrete is plastic under the vibratory action.

A sufficient number of vibrators shall be employed so that, at the required rate of placement, thorough consolidation is secured throughout the entire volume of each layer of concrete. Extra vibrators shall be on hand for emergency use and for use when other vibrators are being serviced.

The use of surface vibrators to supplement internal vibration will be permitted when satisfactory surfaces cannot be obtained by the internal vibrations alone and when the contractor has obtained the approval of the engineer of the equipment to be used. Surface vibrators shall be applied only long enough to embed the coarse aggregate and to bring enough mortar to the surface for satisfactory finishing.

The use of approved form vibrators will be permitted by the engineer only when it is impossible to use internal vibrators. They shall be attached to or held on the forms in such a manner as to effectively transmit the vibration to the concrete and so that the principal path of motion of the vibration is in a horizontal plane.

7. Temperature

Concrete when deposited shall have temperatures within the limits shown in the following table:

Temperatures of Air Degrees F	Temperature of Concrete When Placed Degrees F	
Below 30	70 Minimum	90 Maximum
Between 30 and 45	60 Minimum	90 Maximum
Above 45	50 Minimum	90 Maximum

The method of controlling the temperature of the concrete shall be approved by the engineer.

8. Continuous Depositing

Concrete shall be deposited continuously and as rapidly as practicable until the unit of operation, approved by the engineer, is completed. Construction joints in addition to those provided on the plans will not be allowed unless authorized by the engineer. If so authorized, they shall be made in accordance with Sec. I, Art. 1.

9. Bonding

Immediately before depositing new concrete on or against concrete which has hardened, the forms shall be tightened, the surface of the hardened concrete shall be roughened as required by the engineer, thoroughly cleaned of foreign matter and laitance, and saturated with water. Excess water shall be removed so that the surface is wet but free from pools of water. The new concrete placed in contact with hardened or partially hardened concrete shall contain an excess of mortar to insure bond. To insure this excess mortar at the juncture of the hardened and the newly deposited concrete, the cleaned and saturated surfaces of the hardened concrete, including vertical and inclined surfaces, shall first be slushed with a coating of neat cement grout against which the new concrete shall be placed before the grout has attained its initial set.

10. Placing Cyclopean Concrete

Cyclopean aggregate shall be thoroughly embedded in the concrete. The individual stones shall not be closer than 12 in to any surface or adjacent stones. Stratified stone shall be laid on its natural bed. Cyclopean aggregate shall be carefully placed to avoid injury to forms or adjoining masonry.

11. Placing Rubble Concrete

Rubble aggregate shall be thoroughly embedded in the concrete. The individual stones shall not be closer than 4 in to any surface or adjacent stones. Rubble aggregate shall be carefully placed to avoid injury to forms or adjacent masonry.

12. Water Gain

Water gain is characterized by an accumulation of water at the surface. Whenever water gain appears in the concrete poured, the succeeding batches must be placed sufficiently dry to correct the over-wet condition by the reduction of the amount of water per bag of cement, without changing the proportions of the other ingredients.

M. DEPOSITING CONCRETE UNDER WATER

1. General

The methods specified for Depositing Concrete in Air shall be used except when the space to be filled with concrete contains water which cannot be removed in some practical way. In such cases, and when authorized by the engineer, concrete shall be deposited under water in accordance with the following.

The methods, equipment and materials proposed to be used, shall be submitted first to the engineer for his approval before the work is started. The methods used shall be such as will prevent the washing out of the cement from the concrete mixture, minimize the segregation of materials and the formation of laitance, and prevent the flow of water through or over the new concrete until it has fully hardened. Concrete shall not be placed in water having a temperature below 35 deg F.

2. Capacity of Plant

Sufficient mixing, transporting and placing equipment shall be provided to insure that the depositing of all underwater concrete for each predetermined section or unit of the work to be done, shall be continuous until completion.

3. Standard Specifications

The materials, proportions and methods to be used in making concrete to be deposited under water shall all conform to the requirements of these specifications except as modified or supplemented by the following paragraphs.

4. Cement

Not less than 6 bags of cement per cubic yard of concrete shall be used.

5. Coarse Aggregates

Aggregate for this work shall be of exceptionally good quality, strong and durable. The maximum size of aggregate preferably shall be 2 in and shall not exceed 3 in. The coarse aggregate shall be well graded in such proportions that the weight of the coarse aggregate shall be not less than $1\frac{1}{4}$ nor more than twice that of the fine aggregate.

6. Mixing

The cement and aggregates shall be mixed for a period of 2 min with sufficient water to produce a concrete having a slump of not less than 6 in nor more than 8 in for concrete placed by tremies, and not less than 3 in nor more than 6 in for concrete placed by bottom dump buckets or for concrete placed in sacks.

7. Caissons, Cofferdams or Forms

Caissons, cofferdams or forms shall be sufficiently tight to prevent loss of mortar or flow of water through the space in which the concrete is to be deposited. Pumping will not be permitted while concrete is being deposited, nor until 24 hr thereafter.

8. Leveling and Cleaning the Bottom to Receive Concrete

Before starting to deposit concrete under water, the condition of the bottom shall be examined and reported upon to the engineer by a competent diver, and shall be approved by the engineer.

The surface of the bottom, whether of clay, rock, or other material, shall be leveled as directed by the engineer, before depositing concrete under water.

Where the bottom on which concrete is to be deposited under water is, or is likely to be, covered with silt, such material shall be removed down to solid material before any concrete is placed. The method to be used to clean the bottom of silt or similar material, shall be subject to the approval of the engineer.

9. Continuous Work

Concrete shall be deposited continuously until it is brought up to the required elevation. While depositing, the top surface shall be kept as nearly level as possible, and the formation of laitance planes avoided.

10. Methods of Depositing

One of the following methods of depositing shall be used:

a. *Tremie.* When concrete is to be deposited under water by means of a tremie, the top section of the tremie shall be a hopper large enough to hold one entire batch of the mix or the entire contents of the transporting bucket, when one is used. The tremie pipe shall be not less than 8 in. in diameter and shall be large enough to allow a free flow of concrete and strong enough to withstand the external pressure of the water in which it is suspended, even if a partial vacuum develops inside of the pipe. Preferably, flanged steel pipe should be used, of adequate strength to sustain the greatest length and weight required for the job. A separate lifting device shall be provided for each tremie pipe with its hopper at the upper end. Unless the lower end of the pipe is equipped with an approved automatic check valve, the upper end of the pipe shall be plugged with a wadding of gunnysacking or other approved material, before delivering the concrete to the tremie pipe through the hopper, which plug will be forced to and out of the bottom end of the pipe by filling the pipe with concrete. It will be necessary slowly to raise the tremie in order to cause a uniform flow of the concrete, but the tremie shall not be emptied so that water enters above the concrete in the pipe. At all times after the start of placing the concrete and until all concrete is placed, the lower end of the tremie pipe shall be below the top surface of the plastic concrete. This will cause the concrete to build up from below instead of flowing out over the surface, to avoid formation of laitance layers. If the charge in the tremie is lost while depositing, the tremie shall be raised above the concrete surface, and unless sealed by a check valve it shall be replugged at the top end, as at the beginning, before refilling for depositing concrete.

(Note—Experience has shown that tremie concrete can be placed as above specified, so that it will build up the flow as much as 50 ft horizontally from the discharge end of the tremie with a slope of less than 3 ft in 50 ft.)

b. *Bottom Dump Bucket.* Where concrete is to be deposited under water by means of a bottom dump bucket, the bucket shall be of the type that cannot be dumped until after it has rested, with its load, on the surface upon which the concrete is to be deposited. The bottom doors shall be so equipped as to be automatically unlatched by the release of tension on the supporting line or cable of the bucket, and the bottom doors shall then open downward and outward as the bucket is raised. The top of the bucket shall be fitted with double, overlapping canvas flaps, or other approved covers, to cover the contained concrete and to protect it from wash when it enters the water and as the bucket descends to the bottom. The bucket, preferably, should be so designed that the hinged bottom doors will operate inside of a steel skirt, which skirt will surround the bucket while the bottom doors are shut and will extend below the bucket as the bottom doors open and hence minimize turbulence and motion while the concrete is being deposited. The bucket shall be submerged slowly until it is completely under water. The normal line speed after that shall not exceed 200 ft per min. After the bucket has reached the surface on which the concrete is to be deposited, it shall be raised slowly for the first 6 or 8 ft while the concrete is being deposited.

c. *Placing Sacks of Concrete.* Where a relatively small amount of concrete is to be placed that does not warrant the equipment required for either tremie or open-bottom bucket methods, concrete may be placed under water in sacks or bags. In such case the space shall be filled with sacks of concrete carefully placed by hand in header and stretcher formation, so that the whole mass becomes interlocked. Sacks used for this purpose shall be made of jute or other coarse material, free from deleterious materials, and shall be filled about two-thirds full of concrete and the sack openings securely tied.

d. *Grouted Aggregate.* Installed by placing coarse aggregate in the forms, then injecting cement grout through pipes which extend to the bottom of the forms. The pipes are withdrawn as grouting proceeds. The grout forces the water from the forms and fills interstices in the aggregate.

The coarse aggregate should be placed in horizontal layers of such maximum thickness as will provide a dense fill without segregation and should be well compacted.

The grout mixture should be applied under such pressure and at such consistency as will insure complete filling of voids; and grout pipes should be properly spaced to be consistent with this requirement.

Mineral fillers and admixtures may be added to the grout mixture if approved by the engineer.

The grout mixture required for this class of work necessitates the use of special mixers and agitators to deliver suitable grout in place. This equipment and all grout lines should be maintained in good operating condition. After every shift or work stoppage, they should be cleaned of all grout.

11. Soundings

During the time that concrete is being deposited under water, soundings shall be continuously taken to the surface of the deposited concrete and recorded. The surface of the deposited concrete shall be maintained relatively level over the area being covered.

12. Removing Laitance

Upon completing a unit or section of under-water concrete, any laitance or silt collecting on the upper surface of the same shall be removed and the concrete surface thoroughly cleaned, if additional concrete is to be deposited on that surface.

13. Concrete Seals

Under favorable conditions it will be possible to place under-water concrete of a limited thickness in the bottoms of caissons or cofferdams and so completely seal the structures that after the concrete has set, all water can be pumped out. In such cases, if it is economical to do so, the water shall be pumped out, the exposed surfaces cleaned and the balance of the concrete deposited in air.

N. CONCRETE IN SEA WATER

1. Concrete

Concrete in, or exposed to, sea water shall be air-entrained in accordance with Sec. J, Art. 4, and shall be made with Type II or IIA cement. Concrete in sea water from 2 ft below low water to 2 ft above high water, or from a plane below to a plane above wave action, shall contain a minimum of 7 bags of portland cement per cubic yard in place. Other concrete in sea water or exposed directly along the sea coast shall contain a minimum of 6 bags of portland cement per cubic yard in place. The net amount of mixing water used shall not exceed the quantities shown in Table 1, Sec. J. Porous or weak aggregates shall not be used.

2. Depositing in Sea Water

Sea water shall not be allowed to come in contact with the concrete until it has hardened for at least 4 days. Concrete may be deposited in sea water only when so approved by the engineer.

3. Construction Joints

Concrete shall be placed in such a manner as to minimize the number of construction joints, and all construction joints shall be made as described in Sec. I, Art. 4, and Sec. L, Art. 9.

4. Cover on Reinforcement

Reinforcing steel or other corrodible metal shall be placed not less than 3 in from any plane or curved surface, and at corners shall be not less than 4 in from adjacent surfaces.

5. Protecting Concrete in Sea Water

Where severe climatic conditions or severe abrasions are anticipated, the face of the concrete from 2 ft below low water to 2 ft above high water, or from a plane below to a plane above wave action, shall be protected by stone of suitable quality, dense vitrified shale brick as designated on the plans or as required by the engineer, or in special cases the protection may be creosoted timber.

O. CONCRETE IN ALKALI SOILS OR WATERS

1. Condition of Exposure

In territory where sulfate-bearing soil or sulfate-bearing water are known to occur, concrete of one of the two following proportions of water and cement shall be used, depending upon the severity of conditions. Severity of conditions may be judged by the extent of deterioration which has occurred to concrete previously used in the immediate vicinity or from the sulfate concentrations found in either the soil or the water.

If existing concrete has deteriorated slowly during a period of several years, or if the concentrations of water-soluble sulfates in the soil are 0.1 to 0.3 percent, or if the sulfates in the water are 150 to 3000 ppm, the conditions are considered moderately severe.

If deterioration of existing concrete has occurred rapidly during a period of only a few years, or if the sulfates found in either the soil or the water exceed the values given above, the conditions are considered severe.

2. Concrete for Moderate Exposure

Concrete for moderately severe sulfate exposure shall be made using either a Type II, Type IIA, or a Type V portland cement with not less than 6½ bags per cu yd of concrete in place and not more than 5½ gal of water per bag of cement. The concrete shall be air-entrained in accordance with Sec. J, Art. 4.

3. Concrete for Severe Exposure

Concrete for severe sulfate exposure shall be made using a Type V portland cement with not less than 7 bags per cu yd of concrete in place and not more than 5 gal of water per bag of cement. The concrete shall be air-entrained in accordance with Sec. J, Art. 4.

Note—Type V cement is not regularly stocked by most cement mills. It can usually be obtained from mills which are manufacturing this type of cement on order for the U. S. Bureau of Reclamation. If Type V cement is not obtainable, a Type II or Type IIA cement may be selected which has the lowest calculated tricalcium aluminate content of the brands available.

4. Construction Joints

Concrete shall be placed in such a manner as to minimize the number of construction joints, and all construction joints shall be made as described in Sec. J, Art. 4, and Sec. L, Art. 9.

5. Cover on Reinforcement

Reinforcing steel or other corrodible metal shall be placed not less than 3 in from any plane or curved surface, and at corners shall be not less than 4 in from adjacent surfaces.

P. CURING

1. Cold Weather Curing

In freezing weather, or when there is likelihood of freezing temperatures within the specified curing period, suitable and sufficient means must be provided before concreting, for maintaining all concrete surfaces at a temperature of not less than 50 deg F for a period of not less than 7 days after the concrete is placed when Type I, IA, II or IIA portland cement is used, and not less than 3 days when Type III or IIIA portland cement is used.

The temperature of concrete surfaces shall be determined by thermometers placed against the surface of the concrete. Provision shall be made in form construction to permit the removal of small sections of forms to accommodate the placing of thermometers against concrete surfaces at locations designated by the engineer. After thermometers are placed, the apertures in forms shall be covered in a way to simulate closely the protection afforded by the forms.

In determining the temperatures at angles and corners of a structure, thermometers should be placed not more than 8 in from the angles and corners. In determining temperatures of horizontal surfaces, thermometers shall rest upon the surface under the protection covering normal to section involved.

Temperature readings shall be taken and recorded at intervals to be designated by the engineer, over the entire curing period specified, and the temperatures so recorded shall be interpreted as the temperature of the concrete surfaces where the thermometers were placed.

When protection from cold is needed to insure meeting these specification requirements, all necessary materials for covering or housing must be delivered at the site of the work before concreting is started and must be effectively applied or installed, and such added heat must be furnished as may be necessary without depending in any way upon the heat of hydration during the first 24 hr after concrete is placed when Type I, IA, II or IIA portland cement is used, or the first 18 hr when Type III or IIIA portland cement is used. The methods of heating and protecting the concrete shall be approved by the engineer. Chemicals or other foreign materials shall not be mixed with the concrete for the purpose of preventing freezing, unless approved by the engineer.

When heat is supplied by steam or salamanders, covering or housing of the structure shall be so placed as to permit free circulation of air above and around the concrete within the enclosure, but to the exclusion of air currents from without, excepting that where salamanders are used, sufficient ventilation shall be provided to carry off gases. Special care shall be exercised to maintain the specified temperature continuously and uniformly in all parts of the structure enclosure, and to exclude cold drafts from angles and corners and from all projecting reinforcing steel. All exposed surfaces in the heated enclosure shall be kept continuously wet during the heating period unless heat is supplied in the form of live steam.

2. Hot Weather Curing

The temperature of concrete at time of placement should not exceed 90 deg F. When the temperature of the concrete approaches 90 deg F, special efforts to prevent too rapid drying out must be made.

Continuous wet curing is preferred and should commence as soon as the concrete has hardened sufficiently to resist surface damage. Wet curing should be carried out in accordance with the practice recommended under Sec. P, Art. 4. Curing water should not be much cooler than the concrete to avoid temperature-change stresses resulting in cracking. Exposed, unformed concrete surfaces should be protected from wind and direct sun.

3. Wet Curing

All concrete surfaces when not protected by forms, or membrane curing compounds, must be kept constantly wet for a period of not less than 7 days after concrete is placed when Type I, IA, II or IIA portland cement is used, or not less than 3 days when Type III or IIIA portland cement is used.

The wet curing period for all concrete which will be in contact with brine drip, sea water, salt spray, alkali or sulfate-bearing soils or waters, or similar destructive agents, shall be increased to 50 percent more than the periods specified for normal exposures. Salt water and corrosive waters and soils shall be kept from contact with the concrete during placement and for the curing period.

When wood forms are left in place during the curing period they shall be kept sufficiently damp at all times to prevent openings at the joints and drying of the concrete.

4. Membrane Curing

Membrane curing, or membrane curing following preliminary wet curing, may be substituted for wet curing with approval of the engineer.

Liquid membrane-forming compounds shall meet the requirements of ASTM Specification C 309. The compound may be clear (may contain a fugitive dye), white pigmented, or light-gray pigmented as specified by the engineer.

The compound shall be applied at the coverage rate of 150 or 200 sq ft per gal to all exposed concrete surfaces except areas where concrete or other material is to be bonded, such as construction joints and areas to be dampproofed or waterproofed.

The compound shall be sprayed on finished surfaces as soon as the surface water has disaappeared. Spraying equipment shall be of the pressure-tank type with provision for continual agitation of the contents during application. The compound shall be applied in one coat by spraying one-half of a given area in one direction and the remaining one-half at right angles to this direction. If forms are removed during curing period, concrete shall be sprayed lightly with water and the moistening continued until the surface will not readily absorb more water. The compound shall then be sprayed or brushed on the concrete surface as soon as the moisture film has disappeared.

The continuity of the coating must be until the fourteenth day after the concrete has been placed when Type I, IA, II or IIA portland cement is used, or until the seventh day when Type III or IIIA portland cement is used.

5. Steam Curing

For specifications see Part 17, Sec. U, Art. 5 (b) this Chapter.

Q. FORMED SURFACE FINISH

1. General

The following requirements, in addition to the provisions hereof applying to forms, mixing, conveying, depositing, etc., except as modified by the plans or by the direction of the engineer, shall be applicable to the construction of concrete surfaces exposed upon the completion of the structure:

a. All face forms shall be smooth and watertight. If of wood, the face boards shall be sized to a uniform thickness and all offsets or inequalities dressed to a smooth surface. They shall be tightly placed and all openings and cracks pointed flush, as directed by the engineer, to prevent leakage and the formation of fins. Forms shall be so constructed that they can be removed without hammering or prying against the concrete surface.

b. Exposed surface shall be cast in one continuous operation between prescribed construction limits. Joints not shown on the plans shall be made only as directed by the engineer and shall be true to line with sharp unbroken edges, beveled or rounded as specified.

c. The concrete shall be so mixed, placed and compacted that the aggregate is uniformly distributed and a full surface of mortar brought against the form free from air pockets and void spaces.

d. The forms shall be carefully removed, and any fins or projections neatly removed as directed by the engineer. If there should be found any small pits or openings in the exposed surface of the concrete or if bolts are used for securing the forms, the ends of which on removal leave small holes, the surface shall be thoroughly saturated with water and all such holes, pits, etc., shall be neatly stopped with pointing mortar of cement and fine aggregates in the same proportion as used in the concrete, and smoothed even with the surface with a wooden float. The mortar shall be mixed in small quantities and shall be used while still plastic.

All such work in connection with the correction of damaged sections, voids or honeycomb shall be performed under the direction of the engineer.

e. No mortar or cement shall be applied to the surface except to fill pits or voids, tie bolt holes, etc., as above provided, and not by plastering.

2. Rubbed Finish

The surface shall be rubbed only when called for on the plans or directed by the engineer.

After all voids are filled, the surface shall be thoroughly wetted and rubbed with a carborundum brick, or similar abrasive, to a smooth, even finish of uniform appearance without applying any cement or other coating.

R. UNFORMED SURFACE FINISH

1. Top Surface Subject to Wear

For specifications see Chapter 6, Part 9.

2. Top Surface Not Subject to Wear

Top surfaces not subject to wear shall be smoothed with a float and be kept wet for at least 7 days. Care shall be taken to avoid an excess of water in the concrete and to drain or otherwise promptly remove any water that comes to the surface. Dry cement, or a mixture of cement and sand, shall not be sprinkled directly on the surface.

S. DECORATIVE FINISHES

Any special or decorative finish which may be required will be done as called for on the plans and will be covered by a separate specification.

Pages 8–18–1 and 8–18–2

ASTM SPECIFICATIONS AND DESIGNATIONS

Designation A 408–58T—Add "(Tentative)" after description of specification.
Designation A 431–59T—Add "(Tentative)" after description of specification.
Designation A 432–59T—Add "(Tentative)" after description of specification.
After "Designation C 70–47" add the following new item: "C 76–61T Reinforced Concrete Culvert, Storm Drain and Sewer Pipe (Tentative)."
Revise designation numbers as follows:

```
A    6–60T   to A    6–61T
A    7–58T   to A    7–61T
A   82–58T   to A   82–61T
A  185–58T   to A  185–61T
C   33–59    to C   33–61T Add "(Tentative)" after description of specification.
C   39–59    to C   39–61
C   88–59T   to C   88–61T
C  117–49    to C  117–61T Add "(Tentative)" after description of specification.
C  136–46    to C  136–61T Add "(Tentative) "after description of specification.
C  150–60    to C  150–61
C  175–60    to C  175–61
C  205–60T   to C  205–61T
C  227–60T   to C  227–61T
C  289–57T   to C  289–61T
C  290–57T   to C  290–61T
C  291–57T   to C  291–61T
C  292–57T   to C  292–61T
C  310–57T   to C  310–61T
C  342–55T   to C  342–61T
```

Report on Assignment 8

Waterproofing for Railway Structures
Collaborating with Committees 6, 7 and 15

R. J. Brueske (chairman, subcommittee), A. L. Becker, D. E. Bray, G. W. Clarvoe, J. R. Iwinski, J. M. Williams, K. B. Woods.

Your committee has reviewed Part 3, Chapter 29, and has found that the specification for dampproofing materials is a satisfactory specification but is considered too rigid by the manufacturers, which did not feel justified in keeping material meeting the specification in stock for the small quantities used by railroads. Therefore, your committee offers for adoption the following recommendations in connection therewith:

Pages 29–3–1 to 29–3–4, incl.

SPECIFICATIONS FOR BITUMINOUS EMULSIONS FOR DAMPPROOFING

Delete entire subject matter under this heading and add: "Temporarily withdrawn. Subject matter to be studied and rewritten."

Your commitee also recommends for adoption the following revisions to Parts 2 and 4, Chapter 29.

Pages 29–2–1 to 29–2–8, incl.

SPECIFICATIONS FOR MEMBRANE WATERPROOFING

Page 29–2–3, Sec. B, Art. 4 (d), Asphalt Mastic (2)—Change to read:

"Coarse mineral aggregate shall be well graded crushed stone, crushed air-cooled iron-blast-furnace slag or washed gravel that will meet the requirements of ASTM Specifications, designation D 692–54, Size 8 ($\frac{3}{8}$ in to No. 8). It shall be free from soft particles, organic matter and other deleterious material."

Pages 29–4–1 to 29–4–5, incl.

SPECIFICATIONS FOR WATERPROOFING COATINGS FOR EXPOSED CONCRETE SURFACES

Page 29–4–2, Sec. B, Art. 3 (b) Materials—Change third paragraph to read:

"Coarse aggregate—crushed stone, crushed air-cooled iron-blast-furnace slag, or gravel to be of one size, which passes a $\frac{3}{8}$ in sieve, and all of which is retained on a No. 4 sieve (4760 microns)."

Report of Committee 3—Ties and Wood Preservation

R. B. RADKEY, *Chairman*

W. E. FUHR,
Vice Chairman

C. S. BURT
H. F. KANUTE
W. W. BARGER
L. C. COLLISTER
W. L. KAHLER

P. D. BRENTLINGER
M. J. HUBBARD
G. A. WILLIAMS
W. F. ARKSEY
A. B. BAKER
R. S. BELCHER (E)
R. G. BROHAUGH
WALTER BUEHLER (E)
C. M. BURPEE
W. J. BURTON (E)
G. B. CAMPBELL
R. W. COOK
D. L. DAVIES
C. E. DEGEER
R. F. DREITZLER
L. P. DREW
H. R. DUNCAN
K. C. EDSCORN
T. H. FRIEDLIN (E)
A. K. FROST
F. J. FUDGE
H. M. HARLOW
F. F. HORNIG
M. S. HUDSON

R. P. HUGHES
F. S. HUNTER
H. E. HURST
C. E. JACKMAN
W. R. JACOBSON
H. W. JENSEN
L. W. KISTLER
W. E. LAIRD
R. W. ORR
T. H. PATRICK
C. A. PEEBLES
R. R. POUX
A. P. RICHARDS
H. S. ROSS
N. A. SALZANO
J. T. SLOCOMB
O. W. SMITH
R. B. SMITH
E. F. SNYDER
L. S. STROHL
S. THORVALDSON
H. K. WYANT
R. G. ZIETLOW
Committee

(E) Member Emeritus.
Members listed in bold face are the official representatives of the Engineering Division, AAR.

To the American Railway Engineering Association:
Your committee reports on the following subjects:

1. Revision of Manual.
   Progress report, with recommendations submitted for adoption .... page 316

2. Cross and switch ties.
   Brief progress report, presented as information .... page 321

3. Wood preservatives.
   Brief progress report, presented as information .... page 322

4. Conditioning and preservative treatment of forest products.
   Manual material concerning treatment of timber and bridge ties submitted
   for adoption. Report on retreatment of salvaged cross ties, fire-retardant
   treatment of bridge timbers and definition of refusal treatment presented
   as information .... page 322

5. Service records.
   Data on treated fence posts after 30 years' service presented as information.
   (Information on annual tie renewals and progress of the termite stake test
   has been published in recent Bulletins) .... page 325

6. Methods of prolonging service life of ties.
   Progress report, submitted as information, on AAR laboratory work on
   anti-splitting devices .... page 325

THE COMMITTEE ON TIES AND WOOD PRESERVATION,
R. B. RADKEY, <i>Chairman.</i>

AREA Bulletin 568, December 1961.

Report on Assignment 1
Revision of Manual

C. S. Burt (chairman, subcommittee), W. W. Barger, R. S. Belcher, C. M. Burpee,
W. J. Burton, G. B. Campbell, R. W. Cook, H. R. Duncan, R. P. Hughes, R. R.
Poux, N. A. Salzano, R. B. Smith, E. F. Snyder.

Your committee recommends the following revisions, deletions, and reapprovals of
and additions to Manual material in Chapter 3—Ties.

Page 3-1-6

EXPLANATION OF CROSS TIE DESIGN

Reapprove with the following revisions:

In the first line of Par. 5 delete the colon following the word "recommended" and
add the following words "in the interest of promoting economy in track maintenance:"

Change Par. 5 (b) to read: "The use of 9-ft ties for making renewals in lines of
heavy traffic."

Pages 3-1-8 and 3-1-9

SIZE OF HOLES BORED FOR SPIKES

Reapprove with the following revision:

Change the second sentence of Par. 5 to read: "Under such conditions, it has been
found preferable not to bore through the ties in order to minimize corrosion of the
spikes from electrolysis."

Delete Specifications for Devices to Control the Splitting of Wood Ties, pages 3-1-12
and 3-1-13, substituting therefor the following:

SPECIFICATIONS FOR DEVICES TO CONTROL THE SPLITTING OF WOOD TIES

A. ANTI-SPLITTING DEVICES

Anti-splitting devices may be of (a) the type made from a strip of steel and applied
by driving into the end (cross section) of the tie, or (b) of the steel dowel type, applied
parallel to the wide face of the tie, transverse to is length.

B. MATERIALS

1. Irons

a. If of the strip-steel type, it shall be manufactured in accordance with the following specifications. The strips shall be hot-rolled open-hearth, bessemer or shell steel, as may be stipulated in the order for the irons.

b. *Chemical Requirements:* The steel shall conform to the following chemical composition:

	Open Hearth Steel	Bessemer Steel	Shell Steel
Carbon	0.25–0.35	0.15–0.25	0.45–0.55
Manganese	0.40–0.60	0.30–0.60	0.60–0.90
Phosphorous	0.050 Max	0.110 Max	0.035 Max
Sulfur	0.050 Max	0.060 Max	0.055 Max
Copper	0.20 Min	0.20 Min	0.20 Min

c. *Physical Requirements:* The manufacturer may, at his option, substitute tension tests for the chemical analysis specified above. When this option is exercised:

(1) Phosphorous and sulfur shall not exceed the limits of the chemical requirements.

(2) One tension test shall be made for each lot of 10 tons or fraction thereof.

(3) The finished strip shall conform to the following minimum tensile properties:

Strength, psi	75,000
Yield point, psi	40,000
Elongation in 2 in, percent	20

d. *Design:* Anti-splitting irons shall be of the shape and size stipulated by the purchaser.

e. *Manufacture:* Anti-splitting irons shall have smooth surfaces and be free of distortion, scale, jagged ends, and blunt bevelled edge.

The dimensions of the steel strip in anti-splitting irons shall be not less than:

Thickness	0.083 in
Width	$\frac{1}{4}$ in
Length	as required by the design

f. *Variations:* Variations (over or under) from dimensions specified shall not exceed:

Thickness	0.005 in
Width of strip	$\frac{1}{32}$ in
Width of bevel	$\frac{1}{32}$ in
Length	$\frac{1}{8}$ in

2. Dowels

If the steel dowel type is used, it shall be of the dimensions and thread design specified by the purchaser, and shall be of C–1020 steel, ASTM Specifications, designation A 107, with a minimum of 2 percent copper.

C. INSPECTION

Inspectors representing the purchaser shall have free entry, at all times while work on the contract of the purchaser is being performed, to all parts of the manufacturer's works which concern the manufacture of the material ordered. The manufacturer shall afford the inspectors, without charge, all reasonable facilities to satisfy them that the material is being supplied in accordance with these specifications. Unless otherwise agreed all inspection and tests shall be made at the place of manufacture prior to shipment, and shall be so conducted as to not interfere unnecessarily with the operation of the works.

D. DELIVERY

Accepted devices shall be shipped by the seller in accordance with instructions in the order covering them, securely packed in containers marked with the name, type, grade, and quantity of the material therein, and with the name of the seller and the number of the buyer's contract or order.

Delete Application of Anti-Splitting Devices, page 3–1–4, substituting therefore the following:

APPLICATIONS OF ANTI-SPLITTING DEVICES

All hardwood ties (those from broadleaved trees) are subject to splitting and when so designated by the purchaser shall have anti-splitting devices applied. Either strip irons or dowels may be used for this purpose.

Anti-splitting devices designed to control the splitting of ties should be applied prior to or at the time the ties are delivered to the treating plant.

1. Irons

Anti-splitting strip irons driven into the ends of ties should be so placed as to cross at right angles the greatest possible manner of radial lines of the wood. Irons should be placed far enough from the wide faces to prevent splitting.

2. Dowels

Two dowels should be inserted in each end of ties in prebored holes ⅛ in less in diameter than the diameter of the dowels and parallel to the wide faces.

Purchaser to specify the location of the dowels in relation to the end and the two faces of the tie.

Dowels should be inserted by mechanical machines capable of holding the ties in a clamped position under pressure while the holes are drilled and the dowels inserted.

Pages 3–2–1 to 3–2–4, incl.

SPECIFICATIONS FOR SWITCH TIES

Reapprove with the following revisions:

Page 3–2–1, Sec. A, Art. 1—Delete "Chestnut" as being acceptable.

Delete Sec. B, Art. 3. Resistance to Decay.

Page 3–2–2, Sec. C, Art. 1 (c)—Delete reference to "hewed" switch ties.

Sec. D—Delete reference to "hewed" ties.

Sec. E. Art. 1—Delete the words "steamer lines" and "steamer landing", making the last part of the paragraph read as follows: ". . . while away from rail lines and transfer him from and to a railway station."

Sec. E, Art. 3—Change to read: "The following decay will be allowed: in cedar and in cypress, 'pipe or stump rot' and 'peck', respectively, up to the limitations as to holes; 'blue stain' is not decay and is permissible in wood."

Page 3–2–3, Sec. E, Art. 7—Change to read: "One which is not more than 5 in long will be allowed. The purchaser shall specify the anti-splitting devices to be applied, if any."

Sec. F, Art. 1—Add the following sentence at the beginning of the paragraph: "Ties delivered for inspection shall be stacked at suitable and convenient places."

Page 3–2–4—Delete Sec. F, Art. 4. Class U—Ties Which May Be Used Untreated. Redesignate present Art. 5 as Art. 4.

In present Art. 5 (to be redesignated as Art. 4) delete the word "Sap" wherever it appears. Also delete "Chestnut" in Group Td.

Page 3–3–1

INSTALLATION AND KEEPING RECORDS OF TEST SECTIONS

Reapprove without change.

Page 3–3–2

MARKING TIES FOR SERVICE RECORDS

Reapprove without change.

Pages 3–4–1 to 3–4–3, incl.

ECONOMIC COMPARISONS OF TIES

Reapprove with the following revision:

Page 3–4–2—Correct formula under "Total anual cost" to read:

$$I + A = \frac{CR\,(1 + R)^n}{(1 + R)^n - 1}$$

Page 3–4–4

TRAFFIC UNIT FOR USE IN COMPARING TIE LIFE

Reapprove without change.

BEST PRACTICE FOR TIE RENEWALS

Reapprove without change.

Pages 3–5–1 and 3–5–2

FUNDAMENTALS TO BE CONSIDERED IN DESIGNS
OF SUBSTITUTE TIES

Reapprove without change.

Pages 3–M–1 to 3–M–9, incl.

THE HANDLING OF TIES FROM THE TREE INTO THE TRACK

Reapprove with the following revisions:

Page 3–M–2.Art. 3. Piling in the Woods—Substitute the word "ground" for "forest floor" in last sentence.

Art. 6. Conditioning—Eliminate first sentence of first paragraph. Change second sentence of first paragraph to read: "Ties should be conditioned in accordance with AREA Plant Practice, Part 1, Chapter 17."

Page 3–M–3, Art. 8. Machining—Delete the word "just" from first sentence.

Art. 9. Storage—Delete the words "over 6 in high" from first sentence, first paragraph. Second paragraph, change first sentence to read: "Ties should be stacked for seasoning so as to provide . . ." Delete third paragraph.

Page 3–M–4. Change wording above Figs. 1, 2, and 3 to read: "Open stacking for seasoning ties where they will not dry too rapidly." Change wording above Fig. 4 to read: "Semi-solid stack for seasoning ties where they may dry too rapidly."

Page 3–M–5. Second paragraph—Delete "for prompt insertion in track" from first sentence, and change sentence to read, "Ties in excess of those needed on line and which are to be stored after treatment . . ."

Art. 10. Preservation—Delete from fourth paragraph the third sentence, reading: "Neither the preservative nor the heat . . ."

Page 3–M–6. Art. 12. Distribution—Delete fourth paragraph.

Delete Pars. 6, 7, and 8 and replace with the following:

"Ties will be shipped from treating plants for usage under three general conditions:

"a. A large number of ties for renewal in connection with mechanized maintenance work over a several consecutive mile work location. These ties are best handled in special tie cars designed for this purpose, facilitating rapid unloading with minimum maintenance labor. Ties should be unloaded at the point of usage to avoid labor expense in stacking or rehandling. Time between unloading and insertion should be the minimum practical to avoid damage to ties due to exposure to elements.

"b. A small number of ties to separate locations for use as on-line emergency stock or spot renewals. Handling can be in special tie cars or in such other cars as the railroad has available.

"c. A large number of ties shipped for use in construction work. The loading and handling method should mesh with the construction situation. Banded bundles can be used if ties are to be transferred from railway cars to trucks. Special tie cars are practical for adjoining track construction."

Page 3–M–7, Art. 13. Care During and After Distribution—Delete second paragraph.

Third paragraph, change to read: "Ties on line not needed for immediate use should be stacked as shown in Figs. 5 or 6. Ties should be covered with cinders or earth for protection against weather and sparks from locomotives."

Art. 14. Care During and After Installation, Par. 5—Change first word "creosoted" to "treated."

Delete Pars. 6 and 7.

Page 3–M–8, first paragraph—Delete from third sentence: "provided the total number removed do not exceed the authorized renewals."

Delete from fourth sentence, first paragraph: "as when ballast is strengthened or rail relaid," and change to read: "but when track is given a general out of face overhauling, all ties which appear to be nearing the end of their service life may be removed."

Par. 5—Delete first word, "untreated."

Art. 16. Salvage, second paragraph, second sentence—Delete "as when ballast and rail are being renewed."

Add the word "spot" to last sentence. This sentence will then read, ". . . high-speed lines where spot tie renewals . . ."

Par. 3—Delete the last six words from last sentence, "as a result of mechanical wear."

Page 3–M–9—Delete Par. 2.

Add the following sentence to Par. 3: "Such ties should be re-inserted in track quickly to avoid accelerated deterioration."

Par. 5—Change the word "creosoted" to "treated."

Fence posts.—Delete Pars. 1 and 2. Insert "some ties removed from track will make satisfactory fence posts."

Delete paragraph dealing with *Labor exchange.*

Page 3–M–10

CONSERVATION OF TIMBER SUPPLY

Reapprove with following revisions:

Par. 1—Delete "wherever practicable."

Par. 2—Change last sentence to read:"Items which tend to retard the destruction of ties by mechanical wear are: (1) heavier rail; (2) type of rail joint; (3) continuous welded rail; (4) tie plates of correct type and size securely fastened to the tie; (5) tie pads; (6) adzing and boring of ties; (7) rail anchors, and (8) track maintenance.

Report on Assignment 2

Cross and Switch Ties

H. F. Kanute (chairman, subcommittee), W. F. Arksey, P. D. Brentlinger, R. W. Cook, A. K. Frost, F. J. Fudge, F. F. Hornig, C. E. Jackman, W. R. Jacobson, L. W. Kistler, W. E. Laird, H. S. Ross, N. A. Salzano, J. T. Slocomb, H. K. Wyant.

Your committee submits the following brief progress report, as information, on the development of specifications for dry ties and concrete ties, and specifications on the desirable characteristics of salvaged ties to be reused.

Specifications for dry ties are not practicable because of ramifications in prescribing tolerances for seasoning defects.

Present tests of concrete ties have not shown conclusively that the present design is satisfactory. Longer service life tests should be made before specifications are written.

There is a great deal of variation on the various railroads on the use of second-hand ties. Information will be secured from various railroads in regard to use of second-hand ties, expected life and type of tracks where second-hand ties are used, and reported on next year.

Report on Assignment 3

Wood Preservatives

W. W. Barger (chairman, subcommittee), C. E. DeGeer, K. C. Edscorn, M. S. Hudson, L. W. Kistler, T. H. Patrick, A. P. Richards, D. L. Davies, R. F. Dreitzler, W. Buehler, R. B. Radkey, W. E. Fuhr.

This assignment is divided into three parts as follows:

Part 1—Keep Up To Date Current Specifications for Preservatives

During the last two years, all preservative specifications have been reapproved or revised and brought up to date. This assignment is continually under a study by your committee.

Part 2—Review and Report on New Preservatives

No new preservatives have come to the attention of your committee during the last year. A new creosote-coal tar solution specifications for use in marine piles will be studied.

Part 3—Keep Up To Date Current Specifications for Petroleum as a Carrier for Standard Wood Preservatives

This assignment is continually under study by your committee. New developments will be reported as they come to the attention of the commitee.

Report on Assignment 4

Conditioning and Preservative Treatment of Forest Products

L. C. Collister (chairman, subcommittee), W. F. Arksey, G. B. Campbell, D. L. Davies, R. F. Dreitzler, H. R. Duncan, T. H. Friedlin, M. S. Hudson, F. S. Hunter, H. E. Hurst, R. W. Orr, T. H. Patrick, R. R. Poux, R. B. Smith, R. G. Zietlow.

Your committee submits for adoption the following recommendation with respect to Chapter 17 of the Manual:

Pages 17–4–1 to 17–4–18.1, incl.

SPECIFICATIONS FOR TREATMENT

Page 17–4–9. To the table of specific requirements for preservative treatment of lumber, timbers and bridge ties, add oil-borne preservatives for fir, hemlock, larch and oak, as follows:

	Above Ground	Ground Contact
Oil-borne preservatives Pentachlorophenol***	0.30	0.4
Copper Naphthenate (Copper metal)	0.05	0.1

TREATMENT OF SALVAGED TIES

The committee submits the following report, as information, on the treatment or re-treatment of salvaged ties from an abandoned rack on the B&O. Several members of this committee were fortunate enough to view the operations. The general procedure was as follows:

Ties were removed from the roadbed, inspected and loaded in coal cars for shipment to the treating plant. Here approximately 30 percent of the ties were culled for decay, split and other defects. If the tie was split through the rail base and had any decay showing, the tie was thrown out. If the tie was plate cut as much as $\frac{3}{4}$ in or had soft wood in the tie plate area to a depth of $\frac{3}{4}$ in, it was discarded. If there was any mechanical breakage, the tie was discarded.

The ties when received at the treating plant were unloaded on skids where men straightened them and plugged the spike holes with untreated pine plugs. Ties with spikes or other metal objects were thrown out here. With the reinspection, probably another 5 to 10 percent were discarded for additional defects.

The ties proceeded down a conveyor, and those marked for discard were removed. The remainder were washed under water pressure of 250 lb. The ties were then adzed to level up the rail base and permit the use of 14-in plates. Another 2 to 3 percent of the ties were thrown out at the boring machine for doweling. Ties were then put into the treating cinders where they received 4-lb treatment with 70/30 creosote/coal tar solution. The ties processed here were all sawn hardwood, approximately 80 to 85 percent oak 6 in and 7 in by 8 ft in length. Their original treatment was 7-lb 80/20 creosote/coal tar solution. It was estimated that the age of the salvagable ties varied from 5 to 20 years.

The total cost to the railroad for this entire operation from the time ties are received at the treating plant until they are shipped is $1.30 per tie. It is felt that this treatment will extend the tie life by 10 years.

The following specifications for re-treatment of salvaged ties is presented for information:

TREATMENT
Creosote-type preservatives ....Empty or full-cell
Pressure-psi max   150

RESULTS OF TREATMENT
Retention-lb per cu ft min:
Creosote-type preservatives .... General Use**
     Creosote   4
     Creosote–coal tar   4
     Creosote–petroleum   4

TEST OF FIRE-RETARDANT TREATMENTS

For further information, the Santa Fe has successfully concluded the test burning of two full-scale bridge replicas that were treated with an emulsion of borax/boric acid, penta, petroleum oil, diesel oil and marasperse. The percentages of the preservatives by weight in one replica was as follows:

Seguro oil 32.4
Diesel oil 8.7
Penta 6.5
Borax 8.2
Boric acid 8.2
Water 35.1
Marasperse 0.9

The retention was 11 lb per cu ft.

This replica was burned in a tumbleweed fire on June 6, 1961, with a peak temperature of 1600 F 30 sec after ignition. Fire in the weeds was out in 2 min 55 sec, on the windward side of the replica in 3 min, and on the leeward side in 4 min. The char was less than ⅛ in.

The other replica was treated with a borax/boric acid–creosote/petroleum emulsion with the following percentage of preservatives by weight:

Creosote 30.97
Seguro oil 27.82
Borax 7.45
Boric acid ....:.................. 7.45
Water 25.42
Marasperse .... .89

The retention was 13 lb per cu ft.

This replica was also burned on June 6, 1961, with a peak temperature of 1810 F 45 sec after ignition. Fire in the weeds was out in 2 min 25 sec, on the windward side of the replica in 3 min, and on the leeward side in 4 min 45 sec. The char was less than ⅛ in.

The added cost of this treatment to the regular 50/50 creosote/petroleum mixture is about $6 to $10 per thousand board feet. The tri-cressyl phosphate treatment now costs $60 to $70 per thousand board feet.

From a weatherometer test after 1500 hr of exposure, it was found that the borax/boric acid mixture is relatively two-thirds as permanent as the creosote/petroleum mixture. It is felt that by adding penta to the creosote/oil emulsion there will be adequate toxic protection as well as fire protection.

DEFINITION OF REFUSAL POINT AND REFUSAL TREATMENT

For further information, it is the committee's thought that "refusal point" and "refusal treatment" should be more clearly defined. The committee's recommendation is as follows:

REFUSAL POINT.—The point reached in timber treatment with the pressure and temperature maintained until the quantity of preservative absorbed is not more than 2 percent of the amount already injected in each of any two consecutive half hours for Douglas fir and oak and ½ percent in any half hour for other species.

REFUSAL TREATMENT.—The treatment of timber by the full-cell process for maximum retention and empty-cell process for maximum penetration, with pressure and temperature maintained until refusal point is reached.

METHODS AND PROCESSES FOR CONDITIONING FOREST PRODUCTS BEFORE TREATMENT

No new methods or processes for conditioning forest products before treatment are being studied.

REVIEW RETENTIONS FOR TC–TD TIES

It is the recommendation of the committee that this instruction be dropped.

Report on Assignment 5

Service Records

W. L. Kahler (chairman, subcommittee), A. B. Baker, C. M. Burpee, W. Buehler, C. E. DeGeer, F. J. Fudge, W. E. Fuhr, H. M. Harlow, R. P. Hughes, A. P. Richards, R. B. Radkey, J. T. Slocomb.

Tie Renewals and Cost per Mile of Maintained Track as Furnished by the AAR, Bureau of Railway Economics, Washington, D. C., from Annual Reports of Class I Railways to the Interstate Commerce Commission

These statistics were published, in Bulletin 565, June–July 1961, page 1026.

Service Records for all Forest Products Used for Railway Construction and Maintenance

Your committee submits the following report on service test of 313 fence posts treated by the Lowry process with Grade 1 creosote in 1930 by the New York Central Railroad and installed in right-of-way fence in the vicinity of Rome, N. Y.

The 1940 inspection of these posts was included in AWPA U-5 Report on Fence Posts. The last inspection was made in November 1960 and developed the following:

Initial Test		*Summary of November 1960 Inspection*
50 beech posts	....	None removed
50 birch posts	....	None removed
50 cedar posts	....	15 removed (6 removed 1947 a/c property lease)
		(9 removed a/c badly decayed)
62 chestnut posts	....	36 decayed
51 maple posts	....	1 decayed
50 oak posts	....	36 removed in 1947 account leasing property

The beech, birch, maple and oak posts are in excellent condition; however, the oak posts show signs of checking.

The chestnut posts show heart decay and are badly weathered. They will probably not last much longer.

The cedar posts decayed at the ground line.

Cooperate with the AAR and University of Florida in the Operation and Evaluation of the Termite Stake Test at Gainesville, Fla.

This is a research project. Three progress reports have been made; the latest is printed in Bulletin 566, September–October 1961, page 53.

Destruction by Marine Organisms

No report to submit this year. Information of interest will be submitted as it develops.

Report on Assignment 6

Methods of Prolonging Service Life of Ties

P. D. Brentlinger (chairman, subcommittee), R. S. Belcher, R. G. Brohaugh, C. S. Burt, L. C. Collister, T. H. Friedlin, A. K. Frost, F. F. Hornig, H. E. Hurst, C. E. Jackman, H. W. Jensen, H. F. Kanute, C. A. Peebles, H. S. Ross, O. W. Smith, E. F. Snyder, S. Thorvaldson, G. A. Williams, H. K. Wyant, R. G. Zietlow.

Your committee presents as information a progress report on a field study of anti-splitting devices made by the research staff of the Association of American Railroads.

The research staff also made laboratory tests of ties of Jarrah wood (a product of Australia) and prepared a report which may be obtained from the director of engineering research, Association of American Railroads, 3140 S. Federal St., Chicago 16.

Inspection of Ties for Anti-Splitting Devices

An inspection was made on June 2, 1961, of the 2500 ties which are to be used in the field study of anti-splitting devices conducted in cooperation with the Chicago, Milwaukee, St. Paul & Pacific Railroad. At this time, the seasoning of these ties having been completed, the ties were ready for preservative treatment, after which they will be installed in track for service tests. T. H. Patrick, supervisor tie bureau, of the Chicago, Milwaukee, St. Paul & Pacific Railroad, R. C. Studebaker, superintendent, Timber Preserving Plant of the T. J. Moss Tie Company, and G. M. Magee, director of engineering research, AAR, made the inspection.

These ties are mixed oak from southern Indiana, which started seasoning June 10, 1960, and from northern Wisconsin, which started seasoning June 20, 1960. In all, there are five test groups.

M60–1 are control ties having no anti-splitting devices applied. M60–2 are saw-kerfed ties in which two horizontal saw cuts were made in each end of each tie. M60–3 are ties doweled green out-of-face. M60–4 are to be doweled out-of-face just prior to treatment. M60–5 are to be doweled selectively just prior to treatment, choosing only split ties for doweling which would appear to require or be benefited by the application of dowels. Actually the M60–5 group was adzed, bored, and selectively doweled the afternoon of June 2; 42 ties from the northern group and 73 from the southern group were doweled.

The inspection consisted of examining each end of each tie and classifying the ties with respect to the width of the maximum size split that had developed. A taper gage was used to measure the width of splits or checks, although it was found that for most ties the classification could be made visually with an occasional check made with the taper gage. The accompanying tabulation shows a comparison of the condition of splitting in each of the test sections for the ties as measured June 28, 1960, near the beginning of the seasoning period and as rechecked on June 3, 1961 at the end of the seasoning prior to preservative treatment.

General Remarks

It will be noted in the M60–1 control ties that after the seasoning period considerably more ties were in the $\frac{1}{8}$ to $\frac{1}{4}$ in range and less in the 0 to $\frac{1}{8}$ in than when the seasoning started, and that the southern ties showed this change to a much larger extent than the northern ties. In fact, the southern ties showed evidence of more splitting tendency at the beginning of the seasoning period.

The M60–2 saw-kerfed ties appeared to have received some benefit in prevention of splitting and checking. Although there was some increase in the amount of checking and splitting as indicated by the increase in number in the $\frac{1}{8}$ to $\frac{1}{4}$ in range and decrease in the number in the 0 to $\frac{1}{8}$ in range, this change was not as pronounced as in the M60–1 control ties.

M–60–3 ties doweled green out-of-face also showed benefit in reducing checking and splitting from the use of the dowels. Here again as in the M60–1 and M60–2 ties, the progression of splitting and checking was less in the northern ties than in the southern ties.

Statement of Condition of Checks and Splits Before and After Seasoning

Max. Split	Northern				Southern			
	East End		West End		East End		West End	
	No.	%	No.	%	No.	%	No.	%
			M60-1 Control Ties					
			Condition June 28, 1960 at start of seasoning					
0 – 1/8	235	94.0	241	96.4	203	81.2	224	89.6
1/8-1/4	13	5.2	9	3.6	45	18.0	26	10.4
1/4-38	2	0.8	0	0.0	1	0.4	0	0.0
3/8-1/2	0	0.0	0	0.0	1	0.4	0	0.0
1/2 +	0	0.0	0	0.0	0	0.0	0	0.0
Total	250	100.0	250	100.0	250	100.0	250	100.0
			Condition June 2, 1961 after seasoning					
0 – 1/8	199	79.6	214	85.6	106	42.4	173	69.2
1/8-1/4	49	19.6	36	14.4	132	52.8	76	30.4
1/4-3/8	2	0.8	0	0.0	11	4.4	1	0.4
3/8-1/2	0	0.0	0	0.0	0	0.0	0	0.0
1/2 +	0	0.0	0	0.0	1	0.4	0	0.0
Total	250	100.0	250	100.0	250	100.0	250	100.0

Max. Split	Northern				Southern			
	East End		West End		East End		West End	
	No.	%	No.	%	No.	%	No.	%
			M60-2 Saw Kerfed Ties					
			Condition June 28, 1960 at start of seasoning					
0 – 1/8	210	84.0	223	89.2	202	80.8	193	77.2
1/8-1/4	38	15.2	24	9.6	43	17.2	53	21.2
1/4-3/8	1	0.4	2	0.8	5	2.0	3	1.2
3/8-1/2	1	0.4	0	0.0	0	0.0	1	0.4
1/2 +	0	0.0	1	0.4	0	0.0	0	0.0
Total	250	100.0	250	100.0	250	100.0	250	100.0
			Condition June 2, 1961 after seasoning					
0 – 1/8	205	82.0	204	81.6	174	69.6	181	72.4
1/8-1/4	38	15.2	43	17.2	62	24.8	62	24.8
1/4-3/8	7	2.8	2	0.8	13	5.2	7	2.8
3/8-1/2	0	0.0	0	0.0	0	0.0	0	0.0
1/2 +	0	0.0	1	0.4	1	0.4	0	0.0
Total	250	100.0	250	100.0	250	100.0	250	100.0

Statement of Condition of Checks and Splits Before and After Seasoning

	Northern				Southern			
	East End		West End		East End		West End	
Max. Split	No.	%	No.	%	No.	%	No.	%
M60-3 Doweled Green out-of-face								
Condition June 28, 1960 at start of seasoning								
0 – 1/8	227	90.8	234	93.6	214	85.6	214	85.6
1/8-1/4	21	8.4	16	6.4	35	14.0	35	14.0
1/4-3/8	1	0.4	0	0.0	1	0.4	1	0.4
3/8-1/2	0	0.0	0	0.0	0	0.0	0	0.0
1/2 +	1	0.4	0	0.0	0	0.0	0	0.0
Total	250	100.0	250	100.0	250	100.0	250	100.0
Condition June 2, 1961 after seasoning								
0 – 1/8	215	86.0	225	90.0	182	72.8	193	77.2
1/8-1/4	33	13.2	25	10.0	67	27.8	57	22.8
1/4-3/8	2	0.8	0	0.0	1	0.4	0	0.0
3/8-1/2	0	0.0	0	0.0	0	0.0	0	0.0
1/2 +	0	0.0	0	0.0	0	0.0	0	0.0
Total	250	100.0	250	100.0	250	100.0	250	100.0

	Northern				Southern			
	East End		West End		East End		West End	
Max. Split	No.	%	No.	%	No.	%	No.	%
M60-4 Doweled Seasoned out-of-face								
Condition June 28, 1960 at start of seasoning								
0 – 1/8	224	89.6	230	92.0	198	79.2	215	86.0
1/8-1/4	25	10.0	20	8.0	52	20.8	34	13.6
1/4-3/8	1	0.4	0	0.0	0	0.0	1	0.4
3/8-1/2	0	0.0	0	0.0	0	0.0	0	0.0
1/2 +	0	0.0	0	0.0	0	0.0	0	0.0
Total	250	100.0	250	100.0	250	100.0	250	100.0
Condition June 2, 1961 after seasoning								
0 – 1/8	182	72.8	208	83.2	137	54.8	199	79.6
1/8-1/4	63	25.2	40	16.0	105	42.0	43	17.2
1/4-3/8	5	2.0	1	0.4	8	3.2	8	3.2
3/8-1/2	0	0.0	0	0.0	0	0.0	0	0.0
1/2 +	0	0.0	1	0.4	0	0.0	0	0.0
Total	250	100.0	250	100.0	250	100.0	250	100.0

Statement of Condition of Checks and Splits Before and After Seasoning

	Northern				Southern			
Max. Split	East End		West End		East End		West End	
	No.	%	No.	%	No.	%	No.	%

M60-5 Doweled Selectively

Condition June 28, 1960 at start of seasoning

No measurements taken

Condition June 2, 1961 after seasoning

0 - 1/8	179	71.6	176	70.4	119	47.6	140	56.0
1/8-1/4	65	26.0	69	27.6	111	44.4	102	40.8
1/4-3/8	4	1.6	4	1.6	19	7.6	8	3.2
3/8-1/2	0	0.0	0	0.0	0	0.0	0	0.0
1/2 +	2	0.8	1	0.4	0	0.4	0	0.0
Total	250	100.0	250	100.0	250	100.0	250	100.0

The M60-4 ties to be doweled out-of-face after seasoning and the M60-5 ties to be doweled selectively after seasoning may be compared during the seasoning period with the M60-1 ties. It will be noted that there are some differences in the number of ties in the 0 to 1/8 in and 1/8 to 1/4 in ranges after the seasoning period. However, the agreement generally is probably as good as could be expected. There does seem to be a significant reduction in splitting between the M60-2 saw-kerfed ties and the M60-3 doweled green out-of-face ties, as compared with the M60-1, 4, and 5 ties which were seasoned without any anti-splitting devices or treatment.

Arrangements were made to designate each tie of each test section with two bored holes on the top of the tie approximately 4 in from the midlength of the tie. Standing on the end of the tie towards which end the drilled holes are displaced from the midlength, two holes are used to simulate the center and the end of the hour hand on a clock, so the 12:00 o'clock and 6:00 o'clock positions would be on a line with the center line of the tie. The test section numbers 1, 2, 3, 4, and 5 are designated by corresponding positions of 1, 2, 3, 4, and 5 o'clock with the two drilled holes. For the northern ties the hole representing the center of the hour hand is made with a smaller diameter than the hole representing the end position of the hour hand. For the southern ties both holes are made of the large diameter and of the same size. Mr. Patrick agreed to nail a metal tag on the ties in section 5 that were selectively doweled so they could be readily identified in track.

Report on Assignment 7

Substitutes for Wood Ties

M. J. Hubbard (chairman, subcommittee), W. J. Burton, L. P. Drew, K. C. Edscorn, F. S. Hunter, H. W. Jensen, W. E. Laird, R. W. Orr, C. A. Peebles, O. W. Smith, L. S. Strohl, S. Thorvaldson.

Your committee submits, as information, the following summary of a report prepared by the Research Department, Association of American Railroads, designated ER-20*, covering investigation of prestressed concrete ties.

With respect to the economic comparison between prestressed concrete ties and wood ties installed in new track, as set forth in the summary, the committee would point out that the savings shown are based upon assumptions with respect to wood ties that are too conservative, and that more liberal assumptions would materially alter the indicated annual savings favorable to prestressed concrete ties. The committee also feels that any such economic comparison should also include annual maintenance costs for the two types of installations under consideration.

Summary of Report on AAR Prestressed Concrete Tie Investigation

This report embraces a description and analysis of an investigation of prestressed concrete ties. The investigation covers the design of the tie and rail fastenings to carry operating loads used by the railroads in the United States and Canada, the laboratory testing, installation of experimental ties, field investigation, service performance, economic comparison beween wood and concrete ties and recommendations for future installations.

The investigation was started in 1957 even though it was realized that the supply of wood ties was ample at present and in the foreseeable future, as it was felt desirable to have a few ties installed on several railroads to secure service performance over a period of years in case our timber supply or economic conditions should change.

The first ties were designed on the basis of replacing one wood tie with one concrete tie. Such ties were made and then tested under static and repeated loads at the Research Center. The results of this preliminary investigation were satisfactory from a technical standpoint, but cost estimates indicated such a tie-for-tie replacement would not be economical.

The next phase of the investigation was to consider an increased spacing of the concrete ties and the effect of such an increased spacing on the rail stresses, percent of axle load carried by each tie and width of tie as determined by pressure of tie on the ballast. This study indicated.

1. The rail stresses would only be increased about 10 percent with the ties spaced at 30 in instead of 20 in.

2. Each tie spaced at 30 in would carry about 50 percent of the axle load instead of 40 percent for the ties at 20 in.

3. The pressure of the tie on the ballast for the 30-in spacing could be kept the same as that under the 9-in-wide wood ties at 20-in spacing by making the ties 12 in wide. This pressure is approximately 65 psi.

---

* Copies of the complete report may be obtained from the director of engineering research, Association of American Railroads, 3140 S. Federal St., Chicago 16, Ill.

As a result of the studies on increased tie spacing, a tie 12 in wide was designed to carry the necessary bending moments, and several ties were made in accordance with details prepared by the research center. The ties were subjected to both static and repeated loads. Ties made with lime-rock aggregate were not satisfactory, but those made with granite were able to meet the criteria of a static bending moment of 150,000 in-lb at center line of rail and to sustain 2,000,000 cycles of repeated load, producing a bending moment of 200,000 in-lb at center line of rail without failure.

The laboratory investigation also included the development of suitable rail fastenings, such as rail clips, insulating pads, anchor bolts and thimbles for electrical insulation. The fastenings developed and tested in the tie wear machine provide sufficient pressure on the rail to restrain expansion and contraction movement at the ends of welded rail in about the same distance as the usual rail anchors.

Upon completion of the laboratory investigation, arrangements were made with the Atlantic Coast Line Railroad and Seaboard Air Line Railroad to install a quarter-mile section of the 12 in wide by 7 in deep ties in their welded rail territory near Rocky Mount, N. C. and Tampa, Fla., respectively. The ties were installed with mechanized equipment. At the present time, about five miles of concrete ties have been installed on various railroads, but such installations still are to be considered as experimental. A list of the various installations is shown in Table 5.*

Prior to the installation of the ties on the Seaboard Air Line, six ties were completely instrumented with strain gages. These ties were then installed in the track. Allowabout six months for the consolidation of the ballast, the strains in the concrete were recorded under trains operating at speeds varying from 10 to 80 mph. In addition to the strains in the concrete, the strains in the clips and anchor bolts were determined with various amounts of bolt tension, stresses in the rail base determined and the vertical movements of the rail and tie measured.

A brief summary of the analysis of the data as secured during the field investigation is as follows:

1. The bending moments at the center line of rail, as determined by the recorded concrete strains on the top surface of the tie, were slightly higher than the assumed bending moment of 150,000 in-lb.

2. The bending moments at the center line of rail, as determined by the recorded concrete strains on the bottom surface of the tie, were lower than the assumed bending moment.

3. The recorded concrete strains on the top and bottom surface of the tie at center of track indicates bending moments slightly below that required to crack the tie. As a result of these data, changes were made in the tie details to reduce the possibility of tie cracking.

4. The oscillograms recorded at high train speeds indicate that the ties are subjected to high-frequency vibrations of about 100 to 150 cycles per second, but the stresses resulting from these vibrations are only about 10 to 20 percent of the total stress in the tie.

5. The recorded bending moments along the length of the tie indicate the center portion of the tie is producing some pressure on the ballast. This pressure tends to reduce the bending moment at center of track produced by the overhang of the tie outside the rail.

---

* The table numbers referred to in this summary are the same as those in the complete report.

TABLE 3

PRESTRESSED CONCRETE TIE INVESTIGATION

INSTALLATIONS OF CONCRETE CROSS TIES

Owner	Date Installed	Location	No. of Ties	Type of Ties	Spacing in.	Rail Size	Train Speed M PH	Yearly Tonnage over Track	General Comments
Atlantic Coast Line Railroad	March 1960	Near Four Oaks, N. C.	500	AAR Type "E" Indirect Fixation (1)	30	132 Welded	75	10,000,000	1/2 wedge nut 1/2 channel lock
Seaboard Air Line Railroad	March 1960	Tampa, Florida	600	AAR Type "E" Indirect Fixation	30	115 Welded	80	17,000,000	1/2 wedge nut 1/2 channel lock
St. Louis, San Francisco Railway	March 1960	Industrial Siding in Pensacola, Florida	22	AAR Type "E"	30	100 Jointed	6		
Western Pacific Railroad	March 1960	Ferry Approach Siding San Francisco, California	49	Gerwick	26 24 30	85 Jointed	5		
Duluth, Missabe & Iron Range Ry. Co.	October 1960	Near Saginaw Minn.	99	AAR Type "E" Indirect Fixation	26	115 Welded		20,000,000	
Central Railroad of New Jersey	May 1961	Ashley, Pennsylvania	10	French RS	21	Jointed			Ties shipped from France (No Modif.)
General Portland Cement Co.	June 1961	Lake Charles, La.	450	AAR Type "E" Indirect Fixation	30	90 Jointed	10		Plan three plates 5" x 11" plates tangent track
Ideal Cement Co.	June 1961	Black Pt. Terminal Tampa, Fla.	250	AAR Type "E" Direct Fixation (2)	32	100 Jointed	10		100 ties on 20-degree curve gage to 4'-9"
Seaboard Air Line Railroad	June 1961	Near Macclenny, Fla.	153	AAR Type "E" Direct Fixation	30	100 Jointed	80		Installed with steel tie plates on prestressed concrete trestle
Union Railroad	July 1961	East Pittsburgh, Pa.	2	AAR Type "E" Direct Fixation	30	131			
Canadian National Ry.	October 1961	Drummondville, Que., Canada M. P. 102	800 5	AAR Type "E" Indirect Fixation AAR Type "E" Direct Fixation	30	132 Welded	75	16,000,000	
Canadian National Ry.	October 1961	Drummondville, Que., Canada M. P. 103	1000	French RS	26 30	132 Welded	75 75	16,000,000	
Gen. Portland Cement Co.	October 1961		150	AAR Type "E" Direct Fixation	32	100 Jointed	10		24° 15' curve gage 4'-9-1/4"
Ideal Cement Co.	October 1961	Castle Hayne, North Car.	10108	AAR Type "E" Direct Fixation	30 36	85# ASCE 100# RE Jointed	25		
Atlanta & St. Andrews Bay Line Railroad	November 1961		16	AAR Type "E" Direct Fixation	30	115 Jointed	40		
Foreign, Venezuela	December 1961		1500 500	AAR Type "E" Indirect Fixation AAR Type "E" Direct Fixation	30	132 Jointed	60	20,000,000	
Atlantic Coast Line Railroad	1961	Adjacent to original near Four Oaks, N. C.	500	AAR Type "E" Direct Fixation	30	132 Welded	79	10,000,000	Installed without steel tie plates
United States Steel Co.	1961	Wyoming	577	Gerwick					

(1) The "Indirect Fixation" type of tie has a steel tie plate under the rail with a bearing pad under the tie plate. The rail is held vertically by a clip, the shoulders on the tie plate holding the rail transversely.

(2) The "Direct Fixation" type of tie has a 1:40 cant in the concrete surface under the rails and the rails bear directly on a bearing pad. The rail is held both vertically and transversely by a special clip.

Table as of November 1961

TABLE 13

Estimated Annual Cost of Installation of Concrete Ties vs. Wood Ties
of One Mile of Main Track Laid with 132 RE Continuous Welded Rail

Type	Item	Cost Installed	Years Life	Amortized at 4 percent	Salvage Credit	Net Amort-ized Cost	Annual Renewal Cost
CONCRETE TIES	Concrete Ties with Anchor Nuts	$29,100	50	$1,355	$ 0	$1,355	$582
	Tie Plates, Clips, Washers and Pads	10,095	25	646	40	606	419
	Bolts and Insulating Bushings	3,740 42,935	12-1/2	386 2,387	10 50	376 2,337	369 1,370
	Loss Due to Removing Usable Wood Ties	7,300		340		340	
	ANNUAL COST			$2,727	$50	$2,677	$1,370
WOOD TIES	Creosoted Oak Ties	22,800	25	1,460	0	1,460	900
	Tie Plates	10,600	25	678	80	598	415
	Track Spikes and Rail Anchors	6,400 39,800	12-1/2	661	36	625	550
	ANNUAL COST			$2,799	$116	$2,683	$1,865
Excess: Wood over Concrete				$72	$66	$6	$495

Note: Unit costs of ties, fastenings, store expenses and installation costs shown on Table 14.

6. Simultaneous readings taken on six ties indicate that there is considerable variation in the axle load carried by each tie, with an average value of 43 percent.

7. The recorded compressive concrete strains on the top and bottom surface of the tie at center line of rail indicate no reduction in impact by use of poly-ethylene or fiber-rubber pads under the steel tie plates.

8. The range in stresses in the bolts and clips under the passage of trains was a maximum when the bolt load was about 4,000 lb and a minimum under a bolt load of 10,000 lb.

9. The maximum recorded stress in the base of the 115 RE rail was 15,600 psi.

10. The maximum upward movement of the tie was 0.048 in and occurred as the wheel approached the tie. The maximum downward movement was 0.138 in and occurred when the wheel was over the tie.

The original installations of concrete ties have not been in service long enough to determine any maintenance trends, but the railroads advise that the performance in respect to holding surface and alinement has been good. Some loss in bolt tension has

TABLE 14

Unit Costs Used for Estimated Annual Costs
in Comparison of Concrete Ties vs. Wood Ties

Type	Item	Number of Units	Cost per Mile	Unit Cost				Total
				Price Each	Store Expense	Labor Installing	Payroll Taxes, etc.	
CONCRETE TIES	Concrete Ties	2,168	$26,420	$9.15*	$0.19	$2.25	$0.60	$12.19
	Anchor Nuts	8,672	2,680	0.30	0.01	Cast in Tie		0.31
	Tie Plates	4,336	5,105	0.985	0.065	0.10	0.025	1.175
	Rail Clips	8,672	3,225	0.35	.022	(a)		0.372
	Spring Washers	8,672	530	0.06	.001	(a)		0.061
	Plate Washers	8,672	175	0.02	–	(a)		0.02
	Tie Pads	4,336	1,060	0.24	.005	(a)		0.245
	Clip Bolts	8,672	1,970	0.16	.003	0.05	0.013	0.226
	Insulating Bushings	8,672	1,770	0.20	.004	(b)		0.204
WOOD TIES	Creosoted Oak Ties	3,250	$22,800	$5.00	$.010	$1.50	$0.40	$7.00
	Tie Plates	6,500	10,600	1.48	.09	0.05	0.01	1.63
	Track Spikes	13,000	1,635	0.088	.006	0.025	0.007	0.126
	Rail Anchors	8,272	4,765	0.48	.033	0.05	0.013	0.576

*　Base price $8.75 – Remainder is off line freight
(a)　Labor cost included in placing the plate
(b)　Labor cost included in placing bolts

taken place, the ties installed with the wood pads under the steel tie plates showing a greater loss than those with polyethylene pads.

A few ties have developed cracks through bolt holes, but most of these have been replaced with machine-made ties having greater concrete strengths. Several ties developed hairline cracks across the top of the tie in the center portion. This center cracking has been eliminated by a slight change in the details of the tie.

A study of the economics of the concrete tie versus the wood tie must necessarily be based on assumptions regarding the life of the concrete tie as well as the fastenings which are used with it. Fairly reliable information is available relative to tie life of treated oak ties although there is not much data covering the life of track fastenings.

Table 13 covers the estimated annual cost of one track mile on an amortized basis with 4 percent interest, using 50 years as the life of concrete ties and 25 years for creosoted oak ties. The detailed costs of the ties and fastenings are shown on Table 14. If the concrete tie is properly manufactured with rigid control of the aggregate and prestressing, a useful life of 50 years or even greater does not appear out of reach. The report of Committee 3—Ties and Wood Preservation, page 561, Proceedings, Vol. 59, indicates an average life of 23.2 years for treated oak ties in main track.

The item "Loss Due to Removing Usable Wood Ties" represents the loss of tie life in a complete out-of-face renewal of one mile of track less salvage plus the cost to pick up and reinstall at a new location. As will be noted, this loss is amortized over a period of 50 years, the same as the concrete ties. This cost will not be recurring at the end of the 50 year period.

The final column of Table 13, headed "Annual Renewal Costs," includes labor and material charges as a matter of information to indicate probable charges after normal renewal cycles are reached.

It is quite apparent from Table 13 that a saving of only $6 per track mile per year will be realized in the use of concrete ties over wood ties when they are installed in an existing track because of the loss of serviceable life of the wood ties in their original location and the cost of reinstalling elsewhere. However, it can be seen that a saving of $346 per track mile per year could be made if the concrete ties were installed in new track where such loss in the present wood ties would not occur.

Recommendations for future work on prestressed concrete ties with fastenings include the revised details of the tie, details of the clips, pads and insulating bushings, and acceptance testing.

The ties for the first phase of the investigation were made by Prestressed Concrete of Colorado, Denver, Colo., while the 12-in ties for the second phase were made by The American Concrete Crosstie Corporation of Tampa, Fla. The latter company also furnished the ties, without cost, for the original test installations on the Atlantic Coast Line Railroad and Seaboard Air Line Railroad.

Report on Assignment 8

Making Charcoal From Used Ties

G. A. Williams (chairman, subcommittee), A. B. Baker, R. G. Brohaugh, L. P. Drew, W. E. Fuhr, H. M. Harlow, M. J. Hubbard, W. R. Jacobson, W. L. Kahler, R. B. Radkey, L. S. Strohl.

Your committee submits the following report on the feasibility and economy of producing and marketing charcoal made from old scrap ties. The report is for information only.

For the past several years and prior to the assignment of this subject, inquiries have been made of various charcoal producers as to the feasibility of making charcoal from used ties. Of the ten producers contacted, seven replied that it was not feasible, one replied in the affirmative, and two were uncertain.

Of the seven who replied in the negative, various reasons were given. Among these reasons were that:

1. The cost would be prohibitive.
2. The creosote treatment would render the charcoal unsuitable for cooking purposes or use in the chemical industry.
3. The oil and foreign matter in old ties would make their use impractical.

In addtion to the above-mentioned inquiries, experiments were conducted on Pennsylvania Railroad property in 1959, by the use of a kiln consisting of an open-faced, sheet-steel cube with a chimney and several draft-control doors. These kilns were large enough to cover a stack of 100 ties at one burning. The stacks were burned for 14 days and were reported by the director of engineering research of the AAR, under date of August 18, 1959, as having produced a relatively good grade of charcoal. However, no cost figures were developed nor definite conclusions reached as to the practicability of such production.

It is the intention of your committee to reinaugrate additional experiments along this line, if possible, and to handle the matter further with other charcoal producers, particularly those who indicate there is a possibility of using old ties for the production of charcoal.

Report of Committee 28—Clearance

J. G. GREENLEE,
Chairman

R. L. WILLIAMS,
Vice Chairman

B. BRISTOW
C. W. HAMILTON
J. F. SMITH
E. E. MILLS
W. P. KOBAT
R. A. SKOOGLUN
J. A. CRAWFORD
M. E. VOSSELLER

J. D. BATCHELDER
C. O. BIRD
E. S. BIRKENWALD
D. H. BROWN
S. M. DAHL
R. D. ERHARDT
J. E. FANNING (E)

J. E. GOOD
A. R. HARRIS
W. F. HART
C. F. INTLEKOFER
M. L. KOEHLER
J. R. MOORE
R. C. NISSEN
J. F. PEARCE
C. E. PETERSON
R. C. RANKIN
W. S. RAY
J. W. WAGNER
J. W. WALLENIUS
H. G. WHITTET
M. A. WOHLSCHLAEGER
Committee

(E) Member Emeritus.
Members listed in bold face are the official representatives of the Engineering Division, AAR.

To the American Railway Engineering Association:

Your committee reports on the following subjects:

1. Revision of Manual.
   Progress report, with recommendations submitted for adoption .... page 338

2. Clearances as affected by girders projecting above top of track rails, structures, third rail, signal and train control equipment, collaborating with Committee 18, with Communication and Signal Section, AAR, and with Mechanical and Operating–Transportation Divisions, AAR.
   No report. Assignment completed.

3. Review clearance diagrams for recommended practice, collaborating with AREA Committees concerned, and the AAR Joint Committee on Clearances.
   No report. Committee is giving further study to proposed Fig. 9—Clearance Diagram for Overhead Bridges and Other Structures not Otherwise Provided for, which is being held in abeyance.

4. Compilation of the railroad clearance requirements of the various States.
   Progress report, submitting information on New York State statutes effective April 20, 1961 .... page 338

5. Clearance allowances to provide for vertical and horizontal movements of equipment due to lateral play, wear and spring deflection, collaborating with the Mechanical Division AAR.
   Progress report, submitted as information .... page 339

6. Compilation in table form of offsets for overhanging loads on curves.
   Final report, with recommendation submitted for publication in the Manual page 339

7. Methods of measuring high and wide shipments, collaborating with Mechanical Division, AAR.
No report. Assignment completed.

8. Review present methods of presenting published clearance information to determine how this can be simplified and/or standardized.
Progress report, submitted as information .... page 340

9. Review clearance records of various railroads, looking to developing a standardized method for charting all obstructions.
Progress report, submitted as information .... page 340

THE COMMITTEE ON CLEARANCES,
J. G. GREENLEE, *Chairman.*

AREA Bulletin 568, December 1961.

Report on Assignment 1

Revision of Manual

B. Bristow (chairman, subcommittee), J. G. Greenlee, R. L. Williams, J. D. Batchelder, C. O. Bird, E. S. Birkenwald, S. M. Dahl, J. E. Good, W. F. Hart, M. L. Koehler, E. E. Mills, J. R. Moore, J. F. Pearce, J. F. Smith, J. W. Wallenius, H. G. Whittet, J. E. Fanning.

Your committee submits for adoption the following recommendations with respect to the Manual:

Page 28–1–1

GENERAL INFORMATION

Reapprove with the following revisions:

Change Par. 2 to read:

"2. On curved track the clearances shall be increased to allow for the overhang and tilting of a car 85 ft long, 65 ft between centers of trucks, and 15 ft 1 in high."

Change the Note to read:

"*Note—Pars. 1, 2, 3, and 4 apply to Figs. 1 to 8, incl. Par. 5 applies to Figs. 1 to 8, incl., and also to Fig. 1, Part 3, this Chapter.*"

Report on Assignment 4

Compilation of Railroad Clearance Requirements of the Various States

J. F. Smith (chairman, subcommittee), C. O. Bird, B. Bristow, D. H. Brown, R. D. Erhardt, J. E. Good, J. G. Greenlee, A. R. Harris, W. P. Kobat, E. E. Mills, J. R. Moore, C. E. Peterson, R. A. Skooglun, J. W. Wagner, J. W. Wallenius, R. L. Williams, M. A. Wohlschlaeger.

Your committee submits, as information, the clearance dimensions of the State of New York. These are given by column numbers, for convenience, for addition to the clearance chart dated October 1, 1960, which was published in Bulletin 561, December 1960. The revision date of the chart is July 20, 1961.

The New York statutes became effective on April 20, 1961.

LEGAL CLEARANCE REQUIREMENTS, STATE OF NEW YORK

Column No.	4&5	6	7	8	9, 10, 11	12, 13, 14, 15
Dimension	13' 6"	15' 0"	18' 0"	19' 0"	13' 6"	22' 0"

Column No.	16	17	18	19	20	21
Dimension	18' 0" note 3	18' 0"	8' 6"	8' 0"	8' 6"	8' 0"

Column No.	22	23	24	25	26	27 note 9
Dimension	8' 0" note 3	8' 0"	0' 8"	5' 1"	4' 0"	5' 7"

Column No.	28	29	30	31	32	33
Dimension	4' 0"	8' 6" notes 8, 10	8' 6"	3' 0"	6' 0"	0' 4"

Column No.	34	35	36	37	38	39
Dimension	3' 0"	8' 6"	6' 5" note 11	8' 0"	E	8' 0"

Report on Assignment 5

Clearance Allowances to Provide for Vertical and Horizontal Movements of Equipment Due to Lateral Play, Wear and Spring Deflection

Collaborating with the Mechanical Division, AAR

E. E. Mills (chairman, subcommittee), C. O. Bird, B. Bristow, D. H. Brown, S. M. Dahl, R. D. Erhardt, J. G. Greenlee, C. W. Hamilton, A. R. Harris, W. F. Hart, C. F. Intlekofer, J. R. Moore, J. F. Pearce, C. E. Peterson, W. S. Ray, J. F. Smith, R. L. Williams, M. A. Wohlschlaeger.

Your committee submits the following report of progress in the collection of data on the static and dynamic behavior of flat-car trailer carriers and three-level auto carriers.

Tests were run on the Burlington Railroad on May 24, 1961, from Chicago to Galesburg, Ill. at speeds up to 70 mph on the main line; on May 25, 1961, from Galesburg to Barstow, Ill., and return at speeds up to 63 mph on a branch line; and on May 26, 1961, from Galesburg to Chicago on a passenger train with an 85-ft flat car carrying two fully loaded trailers at speeds up to 83 mph.

On May 30, 1961, a similar car equipped with a three-level auto rack and loaded with 12 new automobiles was run from Chicago to St. Paul, Minn., at speeds up to 70 mph, but predominately around 50 mph because of the length of the train.

The results of these tests will be the subject of a future report.

Report on Assignment 6

Compilation in Table Form of Offsets for Overhanging Loads on Curves

W. P. Kobat (chairman, subcommittee), J. D. Batchelder, J. A. Crawford, J. G. Greenlee, M. L. Koehler, R. C. Nissen, C. E. Peterson, R. C. Rankin, J. F. Smith, M. E. Vosseller, R. L. Williams.

Your committee now submits for adoption and publication in the Manual the two tabulations showing offsets for overhanging loads on curves that were presented as information with last years report (Proceedings, Vol. 62, 1960, page 412).

Sheet 1 gives the offsets for a 1-deg curve at the middle of car or load for distances between truck centers from 20 to 65 ft, with illustrations. Sheet 2 gives the end offsets for a 1-deg curve for end overhang from 6 to 40 ft.

The purpose of the tabulations is to save time in calculating clearances for excessive-dimensioned shipments on curves.

Report on Assignment 8

Review Present Methods of Presenting Published Clearance Information to Determine How This Can Be Simplified and/or Standardized

J. A. Crawford (chairman, subcommittee),E. S. Birkenwald, D. H. Brown, J. G. Green-lee, C. W. Hamilton, C. F. Intlekofer, J. W. McMillen, R. C. Nissen, C. E. Peterson, R. C. Rankin, R. A. Skooglun, J. F. Smith, J. W. Wagner, R. L. Williams, M. A. Wohlschlaeger.

Your committee submits the following report of progress in its study of present methods of presenting clearance information. Analysis and evaluation of the present representation of the different railroads will provide the basis upon which to determine how published clearance information can be simplified and/or standardized and for recommendations by your committee.

Report on Assignment 9

Review Clearance Records of Various Railroads, Looking To Developing a Standardized Method For Charting All Obstructions

M. E. Vosseller (chairman, subcommittee), C. O. Bird, E. S. Birkenwald, D. H. Brown, J. A. Crawford, J. G. Greenlee, C. W. Hamilton, C. F. Intlekofer, R. C. Nissen, J. F. Pearce, C. E. Peterson, R. C. Rankin, R. A. Skooglun, J. W. Wagner, R. L. Williams, M. A. Wohlschlaeger.

Your committee submits its first report of progress on this assignment.

Sample clearance charts and suggestions were requested and received from subcommittee members. A chart containing the best features of the information received is being prepared for submitting to subcommittee members for their review and comments.

After receipt of advice from the subcommittee the chart will be finalized for presentation to the full membership of Committee 28 for review and advice.

Report of Committee 25—Waterways and Harbors

J. G. MILLER	BENJAMIN ELKIND (E)
M. A. MICHEL	OSCAR FISCHER
R. J. CLARKE	DOSWELL GULLATT
J. C. FENNO	L. W. HAYDON
	H. F. KIMBALL
L. E. BATES	SHU-T'IEN LI
G. W. BECKER	W. J. O'CONNELL
E. A. BEEKLEY	M. S. PATTERSON
G. W. BENSON	A. L. SAMS
R. L. BOSTIAN	E. M. SKELTON
C. M. BOWMAN	F. R. SPOFFORD
B. M. DORNBLATT	*Committee*

JOHN F. PIPER, *Chairman*
FRANK J. OLSEN,
    Vice Chairman

(E) Member Emeritus,
Members listed in bold face are the official representatives of the Engineering Division, AAR.

To the American Railway Engineering Association:

Your committee reports on the following subjects:

1. Revision of Manual.
   Progress report, with recommendations submitted for adoption .... page 342

2. Current policies, practices and developments dealing with navigation projects.
   This committee has devoted the work year to reviewing publications concerning waterway matters which were issued by the AAR and various public sources. It has been decided that, since this is not specifically a subject normally within the function of an engineering committee, this subcommittee will be assigned new work in the forthcoming year, and the present course of study will be dropped.

3. Bibliography relating to benefits and cost of inland waterway projects involving navigation.
   Progress report, submitted as information, including six additional references, with annotations .... page 343

4. The use of hydraulic models for the study and resolution of waterway problems.
   The committee has reviewed various publications from the Waterways Experiment Station, U. S. Army Corps of Engineers, Vicksburg, Miss., for extraction of pertinent information and criteria to be used in the design of hydraulic models. It is presently exploring other sources for relevant information on this subject. It is also preparing data which will be incorporated in the Manual, setting forth application and design criteria for hydraulic models.

6. Planning, construction and maintenance of rail-water transfer facilities. This committee is directing its efforts to the study of rail-water terminal facilities for roll-on, roll-off, lift-on, lift-off, and conveyor-type operations.

7. Relative merits and economics of construction materials used in waterfront facilities.

Report on "Service Performance Records of Greenheart in Docks and Harbors of the United Kingdom" .... page 345

Report on "Azobe as a Construction Material and Its Comparison with Greenheart in Waterfront Facilities" .... page 345

The Committee on Waterways and Harbors,
J. F. Piper, <i>Chairman.</i>

AREA Bulletin 568, December 1961.

Report on Assignment 1

Revision of Manual

J. F. Piper (acting chairman, subcommittee), J. G. Miller, A. L. Sams, F. R. Spofford.

Your committee submits for adoption the following recommendations with respect to Chapter 25 of the Manual:

Pages 25–1–1 to 25–1–3, incl.

PUBLIC IMPROVEMENTS—THEIR COSTS AND BENEFITS

Reapprove without change.

Pages 25–5–1 to 25–5–7, incl.

DREDGING SPECIFICATIONS

Reapprove without change.

Pages 25–5–8 and 25–5–9

TYPES OF DREDGES AND THEIR RESPECTIVE USES

Reapprove without change.

Page 25–5–9

ALLOWABLE OVER-DEPTH IN DREDGING OPERATIONS TO OBTAIN THE DESIRED OPERATING DEPTH

Reapprove without change.

Page 25-5-10

USUAL SLOPES TAKEN IN DEEP WATERWAYS

Reapprove without change.

Pages 25-5-10 to 25-5-12, incl.

SOUNDING METHODS

Reapprove without change.

Parts 2, 3 and 4 of Chapter 25 will be reviewed and revised during the coming year.

Report on Assignment 3

Bibliography Relating to Benefits and Costs of Inland Waterway Projects Involving Navigation

M. A. Michel (chairman, subcommittee), G. W. Becker, C. M. Bowman, B. M. Dornblatt, H. F. Kimball, W. J. O'Connell, M. S. Patterson.

Your committee submits the following report of progress which presents six additional references, with annotations.

1960

1. May, 1960—Committee Print No. 11, "Future Needs for Navigation", U. S. Select Senate Committee on National Water Resources Pursuant to S. Res. 48, 86th Congress. U. S. Government Printing Office, Washington, D. C.

Committee Print No. 11 is a 28-page publication, prepared by the Corps of Engineers, designed to inform the Senate Committee as to future navigation needs of the nation. Briefly, it indicates that more than 10,000 miles (about half) of the existing waterway system requires improvement of channels, locks and terminals at a cost of $4.4 billion, and that 3,000 miles of new waterways could be added to the present navigation system at an additional cost of $3.5 billion.

Of considerable interest in the report is the fact that it uses the *average* of 3,000,000 ton miles of freight per mile of road per year, as though it were a *capacity* measurement on a railroad, instead of an overall average for Class I Railroads. This, and other controversial issues contained in the report, resulted in the preparation and filing of the succeeding publication with the Select Senate Committee.

1961

2. January, 1961—Committee Print (No number) "Supplemental Information on Subjects covered by the Committee's Studies", U. S. Select Senate Committee on National Water Resources Pursuant to S. Res. 48—86th Congress. U. S. Government Printing Office, Washington, D. C.

Of specific interest in this "Supplemental Information" is a letter by Daniel P. Loomis, dated Sept. 12, 1960, which gives the Select Committee the views of the AAR with respect to Committee Print 11; wherein, among other things, it is pointed out that certain heavy railroad traffic segments "had exceeded 80 million ton miles per mile of road per year."

This Committee Print also contains a "Future Needs for Transportation" report, by R. L. Banks & Associates, which is a "critical evaluation of Committee Print No. 11". It points out a number of fallacies in the Army Engineers' Report, and contains a chapter entitled "Justifying Waterway Improvements", which deals with (1) Economic Analysis; (2) Cost Estimation; (3) Benefit Measurement.

3. January 30, 1961—Report No. 29, "Report of The Select Committee on National Water Resources—Pursuant to S. Res. 48—86th Congress—Together with Supplemental and Individual Views." Government Printing Office, Washington, D. C.

In this, the Senate Committee's Report, under a section dealing with "Water resources development neds"—NAVIGATION—the Committee mentions both of the foregoing reports, i.e., Committee Print No. 11 and the Banks critical analysis, in a neutral manner, as follows: "The Committee reproduces these two reports in order to make available information on these subjects without endorsing either view."

4. "Waterways of The United States"—Rivers, Harbors, Lakes, Canals—Published by the National Association of River and Harbor Contractors, 15 Park Row, New York 38.

This book stresses the importance of the Nation's waterways system; contains considerable propaganda in favor of inland waterway projects; points out the "16 steps from conception to completion of waterway improvement projects"; and indicates benefits vs costs for 16 of the Nation's major waterways.

5. "The Seaway Story", by Carleton Mabee. Published by the Macmillan Company, New York.

Although this volume does not contain data involved directly in benefit cost matters, it points to a number of prime factors involved in connecion with forwarding the St. Lawrence Seaway Project, politically. It also narrates many allied costs which were not indicated in the Seaway Study, such as deepening and improving Great Lakes harbors; improving the Cal-Sag channel; deepening of connecting channels, etc.

6. March, 1961—"Saint Lawrence Seaway Development Corporation, Annual Report 1960. U. S. Government Printing Office, Washington, D. C.

Of outstanding interest in this report is the fact that when an actual record is kept in a businesslike manner—as provided by law—the interest rate is ". . . the current average rate on current marketable obligations of the United States of comparable maturities as of the last day of the month preceding the issuance of the obligation . . .". (Laws of 83rd Congress, 2nd Session, Public Law 358 approved May 13, 1954, etc.) During the construction of the Seaway this average rate was 3.4 percent. This is of particular interest to those studying benefit cost ratio as applied in the Corps of Engineers' reports, wherein rates in the neighborhood of $2\frac{1}{2}$ percent are in current use.

Report on Assignment 7

Relative Merits and Economics of Construction Materials Used in Waterfront Facilities

Shu-t'ien Li (chairman, subcommittee), C. M. Bowman, B. M. Dornblatt, E. M. Skelton, F. R. Spofford.

Your committee submits its report on this assignment in two parts, as follows:

Part 1—Service Performance Records of Greenheart in Docks and Harbors of the United Kingdom.

Part 2—Azobé as a Construction Material and Its Comparison with Greenheart in Waterfront Facilities.

Part 1

Service Performance Records of Greenheart in Docks and Harbors of the United Kingdom

By Shu-t'ien Li

Chairman, Subcommittee 7, Committee 25—Waterways and Harbors

The outstanding merits of genuine greenheart *(Nectandra Rodioei)* for marine applications have been more convincingly revealed in other parts of the world than in the United States. Here, its use in marine construction is comparatively recent; no conclusive service-life range could be established.

In an effort to ascertain its serviceable life, inquiry was directed to the United Kingdom. Through the courtesy of Alexander McDonald, M.I.C.E., secretary, the Institution of Civil Engineers (London), R. P. Woods, principal scientific officer, Timber Development Association Ltd. (21 College Hill, London, E.C. 4), based on a survey of dock and harbor authorities carried out in the early 1950's, cites the following service performance records of greenheart used in waterfront facilities in the United Kingdom:

Location	Life or Condition	Location	Life or Condition
Ayr	50 years good	Harwich	25 years plus
Dover	30 years plus	Littlehampton	Good
Dundee	over 50 years	Poole	70 years approximately
Grangemouth	40/50 years	River Tees Ports	30/40 years
Granton	80 years life	Southampton	50 years and up
Greenock	Up to 75 years life, excellent	Sunderland	At least 50 years
		Troon	50 years good

It is felt that these service records of greenheart over the past 80 years are of sufficient significance to merit its use in sea water in comparison with treated soft wood.

Part 2

Azobé as a Construction Material and its Comparison with Greenheart in Waterfront Facilities

By Shu-t'ien Li

Chairman, Subcommittee 7, Committee 25—Waterways and Harbors

Nomenclature

"Azobé" is the original French name of the wood and of the tree. In Africa and elsewhere, the wood is also known as "ekki." Its scientific name is *Lophira procera* after A. Chev. It is a tropical hardwood.

Source and Production

Azobé is a large sturdy tree growing only in Central West Africa. The finest stands occur in the forests of the Cameroun. The area had been developed under French mandate after World War I, was a trust territory since World War II, and has gained independence in recent years.

Azobé plank and timber are produced today in large, modern sawmills equipped with American machinery. Production has increased in each of the past 10 years during the 1950's and now attains 20 million board feet per annum. Practically all of this production has been consumed in Europe and South Africa. Production is being further increased at present to provide for requirements in the United States and Canada.

Available Sizes

All sizes of Azobé are sawn as plank or timber, and are available within the following limits:

(1) As boxed heart timber from 14 by 14 in up to 20 by 20 in or slightly larger.
(2) Sizes 10 by 10 in or 12 by 12 in are not readily obtainable except in shorter lengths.
(3) As timber up to and including 8 by 16 in by 20 ft.
(4) As plank from 1 by 6 in up to 6 by 12 in or 7 by 10 in, free of heart.
(5) Most readily available lengths are between 12 and 20 ft, possibly up to 23 ft.
(6) Longer lengths are obtainable only in boxed heart sizes.
(7) Small-dimension sizes are also available.
(8) Not available as poles in long lengths for piling.

Outstanding Merits

As an industrial and structural timber, Azobé has the following outstanding merits:

(1) Superior strength and durability.
(2) Outperforms and outlasts ordinary woods and corrosible metals in installations subject to severe abuse or strain.
(3) Saving in maintenance cost as a result of durability.
(4) Saving in material as the result of higher strength.
(5) Saving in cost in the long run wherever its application is appropriate.
(6) Resistant to:
    (a) Marine borers.
    (b) Fungi and decay.
    (c) Termites.
    (d) Acids.
    (e) Abrasion.
    (f) Flame spread.
(7) Excellent as a non-conductor of electricity.
(8) Readily workable with power tools, though an exceptionally hard wood.
(9) Does not split like many timbers of comparable density.

Marine, Transportation, and Industrial Applications

The first broad commercial use of Azobé was introduced in France during its mandate of the territory. This wood has since been used extensively there for marine, railway, and utility construction. Its use for various kinds of flooring has spread in Europe and in chemical plants in both Europe and the United States. The following accounts describe its applications:

(1) *Marine Applications*

(a) In the French channel ports, where tides of 20 to 30 ft are not uncommon, the jetties and docks built of azobé in the 1930's which survived bomb damage during World War II are up to now as sound and durable as when installed. Since the war, construction in these harbors has been designed primarily in azobé.

(b) Crandall Dry Dock Engineers of Cambridge, Mass., has had experience with azobé in marine construction for the Chamber of Commerce at Boulogne-sur-Mer, France. In a communication dated August 3, 1961, Kenneth M. Childs, Jr., of this firm states:

"I was able to see pieces of azobé, which had been in the water for perhaps 15 years in this port. These pieces showed no signs at all of any attack by marine borers. This could be due in part to the wood's being partially submerged or to the lack of borer infection, since all of the timber used in waterfront structures is azobé and, thus, there is no readily available source of food for any borer which may be brought into the waters by wooden hull vessels. It is significant, however, that there was no sign of attack, and from the reports I have received from the engineers of the Ponts et Chaussées, there is no attack in any of the azobé in the port."

(c) Azobé has been used in Belgium in marine work by Crandall Dry Dock Engineers, at Ostende for the City of Ostende and at Nieuport for the Belgian Navy.

(2) *Transportation Applications*

(a) One line of the Metro trains in Paris has been transformed into a swift, noiseless, smooth-riding model of urban transportation by having the trains equipped with rubber tires running on a track of azobé, providing superior traction with long-lasting life.

(b) In Europe, azobé is widely used now for rail-car floorings and truck floorings to withstand rough abuse without frequent replacement.

(3) *Factory Flooring Applications*

Besides rail-car floorings and truck floorings, azobé has also been widely used in Europe for industrial floorings where severe use and excessive abrasion occur and acid proofing is necessary.

(4) *Utility Applications*

In the French mountains, power-transmission towers with azobé crossarms have withstood severe ice conditions better than steel crossarm towers. On many of these mountain lines, steel crossarm towers are being paralleled with azobé crossarm towers as a precaution against winter service interruptions.

(5) *Process-Industries Applications*

(a) Many chemical plants in Europe use azobé in their process equipment for tanks or filter presses, or wherever a wood is needed that can withstand severe erosion conditions.

(b) In the United States azobé has been found to be of unique value in the tank rooms of electrolytic copper refineries. After considerable experimenting, they have found that, from a cost standpoint, azobé is by far the best material for use as capping boards in their electrolytic tank house.

(c) The use of azobé in the process tanks of a chemical plant producing phosphoric acid has shown that there was no deterioration in this wood after 6 months of operation in a solution of hot, 175 F, 45 percent phosphoric acid subject to erosion from agitation. The wood stood up even better than the 316 stainless-steel plates which were bolted on the tank timbers.

Advantages and Limitations in Marine Use

Due to its inherent strength, durability, resilience, and resistance to marine borers, fungi and decay, termites, abrasion, and flame spread, azobé is well adopted to:

(1) Marine capping.
(2) Marine decking.
(3) Suspended marine fendering sections.

However, on account of its limited lengths, generally not over 23 ft, though longer lengths can be obtained in special cases with restrictive quantity and at a premium, it should not be used for:

(1) Marine bearing piles.

(2) Marine fender piles and dolphins, where usually much longer lengths are required, without advance confirmation from the producers.

Marine-Borer Resistance—Azobé vs. Greenheart

In French channel ports, azobé has remained sound for more than 25 years since the 1930's.

The installations of this wood in Belgian harbors have been comparatively recent; no decisive service performance could yet be concluded.

A report from the Clapp Laboratories states that a sample of azobé had been exposed at Wrightsville Beach, N. C., for the past six months prior to October 1960, with no marine-borer attack, as yet.

Crandall Dry Dock Engineers had experience with a sample exposed 4 months prior to October 1960, in the harbor at Kingston, Jamaica, and noted only two very small penetrations by *Teredo*. It must be borne in mind, however, that *Teredo* always enters timber in minute pin holes during its larvae stage. Such pin holes are almost undetectable macroscopically except by careful inspection. The *Teredo* animals grow inside the timber.

At practically the same and nearby locations of the foregoing, Greenheart has shown the following service performances:

Location	Years of Service up to 1955	Remarks of Condition
France		
Dunkerque	24	Slight attack
Havre	13	Good
Belgium		
Ostende	16	Slight attack
Zeebrugge	6	Slight attack
United States		
Hampton Roads, Va.	10	Good
West Indies		
Trinidad (Br. Is.)	10–15	Failed

The performance of Greenheart at Dunkerque and Havre, France, is comparable with that of azobé in French channel ports. Whether the sample of azobé at Wrightsville Beach, N. C., will last as long as Greenheart at Hampton Roads, Va., and the sample of azobé at Kingston, Jamaica, will perform as well as that of Greenheart at Trinidad, remain to be seen.

Available Lengths—Azobé vs. Greenheart

Azobé is readily available only in sawn timber not over 23 ft in length. While there is no difficulty to obtain azobé timber up to 8 by 16 in by 20 ft, greenheart of 16 by 16 in by 40 ft is readily available.

Azobé is not available as poles in long lengths for piling. On the other hand, round greenheart piles are readily available in lengths up to 75 ft, and to 85 ft on an accumulative basis. Lengths of 80 ft and over are available upon special order. All greenheart piles are graded in accordance with ASTM standards, with butt and tip sizes conforming to Class "A" or "B" stipulations, whichever is required.

Because of the high strength possessed by greenheart, extremely long piles can be built up in the field by splicing.

In the United States, greenheart has been exempted from the "Buy American Act."

Cost—Azobé vs. Greenheart

As of October 13, 1960, azobé was about twice the cost of creosoted southern yellow pine in the United States, according to Kenneth M. Childs, Jr., of Crandall Dry Dock Engineers.

During the past decade, the higher increase in the cost of creosoting and the less increase in the price of greenheart have leveled off the slight higher cost for greenheart. Back in May, 1953, writing in The Dock and Harbor Authority, A. M. Robertson states: ". . . as it is now the same cost as properly creosoted softwood and is vastly superior, . . ."

As of end of September 1959, sawn greenheart timber, f.o.b. dock, port of entry, Gulf or Atlantic coast, had a price structure within the following limits:

Size	Run of Lengths	Price per MBM not over
Up to 4 by 12 in	10 ft and longer	$275
14 by 14 in	10 ft and longer	300

Lengths of Greenheart Piles, Ft		Cost per Lin Ft
50	....	$2.00
75	....	2.50

Mechanical Properties—Azobé vs. Greenheart

Results of tests on azobé conducted by the Imperial Institute, London, and those on greenheart by the Forest Products Research Laboratory, United Kingdom, are tabulated on page 350 for purposes of comparison, with specimens of azobé at 12.2 percent moisture content and of greenheart at 12.0 percent moisture content. This 12.0 percent moisture content represents a standard air-dry moisture content. To derive allowable working stresses, a moisture content under specific service conditions must be used to transform the mechanical properties according to methods given in the 1955 Wood Handbook of the Forest Products Laboratory of U. S. Department of Agriculture, and a factor of safety applied.

Specific Design

In any design in azobé, as in greenheart, the design has to be proportioned according to the specific allowable unit stresses. The greater strength of azobé permits the use of smaller cross sections or larger spacings, or both, to withstand the same load, or absorb the same energy, as larger sections at smaller spacings of ordinary timber.

Alternative designs of treated pine (or fir) and untreated azobé will require different designs. To simply substitute treated softwood design by untreated azobé would lose the economic advantages of the higher strength of the latter.

Comparative Economics

Any comparative economics in the use of azobé versus treated soft wood has to be figured on an annual cost or a capitalized cost basis, taking not only the first cost, maintenance cost, and replacement cost of each into consideration but also the longer life expectancy of azobé.

MECHANICAL PROPERTIES OF AZOBE' AND GREENHEART

Properties	Azobé (Lophira procera)	Greenheart (Nectandra rodioei)
Moisture content, percent	12.2	12.0
Unit weight, lb per cu ft	65.5	62.0
Static bending:		
Fiber stress at elastic limit, psi	15,420	16,200[1]
Modulus of rupture, psi	25,460	26,900
		30,500[2]
Modulus of elasticity, psi	3,000,000	2,970,000
		3,400,000[2]
Compression parallel to grain:		
Fiber stress at elastic limit psi	10,260	10,140[1]
Maximum crushing strength, psi	11,610	14,100
		14,940[2]
Modulus of elasticity, psi	3,388,000	4,140,000[1]
Compression perpendicular to grain:		
Fiber stress at elastic limit, psi	2,870	2,560[3]
Shearing parallel to grain:		
Shearing strength, psi		2,830
Radial, psi	2,650	
Tangential, psi	2,790	
Tension perpendicular to grain:		
Tensile strength, psi		1,020[1]
Radial, psi	1,452	
Tangential, psi	2,315	
Hardness:		
Indenting load on side grain, lb		2,520
		2,650[2]
Radial surface, lb	4,140	
Tangential surface, lb	3,980	

[1] Results of University of Michigan tests on black variety at 14.4 percent moisture content.
[2] Small-size test specimens.
[3] Approximate calculated value.

Further Evaluation of Azobé

Azobé is a promising natural material for many marine, transportation, and indus-trial applications. Further evaluation of its physical and mechanical properties and additional long-duration test boards under different marine environments are basic toward its general introduction into the United States.

Report of Committee 27—Maintenance of Way Work Equipment

R. S. RADSPINNER,
Chairman
R. W. BAILEY,
Vice Chairman
R. M. JOHNSON,
Sec'etary

R. E. BERGGREN
G. L. ZIPPERIAN
W. M. LUTTS
G. E. ROBERTS
R. O.CASSINI
H. D. HAHN
W. GLAVIN

R. M. BALDOCK
T. S. BEAN
C. T. BLUM
R. E. BUSS
L. B. CANN
G. R. COLLIER
L. E. CONNER
B. E. CORS
D. E. COWELL
J. W. CUMMINGS
A. C. DANKS
K. J. DECAMP
J. O. ELLIOTT
E. H. FISHER
W T. FRIEDLINE
S.E. HAINES
W. T. HAMMOND
E. W. HODGKINS
HAYNIE HORNBUCKLE
R. A. HOSTETTER
N. W. HUTCHISON
W. R. JACOBS
R. K. JOHNSON
H.E. KENISTON
M. E. KERNS
S. H. KNIGHT
W. F. KOHL
W. E. KROPP

JACK LARGENT
C. F. LEWIS
H. F. LONGHELT
G. J. LYON
PAUL MARTIN
C. E. MCENTEE
E. L. MIRE
C. F. MONTAGUE
A. W. MUNT
V. W. OSWALT
P. G. PETRI
H. C. POTTSMITH
J. E. REYNOLDS
J. W. RISK
F. E. SHORT
F. N. SNYDER
M. M. STANSBURY
J. E. SUNDERLAND
M. C. TAYLOR
T. H. TAYLOR
H. A. THYNG
S. E. TRACY
C. R. TURNER
L. B. WATERMAN
ALFRED WISMAN, JR.
F. E. YOCKEY

Committee

Members listed in bold face are the official representatives of the Engineering Division, AAR.

To the American Railway Engineering Association:

Your committee reports on the following subjects:

3. Standardization of parts and accessories for work equipment.
   Continuing subject, with work being completed on hydraulic tanks and
   fittings. Other components or accessories of machines will be reported on
   both as information and Manual recommendations.

4. Study of average annual costs of repairing and servicing various widely
   used multi-tool power tampers.
   The committee recommends that this subject be dropped due to lack of
   realistic material and information available to give accurate and satisfactory
   report.

5. Set-offs and set-off attachments for work equipment.
Final report, presented as information .... page 366

6. Procurement and stocking of parts and materials for the repair of work equipment.
Information is being assembled from railroads to study various methods now being used. Report will show most desirable procedure to benefit all departments on average railroad.

7. Vehicles and equipment for routine track inspection.
Final report, presented as information .... page 369

8. Equipment for the control and performance of jacking in track surfacing operations.
Information is being collected to include latest power jacks and surfacing equipment. Report will include operation and specifications of each unit.

THE COMMITTEE ON MAINTENANCE OF WAY WORK EQUIPMENT,
R. S. RADSPINNER, Chairman.

AREA Bulletin 568, December 1961.

Report on Assignment 1
Revision of Manual

R. S. Radspinner (chairman, subcommittee), R. W. Bailey, R. M. Baldock, L. B. Cann, E. H. Fisher, S. E. Haines, W. T. Hammond, R. K. Johnson, R. M. Johnson, M. E. Kerns, S. H. Knight, Jack Largent, H. F. Longhelt, Paul Martin, E. L. Mire, V. W. Oswalt, J. E. Reynolds, J. W. Risk, F. E. Short, L. B. Waterman.

Your committee submits for adoption the following recommendations with respect to Chapter 27 of the Manual:

Pages 27–1–3 and 27–1–4

CARE AND OPERATION OF MAINTENANCE OF WAY EQUIPMENT

Page 27–1–3. Add the words "It is desirable that" to the beginning of Par. 2, making the first sentence of this paragraph read: "It is desirable that adequate instructions for the care and operation of maintenance of way equipment should be issued." Change the first word of the second sentence of Par. 2 from "Such" to "Suitable."

Page 27–1–4. Change the first word of Par. 10 from "Necessary" to "Adequate."

Pages 27–1–5 to 27–1–9, incl.

WIRE ROPE USED WITH WORK EQUIPMENT

Pages 27–1–5 and 27–1–6. Rewrite Arts. 1, 2, 3, and 4 to read as follows:

1. General

While wire rope and manila rope have certain advantages for specific uses, the latter is limited in general to manual operations because it is easier on the hands and because of its flexibility and elastic qualities. For use with work equipment, wire rope, size for size and weight for weight, has the advantage of greater strength, greater resistance to abrasion and crushing, greater reserve strength, and greater stability under all types of weather conditions.

2. Construction

Wire rope is composed of wires, strands, and core. The wires are drawn to a predetermined size and helically laid to a definite pitch in various arrangements to form a strand. Then the required number of strands are helically laid to a definite pitch around a core to form the wire rope. The wire rope core may be a Fiber Core (FC), a Wire Strand Core (WSC) or an Independent Wire Rope Core (IWRC).

The size, number, and arrangement of wires, the number of strands, the pitch, and the type of core in a rope are determined largely by the service for which the rope is to be used.

In general, the greater the number of wires in a strand and the greater the number of strands, the more flexible the rope.

Guy ropes and sash and ball cords do not require flexibility and are usually made of six strands of seven wires each. Hoisting ropes require flexibility and are usually made up of 6 strands of 19 (16–26) wires per strand and 6 strands of 37 (21 to 49) wires per strand. Some elevator and crane hoisting ropes are made of 8 strands of 19 wires per strand.

Depending upon the service conditions, 6-strand ropes may have a Fiber Core (FC), a Wire Strand Core (WSC), or an Independent Wire Rope Core (IWRC), and may be either regular lay or lang lay. Eight strand ropes are usually regular lay with Fiber Core.

Wire rope for use with work equipment is generally furnished in improved plow steel grade in which the wire tensiles range from 220,000 to 250,000 psi minimums. Guy ropes and sash cords may be furnished in iron grade having wire tensiles ranging from 70,000 to 80,000 psi minimums. Improved plow steel grade has the greater strength and is particularly resistant to abrasion, shock, vibration, and fatigue.

During manufacture, the wire rope is thoroughly coated with a suitable lubricant. If the wire rope has a fiber core, the fiber is thoroughly impregnated with a lubricant suitable for the preservation and lubrication of the core fibers. Factory lubrication of a wire rope cannot be expected to last throughout the life of a rope and must be supplemented while in use. Failure -to lubricate wire ropes in the field, and especially those exposed to corrosive elements or under severe stress or bending conditions, will cause premature corrosion, wear, and/or fatigue to the wires and early drying out of the fiber core, all resulting in greatly reduced rope life.

The function of a wire rope core, be it FC, WSC, or IWRC, is to support the strands laid around it. If the core is allowed to deteriorate, then rope life will be affected. The life of the rope is in a great degree measured by the life of the core. In 6-strand ropes a WSC or IWRC adds about $7\frac{1}{2}$ percent to the rope strength, 10 percent to the rope weight, and 20 percent to the rope cost."

3. Lay

Lay refers to the direction of helical twist in a strand or rope. The strand lay refers to the direction of helix of the wires composing the strand. The rope lay refers to the direction of helix of the strands composing the rope. In a right-lay rope, the strand lay corresponds to that of a right-hand screw thread. In a left-lay rope, the strand lay corresponds to that of a left-hand screw thread. A regular-lay rope is a rope in which the lay of the wires in the strand is opposite in direction to the lay of the strands in the rope. Ropes may be either right regular lay or left regular lay.

A lang-lay rope is a rope in which the lay of the wires in the strand is the same direction as the lay of the strands in the rope. Ropes may be either right lang lay or left lang lay. Lang-lay ropes have good flexibility and are highly resistant to abrasion and fatigue but are susceptible to crushing, distortion, rotation under load, and kinking, and must be handled with special care.

An alternate-lay rope is a rope that is a combination of regular lay and lang lay because the lay of the wires in the adjacent strands is alternately right lay and left lay. Ropes of this type are special in their application.

The pitch of a rope (or strand) is sometimes referred to as "length of lay" and is the distance parallel to the axis of the rope (or strand) in which a strand (or wire) makes one complete helical convolution about the core (or center).

4. Various Strand Constructions

Round Strand: This construction consists of a number of round wires helically laid about an axis to form a round strand. The round strands generally fall in the three classifications—7-wire, 19-wire (16 to 26), and 37-wire (27 to 49). Six or eight round strands are usually used to make a round strand wire rope.

Flattened Strand: This construction consists of a number of round wires sometimes used in conjunction with a triangular or oval heart wire so arranged as to form a triangular-shaped strand. When such strands, usually 6 in number, are used to form a rope, the outer contour of the rope more closely approaches a circle, providing for greater wear surface and smoother contact with drums and sheaves.

Non-Rotating: The usual type of round-strand non-rotating wire rope consists of 18 strands of 7 wires each in which the lay of the outer 12 strands is opposite to the lay of the inner 6 strands. This arrangement tends to equalize the tendency for the rope to rotate that is present in the ordinary 6- and 8-strand ropes.

Preformed Wire Rope: In a preformed wire rope, the strands are shaped to the helical form they assume in the rope. Preformed wire ropes do not unravel when they are cut. They are easier to handle, are more resistant to severe banding conditions, are more resistant to kinking, and spool better than non-preformed ropes.

Page 27-1-7. Change the word "must" to "should" in both Par. i. Improper fitting, and Par. k. Kinking.

Page 27-1-7. Rewrite "Shop Cranes" under Art. 7 to read:

Shop Cranes: As a rule, the drums and sheaves are so small that ropes of 6 strands of 37 wires, fiber core, are preferred, but 8 strands of 19 wires, fiber core, are acceptable.

Pages 27-1-8 and 27-1-9. Delete Art. 10, including Table 2, substituting therefor the following:

10. Sockets and Mandrels

When it is necessary to anchor an end of a wire rope in a socket or mandrel, the following procedure should be carried out:

a. Wire Rope Preparation

(1) Seizing—The wire rope should be securely seized or clamped on the end before cutting. Measure from the end of the rope a length equal to the length of the socket basket. Seize or clamp at this point. Use as many seizings as necessary to prevent the rope from unlaying.

(2) Brooming—After the rope is cut, the end seizing should be removed. The wires should then be broomed out and the cores treated as follows:

(a) Fiber Core—Cut back length of socket basket.

(b) Steel Core—Broom out.

(3) Cleaning—The wires should be carefully cleaned with a suitable solvent for the distance they are to be inserted in the socket. They then should be dipped in commercial muriatic acid until thoroughly cleaned. The depth of immersion in acid should not be more than three fourths of the cleaned length.

Note: Fresh acid should be prepared when satisfactory cleaning of the wires requires more than 1 min (prepare new solution—do not merely add new acid to old). Be sure acid surface is free of oil or scum. The wires should be dipped in clean hot water.

b. Attaching Socket or Mandrel

(1) Installing—Preheat the socket to approximately 200 deg F and slip socket or mandrel over ends of wire. Distribute all wires evenly in the basket and flush with top of basket. Be sure socket is in line with axis of rope.

(2) Pouring—Use only zinc not lower in quality than "high grade" per ASTM Specification B 6. Heat zinc to a range of 850 to 1000 deg F. Skim off any dross which may have accumulated on the surface of the zinc bath. Pour molten zinc into the socket basket in one continuous pour if possible.

c. Final Preparation

Remove all seizings. After cooling, apply lubricant to rope adjacent to socket to replace lubricant removed by heat of socketing. Socket is then ready for service.

11. Wire Rope Clips

Where clips are applied to obtain maximum strength, make sure the proper size used. The U-bolt and saddle grooves should fit snugly on the rope. All clips should be applied in the same manner, namely, the U-bolt over the free end and the saddle to that portion of the rope carrying the load. Table 2 gives the number of clips and the spacing of them required for ropes of different sizes.

TABLE 2

NUMBER AND SPACING OF CLIPS FOR ROPES OF VARIOUS SIZES

Dia. of Rope, Inches	Number of Clips	Center to Center Between Clips, Inches	Length of Wrench, Inches
$\frac{1}{4}$	2	$1\frac{1}{2}$	6
$\frac{3}{8}$	2	$2\frac{1}{4}$	9
$\frac{1}{2}$	3	3	12
$\frac{5}{8}$	3	$3\frac{3}{4}$	12
$\frac{3}{4}$	4	$4\frac{1}{2}$	18
$\frac{7}{8}$	4	$5\frac{1}{4}$	18
1	4	6	24
$1\frac{1}{8}$	5	$6\frac{3}{4}$	24
$1\frac{1}{4}$	5	$7\frac{1}{2}$	24
$1\frac{3}{8}$	6	$8\frac{1}{4}$	24
$1\frac{1}{2}$	6	9	24
$1\frac{5}{8}$	6	$9\frac{3}{4}$	30
$1\frac{3}{4}$	7	$10\frac{1}{2}$	30
$1\frac{7}{8}$	8	$11\frac{1}{4}$	30
2	8	12	30

TABLE 3

DATA ON VARIOUS TYPES OF ROPE USED WITH MAINTENANCE OF WAY WORK EQUIPMENT

Rope Dia., Inches	Construction	Approx. Wt. Per Foot, Pounds		Strength, Tons		
				Iron	Improved Plow Steel	
		FC	IWRC	FC	FC	IWRC
¼	6x 7 Galv.	0.094	--------	0.918	2.46	--------
½	6x 7 Galv.	0.38	--------	3.58	9.59	--------
¾	6x 7 Galv.	0.84	--------	7.90	21.2	--------
1	6x 7 Galv.	1.50	--------	13.8	37.0	--------
1¼	6x 7 Galv.	2.34	--------	21.2	56.8	--------
1½	6x 7 Galv.	3.38	--------	28.9	77.6	--------
¼	6x19	0.105	0.116	--------	2.74	2.94
½	6x19	0.42	0.46	--------	10.7	11.5
⅝	6x19	0.66	0.72	--------	16.7	17.9
¾	6x19	0.95	1.04	--------	23.8	25.6
1	6x19	1.68	1.85	--------	41.8	44.9
1¼	6x19	2.63	2.89	--------	64.6	69.4
1½	6x19	3.78	4.16	--------	92.0	98.9
2	6x19	6.72	7.39	--------	160.	172.
¼	6x37	0.105	0.116	--------	2.59	2.78
½	6x37	0.42	0.46	--------	10.2	11.0
⅝	6x37	0.66	0.72	--------	15.8	17.0
¾	6x37	0.95	1.04	--------	22.6	24.3
1	6x37	1.68	1.85	--------	39.8	42.8
1¼	6x37	2.63	2.89	--------	61.5	66.1
1½	6x37	3.78	4.16	--------	87.9	94.5
2	6x37	6.72	7.39	--------	154.	165.
¼	8x19	0.098	--------	--------	2.35	--------
½	8x19	0.39	--------	--------	9.23	--------
⅝	8x19	0.61	--------	--------	14.3	--------
¾	8x19	0.88	--------	--------	20.5	--------
1	8x19	1.57	--------	--------	36.0	--------
1½	8x19	3.53	--------	--------	79.4	--------
½	18x 7	0.43	--------	--------	9.85	--------
⅝	18x 7	0.68	--------	--------	15.3	--------
¾	18x 7	0.97	--------	--------	21.8	--------
1	18x 7	1.73	--------	--------	38.3	--------
1¼	18x 7	2.70	--------	--------	59.2	--------
1½	18x 7	3.89	--------	--------	84.4	--------

Pages 27–2–1 to 27–2–21, incl.

MOTOR CARS, PUSH CARS AND TRAILERS

Page 27–2–1. Change title to read "Specifications for Motor Cars, Push Cars and Trailers."

Page 27–2–3. In the third line from bottom of page, change the word "safety" to "hand."

Page 27–2–4. In Par. (h) *Safety Rails,* change the word "safety" to "hand" wherever it appears.

Page 27–2–15. In the caption tor Fig. 10, change the word "safety" to "hand."

Page 27–2–16. In the caption for Fig. 11, change the word "safety" to "hand."

Report on Assignment 1 (a)

Revision of Handbook of Instructions for Care and Operation of Maintenance of Way Equipment

R. E. Berggren (chairman, subcommittee), D. E. Cowell, A. C. Danks, J. O. Elliott, W. T. Hammond, H. E. Keniston, W. E. Kropp, C. F. Montague, V. W. Oswalt, J. E. Sunderland, T. H. Taylor, S. E. Tracy, C. R. Turner.

In its last report your committee submitted reports on two units of equipment and one additional attachment, with the recommendation that they be included in the next revision of the Handbook (Proceedings, Vol. 62, 1961, pages 455 to 465).

Investigation develops that the three additional machines covered by this year's report have sufficient general distribution and use to justify writing instructions covering their care and operation.

It is the recommendation of the committee that this material also be included in the next revision of the Handbook.

4-TOOL SPOT TAMPER (TYPE 1)
(1962)

DESCRIPTION: A self-propelled, four-wheel, insulated, track-mounted unit consisting of gasoline or diesel power unit and air compressor, with hydraulic propulsion through 3-speed transmission. Tamper consists of four heavy-duty air tamping tools mounted on a hydraulically operated vertical and traversing tamping head to either rail. Machine has hydraulic rail clamps, hydraulic jacks, hydraulic integral center lift and turntable. Controls are for one-man operation. Brakes are four wheel air operated.

USE: Smoothing, spot surfacing, yard and terminal maintenance and tamping ties behind tie renewal gangs.

APPROXIMATE WEIGHT: 11,800 lb with diesel engine.

DIMENSIONS:

Height 7 ft 9 in, Length 13 ft 1 in, Width 10 ft 5½ in (working), 8 ft 0 in (traveling).

TRAVEL SPEED:

5 mph (low speed), 10 mph (intermediate), 20 mph (high speed).

MANUFACTURER'S RECOMMENDED SPEED:

Engine, 1800 rpm

CAPACITIES:

Compressor 125 cfm
Air receiver pressure 115 psi (working pressure)
Hydraulic system 30 gal
Fuel tank   18.5 gal
Air receiver–oil   5 gal

CARE AND OPERATION:

Starting Power Unit

3.32.45 The operator must make daily inspection of:

 a Engine water coolant.
 b Crankcase oil level.
 c Hydraulic oil tank level.
 d Compressor receiver oil level.

Hydraulic System

3.32.46 Reservoir must be kept filled to operating level (indicated on sight glass).

3.32.47 Use only approved hydraulic oil.

3.32.48 Clean line filter every 60 days.

3.32.49 Change hydraulic oil every season or once a year.

3.32.50 Daily inspection must be made of hose and connections, and any defects found must be corrected.

Air Compressor

3.32.51 Air receiver pressure must be set at 115 lb.

3.32.52 Approved oil must be kept at proper operating level in the compressor receiver.

3.32.53 Change compressor receiver oil after every 500 hr of operation.

Transmission

3.32.54 Keep filled to capacity with SAE 90 gear lubricant

Air Tamping Guns

3.32.55 See instruction in Sec. 4 (Tools—Pneumatic).

3.32.56 Set lubricator to feed about 2 qt of proper lubricating oil per working day.

Tamping Head

3.32.57 Safety pin must be known to be latched or otherwise positively secured when in travel position or when men are performing work under the tamping head.

3.32.58 Head should be in center position when traveling or being set off track.

Compressor Air Control Valve (Located on control panel)

3.32.59 When traveling, the air control valve must be positioned on working position in order to maintain full braking power.

Set-Off

3.32.60 Set-off pedestals and turntable must be positioned on level ground or blocking.

3.32.61 When machine is placed on the set-off, make certain that machine is secured to set-off with a chain and lock.

Parking

3.32.62 Emergency hand brake must be set when machine is not attended by operator on or off the track.

Lubrication

3.32.63 Follow manufacturer's recommendations as follows:

a Use open gear grease to lubricate cross-head travel bars.

b Use light oil to lubricate drive chains.

c Fill pressure-gun grease fittings on tamping tool retainers with chassis or graphite grease daily to protect tamping bar shanks.

d Lubricate tamping head guide posts daily with light oil.

e Lubricate cam plate slots with chassis grease daily.

f Add alcohol to (Alcohol drip) for cold weather operation.

Safety Pins

3.32.64 The operator must see that the safety pins and chains are latched when machine is traveling or being set off track.

3.32.65 The operator must always be careful to lock all safety pins in their respective locations to protect himself as well as the machine.

3.32.66 It is possible to bump one of the desk controls and activate any one of the following components, where safety lock pins and safety chains should be installed:

a Center tamping head with lock pin (lateral movement).

b Secure all tamping guns with lock pins.

c Secure tamping head with lock pin (elevation movement).

d Secure rail clamps with lock pins.

e Secure side jacks with safety chain.

f Secure center lift with safety chain.

8-TOOL MULTI-PURPOSE TAMPER (TYPE 1)
(1962)

DESCRIPTION: A self-propelled, four-wheel-drive, track-mounted unit consisting of diesel power unit and rotary air compressor with hydraulic propulsion through 3-speed transmission. Tamper consists of eight heavy-duty air tamping tools mounted four each on independently operated tamping heads. Machine has hydraulic rail clamps, hydraulic jacks, hydraulic integral center lift and turntable. Controls are for one-man operation. Brakes are four-wheel, air operated.

USE: As a production tamper, a spot tamper or a jack-tamper.

WEIGHT: 15,400 lb

DIMENSIONS:

Length 14 ft 5 in, Height 7 ft 9 in, Width 10 ft 5 in (working), 8 ft 0 in (traveling).

TRAVEL SPEED:

5 mph (low speed), 10 mph (intermediate), 20 mph (high speed).

MANUFACTURER'S RECOMMENDED SPEED:

Engine, 1800 rpm.

CAPACITIES:

Compressor 250 cfm
Air receiver 115 psi (working pressure)
Hydraulic system 35 gal
Fuel tank   33 gal
Air receiver–oil 12 gal

CARE AND OPERATION:

Starting Power Unit

3.32.75 The operator must make daily inspection of:

a Engine water coolant.
b Crankcase oil level.
c Hydraulic oil tank level.
d Compressor receiver oil level.

Hydraulic System

3.32.76 Reservoir must be kept filled to operating level with approved hydraulic oil.

3.32.77 Line filter must be cleaned every 60 days.

3.32.78 Oil must be changed every season or once a year.

3.32.79 Daily inspection must be made of all hose connections, and all defects corrected promptly.

Brakes

3.32.80 Emergency hand brake must be set when machine is not attended by the operator, on or off the track.

Air Compressor

3.32.81 Air Receiver pressure must be set at 115 lb.

3.32.82 Approved oil must be kept to operating level in the compressor receiver.

3.32.83 Change compressor receiver oil after every 500 hr of operation.

Transmission

3.32.84 Keep filled to proper level with SAE 90 gear lubricant.

Air Tamping Guns

3.32.85 See instructions in Sec. 4 (Tools—pneumatic)

3.32.86 Set safety gun pins when in travel position.

Tamping Head

3.32.87 Safety pin must be known to be latched or otherwise positively secured when in travel position or when men are performing work under the tamping head.

3.32.88 Heads should be in center position when traveling or being set off track.

Air Control Valve

3.32.89 Valve must be positioned in working position when traveling in order to maintain full braking power.

Set-Off

3.32.90 Safety pins must be latched when traveling or being set off the track.

3.32.91 Set-off rail pedestals must be properly blocked to prevent machine from tipping.

3.32.92 When stored, the machine must be secured to set-off rail with chain and locked.

Lubrication

3.32.93 Set lubricators (2) to feed approximately 2 qt of lubricating oil each per working day.

3.32.94 Use open gear grease to lubricate cross-head travel bars.

3.32.95 Use light oil to lubricate drive chains.

3.32.96 Fill pressure gun grease fittings on tamping tool retainers with chaissis or graphite grease daily to protect tamping bar shanks.

3.32.97 Lubricate tamping head guide posts daily with light oil.

3.32.98 Lubricate cam plate slots with chassis grease daily.

3.32.99 Add alcohol to (Alcohol Drip) for cold-weather operation.

Safety Pins

3.32.100 The operator must see that safety pins are latched when machine is traveling or being set off track.

3.32.101 The operator must always be careful to lock all safety pins in their respective locations to protect himself as well as the machine.

3.32.102 It is possible to bump one of the desk controls and activate any of the following components, where safety lock pins and safety chains should be installed:

> a Center tamping head with lock pin (lateral movement).
>
> b Secure all tamping guns with lock pins.
>
> c Secure tamping head with lock pin (elevation movement).
>
> d Secure rail clamps with lock pins.
>
> e Secure side jacks with safety chain.
>
> f Secure center lift with safety chain.

TIE SPACER (TYPE 1)
(1962)

DESCRIPTION: A self-propelled, four-wheel, insulated, track-mounted unit consisting of an air-cooled gasoline power unit with hydraulic propulsion through a 2-speed transmission. A set of hydraulic tie-shifting devices, one on each side of the machine, are controlled and powered hydraulically. The machine has an integral hydraulic center lift and turntable. Brakes are four-wheel, hydraulic type.

USE: Spacing ties and correcting slewed tie conditions ahead of power jacks or hand jacks in a surfacing gang, where the track has been skeletonized

WEIGHT: 4,000 lb

DIMENSIONS:

Length 6 ft 0, Height 6 ft 5 in, Width 6 ft 0 in.

TRAVEL SPEED:

13 mph (low), 25 mph (high).

MANUFACTURER'S RECOMMENDED SPEED:

Engine, 2100 rpm

CAPACITIES:

Hydraulic system, 33 gal

CARE AND OPERATION:

Starting Power Unit

3.30.11 The operator must make daily inspection of:

　　a Crankcase oil level.
　　b Hydraulic tank oil level.

Hydraulic System

3.30.12 Reservoir must be kept filled to operating level.

3.30.13 Oil level checked by a dip stick located at the top of side of hydraulic tank.

3.30.14 Line filters must be cleaned every 60 days (two filters located just outside of hydraulic tank).

3.30.15 Hydraulic oil must be changed every season or once a year.

3.30.16 Daily inspection must be made of all hose and connections and any defects promptly corrected.

Brakes

3.30.17 Emergency hand brake must be set when machine is not attended by operator on or off the track.

Transmission

3.30.18 Keep filled to proper operating level, use SAE 90 gear lubricant.

Set-Off

3.30.19 Set-off pedestals and turntable must be positioned on level ground or blocking.

3.30.20 When machine is placed on the set-off make certain the machine is secured to set-off with a chain and lock.

Lubrication

3.30.21 Use light oil on the drive chain.

3.30.22 Lubricate pusher frame in cylinder guide rollers daily (use chassis grease or graphite).

3.30.23 Lubricate power take-off clutch assembly with chassis grease several times per month (do not over grease).

3.30.24 Pressure grease axle bearings every four months with chassis grease.

Safety Chains

3.30.25 The operator must always be careful to secure all safety chains to their respective locations to protect himself as well as the machine.

3.30.26 It is possible to bump one of the desk controls and activate any one of the following components, where safety lock pins and safety chains should be installed:

a Front and rear pusher frame assembly (4 chains) at left and right side of machine.

b Center lift or turntable safety chains.

3.30.27 These points are stressed in order that no interference will result with rail and highway crossings and crossovers, or any other conditions that may exist.

Report on Assignment 2

Improvements to be Made to Existing Work Equipment

G. L. Zipperian (chairman, subcommittee), C. T. Blume, K. J. DeCamp, W. T. Friedline, R. A. Hostetter, N. W. Hutchison, W. R. Jacobs, C. F. Lewis, Paul Martin, P. G. Petri, M. C. Taylor, H. A. Thyng, F. E. Yockey.

This is a progress report, submitted for information, and is a continuation of the progress reports submitted by this committee and found in Vol. 53, 1922, page 396; Vol. 54, 1953, page 666; Vol. 55, 1954, page 502; Vol. 56, 1955, page 525; Vol. 57, 1956, page 488; Vol. 58, 1957, page 585; Vol. 59, 1958, page 629; Vol. 60, 1959, page 427, and Vol. 61, 1960, page 500, and covers changes in work equipment that this committee has found to be practical and desirable.

Spike Puller, Mechanical

This is a four-wheel, rail-mounted, self-propelled machine for pulling track spikes in rail gangs.

Suggested improvements to this machine:

1. Provide a guard to cover the opening on the walking arm at the front so no access is possible when the walking arm rises to the top position.
2. Provide a guard to cover the fly wheel so that it will not be exposed to the operator or other personnel when the machine is working.

Tie Adzer

This is a self-propelled machine balanced on one rail, with adjustable bevel guide wheels for controlling the position of the cutter head when adzing. It is used with rail-laying track gangs.

Suggested improvements to this machine:

1. Apply a hand throttle so the operator may control the engine speed without having to leave his normal working position. The throttle at present is mounted on the engine control panel where the operator is not able to reach it when working.
2. Apply a guard under the drive sprockets, which are mounted on the ends of the drive-wheel axles, to prevent the axles from being bent and the chains and sprockets fouled with dirt when setting off.

3. Apply a metal guard to prevent shavings from the cutter head striking the engine and igniting a fire.

Tie Coaters

This is a rail-mounted machine with gasoline-engine-driven pump with tank heating equipment to apply asphaltic material to ties after adzing.

Suggested improvements to this machine:

1. Apply removable inspection or clean-out plates to the tank adjacent to the strainer. At present the strainer cannot be cleaned when baked material plugs the holes without installing removable side plates.

Truck Cranes (with rail-guide-wheel equipment for on- and off-track operation)

This is a rubber-tired truck crane with rail-guide-wheel equipment for on- and off-track use in bridge repair work and rail laying, and regular off-track duty as a truck crane.

Suggested improvement to this machine:

1. The rail guide wheels should have the regular AAR wheel tread contour and tread width from back of flange to outside of tread. Present guide wheels are approximately 5 in across flange and tread face which does not allow proper operation through self-guarded frogs and around curves.

Tie Tamper (multi-tool, hydraulic squeeze, electrical vibratory)

This is a four-wheel, track-mounted, self-propelled, electromagnetic, hydraulically operated tamper of the split-head type. It is equipped with vertical electric motors which impart rapid vibrations to the tamping bars and forces ballast under the ties with squeeze action. The machine can be equipped with automatic track jacking and leveling equipment.

Suggested improvements to this machine:

1. The present hydraulic oil filters are submerged inside the reservoir, and the reservoir must be drained to clean the filters. The present filters are not sufficient to protect the hydraulic system's components, resulting in numerous failures. Only when the suction-line flow rate is reduced can it be determined that the filters need cleaning. It is felt that a submersible type with automatic cut-off and tell-tale indicator filter should be installed.
2. Machine at present uses turntable and rail wheels for setting off. Transverse wheels, power driven, should be applied, as the machine when equipped with auto-jack is so long that insufficient clearance is available, and it must be moved to a siding to clear trains or when tying up at night.
3. Hydraulic fittings to lower end of work-head rams are bursting, and other hydraulic hoses are giving poor service. Swivels and fittings to avoid kinking and twisting of suctionline hoses should be used. Recommend that high-pressure hose and fittings more generally available and standard be used to avoid excessive delay obtaining replacements when repairs are required.

Tie Tamper (multi-tool, electric-vibratory)

This is a four-wheel, track-mounted, self-propelled, split-head machine with an electric vibratory motor which imparts rapid vibrations to tamping bars to consolidate ballast under the ties with squeeze action.

Suggested improvements to this machine:

1. Canopy top has a drain hole in the eave through right and left front corners, directly over the electric tamper motors on the back row, causing the motors to develop electrical trouble and shorting. Drain holes should be moved to another location not over the tamping motors.

Under Track Positioning and Control Carriage for Plow

This is an on-track, eight-wheel, rail-mounted, self-propelled machine with rail jacking equipment for placing plow under track and for removing same. Also incorporates tie ejecting and track-lining equipment.

Suggested improvements to this machine:

1. A single brake band is provided on the main drive shaft between transmission and drive-chain sprocket. If single roller-type chain breaks, no means is provided to stop during self-propelling operation except by lowering jacking rams, with serious damage resulting. Suggest that power-operated brakes, with brake shoes applying directly on the wheels, be added to this machine.

Utility Crane and Tie Inserter

This is a four-wheel, track-mounted, self-propelled machine used to handle and insert ties in tie renewal gangs.

Suggested improvements to this machine:

1. A great deal of difficulty is being experienced with the wire rope on both the main lifting winch and the tie inserter winch. The wire rope overruns on the winch drums and becomes kinked and tangled. Suggest that the diameter of the drums be increased and flanges made higher properly to handle amount of wire rope used. The wire-rope-retaining rollers on the winches should be made more positive to assure smooth, even spooling of the cable on the drums.

2. The vacuum brake cylinders are mounted beneath the machine frame at each wheel and are being frequently damaged by the swinging load being handled by the crane. Suggest that the vacuum brake cylinders be recessed under the frame to provide a protective cover so they will not be damaged by swinging loads.

Power Track Wrench

This is a rail-mounted machine with transverse cross carriage used to remove and apply track bolts.

Suggested improvements to this machine:

1. The drive gear and clutch assemblies are failing after a short period of service in the field. It seems that due to faulty design and construction, clutch and gear asemblies are not adequate for the service intended. Suggest that the gear and clutch assemblies be re-engineered to provide greater strength. The size of the boss on the gear housing for the clutch fork stud should be increased as the present one is breaking off.

Report on Assignment 5

Set-offs and Set-off Attachments for Work Equipment

R. W. Bailey (chairman, subcommittee), R. E. Buss, G. R. Collier, K. J. DeCamp, W. T. Hammond, E. W. Hodgkins, W. R. Jacobs, R. M. Johnson, Paul Martin, J. E. Reynolds, M. C. Taylor, S. E. Tracy.

This is a final report, submitted as information.

Various methods have been developed for the purpose of removing work equipment from the track. The basic unit is the portable set-off illustrated in **Fig. 1**. This unit consists of two I-Beam supports between vertical pipe posts which are interconnected by adjustable rods to maintain a rigid box section. The vertical supports are drilled at uniform intervals to allow level adjustment of the box section in compensation for irregularities of the ground adjacent to the track. The posts are removable and various lengths are furnished.

The posts are supported on the ground by a large plate to give the structure sufficient ground bearing. The entire box section is connected to the base of the rail by an adjustable clamp and bar to give rigidity.

The set-off rails furnished by the manufacturer of the equipment are mounted on this box section. The removable sections of the rail are placed on the track and pinned to these rails. The item of equipment is placed on the removable section of the rail and moved onto the set-off rails. The removable section is then removed and the total weight rests on the set-off.

In reviewing the methods of removal of various machines from the track, it was found that one type uses a turntable and ramp to raise the machine, turn it 90 deg, and run it off onto the set-off; another type uses hydraulic rams or a mechanical means to

Fig. 1

Fig. 2

Fig. 3

Fig. 4

Fig. 5

raise the machine, then the operator places the removable set-off rails under the set-off wheels and moves the machine sideways onto the set-off.

The gages of set-off devices vary from standard track gage (4 ft 8½ in) to as high as 15 ft; the weight of the equipment varies from a few hundred pounds to as high as 40,000 lb; and the wheel height above the rail varies from rail height to 9 in.

One particular unit has a capacity of 20,000 lb and will handle widths up to 7 ft 9 in. A heavy-duty unit is manufactured which has a capacity of 40,000 lb and will accommodate widths up to 15-ft. The design is such that the individual parts making up the set-off can be handled by two men.

This type of set-off may include such refinements as winches to move machines, locking devices to prevent the machines from moving in relation to the set-off, screw adjustments for elevation, etc. The major advantage to this type of set-off is that the components can be handled by a minimum of personnel.

A new arrangement is available which has a self-propelled hydraulic power pack to operate a series of hydraulic-ram-operated portable units, as illustrated in Figs. 2 through 5. The complete rail section of the set-off is illustrated in Fig. 2 allowing a complete machine or group of machines to be run on the side-moving track and "into the clear." The track portion of the set-off is then raised up clear of the track to allow a train to pass.

A complete, portable turnout is available, which is constructed similar to the boxed-section set-off first discussed. Several of these units are attached by lightweight I-Beam rails which are connected to the existing operating track by means of an old style "street car run-off frog." This allows the machines to be run up on a ramp over the frog which sits on top of the rail, and onto the set-off rails, mounted parallel to the track. The set-off frog and ramp is then removed, allowing passage of trains.

Report on Assignment 7

Vehicles and Equipment for Routine Track Inspection

H. D. Hahn (chairman, subcommittee), T. S. Bean, L. B. Cann, E. D. Cowell, J. W, Cummings, E. H. Fisher, W. F. Kohl, H. F. Longhelt, C. E. McEntee, F. N. Snyder, H. A. Thyng, L. B. Waterman, F. E. Yockey.

This is a final report, submitted as information.

In recent years many railroads have made force reductions which, in turn, lengthened the assigned territories of personnel responsible for the inspection and maintenance of track. As a result, the vehicles and equipment used in routine track inspection work have also undergone modification on numerous railroads. The old practice of track walking is virtually a thing of the past, except in yards and in terminal areas. Many track inspectors continue to inspect their track by riding a locomotive or from the rear platform of a passenger or freight train, and these modes of transport along with the motor car are far from being obsolete for the purpose of track inspections.

VEHICLES

Although those individuals who are required to make periodic track inspections as a part of their regular assignments still depend to a large extent on either light or standard inspection motor cars for transportation, there has been a trend on some railroads to furnish these people with highway-rail sedans (Fig. 1) or light pick-up trucks (Fig. 2). This trend is more significant and apparent in the case of track inspection personnel who

Fig. 1—Four-door sedan equipped with road-rail conversion unit.

Fig. 2—Inspection truck equipped with conversion unit.

Fig. 3—Station wagon equipped for rail or highway operation.

make less intensive track inspections less frequently. The popular mode of transportation for these individuals is the six or nine passenger highway-rail station wagon (Fig. 3).

There are two manufacturers which can furnish highway-rail conversion equipment for installation at the railroads' shops or at their own factories. These firms offer a wide range of makes and models of passenger cars and trucks upon which their equipment can be mounted, and, if the railroad prefers, the complete vehicle can be furnished with or without a special body. For routine track inspection work conversion units are usually mounted on vehicles with a gross vehicular weight of 6000 lb or less.

One manufacturer's standard highway-rail conversion unit normally used on this size of vehicle is described as follows: Retractable flanged guide wheels, 12½ in. in diameter, made of machined cast steel with a replaceable fiber insulation set, and machine balanced for high-speed performance. These wheels are on heavy-duty, 5-stud automotive hubs with heavy-duty tapered roller bearings and special 1¾-in spindles. The torsion-type coil springs cushion guide wheel action on the rail, and the spring retainer collar can be adjusted to balance wheel loads. Positive travel stops restrict vertical wheel travel, and rubber pads help reduce rail shock. The mounting brackets are designed for the specific make and model of vehicle to be equipped. These brackets bolt to the guide-wheel unit for proper positioning, following alinement. A steering wheel lock holds the front wheels in the proper track, and, when required, wheel dishing, and/or spacing blocks are furnished to correct vehicle tread to the necessary 59½ to 60½ in. A 36-in steel operating handle is shaped for easy actuation on a particular vehicle. The weight of this conversion unit averages 248 lb for the front assembly and 262 lb for the rear assembly.

Optional equipment which can be furnished on routine track inspection vehicles includes non-slip differential, heavy-duty front and rear springs, outside rear-view mirrors, rubber tread for the guide wheels, hand-operated hydraulic booster, snow-tread-type rear tires for better wheel traction on slippery rail, roof-mounted spotlights, flasher lights on the cab, and, under certain conditions, power steering and brakes. The optional rubber-tread guide wheels are generally used on station wagons and sedans, while the steel tread continues in use on the trucks. Some railroads have recently applied the rubber-tread guide wheels to pick-up trucks used by track inspectors since they provide an easier ride and a lower noise level within the truck. There still are many railroads, however, which believe that truck inspectors can get a better "feel" of the track if they can hear the "sing" of the flanged steel wheels.

This same manufacturer offers a new lightweight conversion unit with 11-in diameter guide wheels on integral hubs for mounting on vehicles under 4800 lb gross vehicular weight. This unit weighs 340 lb and is adaptable to many 1961 vehicles and most "compact" station wagons. Some of the features of this unit include torsion-bar suspension, three-piece wheels, permanently lubricated bearings, replaceable rubber wheel treads and improved insulation. High flotation tires are available as an option for some vehicles to allow the vehicle to mount and leave the rail anywhere. A larger highway-rail conversion unit featuring hydraulic operation and manual locks is available from this manufacturer for installation on vehicles having a gross vehicular weight of from 9,000 to 50,000 lb. Since vehicles of this size are not normally used for track inspection work a detailed description is not given in this report.

The other manufacturer of highway-rail conversion units offers a manually operated unit for installation on station wagons, ½-ton trucks, and ¾-ton trucks. This equipment consists of load-bearing guide wheels with 8½-in diameter tread, insulated, and mounted on tapered roller bearings. These guide wheels have independent action, and the load is applied through a rubber-cushion torque unit. The guide wheels are held in position by a mechanical lock, and a hand operated steering wheel lock is mounted on the steering column. Normally station wagons are equipped with rubber-tread guide wheels, and trucks with steel guide wheels. Wheel spacers are required for most vehicles to give the proper wheel spacing for on-track operation. Heavy-duty springs and supplementary rear supports are recommended for most vehicles. Conversion equipment for larger 1½-, 2-, and 2½-ton trucks is also available from this manufacturer with hydraulic operation of the guide wheels available when required.

A third manufacturer offers a single make of vehicle with special flanged wheels on a complete unit basis only. This vehicle is basically a foreign truck (Fig. 4) or station wagon (Fig. 5) equipped with 14-in flanged, steel guide wheels fitted on roller bearings. During on-track operation the weight of the rear part of the car is distributed half to the rear tires and half to the rear flanged wheels while the weight of the front part of the car rests entirely on the flanged wheels. These flanged wheels are secured by double safety catches while on the track, and both front and back track attachments are insulated. The steering wheel can be locked for on-track operation, and an electric turntable jack enables the operator to take the vehicle on or off the rails at any road crossing or turn it around without leaving the rails. This turntable is powered by a 12-v electrical system separate from the car's 6-v system. Four wheel braking is provided for both on and off-track operation. The front flanged wheels are equipped with hydraulic brakes which, together with the rear wheels on the car, afford four-wheel on-track braking. The retractable guide wheels are manually controlled. Rubber-cushioned flanged rear wheels can be furnished as an option. The driver's seat is over the front wheels, providing an excellent view of the track for inspection work.

Fig. 4—Highway–rail truck equipped with electric turntable jack.

Fig. 5—Highway–rail station wagon for inspecting track or
transporting men.

Normal maintenance of highway-rail conversion equipment requires lubrication at the time of lubricating the vehicle plus periodic inspections of the rail wheel attachments. The principal cost of repair is for flanged wheels.

There are other track inspection cars which detect, locate, and graphically record defects in the track. These cars measure the quality of track by recording such data as surface, alinement, cross level superelevations, and gage. Such cars ar either self-propelled or pulled by passenger trains. Since these units are normally used only on annual or semi-annual inspections, they cannot be considered vehicles which are used for routine track inspections, and therefore detailed descriptions are not covered in this report. There are other large bus-type vehicles (Fig. 6) which operate exclusively on the rails

Fig. 6—Bus-type on-track inspection car.

and are available from either American or foreign manufacturers. These units are designed for transporting inspection parties and not for the recording of track characteristics.

EQUIPMENT

The type and quantity of equipment carried by inspection parties is usually dependent upon the number of men in the party, the mode of transportation, the purpose of the inspection trip, and whether or not corrective measures are to be taken at the time defects are found. The optional equipment which can be obtained for the inspection vehicle is described in the previous section of this report. Some of the major items of track inspection equipment other than those which are a part of the vehicle are described briefly as follows:

1. Combination Level and Gage

(a) *Aluminum Type*—This tool is constructed of a heat-treated aluminum alloy. It is available with either the adjustable $22\frac{1}{2}$-in radius or the AREA $57\frac{1}{2}$-in radius vials. The gage attachment with guard rail spacer is insulated, and it is adjustable through the medium of shims to provide for wear. A 7-in elevation scale graduated in eighths is concealed in the end of gage when not in use. Some general characteristics not found in wooden type units are (1) light weight (5 lb), (2) does not absorb or expel moisture due to constant temperature and humidity changes, and (3) aluminum tube section provides rigidity to withstand abuse.

(b) *Wooden Type*—This tool is made from seasoned, select hardwood with all metal parts being machined bronze. It has built-in features such as (1) adjustable level glass, (2) track gage with guard rail lug, (3) graduated elevation scale, (4) master gage for testing other gages, and (5) scale showing variations of gage on curves.

2. Track Gage Indicator (Type 1)

This is a portable hand tool weighing 29 lb which is used in checking gage. It is constructed of steel tubing, angles, and ball bearing rollers. It is insulated to prevent shunting track circuits, and moves on the rail on eight ball-bearing rollers. Two ball bearings run on the top of each rail, and two on the inside of the rail head at the gage line. A spring inside the frame keeps the rollers in tight contact on each rail. A gage-indicating pointer multiples gage irregularities approximately four times so that small variations can be easily seen on the dial. This dial is calibrated to read gage variations from correct to 1 in wide or 1 in tight. This tool is used primarily in regaging operations although it can be used for routine inspection work.

3. Track Gage Indicator (Type 2)

One railroad has recently developed a track gage indicator which can be pushed at a speed of about 5 mph by a small highway-rail inspection vehicle (Fig. 7). This device consists of a gage-measuring dial mechanism mounted on a unit with four flanged wheels. A spring-loaded steel shaft with a tubular receptacle bears against the gage lines of both running rails by means of steel contact wheels at each end of the shaft, thereby measuring the variations in gage. These variations are transferred through rack and pinion assemblies to an indicator dial which is mounted at eye-level height on the front of the indicator car (Fig. 8). The variations are amplified at a 5 to 1 ratio to read on the dial to the nearest $\frac{1}{8}$ in. This dial is colored green for readings under 57 in and red for readings over that figure. This road is also planning to install a pendulum-potentiometer on this device to detect variations in cross level.

Fig. 7—Road–rail inspection vehicle pushing track gage indicator.

Fig. 8—Indicator dial on indicator car.

4. Track Inspection Machine (Fig. 9)

This instrument was designed to indicate irregularities in the alinement and surface of track, and is commercially manufactured. It consists of a solid brass frame mounted on a wooden block, a horizontal and a vertical flexible steel stem attached to the frame, a pedometer on each stem actuated by shock and registering the number accumulated, and a cyclometer attached to the frame for each stem, to which it is attached by hinged arms, the cyclometers operating through a ratchet. This instrument is located on the floor of a car approximately over the center of the rear truck.

Water spills are obtained by a small instrument (Fig. 10) consisting of a framework holding in place a funnel, the spout of which empties into a glass underenath. A tumbler, kept continuously filled with water, stands in the center of the funnel. The quantity of water spilled into the glass from the funnel constitutes the water spills.

The readings of the cyclometers and pedometers taken in conjunction with the water spills obtained over the distance travelled, adjusted for the average speed in miles per hour, form the basis for comparison among the divisions or subdivisions of a railroad. The following formula is used in making these comparisons:

$$\text{Mark} = 10 - \frac{\text{Readings per Mile}}{\text{Avg. Speed in MPH}}$$

$$\text{Average Speed in MPH} = \frac{60 \times \text{Distance in Miles}}{\text{Minutes}}$$

$$\text{Total Readings per Mile} = \frac{\text{Horizontals} + \text{Verticals} + \text{Pedometers} + \text{Spills}}{\text{Distance in Miles}}$$

5. Track Recorder

This instrument, of foreign design, has been used by many American railroads to get a comparison of ride qualities over a particular section of railroad on successive

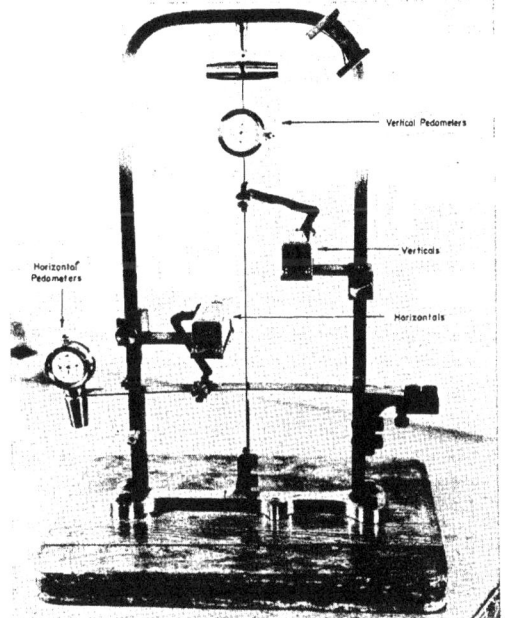

Fig. 9—Track inspec-
tion machine.

Fig. 10—Device for measuring water spills.

trips. It works on a pendulum principle as three series of weights move in directions parallel to the track, transversely to the track, and vertically to the track, respectively. Records of pendulum movements are made on a traveling band of paper by four steel points through a sheet of carbon paper. This instrument is worked in the compartment of a car over one of the axles.

On some railroads today it is thought to be desirable for track inspection parties of 2 or 3 men to be equipped with a small complement of power tools so that more of the repair work can be done on the spot. One road is experimenting with a foreign truck converted into a highway–rail vehicle and equipped with a small 85-cfm compressor and power tools such as tampers and air impact wrenches. Future track inspection parties may have with them a small complement of power-operated tools of either the electric hydraulic or pneumatic type.

CONCLUSIONS

1. The practice of walking track is virtually obsolete except in yards and terminal areas.

2. Track inspections are still being made from locomotive cabs and from the rear end of passenger and freight trains. This practice is likely to continue in the future, since track irregularities which cannot be detected on a motor car can often be detected by this means.

3. Motor cars are still one of the primary modes of transportation for men who make periodic track inspections on a regular schedule.

4. Highway-rail vehicles are replacing motor cars on some railroads as a means of transportation for track inspection personnel making regular periodic inspections. Sedans, light pick-up trucks, and "compacts" are the favorite vehicles used by these individuals. Station wagons of various types are popular vehicles for personnel who make inspections less frequently.

5. The lightweight combination level/gage is a favorite piece of equipment for track inspectors. It has eliminated the necessity of carrying both a track gage and a track level.

6. Portable instruments are available for making comparisons of the ride qualities of a particular section of track on successive trips.

7. The future may bring new developments in small, lightweight power tools such as impact wrenches, tie tampers, and spike hammers which can be used by track inspection parties of from 1 to 4 men. These tools would be either electrically, hydraulically, or pneumatically operated from small portable generator sets or air compressors carried on highway–rail vehicles.

8. Railroads and railroad supply companies continue to experiment and develop equipment which will quickly and efficiently determine the quality of track. New tools and equipment may be utilized in making more extensive on the spot repairs of track defects.

Nothing does it as well as

The **JACKSON YARD TAMPER**

Note the illustrations at the left and we believe you will agree the new JACKSON YARD TAMPER is about to revolutionize yard tamping. It gets into places that other on-track tampers can not tamp and completely fulfills the major portion of all yard tamping requirements. As one user of this equipment has put it: "This equipment increases our production effort in surfacing leads by at least 100%, with half as many men."

The Jackson Yard Tamper has two independent workheads which may be moved back and forth laterally by means of hydraulic rams. Hence the tamping blades may be positioned for tamping in any desired locations. The single blades are unusually long (26") and so mounted that as penetration into the ballast increases the slant becomes progressively greater. Thus thorough consolidation of the ballast is achieved in a range from 18" inside the rail, directly under the rail and to the end of the tie. And the same blade action makes it possible to do a thorough job of tamping under large frogs and in other tough places. Powerful Jackson vibratory motors insure rapid penetration.

Readily convertible to spot tamping, jack tamping and other normal tamping uses ... simply by substituting the regular Jackson double blade tamping units for the single blades. It's a tremendously useful machine which is bound to pay big dividends on its investment cost. Let us give you the complete facts.

As always, direct sales, leases and service to all U.S railroads

JACKSON VIBRATORS, INC.
RI-10
LUDINGTON, MICHIGAN

for effective
weed control...

- **Concentrated BORASCU®**
- **POLYBOR-CHLORATE®**
- **UREABOR®**
- **MONOBOR-CHLORATE®**

These <u>borate weed killers</u> are proving best for roads in every way... *efficiency, safety, economy, convenience, easy application.*

Today's use of borates for maximum control of vegetation began years ago with our pioneer work in the field. Continued research has developed the group of herbicides, listed above, which most roads now favor for every phase of weed control. These four weed killers are nonselective. They are widely used for year-round maintenance of weed-free conditions about trestles, tie piles, yards, signals, switches, and rights of way. Find out how you, too, can do a better job on weeds... write today.

AGRICULTURAL SALES DEPARTMENT

U.S. BORAX

630 SHATTO PLACE · LOS ANGELES 5, CALIFORNIA

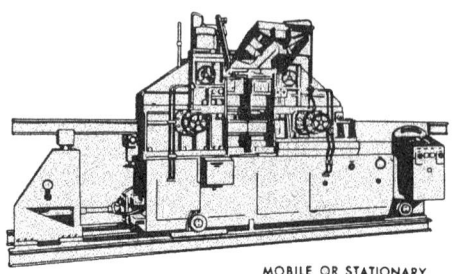

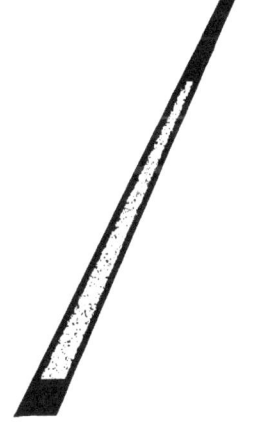

VEGETATION CONTROL
WITH
CHEMICALS

READE MANUFACTURING COMPANY, INC.

Jersey City—Chicago—Minneapolis—Kansas
City—Birmingham—Stockton

SERVING RAILROADS OF AMERICA FOR
MORE THAN FORTY YEARS

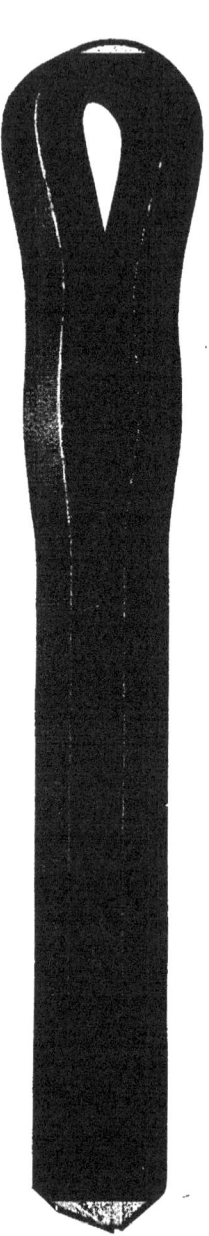

SPEED SWING

MODEL 441

Developed and Built
for Railroad Maintenance

180° BOOM SWING

DOES ALL JOBS!

LAYING STANDARD RAIL

CUTS MAINTENANCE COSTS

12 FAST CHANGE ATTACHMENTS

- Forks
- 1¼ Cu. Yd. Bucket
- Tote Hook
- 18' Boom Extension
- Fork Tie Baler
- Track Cleaning Bucket

- Back Hoe
- Clamshell
- Back Filler Blade
- Pull Drag Bucket
- 4 Cu. Yd. Snow Bucket
- Pile Hammer

Optional Attachment
Flanged Wheels, Hydraulically Controlled

PETTIBONE MULLIKEN CORPORATION

RAILROAD **PMCO** DIVISION

141 W. JACKSON   CHICAGO 4, ILL.

*80 Years of Service
to the Railroad Industry*

HOLD EVERYTHING...

till you have the complete dope on these vastly important improvements that make the 1962 JACKSON TRACK MAINTAINER more than ever the most advantageous machine of its kind:

FASTER PENETRATION . . . FASTER WORK-HEAD ACTION . . . FASTER INDEXING . . . FULL ELECTRIC OPERATION WITH PUSH-BUTTON CONTROL . . . GREATLY INCREASED VERTICAL AND HORIZONTAL VIBRATORY FORCES . . . MOTOR AND GENERATOR PRO-TECTION AGAINST SHORT CIRCUITS, SINGLE-PHASING AND LOW VOLTAGE . . . FASTER, MORE POSITIVE BRAKING.

JACKSON VIBRATORS, INC.
LUDINGTON, MICHIGAN

WRITE, WIRE OR
PHONE FOR
COMPLETE
INFORMATION

R1-9

The only man on his feet...

He's the pin puller—uncoupling cars. He could be in any of the GRS-equipped automatic yards. Once a car is free, its speed is controlled automatically by Class-Matic®, the GRS system of yard automation. At the crest, the conductor pushes a button on his console for each cut, to route the cars automatically to the proper tracks. In the retarder tower, the operator merely monitors the system.

The free-rolling cars are judged individually as to rollability, weight, route, distance to coupling, and other factors which are fed to the analog computer. This controls the retardation automatically so the cars glide gently to coupling.

What a saving in manpower! And consider the safety factor—only one man near the cars at any time. The saving in freight time alone will vindicate your choice of GRS Class-Matic, the system that pays for itself in short order.

3084

GENERAL RAILWAY SIGNAL COMPANY

ROCHESTER 2, NEW YORK

NEW YORK 17, NEW YORK   CHICAGO 1, ILLINOIS   .ST. LOUIS 1, MISSOURI

Kershaw Trackwork Machines

Designed, Tested, and Proven on America's Railroads

MONTGOMERY   ALABAMA

Trackwork Equipment Developed and Proven On the Job

Heavy Duty Ballast Regulator, Scarifier and Plow, Standard Ballast Regulator, Scarifier and Plow, Track Broom, Super Jack-All, Standard Jack-All Kershaw Kribber, Two-Wheel Kribber, Tie Bed Cleaner, Track Undercutter-Skeletonizer, Ballast Cleaner, Multiple (Spot) Tamper, Foreman's Sight Car, Crib-Adze, Rail Re-Layer, Mocar Crane, Track Crane and Tie Inserter, Utility Derrick, Two-Ton Rail Derrick, Chemical Spreader Car, Tie Replacer, and Universal Side Set-Off Assembly.

Kill more weeds *per mile...per dollar*
with *Liquid* **UROX**®!

Liquid Urox Weed Killer is the *first* liquid – substituted urea-type herbicide ever developed for railroads. It's fast-acting . . . withers annual and perennial grasses as well as broadleaved weeds within 12 hours after application, regardless of weather. It's long-lasting . . . just *one* application wipes out weeds and brush for 8 to 18 months. What's more, control can be continued economically each year with small "booster" doses.

Liquid Urox is ideal for railroad spray trains . . . doesn't need continuous agitation . . . won't clog spray nozzles . . . won't settle out . . . can be mixed with fuel oil, diesel oil or ordinary weed oils. Write today for the complete story on railroad-proved liquid Urox Weed Killer.

GENERAL CHEMICAL DIVISION
40 Rector Street, New York 6, N. Y.

Assure <u>lower</u> maintenance costs,
<u>better</u> performance with...

TEXACO

Railroad Lubricants and Systematic Engineering Service

TEXACO INC.
RAILWAY SALES DIVISION
135 East 42nd St., New York 17, N. Y.

NEW YORK • CHICAGO • SAN FRANCISCO • ST. LOUIS • ST. PAUL • ATLANTA

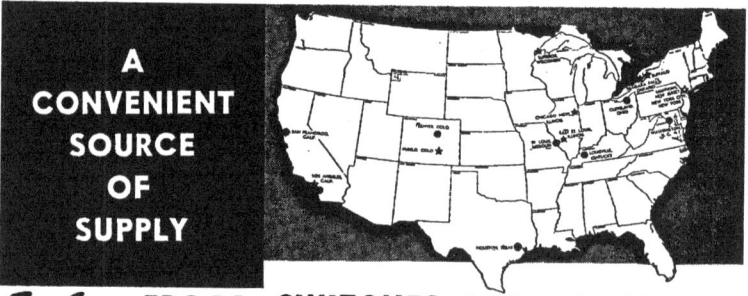

Model N U Tie Cutter

HERE IS THE WINNING TEAM

The Woolery NU Tie Cutter and the Woolery Tie-end Remover preserve the line and surface of the track and at the same time reduce the cost of tie renewals. Ties can be removed without trenching, jacking up track or adzing tops of rail-cut ties. With this team you simply cut both ends of tie, pry out center piece, insert in its place the tie-end remover and out go the tie ends pushed by the double acting, double ended hydraulic cylinder of the Tie-end remover.

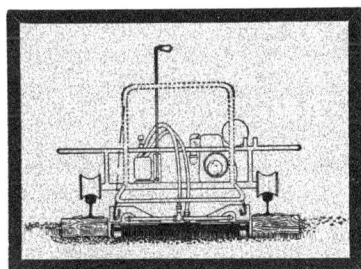 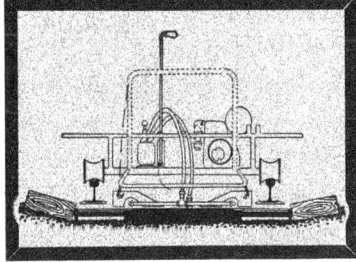

FOR HIGHEST EFFICIENCY USE TWO TIE CUTTERS WITH ONE TIE-END REMOVER

WOOLERY MACHINE COMPANY
MINNEAPOLIS, MINN.

PROGRESS REPORT

The New, Low-Cost

JACKSON UTILITY TAMPERS

Can be tailored specifically to **your** tamping and budget requirements

These machines offer exceptional opportunity to handle all tamping requirements for which large production tampers are not indicated or available . . . to do so with maximum efficiency and at the lowest investment and lowest costs consistent with quality work. To see how they can be fitted perfectly to your requirements note the following:

AVAILABLE WITH EITHER GAS OR DIESEL ENGINES— WITH VERY POWERFUL VIBRATORY TAMPING MOTORS for maximum penetration in hard going and handling any type of ballast in any condition—OR, OUR SOMEWHAT SMALLER VIBRATORY MOTORS AT LESS COST. These are thoroughly suited to smaller railroads and branch lines where utmost power is not a first consideration . . . fine for spot tamping and surfacing work, tightening up behind the tie gang and emergency tamping. OK for general use in small or soft ballasts. Excellent for new construction and high lift ballast insertions anywhere.

As always, direct sales, leases and service to all U. S. railroads.

JACKSON VIBRATORS, INC.
LUDINGTON, MICHIGAN

R1-11

Tops optionally available.

AVAILABLE WITH JACKS. When so equipped, the Jackson Utility Tamper becomes a very fast jack tamper for raising ahead of production tampers, a very versatile machine, indeed. Adaptable to surfacing devices. Tops, optionally available.

Simplified throughout for easy operation and economical maintenance

Let us show you how a Jackson Utility Tamper can be ideally tailored to your needs.

CROSS TAMPING: Jackson Tampers, like no others, are highly efficient in cross tamping because of their unique and powerful vibratory action which uniformly consolidates to maximum compaction a perfect tie bed of large proportions right under the rail base where it belongs.

the BORTUNCO GROUP

BorTunCo has an established record with engineers and contractors for capable sub-contract job performance*

American Railway
Engineering Association—Bulletin

Vol. 63, No. 569   January 1962

REPORTS OF COMMITTEES

The reports in this issue of the Bulletin will be presented to the 1962 convention of the Association at the Conrad Hilton Hotel, Chicago, March 9–10. Comments and discussion with respect to any of the reports are solicited, and should be addressed to the chairman of the committee involved, in writing in advance of the convention, or from the floor of the convention.

Published by the American Railway Engineering Association, Monthly, January, February, March, November and December; Bi-Monthly, June–July, and September–October, at 2211 Fordem Avenue, Madison, Wis.; Editorial and Executive Offices, 59 Van Buren Street, Chicago 5, Ill. Second class postage paid at Madison, Wis. Accepted for mailing at special rate of postage for in Section 1103, Act of October 3, 1917, authorized on June 29, 1918. Subscription $10 per annum.

Report of Committee 15—Iron and Steel Structures

C. Neufeld, *Chairman*

G. W. Salmon,
Vice Chairman

J. M. Hayes, *Secretary*

E. S. Birkenwald
J. E. South
E. T. Franzen
A. R. Harris
R. C. Baker
W. E. Dowling
J. C. King

H. A. Balke
Ethan Ball*
J. L. Beckel
L. S. Beedle
R. S. Bennett
J. E. Bernhardt
E. D. Billmeyer
R. T. Blewitt
H. F. Bober
E. T. Bond, Jr.
R. N. Brodie
E. E. Burch
Abram Clark*
J. G. Clark
R. P. Davis
C. E. Ekberg
T. L. Fuller
G. K. Gillan
R. W. Gustafson
C. D. Hanover, Jr.
Alfred Hedefine
W. C. Howe
D. L. Jerman*
E. A. Johnson
B. G. Johnston
E. W. Kieckers
M. L. Koehler
K. H. Lenzen
Shu-t'ien Li
J. F. Marsh
M. L. McCauley

D. V. Messman
James Michalos
W. G. Mitchell
R. F. Moline
N. W. Morgan*
W. H. Munse
N. M. Newmark
J. C. Nichols
D. L. Nord
R. D. Nordstrom
T. G. O'Neil
E. E. Paul
G. H. Perkins
R. A. Peteritas
A. G. Rankin
C. A. Roberts
W. E. Robey
G. E. Robinson
D. D. Rosen
Henry Seitz
R. I. Simkins
A. E. Smith
H. F. Smith*
R. D. Spellman
H. C. Tammen (E)**
W. M. Thatcher
L. E. Titlow
H. T. Welty (E)
A. R. Wilson (E)
L. T. Wyly
Committeee

* Resigned from committee prior to Dec. 1, 1961.
** Died July 6, 1961.
(E) Member Emeritus.
Members listed in bold face are the official representatives of the Engineering Division, AAR.

To the American Railway Engineering Association:

Your committee reports on the following subjects:

1. Revision of Manual.

2. Composite steel and concrete spans; non-ferrous metal bridges, collaborating with Committees 8 and 30.

Non-ferrous metal bridges is a new subject, and your committee submits
below as information an outline of the steps it proposes to take in carrying
out the assignment:

1. Assemble and study data and references on existing aluminum bridges and their specifications.
2. Correlate the above with reference to use for railway bridges.
3. Assemble and study data on research to date on the use of aluminum for structures.
4. Study the economics of the problem.
5. List future or further problems and study the need for additional testing and research.

Step No. 1 has been carried out and Step No. 2 is underway.

3. Corrosion of deck plates.

The details covering the installation of plate specimens of 12 different metals at the AAR Research Center and on the Huey P. Long Bridge at New Orleans, to determine their relative resistance to corrosion, was reported in the 1960 Proceedings, Vol. 61, page 580. These specimens are being subjected to severe brine exposure. To measure the effect of corrosion, half of the specimens will be left undisturbed for at least three more years and then thoroughly cleaned to remove all rust and then weighed. The other half of the specimens will be left undisturbed for several more years before cleaning and weighing. A visual inspection indicates that all the specimens, except stainless steel and aluminum, are corroding at about the same rate.

4. Stress distribution in bridge frames.

(a) Floorbeam hangers.

Progress report, presented as information .... page 399

(b) Truss bridge research project.

Progress report, presented as information .... page 400

5. Design of steel bridge details.

Your committee is reviewing Arts. 43, 44, 45, 68 and 79 of Part 1, Sec. A, and investigating the use of new materials for expansion bearings and for waterproofing steel bridge decks.

6. Preparation and painting of steel surfaces; synthetic resins and other adhesive materials for protective coating and reinforcement, collaborating with Committee 7.

No tests have been initiated this past year on account of limited research funds. Reports on previous paint tests will be reported as important information develops from these tests. Although the AAR is not now contributing money to the Steel Structures Painting Council, some member roads have made financial contributions to the Council. Your committee has studied the report of Committee 7 (Proceedings, Vol. 62, 1961, pages 526 to 537, incl.) insofar as the use of synthetic resins and adhesive materials apply to iron and steel structures. Some applications have been made but it is too early to report on their behavior.

7. Bibliography and technical explanation of various requirements in AREA specifications relating to iron and steel structures.

Your committee is in the process of reviewing each article of the specifications for iron and steel structures with regard to possible forthcoming recommendations for changes, additions, deletions, rearrangements, etc. Inquiries and helpful suggestions have been received from men not on this committee. Progress is being made on the overall task; however, more time will be needed to complete the assignment. Already your committee has made recommendations to other committees with favorable actions resulting therefrom.

8. Specifications for the design of corrugated metal culverts, including corrugated metal arches.

This assignment was completed with adoption and publication in the Manual of Specifications for Corrugated Structural Plate Pipe, Pipe-Arches and Arches, 1961, pages 1–4–25 to 1–4–32.1, incl.

10. Effect of continuous welded rail on bridges, collaborating with the Special Committee on Continuous Welded Rail.

Progress report, presented as information .... page 400

THE COMMITTEE ON IRON AND STEEL STRUCTURES,

C. NEUFELD, *Chairman.*

AREA Bulletin 569, January 1962.

MEMOIR

Henry C. Tammen

Committee 15 will miss the wisdom, intellectual curiosity, patience and mental ability of Henry Casper Tammen, who at the age of 76 years passed away suddenly on July 6, 1961. During his 17 years as member and 8 years as Member Emeritus of Committee 15, he served the committee well, contributing greatly to the formulation of modern specifications for shop fabrication and for movable bridges and to the rigorous analysis of theory and fundamentals in their application to the committee's specifications.

Mr. Tammen joined the Association in 1929; was a member of the committee from 1933 to 1950; and was elected Member Emeritus in 1953.

His early education was received in the public schools of Yankton, S. D., where he was born on November 15, 1884. He was graduated from the University of California with the degree of Bachelor of Science in Civil Engineering. Following graduation he served two years as rodman and draftsman for the Western Pacific Railroad and the Southern Pacific Company. From 1908 to 1914, he was employed by Waddell & Harrington in Kansas City as draftsman, during which time he assisted in the preparation and accumulation of much of the data for Doctor Waddell's book "Bridge Engineering." When the firm of Harrington, Howard and Ash was formed in 1914, he became assistant engineer and chief designer. From 1928 to 1940 he was a partner of the consulting firm of Ash, Howard, Needles and Tammen, and from 1940 until his death, partner in the firm of Howard, Needles, Tammen & Bergendoff. During his latter years he served as special consultant to Howard, Needles, Tammen & Bergendoff.

Mr. Tammen's specialty was bridge design, in which field he held several patents relative to movable bridges. During his career he was responsible for the design of 16 of the railroad bridges crossing the Welland Canal, 25 bridges for the Kansas City Southern Railroad, the Southern Pacific Neches River Bridge at Beaumont, Tex., and the Seaboard's Savannah River Bridge at Savannah, Ga. Some of the structures in which he was particularly involved included the Harlem River lift span of the Triborough Bridge in New York City, the Delaware Memorial Bridge at Wilmington, Del., the Passaic River lift span on U. S. Route 1 in New Jersey, the Old Lyme–Old Saybrook Bridge in Connecticut, the Edison Bridge over the Raritan River at Perth Amboy, N. J. In addition he was responsible for the design of many movable bridges throughout New Jersey, Virginia and Florida and many large bridges across the Mississippi and Missouri Rivers.

Mr. Tammen was a fellow and life member of the American Society of Civil Engineers, a member of the American Institute of Consulting Engineers, the Canoe Brook County Club in Milburn Township, New Jersey, the New Jersey Seniors Golf Association and the Central Presbyterian Church of Summit, N. J.

Survivors include his wife, Gertrude Armstrong Tammen; a son, John G. Tammen of Newark, Ohio; a daughter, Miss Margaret M. Tammen of New York City; two grandchildren; and three sisters, Mrs. Andrew J. Odgard of Glendale, Calif., Mrs. Beverly T. Ward of Costa Mesa, Calif., and Mrs. J. Manor Howard of Del Mar, Calif.

Report on Assignment 1

Revision of Manual

E. S. Birkenwald (chairman, subcommittee), E. Ball, J. L. Beckel, R. P. Davis, A. R. Harris, J. M. Hayes, W. C. Howe, J. C. King, J. F. Marsh, D. V. Messman, E. E. Paul, D. D. Rosen, G. W. Salmon, R. D. Spellman.

Your committee submits the following revisions of specifications for adoption and publication in the Manual:

Pages 15–1–1 to 15–1–58, incl.

SPECIFICATIONS FOR STEEL RAILWAY BRIDGES

Page 15–1–1, Foreword, 1st paragraph, last sentence: Substitute the words "light-traffic and branch-line" for the words "light and branch."

Page 15–1–3, Art. 3, Sec. A: Replace with the following:

3. Shop Drawings

(a) After the contract has been awarded, the contractor shall submit to the engineer, for review and approval as to conformity to contract requirements, prints from checked plans in the number required, of stress sheets, shop drawings and erection procedures, unless such sheets, drawings and procedures have been prepared by the company.

Shop drawings shall be 24 in by 36 in. in size, including left hand margin 1½ in wide and ½-in margin on other edges. An approved title shall be in the lower right hand corner.

If any changes or corrections are required by the engineer, one print with changes noted thereon shall be returned to the contractor. Prints from corrected plans shall be submitted to the engineer for review and this procedure will continue until each drawing, etc., is approved.

No change shall be made on such approved drawings without the consent of the engineer.

The contractor shall furnish to the company as many prints of the drawings as may be necessary to carry out the work.

(b) The contractor shall be responsible for the correctness and completeness of his drawings, regardless of any approval by the engineer.

(c) Any work performed or material ordered prior to approval by the engineer shall be at the sole risk of the contractor.

(d) The original drawings shall be ink on tracing cloth or legible drawings photographically reproduced on tracing cloth by an approved method. They shall be delivered to and become the property of the company upon completion of the contract.

Page 15-1-4, Art. 8, Sec. A: In the second last line, substitute the words "if special" for the word "where".

Page 15-1-4, Art. 9, Sec. A: Replace with the following:

9. Spacing of Trusses, Girders and Stringers

The distance between centers of outside trusses or girders shall be sufficient to prevent overturning by the specified latrai forces. In no case shall it be less than 1/20 of the span for through spans, nor 1/15 of the span for deck spans.

Where the track is supported by a pair of deck girders or stringers, the distance center to center shall be not less than 6 ft 6 in. If multiple girders or stringers are used, they shall be arranged as nearly as possible to distribute the track load uniformly to all members.

Page 15-1-32, Art. 1, Sec. B: Insert in the second paragraph after the words "open-hearth", the following, "basic-oxygen."

Page 15-1-33, Art. 2 (c), Sec. B: Replace with:

(c) The steel shall be made by the open-hearth, basic-oxygen or the electric-furnace process.

Page 15-1-40, Art. 1: Revise the formulas for bearing on expansion rollers and rockers to read:

$$\textit{Lb per Lin In}$$

For diameters up to 25 in .... $\dfrac{p-13{,}000}{20{,}000} \, 600d$

For diameters from 25 in to 125 in .... $\dfrac{p-13{,}000}{20{,}000} \, 3{,}000 \, \sqrt{d}$

Pages 15-2-1 to 15-2-73, incl.

SPECIFICATIONS FOR MOVABLE RAILWAY BRIDGES

Page 15-2-1, Foreword, 1st paragraph, last sentence: Substitute the words "light-traffic and branch-line" for the words "light and branch".

Page 15-2-14, Art. 6, Sec. C: Substitute the following for the last paragraph:

When the moving span is normally left in the closed position, the machinery for bascule and swing bridges shall also be proportioned to hold the span in the fully open position against a wind load of 20 lb per sq ft on any vertical projection of the open bridge. For swing bridges, provision shall be made for a wind load of 25 lb per sq ft on the other arm. In proportioning the machinery for these conditions, 1.5 times the normal unit stresses may be used.

When the moving span is normally left in the open position, the machinery for bascule and swing bridges shall also be proportioned to hold the span in the fully open position against the wind loads specified in Art. 5 (d), Sec. C. If desired, the machinery may be proportioned as specified in the preceding paragraph and the difference between the wind loads specified in Art. 5 (d), Sec. C, and those specified in the preceding paragraph shall then be cared for by separate holding devices. In proportioning the machinery for these conditions, 1.5 times the normal unit stresses shall be used.

Page 15–2–15, Art. 9, Sec. C: In the fourth paragraph, delete, in the first line concerning capacity of machinery brakes, the words "and equal arm swing bridges"; also, at the end of the third and beginning of the fourth line, the words "unequal arm".

Pages 15–2–16 and 15–2–17, Art. 13, Sec. C: Substitute the following:

13. Unit Stresses in Machinery Parts

The following allowable unit stresses in pounds per square inch shall be used for machinery and similar parts:

Material	Tension	Compression	Fixed Bearing	Shear
Structural carbon steel	12,000	$12,000 - 55\frac{l}{r}$	16,000	6,000
Class C_1 forged carbon steel	13,000	$13,000 - 55\frac{l}{r}$	16,000	6,500
Class E forged carbon steel—except keys	15,000	$15,000 - 65\frac{l}{r}$	18,000	7,500
Class E forged carbon steel—keys			15,000	7,500
Class A forged alloy steel	16,000	$16\,000 - 70\frac{l}{r}$	21,000	8,000
Hot-rolled shafting and cold-finished shafting (equal to Class E forged) ..	15,000	15.000		7,500
Cast steel	9,000	$10,000 - 45\frac{l}{r}$	13,000	5,000
Cast iron—Class No. 25	2,000	10,000*		
Bronze—Alloy D	7,000	7,000		

* For struts whose $\frac{l}{r}$ is 20 or less.

STRESS IN EXTREME FIBERS OF TRUNNIONS

	Rotation More Than 180 Deg	Rotation 90 Deg or Less	Fixed Trunnions
Class C_1 forged carbon steel	13,000	13,000	15,000
Class E forged carbon steel	15,000	15,000	17,000
Class A forged alloy steel	16,000	20,000	22,000

For stresses in rotating parts, and in frames, pedestals, and other units which support rotating parts, the computed stresses shall be multiplied by the impact factor K.

$K = 1.0$ for trunnions and for counterweight sheaves and their shafts.

$K = 1.0 + 0.03\sqrt{n}$ for other parts where $n = $ rpm of rotating part.

All of the unit stresses specified in this article provide appropriate safety factors against static failure and against failure by fatigue with and without reversal of stresses. In the determination of the safety factor against fatigue failure, provision was made for stress-raisers by assuming that the stress-raiser increases by 40 percent of the computed stress which is used for design at the unit stress specified. For trunnions and counterweight sheave shafts, this provides for the customary increase in diameter inside the bearings with the two diameters connected by fillets of reasonable radius; for gear arms, it provided for the increase in stress at the faces of a pinion forged into one piece with the shaft or, where the gears and pinions are keyed to the shaft, for the increase in stress caused by the keyways. The increase in stress provides for one or two keyways, 120 deg apart, each having a width not more than $\frac{1}{4}$ and a depth not more than $\frac{1}{8}$ the shaft diameter. In the absence of keyways or other stress-raisers in a shaft, the unit stresses for torsion and flexure in a shaft may be increased 20 percent.

Page 15-2-17, Art. 14 (a), Sec. C: Delete the line reading "Trunnion bearings and counterweight sheave bearings, rolled or forged steel on Alloy B bronze .... 1,500" and substitute therefor:

Trunnion bearings and counterweight sheave bearings, rolled or forged steel on Alloy B bronze:

For loads while in motion 1,500
For loads while at rest 2,000

Page 15-2-19, Art. 17, Sec. C: Revise the two formulas to read:

$$f = \frac{16\,K}{\pi\,d^3}\ (M + \sqrt{M^2 + T^2})$$

$$S = \frac{16\,K}{\pi\,d^3}\ \sqrt{M^2 + T^2}$$

Delete the explanation for K_2, K_3 and n below these formulas and substitute: "For values of K and allowable unit stresses with and without keyways, or other stress-raisers, see Art. 13, Sec. C."

Pages 15-2-19 and 15-2-20, Art. 19, Sec. C: Delete the present article and substitute the following:

19. Machinery Design

The machinery for moving the span shall be designed at normal unit stresses for the following percentages of full-load rated torque of the prime mover at Condition A, Art. 6, Sec. C speed:

Electric motors ....150 percent
Internal combustion engines ....100 percent

For manual operation, the machinery shall be designed as specified in Art. 2, Sec. F.

The machinery shall also be designed for the braking forces, and at the unit stresses, specified in Art. 6 and 9, Sec. C.

Pages 15-6-9 to 15-6-14, incl.

INSTRUCTIONS FOR THE MAINTENANCE INSPECTION OF STEEL BRIDGES

Page 15-6-9: Revise Par. 1 to read:

1. The inspection of steel bridges should be under the general supervision of the chief engineer in charge of maintenance, the bridge engineer, or other general officer in charge of bridge maintenance. Direct supervision on each maintenance unit like a division, district or region, should be under some designated officer herein called general bridge inspector.

Page 15-6-9, Par. 2, 2nd line: Change the word "shall" to "should."
Par. 3, 1st line: Change the word "shall" to "should."
Par. 4, 1st and 2nd lines: Change the word "shall" to "should."
Par. 5, 2nd and 7th lines: Change the word "shall" to "should."

Par. 5, 4th line: Change the word "must" to "should."
Par. 6, 1st, 2nd and 4th lines: Change the word "shall" to "should."
Par. 7, 3rd line: Change the word "shall" to "should."
Page 15–6–10, Par. 9, 1st line: Change the word "shall" to "should."
Page 15–6–14, Par. 10, 1st and 2nd lines: Change the word "shall" to "should."
Par. 11, 1st line: Change the word "shall" to "should."

Your committee submits for adoption the following further recommendations with respect to material in Chapter 15 of the Manual:

Pages 15–3–1 to 15–3–8, incl.   –

SPECIFICATIONS FOR STEEL RAILWAY TURNTABLES

Reapprove without change.

Pages 15–4–1 to 15–4–3, incl.

SPECIFICATIONS FOR THE DESIGN OF RIGID-FRAME STEEL BRIDGES

Reapprove without change.

Pages 15–4–4 to 15–4–6, incl.

SPECIFICATIONS FOR THE DESIGN OF CONTINUOUS STEEL RAILWAY BRIDGES

Reapprove without change.

Pages 15–5–1 to 15–5–6, incl.

SPECIFICATIONS FOR THE ERECTION OF STEEL RAILWAY BRIDGES

Reapprove without change.

Pages 15–6–1 and 15–6–2

INSTRUCTIONS FOR THE MILL INSPECTION OF STRUCTURAL STEEL

Reapprove without change.

Pages 15–6–2 to 15–6–4, incl.

INSTRUCTIONS FOR THE INSPECTION OF THE FABRICATION OF STEEL BRIDGES

Reapprove without change.

Pages 15–6–5 to 15–6–9, incl.

INSTRUCTIONS FOR THE INSPECTION OF BRIDGE ERECTION

Reapprove without change.

Pages 15-7-1 to 15-7-3, incl.

CLASSIFICATION OF RAILWAY BRIDGES

Reapprove without change.

Pages 15-7-3 to 15-7-8, incl.

RULES FOR RATING EXISTING IRON AND STEEL BRIDGES

Reapprove without change.

Pages 15-7-8 to 15-7-14, incl.

METHODS OF STRENGTHENING EXISTING BRIDGES

Reapprove without change.

Page 15-M-3

FUSION WELDING

Reapprove without change.

Pages 15-M-20 and 15-M-21

TRACK ANCHORAGE ON BRIDGES AND SIMILAR STRUCTURES

Reapprove without change.

Pages 15-M-27 to 15-M-29, incl.

SPECIFICATIONS FOR ASSEMBLY OF STRUCTURAL JOINTS USING HIGH-STRENGTH STEEL BOLTS IN STEEL RAILWAY BRIDGES

Reapprove without change.

———————

Your committee presents as information the following recommendations with respect to the substitution of A 36 for A 7 steel in the Specifications for Steel Railway Bridges, to be considered for adoption one year hence:

Pages 15-1-1 to 15-1-58, incl.

SPECIFICATIONS FOR STEEL RAILWAY BRIDGES

Page 15-1-4, Art. 10, Sec. A: Revise the second paragraph to read:
Where shallower depth spans are required, they shall be proportioned so that the deflection from the combined effect of live load and impact shall not exceed a deflection-span ratio of 1/640.

Page 15-1-11, Art. 31 (a), Sec. A: Revise to read:

31. Unit Stresses

The allowable unit stresses to be used in proportioning the parts of a bridge shall be as follows:

Lb per Sq in

(a) *Structural and Rivet Steel:*

Axial tension, structural steel, net section 20,000

Tension in floorbeam hangers, including bending:
Using rivets in end connections 14,000
Using high-strength bolts in end connections 20,000

Tension in extreme fibers of rolled shapes, girders and built sections, subject
to bending 20,000

Axial compression, gross section:
For stiffeners of plate girders 20,000

For compression members centrally loaded and with values of l/r not
greater than 130:

Riveted ends ....17,000$-$0.31$\dfrac{l^2}{r^2}$

Pin ends ....17,000$-$0.43$\dfrac{l^2}{r^2}$

$l =$ length of member in inches.
$r =$ least radius of gyration of member in inches.

For compression members with values greater than 130 and for compression
members of known eccentricity, see Appendix A.
Compression in extreme fibers of rolled shapes, girders and built sections

subject to bending (for values of l/b not greater than 40) ....20,000$-$6$\dfrac{l^2}{b^2}$

$l =$ length in inches of unsupported flange between lateral connections
or knee braces.
$b =$ flange width in inches.

Diagonal tension in webs of girders and rolled beams at sections where
maximum shear and bending occur simultaneously 20,000
Stress in extreme fibers of pins 30,000
Shear in plate girder webs, gross section 12,500
Shear in power driven rivets 13,500
Shear in pins 15,000
Shear in bolts and hand-driven rivets 11,000
Bearing on pins and power driven rivets 27,000
Bearing on milled stiffeners and other steel parts in contact 30,000
(Rivets driven by pneumatically or electrically operated hammers are
considered power-driven).
Bearing between rockers and rocker pins 13,500
Bearing on turned bolts 22,500
Bearing on hand-driven rivets 20,000
Bearing on expansion rollers and rockers, lb per lin in:
For diameters up to 25 in .... $\dfrac{P-13,000}{20,000}$ 600 d

For diameters from 25 in to 125 in .... $\dfrac{P-13,000}{20,000}$ 3,000 $\sqrt{d}$

Page 15–1–12, Art. 31 (b), Sec. A: Revise to read:

(b) *Cast Steel*

For cast steel, the allowable unit stresses in compression and bearing shall be 0.9 of
those for structural steel. Other allowable unit stresses shall be 2/3 of those for
structural steel.

Pages 15–1–32 and 15–1–33, Art. 1, Sec. B: Revise to read:

1. Structural and Rivet Steel

Except as otherwise specified in this section, structural and rivet steel shall conform to the requirements of the current ASTM Specifications as follows: for structural steel, designation A 36; for rivet steel, designation A 141.

The steel shall be made by the open-hearth, basic-oxygen, or the electric-furnace process.

The chemical composition shall conform to the following table:

	Ladle Analysis	Check Analysis
Carbon, max percent	0.28	0.32
Phosphorus, max percent	0.04	0.05
Sulfur, max percent	0.05	0.063
Copper, when copper steel is specified, min percent	0.20	0.18

The tensile test requirements shall be as follows:

	Structural Steel	Rivet Steel
Tensile strength, psi	60,000–80,000	52,000–62,000
Yield point, min psi	36,000	28,000
Elongation in 8 in, min percent	20[a]	24
Elongation in 2 in, min percent	23[b]	

[a] Except as modified in Secs. 5 (c) and 5 (d) of ASTM Specifications, designation A 36.
[b] Except as modified in Sec. 5 (e) of ASTM Specifications, designation A 36.

The crosshead speed of the testing machine, after a stress of $\frac{1}{2}$ the specified yield point is attained and during determination of yield point, shall not exeed 1/16 in per min per in of gage lengh; and during determination of ultimate strength shall not exceed $\frac{1}{2}$ in per min per in of gage length.

The bend test specimen shall stand being bent cold through 180 deg without cracking on the outside of the bent portion to an inside diameter which shall bear the following ratio to the thickness of the specimen:

Thickness	Ratio
Structural steel, $\frac{3}{4}$ in and under	$\frac{1}{2}$
Structural steel, over $\frac{3}{4}$ in to 1 in, incl	1
Structural steel, over 1 in to $1\frac{1}{2}$ in, incl	$1\frac{1}{2}$
Structural steel, over $1\frac{1}{2}$ in to 2 in, incl	$2\frac{1}{2}$
Structural steel, over 2 in	3
Rivet steel	flat on itself

In order to meet the requirements of Art. 134, Sec. A for the properties of full-size annealed forged-head eyebars, the manufacturer of the eyebars may determine the physical properties required for the specimen tests. The steel shall conform to the above requirements as to physical properties other than tensile strength.

Page 15-1-31, Art. 134, Sec. A: Change yield point from 33,000 to 36,000.

Page 15-1-42, Appendix A: Revise as follows·

Under y, add after structural steel "(A7)" and between this and the next line, insert another line reading: "36,000 for structural steel (A36)."

In the paragraph for values of p with $\dfrac{l}{r} = 0$, after the words "structural steel", add "(A7)"; also between this and the next line, insert another line reading: "= 17,000 for structural steel (A36)"

For values of f, insert a line between high-strength steel and structural steel to read: "$= 1.69$ for structural steel (A36)", and add "(A7)" after structural steel in what will become the third line.

Your committee presents as information the following recommendation with respect to the current specifications covering the use of high-strength bolts, to be considered for adoption one year hence:

Pages 15–M–27 to 15–M–29, incl.

SPECIFICATIONS FOR ASSEMBLY OF STRUCTURAL JOINTS USING HIGH-STRENGTH STEEL BOLTS IN STEEL RAILWAY BRIDGES

Delete, substituting the following:

SPECIFICATIONS FOR STRUCTURAL JOINTS USING HIGH-STRENGTH STEEL BOLTS IN STEEL RAILWAY BRIDGES

A. SCOPE

1. These specifications cover recommended practice for the design and assembly of structural steel joints using high-strength steel bolts tightened to a high tension.

2. Design and construction shall conform to AREA Specifications for Steel Railway Bridges, Part 1, this Chapter, except as provided herein.

3. Joints required to resist shear between their connected parts are designated as either friction-type or bearing-type connctions. Shear connections shall be friction-type when subjected to stress reversal or severe stress fluctuation.

4. The Commentary following these specifications provides guidance in their application.

B. BOLTS, NUTS AND WASHERS

1. Bolts, nuts and washers shall conform to requirements of the current edition of the Specifications for High-Strength Steel Bolts for Structural Steel Joints of the American Society for Testing and Materials (ASTM Specifications, designation A 325), except as hereinafter provided.

2. Bolt dimensions shall conform to the current requirements for heavy hexagon structural bolts of the American Standards Association (ASA Standard B18.2). See Fig. 1 for details.

3. Nut dimensions shall conform to current requirements for heavy semi-finished hexagon nuts of the American Standards Association (ASA Standard B18.2). See Fig. 1 for details.

4. Circular washers shall be flat and smooth, and their nominal dimensions shall conform to the current requirements for Type A washers of the American Standards Association (ASA B27.2) with nominal dimensions as given in Table 1. Beveled washers shall be square or rectangular, taper in thickness, and conform to the dimensions given in Table 1.

Where necessary, washers may be clipped on one side to a point not closer than seven-eighths of the bolt diameter from the center of the washer.

Fig. 1.

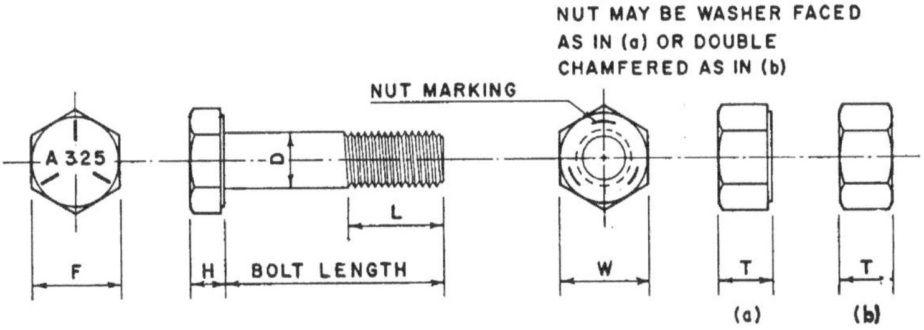
NUT MAY BE WASHER FACED
AS IN (a) OR DOUBLE
CHAMFERED AS IN (b)
NUT MARKING
A 325
D
L
F
H
BOLT LENGTH
W
T
T
(a)
(b)

Bolts manufactured to ASTM Specifications, designation A 325, are identified on the top of the head by 3 radial lines, the legend "A 325", and the manufacturer's mark. Heavy semi-finished hexagon nuts manufactured to ASTM Specifications, designation A 325, are identified on at least one face by 3 circumferential marks, or by the number "2" and the manufacturer's mark.

BOLT AND NUT DIMENSIONS (INCHES)

Nominal Bolt Size D	*Heavy Hexagon Structural Bolts*			*Heavy Semi-Finished Hexagon Nuts*		*To Determine* Required Bolt Length, Add to Grip (Inches)*
	Width Across Flats F	*Height H*	*Thread Length L*	*Width Across Flats W*	*Height T*	
3/4	1 1/4	15/32	1 3/8	1 1/4	47/64	1
7/8	1 7/16	35/64	1 1/2	1 7/16	55/64	1 1/8
1	1 5/8	39/64	1 3/4	1 5/8	63/64	1 1/4
1 1/8	1 13/16	11/16	2	1 13/16	1 7/64	1 1/2
1 1/4	2	25/32	2	2	1 7/32	1 5/8

*In order to determine the required bolt length, the values shown in the right-hand column of the above table should be added to the grip, i.e., the total thickness of all connected material exclusive of washers. These values are generalized, with due allowance for manufacturing tolerances, to provide for the use of a heavy nut with adequate "stick through" at the end of the bolt. For each hardened flat washer that is used, add 3/16 in and for each beveled washer, add 5/16 in. The length determined by the use of the above table should be adjusted to the next longer 1/4-in length increment.

Heavy hexagon structural bolts have slightly shorter thread lengths than other standard bolts. By making the body length of the bolt a basic dimension, it has been made possible to exclude the thread from all shear planes, except in the case of thin outside parts adjacent to the nut. Depending on the amount of bolt length added to adjust for incremental stock lengths, the full thread might extend into the grip as much as 3/8 in for 3/4- 7/8- and 1 1/4-in bolts and as much as 1/2 in for 1- and 1 1/8-in bolts. (Inclusion of some of the thread run-out into the plane of shear is permissible). When the thickness of an outside part adjacent to the nut is less than these values, it may be necessary to use the next increment of bolt length together with a sufficient number of flat circular washers to insure full seating of the nut. The higher working value in shear permitted in bearing-type joints can then still be the basis for determining the number of bolts in the connection.

TABLE 1—NOMINAL WASHER DIMENSIONS (INCHES)

Bolt Size	Circular Washers				Square or Rectangular Washers for American Standard Beams and Channels		
	Inside Dia.	Outside Dia.	Thickness Gage No.	Thickness Inches	Minimum Width	Mean Thickness	Slope or Taper in Thickness
¾	13⁄16	1¾	9	0.148	1¾	5⁄16	1:6
⅞	15⁄16	2	8	0.165	1¾	5⁄16	1:6
1	1 1⁄16	2¼	8	0.165	1¾	5⁄16	1:6
1⅛	1¼	2½	8	0.165	2¼	5⁄16	1:6
1¼	1⅜	2¾	8	0.165	2¼	5⁄16	1:6

C. BOLTED PARTS

1. Surfaces of bolted parts in contact with the bolt head and nut shall not have a slope of more than 1:20 with respect to a plane normal to the bolt axis. Bolted parts shall fit solidly together when assembled without interposition of gaskets or other compressible material. Holes may be punched, sub-punched and reamed, or drilled as required by the AREA Specifications and shall be of a diameter not more than 1/16 in. in excess of the nominal bolt diameter.

2. When assembled, all joint surfaces, including those adjacent to the bolt heads, nuts or washers, shall be free of scale except normal mill scale. Surfaces shall be free of dirt, loose scale, burrs and other defects that would prevent solid bearing of the parts.

3. Contact surfaces shall be free of oil, paint, lacquer or galvanizing.

D. ALLOWABLE WORKING STRESSES

1. *Tension.* Bolts required to support applied loads by means of direct tension shall be so proportioned that their average tensile stress, using the area based on the nominal bolt diameter and independent of any initial tightening force, will not exceed 36,000 psi.

2. *Shear.* Bolts in friction-type connections assembled in accordance with the requirements of Sec. C–3 and bolts in bearing-type connections having threading in plane of contact surfaces of the connected parts, shall be proportioned on the basis of a shear stress of 13,500 psi. In friction-type connections this shear stress may be used to proportion high-strength bolts used in combination with rivets designed in accordance with provisions of Part 1, this Chapter.

Bolts in bearing-type connections, where bolt threads are excluded from the shear planes of the contact surfaces between the connected parts, shall be proportioned on the basis of a shear stress of 20,000 psi.

3. *Bearing.* In friction-type connections, there need be no consideration of bearing, fillers need not be developed, and no extra bolts required because of long grip. In bearing-type connections, the computed bearing pressure, assumed to be distributed over an area equal to the nominal bolt diameter times the thickness of the connected piece, shall not exceed 36,000 psi.

In connections of this type having no more than two bolts in a line parallel to the direction of stress, the distance from the end of the member to the center of the nearest bolt shall be not less than the nominal shearing area of the bolt (single or double shear, as the case may be) divided by two-thirds of the plate thickness. This end distance may be proportionately less where the stress per bolt is less than that permitted in this section, but not less than 1.5 times the bolt diameter.

4. Increases in working stresses allowed in other parts of Chapter 15 may be applied to the stresses given in this section.

5. The foregoing working stresses for fasteners shall be used in proportioning the connections for members designed and fabricated of higher strength steels.

E. ASSEMBLY

1. Heavy hexagon structural bolts with heavy semi-finished hexagon nuts may be installed without hardened washers, except that washers shall be used under the following conditions:

    a. where the bolt head or nut is tightened against a surface with a slope greater than 1:20 with respect to the bolt axis;

    b. where the bolt holes are oversize; and

    c. where the bolts are subject to tensile loads.

When an outer face of the bolted parts has a slope of more than 1:20, a smooth beveled washer shall be used to compensate for the lack of parallelism. In the other exceptions noted, flat circular washers shall be used under both the bolt head and nut.

2. Tightening shall be done by the turn-of-nut method. In the turn-of-nut method there shall first be enough bolts brought to a snug-tight condition to insure that the several parts of the joint are properly compacted, i.e,, brought into *full* contact with each other. Snug-tight is the tightness attained by a few impacts of an impact wrench or the full effort of a man using an ordinary spud wrench. Following the initial step, bolts shall be placed in any remaining holes in the connections and brought to snug tightness. Then all bolts in the joint shall be tightened additionally by the applicable amount of nut rotation specified in Table 2, with tightening progressing systematically from the most rigid part of the joint to its free edges.

TABLE 2—NUT ROTATION* FROM SNUG-TIGHT CONDITION

Bolt Diameter	Grip	Both Faces Normal to Bolt Axis	One Face Normal to Bolt Axis, One Sloped 1:20	Both Faces Sloped 1:20 from Normal to Bolt Axis
¾, ⅞ 1, 1⅛, 1¼	to 5″ incl. to 8″ incl.	½ turn	¾ turn	1 turn
¾, ⅞ 1, 1⅛, 1¼	over 5″ over 8″	¾ turn	1 turn	1 ¼ turn

*Nut rotation is rotation relative to bolt regardless of the element (nut or bolt) being turned. Tolerance on rotation: ¼ turn over; nothing under.

3. If required because of bolt-entering and wrench-operation clearances, tightening may be done by turning the bolt while the nut is prevented from rotating.

F. INSPECTION

1. The inspector shall satisfy himself that all requirements of these specifications are met.

2. The inspector shall approve the procedure for the installation of bolts and shall further observe the field installation to determine that the bolts are brought to a snug-tight condition and then given the specified turn.

3. Bolts, nuts and washers are normally received with a light residual coating of oil. This coating is not detrimental even to friction-type connections and need not be removed.

4. Bolts tightened by the turn-of-nut method may have the outer face of the nut match-marked with the protruding bolt point before final tightening, thus affording the inspector visual means of noting the actual nut rotation. Such marks can be made by the wrench operator with a crayon or dab of paint, after the bolts have been brought up snug tight.

COMMENTARY

Scope

When first approved by the Research Council on Riveted and Bolted Structural Joints of the Engineering Foundation, January 1951, the Specifications for Assembly of Structural Joints Using High-Strength Steel Bolts merely permitted the substitution of a like number of high-strength bolts for hot-driven rivets of the same nominal diameter. In succeeding years, tests on bolted structural joints developed the fact that in a great many cases movement of the connected parts that bring the bolts into bearing against the sides of their holes was in no way detrimental. When the nature of the loading is such that fatigue-type failure or reversal of movement will not occur, the high clamping force in the bolts provides a rigid assembly in the "slipped" position, and the shear strength of the high-strength bolts, when threads are excluded from contact surface shear planes, is even greater than that of hot-driven rivets required to function under similar circumstances. Accordingly, the present specifications recognize two kinds of shear connections, designated as friction-type and bearing-type respectively.

Just how much stronger the high-strength bolts are in resisting actual shearing forces and what effect the higher stresses in the bolts have upon the strength of the connected parts have been the subjects of extensive study sponsored by the Research Council since 1954. The results of these studies, together with improvements in installation practices which are the outgrowth of extensive experience in the use of high-strength bolts, form the background for the current edition.

Bolted Parts

High initial bolt tension provides worthwhile advantages; therefore the same initial tensioning is recommended for bearing-type connections as for the friction-type. Among these benefits are overall joint rigidity, a better stress pattern and security against nut loosening.

Allowable Working Stresses

Tension

The working stress recommended is intended to include as part of the calculated bolt load the tension resulting from prying action produced by deformation of the connected parts. When subjected in tension to the recommended working value, approximately two-thirds the initial tightening force, high-strength bolts will experience little if any actual change in stress. Tests (1)* have demonstrated that their fatigue strength under this condition of loading is not adversely affected.

---

* The numbers in parenthesis refer to items in the References.

Shear: Friction-Type Connections

No change has been made in the recommended working value for bolts used in friction-type joints. They are, as heretofore, given the "shear" value recommended in these specifications for hot-driven rivets of the same nominal diameter.

Resistance to slip is determined by the amount of bolt tension and the nature of the contact surfaces in a given connection, and is independent of the basic stress under which the connected parts are designed.

Conections having contact surfaces of unrusted mill scale offer the least resistance to slip of any unpainted joints; rusted surfaces which have been well cleaned may provide up to two times as much resistance. The recommended "shear" value, equal to 13,500 psi, is based on numerous tests, (2) (3) (4) (5), and can be correlated with a coefficient of friction of 0.35. While lower coefficients have been observed in some laboratory tests of joints having contact surfaces of tight unrusted mill scale, or surfaces made smooth by grinding, a coefficient of friction of 0.35 is more representative of values likely to be encountered in actual construction.

Applying this value to the recommended minimum bolt tension, the factor of safety against slip can be computed as

$$N = \frac{0.35 \times \text{bolt proof load}}{13,500 \text{ psi} \times \text{nominal bolt area}}$$

For $\frac{7}{8}$- and 1-in bolts N equals 1.55 for bridges designed in accordance with these specifications. Since the proof load for various size bolts is not exactly proportional to their nominal area, the values for N would be slightly higher for bolts smaller than $\frac{7}{8}$ in and slightly lower (2 to 5 percent) for bolts larger than 1 in. in diameter. This factor of safety against slip compares with a factor of safety of 1.83 against yielding of current ASTM Specifications, designation A 7, steel.

Under repeated loading the factor of safety against slip indicates the margin against the condition where a reduced fatigue strength may develop. Under static load conditions it may represent the margin against a one-time displacement movement, as under lateral shock or maximum wind loading, which is seldom likely to be reversed. A factor of safety against slip less than 1.83 is acceptable, except where there must not be movement under overloads.

When the allowable "shear" value is increased one-third for wind the value of N in the above equation approaches unity. If the satisfactory performance of the structure depends upon joints which must not move, the designer should so proportion these joints as to satisfy himself that the margin against slip is adequate.

Connections of the type shown in Fig. 1 (A) in which some of the bolts (A) lose a part of their clamping force due to applied tension, suffer no overall loss of frictional shear resistance. The bolt tension produced by the moment is coupled with a compensating compressive force (C) on the other side of the axis of bending. In a connection of the type shown in Fig. 1 (B), however, all of the fasteners (B) receive applied tension which reduces the initial compression at the contact surface. If bolts are used, and slip under load cannot be tolerated, the working value of the bolts in shear should be reduced in proportion to the ratio of residual tension to initial tension.

Because bolts in friction-type connections do not depend upon bearing against the sides of their holes, those provisions of the general design specifications intended to guard against high bearing stresses, and bending of the bolt due to bearing, are waived.

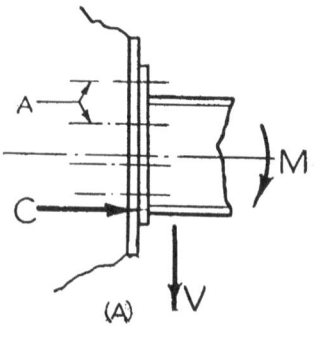
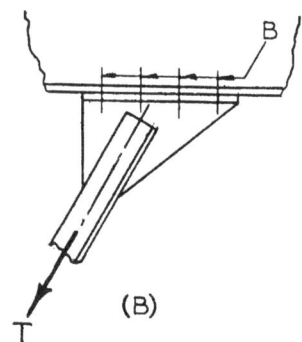

Fig. 1

Shear: Bearing-Type Connections

In connections where the bolts may bear against the holes in the connected parts, the allowable stress of bolts is dependent upon the presence or absence of bolt threading at the plane of contact surfaces where shearing occurs. If the unthreaded shank of the bolt is available to resist this shear at all planes where it occurs, it has been shown (4) (5) that a shear stress equal to 20,000 psi affords as large a factor of safety against high-strength bolt shear failure as that provided in these specifications for rivets. On the other hand, it was found that failure occurs at 15 percent less load when threading is present at one of the two shear planes of an enclosed part, and at 30 percent less load when threads are present in both shear planes. This latter load, giving a working value of 13,500 psi, could be expected also for single-shear joints with threads in the shear plane. Similar observations have been made from tests using ordinary bolts. They merely reflect the ratio of area at the root of thread to the nominal bolt area.

For convenience, different working values (20,000 psi and 13,500 psi) are given, applicable to the nominal bolt area; a single value (the higher one cited) could have been recommended, leaving to the designer the determination of actual area available to resist shear. However, such a determination might require prior knowledge of the side of the joint from which the bolts would be installed. Such information is seldom available at the time shop drawings are made.

In many double-shear connections it would be possible to present the unthreaded shank at one shear plane without taking any precaution to exclude threading from the other plane. In such cases a working value intermediate between 13,500 psi and 20,000 psi could be justified, provided the outside parts of the connection were integral to a single member. Such would not be the case, however, with the outstanding legs of the framing angles for two beams on opposite sides of a supporting girder.

Bearing

Tests (6) (7) have shown that bearing pressure on rivets in double or single shear, computed on the basis of an area equal to the product of the part thickness and nominal rivet diameter, has no significant effect on the strength of the connected parts when this pressure is not more than 2.25 times the tensile stress applied to the net area of these parts. A factor of 2.0 is used in these specifications. It would appear that the ratio of fastener spacing normal to the line of force, to fastener diameter, rather than unit pressure as such, is the critical factor, and that computed bearing stress is a convenient index of safe net section.

When there are not more than two bolts in the line of stress and the pressure from the bolt is directed toward the end of a connected part, a 50 percent increase in end distance, above that required for rivets under similar circumstances, is recommended, since the working value of the bolt has been increased in the ratio of 20,000 psi to 13,500 psi, or nearly 50 percent. Otherwise the end bolt conceivably might push out of the connected part before the full tensile strength of the net section is attained.

Installation

Tests (8) have shown that a hardened washer is not needed to prevent minor bolt relaxation resulting from the high stress concentration under the bolt head or nut. Such relaxations were less than 5 percent of the initial tension; took place within hours of bolt tightening, after which further loss of tension was negligible; and were substantially the same with and without the use of washers. Tests have also shown that any galling which may take place where nuts are tightened directly against the connected parts is not detrimental to the static or fatigue strength of the joint.

Instead of suggesting one full turn of the nut from a finger-tight position, a somewhat smaller rotation, from a snug-tight condition, is now specified. On an average, the bolt tension provided by either prescription is approximately the same (9). However, measuring the nut rotation from a snug-tight condition, which necessitates first drawing the several parts of the connection tightly together, has been found to produce more uniform bolt tension (10).

The percentage of bolts in a given connection which must be made snug-tight in order to compact the joint will depend upon the stiffness of the several connected parts and their initial straightness. In extreme cases it may be necessary to snug-up bolts in all of the holes not used for pinning, in order to seat the parts.

After the parts are suitably drawn together bolts are installed in any remaining open holes, tightened to a snug-tight condition, and all nuts are then rotated by the prescribed amount, after which bolts are installed in the holes originally pinned, and tightened using the same procedure.

Tightening of the bolts in a joint should commence at the most rigidly fixed or stiffest point, and progress toward the free edges, both in the initial snugging up and in the final tightening. During tightening, the bolt head should be held by a hand wrench to prevent turning

REFERENCES

(1) Munse, W. H.—Research on Bolted Connections, ASCE Transactions Paper 2839, Vol. 121, 1956, page 1255.

(2) Hechtman, R. A.; Young, D. R.; Chin, A. G.; and Sarikko, E. R.—Slip of Joints Under Static Loads, ASCE Transactions Paper 2778, Vol. 120, 1955, page 1335.

(3) Vasarhelyi, D. D.; Beano, S. Y.; Madison, R. B.; Lu, Z. A. and Vasishth, U. C.— Effects of Fabrication Techniques on Bolted Joints, ASCE Proceedings Paper 1973, Vol. 85, 1959, No. ST3, page 71.

(4) Foreman, R. T.; and Rumpf, J. L.—Static Tension Tests of Compact Bolted Joints, ASCE Proceedings Paper 2523, Vol. 86, 1960, No. ST6, page 73.

(5) Bendigo, R. A.; and Rumpf, J. L.—Static Tension Tests of Long Bolted Joints, Lehigh University, Fritz Engineering Laboratory Report 271.8, 1960.

(6) Jones, J.—Effect of Bearing Ratio on Static Strength of Riveted Joints, ASCE Transactions Paper 2949, Vol. 123, 1958, page 964.
(7) Munse, W. H.—The Effect of Bearing Pressure on the Static Strength of Riveted Connections, University of Illinois, Engineering Experiment Station Bulletin 454, 1959.
(8) Lewitt, C. W.; Chesson, E., Jr.; and Munse, W. H.—Studies of the Effect of Washers on the Clamping Force in High-Strength Bolts, University of Illinois, SRS No. 191, 1960.
(9) American Railway Engineering Association—Tightening High-Strength Bolts, Proceedings, Vol. 56, page 599.
(10) Bendigo, R. A.; and Rumpf, J. L.—Calibration and Installation of High-Strength Bolts, Lehigh University, Fritz Engineering Laboratory Report 271.11, 1960.
(11) Research Council on Riveted and Bolted Structural Joints of the Engineering Foundation—Specifications for Structural Joints Using ASTM A 325 Bolts, 1960.

Report on Assignment 2

Composite Steel and Concrete Spans; Non-Ferrous Metal Bridges

Collaborating with Committees 8 and 30

James Michalos (chairman, subcommittee), H. F. Bober, R. W. Gustafson, D. L. Jerman, K. H. Lenzen, W. H. Munse, R. D. Nordstrom, T. G. O'Neil, D. D. Rosen.
Your committee submits the following progress report as information, to be considered for adoption one year hence:

SPECIFICATIONS FOR COMPOSITE STEEL AND CONCRETE SPANS

Simple-span bridges consisting of steel beams and concrete slab working integrally, shall have their composite beams proportioned by the moment-of-inertia method, using the net composite section.

The effective width of flange on either side of any beam web shall not exceed the following:

(a) One-half the distance to the center line of the adjacent beam.
(b) One-eighth of the span length of the beam.
(c) Six times the thickness of the slab.

For exterior beams the effective width of flange on the exterior side shall not exceed the actual overhang.

When no temporary intermediate supports are provided for the beams during casting and curing of the concrete slab, then the imposed dead loads shall be considered as acting on the steel beams alone, and all subsequent loads as acting on the composite section. If the beams are provided with effective temporary intermediate supports which are kept in place until the concrete has attained 75 percent of its required 28-day strength, then all loads shall be assumed as acting on the composite section.

The effect of creep shall be considered in the design of composite beams which have dead loads acting on the composite section. Stresses and horizontal shear produced by such dead loads shall be taken as the greater of those computed for the value of n

given in AREA Specifications for Design of Plain and Reinforced Concrete Members, Chapter 8, Part 2, Sec. D, Art. 1, or for 3 times that value.

Resistance to horizontal shear at the junction of the slab and the beam shall be provided by mechanical means.

Shear shall be computed by the following formula:

$$S = \frac{VQ}{I}$$

wherein

 S is the horizontal shear per linear inch at the junction of slab and beam at the section of the beam in question.

 V is the vertical shear.

 Q is the statical moment of the effective area of the slab about the neutral axis of the composite section.

 I is the moment of inertia of the composite section.

The spacing of the shear transfer devices shall be determined by dividing the resistance value of the individual device by the shear per linear inch, S. The maximum spacing shall be 2 ft.

Shear transfer devices shall be of such construction as will permit thorough compaction of the concrete and will insure that their exposed surface is in contact with the surrounding slab. They shall be capable of resisting both horizontal and vertical movement between the concrete and steel.

The flange of the composite beam shall not be considered effective in resisting vertical shear. This force shall be considered to be resisted entirely by the web of the steel beam.

Report on Assignment 4

Stress Distribution in Bridge Frames

(a) Floorbeam Hangers

E. T. Franzen (chairman, subcommittee), L. S. Beedle, E. S. Birkenwald, J. E. Bernhardt, R. P. Davis, G. K. Gillan, J. M. Hayes, K. H. Lenzen, J. Michalos, W. H. Munse, L. T. Wyly.

Your committee submits the following report of progress:

The work on Assignment 4 (a)—Floorbeam Hangers, was essentially completed with the revisions of Chapter 15 of the Manual recommended and approved last year.

Clarification will be made of the portions of Chapter 15 which specify unit stresses to be used for the design and rating of floorbeam hangers.

The following amendment, intended to become new Par. 3 of Art. 1, page 15-7-12, Sec. D. Trusses, of Methods of Strengthening Existing Bridges, Part 7, Chapter 15, is submitted as information:

"Floorbeam hangers are frequently highly stressed from a combination of bending and direct axial tension. To reduce the possibility of fatigue cracking in these highly stressed hangers, sharp copes or re-entrant cuts should be eliminated or modified. The use of high-strength bolts at the top connections of the floorbeam hangers to replace all rivets should also be considered to improve the load transfer to the gusset plates."

Report on Assignment 4 (b)

Truss Bridge Research Project

E. T. Franzen (chairman, subcommittee), L. S. Beedle, E. S. Birkenwald, R. P. Davis, G. K. Gillan, J. M. Hayes, K. H. Lenzen, J. Michalos, W. H. Munse, L. T. Wyly.

Your committee submits the following report of progress:

During the past year six separate investigations were conducted to determine the ultimate carrying capacity of the truss span with a damaged end post. The amount of damage was successively increased from a bend of 1 in downward and ½ in outward in the first investigation to nearly 6 in downward and 3 in outward in the sixth investigation. Whereas the carrying capacity diminished as the amount of damage increased, the total load on the bridge in the condition of greatest damage produced average calculated axial stresses in the end post equal to 0.52 of the yield stress. It was apparent that with increased damage to the end post there was a redistribution of load in the end panel to the remainder of the bridge. None of the measured stresses in other members of the truss span exceeded the yield point of current ASTM Specifications, designation A 7.

The damaged end post was removed and replaced with a new end post.

The above series of investigations were conducted with the damaged end post in Truss A (with lacing bars and angles turned out). A similar series will be conducted next on Truss B (with perforated cover plates and angles turned in).

A complete report, including all preliminary loads and the loading of the damaged end post in Truss A, is in preparation.

Report on Assignment 10

Effect of Continuous Welded Rail on Bridges

Collaborating with the Special Committee on Continuous Welded Rail

J. C. King (chairman, subcommittee), R. T. Blewitt, A. Hedefine, E. W. Kieckers, K. H. Lenzen, D. V. Messman, J. C. Nichols, D. L. Nord, R. D. Nordstrom, T. G. O'Neil, W. E. Robey, J. E. South.

This is a progress report submitted as information.

Your committee in collaboration with the Special Committee on Continuous Welded Rail has been working toward specific recommendations for the use of continuous welded rail on bridges. It is believed that such recommendations must be based on field experience to a greater extent than on theory. This report is, therefore, a consensus of the practice of roads represented by your committee and by the chairman and subcommittee chairmen of the Special Committee on Continuous Welded Rail.

Some roads use welded rail only on bridges of short length and on tangent track. Others extend the use of welded rail to track over bridges of long length or having considerable curvature with the precautions outlined herein.

Expansion Joints: Lateral forces change in direct proportion to temperature change, degree of curvature and weight of rail. Most bridges will not be hurt by welded rail with curvature of 1 deg or less; some, with special precautions, are considered satisfactory with rail up to 4 deg curvature; but bridges of 4-deg curvature or more are customarily protected by special expansion joints in the welded rail. These expansion joints are also used adjacent to or on long open-deck bridges where joints between strings cannot be adequately protected by box anchorage of ties.

Ballasted-Deck Bridges: No special consideration is given to length of string, location of joints or method and number of anchors except when a joint occurs on a bridge with curvature exceeding 2 to 3 deg, in which case special precautions are taken either to change the location of the joint and/or to provide additional anchorage and ballast section to prevent lateral movement.

Open-Deck Bridges: Except for bridges of very short total length no bolted joint is permitted on or within 6 rail lengths of the bridge. Rails are box anchored to every other track tie for two rail lengths next to the bridge and to every tie for a minimum of 4 rail lengths beyond that joint. Some roads apply box anchors to every bridge tie which is hook bolted or otherwise fastened to the structure. Others apply no box anchors to bridge ties except near the center of a section of rail having expansion joints on each end of the section.

The foregoing is a conservative consensus of the practice presently in use by a number of roads and is submitted as information. Suggestions and experience records are solicited.

Report of Committee 24—Cooperative Relations with Universities

W. W. HAY, *Chairman*

H. E. KIRBY,
Vice Chairman

W. S. KERR, *Secretary*
J. F. DAVISON

T. D. WOFFORD, JR.
H. E. HURST
C. E. R. HAIGHT
J. L. ALVORD
B. G. ANDERSON
H. C. ARCHDEACON
W. S. AUTRY
J. B. BABCOCK (E)
GEORGE BAYLOR
R. H. BEEDER
J. B. CLARK
R. P. DAVIS (E)
G. H. ECHOLS
E. I. FIESENHEISER
R. J. FISHER
R. T. FORTIN
R. R. GUNDERSON
C. L. HEIMBACH
L. J. HOFFMAN
R. P. HOWELL
W. H. HUFFMAN
S. R. HURSH (E)
A. V. JOHNSTON
CLAUDE JOHNSTON

FRANK KEREKES
E. C. LAWSON
B. B. LEWIS
R. E. LOOMIS
R. W. MIDDLETON
G. W. MILLER
JERRY NEBEN
R. C. NISSEN
W. A. OLIVER
J. F. PEARCE
J. E. PERRY
G. B. PRUDEN
R. B. RICE
R. W. RIPLEY
V. J. ROGGEVEEN
J. A. RUST
P. S. SETTLE, JR.
H. O. SHARP
E. R. SHULTZ
JOHN STANG
R. J. STONE
D. W. TILMAN
EGONS TONS

Committee

(E) Member Emeritus.
Members listed in bold face are the official representatives of the Engineering Division, AAR.

To the American Railway Engineering Association:

Your committee reports on the following subjects:

1. Stimulate greater appreciation on the part of railway managements of
   (a) the importance of bringing into the service selected graduates of colleges and universities, and
   (b) the necessity for providing adequate means for recruiting such graduates and of retaining them in the service.

2. Stimulate among college and university students a greater interest in the science of transportation and its importance in the national economic structure by
   (a) cooperating with and contributing to the activities of student organizations in colleges and universities, and
   (b) presenting to students and their counselors a positive approach to the attractive and interesting features of the railroad industry and the advantages of choosing railroading as a career.

3. The cooperative system of education, including summer employment in railway service.

Progress report, submitted as information .... page 408

4. What constitutes a desirable curriculum for students to pursue in preparation for a career in railroad engineering?

Final report, submitted as information .... page 409

5. Ways in which railroads can cooperate with universities in developing research, including the revising of "Suggested Topics for Theses on Railroad Subjects".

Progress report, presented as information .... page 413

6. Procedures for orienting and developing newly employed engineering personnel.

Work is in progress and the study will be continued in 1962.

7. Stimulate an interest by college and university staff members in current railroad problems and practices including AREA membership.

Work is in progress and the study will be continued in 1962.

THE COMMITTEE ON COOPERATIVE RELATIONS WTH UNIVERSITIES,

W. W. HAY, Chairman.

AREA Bulletin 569, January 1962.

Report on Assignment 1

Stimulate Greater Appreciation on the Part of Railway Management of

(a) the importance of bringing into the service selected graduates of colleges and universities, and

(b) the necessity of providing adequate means for recruiting such graduates and of retaining them in the service

J. F. Davison (chairman, subcommittee), H. C. Archdeacon, R. H. Beeder, G. H. Echols, R. J. Fisher, R. T. Fortin, C. E. R. Haight, W. W. Hay, W. H. Huffman, Claude Johnston, H. E.. Kirby R. W. Middleton, J. E. Perry, G. B. Pruden, R. W. Ripley.

The current activity of your committee on this assignment is directed towards the establishment of the type and interests of graduates most likely to be attracted to and remain in railway employment. To this end, a questionnaire is being developed, the replies to which would indicate the profile of the present-day railway engineer.

A preliminary questionnaire has been developed and tested cn a sample group of graduate engineers employed on a few representative railroads. The replies received indicate that much useful information can be obtained in this manner, but some refinements in certain questions are required. It is now proposed to revise the questionnaire and rephrase the questions so that the replies can be recorded on punch cards for easy sorting and tabulation. Upon completion, the questionnaires will be circulated to railway engineering staff members through their respective chief engineers.

An Appendix to this report contains a review by G. B. Pruden, member of Subcommittee 1, of a recently published study on college recruiting conducted by the

Bureau of Industrial Relations, the University of Michigan. Railroads interested in improving their recruiting techniques will find much of value in the report of this study. This progress report on Assignment 1 is presented as information.

Appendix to Report on Assignment 1

Effective College Recruiting*

This is a comprehensive study prepared on the campus of the University of Michigan covering the "million-dollar manhunt" by industry seeking new employees and specifically a study of the on-campus interview between the industry recruiter and the graduating student. Some of the facts presented are:

1. Today's tight manpower situation that makes campus recruiting vital.
2. The student—his qualifications for employment and his reactions to the recruiters' interviewing methods.
3. The recruiter—his tactics and performance.
4. The company—its planning and follow through.
5. The college placement office—its operation and problems.

Chapters in the book are:

1. Why College Recruiting?
2. Research into the Recruiting Process
3. How the Michigan Research Was Conducted
4. How the College Placement Office Operates in Campus Recruiting
5. How Do Companies Prepare To Recruit?
6. Tactics of Campus Interviewing
7. Strong and Weak Students—the Recruiter's Viewpoint
8. How Students Rate Recruiters
9. Interviews That Students Like
10. The Recruiting Brochure and the Company Image
11. How General Are the Results of the Research?
12. What Every Recruiter Should Know About Students

From the point of view of the company interested in improving its recruiting program, the student who will soon graduate and has no idea of the steps in obtaining employment, or the college desiring to improve its placement office, this book offers a wealth of information. It might well be read by management of industry in order to become aware of the magnitude of college recruiting, by college students to realize the importance of scheduled planning in seeking employment, and by college administrators to remind them of the importance of their placement office. Each of these three groups has an important role to play in the story of successful employment, which is the future of industry and education. Each needs to realize fully the importance of this role. This book gives him such insight.

Interviews and questionnaires were participated in by 94 recruiters, 1134 students and 99 companies. After the information was collected and assembled, it was tabulated and programmed for use on computers available on the Michigan campus. These computers made it possible for the authors to explore many interesting cross relations in such data. Charts are presented to show these relationships.

* By George S. Odiorne and Arthur S. Hann. Published by the Bureau of Industrial Relations, University of Michigan, Ann Arbor, Mich., as report No. 13. Hard back, 288 pages. Price $5.

Whereas the report originated in the Michigan School of Business Administration, Chapter 11 specifically relates the findings to the field of engineering. Jack Young, director of the Placement Center, University of Michigan College of Engineering, says "Your basic results would apply in any college recruiting situation. The specific data regarding accountants, actuaries, and MBA's would differ, but the guides to companies on how to prepare, how to interview, and how to follow up are perfectly valid in engineering as elsewhere". The report suggests that future studies to test these results in engineering colleges can be made on a comparable scale.

It is noted that reference is made to the following published information concerning college recruiting. These references are listed for further investigation:

1. Stephen Habbe, "Employment of the College Graduate", National Industrial Conference Board Studies in Personnel Policy, No. 152, New York City, 1956.
2. Stephen Habbe, "College Graduates in Industry", National Industrial Conference Board Studies in Personnel Policy, No. 9, New York City, 1948.
3. "Techniques of College Recruiting", The Bureau of National Affairs, Inc., Washington, D. C., 1941.
4. "Recruiting and Placing College Graduates in Business", Metropolitan Life Insurance Company, New York City, 1950.
5. "A Guide to College Recruitment", Society for Personnel Administration, Booklet No. 12, Washington, D. C., 1956.
6. Frank S. Endicott, "Trends in the Employment of College and University Graduates in Business and Industry", Northwestern University, 1956.

To those in the railroad industry interested in acquiring college graduates for employment in their companies, this book gives a review of the scope of college recruiting which will, if properly applied by qualified personnel, certainly help to obtain more easily the men sought. Substantial savings in time, effort, and expense can be realized by applying the findings presented in this report.

Report on Assignment 2

Stimulate Among College and University Students a Greater Interest in the Science of Transportation and Its Importance in the National Economic Structure By

(a) cooperating with and contributing to the activities of student organizations in colleges and universities, and
(b) presenting to students and their counselors a positive approach to the attrative and interesting features of the railroad industry and the advantages of choosing railroading as a career

B. B. Lewis (chairman, subcommittee), B. G. Anderson, J. B. Clark, R. P. Davis, G. H. Echols, W. W. Hay, S. R. Hursh, A. V. Johnston, H. E. Kirby, G .W. Miller, Jerry Neben, R. B. Rice, J. A. Rust, John Stang, Egons Tons.

The entire membership of Committee 24 has been active in contributing to this assignment. The following are typical of the activities of the members in promoting the railroad industry before student groups.

R. H. Beeder, chief engineer—system, Atchison, Topeka & Santa Fe Railway, was the principal speaker at a meeting of the Kansas Engineering Society at Lawrence, Kans. He spoke on the Williams-Crookston line change in Northern Arizona. He also showed the Morrison-Knudson film of this project. The meeting was attended by the Topeka and Eastern Chapters of the Kansas Engineering Society, along with engineering students and faculty of the University of Kansas.

C. J. Code, assistant chief engineer—staff, Pennsylvania Railroad, gave an address at the University of Illinois to seniors of civil engineering on November 9, 1960, on the subject, "Some Applications of Engineering Technology to Railroad Track Problems." Mr. Code also provided mimeographed copies of his address for distribution and use in railroad engineering classes.

H. E. Kirby, cost engineer—system, Chesapeake & Ohio Railway, gave a talk to high school exhibitors at Area Science Fair at Marshall University, Huntington, W. Va.

Professor V. J. Roggeveen, Stanford University, reports the following railroad-oriented activities in the Department of Civil Engineering at Stanford in the past academic year.

1. A senior student did his senior report on a topic pertaining to decisions about bridge replacement on the Southern Pacific and received worthwhile cooperation and help from the engineering management of the Southern Pacific Company.

2. A graduate student did a special study related to decisions on capital investments in general, and also received most worthwhile cooperation and help from the engineering management of the Southern Pacific Company.

3. Another graduate student has almost completed an analysis of the decision to make the line change in Arizona, recently completed by the Santa Fe, as a special report topic. For this he received most invaluable and friendly help from Mr. Beeder, chief engineer of the Santa Fe.

4. The AREA slide set on railways was shown to the transportation engineering class of 35 students.

E. R. Shultz, engineer—maintenance of way-system, Pennsylvania Railroad, spent two days at Penn State and showed slides and pictures to students individually, in connection with explaining job opportunities on the Pennsylvania Railroad. He also visited Penn State Center at Wilkes-Barre and interviewed 12 students of the two-year courses in drafting and surveying.

The seniors in the School of Civil Engineering, Purdue University, in their annual inspection trip visited Markham Yard of the Illinois Central Railway and had an opportunity to observe and discuss its operation.

Wm. J. Hedley, chief engineer of the Wabash Railroad, spoke before the Civil Engineering Seminar at Purdue University on March 1, 1961. Mr. Hedley also was the principal speaker at the Indiana Section of the American Society of Civil Engineers at Indianapolis. Prior to this meeting, an inspection was made of the New York Central Yard at Indianapolis. Three Purdue students made the inspection of the yard and attended the meeting at the invitation of Chairman Lewis.

Dr. L. K. Sillcox, honorary vice chairman of the board, New York Air Brake Company, gave an address to the senior civil engineering students and others at the University of Illinois on October 11, 1961. His subject was "Stamp of Success."

At the Engineering Open House held on March 10 and 11 at the University of Illinois, exhibits were made available to the transportation and railroad engineering area of civil engineering by the Illinois Central Railroad, New York Central Railroad and the

Union Switch and Signal Division, Westinghouse Air Brake Company. Pictures and drawings of the Robert E. Young Yard at Elkhart, Ind., were furnished by the New York Central. The Illinois Central placed a locomotive, airbrake instruction car, dining car, coach, caboose, rail detector car, and a ballast regulator on the University side track. Union Switch and Signal provided descriptive pamphlets, signal models, and a model train identification system.

During the 1960–1961 school year there were 52 Student Affiliates enrolled with the AREA. Several of these have since taken railroad jobs and applied for AREA junior membership.

This report of progress is presented as information.

Report on Assignment 3

The Cooperative System of Education, Including Summer Employment in Railway Service

W. A. Oliver (chairman, subcommittee), J. L. Alvord, George Baylor, J. B. Clark, E. I. Fiesenheiser, R. R. Gunderson, W. W. Hay, S. R. Hursh, H. E. Kirby, R. C. Nissen, J. F. Pearce, R. B. Rice, E. R. Shultz, R. J. Stone, D. W. Tilman.

In fulfilling its assignment in 1961 the committee followed essentially the same procedures used in the preceding years. A questionnaire was sent to the chief engineering and maintenance officers of the railroads in late February requesting information concerning their summer employment needs. The replies to the questionnaire were returned to the subcommittee chairman, the information was tabulated, reproduced in the secretary's office, and sent to some 125 engineering colleges on March 23.

As a result of this annual cooperative venture many students have obtained their first introduction to railroad work and now some have become permanent employees. Furthermore, the project has shown these students that there are excellent opportunities for an engineering career in the railway field, thus increasing the interest of the engineering colleges in encouraging this type of employment among their students.

As indicated in the following tabulation covering the years 1959, 1960, and 1961, it will be seen that there was a smaller number of opportunities offered in 1961 than in 1960, the peak year, although the number of railroads returning the questionnaire was comparable. As in previous years, it was apparent that the number obtaining employment through the efforts of the committee was small in comparison to the number employed through the direct recruiting efforts of the railroads. This last statement would seem to raise a question concerning the value of the work of the subcommittee as far as being of worthwhile assistance to the railroads is concerned. Furthermore, two railroad officials stated that the time consumed in handling the resulting correspondence was far more than was justified by the results obtained. Consequently, they did not return the questionnaire in 1961.

COMPARISON OF THE RESULTS OF QUESTIONNAIRE FOR THE YEARS 1961, 1960, and 1959

1961 Offering summer employment through response to questionnaire ..... 6 railroads
 (49 jobs)
 *Summer employment by direct recruiting 12
 Negative replies 40

 Total Return 58

* Including railroads having arrangements with colleges with cooperative programs.

1960 Offering summer employment through response to questionnaire .....12 railroads
      (151 jobs)
      Summer employment by direct recruiting 14
      Negative replies 37
                                                                      —
      Total Return 63   "

1959 Offering summer employment through response to questionnaire ......12 railroads
      (69 jobs)
      Summer employment by direct recruiting 10   "
      Negative replies 46
                                                                      —
      Total Return 68

The committee believes that the efforts put forth in the past have been worth while, but it now realizes that the correspondence involved in answering the many applications for summer employment has put an impossible burden on some railroad personnel. In order to alleviate this situation and at the same time maintain at the colleges the interest which the project has stimulated, it now proposes a modified program. It intends to take advantage of the previously stated fact that in addition to cooperating with Committee 24 in its summer employment program, many railroads have continued their established custom of recruiting for summer employment along their own right-of-way. A simple four-part questionnaire is now proposed which will be submitted with the request that if it is not considered desirable to answer all of the questions, the addressee should return whatever part of the requested information he is willing to have distributed among the engineering colleges.

The four proposed questions are as follows:

1. Did you employ engineering students during the summer of 1961?
2. If you did, how many?
3. Are you willing to have students apply for summer employment along your railroad?
4. If your answer to question 3 is yes, and you are willing to have published the names to whom such applications should be sent, please include such a list.

This information will be distributed to the engineering colleges in late March, with the stipulation that the student should apply to the railroads in his immediate vicinity.

The committee requests the continued cooperation of the railroads.

Report on Assignment 4

What Constitutes a Desirable Curriculum for Students to Pursue in Preparation for a Career in Railroad Engineering

T. D. Wofford, Jr. (chairman, subcommittee), J. B. Babcock, George Baylor, E. I. Fiesenheiser, W. W. Hay, C. L. Heimbach, L. J. Hoffman, W. H. Huffman, A. V. Johnston, Claude Johnston, H. E. Kirby, Jerry Neben, R. C. Nissen, J. F Pearce, R. W. Ripley, H. O. Sharp.

Your committee has been asked to address itself to the question of "What Constitutes a Desirable Curriculum for Students to Pursue in Preparation for a Career in Railroad Engineering." A curriculum is broadly defined as "a specified, fixed course or a body of courses offered in an educational institution or by a department thereof leading to a degree."

Today's college undergraduate, for the most part, finds it necessary to select and pursue a degree in one of the traditional engineering curricula, i.e., civil, electrical, mechanical, industrial, etc., as preparation for a career in railroad engineering. For most students, the selection of a specific curriculum is in order before he has become career conscious or industry orientated, and selection is made difficult by the student's usual lack of understanding of an industry's needs and opportunities. Several of the major fields of railroad engineering and the more traditionally recommended curricula are listed as follows:

Field of Railroad Engineering	*Recommended Curriculum*
Maintenance of way	Engineering—civil and/or industrial
Design and construction	Engineering—civil, electrical, mechanical, architectural, industrial
Communication and signals	Engineering—electrical
Maintenance of equipment	Engineering—mechanical, electrical, industrial

What constitutes a desirable curriculum for all types of engineering careers, not only railroad engineering, is the subject of much active and continuing discussion among educators and industry. Professor K. B. Woods of Purdue University referred in his address before the 1961 annual meeting of the AREA to the changes under way and being contemplated in the content of engineering education; more particularly, to the curriculum of civil engineering. Recent trends in engineering education are away from the traditional teaching by process of synthesis in which component principles of science are put together to produce a desired result, i.e., education useful for a specific industry. In its place is evolving a curriculum based on the teaching of engineering principles broken down into its component parts, but not orientated to specific industries. Engineering educators almost unanimously agree that henceforth the four-year curriculum will contain time only for the presentation of basic principles of mathematics and science, with a necessary minimum attention to language, humanities, social science and other disciplines to broaden the education of engineers. The time element in engineering education at the college level emphasizes the importance of high school and preparatory school training, particularly in the fields of language and basic sciences.

At most colleges the teaching of surveying, drafting, and design courses in the undergraduate curriculum, with specific reference to railroad construction, operation and maintenance, will be soon a thing of the past. At a few institutions undergraduate courses are being offered in transportation engineering which encompass the basic principles of economics and regulation, and cover in a general way the facilities and operations relevant to the major modes of transportation. Courses of this nature will be beneficial in the curriculum of engineering students preparing for a career in railroad engineering, but likely will be available for undergraduates at only a limited number of larger institutions.

Graduate study immediately upon receipt of the bachelor's degree is becoming increasingly prevalent among engineering students at the present time. Much of this work is being done in preparation for teaching careers, but increasing numbers of engineers with advanced degrees are finding their way into industry, government service and consulting work. Current requirements in the field of railroad engineering for education above the bachelor's level mostly relate to specialized areas, including but not limited to, soil mechanics, geology, research and development, and computer applications.

Among engineering educators there are many proponents of cooperative education, five-year undergraduate curriculums, and other variations in the educational program for engineers, all of which lengthen the period of college leading to the bachelor's degree. Suggestions are being made that the undergraduate curriculum contain only basic engineering sciences and mathematics, language, social studies, etc., with all engineering specialization concentrated in post-graduate study. The reasons for these suggested changes are the greatly increased store of knowledge which engineers need to master, together with the greatly expanding horizon of scientific and engineering problems to be met by engineers in the future. A gradual transition to a "common core curriculum" in engineering education at the undergraduate level is under way to some extent in practically all colleges, although there is great variation in the practice among various institutions. Among the aims in revising engineering curricula is better preparation of the undergraduate for writing. speaking and the cultivation of interest beyond strictly technical fields.

At a special conference held in July 1960, delegates from the civil engineering staffs of engineering colleges and representatives from ASCE, ASEE, ECPD and the National Science Foundation met to discuss the long-range goals and to develop a sense of future direction for civil engineering education. Certain resolutions were offered which are indicative of the long-range trends that may develop in all branches of engineering. A condensed report on the voting, giving a brief synopsis of each resolution for which the majority vote was favorable, is as follows:

(1) *Move toward* a pre-engineering, undergraduate, degree-eligible program for *all* engineers, to be followed by a professional or graduate civil engineering curriculum leading to the *first* civil engineering degree, and request ECPD and EJC support..

(3) *Abandon* the C.E. degree, where this is a non-resident, non-academic degree.

(4) *Establish* graduate professional schools offering programs leading to the degrees *Master of Engineering* and *Doctor of Engineering* in the several engineering specialties, and continue current programs leading to the degrees *Master of Science* and *Doctor of Philosophy* in the several engineering specialties.

(5) Do *not* develop 3-year pre-engineering programs in colleges of *Arts and Sciences,* followed by 3-year engineering programs in engineering colleges.

(7) *Encourage* individual and joint action by ASCE and ECPD to achieve and maintain truly professional standards of performance and ethics.

(8) *Promote* increased recognition by civil engineering faculties of education in basic sciences and general culture.

(9) *Perfect* programs for the more gifted students on a school-to-school basis with experimentation and free exchange of experience, in order to meet the broad spectrum of needs of the civil engineering profession.

(10) *Extend* the curriculum from 4 to 5 years for the award of the *first* civil engineering degree.

Among railroad officers, there is wide variance of opinion concerning the adequacy of present engineering education and the extent to which changes need to be made for the purpose of providing a desirable curriculum for students to pursue in preparation for a career in railroad engineering. To quote the chief engineer of one large railroad, "My opinion is that . . . engineering education, leading to a bachelor's degree, should be broad with more emphasis on pure science and those courses leading to a fine command of the English language, written and oral. I do not believe it necessary to school the young engineers in specifics at this level." By way of contrast, many railroad officers feel that it would be a mistake to decrease the teaching of basic mathematics, science or practical engineering courses to make room for increased education in language or social sciences, the thought being that these latter fields should be covered in high school or by outside study.

There are many railroad engineering officers and supervisory employees who are sincerely concerned over the de-emphasis on teaching by the method of synthesis, especially to the extent that it may no longer produce an engineering graduate with an understaanding of railroad engineering. Many ask this question: "How will we secure men who are qualified to handle surveying and track layout, preparation of plans, design of railroad bridges and buildings, and how can we have competent officers for the future when men do not have these basic skills?" It now becomes apparent that skills and knowledge peculiar to the railroad industry will have to be obtained by engineering graduates through training programs, on-the-job experience, and perhaps post-graduate study. It is becoming quite common for engineering graduates with the bachelor's degree to take post-graduate work in economics and business administration while working in business. Training of engineers on the job involves considerable expense to industry, and the pursuit of post-graduate study requires the expenditure of considerable time and funds by students. These factors are of great importance to both employers and to students and must be considered in evaluating changes in engineering curriculum.

With the increasing trend in engineering education toward analysis, i.e., the breaking into component parts of the principles of engineering science, increasing importance will need to be given to the technical school as a source of technically trained assistants. Several years ago Committee 24 reported on the use of technicians and trends in this direction. The National Society of Professional Engineers recently authorized the establishment of an Institute for the Certification of Engineering Technicians. The purpose of this step was to elevate the performance standards of engineering technicians as an important part of the engineering team and to determine the competence of engineering technicians through investigation and examination.

In concluding the report, it is felt that the attention of railroad management can be called to the changes that are being made in the "common core curriculum" at many engineering colleges. Many of these changes are in the direction of improvement which industry has said it wants. These changes, however, place increasing importance to on-the-job training programs and even to additional education past the bachelor-of-science-degree level. For the same reason, use of undergraduates for summer employment becomes a more important source of training and experience for the undergraduate engineering student. There still remains a great deal of variation in the engineering curriculum from one school to another, and some are definitely better oriented to the needs of the railroad engineer. In general, it is concluded that the engineering curriculum at most colleges continues to provide the railroads with adequately trained men for their engineering organization. It is, however, necessary for railroad management to keep fully abreast of further changes in the engineering curriculum that are being discussed and to exert their influence where possible for the best interest of the industry and the profession.

This is submitted as a final report with the recommendation that the assignment be reactivated within the next five years as a means of keeping informed on future trends in engineering education and to maintain contact between the industry and the field of education.

Report on Assignment 5

Ways In Which Railroads Can Cooperate With Universities In Developing Research, Including The Revision of "Suggested Topics For Theses on Railroad Subjects"

H. E. Hurst (chairman, subcommittee), W. S. Autrey, R. J. Fisher, R. T. Fortin, W. W. Hay, W. S. Kerr, H. E. Kirby, E. C. Lawson, R. E. Loomis, V. J. Roggeveen, H. O. Sharp, John Stang, P. S. Settle, Egons Tons.

Your committee submits the following report of progress as information.

Research offers an effective way for railroads to develop solutions to problems confronting them. In addition, it affords an excellent opportunity to create and develop interest among university and college faculties and students in the railroad industry and railroad engineering.

The inherent value of engineering research is well known. The development of new knowledge, new techniques, and new devices has become increasingly important in the modern age, and the direct rewards are generally recognized. However, there are other important benefits to be derived from engineering research, although perhaps less obvious. The values gained for educational programs are more difficult to evaluate but may be realized in the careers of engineers educated in stimulating research environments.

Of increasing importance in engineering, as well as in other fields, is the substantial increase in graduate activity leading to higher degrees, such as masters and doctorates.

Basically, an active research program in an educational institution is a necessary function to assist in maintaining an interested, creative faculty developing capabilities which will more adequately promote a higher caliber of engineering teaching in an active, changing field. Such programs create that atmosphere which best develops student inquiry and intellectual activity. Further, a graduate study program requires research as one of its major components.

Universities form advantageous environments for research because the efforts from a group of researchers may combine the knowledge of several fields of engineering and science in a single project. In a similar manner, expensive equipment which may be required in engineering research is readily available. Further, the size and diversity of a large university makes possible a better quality and flexibility of engineering research.

Your committee is considering ways in which railroads can cooperate in developing research. Some of those under immediate consideration are as follows:

(1) Promote increased activity, particularly in the railroad field, by directly sponsoring research projects; providing support by monetary contributions. (Sponsorship may be covered by formal contract or agreement.)

(2) Sponsor research projects through faculty members employed by the sponsoring agency, thus bypassing direct formal contract with the university or college. Students handling routine aspects of such research to be compensated by the sponsor.

(3) Sponsor directed research by employment of faculty members during summer months either on the railroad proper or at the university.

(4) Utilize grants in aid to selected and qualified graduate students at colleges and universities to work on graduate programs and research in fields of railroad interest. This could be promoted through the establishment of fellowships.

(5) Establish a fund of $5000 or more by which the railroad industry can pro-
vide assistance to faculty members and students in their study and research
activities in connection with current railroad problems; and, thereby, promote
the interest of college and university students and faculty members in these
problems and the railroad story, as well as fostering closer relationships
between the schools and railroads.

(6) Promote interest and research activities by making data available and providing access by faculty staff members and students to railroad properties
and records under mutually agreeable arrangements.

(7) Make available for research programs, by loan or use, items of railroad
material or equipment.

(8) Develop interest by supporting activities of Student affiliates, such as article
writing, publishing periodicals, etc.

(9) Promote the voluntary support of faculties and faculty members interested
and sympathetic to railroads to encourage graduate students to do research
and theses in the railroad transportation field.

In consideration of the foregoing methods or ways of developing research, it appears
desirable to state that experience indicates that research programs are determined primarily by the interests and activities of the faculty members.

*Most projects develop from ideas and the initiative of individual faculty members.
Experience reveals that the interest of students follow those of the instructor. It is,
therefore, important to interest the instructor in the area of railroad activity.*

The 1960 report on assignment 7 by your committee largely confirmed the fact
that railroads, as compared with other industries, have shown little interest in the facilities or research abilities of our universities and colleges. Thereby, railroads have reduced
their prestige and impaired interest in their problems and activities. Perhaps one of the
most important means of developing research activity is to expand railroad interest by
indicating to them the desirability of such interest in promoting desired research.

At least five general categories or arrangements for cooperative research programs
can be identified. The first is contract research in which the institution, for payment
of the costs of a project, agrees to undertake a study, furnish a report or reports of the
findings, and render an accounting for funds. A second type or arrangement involves
the establishment of industrial laboratories on the campus of a college or university.
A third type includes those arrangements for financial assistance of graduate students.
In the fourth type are the various forms of grant-in-aid which vary immensely in their
form and content. Fifth, and finally, some institutions establish industrial cooperative
plans which provide for annual distribution or payment by industrial concerns for support of research in exchange for advisory services.

Successful institution-industry research depends upon a competent and interested
research leader, mutual confidence, and satisfactory communications. Since there is
usually no physical product resulting from the research efforts of the institution, it is
important that industrial administrators or business officers be adequately informed of
the objectives of the research and recognize that success is not measured by the same
standards as when a physical object is produced. It is also desirable to have those persons in a company who might be affected by the results of the research efforts of an
institution participate in conferences or seminars as the work progresses.

Regardless of the type of research agreed upon with an educational institution, it is
essential that a representative of the university's staff be brought into the negotiations
before any final arrangements are agreed upon with scientific and engineering personnel.

Two publications which deal with research policies at colleges and universities and university patent policies are as follows:

(1) Engineering College Research Review, published annually by the Engineering College Research Council (ECRC) of the American Society for Engineering Education. This publication presents research officers, personnel, expenditures, active projects, and research policies for all the accredited engineering schools and colleges in the United States.

(2) University Patent Policies and Practices is Publication No. 257 of the National Research Council of the National Academy of Science, 2101 Constitution Avenue, Washington 25, D. C. This publication gives formalized policies, general situation, and institutional policies and practices for the major United States colleges and universities.

In general, universities normally reserve the right to publish worthwhile results and to direct the use of patentable discoveries for the best interests of the public. However, the details of publication and patent rights may be tailored to the particular project within the scope of the general policy.

Research and development surveys analyzing the cost of such work in universities and colleges have separated operating costs into three major components, as follows:

(1) The first component includes research and development sponsored by sources such as the Federal Government, industry, private foundations, voluntary health agencies, or by the schools themselves.

(2) The second component consists of the institution's contribution to the indirect costs of separately budgeted research and development.

(3) The third component of the total costs of research—departmental research—i.e., research allied to the instructional function of the departments, covers the assistance rendered by faculty members to outside—sponsored projects for which no reimbursement is made to an academic institution.

A study for the fiscal year 1958 indicated that operating expenditures for separately budgeted scientific research and development in engineering schools amounted to $70.9 million of a total $327.5 million, of which $105.5 million represented research and development in medical schools. It will be noted that the engineering and medical schools together accounted for slightly more than one-half of such expenditures in colleges and universities proper. Further, that such expenditures do not include the third "cost component" referred to, since this cannot readily be determined.

A recent survey listed 11,000 projects in progress in 121 leading engineering colleges with a total of expenditures in excess of $180 million for the year 1960.

The foregoing cost consideration of research and development may assist in correlating the increasing scope of research and factors under consideration. However, consideration should be given to the fact that federal funds for research and development in 1957–58 were twice that of the total from other sources. Federal support amounted to $48.6 million, and the funds from other sources $22.3 million. Most of the federal funds represented contracts with the Defense Department.

One factor which should be pointed out in particular is that there has been a decreasing trend in the enrollment of undergraduate students and a rapidly increasing enrollment of graduate students in engineering schools. At some universities, the number of graduate students approaches approximately one-third of the total undergraduate engineering student enrollment.

Graduate students supplement their formal classroom study with research experience and graduate research assistantships provide financial support which is often necessary to make graduate study possible.

The undergraduate teaching program benefits in that the participation of the faculty in research helps to maintain professional competence and keeps them alert to the latest advances in engineering. A "research atmosphere" keeps students aware that engineering is a progressive, developing field which requires continued learning and development throughout the engineer's life.

The participation of graduate students in the research program makes their talents and training available to the program. In addition, research experience gained insures a future supply of men with advanced degrees who will, in turn, be able to contribute to engineering progress.

Some general conclusions concerning the present state of graduate study in engineering schools are offered as a matter of information and interest:

In general, graduate education must accommodate itself to both the pressure of expanding science and knowledge as well as the increasing number of students. The increasing body of knowledge handled by the graduate school will continue to grow and, thereby, add problems of both complexity and specialization.

Further, graduate education will continue to grow in degrees, students, faculty, departments, institutions, and support. Likewise, the demand for the graduate school product, while already substantial, will continue to grow.

The growth of the professional fields has developed an increasing demand for professional work within them.

The graduate school serves as a career ladder for able students, particularly in the engineering fields. It becomes a central channel not only for training, but also for recognition and selection.

Most students obtain support in some manner through the graduate school, and almost all doctoral students receive support from the universities or, through them, from the large fellowship programs.

A particularly interesting and recent review of the present state of graduate study is covered by Dr. Bernard Berelson in his book, "Graduate Education in the United States," published in 1960 by McGraw-Hill Book Co., Inc. The book is a result of one of a number of studies supported by grants of the Carnegie Corporation of New York.

———

Your committee was charged with the preparation of a revised compilation of railroad problems for research and development by students, primarily for Master and Doctoral postgraduate research study. Initially titled, "Suggested Topics for Theses on Railroad Subjects," it has appeared desirable to modify the title of the listing which, is here again offered with the realization that it could be a more comprehensive listing of problems to be found in railroad engineering and operation, and will require further revision at intervals in the future. By the same token, the present listing will undoubtedly suggest other and related topics.

SUGGESTED TOPICS FOR STUDY AND RESEARCH ON RAILROAD SUBJECTS

Accounts and Statistics

1. Develop a set of standard methods, possibly with the use of computers, which would produce figures to give a uniform means of cost comparison. One of the big difficulties in deciding on the type of equipment or the theoretical advantage of one system of maintenance over another is the problem of getting good cost figures on which to base an opinion. The ICC standards of accounting make it difficult to get actual cost figures on new methods. Each railroad tends to present its own ideas in the most favorable light. On many railroads, the efficiency of their particular operations is judged by a time study taken in one day or a half day, and does not take in account the long-range results or a monthly or yearly average.

2. Modify income accounting to include capital expenditures.

3. Develop depreciation reserve accounting for the Internal Revenue Service.

4. Modify Interstate Commerce Commission reports to an I.R.S. basis.

5. Study and utilization of machine accounting to provide up-to-date cost data for maintenance and construction, etc.

Ballast

1. Evaluate desirable characteristics and economics involved, i.e, stability, drainage, load-carrying abilities, durability, frictional characteristics, etc.

2. Analysis of materials and gradation to obtain favorable characteristics for railroad ballast.

3. Technical and economic considerations in making ballast selection for various railroad traffic conditions.

See also Bridges.

Bridges

1. Economics of design and fabrication of welded wide-flange beams and welded girders fabricated with steel plates. This study to determine the economical thickness of steel plates to be used in the webs and flanges of such beams and girders of various span lengths to accommodate railroad loading. The topic might include some discussion on the possibility of using horizontal stiffeners on the outside of webs to improve the appearance of girder spans.

2. Economics of various types of ballast troughs for supporting railway track ballast on railway bridges, taking into consideration service life, cost of maintenance, and effect of the dead load of the troughs. The following types of ballast troughs are suggested for study:

  a. Wrought-iron plates provided with waterproof covering protected by asphalt plank to resist abrasion by the track ballast.
  b. Creosote-treated timber with galvanized steel fastenings.
  c. Reinforced concrete slabs:
    (1) Precast slabs
    (2) Poured-in-place slabs

This topic could also include a discussion of methods for installing ballast troughs on existing railway bridges so as to achieve minimum interruption to railway traffic.

3. Fireproofing railroad bridges, methods and evaluation of economic factors involved.

4. The use of timber in railway bridges—including economics and development of chemical treatments to increase service life.

5. Develop impact formula for the design of ballasted and non-ballasted railway bridge structures of steel and reinforced concrete for diesel engines and modern loading.

6. The participation of floor and lateral systems in the stress-carrying capacity of the chords of steel railway bridges.

7. The design and adaptation of post-tensioned reinforced concrete for railroad bridges.

8. The effect of creep and relaxation of the tendons in the design of post-tensioned precast concrete slab and beam units for medium span railway bridges.

Buildings

1. Study design, materials, and economies in using portable and semi-portable buildings, particularly prefabricated, at terminals and stations where flexibility due to changing needs is paramount.

2. Use of infra-red rays for heating shops, warehouses, and other miscellaneous railroad buildings.

Centralized Traffic Control

1. Develop economics of siding spacing and relation to traffic densities, etc.

2. Developments and adaptation of TV monitoring to CTC operations, automatic train operations, and automation potentials of TV monitoring.

Clearance

1. Design of a railway clearance car and/or equipment.

2. Study to develop adaptation of electronics equipment and other devices to replace or supplement mechanical measuring devices on clearance cars and/or equipment.

Communications

1. The application of microwave communications on railroads, both for use in special areas and eventually for general use to replace open-wire pole lines. The use of microwave as a means of communication on the railroads has barely been started, there being installations on the Santa Fe, Southern Pacific, Rock Island, and other railroads. During the next 10 years the use of microwave will no doubt be gradually enlarged; however, there are many unsolved problems.

2. The role of radio in railroad communication.

Competition

1. The position and role of the railroads in the current highway and waterway development programs, and in the tremendous development of cities that is going on at this time.

2. Truck and rail freight-haul costs. Determine the zones of distance where truck haul is cheaper to the shipper on the basis of tons shipped. It is generally considered that, for short haul, the truck is cheaper; and also for less than car loads. However, for a given tonnage to be hauled there must be a "break even" distance beyond which the advantage in costs is in favor of the railroads. It is believed that a determination of this point would be of value to the shipper, if determined by a neutral, factual study of the problem.

3. Study to determine additional legislation necessary to permit operation of railroads on a more equitable competitive basis with subsidized and other forms of common carriers.

See also: Costs; Management

Computers, Electronic

1. Review the savings to be realized by the railroad industry from the more general use of electronic computers, including the adoption of such equipment in railroad engineering departments.

2. Universal computer-oriented language for railroad engineering computer programming—writing computer programs for typical engineering problems suitable for any computer and any railroad.

3. A proposed universally acceptable waybill brought up to date in view of modern electronic business machines:

    a. State of the art in "Character Recognition".

    b. Machine identification of magnetic or similar ink, type, face, etc.

4. Factors that will enter into computer solution of the question of long, slow, few freights vs. short, fast, frequent trains.

    a. When is the optimum time to run a freight?

    b. We lose business to trucks because they leave their terminals shortly after loading—on 3-min headway. Railroads frequently hold freight for a day "because Grandpa did." This is basically a topic in linear programming.

See also: Operations.

Costs

1. A restudy and revision of the "Yager Formula" of the AREA for determining the relationship and changes in annual costs of maintenance of way expense, due to changes in volume of traffic. This formula was originally determined during the federal control of railroads, and is now being revised to meet present methods and maintenance costs.

2. Make an up-to-date and more accurate determination of the relation between fixed and variable costs in the several systems of transport.

3. Application of work-study techniques to operation and maintenance. Development of potential savings to railroad activities.

4. Advantages and disadvantages of leased or rented facilities versus railroad ownership.

5. Effects of automation and mechanization upon railroad maintenance costs.

6. Economics of standardization in railroad practices.

See also: Accounts and Statistics; Competition; Operation, Track.

Crossings, Grade

1. Develop responsibility relationship of public versus railroads.

2. Economics study of grade crossing construction and maintenance.

3. Relative advantages and disadvantages of various types of grade crossing construction.

Drainage

1. Effects of changes in land use on water runoff and culvert design.

2. Effects upon ballast and roadbed.

Freight Stations and Freight Handling

1. Study materials and design of pallets and/or containers which will provide handling of less-carload lots and carload lots by fork-lift trucks. Most of the freight is presently loaded into and out of cars by hand, which results in high labor costs. This could be expanded into a study of economics resulting from manufacturers and processors packaging their goods on pallets or containers which can be handled as a unit through the wholesale warehouses to the retail outlet.

2. Study to determine what measures could be taken to reduce damage to commodities transported by rail and the cost of claims for such damage.

See also: Piggy-Back.

Freight Traffic

Development of freight traffic trends.

General

1. Relationship of transportation facilities to city planning, development, and redevelopment.

2. Opportunities for solving railroad problems through research.

3. Automatic and non-destructive testing and inspection devices.

Legislation

1. Analysis of restrictive railroad legislation passed in the various states:

   a. Justification.

   b. Arbitrary character.

   c. Need for constructive bills.

2. Consolidation of all forms of transportation under one Secretary of Transportation or formulation of a uniform national policy.

3. Study of railroad taxation inequities, and relation to the community life.

4. History of, and character of government control of railroads and other forms of transportation, proposed legislative changes.

5. Mergers (including techno-economic implications).

Location

1. Review, revise, and develop more accuracy in all railroad engineering design and location criteria.

2. Economics of grade and line revision—objectives: reductions in operating and maintenance costs.

3. Utilization of photogrammetry in construction and relocation of railroads.

4. Utilization of computor programs in evaluation of alternative locations.

Management

1. Planned allocations of resources for maintenance of way—budget forecast, revenue, and traffic factors, plannning for steady base loads of work.

2. Investigate the advantages and disadvantages of broadening the deferred-payment method to cover large expenditures for additional railrcad plant and equipment. Conditional sales agreements with manufacturers are used to cover large expenditures for furnishing and installing various types of permanent facilities, such as signals, interlocking, car retarders, etc. Also, agreements with finance companies are used to cover lease of more expendable items such as trucks and roadway maintenance equipment.

3. Study of the potential consolidation of railroad systems on a national basis and the problems involved, with suggestions for solving these problems.

See also: Competition; Legislation.

Materials

1. Economics of standardization.
2. Dynamics of standardization.

See also: Costs.

Operations

1. Study and development of ways and means to effect transit-time reduction between origin and destination of freight rail shipments.

2. What are the economic and operational factors affecting fixed-consist, light-tonnage, through-freight-train operation between major terminals, i.e., Chicago to the West Coast, New York through St. Louis or Chicago to the West Coast, etc.

3. What are the reasons for not advancing automatic remote operation of trains?
   a. Technical developments needed.
   b. Operational changes required.
   c. Labor contracts.
   d. Regulation by state bodies.

4. Can a railroad be run from a centralized point?
   a. CTC.
   b. Car dispatching or distributing.
   c. Freight house.
   d. Billing.
   e. "Personal contact."

5. Determine the effects of railroad capacity and operation of varying geographical areas of the United States (for example, the short-haul, multi-terminal operation of eastern railroads as contrasted with the long-haul, and few terminals encountered, by railroads of the West; also mountainous terrain as contrasted to flat terrain, etc.)

6. Develop more nearly exact measures of track and traffic capacity for railroads. The increasing traffic load on line-haul and rapid-transit facilities and the need for increased efficiency in utilization of the minimum plant requirements make desirable an improvement in existing and cumbersome methods of computation. The adaptation of such new methods to computer solution should be included.

7. Evaluate various types of transport as to their relative engineering and economic efficiencies for various specific functions.

8. Commuter traffic problems.

See also: Computers; Costs; Freight Houses and Freight Handling; Freight Traffic; Passenger Traffic; Piggy-Back; Rolling Stock; Weather; Yards.

Passenger Traffic

1. The role of railroads in commuter service as a result of suburban living.

2. Investigate what would be necessary to make passenger business a profitable operation.

3. A railroad with $1 billion invested in passenger facilities and equipment has an average revenue per passenger-train mile of $7. The full cost to the railroad is $9. If one train mile is taken off, the railroad saves $5. The railroad cannot increase the earnings per train mile to reduce the cost. At the same time, it cannot very well take off a $7-

per-mile train to save only $5 even though the full cost is $9. What is the answer? Note: The railroad proposing this question has offered full cooperation to the investigator in securing all pertinent data.

4. Development of optimum usage and coordination of all forms of passenger service, including possible combination of terminals.

5. Location of passenger terminals.

6. Role of rail transportation in urban development.

See also: Operations; Costs.

Personnel

1. The advantages to the individual, it any, for engineering students to become trained in railroad work by becoming student trainees with the railroad during each vacation period, with a job guarantee upon completion of the training and graduation, such as has been followed by the Southern Pacific Company; compared with enrolling students in trainee programs after graduation.

2. Industrial versus academic training and education. Relationship, value, and limitations.

3. Role of the engineer in railroad transportation.

4. Role of the graduate student in railroad transportation.

Piggy-Back

1. The advantages to truck transportation in costs by using the so-called piggy-back rail service for trailers, as against all highway truck operation between similar origin and destination points.

2. Critical review of various methods for handling "Trailers on Flat Cars" (TOFC).

3. Probability of replacement of freight box cars by flat cars suitable for TOFC— influence on turn-around time, freight rates, operating costs, etc.

4. Investigate the various types of "piggy-back" and container transportation equipment and services now in use on railroads and determine which offers the greatest economic advantage, or recommend a new system which would have a greater economic advantage.

See also: Freight Stations and Freight Handling; Operations.

Piping

1. Study the feasibility of using plastic pipe to transport air, oil, and other materials around terminals.

2. Cathodic protection of underground steel pipelines and steel tanks, including methods for determining soil conductivity before making installations and including instructions for making tests after the completion of installations to ascertain whether proper protection has been achieved.

Rail

1. Study the flow of metal in rail. This could be an extension of the field of study opened up by C. J. Code's recent experiments of inserting brass plugs in the head of a rail and observing over a period of time the change in size and shape of these plugs due to the flow of metal in the head of the rail.

2. Program to determine economical rail size. This involves developing a guide to judgment in the economic selection of rail under different traffic conditions.

3. The theoretical and practical determination of rail contact stresses.

4. Rail grinding to control rail joint batter.

5. Rail stresses.

See also: Track.

Roadbed

1. Evaluation of stabilization methods (mechanical, chemical, etc.)

2. Protection by moisture barriers (bitumens, synthetic polymers, etc.)

3. Applications of modern soil mechanics techniques to roadbed design, and construction.

4. Determination of the passive resistance pressure of soils and application to the design of flexible culvert structures.

Rolling Stock

1. Study the need for specialized freight equipment.

2. Plan to increase car ownership. One of the very pressing problems for the entire railroad industry is the freight-car shortage which occurs periodically during high levels of business. The Association of American Railroads has made frequent studies of this problem in conjunction with the officers of the railroads. Many suggestions have been offered to increase the ownership of cars, including higher per diem, national car pool, formula for individual railroad ownership, and other items. There is no plan so far submitted that has had any appeal to a majority of the railroads in any part of the country.

3. Ultimate effect, including cost analysis, to be gained by equipping all railroad cars and equipment with roller bearings.

4. Passenger comfort. Study designs of reclining seats which will allow passengers to relax comfortably. Most present seats do not provide satisfactory rest for the legs and head.

5. A study to determine what it is costing the railroads to provide specialized freight equipment versus freight equipment of standard design.

6. Study of the relative advantages or disadvantages—economic and mechanical— resulting from the use of roller bearings vs. friction bearings on railroad freight equipment.

7. Analyze the railroad journal-hot-box problem and develop recommendations for a practical and economically feasible solution.

See also: Operations; Passenger Traffic; Piggy-Back.

Signals

1. Evaluate the actual annual costs as between a section of double track compared with centralized traffic control on single track for the same volume of business. Several railroads have abandoned sections of double track by installing CTC on a single track, claiming large savings. These savings occur principally in maintenance.

2. Standardization potentials related to railroad requirements and economics.

Statistics

See: Accounts and Statistics

Ties

1. Economics of tie spacing. The present practice is to place as many ties in track as possible and still have room to tamp around them. If the number of ties were reduced,

it is possible that they might have to be tamped more frequently, but tending to offset this would be the fact that there would be fewer ties to tamp, fewer ties in track to be maintained, and less investment in ties, tie plates and fastenings.

2. Tie renewal program. Tie renewals by the out-of-face method compared with individual renewals when the remaining life of individual ties is exhausted. Determine what percentage of a tie's total expected life can be economically thrown away in order to make renewals with specialized tie gangs, as against individual renewal when a tie is worn out.

3. Economics of replacing timber ties with reinforced concrete ties.

See also: Track

Terminals

1. Developments in highway-railroad transfer service.
2. Potential effects of intermodal integration of terminal facilities.
3. Location of terminal facilities.

Track

1. Study effects of curvature on location and operation, and costs of railroad curvature. The increased resistance due to curvature, the causes thereof, the costs, and the effects of curvature on operation under modern conditions have been the subject of much contradictory evidence and argument. An impartial study could aid in establishing acceptable principles.

2. Investigate the status of the railroad track structure in the light of current and future types of motive power, speeds, axle loadings, equipment, and materials.

3. Equation of track for maintenance. Top maintenance officers consider it quite a problem to equate track values so that each section foreman would be assigned approximately the same amount of work. It is general experience that some sections are excellently maintained and others poorly maintained, and the question to be determined is whether in the first case this is due to the efficiency of the foreman or the fact that he is assigned less work to be done.

4. Economics of deferred maintenance practices.

See also: Ballast; Cost; Operations; Rail; Ties.

Tunnels

1. Ventilation of long railroad tunnels.
2. Modernization requirements and methods of obtaining increased clearances.

Vegetation Control

1. Evaluation of control methods, i.e., mechanical, chemicals.
2. Ballast fouling problems resulting from inadequate weed control measures.

Weather

1. Sub-zero operation. Study the effects of sub-zero weather, snow, and ice on lubrication and on steam and air equipment on coaches and cars.

2. Effects of temperature on rail, joint bar, wheel, and axle breakage and the problems arising therefrom.

3. Effects of weather on overall speed of trains and on track and traffic capacity.

4. Effects of weather on speed of car movements through yards and on yard locomotive capacity; this would involve a study of tonnage ratings with the new types of motive power.

5. Variations in regional operating costs and problems due to weather.

6. Problems of weather imposed on specific commodities; for example, how much extra tonnage does a locomotive have to pull when cars of coal, sand, or ore are water-soaked or iced and how does this affect locomotive and traffic capacity? What are the effects of temperature on shipments of liquids and the effects on costs and capacities, etc.?

See also: Operation; Location.

Yards

1. Factors influencing location and design of freight classification yards—mathematical models, systems design, instrumentation.

2. Interior illumination of railroad-yard control towers, with emphasis on elimination of glare or reflected light from windows.

3. Illumination of large railroad yards.

4. Developments in railroad yarding methods.

5. Relative effects of factors involved in developing efficient hump yard operations.

6. Yard locations.

See also: Buildings; Operations; Weather.

Report of Committee 11—Engineering and Valuation Records

L. W. Howard,
Chairman

M. C. Wolf,
Vice Chairman

W. S. Gates, Jr.,
Secretary

W. A. Krauska
J. Bert Byars
E. W. Smith

W. J. Pease
C. R. Dolan
M. M. Gerber
C. F. Olson
H. N. Halper
R. B. Aldridge
F. B. Baldwin (E)
J. L. Becker
B. A. Bertenshaw (E)
G. R. Berquist
H. C. Boley
H. T. Bradley
C. E. Bynane
J. R. Clayton
C. E. Clonts
B. J. Cook
F. O. Crosgrove
Spencer Danby
R. S. Danis
F. H. DeMoyer
R. L. Ealy
A. L. Engwall
M. Freidman
R. F. Garner
E. W. Gibson
W. M. Hager
C. C. Haire (E)
N. Hammond
J. H. Hande (E)
G. J. Harris

M. J. Hebert
J. W. Higgins
J. A. L. Houston
R. D. Igou
W. H. Kiehl
C. E. Lex, Jr.
W. M. Ludolph
C. B. Martin*
C. W. Meyer
B. H. Moore
B. F. Nauert
F. H. Neely
D. G. Paeske
D. E. Pergrin
C. H. Rapp
H. L. Restall
F. A. Roberts
C. S. Robey
E. J. Rockefeller
H. B. Sampson
R. L. Samuell
R. S. Shaw, Jr.
J. N. Smeaton
J. E. Stein
J. B. Styles
J. R. Traylor
R. C. Watkins*
W. C. Wieters
H. R. Williams
Louis Wolf (E)

Committee

---

* Deceased.
(E) Member Emeritus.
Members listed in bold face are the official representatives of the Engineering Division, AAR.

To the American Railway Engineering Association:

Your committee reports on the following subjects:

4. Use of statistics in railway engineering.
   Progress in study, but no report.

5. Construction reports and property records
   Committee is keeping abreast of current happenings which affect this assignment but has no report.

6. Valuation and depreciation:
   (a) Current developments in connection with regulatory bodies and courts.
       Progress report, submitted as information .... page 447
   (b) ICC valuation orders and reports.
       No report.
   (c) Development of depreciation data.
       No report.

7. Revisions and interpretations of ICC accounting classilfications.
   Progress report, submitted as information .... page 449

8. Simplification of records to determine original costs of tracks to be used in their retirements from the investment account. Committee is keeping abreast of current happenings which affect this assignment but has no report.

9. Simplification of annual reports on Form 588 to the Interstate Commerce Commission, and underlying Completion Reports.
   No report because this assignment is dependent on disposition of Valuation Order 30 of the Interstate Commerce Commission.

THE COMMITTEE ON ENGINEERING AND VALUATION RECORDS,
L. W. HOWARD, Chairman.

AREA Bulletin 569, January 1962.

MEMOIR

Charles B. Martin

Charles B. Martin, chief valuation engineer of the New York Central System, died suddenly on November 14, 1961, while on a short vacation in Escanaba, Mich.

Mr. Martin was born on November 19, 1908, in Brooklyn, N. Y., a son of Adelaide Ketchum and the late Bertram C. Martin. Surviving are his wife, the former Elizabeth Barker; two sons, Charles, Jr. and Clifford; a brother, George N. Martin; and his mother.

While still attending college in 1929, Mr. Martin entered railroad service in a temporary position as chainman with the New York Central Railroad at Albany, N. Y. He went on to complete his Bachelor of Science degree in civil engineering at the University of Maine in 1930. Returning permanently to railroad service, Mr. Martin held the positions of junior engineer and detailer with the engineering department of the New York Central until 1941, at which time he was assigned to the Pittsburgh & Lake Erie Railroad in Pittsburgh. Returning to New York and the New York Central in 1942, Mr. Martin served as assistant engineer with the engineering department until 1944, when he joined the valuation department for a two-year period. Returning again to the engineering department in 1946, Mr. Martin served as designer until 1950, when he was asked to return to the valuation department. Mr. Martin rose to the position of valuation engineer in 1951 and was appointed chief valuation engineer for the entire New York Central System on January 1, 1960.

Mr. Martin joined the American Railway Engineering Association in 1945. He served conscientiously and ably as a member of Committee 11—Engineering and Valuation Records, holding the position of subcommittee chairman and, most recently, was elected to the position of vice chairman—designate. Mr. Martin also served as a member of the Joint Equipment Subcommittee of the Joint Equipment Committee in Washington, D. C.

Mr. Martin was a member of a family well steeped in Americana, railroading, and engineering. His mother is a present member of the Daughters of the American Revolution; his father was resident and district engineer in the construction of the Hudson River Connecting Railroad, and his brother, formerly a member of the engineering staff of the New York Central, is now president of a leading construction company in Chicago.

Mr. Martin devoted his life to railroad service. His ability and experience earned him the respect of all his associates. His good nature and fine humor won him the friendship of all who knew him.

J. E. STEIN, *Chairman,*
L. W. HOWARD,
M. C. WOLF,
Committee on Memoir.

MEMOIR

James H. Roach

James H. Roach, retired chief valuation engineer of the New York Central System, died at his home in Bronxville, N. Y., April 21, 1961, after an extended illness.

Mr. Roach was born on March 3, 1883, at St. Louis, Mo., a son of the late James H. and Virginia Donne' Roach. Surviving are his wife, the former Estelle Pearl Freeman; a son, James Rodney Roach; a daughter, Mrs. Albert C. (Estelle Marie) Martin; a granddaughter and two great granddaughters.

A graduate of Washington University in 1901, Mr. Roach entered railroad service the same year, as structural draftsman and designer with the Missouri Pacific Railroad at St. Louis. He joined the New York Central System in the engineering department of the Lake Shore & Michigan Southern Railway in Cleveland, Ohio, in 1907. In 1914 he entered the valuation department as assistant valuation engineer, Lines West, was transferred to New York as valuation engineer in 1918 and was appointed chief valuation engineer of the System in 1925, and continued in this position until he retired on December 31, 1950. After his retirement he continud as consulting valuation engineer rounding out 51 years of railroad service in this capacity in 1952.

Mr. Roach joined the American Railway Engineering Association in 1910 and became a Life Member in 1945. He served conscientiously and ably on the Committee on Contract Forms, the Tie Committee, and in 1941 became a member of Committee 11— Engineering and Valuation Records. He was a member of the American Society of Civil Engineers, and a licensed professional engineer, New York State. Mr. Roach also served on many engineering and railroad valuation committees and as chairman of the Engineering Committee on Valuation of the Eastern Group of the Presidents' Conference Committee. Upon the formation of the Association of American Railroads in 1934, Mr. Roach was appointed a member of the Valuation Advisory Committee, later chairman of the Engineering Valuation Committee, and finally General Chairman of the Valuation Division of the Association of American Railroads. He was also a member of the Masonic order.

Mr. Roach's ability and long experience in the Federal Valuation of the American Railroads, his untiring efforts in behalf of the Railroads in his own work and in valuation and engineering committees established him as one of the deans of the profession, and earned him the respect of those who knew him.

MEMOIR

Robert C. Watkins

Robert C. Watkins, engineer—valuation, of Chesapeake & Ohio Railway, died at his home in Richmond, Va., August 29, 1961, after an extended illness.

Mr. Watkins was born on January 30, 1904, at Gordonsville, Va. Surviving are his wife, the former Margaret Wiley; a daughter, Mrs. Louis J. Read; and two grandchildren.

Mr. Watkins attended public schools at Gordonsville and the Virginia Polytechnic Institute at Blacksburg, Va. He first entered C&O Railway service in 1922 as a rodman. He advanced from that position to inspector, ballast inspector, assistant cost engineer, assistant supervisor bridges and buildings, supervisor of track, and assistant division engineer. In June 1944, he was promoted to division engineer, Hocking Division, Peru, Ind. and continued in this position until January 1948, at which time he was promoted to division engineer, Columbus, Ohio. He entered the valuation department at Richmond in 1949 as engineer, and continued in this position until his death.

Mr. Watkins joined the American Railway Engineering Association in 1955 and in 1957 became a member of Committee 11—Engineering and Valuation Records on which he served on Subcommittees 5 and 9. He was a licensed professional engineer, Commonwealth of Virginia and a member of the Masonic Order.

Mr. Watkins took an active part in the affairs of Committee 11. His many friends and associates are saddened by his passing and will miss him and his valued counsel during the coming years.

Report on Assignment 1

Revision of Manual

W. A. Krauska (chairman, subcommittee), G. R. Berquist, J. A. Houston, W. M. Ludolph, F. A. Roberts, R. L. Samuell, J. N. Smeaton, E. W. Smith, H. R. Williams, M. C. Wolf.

During the past year your committee has reviewed the following Manual material and recommends adoption of the revisions thereof indicated.

Pages 11–2–1 to 11–2–4, incl.

CONSTRUCTION REPORTS AND PROPERTY RECORDS— RELATION TO CURRENT PROBLEMS

Reapprove with the following revisions:

Page 11–2–1. In Par. 1. Accounting for Cost of a Railway . . .,, change the words "ICC accounting classification" to "ICC Uniform System of Accounts for Railroad Companies." Also, change "Account 701" to "Account 731."

Page 11–2–2. In the second and fifth lines from the top of the page, change the words "ICC Bureau of Valuation" to "ICC Bureau of Accounts."

Change Par. 3 to read as follows:

3. *Federal Income and Excess Profits Taxes.*—The Internal Revenue Service permits, in computing the annual tax payment due, a charge against income for the depreciation incurred during the year. Provided the carrier has sought and obtained permission to do so, this depreciation covers certain classes of roadway property and equipment. The basis of cost to which approved rates are applied is based on the carrier's Property Records. Unlike the ICC Accounting Regulations, several methods of depreciation are permitted (Straight Line, Declining Balance, Sum-of-the-years digits) as well as special adaptations to meet special conditions. In each case the conditions, rates and bases are arrived at after conferences with the Internal Revenue Service representatives."

In Pars. 4 and 10, change the words "ICC Bureau of Valuation" to "ICC Bureau of Accounts."

Page 11–2–3. In Par. 2 change the words "ICC Bureau of Valuation" to "ICC Bureau of Accounts."

Change Pars. 4, 8, and 9 to read as follows:

"4. For computation of rates of depreciation and depreciation bases to be used by railroads in accounting for depreciation in accordance with ICC or IRS requirements."

"8. For making estimates as to the cost or value of trackage rights to be used or leased."

"9. In setting up switching charges to determine the value of the facilities that are to be used in the operation."

Page 11–2–4. In Pars. 19 and 20 change the words "Bureau of Internal Revenue" to "Internal Revenue Service."

Pages 11–2–4 to 11–2–7, incl.

AUTHORITY FOR EXPENDITURE

Reapprove with the following revisions:

Page 11–2–4. Under Instructions for Use of Form, change the first sentence to read "ICC Valuation Order 3 requires that request for Authority for Expenditure . . ."

Page 11–2–5. In the fifth paragraph change the words "ICC Classification of Accounts" to "ICC Uniform System of Accounts."

Page 11–2–6. Delete the last four paragraphs, substituting therefor the following single paragraph:

"Credits: When temporary construction is necessary a credit should be shown for the salvage recovered from this temporary work. Care should be taken to keep this credit separate from salvage from retired property."

Pages 11–2–8 and 11–2–9

DETAILED ESTIMATE

Reapprove with the following revisions:

Page 11–2–8. Delete the first two paragraphs under Instructions for Use of Form, substituting therefor the following single paragraph:

"Each request for authority foɪ expenditure should be accompanied by a detailed estimate of the project, unless the proposed change in the property is so small and so simple that it is deemed unnecessary to show the details of the estimate. A suggested form for this purpose is the Detailed Estimate form. It should be signed by the maker and approved by various officers in accordance with the prevailing organization, one copy to be attached to each copy of the AFE."

Pages 11–2–10 and 11–2–11

REGISTER OF AUTHORITIES FOR EXPENDITURE

Reapprove without change.

Pages 11–2–12 to 11–2–14, incl.

TIME ROLL—LABOR—MONTHLY

Reapprove without change.

Page 11–2–15

TRACK FOREMAN'S DAILY MATERIAL REPORT

Reapprove without change.

Page 11–2–16

FOREMAN'S MONTHLY MATERIAL REPORT—BRIDGES

Reapprove without change.

Page 11–2–17

FOREMAN'S BRIDGE SECTION TOOL REPORT

Reapprove without change.

Pages 11–2–18 and 11–2–19

FOREMAN'S DAILY REPORT OF WORK TRAIN PERFORMANCE

Reapprove without change.

Pages 11–2–20 and 11–2–21

CONDUCTOR'S DAILY REPORT OF WORK TRAIN PERFORMANCE

Reapprove without change.

Pages 11–2–22 to 11–2–29, incl.

ROADWAY AND STRUCTURES—RECORDS AND REPORTS

Reapprove with the following revisions:

Page 11–2–22. In the first paragraph change the words "ICC Classification of Accounts" to "ICC Uniform System of Accounts for Railroad Companies."

Page 11–2–28. Change last word in heading from "Bridges" and "Buildings."

Pages 11-2-30 to 11-2-34, incl.

BRIDGE CONSTRUCTION REPORTS

Reapprove without change.

Pages 11-2-35 and 11-2-36

MONTHLY TRACK MATERIAL REPORT

Reapprove without change.

Pages 11-2-37 and 11-2-38

REPORT OF QUANTITIES IN COMPLETED WORK

Reapprove without change.

Pages 11-2-39 to 11-2-41, incl.

COST OF PROPERTY RETIRED

Reapprove without change.

Pages 11-2-42 and 11-2-43

ROADWAY COMPLETION REPORT

Reapprove without change.

Pages 11-2-44 and 11-2-45

EQUIPMENT COMPLETION REPORT

Reapprove without change.

Pages 11-2-46 and 11-2-47

RECORD OF BALLAST CHANGES

Reapprove without change.

Page 11-2-48

PROGRESS PROFILE

Reapprove without change.

Pages 11-3-1 to 11-3-19, incl.

COST-KEEPING METHODS, STATISTICAL RECORDS, FORMS FOR ANALYZING EXPENDITURES FOR ASSISTANCE IN CONTROLLING EXPENDITURES

Reapprove with the following revisions:

Page 11-3-1. Change headings to read: "Cost Accounting Methods, Statistical Records and Forms for Analyzing Expenditures for Assistance in Controlling Expenditures."

Page 11–3–4. In the second line change the words "ICC accounting classifications" to "ICC Uniform System of Accounts for Railroad Companies."

Change the last sentence of the first paragraph to read as follows: "He should understand that the accounting dollars, which are accumulated monthly, may contain adjustments in prior month's figures, non-cash items, accruals, improper distribution or classification, etc."

In the third paragraph, second line, change the word "seven" to "eight." Add an eighth item to the list following this paragraph, as follows:

"8. Maintenance of Way Power Work Equipment." ·

Page 11–3–13. Under Art. 1. The Budget—Its Function, add the following sentence to the first paragraph: "The amounts to be expended for upkeep and betterment of the property are established through the medium of a budget, or program, as determined by the condition of the units of the property."

Page 11–3–14. Delete the paragraph beginning "The amounts to be expended . . ." (first full paragraph at top of page).

Under B. Budget—Maintenance of way and structures, third line, change the words "ICC classification of accounts" to "ICC Uniform System of Accounts for Railroad Companies."

Page 11–3–18. Add to line 39: "(See Item 60)". Add after line 59 the following:

"M. of W. Power Work Equipment
60.   269   Power Work Equipment"
Change present line 60 to 61.

Pages 11–4–1 to 11–4–9, incl.

GENERAL AND TEXT

Reapprove with the following revisions:

Pages 11–4–1 and 11–4–2. Delete all the material on these pages, substituting therefor the following rewritten version:

Purpose

1. To establish standard drafting practice.
2. Obtain enough information.
3. Omit unnecessary information.
4. Raise quality and accuracy of drawings.

Clearance.—All less than standard clearances should be shown on maps and where necessary supported with memorandum and/or sketch showing details.

Dimensioning.—Dimensions for architectual and bridge drawings should be expressed in inches up to and excluding 12 in; feet and inches over 12 in; pipe diameter in inches only—length in feet and inches; maps and profiles in feet and tenths.

Direction of Line.—Generally, except in some cases of diverging lines, eastbound and southbound should be shown toward the right side of the sheet and westbound and northbound towards the left side of the sheet; terminals should be indicated as "To Chicago" (or name of terminal) with small arrow pointer.

Forms.—Forms are used to assemble information for records and comparative pur_ poses. The weight and spacing of lines should be such as to make the form easily read_ able. For example, by making the third or fifth lines heavier. If the form is of a nature that calls for explanation or will give rise to questions, each column should be lettered and each line numbered.

Graphs and Charts.—Graphs are used where they can illustrate the relation between variable factors to better advantage than they can be shown in tables. The type of graph to be used depends on the requirements. The graph should be executed boldly so as to depict the relationships clearly.

Lettering.—The objective should be to provide distinct, uniform letters and figures with reasonable rapidity. Single-stroke commercial gothic letters and figures should be used, made freehand or by use of mechanical lettering devices, using guide lines for freehand lettering. Vertical style is preferable for standard drawings, general tables, forms, charts and work which requires a more finished appearance. The inclined style is preferable for working drawings, field work and general lettering where speed is an important requisite. Lettering should be read from bottom or the right hand edge of the sheet as determined by the position of the title, which is always in the lower right corner. The height of letters and figures should be specified in twentieths so as to provide for the use of the horizontal lines of Plate A profile paper.

Maps and Profiles.—These shall show the nearest mile post or station, valuation station, degree of curve and all information necessary to identify the location.

Reduction.—If the drawing is to be reproduced to a smaller size, the thickness of the lines and the weight and size of the lettering shall be proportionally increased to avoid loss of detail or legibility in the copy.

Right-of-Way Maps.—Right-of-way maps of new lines and extensions of existing lines shall be made on size C drawings and shall be platted continuously from left to right and shall otherwise conform with ICC specifications.

Scales.—Drawings shall be made on a sufficiently large scale so that they may be properly dimensioned. Determine what information is necessary to obtain a full and complete picture, orient the map to best possible advantage to conserve space, and avoid showing unnecessary information. See Art. 4, Sec. B of Specifications for Preparation of Maps and Profiles, Part 4, this Chapter.

Sheet Number.—When a drawing covers several sheets, each sheet shall be designated "sheet .... of ...." It is preferable that all sheets of a series be of the same size.

Specifications.—Drawings shall refer to the latest revisions of specifications, so as to obviate a revision of the drawing every time a specification is revised.

Symbols and Material Sections.—These should conform with AREA recommended practice.

Tables, Notes and Material Lists.—Should be placed on right-hand side of sheet if practicable, otherwise where space is available. If desirable material lists can be confined to 8½ by 11- in or letter-size paper so they can be printed and attached to requisition.

Terms.—For the designation of steel and iron wire gage use Aswg. For copper wire gage use Awg. Use the term "switch point" instead of "switch rail."

General.—Leave sufficient tracing cloth or paper around the drawing to provide for taping to table or board. On rolled drawings leave sufficient cloth on the outside end to serve as a protection in handling and to provide a place for showing the file number and drawing number. If approval of the drawing is required, the blank for the signature shall be placed immediately to the left of the title block and just above the border line.

Lines drawn for sectioning shall be at an angle of 45 deg. In general, no shade lines are to be used.

Page 11-4-5. Add item 18 to Fig. 2 as shown in revised Fig. 2 presented herewith.

Pages 11-4-10 to 11-4-24, incl.

GRAPHICAL SYMBOLS

Reapprove with the following revisions:

Page 11-4-13. In Fig. 3 delete "Magnetic Meridian" and symbol; substitute "North Point" and symbol. (Revised Fig. 3 presented herewith.)

Page 11-4-14. In Fig. 4 add "Sanding Tower" and symbol therefor. (Revised Fig. 4 presented herewith.)

Page 11-4-15. In Fig. 5 add "Hot Box Detector" and symbol therefor; also, "Automatic Car Journal Oiler" and symbol therefor. (Revised Fig. 5 presented herewith.)

Page 11-4-16. In Fig. 6 add the following to the symbol for Masonry Arch or Flat Top Culvert: "Size: Hor., Vert., Lgth." After "Pipe" delete the words "over 36" dia." After the words "Draw Span", add "or Swing Bridge." (Revised Fig. 6 presented herewith.)

Page 11-4-18. In Fig. 8, add the word "Telltale" after the words "Bridge and Tunnel Warning." (Revised Fig. 8 presented herewith.)

Page 11-4-22. Delete Fig. 12.

Pages 11-4-25 and 11-4-26

METHODS OF FOLDING DRAWINGS

Reapprove without change.

Pages 11-4-27 to 11-4-31, incl.

SPECIFICATIONS FOR PREPARATION OF MAPS AND PROFILES

Reapprove without change.

This completes the committee's review of Chapter 11 of the Manual in connection with the proposed 1962 reprinting of the Manual.

SIZES OF SHEETS FOR ENGINEERING DRAWINGS, FORMS AND CHARTS

SIZE NO.	SHAPE	A	B	C	D	E	F	G	H	J	TO BE USED FOR	REMARKS (All Dimensions in Inches)
1	A	11	8½	1½	¼	2¼	3¼	1			REGULAR DWGS. FORMS AND CHARTS	LENGTHEN BY MULTIPLES OF 7
2	C	11	17		¼	2¼	3¼	1			FORMS AND CHARTS	FOR AREA REDUCE 11 HEIGHT TO 7
3	B	11	8½	1½	¼						REGULAR DWGS. R. OF W. AND CONTRACT	TITLE MAY BE CHANGED AS INDICATED
4	A	10½	8	1½	¼						I.C.C. APPLICATION MAP	NO TITLE BLOCK
5	A	21	16	1½	½						DO	DO
6	D	17	22	1½	½	3	4½	1¼			REGULAR DRAWINGS	LENGTHEN BY MULTIPLES OF 14
7	D	22	34	1½	½	3	4½	1¼			DO	DO
8	D	34	44	1½	½	3	4½	1¼			DO	DO
9	C	24	56		½	6	9				I.C.C. RIGHT OF WAY MAPS	REDUCTION OF SIZES - 11-12
10	A	7	4		½						CUTS FOR AREA REPORTS	WORKING SIZE
11	A	14	8		½						DETAIL DWGS. FOR AREA BULLETINS	DO
12	A	21	12		½						DO	
13	D	9	12½	1½	½						AREA TRACKWORK PLANS	LITHOGRAPHIC SIZE
14	D	18	25	3	1						DO	WORKING SIZE. INCREASE B TO SUIT
15	D	8½	17	1½	½	2½	3½	1¼			FOR STANDARD DWGS. ONLY	LITHOGRAPHIC SIZE
16	D	17	34	3	1	5	7	2½			DO	FOR DETAILING. INCREASE B TO SUIT
17	D	11	17	1½	½	3	4½	1¼			REGULAR DRAWINGS	
18	C	24	36	1½	½	3	4½	1½			FOR STATE HIGHWAY DRAWINGS	

Fig. 2.

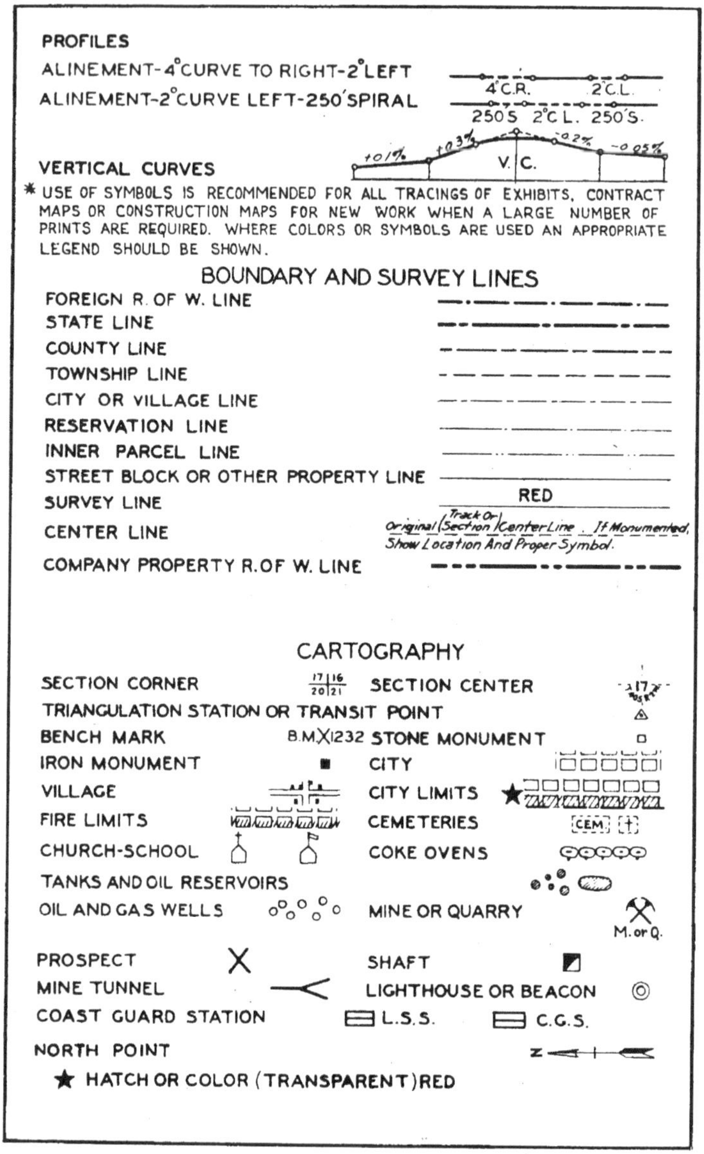

PROFILES

ALINEMENT-4°CURVE TO RIGHT-2°LEFT

ALINEMENT-2°CURVE LEFT-250´SPIRAL

4°C.R.   2°C.L.

250´S.   2°C.L.   250´S.

VERTICAL CURVES

+0.1%   +0.3%   +0.2%   -0.05%

V. C.

✱ USE OF SYMBOLS IS RECOMMENDED FOR ALL TRACINGS OF EXHIBITS, CONTRACT MAPS OR CONSTRUCTION MAPS FOR NEW WORK WHEN A LARGE NUMBER OF PRINTS ARE REQUIRED. WHERE COLORS OR SYMBOLS ARE USED AN APPROPRIATE LEGEND SHOULD BE SHOWN.

BOUNDARY AND SURVEY LINES

FOREIGN R. OF W. LINE

STATE LINE

COUNTY LINE

TOWNSHIP LINE

CITY OR VILLAGE LINE

RESERVATION LINE

INNER PARCEL LINE

STREET BLOCK OR OTHER PROPERTY LINE

SURVEY LINE   RED

CENTER LINE   Track Or-/ Original (Section /Center-Line ... If Monumented, Show Location And Proper Symbol.

COMPANY PROPERTY R. OF W. LINE

CARTOGRAPHY

SECTION CORNER $\frac{17|16}{20|21}$ SECTION CENTER $\frac{17}{20}$

TRIANGULATION STATION OR TRANSIT POINT   △

BENCH MARK   B.M.X1232   STONE MONUMENT   ▫

IRON MONUMENT   ▪   CITY

VILLAGE   CITY LIMITS   ★

FIRE LIMITS   CEMETERIES   CEM [✝]

CHURCH-SCHOOL   COKE OVENS   ⊖⊖⊖⊖⊖

TANKS AND OIL RESERVOIRS   ● ∴ ◯

OIL AND GAS WELLS   ₀°₀° ₀° ₀   MINE OR QUARRY   ⚒   M. or Q.

PROSPECT   X   SHAFT   ◢

MINE TUNNEL   ⟨   LIGHTHOUSE OR BEACON   ◎

COAST GUARD STATION   ⊟ L.S.S.   ⊟ C.G.S.

NORTH POINT   z ⟨─┼ ⟨

★ HATCH OR COLOR (TRANSPARENT) RED

Fig. 3.

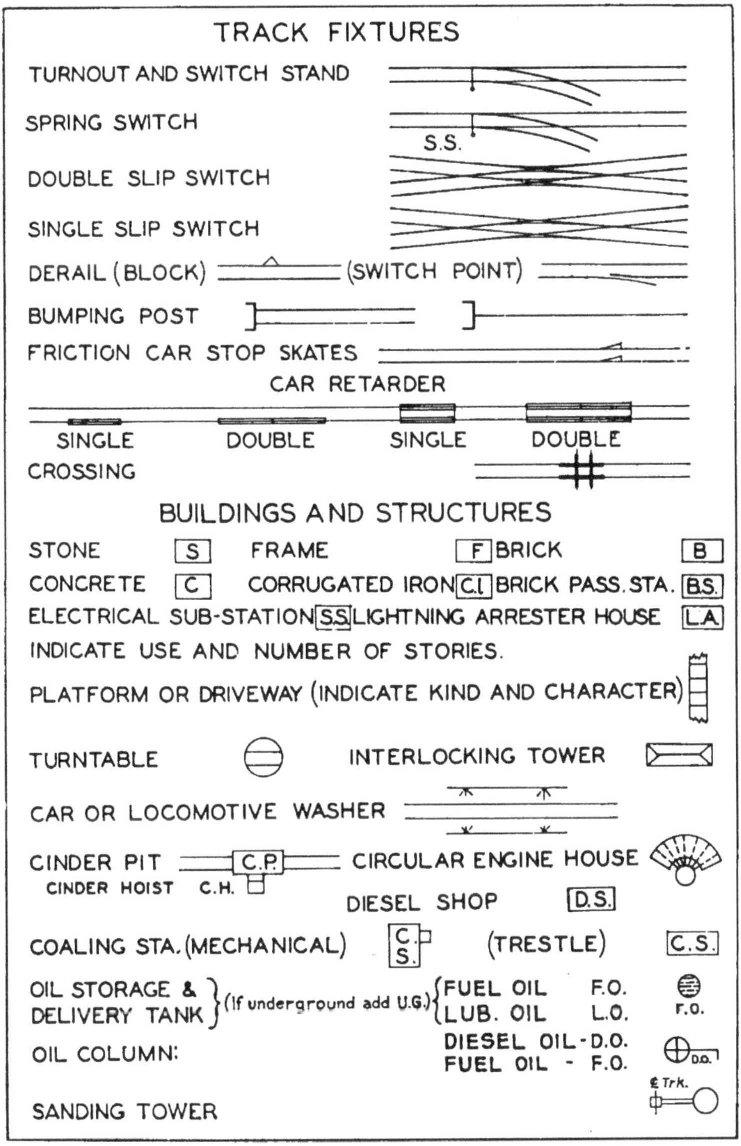

TRACK FIXTURES

TURNOUT AND SWITCH STAND

SPRING SWITCH
S.S.

DOUBLE SLIP SWITCH

SINGLE SLIP SWITCH

DERAIL (BLOCK) ═══════ (SWITCH POINT) ═══════

BUMPING POST

FRICTION CAR STOP SKATES

CAR RETARDER

SINGLE   DOUBLE   SINGLE   DOUBLE

CROSSING

BUILDINGS AND STRUCTURES

STONE   S   FRAME   F BRICK   B

CONCRETE   C   CORRUGATED IRON C.I BRICK PASS.STA. B.S.

ELECTRICAL SUB-STATION S.S LIGHTNING ARRESTER HOUSE L.A

INDICATE USE AND NUMBER OF STORIES.

PLATFORM OR DRIVEWAY (INDICATE KIND AND CHARACTER)

TURNTABLE   INTERLOCKING TOWER

CAR OR LOCOMOTIVE WASHER

CINDER PIT ═══ C.P ═══ CIRCULAR ENGINE HOUSE

CINDER HOIST   C.H.

DIESEL SHOP   D.S.

COALING STA.(MECHANICAL)   C S.   (TRESTLE)   C.S.

OIL STORAGE &
DELIVERY TANK } (If underground add U.G.) { FUEL OIL   F.O.
                                            LUB. OIL   L.O.   F.O.

OIL COLUMN: DIESEL OIL-D.O.
                 FUEL OIL - F.O.   D.O.

                                      ₡ Trk.

SANDING TOWER

Fig. 4.

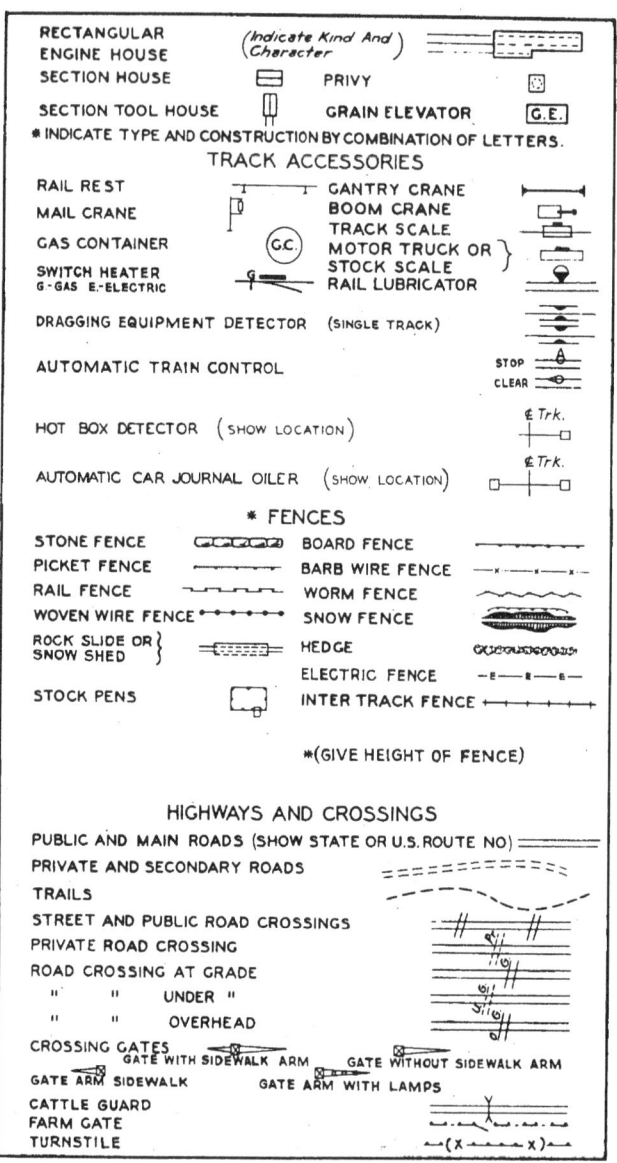

RECTANGULAR   (*Indicate Kind And*)
ENGINE HOUSE   (*Character*)
SECTION HOUSE   PRIVY
SECTION TOOL HOUSE   GRAIN ELEVATOR   G.E.
* INDICATE TYPE AND CONSTRUCTION BY COMBINATION OF LETTERS.

TRACK ACCESSORIES

RAIL REST   GANTRY CRANE
MAIL CRANE   BOOM CRANE
                             TRACK SCALE
GAS CONTAINER   MOTOR TRUCK OR }
                             STOCK SCALE
SWITCH HEATER   RAIL LUBRICATOR
G - GAS E.-ELECTRIC

DRAGGING EQUIPMENT DETECTOR   (SINGLE TRACK)

AUTOMATIC TRAIN CONTROL   STOP
                                             CLEAR

HOT BOX DETECTOR (SHOW LOCATION)

AUTOMATIC CAR JOURNAL OILER   (SHOW LOCATION)

* FENCES

STONE FENCE   BOARD FENCE
PICKET FENCE   BARB WIRE FENCE
RAIL FENCE   WORM FENCE
WOVEN WIRE FENCE   SNOW FENCE
ROCK SLIDE OR }   HEDGE
SNOW SHED   }
                         ELECTRIC FENCE
STOCK PENS   INTER TRACK FENCE

*(GIVE HEIGHT OF FENCE)

HIGHWAYS AND CROSSINGS

PUBLIC AND MAIN ROADS (SHOW STATE OR U.S. ROUTE NO)
PRIVATE AND SECONDARY ROADS
TRAILS
STREET AND PUBLIC ROAD CROSSINGS
PRIVATE ROAD CROSSING
ROAD CROSSING AT GRADE
   "   "   UNDER "
   "   "   OVERHEAD
CROSSING GATES
        GATE WITH SIDEWALK ARM   GATE WITHOUT SIDEWALK ARM
GATE ARM SIDEWALK   GATE ARM WITH LAMPS
CATTLE GUARD
FARM GATE
TURNSTILE

Fig. 5.

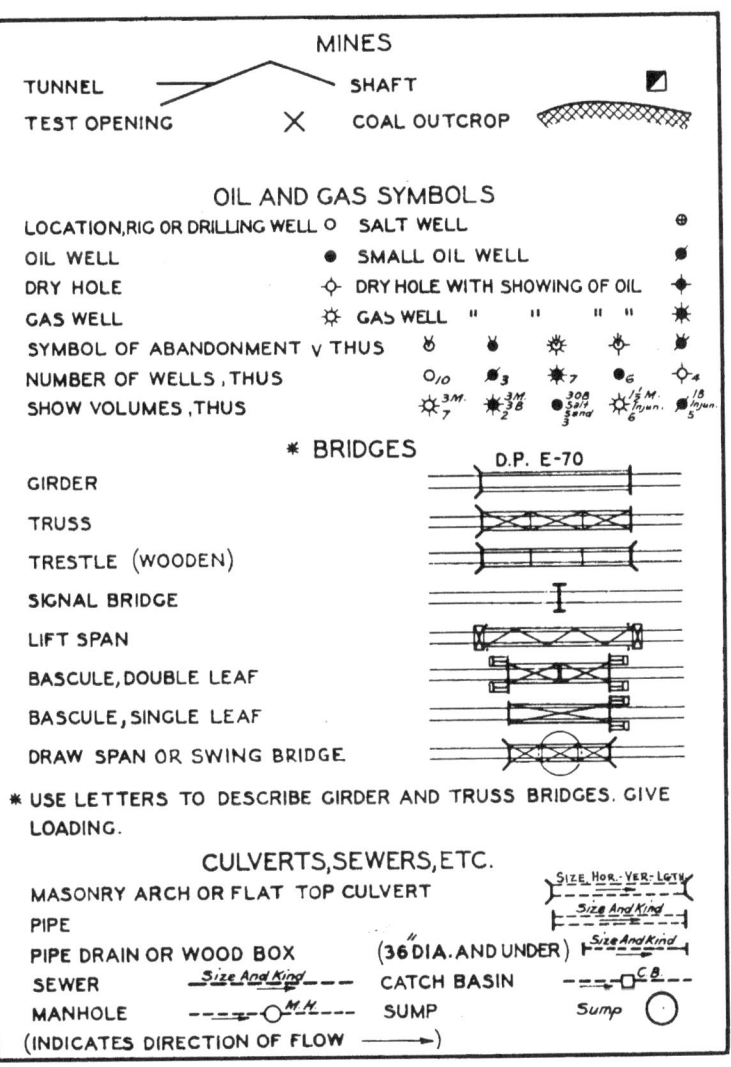

MINES

TUNNEL   SHAFT

TEST OPENING   X   COAL OUTCROP

OIL AND GAS SYMBOLS

LOCATION,RIG OR DRILLING WELL O   SALT WELL
OIL WELL   ●   SMALL OIL WELL
DRY HOLE   ◇   DRY HOLE WITH SHOWING OF OIL
GAS WELL   ☼   GAS WELL "   "   "   "
SYMBOL OF ABANDONMENT ᵥ THUS
NUMBER OF WELLS , THUS
SHOW VOLUMES ,THUS

* BRIDGES   D.P. E-70

GIRDER

TRUSS

TRESTLE (WOODEN)

SIGNAL BRIDGE

LIFT SPAN

BASCULE,DOUBLE LEAF

BASCULE , SINGLE LEAF

DRAW SPAN OR SWING BRIDGE

* USE LETTERS TO DESCRIBE GIRDER AND TRUSS BRIDGES. GIVE
LOADING.

CULVERTS,SEWERS,ETC.

MASONRY ARCH OR FLAT TOP CULVERT   Size Hor.-Ver-Lgth
PIPE   Size And Kind
PIPE DRAIN OR WOOD BOX   (36″DIA. AND UNDER)   Size And Kind
SEWER   Size And Kind   CATCH BASIN   CB
MANHOLE   M.H.   SUMP   Sump
(INDICATES DIRECTION OF FLOW ────►)

Fig. 6.

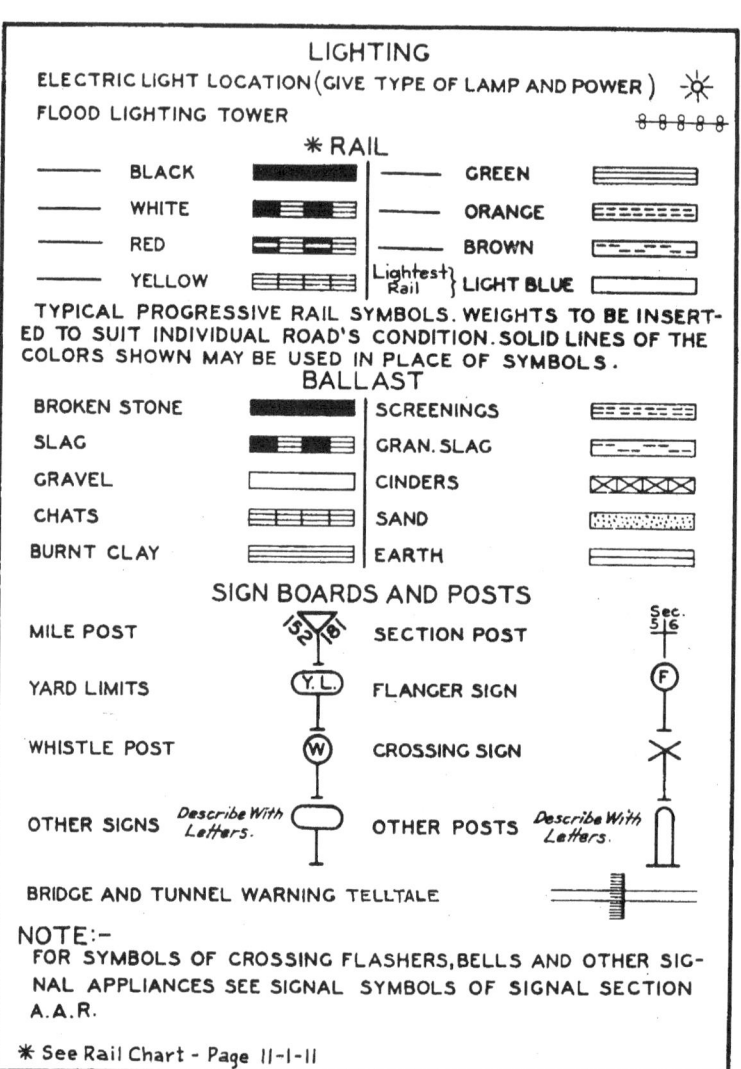

Fig. 8.

Report on Assignment 2

Bibliography on Subjects Pertaining to Engineering and Valuation Records

J. B. Byars (chairman, subcommittee), C. E. Clonts, C. R. Dolan, Morton Friedman, R. F. Garner, H. N. Halper, L. W. Howard, J. A. Houston, R. D. Igou, C. E. Lex, Jr., B. H. Moore, F. H. Neely, C. F. Olson, W. J. Pease, E. J. Rockefeller, W. C. Wieters, H. R. Williams, M. C. Wolf.

Valuation

"Do Railroads Face Heavy Diesel Replacements?", Modern Railroads, June 1960, page 172.

In a paper given at the ASME—AIEE railroad meeting, H. F. Brown of Gibbs and Hill, Inc., declared that railroads will be faced with a serious motive power replacement program in the next five or six years. "Within this period nearly 80 percent of the road motive power now in use will become due for replacement." The "economic life" of road diesel locomotives, based on a combination of average annual depreciation charges plus repair costs, is about 15 years, Mr. Brown declared. This is less than the economic life of either steam or straight electric motive power. "The necessity for more economical types of motive power is a challenge to the locomotive manufacturers and is a problem for serious study on the part of the railroads."

"Uniform System of Accounts for Railroad Companies Proposed Regulations," Part 2, Federal Register, Vol. 26, No. 163, Washington, D. C., Thursday, August 24, 1961.

"Special Office Report: the Decision that Can't Wait", Dun's Review and Modern Industry, September 1961, beginning on page 38.

Struggling in a sea of paper work, battered by mounting clerical costs, business today has no choice but to turn to new advanced tools and techniques of data handling. Help is at hand—but in perplexing variety. Which way should management turn for rescue?

Contents of above article:

Introduction—Why the rush?

  I. March of the "Marvel Machines"
    To streamline office work manufacturers are turning out a glittering array of efficient new equipment.

  II. The Customer's Dilemma
    Bogged down in obsolete procedures, he needs help in a hurry But how does he pick the tools for the job.

  III. It's the Results that Count
    These five companies knew what they needed, invested in the right system, and realized a solid payoff.

  IV. The Invasion from Abroad
    Once a rarity on the American office scene, foreign equipment has established a broadening beach head.

  V. The Fight for Inner Space
    With costs of office space mounting, management must battle to get more production from every square foot.

"The Tax 'Cannibalization' of Our Railroads", Modern Railroads, February 1961, page 65.

With wholly unrealistic tax write-offs and inflation devouring plant faster than it can be replaced, the railroad industry's need for depreciation tax reform is desperate.

"What Support for Appraisers' Opinion?" Clients Service Bulletin of the American Appraisal Company, page 2, Vol. 37, No. 6, of Christmas 1960 issue.

Where property values are in dispute, the court will consider the testimony of qualified expert witnesses, but weight accorded the testimony will hinge largely on the supporting evidence which each submits.

Depreciation

"Depreciation and the Railroad Industry," Railway Materials and Equipment, May–June 1961, page 15.

"The railroad industry has addressed a four-part plea to Congress and government authorities to remedy long-standing legislative inequities in the transportation industry. The request covered: freedom from discriminatory regulation; from discriminatory taxation; freedom from subsidized taxation and freedom to provide a diversified transportation service."

"A Bill to Amend the Internal Revenue Code of 1954 as Amended," Senate Bill S–1370; House of Representative Bill 231, January 3, 1961.

"Why current Values?" Clients' Service Bulletin of the American Appraisal Company, page 2, Vol. 37.

In that long-ago period prior to the income tax, property records were generally very loosely handled, and charges to the depreciation reserve were often based on expediency rather than on fact . . .

"The Engineer's Role in Property Accounting," Clients' Service Bulletin, American Appraisal Company, October 1961, page 2; also, the American Engineer, September 1960.

The engineer is in a position to play an important part in the field of property and depreciation accounting.

Office Procedures

"The Computer", The Saturday Evening Post, November 4, 1961, page 25 and 72 through 73, based on an article by J. R. Pierce, Ph.D., executive director of research for the Communications Principles Division of the Bell Telephone Laboratories.

A computer can be compared to an adding machine in an elementary way. Numbers are represented by means of little wheels. Each wheel can be in any of ten positions, exposing one of the digits, 0 through 9. A computer, however, makes use of electronic devices—wire passing through magnetic rings, or "cores" which can be magnetized in the direction of one pole or the other; magnetic tapes on which a small region can be similarly magnetized; transistor or vacuum tube circuits which can be either "off" or "on"; wires in which a current can flow in one direction or the other; and perforated cards or tapes which activate the machine.

A computer is equipped with a "memory" made up of groups of magnetic cores. Each core of a group can store one digit of a binary number, by being magnetized in one direction for a 0 or in the other direction for a 1. The digit can be read into a read-out of the core by passing a current through a wire which threads the core. The digit remains in the core (that is, it is memorized) until expunged by a subsequent suitable current.

The first requirement in using a computer is to put appropriate binary numbers into the core memory. This is usually done in a sequence of steps. The numbers are first written on a sheet of paper, punched into cards, and transcribed to magnetic tape, from which the computer reads them into its memory. At the end of a computation the results, having been stored in the computer's memory, are printed out on magnetic tape.

The computer can use the numbers in its memory in two ways. Some numbers represent the data to be processed, and the computer stores the intermediate and the final results of its computations in its memory. Other memory positions are reserved for numbers representing commands to act upon data.

One might think that a computer would be limited because it deals only with numbers. The number, however, can stand for letters, words, mathematical signs, musical notes.

The principal limiting factor of the computer is human direction in the form of a program which will guide the machine in a given task. Thus programming is the greatest intellectual challenge in the world of computers. Without a program, a computer is a senseless heap of complexity. The 'how" must come from the human mind.

"The Engineer's Rule in Property Accounting", Clients Service Bulletin, American Appraisal Co., October 1961, page 2, based on an article by Maurice E. Peloubet, published in the September 1960 issue of the Amreican Engineer.

"The engineer is in a position to play an important part in the field of property and depreciation accounting." The author stresses the qualifications of the engineer to estimate the effects of technological improvements and the probable time when the machinery and equipment will become obsolete, and points out that the engineer is best suited to determine the extent of physical wear and tear and the type of machine which should be purchased to replace the one which is not obsolete or worn out.

It is pointed out that as general rule the financial and accounting officials and consultants administer depreciation policies and methods. This has a tendency to obscure the facts, so far as the technical and operating features of depreciation are concerned. They are laymen and should rely on engineering advice for the basic information concerning the determination of useful lives, choice of depreciation methods and the replacing of worn-out or obsolete machinery and equipment.

The author also adds that "Statements by Administrative or Accounting Executives . . . will not have the weight with the Treasury (or) Internal Revenue Service . . . that a soundly supported engineering estimate will have."

Such items as, The Problem of Property Control, Physical Property Checks and Determination of Current Value are also discussed by the author in his article.

"Topographic Symbols", The Department of the Army Field Symbols, 1961, 102 pages, illustrated, Catalog No. D–101.20:21–31½.

This Department of Army Field Manual contains general information on topographic maps and symbols; gives examples and illustrations of topographic symbols arranged by categories, such as drainage features, relief features, and roads; lists topographic abbreviations; and discusses marginal information.

Report on Assignment 3

Office and Drafting Practices

E. W. Smith (chairman, subcommittee), B. J. Cook, W. A. Krauska, H. B. Sampson, R. L. Samuell, R. S. Shaw, Jr., J. R. Traylor, W. C. Wieters.

Your committee has carefully reviewed all Manual material in Part 4—Office and Drafting Room Practices, of Chapter 11, and made recommendations for changes to Subcommittee 1 on Revision of Manual, which is reporting the changes.

The following is presented for information:

Reproduction of Drawings, Statements and Correspondence

While there is a wide variation in the type of equipment used to reproduce drawings, statements and letters, as well as considerable difference in the purpose for which it is used, there is a trend toward more extended use of this type of equipment in both chief engineers' and division or district engineers' offices. Various manufacturers are endeavoring to improve the equipment to make it more versatile, and ball-point pen manufacturers are being urged to use ink that will reproduce.

Other reproducing processes, principally of the black-or blue-line chemical type, are being used to reproduce drawings to a greater extent than blue printing. Some roads no longer use blue prints. Also, some use is being made of roll film, photographic and microfilm equipment for both maps and documents.

Paster Title Block and Legend

Some roads use a rubber stamp or paster title legend such as shown below:

LEGEND

For Company Owned Tracks

	Symbol	Color
Existing tracks to remain	————	Black, solid
Existing tracks to be removed	x-x-x-x	Yellow, solid
Existing tracks to be lined	- - - -	Yellow, dashed
New location of tracks to be lined	▪ ▪ ▪ ▪	Red, dashed
New tracks to be constructed	————	Red, solid
Proposed future tracks	o • • • o	Black symbol

For Foreign Line or Private Tracks

	Symbol	Color
Existing tracks to remain	+-+-+-+	Black, hatched
Existing tracks to be removed	x-+-x-+-x	Yellow, hatched
Existing tracks to be lined	+ + + + +	Yellow symbol
New location of tracks to be lined	✦ ✦ ✦ ✦	Red symbol
New tracks to be constructed	⊢+⊢+⊢+⊢	Red symbol
Proposed future tracks	+-+-+-+	Black symbol

Some roads find it more economical to use a paster title block when a portion of existing map or tracing is used—similar to the example shown below—which can be revised to suit any requirement. For complete identification, it may be desirable to provide space for valuation section, map number and/or date; or this information as well as the mile post location and file reference, may be shown in blank spaces.

NORTH AND SOUTH RAILROAD

——————————— Division ——————————————— Sub Division

PRINT SHOWING

_____

_____

AT

_____

Scale 1" = _____ _____ 19 ____

Office of _____ ___ ___ _____ Chicago, Ill.

Report on Assignment 6

Valuation and Depreciation

(a) Current Developments in Connections with Regulatory Bodies and Courts

C. R. Dolan (chairman, subcommittee), R. B. Aldridge, G. R. Berquist, H. T. Bradley, J. B. Byars, J. R. Clayton, C. E. Clonts, S. Danby, F. H. DeMoyer, R. L. Ealy, A. L. Engwall, R. F. Garner, W. S. Gates, Jr., E. W. Gibson, H. N. Halper, G. J. Harris, M. J. Hebert, J. W. Higgins, L. W. Howard, R. D. Igou, C. B. Martin, B. H. Moore, C. F. Olson, C. H. Rapp, Jr., H. L. Restall, C. S. Robey, E. J. Rockefeller, J. B. Styles, H. R. Williams, M. C. Wolf.

ICC Bureau of Accounts

The Section of Valuation was engaged principally during the year in railroad and pipeline work, preparing tentative and final valuations for pipeline carriers. During October statements were completed showing the elements of value for all Class I line-haul carriers and switching and terminal companies as of December 31, 1960.

During the year 1960, Class I line-haul carriers charged Account 459—Valuation Expenses, $824,352 contrasted with $850,565 for the year 1959.

As of October 1, 1961, pipeline and rail carriers were delinquent in the filing of valuation reports with the Bureau, as follows: 1 switching and terminal company for 1957; 8 line-haul and 1 switching and terminal company for 1958; 24 line-haul and 5 switching and terminal companies for 1959. Valuation reports covering property changes for the year 1960 have been received from 34 line-haul and switching and terminal companies and 68 pipeline carriers.

The backlog of valuation work (rail and pipeline carriers) in the Bureau as of October 1, 1961, was as follows:

	Annual Valuation Reports
Engineering inventories	1,756
Present value of land and rights	810
Original cost summaries	18

The total authorized personnel for valuation work in the Section of Valuation on October 1, 1961 was 32.

During the year the Commission released the Schedule of Annual Indices for Carriers by Railroad, and the Schedule of Annual and Period Indices for Carriers by Pipeline for the year 1960.

Report of the Committee on Valuation,
National Association of Railroad and Utilities Commissioners

The 1961 report of the committee on valuation was presented to the annual convention of the Association in September. The complete report consists of 10 printed pages and may be secured from the Association offices, P. O. Box 684, Washington 4, D. C., at 50 cents a copy. A summary of the report follows:

The introduction reviews the 1961 condition of business activity, the trend of prices, employment levels and states:

"The difficult and imponderable circle of valuation, rates, and rates of return continues to be the problem of utilities and commissions alike. Albeit, the daring and resourcefulness of the parties to rate base determination continues apace. Attrition of value probably was not the same problem in 1960–1961 as in past years, but the updating of rate bases still remains as a problem in rate base determination."

A summary is given of the uses of original cost versus reproduction cost in determining rate bases in cases arising during the year. The report finds:

"The pressures of inflation and attrition influenced more commissions to consider reproduction cost as a basis of valuation than in recent years. Although available information on cases in 1961 would indicate that original cost still dominates the rate base picture. Again, fair value was emphasized as the base for valuation. On the other hand, more commissions have considered the problems of attrition in more pragmatic terms. Commissions in California, Colorado, Hawaii, Idaho, Kansas, Oregon and Washington utilized original cost. Investment in the sense of appraised value was not accepted. (Re: Fresno City Lines, Decision No. 59791, Application No. 41559, March 15, 1960.) In a Colorado case, original cost less depreciation was the primary means of valuation, but areas of judgment were implicitly recognized. (Re: Public Service of Colorado (1960) 34 PUR3d 181). Again, references were made to the value of historical figures derived from regulated accounting procedures.

"Reproduction cost as a measure of value was denied in eight of nine cases brought before federal and state commissions. Specifically, the Federal Power Commission turned down the application of the Panhandle Eastern Pipeline Company for consideration of "replacement cost" of the properties of a natural gas pipeline company since there is no requirement for the Federal Power Commission to consider this approach in determining fair value (Re: Panhandle Eastern Pipeline Co. (1960) 23 FPC 352.) In a Kansas case, the reasons for not allowing reproduction cost were traditional, including the conjectural nature of reproduction cost, the opinions and observations of individuals, and the fallaciousness of assuming that the plant could be reproduced as it presently exists. (Re: Southwestern Bell Telephone Co. (1960) 34 PUR3d 257.) Other commissions which turned down reproduction cost included New Jersey, Louisiana, North Carolina and Oregon. Reflecting a historical affinity for reproduction cost, the Indiana Public Service Commission accepted reproduction new less accrued depreciation as an element in determining a fair value rate base. (Re: Hoosier Water Co. (1960) 34 PUR3d 348; Re: Indiana Gas and Water Co. (1960) 35 PUR3d 43; Re: Indiana Water Corp. No. 28233, March 28, 1960.) Reproduction cost in valuation proceedings still seems to be generally unacceptable except where cost factors are easily agreed upon."

Continuing, the report points out that the use of trended cost estimates were rejected by most commissions in cases arising during 1961. This device has been adopted by a number of state commissions using the original cost approach to allow something for the effects of inflation. It is concluded:

"Apparently, the use of index numbers and trended original cost for valuation to soften the effects of inflation and attrition will be used judiciously by the commissions that have allowed its use in the past. The problems of measuring what might have been done, agreement on the base period, the items to be averaged, and the problem of measuring operating efficiency still remain and the commissions are well aware of them."

The report continues, outlining the difficulties in determining a proper rate base and income account, caused by the use of accelerated depreciation and liberalized depreciation for federal income tax purposes and lists a number of cases in which adjustments had been made for these items.

The report concludes with the following summary:

"The baffling problems of cost recovery for the rendering of service still remains with the regulated industry and the commissions that supervise their activity. The state of flux in valuation regulation has existed from Smyth v. Ames (1898) to the Federal Power Commission v. Hope Natural Gas Co. (1944) and continues to be a problem today. At the time of the Hope case, the commissions were unleashed and at the same time the regulatory job became more difficult since the conundrum of rates, rate-making, and valuation had to be analyzed by regulatory commissions to maintain the financial integrity of utilities, to attract capital, and to compensate investors for risk. The 'fairness' test may be dead on the federal level, but it is very much a part of the analysis of state commissions in valuation. Original cost less depreciation dominates the valuation picture today in most state commissions. Reproduction cost and/or replacement cost are rarely accepted as rate bases since there is great uncertainty in the valuation process. Original cost always has the advantages in rate hearings today.

"Much work has been done in the area of index numbers and the establishment of trends in original cost data. A few state commissions have indicated limited acceptance of the concept but the problems of averages, the data chosen for the base, the lack of actual investment, and the approximations of reproduction cost, new, have been cited as reasons for refusing this method of valuation. Attrition has been recognized by a few commissions, and rates of return have been the primary factor adjusted.

"A myriad of problems still confront commissions, including accelerated depreciation, working capital requirements, and construction and acquisition values. As stated, these problems are met on a pragmatic basis by companies and commissions. Inflation and attrition still confront the companies and commissions. The constancy of the price level and the close approximation of wage increases and productivity will contribute the slowdown in the decline in real terms of the rate base. Although commissions must and do implicitly recognize the effects of inflation on the rate base, temporary means of adjustment in rates provides an answer which does not commit the commissions to deep-seated changes in determination of value. Increased efficiency, outside competition in energy production, increased scale of plant, and more research and development should be taken into consideration in valuation. The existence of the utility has been based on normalcy and monopoly, but this does not mean that rapidly changing conditions cannot be met and solutions found by the companies and the commissions."

Report on Assignment 7

Revisions and Interpretations of ICC Accounting Classifications

M. M. Gerber (chairman, subcommittee), C. E. Clonts, C. R. Dolan, A. L. Engwall, W. S. Gates, Jr., W. M. Hager, C. B. Martin, C. W. Meyer, B. H. Moore, W. J. Pease, C. H. Rapp, F. A. Roberts, C. S. Robey, H. B. Sampson, J. R. Traylor.

This is a progress report, presented as information.

The following ICC subjects which were reported in the 1960 and 1961 Proceedings have been further revised, and the consideration by the ICC and the General Committee, Accounting Division, AAR of subjects 467 and 471 was concluded and the changes incorporated in the revision of the Uniform System of Accounts.

Subject No. 467—Acquisition of Railway Operating Property, was last revised May 29, 1961, so that the proposed changes cover only the following:

    Item   I. Acquisition of a Railway Operating Entity or System.
             (a) Merger or consolidation constituting a pooling of interests.
             (b) Acquisition resulting from a purchase other than from subsidiaries.
    Item II. Reorganization.
    Item III. New account 80—Other Elements of Investment.
    Item IV. Cancellation of Account 733—Acquisition Adjustment.

ICC Subject No. 468—Redistribution of Amounts to Primary Road and Equipment Accounts, will receive further consideration from the Commission with respect to differences between balances in the property accounts of the carriers and the Commission's cost records.

ICC Subject No. 471—Rearrangement and Revision of the General and Special Instructions in the Uniform System of Accounts for Railroad Companies.

These instructions were made effective January 1, 1961, under a proposed Order of Rule Making dated October 21, 1960, Docket 32153: This was later cancelled by ICC Order of December 2, 1960.

There appeared in the Federal Register, Part II, dated August 24, 1961, an Interstate Commerce Commission Notice of Proposed Rule Making, covering a revised issue of the Uniform System of Accounts for Railroad Companies (to be known as the Issue of 1962).

ICC proposes tentatively that this revised issue of the Uniform System of Accounts be effective January 1, 1962, subject to further action after consideration of the views and comments of interested parties, submitted to the Commission not later than October 9, 1961.

No objections being filed, the ICC by its order of November 8, 1961, adopted a revised issue and published it in the Federal Register of November 25, 1961. Copies were served on each railroad company. The ICC is arranging for printed bound copies which may be obtained from the Superintendent of Documents, Government Printing Office, Washington, D. C., at a nominal cost.

Report of Committee 7—Wood Bridges and Trestles

K. L. DeBlois, *Chairman*

B. E. Daniels,
Vice Chairman

D. V. Sartore

R. E. Kuehner
W. L. Anderson
A. L. Leach
J. A. Gustafson
L. R. Kubacki
C. V. Lund
J. F. Holmberg
W. A. Genereux
R. E. Anderson
C. E. Atwater
C. J. Barhydt
W. W. Boyer
T. P. Burgess
A. W. Carlson
J. W. Chambers
H. M. Church (E)
F. H. Cramer (E)
E. M. Cummings
D. J. Engle
J. T. Evans
S. L. Goldberg
S. F. Grear (E)
R. W. Gunther

W. A. Hamilton
F. J. Hanrahan
W. C. Howe
R. H. Hunsinger
Milton Jarrell
W. D. Keeney
W. B. MacKenzie
F. W. Madison
L. J. Markwardt
W. H. Martin
E. A. Matney
T. K. May
J. W. N. Mays
D. H. McKibben
C. H. Newlin
W. H. O'Brien
W. A. Oliver
F. E. Schneider
J. G. Shope
J. D. Tapp, Jr.
D. L. Walker
Lyman W. Wood
Committee

(E) Member Emeritus
Members listed in bold face are the official representatives of the Engineering Division, AAR.

To the American Railway Engineering Association:

Your committee reports on the following subjects:

1. Revision of Manual.

   Progress report, submitted for adoption .... page 452

2. Grading rules and classification of lumber for railway uses; specifications for structural timber, collaborating with other organizations interested.

   A new set of grading rules for hardwood structural timbers is being written. The committee expects to have these rules completed and the tables referring to hardwoods now in the Manual revised so that recommendations can be submitted for adoption and publication in the Manual next year.

3. Specifications for design of wood bridges and trestles.

   No report. This committee is working with Subcommittee 1 to review existing Manual documents on design.

4. Methods of fireproofing wood bridges and trestles. including fire-retardant paints, collaborating with Committees 3 and 6.

   No report due to lack of funds for research.

5. Design of structural glued laminated wood bridges and trestles. No report.

   The committee is making progress on plans and tables for laminated girders and trestle members.

6. Applications of synthetic resins and adhesives to wood bridges and trestles, collaborating with Committees 8 and 15.

   The report published as information in Bulletin 562, January 1961, is being revised.

7. Repeated loading of timber structures.

Laboratory tests were completed during the past year on 24 glued laminated Douglas fir bridge stringers in static and repeated loading. These tests
were conducted principally to explore the relationship of the position of
wheel loads to strength in horizontal shear and constitute part of a more
comprehensive program of tests on sawn and glued laminated stringers
being undertaken in cooperation with the American Institute of Timber
Construction, the National Lumber Manufacturers Association, the American Wood Preserver's Institute and the Forest Products Laboratory,
United States Department of Agriculture. A second series of tests using
southern pine stringers is in progress.

8. Protection of pile cut-offs.

The committee is continuing to investigate various physical tests as conducted by others. It is also considering the use of epoxy resins, but is not
prepared to make a report at this time.

9. Hardening of timber bearing surfaces.

The committee is investigating previously developed information on this
assignment. It is also considering the use of epoxy resins, but is not prepared to present a report.

10. Rules for rating existing wood bridges and trestles.

Progress report, submitted for adoption .... page 456

The Committee on Wood Bridges and Trestles,

K. L. DeBlois, Chairman.

AREA Bulletin 569, January 1962.

Report on Assignment 1
Revision of Manual

D. V. Sartore (chairman, subcommittee), C. E. Atwater, C. J. Barhydt, W. W. Boyer,
J. W. Chambers, B. E. Daniels, W. A. Hamilton, R. H. Hunsinger, Milton Jarrell,
C. V. Lund, W. B. Mackenzie, F. W. Madison, W. H. Martin, E. A. Matney,
J. W. N. Mays, W. H. O'Brien, W. A. Oliver, J. G. Shope, J. D. Tapp, Jr.

In accordance with the request made by the secretary's office two years ago that
all Manual material be reviewed prior to reprinting the Manual in 1962, your committee has continued its review of Chapter 7, and offers the following recommended action
with respect to the various documents.

This completes the review of all documents except those that will require revision
when information from studies now in progress becomes available.

GLOSSARY

Your committee offers the following recommendations with respect to certain terms
in the Glossary that are the responsibility of Committee 7.

Page 10. Delete "Disk Pile" and its definition.

Page 14. Under definition of "Follower", delete the words "when below the foot
of the leads."

Page. 17. Revise the definition of "Hammer, Pile" to read: "A weight used to drive piles. It may be designated as a steam hammer, diesel hammer or drop hammer, depending on the source of the energy."

Page 17. Revise the definition of "Helper Stringer" to read: "A stringer added to an existing panel of stringers."

Page 23. Revise definition of "Pile Hammer" to read: "A weight used to drive piles. It may be designated as a steam hammer, diesel hammer or drop hammer, depending on the source of energy."

Page 26. Delete "Ring" and its definition.

Page 30. Delete "Steam Hammer" and its definition.

Page 30. Delete "Straight" and its definition.

Page 37. Revise the definition of "Wood Trestle" to read: "A wood structure composed of bents supporting stringers, the whole forming a support for loads applied to the stringers through the deck."

Pages 7–3–10 to 7–3–13, incl.

INSTRUCTIONS FOR INSPECTION OF TIMBER TRESTLE RAILWAY BRIDGES

Reapprove with the following revisions:

Page 7–3–13. Under Sec. D. Notes on recommended practices, delete "(a) Safety should be the first consideration," and reletter pars. (b) through (h) as (a) through (g).

Pages 7–3–13 to 7–3–16, incl.

METHODS OF FIREPROOFING WOOD BRIDGES AND TRESTLES

Reapprove with the following revisions:

Page 7–3–14. In the second paragraph under Item 4. *Fire Alarm Systems* replace the words "should be" with the word "are".

Rewrite Item 5. Housekeeping to read as follows:

5. *Housekeeping*—(The following practices, applicable to both open- and ballasted-deck bridges and trestles, are being employed where conditions warrant):

    a. Locomotives are equipped with adequate and well-maintained spark arrestors and tight-fitted ash pans.

    b. Decks are kept clear of all combustible material, and decayed spots in exposed ties or timbers kept trimmed.

    c. Brush and weeds are kept down for a distance of at least 25 ft from the bridge, both underneath and on the embankment at the ends of the bridge or trestle. Also, all sod is removed from under timber bridges and for a distance of 3 ft outside the timbers. This is accomplished by scalping or by the use of a soil sterilant.

    d. Water barrels with buckets are installed on timber bridges, 1 barrel each for structures up to 50 ft long and 1 additional barrel for each additional 150 ft or fraction thereof. For creosoted structures, sand boxes with water-tight covers for keeping the sand dry are used, dry sand being more effective than water in extinguishing small fires on creosoted structures.

Rewrite Item 6. Fire Barriers, to read as follows:

6. *Fire Barriers*—(Applicable to both open and ballasted-deck bridges and trestles) Under this method long bridges and trestles are protected by introducing fire barriers at intervals of about 400 ft. This reduces the hazard by preventing loss of the entire structure in case of fire. Such barriers may be grouped by types of construction, as follows:

 a. Earth fill (see Fig. 1).
 b. Reinforced concrete piers or concrete pile bents (see Fig. 2).
 c. Facing bents with fire-resisting materials (see Fig. 3).
 d. Application of mastic materials to open-deck structures (see Fig. 4).

Page 7–3–17

USE OF GUARD RAILS AND GUARD TIMBERS

Rewrite as follows:

1. On all open-floor railway bridges, the ties should be held securely in their proper spacing; guard or spacer timbers fastened to every tie near its ends are effective. If such continuous timbers are not placed, blocks or other suitable fastenings should be used for spacer timber attachment; on track where speed or other circumstances so indicate it may be advisable also to embed clamping plates or timber connectors between the timbers and ties. Such metal fastenings are more effective than dapping of the spacer timbers, because of the tendency of the wood to split off between daps.

2. Consideration should be given to the use of metal inner guard rails at the following locations, taking into account the alinement, train speed, density and type of traffic, as well as height and length of bridge.

 a. On all through-span bridges.
 b. On all open-floor-deck bridges.
 b. On all tracks of solid-floor-deck bridges for single or double tracks.
 d. On outside tracks only of solid-floor-deck bridges with three or more tracks, unless type of deck construction indicates that special consideration should be given to use on inside tracks also.
 e. On any track the nearer rail of which is within 10 ft of support of an overhead or adjacent structure.
 f. In tunnels.

It is recommended that the inner guard rails, when used, be steel track rails not higher than the running rails. If 5 in or more in height they should not be more than 2 in lower than the running rails. If less than 5 in. in height they should not be more than 1 in lower than the running rails. Normally, they will consist of two rails, spaced about 10 in inside the running rails (measured between near sides of heads) spiked to every tie and spliced with joint bars, fully bolted. The inner guard rails may be tie plated when deemed advisable. They must not contact tie plates of tracks carrying electric signal circuits. Where they protect against a hazard on one side only, a single line of rails may be used, adjacent to the running rail further from the hazard.

It is further recommended that where inner guard rails are used, they extend at least 50 ft beyond the end of the bridge or other structure. This distance may be increased where train speed, curves or other factors warrant the increase, and may be decreased on the leaving end where traffic is in one direction. The ends should run to the center of track and be beveled, bent down or otherwise protected against direct impact. A filler block or plate should be provided at the meeting of the converging rails.

3. Where both guard timbers and inner guard rails are used they should be so spaced that a derailed truck will strike the inner guard rail and not the timber.

Pages 7–4–1 to 7–4–20, incl.

PLANS FOR OPEN-DECK PILE AND FRAMED TRESTLES, MULTIPLE-STORY TRESTLES, AND BALLASTED DECK PILE AND FRAMED TRESTLES

Reapprove with the following revisions:

Page 7–4–2. Under "General Notes" replace the word "shall" with "should" wherever it appears in these notes.

Pages 7–4–3 through 7–4–20. Replace the word "shall" with "should" wherever it appears on Figs. 3, 7, 12, 13 and 15.

Page 7–4–12. Change "Fig. 19" to read "Fig. 10."

Pages 7–4–13 through 7–4–18. Add the following note to Figs. 11, 12, 13, 14, 15 and 16: "To be used where required."

Pages 7–M–4 to 7–M–5.1, incl.

RECOMMENDED PRACTICE FOR OVERHEAD WOOD HIGHWAY BRIDGES

Rewrite as follows:

It is recommended that the current issue of the U. S. Department of Commerce, Bureau of Public Roads, publication entitled "Standard Plans for Highway Bridge Superstructures" be used as a guide for overhead wood highway bridges. Clearances, substructures and details should be in accordance with individual railroad practice.

Delete Figs. 1 and 2 on Pages 7–M–5 and 7–M–5.1, respectively.

Pages 7–M–7 and 7–M–8

RECOMMENDED PRACTICE FOR DESIGN OF WOOD CULVERTS

Reapprove with the following revision:

Page 7–M–8. Replace the word "shall" with "should" wherever it appears in the General Notes on the drawing.

Report on Assignment 10

Rules for Rating Existing Wood Bridges and Trestles

W. A. Genereux (chairman, subcommittee), W. L. Anderson, R. W. Gunther, F. J. Hanrahan, M. Jarrell, J. R. Kelly, L. R. Kubacki, R. E. Keuhner, A. L. Leach, C. V. Lund, W. B. MacKenzie, F. W. Madison, W. H. Martin, E. A. Matney, C. H. Newlin, D. V. Sartore, J. D. Tapp.

Last year your committee presented as information a tentative specification to cover this subject (Proceedings, Vol. 62, 1961, pages 539–542) and invited comments and criticisms thereon. The specification has been reviewed in accordance with comments received and is now presented for adoption and publication in the Manual, at the end of Part 2, Chapter 7, with the following corrections:

On Proceedings page 541, change the tabulated material under Art. 13. Unit Stresses, to read as follows:

	For Equipment or Locomotives Not Regularly Assigned	For Regularly Assigned Equipment or Locomotives
E = modulus of elasticity, in thousands of pounds per square inch.	As shown in Table 1, Specifications for Design of Wood Bridges and Trestles, Part 2, this Chapter.	
f = unit stress in extreme fiber in bending, in pounds per square inch	$1.3\,k\,F_b$	$1.1\,k\,F_b$
All other unit stresses	$1.3\,k$	$1.1\,k$

Change the third paragraph from the bottom of the page to read as follows:

The unit stress in horizontal shear shall not exceed the maximum shown in Table 1, Specifications for Design of Wood Bridges and Trestles, Part 2, this Chapter, for the highest grade of the species under consideration.

Report of Committee 18—Electricity

P. B. Burley, *Chairman*

J. J. Schmidt,
Vice Chairman

W. O. Muller
R. C. Welsh, Jr.°
W. R. Preece, Jr.
E. B. Hager
F. T. Snider
L. B. Curtis
E. M. Hastings, Jr.
B. D. Allison
B. Anderhous
R. J. Berti
L. W. Birch
W. F. Bowers
H. F. Brown
K. A. Browne
Robert Burn
F. J. Corporon
A. B. Costic
H. C. Cross
H. H. Duehne
H. W. Dunn
D. F. Dunsmore
E. D. Feak
W. B. Grimes

B. C. Hallowell
G. B. Hauser
R. E. Hauss
R. B. Hendrickson
W. W. Holloway
R. H. Holmes
T. F. Jelnick
R. L. Kimball
P. O. Lautz
D. R. MacLeod
F. B. McConnel
B. F. McGowan
A. B. Miller
J. J. Miller
H. R. Morgan
R. F. Pownall
E. B. Shew
R. H. Stocksdale
C. A. Stokes
C. M. Summers
E. H. Werner
M. I. Yasuna

Committee

° Died October 15, 1961.

Members listed in bold face are the official representatives of the Engineering Division, AAR.

To the American Railway Engineering Association:

Your committee reports on the following subjects:

1. Revision of Manual, collaborating with the Mechanical Division, AAR.

   Recommendations with respect to the AAR Electrical Manual, submitted for adoption .... page 458

4. and 8. Power supply, motors and controls, collaborating with the Mechanical Division, AAR.

   Due to the death of R. C. Welsh, Jr., chairman of the subcommittee, on October 15, 1961, there is no report. The work of the subcommittee is being progressed under a new subcommittee chairman.

5. Illumination, collaborating with Committee 6 and the Mechanical Division, AAR.

   An investigation of outdoor area lighting has been progressed, dealing primarily with railroad yards. It is expected that a report on this subject will be completed this year and offered as Manual material.

9. Electrolysis and electrolytic corrosion.

   Work is in progress on: (1) methods in use to protect metal structures against corrosion, (2) developments in corrosion-resisting materials, and (3) development of possible effects of cathodic protection installations for underground structures on adjacent railroad signal systems.

10. Wire, cable and insulating materials, collaborating with the Mechanical
    Division, AAR.

    Progress report, submitted as information .... page 460

13. Railway electrification, collaborating with the Mechanical Division, AAR.

    Progress report, submitted as information .... page 464

15. Relations with public utilities, collaborating with Committee 20.

    Considerable progress has been made on the preparation of a schedule of
    fees and rentals for occupancy of railway property by lines of 7500 volts
    and less. It is expected that the schedule will be offered for consideration
    as Manual material next year.

                                        THE COMMITTEE ON ELECTRICITY,
                                            P. B. BURLEY, *Chairman.*

    ─────────
    AREA Bulletin 569, January 1962.

MEMOIR

Robert C. Welsh, Jr.

Robert C. Welsh passed suddenly away on October 15, 1961, after a short period
of illness, as a result of a cerebral hemorrhage caused by a duodenal ulcer. He is survived by his wife, Mary Emma; daughters Patricia A. and Nancy L.; and sons Robert
C. III and John F. He also had one grandchild. To them, the members of Committee
18 and the AREA express their sincere sympathy.

Mr. Welsh was born in Harrisburg, Pa., on January 20, 1907. He was a graduate
of Lehigh University with the degree of B.S. in E.E. He entered Pennsylvania Railroad
service on February 1, 1930, and was general electrician in the Office of the Electrical
Engineer at the time of his death.

In 1943 he became active in the AAR Electrical Section of the Mechanical Division.
In 1952 he continued his activity as a member of the reorganized Electrical Section of
the Engineering and Mechanical Divisions. In 1955 he was appointed Chairman of Committee 8—Power Supply. When the Electrical Section was discontinued at the end of
1960, he became a member of Committee 18—Electricity of the AREA and was appointed Chairman of Subcommittee 4 and 8—Power supply, motors and controls. He
was in the process of completing his first report in his new capacity at the time of
his death.

Bob Welsh will be keenly missed and remembered with great affection by his host
of friends both within and without the railroad industry.

Report on Assignment 1

Revision of Manual

W. O. Muller (chairman, subcommittee), H. W. Dunn, T. F. Jelnick, J. J. Schmidt.

Your committee submits the following recommendations with respect to the AAR
Electrical Manual:

SECTION 14—SAFETY

Delete in its entirety,

SECTION 11—ELECTRIC HEATING

Revise as follows:

Chapter 2, Part 4, Specification for Tubular
Type Electric Heaters for Track Switches

Page 2–4–1, Par. 2. Basis of Purchase: Add new item reading, "(5) Type of terminal. (Par. 6)." Renumber present items (5) to (9), incl., to (6) to (10), incl.

Page 2–4–2, Par. 6. Terminals: Add sentence reading, "Nonseparable type terminal will be provided if so ordered."

Page 2–4–4, Par. 10. Tests A: Change Par. (e) to (d) and Par. (d) to (e). Par. (f) which was inadvertently omitted from the 1960 edition should stay omitted.

Tests B: In Pars. (a) and (b), change the words "test A (d)" to "test A (e)."

Chapter 2, Part 5, Specification for Installation
of Tubular Type Electric Heaters for Track Switches

Page 2–5–1, Par. 1. Purpose: Change first sentence to read, "This specification applies to the application of tubular electric heaters, supports, and clamps. . . ."

Page 2–5–1, Par. 3. Application of Heaters: In the sixth line of (a) change the word "connector" to "terminal."

Page 2–5–2: In diagram change length of bolt to 1¾ in to agree with text and show heater round with a maximum diameter of 0.5 in. Delete vertical dimension reading, "0.625" max."

Chapter 2, Part 6—Specification for Tubular
Hairpin Type Electric Heaters for Track Switches

Page 2–6–3: Change Par. 9 (b) to read, "The attachment bolts shall be ⅜-16 NC by 1¾ inches long and shall be provided with lockwasher and one suitable all-metallic self-locking nut. The material of the bolt, lockwasher, and nut shall be Everdur 1015 bronze or other approved corrosion-resistant material."

Tests A: Change Par. (e) to (d) and Par. (d) to (e). Delete Par. (f).

Tests B: In Pars. (a) and (b) change the words "test A (d)" to "test A (e)."

Chapter 2, Part 8—Specification for Plate
Type Electric Heaters for Track Switches

Page 2–8–2, Par. 9, Tests A: Change (b) to read: "Before receiving current, the heater must withstand a potential of 900 volts, 60 cycles, applied between conductor and sheath for one minute."

Change (d) to read: "After the sheath temperature of this heater has been raised by the passage of current through its element for 10 minutes at rated voltage to equilibrium temperature or not more than 1,000 F, the heater must withstand a potential of 900 volts, 60 cycles, applied between conductor and sheath for one minute and must have insulation resistance of not less than one megohm as measured with a 1000-volt ohmmeter.

Page 2–8–3, Par. 9, Tests A: Change (e) to read: "The heater shall be submerged in water to a line half way up the stud for one hour, at a maximum temperature of 60 F. Immediately upon removal from the water, the heater shall withstand a potential of 900 volts, 60 cycles, applied between conductor and sheath for one minute and must have insulation resistance of at least one megohm measured with a 1000-volt ohmmeter.

Add Par. 13 reading as follows:

13. Note

Due to the construction of these electric heaters, the tests are not as severe as those on other types of electric heaters, and, therefore, plate type heaters are not recommended for heavy main track service.

Pages 2-8-2 and 2-8-3, Tests A: Change Par. (e) to (d) and Par. (d) to (e). Delete Par. (f).

Tests B: In (a) and (b) change the words "test A (d)" to "test A (c)."

Chapter 2, Part 10—Specification for Electric Heaters for Ballast Under Track Switches

Page 2-10-1, Par. 2. Basis of Purchase:

Add new item reading, "(4) Type of terminal. (Par. 5)."
Renumber present items (4) and (5) to (5) and (6), respectively.
Page 2-10-1, Par. 5, Terminals:

Add sentence reading, "A nonseparable type terminal will be provided if so ordered."
Page 2-10-2, Tests A: Change Par. (e) to (d) and Par. (d) to (e). Delete Par. (f).
Tests B: In (a) and (b) change the words. "test A (d)" to "test A (e)."

Report on Assignment 10

Wire, Cable and Insulating Materials

Collaborating with Mechanical Division, AAR

F. T. Snider (chairman, subcommittee), B. Anderhous, R. Burn, W. B. Grimes, R. H. Stocksdale.

Your committee submits as information the following report pertaining to wire, cable and insulating materials:

A. WIRE, CABLE AND INSULATION MATERIAL STANDARDS OF INTEREST TO THE AREA

During the past year the most useful new standard to be published was the IPCEA–NEMA standards publication on Thermoplastic-Insulated Wire and Cable for the Transmission and Distribution of Electrical Energy, designated IPCEA Pub. No. S–61–402, NEMA Pub. No. WC5–1961. These standards apply to materials, constructions and testing of thermoplastic-insulated wires and cables which are used for the transmission and distribution of electrical energy for normal conditions of installation and service, either indoors, aerial, underground or submarine.

Major revisions were made during 1961 to the IPCEA–NEMA Standards Publication for Rubber-Insulated Wire and Cable for the Transmission and Distribution of Electrical Energy, IPCEA No. S–19–81, Third Edition, NEMA No. WC3–1959.

B. THE USE OF ALUMINUM CONDUCTORS FOR INSULATED POWER CABLES

Aluminum has been generally thought of as an unproved substitute metal for insulated cable conductor rather than one that could stand on its own as an alternate. This ambiguous situation has been due to heavy emphasis on the long, successful use of ACSR conductor in high voltage transmission lines, because a really broad picture of installations involving insulated aluminum cable has not been available. As our native copper resources become increasingly scarce, aluminum conductors will become more attractive—perhaps commonplace—far sooner than most engineers suspect. In many cases it will be economically advantageous to use aluminum in power cables even in times when defense production is not absorbing the lion's share of the available copper.

The determining factor as to whether aluminum or copper will be chosen for a particular installation will be one of economics. A summation of the controlling design factors is therefore justified. Since the alternating-current-carrying capacity of electrical-conductor-grade aluminum is less than for a similar size copper conductor, a larger cross section is required for the aluminum to handle the same load and still retain the same conductor temperature. This means a slightly larger overall diameter and hence more cable insulations and coverings. Not only must the resulting increase in the cost of insulation and coverings be offset by a lower conductor cost, but also any extra installation costs arising primarily from need of larger conduits, in some instances, and the stocking of special connectors and terminals.

Larger ducts are not always needed, especially where the conductor diameter is small compared to the overall diameter as in the case of high-voltage cables, or when the copper cable diameter is at the beginning of a given range of particular duct size.

It is hard to justify anything less than No. 6 AWG aluminum, and significant economic advantages do not show up usually until size No. 4 aluminum is reached. This depends upon the cost of the coverings relative to that of the conductor. Wall thicknesses and methods also influence the cost differential.

Blind substitution of aluminum for copper can lead to trouble. On the other hand, if each installation is given some thought, the aluminum can be selected for those applications where it will do the job. Whether or not aluminum can hold these new fields as time goes on will be decided largely by the economics of the situation.

Manufacturing experience to date shows there are no problems from the insulating standpoint provided the conductor size is not less than No. 12 AWG (7×0.0306 in). Being lighter in weight, aluminum conductors bring favorable comment from the workmen. The light weight naturally leads to better handling, particularly of the large CM sizes.

As far as design is concerned, any power cable construction now specifying standard Class B or C stranded annealed copper in sizes No. 6 and larger can be manufactured with aluminum conductors. Control cables as small as No. 12 AWG (7×0.0306 in) are used by some industrials.

From the mechanical standpoint, the medium-hard or $\frac{3}{4}$-hard aluminum conductor will probably be preferred for most applications because it is more easily handled. It has proved to be better suited for compression-applied fittings, and is somewhat less sensitive to "notch effect" and breakage at terminals, an occurrence which has plagued small hard-drawn conductors.

For all practical purposes, the same maximum pulling stress allowed for copper, namely, 10,000 psi or about 0.008 lb per circular mil, can be used in calculations if uniform pulling tensions can be counted upon. It has been general practice in the wire and cable industry to limit the normal stress such as experienced by cables in vertical suspension on their own conductors to about 5000 psi or 0.004 lb per circular mil for copper. This appears to be conservative for both hard and medium-hard aluminum, but is out of bounds for the soft.

There is no question that annealed copper is supreme in its ability to absorb excessive pulling loads or abuse without apparent harm. Service experience has shown that on the small-size hard-drawn aluminum conductors, excessive breakage has occurred during the pulling. In sizes No. 6 and larger, breakage is not a problem. The endurance limit, a measure of cyclic bendability, is approximately the same for medium-hard aluminum and annealed copper. Considering the flex-life of aluminum conductors, indications are that minimum strand size is 14 mils.

There has been interest in the effect of the somewhat greater coefficient of linear expansion which is characteristic of aluminum, namely, 23×10^{-6} per degree C, in contrast to 16.4×10^{-6} for copper. It is not quite as great as that of lead and, therefore, should cause no trouble when used in lead-sheathed cables. Experimental heat runs on single-conductor, 500 MCM aluminum, 15-kv paper-insulated, lead-sheathed cable have failed to disclose any detrimental behavior.

The heat dissipation of aluminum, as indicated by the specific heat, is less than that of copper. However, the increased size of aluminum required to give similar impedance and current-carrying capacity results in almost identical heat values for aluminum and copper. Temperature rises due to short-circuit currents are also similar. In addition, the lower melting point of aluminum will result in a quicker "fusing" with the same currents and this can be an advantage in network operation.

Electrical considerations stem from the fact that the d-c conductivity of aluminum is 61 percent of that of the copper standard. This lower conductivity means that a larger cross-section area will be required with aluminum than with copper. The direct-current resistance of aluminum, size for size, is 1.64 times the bare annealed copper value, and about 1.59 times the tinned copper resistance. It follows that for equal d–c resistance then, the circular mil size of an aluminum conductor will be 1.64 or 1.59 times the respective copper figures. Use of the 1.59 factor results in the aluminum conductor being exactly two AWG sizes larger than the copper, a rule-of-thumb that is sound and can be applied up to 2/0 copper.

Excessive voltage drop has been encountered in some aluminum cable installations where its effect has been overlooked, the size having been chosen only on the basis of load-carrying ability. For similar voltage drop in d-c systems the aluminum circular mils should also be 1.59 times those for copper, and this is also true in 60-cycle a–c circuits up to copper size 2/0. The rule of two AWG sizes larger up to 2/0 copper is also correct to obtain the same voltage drop.

The significance of these points to equipment manufacturers is that the aluminum conductor to handle a particular ampere load with an acceptable voltage drop will certainly be larger than the copper. It could be two AWG sizes larger up to 2/0 copper, and in the larger sizes could have about 50 percent more circular mils, but will vary according to several conditions. Service experience has shown that apparatus terminal fittings should be made big enough to handle the larger aluminum conductors required for a given ampere load. Larger junction boxes may be in order, and sometimes bigger ducts or conduits are necessary.

Aluminum has an advantage over copper in that it is not affected by the sulfur in the rubber insulation nor does it deteriorate the insulation. In paper-insulated cables aluminum does not form reaction products, known as metallic soaps, that are produced by the heating of the impregnating oil in the presence of copper. Aluminum has excellent corrosion resistance except in hydrochloric, oxalic, and hydrofluoric acids, and strong alkalies. Aluminum should not be used for bare neutrals or for bare direct-burial grounding cable since it is anodic to most metals, and chances of galvanic or electrolytic corrosion are high. Alkaline earth will cause rapid chemical corrosion. The inherent oxide film which forms instantaneously on a fresh aluminum surface as soon as it is exposed to air explains its excellent resistance to corrosive atmospheres in some industrial areas, or along the sea coast. Unfortunately, this tight, impervious, oxide film gives rise to the only major problem encountered with insulated aluminum cables, namely, that of splicing and terminating.

The only complaint which can be found concerning the use of aluminum conductors in power cables involves the electrical connections. Aluminum conductors can be readily spliced and terminated if simple precautions are followed, and satisfactory operation guaranteed. Aluminum can be soldered and it can also be joined very successfully by the use of the inert-gas welding technique, and this has found wide application where exteremely large conductors and numerous connections are involved.

Much can be said about the use of splices and terminations of aluminum conductors but, unless a proper splice or termination is made in accordance with recommendations made by the manufacturer, satisfactory operation cannot be obtained.

Aluminum conductors have been used with success in all types of power installations—aerial, in conduits, underground ducts, and direct burial. In a five-month period, a well-known manufacturer shipped from one of its plants half a million feet of finished aluminum conductor cable consisting of over 200 individual items to 70 industrial plants and utilities. The bulk was in sizes from 1/0 to 1,000,000 CM size. The voltages ranged between 600 and 15,000, and the insulations were, in general, oil-base type with some varnished cambric and polyvinyl chloride.

A survey of the above installations after four years of operation disclosed no difficulties encountered in the cable itself. Only 4 of the 70 plants had any trouble, and this was limited to the connections, and this was probably due to their not being informed on the proper method of splicing and terminating.

Typical case histories disclose that, after some initial difficulty in learning the technique of soldering joints and terminals, cable has been placed in service and has operated with no difficulty whatsoever. Overheating joints in another installation was due to poor workmanship resulting from lack of experience. Independent surveys and reports from contractor group meetings indicate that it is in sizes smaller than No. 6 that trouble was encountered more often than not. One of the causes of this trouble was the idea that, by substituting a 75-deg type RH insulation for a 60 deg R insulation, there need be no change from the copper size. While this is true from the standpoint of current-carrying capacity, it leads many people to overlook voltage drop requirements.

Another cause of trouble in sizes smaller than No. 6 has been mechanical breakage at ends, aluminum being more susceptible than copper. This is accentuated by comparing hard or medium-hard aluminum with fully annealed copper. It can be readily observed that aluminum tends to bend in a smaller radius than copper, this being a function of hardness and, when bent with pliers, the bend become highly local, and usually in one or two reversals it fractures completely. Since aluminum is more "notch-sensitive" than copper, any nicks or local flattening increases the tendency unduly. This could add up to terminal block trouble with small conductors, more so if vibration is present.

Actual service records indicate that insulated aluminum conductors, size No. 6 and larger, have better possibilities of being used than the smaller sizes.

Summary

Aluminum's place in the wire and cable field will be decided largely by the overall economics of the installation. Aerial installations constitute the most desirable applications of aluminum cable. The service record of insulated aluminum power cable in sizes AWG No. 6 and larger is excellent. There are millions of feet of insulated aluminum cable giving satisfactory service today. There will be no connection problems if proper fittings are used. In most instances, aluminum conductors can be used in the same construction that now specifies copper for power cable applications.

Report on Assignment 13

Railway Electrification

L. B. Curtis (chairman, subcommittee), R. J. Berti, L. W. Birch, W. F. Bowers, H. F. Brown, K. A. Browne, R. Burn, F. J. Corporon, A. B. Costic, H. H. Duehne, B. C. Hallowell, G. B. Hauser, R. B. Hendrickson, W. W. Holloway, R. L. Kimball, D. R. MacLeod, F. B. McConnel, H. R. Morgan, R. F. Pownall, E. B. Shew, E. H. Werner.

Your committee reports, as information, on the general subject of railway electrification. This subject is so broad that it was necessary to divide it into six subdivisions, several of which were further subdivided in order to cover the field. In addition your committee has 11 electrification problems that it develops in collaboration with the International Union of Railroads whose headquarters are in France. There is no progress to report on these latter assignments this year due in part to a change in secretariat of the UIC during the year. The individual reports follow.

Report on Assignment 13B

Study and Report on Materials and Methods of Electrification

E. H. WERNER, Chairman

ADVANTAGES AND DISADVANTAGES OF CARBON SHOES

Investigation of various railroad experiences with regard to carbon pantograph shoes indicates that on the Pennsylvania Railroad, using imported carbons, considerable trouble was experienced during ice and sleet conditions. In fact, it was necessary to use a copper ice-breaker bar ahead of the carbon strips. It was found that wear was much more rapid in sleet, and since the carbon strips cost more with no more mileage, they are being taken out of service. The road also reported that the carbons were much more satisfactory in freight service, at slower speeds.

During the past year the Erie–Lackawanna started using the carbon shoes. The road reports that four carbon shoes were installed along with four dated copper strip shoes, with the following results:

1. Carbon life matched that of the copper strips within 1,000 pantograph miles, the total being about 41,000 miles; both ran for the same period.
2. There was no burning of the carbons.
3. The test included two minor ice conditions which did not cause any operating difficulty.
4. No extra precautions or special maintenance of the carbons were exercised, nor were any found to be necessary.
5. Carbons tended to groove and valley during the first few runs but eventually smoothed out sufficiently so as not to impair pantograph operation.

STUDY USE OF ALUMINUM IN ELECTRIFICATION

Information continues to arrive in the form of press releases concerning new applications of aluminum for extra-heavy transmission wire, and extra-large aluminum transmission towers. One release was received with photograph of an all-aluminum 20,000-kva transformer, except for the core and certain accessories. This transformer was manufactured for Alcoa for use at its Massena, N. Y., fabricating shop. No significant developments have been made in railroad electrification as such, but the committee is indebted to G. B. Hauser for keeping it informed of all important new developments in the electrical industry.

INVESTIGATE DEVELOPMENT OF ALUMINUM AND PRESTRESSED CONCRETE POLES

After discussing the use of prestressed concrete poles for catenary construction where experience with such is non-existent, contact was made with the engineers of a construction firm of London, England, which had looked into the possibility of prestressed concrete poles as long ago as 1950. But since there was little difference in the cost and the conventional broad-flange beam was of known reliability, there seemed to be no reason to change. Some years later on a job in West Africa, the matter was again looked into, and it was found that the price margin had moved to a minor extent in favor of the prestressed concrete. The construction firm engineers were not completely satisfied with the manufacturing process, as the secret of success was a puddling or agitation process which had to be done at a certain stage while the concrete was wet. This had to be timed accurately by a senior technician who must know what he is doing. Again they found that they could expect high damage in handling or unloading the poles from the flat car.

Further investigation as to what the French are doing on the continent will be investigated in the near future.

Report on Assignment 13C

Investigate and Report on New Types of Bonding

B. C. HALLOWELL, *Chairman*

PREPARE REPORT AND SPECIFICATIONS FOR THERMIT WELDING OF ALUMINUM

Progress to date in the field of protective coatings for aluminum-to-aluminum and aluminum-to-copper welds:

1. A potential difference exists between the aluminum weld metal and an aluminum conductor, with the weld metal more cathodic.

2. A greater potential difference exists between the aluminum weld metal and a copper conductor, with the copper being more cathodic.

3. If a conducting path is established between the components mentioned in Pars. 1 and 2 above, due to moisture from humid air, spray, etc., current can flow and rapid galvanic corrosion will occur, eventually destroying the more anodic component (the aluminum conductor or weld metal). From these conditions, one can readily see the need for establishing a moisture barrier around the connection as soon as possible after welding. Such a barrier should be capable of fulfilling certain other requirements, inherent in the field use of welding material. Some of these requirements are as follows:

(a) Penetration into stranded cable.
(b) Good adherence to both copper and aluminum (materials which are not easily coated.)
(c) Good moisture resistance coupled with good weather resistance.

(d) Relatively quick drying to allow recoating.

(e) Drying by solvent evaporation, rather than catalysis due to the possibility of low-temperature applications.

Summary of Results

Many systems of coatings were evaluated before a brush-coat system was approved for use. Evaluation tests used included salt spray testing, immersion testing, exposure to outside atmospheres, and others.

One of the conclusions reached was that a one-coat system would not meet all of the listed requirements.

Brush-coat system includes a first coat of zinc-chromate-base primer for penetration and adherence and a finish coat of neoprene-base paint for water and weather resistance. To provide the maximum in protection, at least two finish ·coats are recommended. Sample connections of aluminum-to-aluminum welds brush coated have been exposed to 20 percent salt spray for more than one year and show no signs of corrosion.

Sample connections of copper-to-aluminum welds, brush coated, have been exposed to 20 percent salt spray for 800–1000 hr and show only superficial corrosion. These same connections have been exposed to outside weather conditions for more than one year and show no corrosion. For the condition of welding aluminum IPS tubing, rubber plugs coated with a special moisture-resistant grease are used to seal off the weld internally, together with the brush coat applied to the outer surfaces.

Manufacturers of electrical welded terminals have drawings and instructions for application of protective coatings, showing details on how they should be˙ applied.

Future Work

Several developments in weld protection are being investigated. Among these are:

1. The use of zinc-plating on copper lugs and sleeves to reduce the potential between the copper and aluminum conductors. .

2. The use of zinc-pigmented finish coats for the same purpose.

3. The use of an inhibited wash primer for better adhesion and penetration.

4. The use of a modified neoprene with even better weather and moisture resistance.

Report on Assignment 13D

Developments in the Field of Electrification (Domestic and Foreign)

L. W. BIRCH, *Chairman*

UNITED STATES DEVELOPMENTS

The Pennsylvania Railroad has placed in service about 20 of the new E–44 ignitron rectifier-type electric locomotives. Sixty-six of these units are being acquired to replace 90 older, less powerful P5A freight electrics. The new units are rated at 4400 hp and can pull full-length freight trains at 70 .mph. All 12 of its 40-in wheels are powered by 6 traction motors. The electric motors are a standard type, and the Pennsylvania has more than 1000 in service in its diesel-electric locomotives. The E–44 locomotive utilizes the recently proved rectifier-tube method of transforming alternating current into direct current. The 11,000-v, 25-cycle, single-phase alternating current is collected from the overhead catenary by one of two new-type pantographs atop the control cab. These pantographs, developed in France, are now made under license here in the United States.

During April of this year the New Haven Railroad placed in service a silicon-rectifier-equipped commuter car, the silicon rectifiers having replaced ignitron rectifiers. This silicon equipment weighs about one-third as much as the mercury-arc rectifier equipment. The diodes were developed after many years of evaluation under simulated multiple-unit car operating conditions. Performance has been entirely successful.

New York's automated train has been test-operated for some time between 18th Ave. and New Utrecht Ave. on the BMT Sea Beach Line. Train orders are issued by tape to the nearby running rails where the three-car automated train, operating on a section of express track, picks up the coded signals and is started, operated, and stopped automatically. Even the doors are operated with a separate code.

A receiver coil on the truck of the lead car picks up the signals and carries them to an amplifier in the motorman's cab. The amplifier feeds the orders to a decoder— a sort of electronic motorman and conductor combined.

A typical cycle of operation is as follows: after a brief pause, the electronic conductor closes the doors, the electronic motorman gets the "go ahead" and applies the power. The train moves forward very smoothly and picks up speed to approximately 30 mph. Near the end of the experimental track, the electronic impulses flowing into the amplifier cut off the power, operate the brakes, and bring the train to a gentle stop. The doors are opened automatically, and passengers can leave or enter the cars. The experimental stretch is approximately the same length as Times Square to Pennsylvania Station where it is hoped the first commercial operation will be installed

"Trains", said Chairman Patterson of the New York City Transit Authority, "have operated in various parts of the world on remote control, but this is the first time, so far as we know, that one has operated back and forth on a given stretch of track on a fully automatic basis."

FOREIGN DEVELOPMENTS

The Kent Coast electrification (English, second phase) involves 132 route miles; also 214 new electric coaches, 180 of which were built as four-coach corridor express units. These new coaches replace a similar number of steam-hauled coaches.

The British electrification of railroads is predominantly a 25-kv, 50-cycle system, the advantages of which are claimed to be:

1. Power supplies can be drawn direct from national grids.
2. Less copper and steel needed for lighter overhead and supporting structures.
3. Fewer and simpler substations.
4. The motive power units are ac and capable of superior standards of performance.

At the present time, the French National Railways have electrified 4400 miles of lines (Figs. 17 and 18) representing only 17.5 percent of the entire network. However, trains on the electrified sections now handle 55 percent of the total traffic. Included in the 4400 miles there are 1250 miles which have been electrified since 1955 with 25-kv, 50-cycle alternating current, located principally in the northern and eastern sections of France.

It is claimed by French railway engineers that this new high-voltage, commercial-frequency system has been so satisfactory that the French National Railways will continue to electrify lines at the rate of about 200 miles per year. The next electrification will be on the Paris–LeHavre line.

The French National Railways' fleet of locomotives now includes more than 250 "rectifier-type" locomotives with 200 additional units on order. When the first ignitron locomotive was delivered to the railroads ten years ago by American manufacturers, the French engineers did not feel this equipment would find wide application in motive power; however, the thinking changed. Now, the same engineers seem to have more confidence in the silicon rectifier than in other types. For the past two years they have been testing a 5000-hp silicon-rectifier locomotive and claim it has given complete satisfaction. The French feel confident that it is possible to build mass-produced locomotives with silicon rectifiers. To prove their contention they have ordered 15 silicon-rectifier locomotives of 3600-hp capacity.

Barcelona, Spain, has extended its underground railway, principally by means of the "cut-and-cover" method. This section is 1¾ miles long and will be served with five stations.

In South Africa, the end of the approved electrification program is now in sight. The cost of their recent electrifications is in the vicinity of $100 million. During the past year 239 route miles of electrified track were added, bringing the total to 1199. It is interesting to note that the electrification of Capetown suburban lines, a separate project, is scheduled for completion by the end of this year.

By means of an $80 million World Bank loan, the Japanese will soon construct a 311-mile high-speed line between Tokyo and Osaka. This will be a 25-kv, 50-cycle undertaking.

STUDY POSSIBILITY OF LESSENING COST OF THE DISTRIBUTION SYSTEM

During the past year the task group for this subject has considered the use of aluminum structures for both portal and single-track construction and has also considered aluminum for some of the attachments, particularly for single track, the bracket arm being the largest part under consideration.

The preliminary consideration has been carried on with the Chicago South Shore & South Bend Railroad.

At this date further investigation will require some testing, however this will not be undertaken until the economic comparisons have been completed.

Reduced messenger tensions reduce weight of structures. However, ice and snow loads may counteract the anticipated gains. The following discussion compares the merits of standard, anchored catenary spans and counterbalanced, constant-tension catenary spans.

If the overhead line is operated at 25 kv and the commercial frequency of 60 cps, the catenary cables can be of small cross section since the current will be fairly low. However, at high speeds, it is necessary to maintain sufficient mechanical strength in the cables to resist mechanical loads encountered, and also the contact wire should have sufficient mass to afford smooth contact.

In selecting the catenary make-up for this study an approximate equivalent conductivity of 300,000 circular mils was considered desirable, including a messenger that would have sufficient tensile strength to allow a normal sag of 5 ft in a 300-ft span without being over stressed under maximum loading conditions.

Various types of messengers were considered, including bronzes, high-strength and extra-high-strength composite, and aluminum, steel-reinforced. The information shown in this report is based on a 7/16-in high-strength composite cable with 40 percent equivalent conductivity. This, together with a 300,000-circular mil, grooved, 80 percent equivalent conductivity bronze contact wire, gives a total conductivity of 298,300 circular mils.

The total weight of this catenary span, including hangers, is 1.376 lb per ft.

In calculating maximum loadings, ice load was considered only on the messenger cable and not on the contact wire, based on the assumption that trains would not be operating if there were $\frac{1}{2}$ in of ice on the contact wire.

The normal sag of 5 ft in a 300-ft span is the equivalent of approximately 9 ft in a 400-ft span, or 2.25 ft in a 200-ft span.

Fig. 19 shows the variation in sag with different loadings and temperatures when the messenger is anchored at both ends as is customary in North and South America.

Fig. 20 shows the variation in sag under different loadings when the messenger is counterbalanced to maintain a constant tension. With this type of construction, changes in temperature do not affect either tension or sag. Only changes in loading (due to ice and/or wind) affect the sag.

It will be noted that the variation in sag, or vertical movement, is much greater with the constant-tension (or counterbalanced) spans than with the standard, anchored spans. Therefore, all other things being equal, the allowable span length would be less with the counterbalanced span than with the regular anchored span.

On the other hand, with the counterbalanced span, with tension at a constant value of about 60% of the maximum attained with the anchored span, the side loading on curves would be less, this permitting some reduction in structure weights on curves.

Considering vertical movement, and allowing approximately 1 ft above and 1 ft below normal sag, the maximum allowable span length for anchored spans would be 300 ft, and for constant tension spans, 200 ft.

Considering horizontal movement at the center of the span due to a 4-lb wind on $\frac{1}{2}$-in ice, and allowing approximately 1 ft horizontal movement on either side of the normal position, the maximum allowable span length for anchored spans would be about 300 ft, and for constant tension spans, about 225 ft.

Conclusions

There is little, if any, advantage in using a counterbalanced catenary messenger in regions where maximum design loadings include ice.

In fact, there may be a distinct disadvantage, since the maximum allowable span will be less; thus requiring more catenary supporting structures per mile.

There would be more advantage in maintaining a constant tension in the contact wire, since ice loading could be ignored as long as the line is in use.

GENERAL ELECTRIFICATION ECONOMICS

The French National Railways, although having considered other modes of signaling, still retain the track circuit principle. By persistent attention to the quality of maintenance and improvements of the track circuit, the French engineers claim reduction in track circuit losses and some reduction in maintenance expense, particularly in the 25-kv, 50-cycle territory.

A similar statement has not been made by the English as regards their "modernization plan." The newly electrified sections have retained track circuits for signaling.

THIRD RAIL

1. General

Two systems for the distribution of electrical energy to rapid-transit trains are in use in the United States and Canada. These are:

(1) The overhead system.
(2) The third-rail system.

Both systems have their advantages and limitations, and both systems have been highly successful in performing the transfer of electrical energy from cable or rail to the traction motor. The principal reason for selection of a particular system is economic rather than operational since the speeds of most trains are not of sufficient magnitude to eliminate either the overhead system or the third-rail system.

2. Advantages and Disadvantages of Third-Rail over Aerial Distribution

(1) Advantages

  a. Better appearance from an aesthetic viewpoint.
  b. Better visibility for operating personnel and passengers.
  c. For moderate voltages (up to 600 v dc) conductor rail is a much sounder and more lasting mechanical construction than any other form of line conductor.
  d. Maintenance cost is much less than for overhead construction.

(2) Disadvantages

  a. Additional and constant source of danger to track maintenance men.
  b. At complicated junctions and crossovers it is difficult to find room for the third rail.
  c. In case of derailment there is danger of fire, electric shock to passengers, longer interruption to service.
  d. Cost is about double that of overhead construction.
  e. Should not be used for voltages over 600 v dc, because of difficulty in insulating, and increased danger to human life.
  f. Insulation is exposed to mechanical breakage and to the action of corrosive brine dripping from cars.

3. Types of Third-Rail Systems

The danger of the third rail to human life restricts its use to private right-of-way on surface lines, and to elevated, tunnel or subway installation in cities.

It is more suitable for short electrifications than long, because the system voltage is usually limited to 600 v (Fig. 16). Higher voltage (up to 2400 v dc) have been tried, but have been reduced to 750 v or less because of insulation difficulties. Also it is more suitable for direct current than for alternating current due to the increase of resistance by skin effect.

The contact surface may be on top of the rail, on the bottom, or on the side. There are only two installations of under-running (or bottom contact) third rail in the United States and Canada (Figs. 2 and 7) and no side-contact installations. All others are of the over-running (or top-contact) type (Figs. 1 and 6).

The advantages of the over-running third rail are:

  a. More easily designed to suit the available space.
  b. Somewhat easier to install.

c. Does not encroach as much on the clearance limits of ordinary rolling stock as the under-running third rail.

d. A simpler type of insulator and support can be used.

The advantages of the under-running third rail are:

a. Rail is better protected from weather than the over-running contact type—especially for keeping contact surface free of ice.

b. Rail is also more easily and completely protected from accidental contact with track tools.

4. Conductor Rail

Composition

Soft steel—low in carbon and manganese to give high conductivity.

Conductivity

Resistance usually about 6 to $6\frac{1}{2}$ times that of an equal cross section of copper. Overall resistance, including bonded joints, is usually 7 to 8 times that of copper. Conductivity is from $\frac{1}{3}$ to $\frac{1}{2}$ above that of equal cross section of running rails.

Shape

Cross section of rail may have any shape convenient for installation and current collection. If there is no especial advantage in departing from standard rail sections, it is probably better to use them, as they are more readily available.

5. Insulators and Supports

The conductor rail is usually located alongside the track at a fixed distance from the running rail and slightly above it.

Every fourth or fifth tie is extended to hold the third-rail insulator. Spacing between supports is usually 8 or 10 ft.

For over-running third rails (Figs. 1, 5, 11, and 12) the support consists of a porcelain block with cap which holds the rail in lateral position but allows longitudinal movement.

For under-running third rail (Figs. 2 and 7) a bracket is necessary to furnish support from above, and the rail lies in the guide of a split-porcelain insulator.

6. Bonding

It is desirable to bond the joints in the conductor rail to the full current-carrying capacity of the rail.

A type of bond that is widely used for bonding the joints of 150-lb third rail in heavy service consists of two tinned copper bars and two heat-treated bolts. When installed the joint is welded, bolted, and soldered and provides an average of 110 percent conductivity (Fig. 16).

Gaps in the third rail at special-work in the tracks are bridged by cables which are bonded to the ends of the rails. The distance between front and rear collector shoes is greater than the longest gap in order to supply current continuously.

7. Miscellaneous Requirements

Expansion Joints

Expansion joints, similar to Fig. 9, are usually installed, underground, in the third rail at intervals (usually about 10 per mile) to allow for longitudinal movement of the rail due to temperature changes. These joints are constructed to permit longitudinal movement of the rail but restrict lateral movement of rail ends. Fexible bonds of copper of a conductivity equal to that of the third rail are installed around the joint.

Anchors

Anchors, similar to Fig. 5, are usually installed on each section of the third rail between expansion joints to prevent rail creepage. Insulation may be secured by use of wood strain insulators and porcelain pedestal insulators.

Guards

Third-rail guards are provided to prevent accidental contact by workers, and to prevent the formation of ice, sleet, and snow on the third rail.

For over-running third-rail systems the guards usually consist of wooden planks held in position by brackets attached to the track ties, as shown in Fig. 1, while for under-running third rail, the planks are supported by the third rail itself and interlock around the third rail, leaving only the underside of the rail exposed, as shown in Figs. 2 and 7.

8. Current-Collection Apparatus

There are two general types of third-rail shoes:

    a. The gravity type, shown in Fig. 10, which maintains contact pressure by the weight of the shoe. This is essentially an over-running type. The slipper portion of these shoes weighs about 20 lb.

    b. The spring type, shown in Figs. 3, 4, and 8, maintains contact pressure by means of springs. This type of shoe is suitable for over-running, under-running, or side contact.

The spring type allows more complete protection at the third rail, and maintains better contact with the rail at high speed.

Shoes are usually designed with a point of comparative weakness, so that slipper portion alone may be broken off if any obstruction is encountered.

Shoes are usually mounted on a shoe-beam which is made of an insulating material, generally wood. This beam is kept clean and properly painted with an insulating paint to prevent electrical "creepage" or "tracking."

Braided copper shunts carry the current around joints and bearings.

Special folding third-rail shoes are used on the New York, New Haven & Hartford Railroad. These fold up within the equipment clearance line when not in use. They may be raised or lowered from inside the locomotive cab by means of air pressure.

There are generally four shoes on each locomotive or motor car, two on each side.

Power is taken from the shoes to the locomotive by a cable and into a fuse box. Generally each shoe is connected to a separate fuse with a magnetic-blowout arrangement.

9. Estimated Costs Per Mile

The first cost of a third-rail distribution system will vary considerably, depending on whether it is installed in a subway or in the open, how much feeder cable is included, and whether the feeder is carried overhead or underground.

To secure a fair comparison of the cost of third rail and overhead catenary, approximately the same overall conductivity has been used for each system, and all supplementary feeder cable has been considered to be overhead rather than underground.

(Text continues on page 480)

RAILWAY COMMUTER SERVICE

Fig. 2 - Underrunning Rail

Fig. 1 - Overrunning Rail

Fig. 3 - Underrunning Collector

Fig. 5 - Third-Rail Anchor

Fig. 4 - Overrunning Collector

RAPID TRANSIT

Fig. 6 - Overrunning Rail

Fig. 7 - Underrunning
        Rail

Fig. 8 - Underrunning Collector

Fig. 9 - Third-Rail
        Expansion
        Joint

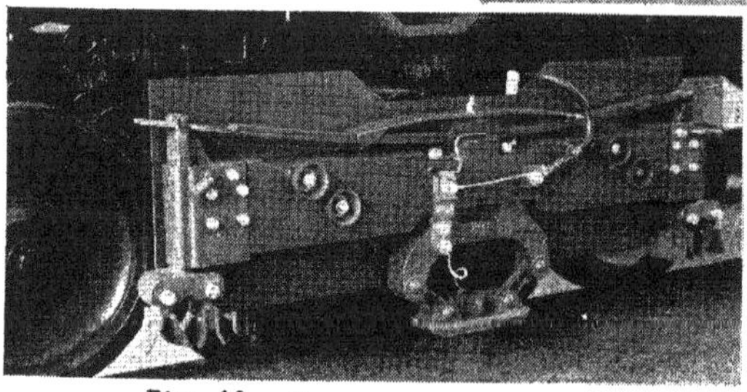

Fig. 10 - Overrunning Collector

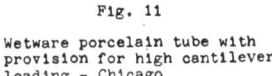

Fig. 11

Wetware porcelain tube with
provision for high cantilever
loading - Chicago

Fig. 12

Wetware porcelain
tube for elevated
track. Assembly is
attached to both
third rail and tie.

Fig. 13

Cleveland Rapid Transit

Fig. 14

Boston

Fig. 15

Third-rail bond applied to
150-lb. rail.

Fig. 16

Early attempts
to use 2400 volts
d-c were not
successful on
Michigan Railways

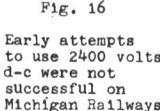

The French National
Railways operated a test
train in May 1961 with
this locomotive and suc-
cessfully collected elec-
trical energy over a 67-
mile section at 25 kv.
from this standard cat-
enary system at an aver-
age speed of 100 mph and
a maximum speed of 118
mph.

Fig. 17

Fig. 18

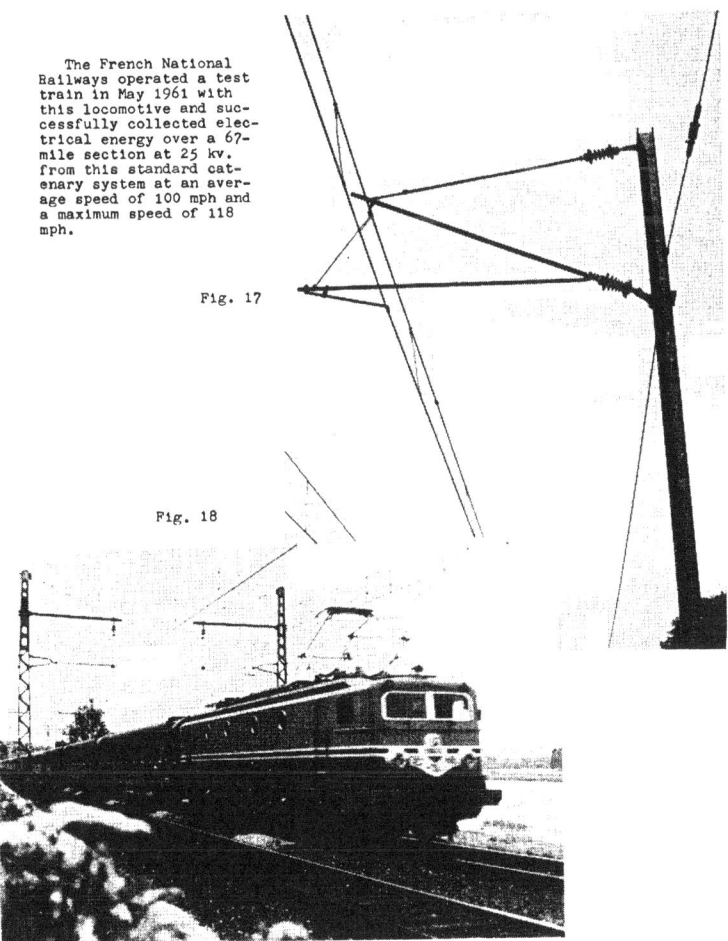

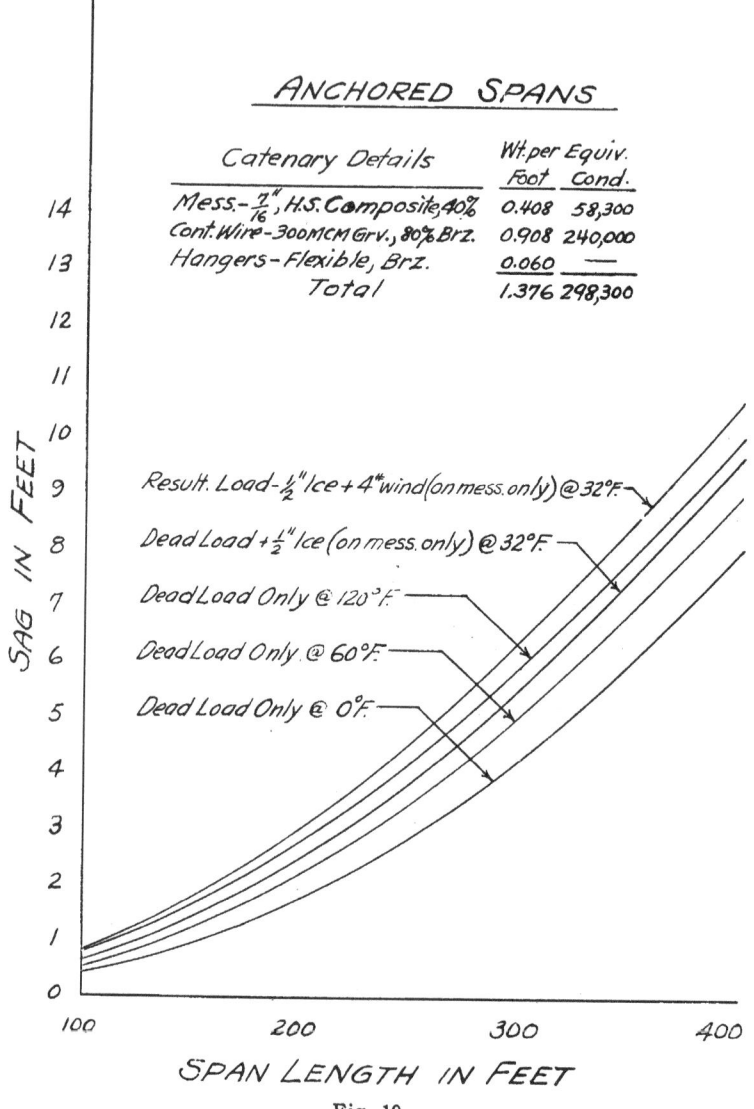

ANCHORED SPANS

Catenary Details	Wt. per Foot	Equiv. Cond.
Mess.-$\frac{7}{16}$", H.S. Composite, 40%	0.408	58,300
Cont. Wire-300 MCM Grv., 80% Brz.	0.908	240,000
Hangers-Flexible, Brz.	0.060	—
Total	1.376	298,300

Result. Load-$\frac{1}{2}$" Ice + 4" wind (on mess. only) @ 32°F.

Dead Load + $\frac{1}{2}$" Ice (on mess. only) @ 32°F.

Dead Load Only @ 120°F.

Dead Load Only @ 60°F.

Dead Load Only @ 0°F.

SAG IN FEET

SPAN LENGTH IN FEET

Fig. 19.

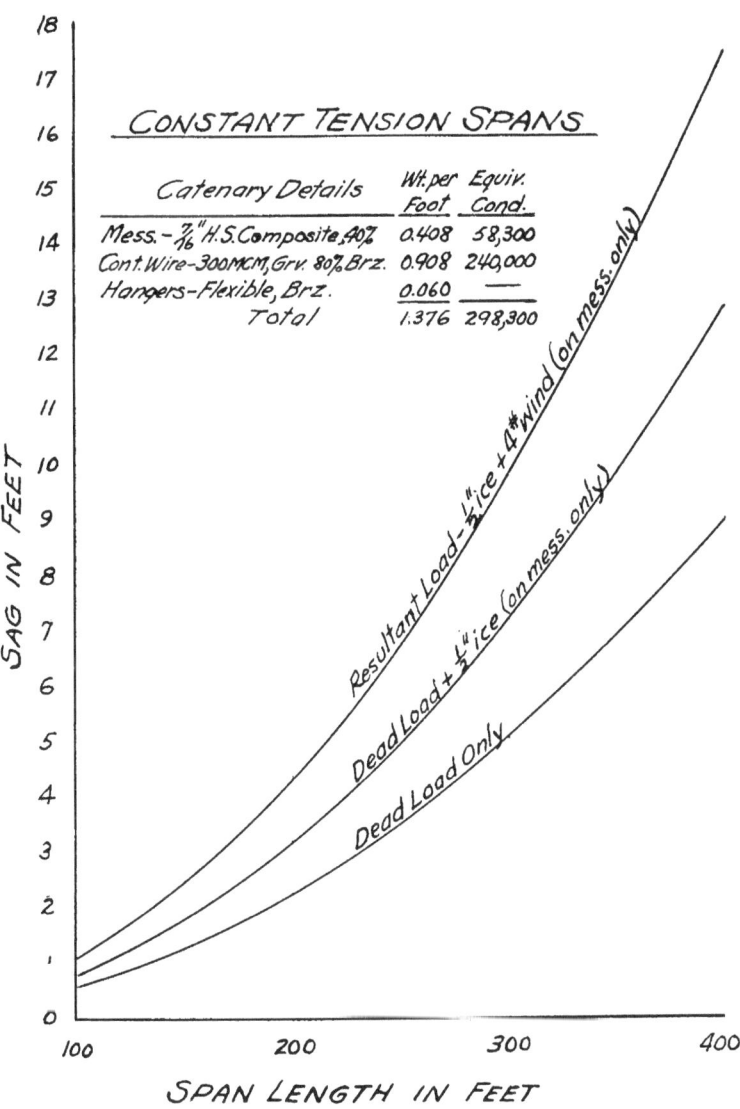

Fig. 20.

A. Conventional 150-lb third rail, 600 v, dc, equiv. cond., 2,590,000
   CM ....$82,000/track mile
B. Variable conductivity third rail, 600 v, dc, runner bar, 6 x 2½-in
   steel channel @ 13.3 lb plus four 500,000 CM bare copper feeders,
   equiv. cond. 2,400,000 CM ....$65,000/track mile
C. Simple overhead catenary, 600 v, dc, 200 ft structure spacing, plus
   two 1,000,000 CM bare copper feeders, equiv. cond. (Figs. 14 and
   15) 2,410,000 CM ....$43,000/track mile

The above estimates include material and labor but do not include the usual percentages which cover insurance, taxes, etc., as well as contractor's profit.

Comparing third-rail maintenance to overhead catenary maintenance, the following figures for the distribution systems may be used. Transformers, tie stations, etc., are not included.

A. Annual maintenance cost for conventional third rail, equivalent conductivity, 2,590,000 CM ....$ 600/track mile
B. Annual maintenance cost for simple overhead catenary and feeders,
   equivalent conductivity, 2,410,000 CM ....$1,400/track mile

SEMI-CONDUCTOR RECTIFIERS FOR RAILWAY ELECTRIC SUPPLY

The Chicago Transit Authority has ordered three 2500-kw, 600-v silicon-rectifier substations. All three rectifiers, to be installed late in 1961, will be located at CTA's new Princeton substation which will service the Englewood line of the rapid-transit system.

The efficiency of the silicon rectifiers will exceed 97 percent, approximately 2 percent more than the efficiency of conventional rectifying equipment. The units will be smaller and should require less maintenance than other types of converters used in substations.

Three multiple-unit commuter cars using silicon rectifiers were placed in revenue service during April 1961. Two of the silicon-rectifier cars are operating on the Pennsylvania Railroad in the Philadelphia area, while the other silicon-rectifier car is operating on the New Haven Railroad in the New York–Stamford area.

Silicon rectifiers have the advantage of being highly efficient, lightweight, and trouble free. The air-cooled silicon equipment can withstand high temperatures, and with no moving parts the rectifier has a long service life. Adequate operating data are not available, but it is expected that the a–c to d–c conversion losses of the silicon rectifier will be approximately one-half the losses encountered with conventional equipment.

Report on Assignment 13E

Develop Tri-annually a Report on Physical, Operating, and Economic Statistics on Railroad Electrifications Systems (Domestic and Foreign)

H. F. BROWN, *Chairman*

TABLE I—ELECTRIFIED RAILROADS THROUGHOUT THE WORLD—1958
(Revised to approximately July 1960)

Country and Railway System	Route Miles	Track Miles	Voltage	System (A)	Contact (B)
Algeria					
Algerian Railways	193	242	3,000	DC	OH
Argentina (I)					
State Railways					
General Mitre	46	94	800	DC	3R
General Urquiza	16	29	550	DC	OH
Sarmiento	26	85	800	DC	3R
Australia					
New So. Wales Gov't. Railway	246	703	1,500	DC	OH
Victorian Gov't. Railway	257	596	1,500	DC	OH
Austria					
State Railways	1,046	2,430	15,000	1/16.7	OH
State Railways	57	64	6,600	1/25	OH
Belgium (H)					
State Railways (SNCB)	547	1,605	3,000	DC	OH
Bolivia					
State Railways	14	19	550	DC	OH
Brazil					
Paulista	307	--------	3,000	DC	OH
Rede Mineiro de Viacao	234	--------	1,500 & 3,000	DC	OH
Central	146	--------	3,000	DC	OH
Sorocabana	304	--------	3,000	DC	OH
Santos Jundiai	68	136	3,000	DC	OH
Parana–Santa Catarina	32	--------	3,000	DC	OH
Ceste Brasileiro	121	--------	3,000	DC	OH
Canada (N)					
Canadian National Railway					
Montreal–Cartierville	30	78	2,700	DC	OH
Chile					
Anglo Chilean Nitrate (C)	24	34	1,500	DC	OH
Bethlehem–Chile Iron Mines (C)	15	24	2,400	DC	OH
Chile Exploration (C)	5	8	660	DC	OH & 3R
Chilean State Railways (H)	183	307	3,000	DC	OH
Chilean Trans Andine	44	51	3,000	DC	OH
Santiago–Puente Alto	13	15	700	DC	OH
Belgian Congo					
BCK (H)	278	--------	25,000	1/50	OH
Costa Rica					
Pacific Railway	77	103	15,000	1/20	OH
Cuba					
Hershey Railway (C)	54	110	1,200	DC	OH
Western Railways of Cuba	25	41	650	DC	OH
Czechoslovakia					
State Railway	4	4	1,200	DC	OH
State Railway	64	88	1,500	DC	OH
State Railway	377	1,463	3,000	DC	OH
Denmark (State Railways)					
Copenhagen Suburban	37	81	1,500	DC	OH

TABLE I (Continued)

Country and Railway System	Route Miles	Track Miles	Voltage	System (A)	Contact (B)
Egypt					
Cairo–Helwan	15	--------	1,500	DC	OH
France					
National Railway (SNCF) (H)	2,879	5,585	1,500	DC	OH
National Railway	22	24	600	DC	3R
National Railway	64	130	650	DC	3R
National Railway	39	39	850	DC	3R
National Railway (I)	1,166	2,226	25,000	1/50	OH
National Railway	29	29	12,000	1/16.7	OH
National Railway	7	14	15,000	1/16.7	OH
Germany					
Eastern Region					
(Russian Occupied Territory)					
Silesian Lines (E)	242	580	15,000	1/16.7	OH
Central Lines (E)	308	923	15,000	1/16.7	OH
Berlin Suburban (D)	181	424	800	DC	3R
Western Region					
German State Railway	3	4	750	DC	OH
State Railways (DBB) (H)	2,097	6,490	15,000	1/16.7	OH
State Railways (DBB) (P)	35	55	20,000	1/50	OH
Privately Owned Lines	43	83	1,200	DC	OH
Great Britain					
British Railways (I)					
London Midland Region	44	217	25,000	1/50	OH
	10	25	1,200	DC	3R
London Midland Region	41	170	1,500	DC	OH
London Midland Region	10	19	6,600	1/50	OH
London Midland Region	33	83	630	DC	4R
London Midland Region	53	130	630	DC	3R
Eastern Region	12	34	630	DC	4R
Eastern Region	94	319	1,500	DC	OH
Eastern Region (I)	69	159	6,250/2,500	1/50	OH
Northeastern Region	43	106	630	DC	3R
Southern & Western Region	798	2,010	660	DC	3R
Hungary					
State Railways (Q)	182	571	16,000	1/50	OH
Budapest Suburban (D)	91	178	1,000	DC	OH
India (I)					
Eastern	88	270	3,000	DC	OH
Western	37	130	1,500	DC	OH
Central	185	430	1,500	DC	OH
Southern	18	36	1,500	DC	OH
Indonesia State Railways	55	77	1,500	DC	OH
Italy					
State Railways (H)	4,090	7,268(R)	3,000	DC	OH
State Railways (J)	820	1,497	3,700	3/16.7	OH
Privately Owned Lines	23	25	800	DC	OH
Privately Owned Lines	81	131	3,000	DC	OH
Privately Owned Lines	16	25	4,000	DC	OH
Privately Owned Lines (G)	594	626	Various	AC & DC	--------
Japan					
State Railways	1,306	3,331	1,500	DC	OH
State Railways	7	15	600	DC	3R
State Railways	6	12	600	DC	OH
State Railways	26	30	750	DC	OH
State Railways (H)	44	105	20,000	1/60	OH
Privately Owned Railways (D)	27	56	600–750	DC	3R
Privately Owned Railways (D)	1,702	3,079	500–600	DC	OH
Privately Owned Railways (D)	170	201	750	DC	OH
Privately Owned Railways (D)	1,703	3,025	1,500	DC	OH
Jugoslavia State Railways (L)	117	217	3,000	DC	OH
Jugoslavia State Railways	5	7	15,000	1/16.7	OH
Luxembourg	12	23	3,000	DC	OH
Luxembourg	42	81	25,000	1/50	OH

TABLE I (Continued)

Country and Railway System	Route Miles	Track Miles	Voltage	System (A)	Contact (B)
Manchuria	78	165	·1,200	DC	OH
Mexico State Railway	64	81	3,000	DC	OH
Morocco State Railway	474	564	3,000	DC	OH
Netherlands State Railway	1,009	2,339	1,500	DC	OH
New Zealand	70	174	1,500	DC	OH
Norway State Railway	985	1,315	15,000	1/16.7	OH
Privately Owned	10	13	10,000	1/16.7	OH
Privately Owned (Meter Gage)	15	19	6,000	1/25	OH
Poland	344	812	3,000	DC	OH
Poland	33	75	800	DC	OH
Poland	25	44	650	DC	OH
Portugal (I)	89	248	25,000	1/50	OH
Portugal	16	38	1,500	DC	OH
Spain State Railways (RENFE)	417	810	1,500	DC	OH
State Railways (RENFE)	19	23	6,000	3/25	OH
State Railways (RENFE) (Meter gage)	7	8	1,350	DC	OH
State Railways (RENFE) (H)	768	1,224	3,000	DC	OH
State Railways (RENFE) (Meter gage)	86	91	1,500	DC	OH
State Railways (RENFE) (Meter gage)	13	13	1,200	DC	OH
State Railways (RENFE) (Meter gage)	58	61	3,000	DC	OH
State Railways (RENFE) (Meter gage)	17	19	600	DC	OH
Privately Owned	69	88	1,500	DC	OH
Privately Owned	20	22	1,200	DC	OH
Privately Owned	126	162	500–650	DC	OH
Privately Owned	24	37	1,300	DC	OH
Privately Owned	105	137	1,750	DC	OH
Sweden State Railways	4,208	6,758	15,000	1/16.7	OH
State Railways (Narrow gage)	68	87	1,500	DC	OH
Privately Owned (Std. gage)	178	255	15,000	1/16.7	OH
Privately Owned (Std. gage)	12	17	1,350	DC	OH
Privately Owned (Narrow gage)	104	126	15,000	1/16.7	OH
Switzerland State Railways (SBB) Std. Gage	1,760	3,640	15,000	1/16.7	OH
Meter Gage	46	59	15,000	1/16.7	OH
Std. Gage	9	20	1,500	DC	OH
Private Lines Std. gage	343	465	15,000	1/16.7	OH
Private Lines Std. gage (G)	81	102	— Various —		
Private Lines Meter gage	18	20	15,000	1/16.7	OH
Private Lines Meter gage	330	394	11,000	1/16.7	OH
Private Lines Meter gage	17	22	1,700	DC	OH
Private Lines Meter gage	27	30	1,200	DC	OH
Private Lines Meter gage	74	86	1,500	DC	OH
Private Lines Meter gage	77	91	1,000	DC	OH
Private Lines Meter gage	17	19	2,200	DC	OH
Private Lines Meter gage (G)	349	417	— Various —		
Turkey	17	35	25,000	1/50	OH
USSR State Railways (D)	19	32	1,200	DC	OH
State Railways	200	433	1,500	DC	OH
State Railways	828	1,100	3,000	DC	OH
State Railways (G)	4,765	----------	3,000	DC	OH
State Railways	85	----------	20,000	1/50	OH

TABLE I (Continued)

Country and Railway System	Route Miles	Track Miles	Voltage	System (A)	Contact (B)
Union South African Republic					
State Railways	1,178	2,672	3,000	DC	OH
United States of America (N)					
CMStP&P	662	921	3,300	DC	OH
CNS&M	105	209	625	DC	OH
CSS&SB	76	90	1,500	DC	OH
E-L	68	157	3,000	DC	OH
IC	10	129	1,500	DC	OH
Long Island	105	256	700	DC	3R
NYC	60	340	666	DC	3R
NY Conn & Long Island	21	65	11,000	1/25	OH
NYNH&H	105	581	11,000	1/25	OH
PRR	644	2,189	11,000	1/25	OH
Reading	84	198	11,000	1/25	OH
Staten Island (B&O)	14	29	600	DC	3R
N&W	130	229	11,000	1/25	OH

Notes

(A) If alternating current, phase and frequency are shown, i. e., 1/16.7 means single phase 16⅔ cycles; 1/25 means single phase, 25 cycles, etc.
(B) OH means overhead contact; 3R, third rail; 4R, positive and negative contact rail plus running rails.
(C) These may be industrial railways not for regular revenue traffic. These are included in some countries and excluded in the U. S. A.
(D) It is not clear whether some of these are city rapid transit lines.
(E) It is possible that electrification has been removed from these lines.
(G) This additional mileage is reported as electrified but data as to voltage and system are lacking.
(H) Additional mileage is being electrified.
(I) Additional electrification is being planned at 25kv, 1/50.
(J) Some of this mileage is being converted to the prevailing national system (3000v, dc).
(K) Additional miles now under construction.
(L) Originally part of Italian State Railway electrification out of Trieste.
(N) Class I Railways only with exception of CNS&M, and CSS&SB.
(P) This is the "Hollenthal" line (has been or will be made 15kv, 1/16.7).
(Q) The earliest line to use "commercial frequency." (Kando System)
(R) Track mileage is incomplete.

for effective
weed control...

- Concentrated **BORASCU**®
- **POLYBOR-CHLORATE**®
- **UREABOR**®
- **MONOBOR-CHLORATE**®

These <u>borate weed killers</u> are proving best for roads in every way... *efficiency, safety, economy, convenience, easy application.*

Today's use of borates for maximum control of vegetation began years ago with our pioneer work in the field. Continued research has developed the group of herbicides, listed above, which most roads now favor for every phase of weed control. These four weed killers are nonselective. They are widely used for year-round maintenance of weed-free conditions about trestles, tie piles, yards, signals, switches, and rights of way. Find out how you, too, can do a better job on weeds... write today.

AGRICULTURAL SALES DEPARTMENT

U.S.BORAX

630 SHATTO PLACE · LOS ANGELES 5, CALIFORNIA

PROGRESS REPORT

SPENO

The only man on his feet...

He's the pin puller—uncoupling cars. He could be in any of the GRS-equipped automatic yards. Once a car is free, its speed is controlled automatically by Class-Matic®, the GRS system of yard automation. At the crest, the conductor pushes a button on his console for each cut, to route the cars automatically to the proper tracks. In the retarder tower, the operator merely monitors the system.

The free-rolling cars are judged individ-ually as to rollability, weight, route, dis-tance to coupling, and other factors which are fed to the analog computer. This con-trols the retardation automatically so the cars glide gently to coupling.

What a saving in manpower! And con-sider the safety factor—only one man near the cars at any time. The saving in freight time alone will vindicate your choice of GRS Class-Matic, the system that pays for itself in short order.

3084

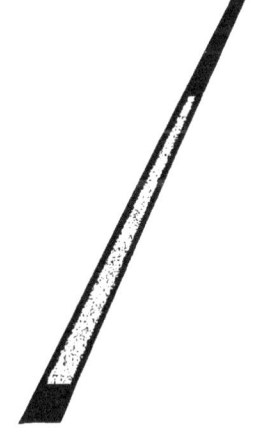

VEGETATION CONTROL
WITH
CHEMICALS

READE MANUFACTURING COMPANY, INC.

Jersey City—Chicago—Minneapolis—Kansas
City—Birmingham—Stockton

SERVING RAILROADS OF AMERICA FOR
MORE THAN FORTY YEARS

WEED AND BRUSH CONTROL

AREA Publications—Price List

The following include some of the Association publications available from the secretary's office on order. Prices shown are for Members only:

	Member Price
Manual of Recommended Practice, complete in 2 volumes, including binders (first copy)	$15.00
Extra binders, each	4.50
Annual Supplements, each	1.00

Separate Chapters

6—Buildings	1.50
7—Wood Bridges and Trestles	1.00
9—Highways	0.50
11—Engineering and Valuation Records	1.25
13—Water, Oil and Sanitation Services	1.00
14—Yards and Terminals	1.00
15—Iron and Steel Structures	1.25
16—Economics of Railway Location and Operation	0.75
20—Contract Forms	1.25
22—Economics of Railway Labor	0.50
25—Waterways and Harbors	0.25
27—Maintenance of Way Work Equipment	0.50
28—Clearances	0.25
29—Waterproofing	0.25
Flexible-cover, loose-leaf binder for separate chapters, each	0.40

Portfolio of Trackwork Plans—119 plans, 8 sheets of specifications, 5 sheets definitions of terms, complete with leatherette cover	$12.50
Track Scale Pamphlet—109 pages, flexible cover	0.80
Federal Valuation of Railroads—87 pages, flexible cover	1.00
Instructions for Mixing and Placing Concrete—24 pages, flexible cover	0.40
Notes on Railroad Location and Construction Procedures from the School of Experience—43 pages, flexible cover	0.50
Handbook of Instructions for the Care and Operation of Maintenance of Way Equipment—149 pages, hard cover	0.85
Instructions for Care and Safe Operation of Welding and Grinding Equipment—23 pages, flexible cover	0.30
Specifications for Steel Railway Bridges (fixed spans)—70 pages, flexible cover	0.75
Specifications for Movable Railway Bridges—73 pages, punched sheets	1.00

Chapters 1, 3, 4, 5, 8, and 17 are not available.

FOR MORE ON-TRACK PRODUCTION TIME
MECHANIZE WITH A
MULTI-GANG*
PACKAGE UNIT

This versatile unit Tamps—Jacks—Lines—Removes and Inserts Ties— Pulls Spikes—Drives Spikes—Bolts and Drills Rails—Transports Men, Tools and Material.

For complete information contact:

Model N U Tie Cutter

HERE IS THE WINNING TEAM

The Woolery NU Tie Cutter and the Woolery Tie-end Remover preserve the line and surface of the track and at the same time reduce the cost of tie renewals. Ties can be removed without trenching, jacking up track or adzing tops of rail-cut ties. With this team you simply cut both ends of tie, pry out center piece, insert in its place the tie-end remover and out go the tie ends pushed by the double acting, double ended hydraulic cylinder of the Tie-end remover.

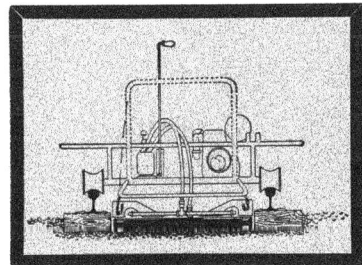 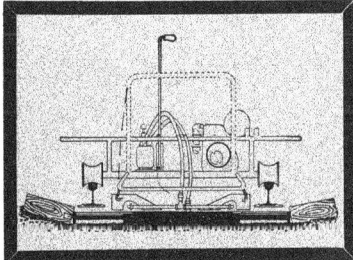

FOR HIGHEST EFFICIENCY USE TWO TIE CUTTERS WITH ONE TIE-END REMOVER

WOOLERY MACHINE COMPANY
MINNEAPOLIS, MINN.

Kill more weeds *per mile*...*per dollar*
with *Liquid* **UROX**®!

Liquid Urox Weed Killer is the *first* liquid — substituted urea-type herbicide ever developed for railroads. It's fast-acting ... withers annual and perennial grasses as well as broadleaved weeds within 12 hours after application, regardless of weather. It's long-lasting ... just *one* application wipes out weeds and brush for 8 to 18 months. What's more, control can be continued economically each year with small "booster" doses.

Liquid Urox is ideal for railroad spray trains ... doesn't need continuous agitation ... won't clog spray nozzles ... won't settle out ... can be mixed with fuel oil, diesel oil or ordinary weed oils. Write today for the complete story on railroad-proved liquid Urox Weed Killer.

Allied Chemical

GENERAL CHEMICAL DIVISION
40 Rector Street, New York 6, N. Y.

REPORTS OF COMMITTEES

The reports in this issue of the Bulletin will be presented to the 1962 convention of the Association at the Conrad Hilton Hotel, Chicago, March 9–10. Comments and discussion with respect to any of the reports are solicited, and should be addressed to the chairman of the committee involved, in writing in advance of the convention, or from the floor of the convention.

Published by the American Railway Engineering Association, Monthly, January, February, March, November and December; Bi-Monthly, June–July, and September–October, at 2211 Fordem Avenue, Madison, Wis.; Editorial and Executive Offices, 59 Van Buren Street, Chicago 5, Ill. Second class postage paid at Madison, Wis. Accepted for mailing at special rate of postage for in Section 1103, Act of October 3, 1917, authorized on June 29, 1918. Subscription $10 per annum.

Report of Committee 5—Track

S. H. Poore, *Chairman*
J. B. Wilson,
Vice Chairman
J. P. Barker, *Secretary*
R. J. Hollingsworth
C. E. Peterson
C. J. McConaughy
L. W. Leitze
L. A. Pelton
N. C. Kieffer, Jr.
R. G. Garland
V. M. Schwing
L. H. Jentoft
J. M. Salmon, Jr.

J. E. Armstrong, Jr.
M. C. Bitner
T. L. Biggar
W. R. Bjorklund
E. E. Brady
J. H. Brown
E. W. Caruthers (E)
R. J. Bruce
G. P. Chandler
J. T. Collinson
W. E. Cornell
E. D. Cowlin
F. W. Creedle
P. H. Croft
A. D. DeMoss
K. E. Dunn
H. C. Fox
R. M. Frey
J. W. Fulmer
L. W. Green
W. E. Griffiths
L. R. Hall
M. J. Hassan
A. E. Haywood
A. B. Hillman, Jr.
A. E. Hinson
E. C. Honath
A. F. Huber (E)
H. W. Jensen
C. H. Johnson

R. J. D. Kelly
C. N. King
R. E. Kleist
R. E. Kuston
E. J. Lisy, Jr.
J. E. Martin
R. E. Misner
W. L. O'Dell
E. J. Osterman
A. C. Parker, Jr.
B. E. Pearson
A. D. Quackenbush
R. P. Roden
M. K. Ruppert
R. N. Schmidt
R. D. Simpson
J. F. Smith
T. R. Snodgrass
G. R. Sproles
J. R. Talbott, Jr.
R. E. Tew
K. H. Von Kampen
C. W. Wagner
S. J. Watson
Troy West (E)
I. V. Wiley
B. J. Worley
M. J. Zeeman (E)
Committee

(E) Member Emeritus.
Members listed in bold face are the official representatives of the Engineering Division, AAR.

To the American Railway Engineering Association:

Your committee reports on the following subjects:

1. Revision of Manual.

   Certain revisions suggested by the AAR are offered for adoption, together with other revisions or reapprovals in view of the proposed reprinting of

2. Track tools, collaborating with Purchases and Stores Division AAR.

   Revisions of the Manual relating to this assignment are being offered.

3. Standardization of trackwork plans, collaborating with Communication and Signal Section, AAR.

   New plans for No. 10, 15 and 20 turnouts for main-line use, and plans for No. 6 and 8 turnouts for yard and switch movements are being offered

485

4. Prevention of damage resulting from brine drippings on track and structures, collaborating with Committee 15 and Mechanical Division AAR.

This assignment covers the laboratory phase of the investigation of brine
corrosion which has been interrupted until funds are available for further
research.

5. Design of tie plates and the use of rubber or composition plates under
insulated joints, collaborating with Committees 3 and 4 and the Communication and Signal Section, AAR.

In 1962 it is planned to inspect the service test at Oakdale, Pa., related to
insulated joints. Your committee recommends this subject be continued.

6. Hold down fastenings for tie plates, including pads under plates; their
effect on tie wear, collaborating with Committee 3.

Progress report, submitted as information .... page 494

7. Effect of lubrication in preventing frozen rail joints and retarding corrosion
of rail and fastenings.

Service tests have been started on the C&NW Railway near Low Moor,
Iowa. Description of these tests may be obtained from AAR director
of engineering research. Your committee recommends this subject be continued.

8. Laying rail tight with frozen joints.

No definitive work done on this project in 1961 and no research budget
is contemplated for 1962. However, installations as outlined in Vol. 62 of
the Proceedings are still in service and it is hoped that results can be
reported in 1963. It is recommended that this subject be continued.

9. Critical review of subject of speed on curves as affected by present day
equipment, collaborating with AAR Joint Committee on Relation Between
Track and Equipment.

This project has been delayed because of the inability to secure equipment
to run the test. The test is to be continued as soon as equipment is available. An advance report on the speed of trains through turnouts appears
on page 67 of Bulletin 566, September–October 1961.

10. Methods of heat treatment, including flame hardening, of bolted rail frogs
and split switches, together with methods of repair by welding; explosive
hardening of manganese track work.

Progress report, submitted as information .... page 495

THE COMMITTEE ON TRACK,
STUART H. POORE, *Chairman.*

Report on Assignment 1

Revision of Manual

R. J. Hollingsworth (chairman, subcommittee), J. P. Barker, R. J. Bruce, P. H. Croft, L. R. Hall, M. J. Hassan, A. F. Huber, H. W. Jensen, L. A. Pelton, S. H. Poore, A. D. Quackenbush, G. R. Sproles, J. B. Wilson, M. J. Zeeman.

Your committee submits for adoption the following recommendations with respect to Chapter 5 of the Manual:

Page 5-2-5

DESIGN OF CUT TRACK SPIKE

Fig. 1—For Use at Rail Joints and Alternate Design for General Use

Delete. Reports from steel suppliers indicate that production of this spike has been negligible.

Page 5-2-6

DESIGN OF CUT TRACK SPIKE

Fig. 2—⅝- and 9/16-In Reinforced Throat Track Spike

Reapprove without change.

Pages 5-3-1 to 5-3-4, incl.

SPIRALS

Reapprove with the following revision:

Page 5-3-1, Art. 1—Purpose, line 1. Add the words "if practicable" after the phrase "main line track."

Pages 5-3-5 to 5-3-9, incl.

STRING LINING OF CURVES BY THE CHORD METHOD

Reapprove with the following revisions:

Page 5-3-5, third paragraph. Substitute "should" for "must" in line 3 and delete the words "and safe" in line 4.

Pages 5-3-9 to 5-3-11, incl.

ELEVATIONS AND SPEEDS FOR CURVES

Page 5-3-9, fifth line from bottom of page. Substitute the words "more than normal" for the word "excessive", and in the next line substitute the word "more" for the words "causing accelerated wear."

Page 5-3-13

VERTICAL CURVES

Reapprove without change.

Page 5-3-14

PERMANENT MONUMENTS

Reapprove with the following revisions:

Change first line to read "Where permanent alinement monuments are used, they should be placed . . .", and delete the word "identically" in the last line.

Pages 5–4–1 to 5–4–4, incl.

SPECIFICATIONS FOR THE LAYING OF NEW TRACK

Reapprove with the following revisions:

Pages 5–4–2 and 5–4–3. Substitute the word "shall" for the word "must" wherever it appears, as being a better word to use in a specification.

Page 5–4–2, Par. 13, third line. Substitute the word "recommendations" for the word "standards", and in Par. 14 substitute the word "shall" for "should."

Pages 5–5–1 to 5–5–3, incl.

SPECIFICATIONS FOR LAYING RAIL

Reapprove with the following revisions:

Page 5–5–2. Delete second paragraph from top of page; revise rail temperature expansion table in Sec. C to read as follows:

33-Ft Rail 160 Joints per Mile		39-Ft Rail 135 Joints per Mile		78-Ft Rail 68 Joints per Mile	
Rail Temperature Deg F	Expansion Inches	Rail Temperature Deg F	Expansion Inches	Rail Temperature Deg F	Expansion Inches
Below −10	$5/16$	Below 6	$5/16$	Below 35	$5/16$
−10 to 14	$1/4$	6 to 25	$1/4$	35 to 47	$1/4$
15 to 34	$3/16$	26 to 45	$3/16$	48 to 60	$3/16$
35 to 59	$1/8$	46 to 65	$1/8$	61 to 73	$1/8$
60 to 85	$1/16$	66 to 85	$1/16$	74 to 85	$1/16$
Over 85	None	Over 85	None	Over 85	None

Page 5–5–3, Sec. D. Change second paragraph to read:

"After rail has been laid, the tops of adjacent rail ends may be ground to a level surface, or low connecting rail built up by welding to proper height."

Page 5–5–4

TEMPERATURE EXPANSION FOR LAYING RAILS

Reapprove with the following revisions:

Change the rail expansion table under Par. 2 to read as follows:

33-Ft Rail 160 Joints per Mile		39-Ft Rail 135 Joints per Mile		78-Ft Rail 68 Joints per Mile	
Rail Temperature Deg F	Expansion Inches	Rail Temperature Deg F	Expansion Inches	Rail Temperature Deg F	Expansion Inches
Below −10	$5/16$	Below 6	$5/16$	Below 35	$5/16$
−10 to 14	$1/4$	6 to 25	$1/4$	35 to 47	$1/4$
15 to 34	$3/16$	26 to 45	$3/16$	48 to 60	$3/16$
35 to 59	$1/8$	46 to 65	$1/8$	61 to 73	$1/8$
60 to 85	$1/16$	66 to 85	$1/16$	74 to 85	$1/16$
Over 85	None	Over 85	None	Over 85	None

Pages 5-5-4 and 5-5-4.1

RAIL CREEPAGE—NUMBER AND POSITION OF RAIL ANCHORS

Page 5-5-4. Change title to read "Rail Creepage—Number and Position of Rail Anchors for Jointed Track and Where Temperature Expansion is Provided."

In the title of Art. 1 add the word "Main" before the word "Track."

Add the word "generally" after the word "is" in line 3 of the first paragraph.

In the last paragraph, line 2, delete the words "at least 2", add the words "as necessary" after the word 'anchors", and delete the words "per rail length." In line 3 of the last paragraph delete the words "the tie with 2 of."

Page 5-5-4.1. Add the word "Main" before the word "Track" in the heading of each of the two diagrams illustrating rail anchorage. Add the word "Main" before the word "Track" in the title of Art. 2.

Page 5-5-6

TRACK BOLT TENSION PRACTICE

Reapprove without change.

Page 5-5-7

GAGE

Reapprove with the following revisions:

Change Par. (c) to read as follows:

"(c) Wide gage due to worn rail, within the permissible limits of wear, should be corrected by closing in or by interchanging the low and high rail."

Delete Par. (e) as this is covered elsewhere in Chapter 5.

Reletter Par. (f) to (e), substitute the word "are" for the word "shall be" after the words "outside spikes" in the third sentence; substitute the word "should" for "shall" and "must" throughout the remainder of the paragraph; change the fourth or last sentence thereof to read 'The old spike holes should be plugged when regaging."

Reletter Par. (g) to (f).

Page 5-5-10

LUBRICATION OF RAIL ON CURVES

Reapprove with the following revisions:

Change the word "coal" to "fuel" in second Par. (c).

Report on Assignment 2

Track Tools

Collaborating with Committee 1 and with Purchases and
Stores Division AAR

C. E. Peterson (chairman, subcommittee), T. L. Biggar, W. R. Bjorklund, E. E. Brady, R. J. Bruce, W. E. Cornell, H. C. Fox, A. E. Haywood, C. N. King, J. E. Martin, C. J. McConaughy, M. P. Oviatt, S. H. Poore, R. E. Sampson, V. M. Schwing, J. R. Talbott, Jr., J. B. Wilson, B. J. Worley.

Your committee offers the following recommendations with respect to specifications and plans for track tools appearing in the Manual, together with specifications and plans for steel drive spikes:

Pages 5–6–1 to 5–6–5, incl.

SPECIFICATIONS FOR TRACK TOOLS

Reapprove without change.

Pages 5–6–5 to 5–6–8, incl.

SPECIFICATIONS FOR ASH AND HICKORY HANDLES FOR TRACK TOOLS

Reapprove without change.

Page 5–6–8

RECOMMENDED LIMITS OF WEAR FOR TOOLS TO BE RECLAIMED

Reapprove without change.

Pages 5–6–9 to 5–6–26, incl.

PLANS FOR TRACK TOOLS

Page 5–6–9. Insert "Metal Specifications" above the headings for the list of plans. Plan 33–61. Drive Spike Extractor Socket Wrench, under "Hardness" change "See Plan" to "300–350".

Page 5–6–25. Plan 30–53—AREA spot board. Change "cast iron plate" to "malleable iron," for use with gage stop. Also change "C–9–4 X 1⅝ in–7¼ lb channel" to "4-in 7.25-lb, channel."

Pages 5–M–1 and 5–M–2

SPECIFICATIONS FOR STEEL DRIVE SPIKES

Reapprove without change.

Pages 5–M–3 and 5–M–4

PLANS FOR DRIVE SPIKES

Reapprove without change.

The following is a progress report, submitted as information:

Claw Bar

An investigation is being conducted to consider the possibility of including the former 1939 AREA claw bar as an alternate plan because a number of railroads are having difficulty with the present AREA claw bar due to the rectangular-shaped. open-ended slot at the spike-pulling end of the bar; the jaw's square corners and blunt end on the open end, and the shortness of the jaw opening seriously hamper engaging the underside of a spike head. When the bar engages the spike before pulling it, the contour of the foot of the bar requires that the bar be almost in a vertical position, which reduces the man-applied leverage; at the same time, it places an average-size man in an unbalanced position.

The 1954 and 1939 claw bars are in test on the Milwaukee Railroad at the present time for the purpose of determining what corrections are necessary in design to overcome the present objections.

Snap-on Ratchet Track Wrench

The snap-on ratchet track wrench is being tested by the Milwaukee and other railroads to determine if it is satisfactory. One railroad reported it has performed admirably in all cases, because of the light weight, good nut removal capacity, fast operation and entirely safe operation.

There were several objections reported, e.g., difficulty with frog bolts where there is insufficient room. It has also been suggested that the length of socket be reduced by reducing space allowed for excess bolt length and by other reductions which will result in better force application.

After the tests are completed, it will be determined if any changes in design are required before adopting it as a standard tool.

Report on Assignment 3

Standardization of Trackwork Plans

Collaborating with Communication and Signal Section, AAR

C. J. McConaughy (chairman, subcommittee), J. E. Armstrong, Jr., J. P. Barker, M. C. Bitner, W. R. Bjorklund, R. J. Bruce, E. W. Caruthers, E. D. Cowlin, W. E. Cornell, F. W. Creedle, A. D. DeMoss, K. E. Dunn, H. C. Fox, R. M. Frey, J. W. Fulmer, M. J. Hassan, A. E. Haywood, R. J. Hollingsworth, E. C. Honath, A. F. Huber, H. W. Jensen, C. H. Johnson, R. J. D. Kelly, N. C. Kieffer, Jr., C. N. King, R. E. Kuston, L. W. Leitze, E. J. Lisy, Jr., R. E. Misner, W. L. O'Dell, E. J. Osterman, A. C. Parker, Jr., B. E. Pearson, C. E. Peterson, S. H. Poore, A. D. Quakenbush, R. N. Schmidt, R. D. Simpson, T. R. Snodgrass, G. R. Sproles, K. H. Von Kampen, S. J. Watson Troy West, I. V. Wiley, J. B. Wilson, B. J. Worley, M. J. Zeeman.

Your committee submits for approval as recommended practice and publication in the Manual (Portfolio of Trackwork Plans) Plans Nos. 221–62, 223–62, 111–62, 113–62, 115–62, 117–62, 121–62, 123–62, 125–62, and 127–62, which are revisions of the following current plans recommended to be withdrawn: Plans Nos. 221–55, 223–55, 111–55, 113–55, 115–55, 117–55, 121–55, 123–55, 125–55, and 127–55. Also recommended is the revision of Appendix A—Specifications for Special Trackwork, by the addition of a new paragraph to page 1. The specific revisions proposed to these plans and specifications are as follows:

PLAN NO. 221–62, REVISION OF PLAN NO. 221–55, DETAILS OF SWITCH POINTS

(Plan No. 221–55 with proposed changes indicated thereon is presented in Part 2 of this issue of the Bulletin)

Add sketch showing machined flange of point rail with note reading "Remove Pointed Edges and Chamfer off Sharp Corners." On three detail sections of points add second arrow to edge on flange from existing note reading "Remove Sharp Edge". This revision is recommended to eliminate the danger of cracks starting in the sharp and jagged edges sometimes left on these rails after machining them.

APPENDIX A—SPECIFICATION FOR SPECIAL TRACKWORK

Add under Item 35, Planing, Division II, Workmanship, the following paragraph:

"Wherever the rail section is altered by machining, all sharp edges and corners must be eliminated by grinding or machining to a suitable chamfer or radius. At the gage corner where wheel contact is made, a minimum $\frac{3}{8}$" radius must be provided."

PLAN NO. 223–62, REVISION OF PLAN NO. 223–55, SWITCH PLATES AND RIGID RAIL BRACES

(Plan No. 223–55 with proposed changes indicated thereon is presented in Part 2 of this issue of the Bulletin)

Increase the shoulder lengths of the slide and heel plates to 4 in beyond the base of rail on the field side and 3 in on the gage side. This is recommended in order to reduce the plate cutting of the ties and bring the shoulder lengths in line with general practice, conforming to lengths used on other switch and frog plates.

PLANS NOS. 111–62, 113–62, 115–62, 117–62, 121–62, 123–62, 125–62 AND 127–62;
REVISIONS OF PLAN NOS. 111–55, 113–55, 115–55, 117–55, 121–55, 123–55,
125–55 AND 127–55; ALL LENGTHS OF SWITCHES HAVING UNIFORM
RISERS AND SINGLE MACHINED TURNOUT PLATES

(Plan No. 111–55 only, with proposed changes indicated thereon, is presented in Part 2 of this issue of the Bulletin. Changes on the other plans are similar)

Increase the shoulder lengths of the turnout plates to 4 in on the field side and 3 in on the gage side; except for those plates with level rail seats increase both shoulder lengths to 4 in to avoid having right and left hand plates.

Your committee has been concentrating its efforts on developing new plans for turnouts incorporating the recommendations previously submitted and approved at the 1960 March Convention. At that time five turnouts with specified lengths and types of switch points and frogs, etc., were approved. This is covered fully in the Proceedings of the 59th Annual Convention, Vol. 61, 1960, and Vol. 62, Bulletin 565, June–July 1961. Preparation of these plans has involved considerable work and investigation. However, they are now completed for the five turnouts as originally approved. The details generally are similar to those now included in the present plans with some revisions in order to cover requirements of the various railroads. The numerous alternate details of construction previously available have been eliminated in the interest of standardization. The five turnouts and their recommended use are as follows:

TURNOUTS FOR MAIN TRACK, THROUGH-TRAIN MOVEMENTS

Turnout No.	Switch	Designation
20	39' curved	High speed
15	26' curved	Medium speed
10	16'6" straight	Low speed

TURNOUTS FOR YARD AND SWITCH MOVEMENTS

Turnout No.	Switch	Recommended Use
8	16'6" straight	For movement at slow speeds. In heavy-duty locations, No. 10 main-track turnout may be used.
6	11' straight	Where limited space prevents using No. 8 turnout and 85-ft or other long cars are not likely to be used.

To eliminate the necessity of consulting numerous plans in order to get data for a complete turnout or crossover, as required under the present system, each of the five turnouts has a layout plan (Sheet 1) showing data generally used, including details for ordering the material and installing same in the field. Then each of these layout plans is followed by other sheets covering details of the switch, frog, and guard rails basically needed for manufacturing purposes.

Accordingly, your committee submits for approval as recommended practice and publication in the Manual (Portfolio of Trackwork Plans) the following new plans (all of which are presented in Part 2 of this issue of the Bulletin):

Turnout No.	Sheet No.	Title
6	1	No. 6 Turnout
6	2	11' 0" Straight Switch
6	3	No. 6 Railbound Mang. Frog and Guard Rails
6	4	No. 6 Solid Self-Guarded Mang. Steel Frog.
8	1	No. 8 Turnout
8	2	16' 6" Straight Switch
8	3	No. 8 Railbound Mang. Steel Frog and Guard Rails
8	4	No. 8 Solid Self-Guarded Mang. Steel Frog
10	1	No. 10 Turnout and Crossover
10	2	16' 6" Straight Split Switch
10	3	No. 10 Railbound Mang. Steel Frog and Guard Rails
10	4	No. 10 Spring Rail Frog
15	1	No. 15 Turnout and Crossover
15	2	26' 0" Curved Split Switch for Rails 115 to 140 lb.
15	3	No. 15 Railbound Mang. Steel Frog and Guard Rails
20	1	No. 20 Turnout and Crossover
20	2	39' 0" Curved Split Switch
20	3	No. 20 Railbound Mang. Steel Frog and Guard Rails

The plans for the Nos. 6, 8 and 10 turnouts are drawn for use with AREA recommended rail sections from 90 RA–A to and including 140 RE rails, but the plans for the Nos. 15 and 20 turnouts cover only the rail sections from 115 RE to and including 140 RE rails, since it is considered that lighter weights of rail will not be generally used for these two turnouts. All plans have been drawn to show the basic principles of construction. If they are approved at the convention all the plans will be redrawn so that they will be uniform in arrangement and set-up, making editorial changes as required for accuracy, clarity and uniformity.

We propose to place these five turnouts with the supporting details in the front of the Portfolio, suitably designated as AREA Recommended Plans and to place all present turnout plans in the back of the Portfolio, as information and use if required by various industries and smaller railroads.

Appendix 3–a

Service Test of Solid Manganese Steel Crossing Frogs with Prestressed Concrete Support versus Timbers

The prestressed concrete crossing support at the intersection of the Indian Harbor Belt Railroad and the Chicago and Western Indiana Railroad at Dolton, Ill., reported in the Proceedings, Vol. 62, page 653, has been in service approximately 17 months.

During the service period to date, the two test crossings have been given normal maintenance, with such surfacing as was needed in the fall and the spring of the year. The castings have also been given periodic bolt tightening and welding to eliminate excessive batter. At the end of the first winter of service, the concrete support was performing much better than the timbers with very little vertical movement under traffic. This situation prevailed until through July 1961, but heavy rainfall in August and September created a subgrade condition so wet that there is considerable pumping under both crossings with vertical movement of the timbers being approximately double that of the concrete.

The check of the batter and flangeway cracks of the castings made in November 1961 indicated no significant change.

An inspection on November 7, 1961, developed that a failure had occurred in one of the post-tensioning rods in the east and west run under the north rail of the IHB. The failure was just inside the anchor plate at the east end. The concrete corner was broken off at the point of failure, but there was no evidence of further distress in the concrete due to reduction of effective rods from six to five. Examination to date has not developed the cause of the break in the rod.

Because of a misunderstanding, the IHB failed to record the maintenance cost of each crossing separately during the first year of service. The cost data should be available for the second year of service. Inspections will be continued and reported to the subcommittee.

Report on Assignment 6

Hold-down Fastenings for Tie Plates, Including Pads Under Plates; Their Effect on Tear Wear

Collaborating with Committee 3

N. C. Kieffer, Jr. (chairman, subcommittee), J. P. Barker, M. C. Bitner, G. P. Chandler, E. D. Cowlin, F. W. Creedle, P. H. Croft, J. W. Fulmer, L. W. Green, L. R. Hall, A. E. Hinson, L. H. Jentoft, C. H. Johnson, R. J. D. Kelly, R. E. Kleist, L. W. Leitze, E. J. Lisy, Jr., J. E. Martin, C. J. McConaughy, W. L. O'Dell, S. H. Poore, M. K. Ruppert J. M. Salmon, Jr., R. N. Schmidt, T. R. Snodgrass, R. E. Tew, C. W. Wagner, Troy West, J. B. Wilson, I. V. Wiley.

The work under this assignment during the past year has been in the nature of laboratory investigations of tie pads and hold-down fastenings at the AAR Research Center. Although somewhat limited by available funds 20 such studies have been made involving approximately 44 million cycles of tie wear machine operation, including tests on both wood ties and concrete ties.

Of 13 investigations with wood ties, three were for and at the expense of the suppliers of the material tested. The remaining 10 were: one control test, two with plates glued to creosoted oak ties, six with tie pads and one hold-down fastening. These were mainly repeat tests for the purpose of checking or confirming similar tests made at an earlier date. The tie plates glued to the creosoted oak ties indicated failure of bond with the wood. It is believed that the failure of the bond with the wood is influenced by the creosote and water making the top of tie soft. Further investigation will be made.

The seven tests with concrete ties involved three with anchor nuts in the ties for holding clip bolts and four with plastic insulating bushings for clip bolts. The tie wear machines have played an important role in the development of pads and fastenings suitable for field use with the concrete tie.

The laboratory investigations are continuing as rapidly as funds will permit, and it is planned to publish a more complete report in 1962 giving the pertinent data in connection with various items studied. It is expected to include in the report sufficient information to permit comparison and ranking of pads, fastenings, etc.

In 1962 the AAR staff will assist the Track committee in inspecting the service tests of tie pads and tie plate fastenings on the Louisville & Nashville Railroad near London, Ky.

Report on Assignment 10

Methods of Heat Treatment, Including Flame Hardening, of Bolted Rail Frogs and Split Switches, Together with Methods of Repair by Welding

J. M. Salmon, Jr. (chairman, subcommittee), J. P. Barker, M. C. Bitner, F. W. Creedle, R. M. Frey, W. E. Griffiths, M. J. Hassan, A. E. Haywood, A. B. Hillman, Jr., A. E. Hinson, L. H. Jentoft, C. H. Johnson, R. E. Kuston, L. W. Leitze, E. J. Lisy, Jr., E. J. Osterman, A. C. Parker, S. H. Poore, Ross P. Roden, R. D. Simpson, W. F. Smith, T. R. Snodgrass, R. E. Tew, K. H. Von Kampen, C. W. Wagner, I. V. Wiley, J. B. Wilson.

The service test of 24 simulated flangeway intersections in bolted rail construction in the Milwaukee Road at Mannheim, Ill., was inspected periodically during the last service period. In November 1961 the metal flow was ground off the tread corners. No welding of the tread corner batter was required. Some repairs by welding may be necessary in 1962. Prior to welding all the units in September 1957 the traffic amounted to 91 million gross tons compared with 131 since welding.

An investigation of the explosive hardening process will be started next year in two crossings at McCook, Ill., in a service test to develop information as to the retardation of flangeway fillet cracks in the depth-hardened flangeways. Supplementary repeated load tests will be made on laboratory specimens.

Report of Committee 4—Rail

W. J. CRUSE, *Chairman*
J. A. BUNJER,
Vice Chairman
O. E. FORT, *Secretary*
C. J. CODE
K. K. KESSLER
J. C. JACOBS
EMBERT OSLAND
L. S. CRANE

T. B. HUTCHESON
A. P. TALBOT
W. D. ALMY
JOHN AYER, JR.
S. H. BARLOW
G. V. BEGANY, JR.
J. M. BENTHAM
H. B. BERKSHIRE
T. A. BLAIR (E)
BLAIR BLOWERS
B. BRISTOW
C. B. BRONSON (E)
R. M. BROWN
T. F. BURRIS
R. E. CATLETT, JR.
J. B. CLARK
M. W. CLARK
C. A. COLPITTS
C. O. CONATSER
G. H. ECHOLS
F. L. ETCHISON
D. T. FARIES
J. L. GRESSITT (E)
C. E. R. HAIGHT
V. E. HALL
C. J. HENRY
C. C. HERRICK

W. H. HOBBS
S. R. HURSH
H. W. JENKINS
R. F. LAWSON
R. R. LAWTON
LEE MAYFIELD
RAY MCBRIAN
B. R. MEYERS
F. R. MICHEAL
C. E. MORGAN
J. S. PARSONS
R. H. PATTERSON
C. F. PARVIN
B. R. PERKINS
G. L. P. PLOW
R. C. POSTELS
R. B. RHODE
C. R. RILEY
J. G. RONEY
W. D. SIMPSON
H. F. SMITH
V. R. TERRILL
J. S. WEARN
D. J. WHITE
H. M. WILLIAMSON
R. P. WINTON (E)
W. L. YOUNG
Committee

(E) Member Emeritus.
Members listed in bold face are the official representatives of the Engineering Division, AAR.

To the American Railway Engineering Association:

Your committee reports on the following subjects:

1. Revision of Manual.

   Progress report, including recommendations submitted for adoption ..... page 498

2. Collaborate with AISI Technical Committee on Rail and Joint Bars in research and other matters of mutual interest.

   Progress report, including as Appendix 2-a, Report on Investigation of Failures in Control-Cooled Rails .... page 503

3. Rail failure statistics, covering (a) all failures; (b) transverse fissures; (c) performance of control-cooled rail.

   Progress report, including statistics on rail failures reported up to December 31, 1960 (on net ton basis) .... page 511

4. Rail end batter; causes and remedies.

   No report. The investigation of the welding of battered rail ends using different welding procedures, together with the evaluation of welding rods and electrodes, is being continued.

5. Economic value of various sizes of rail.

   Progress report, presented as information .... page 528

THE COMMITTEE ON RAIL,

W. J. CRUSE, *Chairman.*

AREA Bulletin 570, February 1962.

Report on Assignment 1

Revision of Manual

J. A. Bunjer (chairman, subcommittee), H. B. Berkshire, J. B. Clark, W. J. Cruse, F. L. Etchison, O. E. Fort, V. E. Hall, C. C. Herrick, R. R. Lawton, Lee Mayfield, Ray McBrian, J. S. Parsons, R. H. Patterson, B. R. Perkins, G. L. P. Plow, R. C. Postels, J. S. Wearn.

Your committee submits the following recommendations with respect to Chapter 4 of the Manual:

Page 4–1–1

Fig. 1—90 RA–A Rail Section

Reapprove without change.

Page 4–1–2

Fig. 2—100 RE Rail Section

Reapprove without change.

Page 4–1–3

Fig. 3—115 RE Rail Section

Reapprove without change.

Page 4–1–4

Fig. 4—132 RE Rail Section

Reapprove without change.

Page 4–1–5

Fig. 5—133 RE Rail Section

Delete and substitute 136 CF&I section to be designated 136 RE rail section. (Drawing of new section presented herewith.)

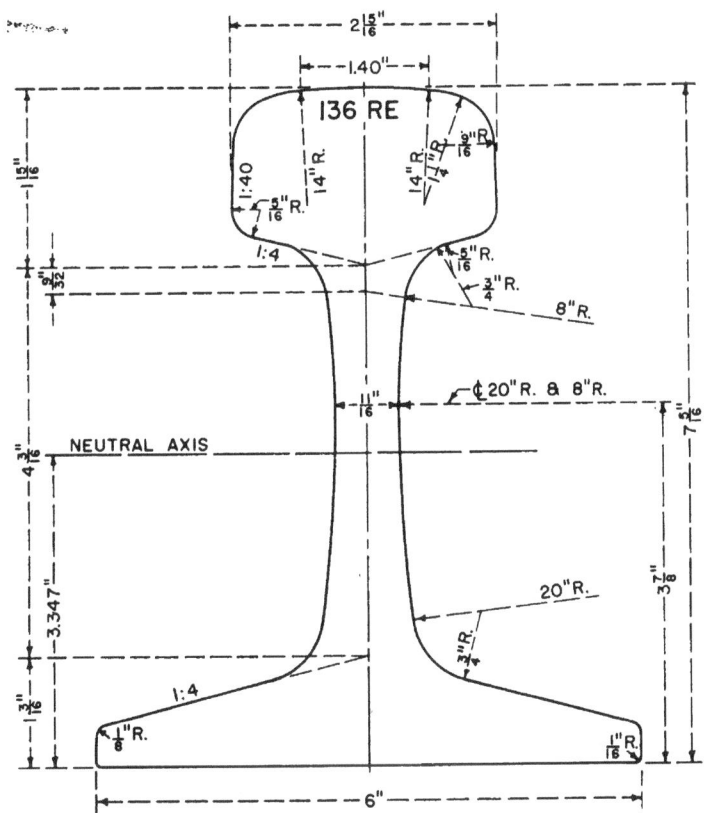

	Area Sq In	Percent
Head	4.86	36.4
Web	3.62	27.1
Base	4.87	36.5
Total	13.35	100.0

Moment of inertia 94.9
Section modulus, head 23.9
Section modulus, base 28.3
Ratio m.i. to area 7.11
Ratio s.m. head to area 1.79
Ratio height to base 1.22
Calculated weight, lb per yd 136.2

Fig. 5—136 RE rail section.

Page 4-1-6
Fig. 6—140 RE Rail Section
Reapprove without change.

Page 4-1-6.2
Fig. 1—Joint Bar and Assembly for 90 RA-A Rail
Reapprove without change.

Page 4-1-7
Fig. 2—Joint Bar and Assembly for 100 RE Rail
Reapprove without change.

Page 4-1-8
Fig. 3—Joint Bar and Assembly for 115 RE Rail
Reapprove without change.

Page 4-1-9
Fig. 4—Headfree Joint Bar and Assembly for 132 RE Rail
Reapprove with the following revision: Change title to read:
"Fig. 4—Headfree Joint Bar and Assembly for 132 RE and 136 RE rail."

Page 4-1-10
Fig. 5—Head Contact Joint Bar and Assembly for 132 RE Rail
Reapprove without change.

Page 4-1-11
Fig. 6—Head Contact Joint Bar and Assembly for 133 RE Rail
Delete.

Page 4-1-12
Fig. 7—Headfree Joint Bar and Assembly for 133 RE Rail
Delete.

Page 4-1-12.1
Fig. 7a—Head Contact Joint Bar and Assembly for 140 RE Rail
Reapprove without change.

Page 4-1-12.2
Fig. 7b—Headfree Joint Bar and Assembly for 140 RE Rail
Reapprove without change.

Page 4-1-13
Fig. 8—Recommended Head Easement for Joint Bars
Reapprove without change.

Page 4-1-14 and 4-1-15
RAIL DRILLINGS, BAR PUNCHINGS AND TRACK BOLTS
Reapprove with the following revisions:
Eliminate reference to 133 RE rails and joint bars in both Table 1, page 4-1-14, and Fig. 1, page 4-1-15, and in its place show 136 RE rails and joint bars. In Table 1, eliminate column for 133 RE, and under column for 132 RE show heading to read 132 RE and 136 RE.

Pages 4–3–16 to 4–3–19, incl.

DESIGN FOR TRACK BOLTS AND NUTS

Reapprove with the following revisions:

Page 4–1–19, Table 3—American Standard Track Bolt Nuts: Correct "Minimum" columns for bolt thickness as shown in Exhibit "A", page 502.

Pages 4–2–1 to 4–2–6, incl.

SPECIFICATIONS FOR OPEN-HEARTH STEEL RAILS

Reapprove without change.

Pages 4–2–9 to 4–2–11, incl.

SPECIFICATIONS FOR HIGH-CARBON-STEEL JOINT BARS

Reapprove without change.

Pages 4–2–12 to 4–2–14, incl.

SPECIFICATIONS FOR QUENCHED CARBON-STEEL JOINT BARS

Reapprove without change.

Page 4–3–7

FORM 402–A. REPORT OF RAIL FAILURES IN MAIN TRACK

Reapprove with the following revision:

Change Par. A under "Instructions" to read as follows:

A. Foreman must fill in this form for each rail removed from main tracks account of being broken, defective or damaged, as shown in the "Description of Rail Failures" on back of form, and send it to the Roadmaster or Supervisor on the day a broken rail is found, or on the day a damaged or defective rail is removed. Foreman will check type of failure from descriptions on back of form. Do not fill in this form for rails which are worn out, damaged, or defective in stretches and are ready for replacement, even though the failure may be a shell, crushed head, or other type of defect, whether or not the result of poor material.

Page 4–3–9

Form 402–C. Yearly Summary of Rail Failures

Reapprove without change.

Pages 4–3–10 and 4–3–11

Form 402–C (a). Instructions for Filling in Forms 402–C and 402–L

Reapprove with the following revisions:

Change Pars. 5 and 7, page 4–3–10, to read as follows, respectively:

5. Under "Other Head" list crushed heads and shelly spots (not including those from which detail fracture has developed).

7. Under "Web-In Joint" list all failures where the cause of rail removal is a bolt hole, fillet crack, or a pipe of fishtail regardless of whether it is a service failure or was detected by visual inspection or by inspection with any of the various types of defect-detecting instruments or equipment, including the supersonic type.

Page 4–3–12

Form 402–L. Annual Report of Progressive Type Head Failures

Reapprove without change.

Fig. 3—Track bolt nuts.

Standard Nut

Optional Nut

EXHIBIT "A"
PROPOSED REVISIONS
0.710
0.833
0.956
1.079
1.187
1.187

Table 3—American Standard Track Bolt Nuts

Nominal Diameter	WIDTH ACROSS FLATS W			THICKNESS U						Chamfer (c) (Optional Nut Only)
				Recommended for Medium Carbon Nuts			Recommended for Low Carbon Nuts			
D	Nominal	Maximum	Minimum	Nominal	Maximum	Minimum	Nominal	Maximum	Minimum	E
¾	1¼	1.250	1.212	¾	0.774	0.726	⅞	0.901	0.849	¼
⅞	1 7/16	1.4375	1.394	⅞	0.901	0.849	1	1.028	0.972	¼
1	1⅝	1.6250	1.575	1	1.028	0.972	1⅛	1.155	1.095	⅜
1⅛	1 13/16	1.8125	1.756	1⅛	1.155	1.095	1¼	1.282	1.218	½
Additional Sizes—Now in Use But Not Recommended for New Design										
13/16	1¼	1.2500	1.212				⅞	0.901	0.849	¼
15/16	1½	1.5000	1.450				1⅛	1.155	1.095	⅜
1	1½	1.5000	1.450				1⅛	1.155	1.095	⅜
1 1/16	1⅝	1.6250	1.575				1⅛	1.155	1.095	⅜
1⅛	1 11/16	1.6875	1.631				1¼	1.282	1.218	½

Notes: (a) All dimensions given in inches.
(b) 60 deg chamfer is optional when specified. (Dimensions for medium carbon nut are same as American Standard Heavy Nut, ASA B-18.2 for sizes shown.)
(c) This dimension not specified on ASA Standard B-18.2, which specifies diameter of top circle instead.

Page 4-3-14

Form 403-B. Diagram of Lines of Wear

Reapprove without change.

Page 4-M-7

END HARDENING OF RAILS AT MILLS

Reapprove with the following revisions:

Preface "End Hardening of Rails at Mills" with the following statement: "The following may be used as a supplement to the Specifications for Open-Hearth Steel Rails."

Eliminate the first sentence which reads as follows: "Control-cooled rails shall be used for end-hardened product."

Report on Assignment 2

Collaborate with AISI Technical Committee on Rail and Joint Bars In Research and Other Matters of Mutual Interest

W. J. Cruse (chairman, subcommittee), W. D. Almy, John Ayer, Jr., G. V. Begany, Jr., J. A. Bunjer, T. F. Burris, C. J. Code, C. A. Colpitts, L. S. Crane, D. T. Faries, O. E. Fort, C. J. Henry, T. B. Hutcheson, J. C. Jacobs, K. K. Kessler, Ray McBrian, B. R. Meyers, Embert Osland, H. M. Williamson.

This committee sponsors two research projects at the University of Illinois, both of which are under the direction of Professor R. E. Cramer. The report on the first project, entitled "Investigation of Failures in Control-Cooled Rails", is presented below as Appendix 2-a.

The report on the second project, entitled "Shelly Rail Studies at the University of Illinois", is presented as Appendix 8-b under Assignment 8.

Appendix 2-a

Investigation of Failures in Control-Cooled Rails

By R. E. Cramer

Research Associate Professor of Engineering Materials, University of Illinois

Organization and Acknowledgement

This investigation is financed by the Research Department of the Association of American Railroads. Student Assistant Jerry Crum has worked on this investigation on a part-time basis.

Control-Cooled Rails Which Failed in Service

Since our last report of October 1, 1960, reports have been prepared on 12 control-cooled rails sent to this laboratory as failed rails. These reports are sent to the railroad engineers supplying the failed rails, and copies go to the rail manufacturer and the director of engineering research, AAR, for the Association's rail failures statistics.

Table 1 lists 11 of these failures. It will be noted that each of the failures is classified differently.

Table 1

FAILED CONTROL-COOLED RAILS EXAMINED BETWEEN OCTOBER 1, 1960 and Oct. 1, 1961

T. F. = Transverse Fissure   D. F. = Detailed Fracture   V. S. H. = Vertical Split Head   H. S. H. = Horizontal Split Head
C. F. = Compound Fissure

Source of Failed Rail	Lab. Failed Rail Number	Size of Rail	Mill	Heat Number Rail Letter Ingot Number	Date Rolled	Classification of Failure
N & W	1055	136	E. Thomson	211152-D-31	4-1943	T. F. from Inclusion
Q. N. S. & L.	1056	132	Dominion		1952	D. F. from Bondwire hole
C. of Ga.	1057	115	Tennessee	89637-A	7-1960	Split Web at Elec. Weld
N. P.	1058	115	Inland	27001-G-24	4-1951	C. F. from Fishtail
I. C.	1059	112	Tennessee	84357 9-B-5	5-1945	Fatigue of Switch Point
G N	1060	112	Gary	49187	5-1939	D. F. at Bond wire weld
I. C.	1061	132	Gary	729177-D-9 .	10-1957	Fracture from Bent tie plate
I C	1062	112	Tennessee	862361-G-14	2-1947	V. S. H. from Fishtail
I C	1063	115	Inland	24050-H-14	5-1956	H. S. H. & V. S. H. from Fishtail
B & A	1 64	100	E. Thomson	204072-D	4-1935	Fracture from Derailment
C. P.	1 66	100	Algoma	24854-F-10	3-1944	D. F. from Shelling

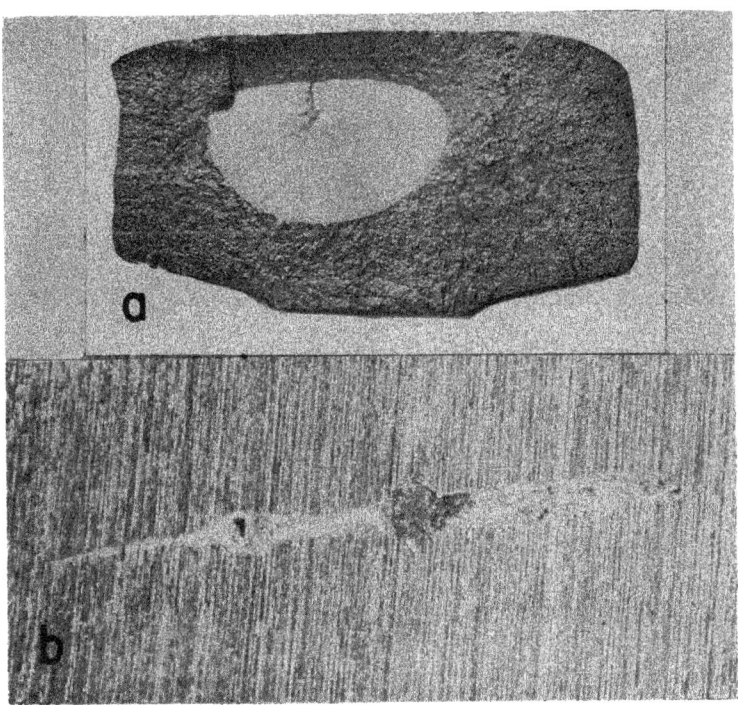

Fig. 1—Rail 1055, transverse fissure from inclusion.
a. Fissure with round nucleus.
b. Sawed slice from rail head magnified 6×. No etch.

Transverse Fissure From Inclusion

Fig. 1a shows the transverse fissure in rail 1055, from a large, round inclusion. Fig. 1b shows another inclusion found in sawing the rail head. This is magnified only 6 times, so it can be seen that it was a large piece of refractory which became trapped in the molten steel.

Split Web at Electric Flash Weld

Fig. 2 shows a service failure in an electric flash weld after only three months in service. This electric weld was made at a commercial welding plant near the Tennessee rail mill. The horizontal crack in the web was the original failure. Fig. 2b shows one side of this web fracture. There are fatigue failures starting at the surfaces of the flash metal on both sides of the web. Fig. 2c shows this same surface polished and etched in ammonium persulfate to show the heated zones. It will be noted that there are dark heat zones at each side of the weld. These areas have been heated above the critical temperature when the weld flash was ground off, apparently while the main body of

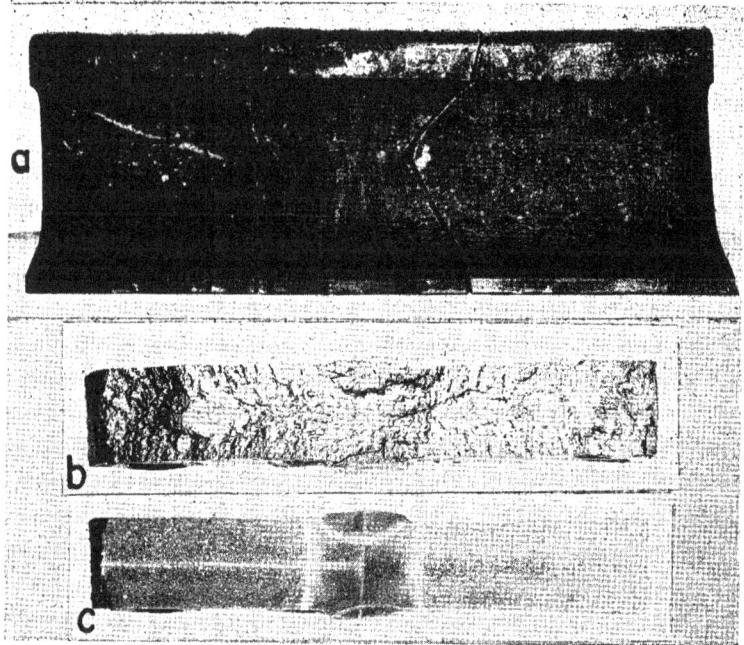

Fig. 2—Rail 1057, split web at electric weld.

a. Rail as received.

b. Horizontal fracture of web starting at weld.

c. Same piece as in b, polished and etched in ammonium persulfate to show heating of weld flashes on both sides during grinding.

the weld was cold. The latter is assumed because Rockwell hardness tests showed the reheated areas to be hard areas of 400 Brinell hardness.

Such a hard area on the side of the rail web of softer steel makes a zone where fatigue cracks would be likely to start. After these started on both sides of the rail web they penetrated across the web and the rail then fractured in service. All weld flashes should be ground off while the weld is still hot. If the weld does cool without grinding, it would be good practice to reheat the welds with torches before grinding off the flashes or to take special precautions in grinding a cold welding flash.

Fatigue of Switch Point

Fig. 3a shows a switch point which developed a fatigue area at the thin edge of the machined rail base. Fig. 3b shows more detail of the thin machined edge and the fatigue area at about $1\frac{1}{3} \times$ magnification. It is poor design to have this sharp point on the side of the rail base that is subjected to high stresses. This has been reported to

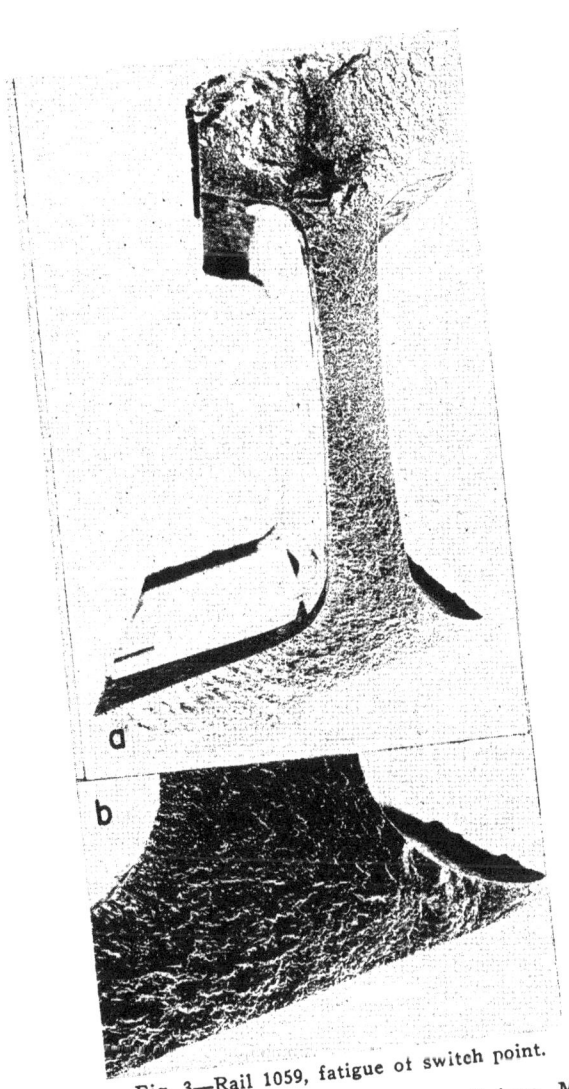

Fig. 3—Rail 1059, fatigue of switch point.
a. Fracture as received.
b. Fatigue area at thin edge of machined rail base. Magnified about 1⅛×.

the AREA Track committee, which has proposed changes in its trackwork plans to eliminate all sharp edges and corners from all places where the rail section is altered by machining.

Fig. 4—Rail 1060, detail fracture at bond wire weld.
a. Fracture and weld on side of rail head.
b. Opposite face polished and etched, showing heated area at
   bond weld. Etched in ammonium persulfate.

Detail Fracture at Bond Wire Weld

Fig. 4a shows a detail fracture covering the complete head area of rail 1060. The
picture also shows the bond wire weld on the side of the rail head where this detail fracture started. Fig. 4b shows the opposite face of the fracture ground smooth and etched

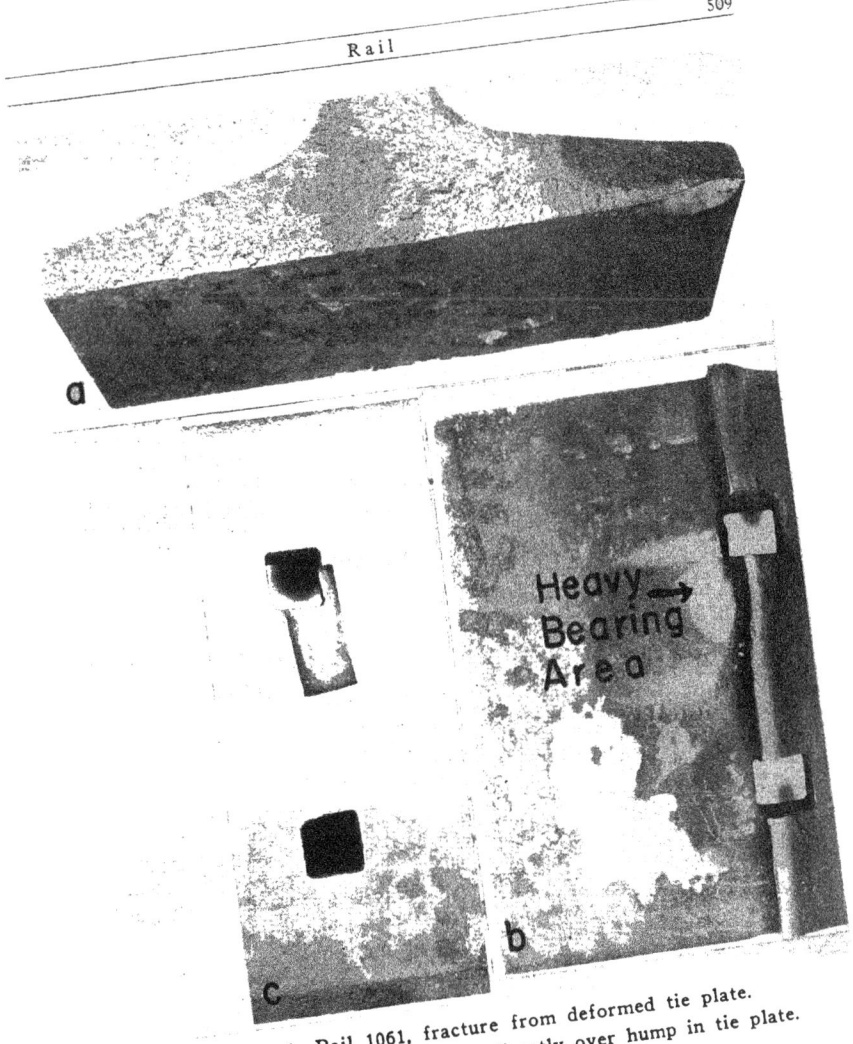

Fig. 5—Rail 1061, fracture from deformed tie plate.
a. Fatigue fracture in rail base directly over hump in tie plate.
b. Top of tie plate showing hump.
c. Under side of tie plate opposite hump.

with ammonium persulfate to show the area of the head heated during the welding of the bond wire. It has been reported that this bond wire weld was not a standard type weld but apparently was a temporary weld made by hand welding. These reheated spots become rather hard when they cool from the welding temperature, and the hard areas on the side of the rail heads may be the starting points for fatigue cracks if there are high bending stresses on the rail in service.

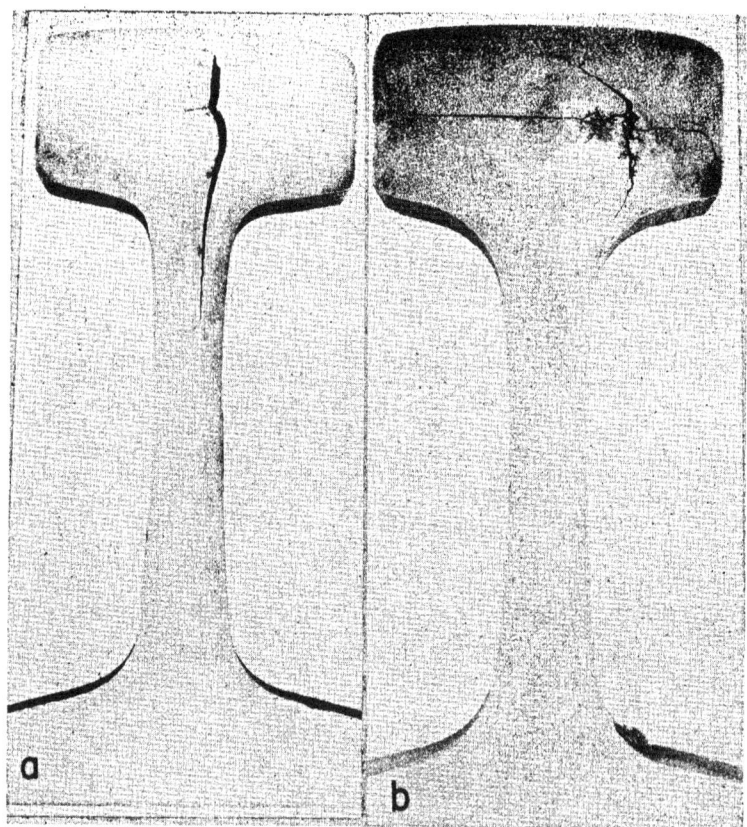

Fig. 6—Rails 1062 and 1063, failures from fishtails.
a. V.S.H. from long narrow fishtail.
b. Both V.S.H. and H.S.H. round fishtail.

Fracture from Deformed Tie Plate

Fig. 5a, rail 1061, shows a service failure in continuous welded rail that was caused by a hump in the tie plate. The dark area of the fracture was a fatigue crack, and the white area on the bottom of the rail is the bearing area on the hump in the tie plate shown in Fig. 5b. Fig. 5c shows the bottom of the tie plate opposite the bearing area; there is a deep punched mark that apparently was made when the plate was hot. It is assumed that one of the punches broke, and the broken-off end caused this defect. The hump on the top of the plate was not discovered before the plate was put in track, and it took about three years for the failure to develop in service.

Rail Failures from Fishtails

Fig. 6 shows two failures from fishtails in the bottom rails of the ingots from two different mills. Fig. 6a was a wide vertical split head at the end of the rail. The etched slice pictured shows a thin white vertical streak in the rail steel. This was the starting point for the vertical split head failure.

Fig. 6b shows another rail which developed both a horizontal split head and vertical split head at a round segregation spot in the rail head. This was also the bottom rail of the ingot, and the defect is in a fishtail area that should have been removed in cropping the rail.

Summary

The foregoing pictures and discussions illustrate the following types of rail failures:

Transverse fissure from inclusion (Fig. 1.
Split web at electric flash weld (Fig. 2).
Fatigue of switch point (Fig. 3).
Detail fracture at bond wire weld (Fig. 4).
Fracture from deformed tie plate (Fig. 5).
V.S.H. and H.S.H. from fishtail defects (Fig. 6).

Report on Assignment 3

Rail Failures Statistics Covering (a) All Failures, (b) Transverse Fissures, (c) Performance of Control-Cooled Rail

C. J. Code (chairman, subcommittee), S. H. Barlow, G. V. Begany, B. Bristow, J. A. Bunjer, J. B. Clark, M. W. Clark, W. J. Cruse, C. O. Conatser, C. E. R. Haight, H. W. Jenkins, R. F. Lawson, F. R. Micheal, Embert Osland, C. F. Parvin, B. R. Perkins, G. L. P. Plow, J. G. Roney, H. F. Smith, J. S. Wearn, D. J. White.

These statistics are based on the rail failures reported to December 31, 1960, and are submitted as information. They include the service and detected failures reported by 54 railroads on all of their main track mileage, which constitutes approximately 90 percent of the main track in the United States and Canada. This report is one of the research activities of the Association of American Railroads Research Department, W. M. Keller, vice president, and was prepared by Kurt Kannowski, metallurgical engineer, under the general direction of G. M. Magee, director of engineering research.

The accompanying tables and diagrams indicate the extent of control of the transverse fissure problem that has been obtained by the use of control-cooled rail and detector car testing, give data on the quality of each year's rollings for the various mills, and show the types of failures that are occurring on the various railroads as related to the mill producing the rail.

Transverse Fissure Failures

Data on service tranverse fissure failures and detected transverse defects are given in Table 1 and Fig. 1. Table 1 shows this information for individual roads for the 10-year period, 1951 to 1960, incl. Comparing the failures for 1960 with those for 1959 it will be noted that there was a slight increase in service failures and a decrease of about 4 percent in detected failures. Making this same comparison for individual roads shows in some cases a slight increase and in others a slight decrease, but no change that is sufficiently large to be significant.

The complete story on service and detected failures is given in Fig. 1. The significance of this figure was explained in detail in a previous report (Vol. 61, page 845) and will not be repeated here. Lines "C" and "D" were discontinued in 1958 because they had served their purpose. The most important line in Fig. 1—"A" for service transverse fissures—shows the fine improvement that has been effected by the use of control-cooled rail, detector car testing, and improved rail sections.

The reduction in detected transverse defects since 1948 is due primarily to the introduction of control-cooled rail in 1937 and the introduction of the new rail sections in 1948. It will be shown later there has been no significant reduction in detector car testing, and the decline in gross ton miles of traffic from 1948 to 1960 is only 16 percent whereas the reduction in detected failures is 50 percent. This reflects a substantial economy in the cost of replacing rails because of detected defects.

Because of the importance of detector car testing in the control of service rail failures, it has been the practice for the past several years to request data from the reporting roads on the number of track miles tested by detector cars. These data are presented below.

Year Tested	No. of Roads Reporting	Track Miles Tested by Detector Cars
1953	59	212,280
1954	56	201,134
1955	56	186,322
1956	50	196,882
1957	57	212,082
1958	54	216,731
1959	53	212,833
1960	53	206,731

It can be noted that there is a decrease of only 6,102 track miles tested by detector cars as compared to last year. This is a small decrease, and it is evident that the unfavorable business conditions in the last few years has not resulted in a corresponding reduction in detector car testing.

Mill Performance

The number of service and detected rail failures that occur during the first five years of service may be considered a good criterion of mill performance and the quality of rail as manufactured. Fig. 2 shows these failures for 1908 rollings to 1955 rollings, incl. The large decrease in failures in the rollings from 1908 to 1914 occurred during the change over from Bessemer to open-hearth production; the further decrease in the rollings to 1937 reflects improvements in mill practice and probably increase in rail size; the decrease after 1937 rollings reflects the benefits from control-cooling, and since the 1948 rollings, from improved design incorporated in the new rail sections. The low failure rates in the 1954 rollings and still lower in the 1955 rollings are believed to be due to improvements in open-hearth practices that have been put in effect in recent years.

Fig. 3 shows the control-cooled rail failure rates cumulatively for the first 10 years of service of rail rolled by the different mills. Because of the difference in service conditions on roads served by the various mills, these failure data should not be taken as necessarily indicative of the rail quality. It is interesting to note the reasons for the relatively high failure rates for some rollings. The failure rate for the Algoma mill was largely due to VSH, Other Head, Web-in-Joint, and Base failures (see Table 6). The rate of 11.5 for the 1959 rollings from this mill was due largely to 7 VSH, 4 Other Head, and 20 Web-in Joint failures that occurred in 100 REHF rail on the Canadian Pacific. Web-in-Joint failures are considered to be due to service conditions rather than rail quality. The failure rate for Carnegie–E. T. rollings was influenced by the relatively high number of CF and DF failures on the Norfolk & Western, which types of failure are also considered to be due to service conditions, particularly heavy traffic and wheel loads and sharp curvature. The Colorado rollings since 1955 show a low failure rate, and the rate for previous years was influenced to a large extent by the large number of CF and DF failures that occurred on the Union Pacific due to service conditions. The failure rate for the Dominion rollings to 1955 was due to the relatively high number of Web-in-Joint failures on the Canadian Pacific. The report for two years ago contained a statement from C. A. Colpitts, chief engineer of the Canadian Pacific, giving the difference in service and defect detection conditions of rail from the Algoma and Dominion mills. The failure rates for the rollings from the other mills require no comments.

Carnegie–E. T. stopped rolling rail in 1958 and Inland in 1957. Dominion rolled no rail in which failures were reported for the railroads included in these statistics in 1955, 1958, and 1959, and Lackawanna in 1959. A comparison of Fig. 3 with corresponding data in the report of 11 years ago shows a marked reduction in failure rate, most of which is due to the new rail sections introduced in 1947. This is further shown in Fig. 4 which gives the service and detected failures per 100 track miles that have occurred to December 31, 1960, for each year's rollings from 1950 to 1959, incl. The dashed line shows the corresponding data for the old sections, 1938–1947 rollings, incl., also control-cooled rail. The dropoff in failure rate for the old sections for the tenth year of service, and to some extent, for the ninth year, should be disregarded because the decrease was due to the fact that considerable tonnage of the original rollings having the highest failure rate was removed from track. It should also be pointed out that the record for the new sections is actually better than shown because many of the failures reported actually occurred in the old rail sections rolled after 1947, as will be discussed more fully later.

Types of Failures

Table 5 shows the accumulated service and detected failures per 100 track miles in rollings 1950 to 1959, incl., that have occurred to December 31, 1960, by types of failures and by mills. This table is helpful in assessing the relative importance of the different types of failures. For example, the CF and DF classification comprises 40 percent of the total failures and Web-in-Joint failures, 25 percent. Both of these types of failures are considered to be due to service conditions and to some extent rail design rather than to mill quality.

To give some indication of the extent to which the "new rail sections" adopted in 1947 have affected the number of failures of each type, the following tabulation shows the accumulated failures in the "old sections" in the 1938 to 1947 rollings, incl.; in the 1950 to 1959 rollings, incl., which include mostly new but some of the old sections; and in the 1950 to 1959 rollings, incl., of the new sections only. These new sections include the 106 CF&I, 112 TR, 115 RE, 119 CF&I, 127 NYC (Mod), 129 TR, 132 RE, 133 RE 136 CF&I 140 RE and 155 PS.

	Accumulated Failures Per 100 Track Mile Years		
	Old Sections (1938–1947)	All Sections (1950–1959)	New Sections (1950–1959)
TF—Verified by University of Illinois	0.02	0.01	0.01
CF and DF	1.73	1.61	1.85
VSH	0.58	0.25	0.09
HSH	0.53	0.12	0.07
Other Head	0.52	0.62	0.41
Broken	0.75	0.06	0.03
Web-in-Joint	3.34	1.02	0.13
Web—Other	1.47	0.19	0.06
Base	0.30	0.15	0.02
All Types	9.24	4.03	2.67

It is evident from the above comparison that no improvement has been effected in the CF and DF classification with the new sections. This does not necessarily mean that the new rail sections are not equally or even more resistant to shelling, because there has been a considerable increase in wheel loads between these two periods, and wheel load is thought to be the most important factor in producing shelling. There has been a large and gratifying reduction in the number of web failures. In this connection it should be noted that the substantial tonnage of welded rail being laid in recent years may be expected to reduce still further the number of Web-in-Joint failures. The change over to diesel power has probably also been a factor in the reduction of Web–Other failures. The substantial reduction in VSH and HSH failures is probably to be attributed to improved mill practices, and in Broken to the change over to diesel power, use of heavier rail, and improved maintenance and operating practices. The gross ton miles of traffic has been only about 6 percent less in the 10 years period for the new sections than for the old sections.

The rail failure reports were put on punch cards for the first time this year which has permitted a comparison of failures reported by rail sections. The results so obtained are shown in Table 5A. Generally, the failure rates per 100 track mile years are quite low. The rates for the CF&I sections are not too significant as yet because of the relatively few years of service. The very high failure rate in the 133 RE section is due to the large number of CF and DF and Other Head failures on the Union Pacific and it is believed that these are due to service conditions and perhaps mill quality rather than rail design.

Table 6 is of interest in comparing the failures by type, by mill, and by individual railroads. It will be noted that most of the failures reported in this table have occurred on a relatively few railroads. It is suggested that study of service conditions on these railroads might be worthwhile in an effort to determine whether practical measures might be found to reduce the failure rate.

Table 7 shows the service and detected failures in the rail web within joint bar limits. Comparing these results with those reported last year it will be noted that the number of joints reported inspected with defect detecting instruments increased from 10,985,340 in 1959 to 13,208,822 in 1960. As a result of this increase in testing the number of detected defects in control-cooled rail (rolled after 1937) increased from 9,013 in 1959 to 9,716 in 1960, and the number of service failures decreased from 3,769 in 1959 to 3,206 in 1960.

Professor R. E. Cramer at the University of Illinois examines every year rail failures submitted by some of the railroads which are thought to be transverse fissures.

These are reported in Table 8 as Accumulated Transverse Fissure Failures in Control-Cooled Rail by mill and year rolled. These failures are classified as transverse fissure failures from shatter cracks, from inclusions and from hot torn steel. Transverse fissures from shatter cracks have been practically eliminated by proper control-cooling. No transverse fissure from shatter cracks has been reported in the rollings since 1951. The transverse fissures from inclusions and hot torn steel are due to mill practices and can be kept to a minimum by proper quality control.

A Table 9 has been added this year because Assignment 11—Rail Damage Resulting from Engine Burns; Prevalence; Means of Prevention; Repair by Welding, has been discontinued. It was felt by the committee that no satisfactory substitute has been developed for the repair of engine burns by welding and that the continued use of the process, properly supervised, is recommended. Even though this table does not give a complete picture of the welded engine burn failures it is felt there is a sufficient sampling of the railroads using the process to indicate that satisfactory results may be obtained.

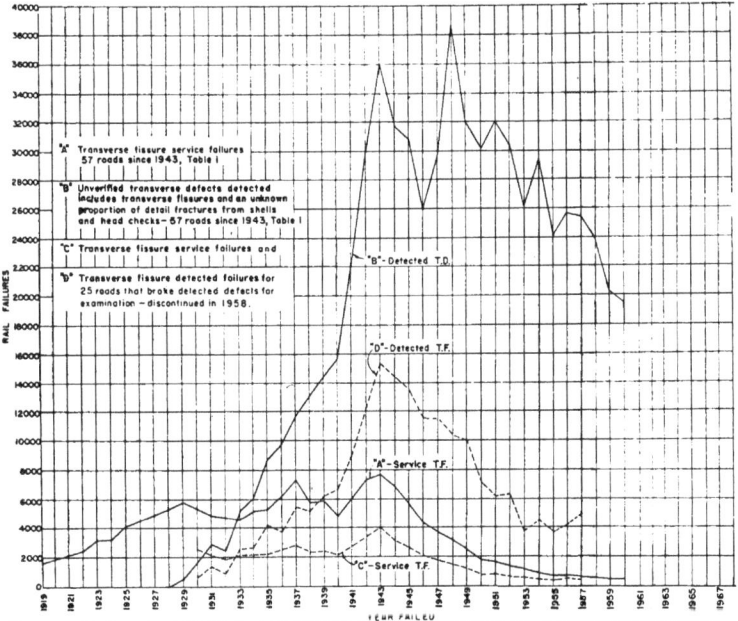

FIG. I - ANNUAL SERVICE RAIL FAILURES DUE TO TRANSVERSE FISSURES AND TO DETECTED TRANSVERSE DEFECTS AS REPORTED BY ALL RAILROADS

Rail

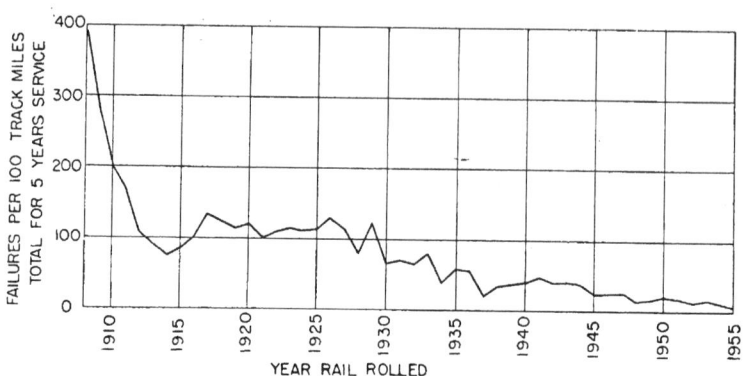

FIG. 2- SERVICE AND DETECTED FAILURES IN UNITED STATES AND CANADA

MILL	YEAR ROLLED	CONTROL COOLED RAIL FAILURES PER 100 TRACK MILE YEARS - ALL TYPES EXCEPT ENGINE BURN FAILURES	MILL	YEAR ROLLED	CONTROL COOLED RAIL FAILURES PER 100 TRACK MILE YEARS - ALL TYPES EXCEPT ENGINE BURN FAILURES
ALGOMA	1950	15.1	INLAND	1950	1.5
	1951	8.5		1951	1.4
	1952	8.7		1952	1.2
	1953	12.4		1953	1.4
	1954	9.5		1954	0.5
	1955	13.5		1955	0.5
	1956	3.7		1956	1.0
	1957	3.0		1957	0.5
	1958	3.9		1958	-
	1959	11.5		1959	-
CARNEGIE-E.T.	1950	8.2	LACKAWANNA	1950	5.8
	1951	2.7		1951	1.3
	1952	6.4		1952	2.8
	1953	13.0		1953	1.1
	1954	11.0		1954	0.8
	1955	4.7		1955	0.6
	1956	12.0		1956	0.3
	1957	1.1		1957	1.8
	1958	-		1958	2.1
	1959	2.9		1959	-
COLORADO	1950	11.1	STEELTON	1950	1.1
	1951	13.2		1951	0.7
	1952	5.3		1952	1.9
	1953	8.3		1953	0.8
	1954	4.0		1954	1.3
	1955	1.3		1955	0.5
	1956	1.4		1956	0.5
	1957	0.8		1957	0.4
	1958	0.8		1958	0.3
	1959	0.6		1959	0.5
DOMINION	1950	30.4	TENNESSEE	1950	1.9
	1951	17.1		1951	0.9
	1952	21.2		1952	0.5
	1953	32.8		1953	0.9
	1954	16.4		1954	0.3
	1955	-		1955	0.3
	1956	0.9		1956	0.1
	1957	1.2		1957	0.4
	1958	-		1958	0.3
	1959	-		1959	0.5
GARY	1950	3.4	ALL MILLS	1950	5.8
	1951	1.4		1951	4.7
	1952	1.2		1952	4.0
	1953	0.9		1953	4.4
	1954	0.5		1954	2.7
	1955	0.3		1955	1.4
	1956	0.3		1956	1.5
	1957	0.3		1957	0.8
	1958	0.1		1958	0.9
	1959	0.4		1959	2.0

FIG 3 - CONTROL COOLED RAIL FAILURE RATES TO DECEMBER 31, 1960 BY MILLS - ALL TYPES EXCEPT ENGINE BURN FAILURES - SERVICE AND DETECTED.

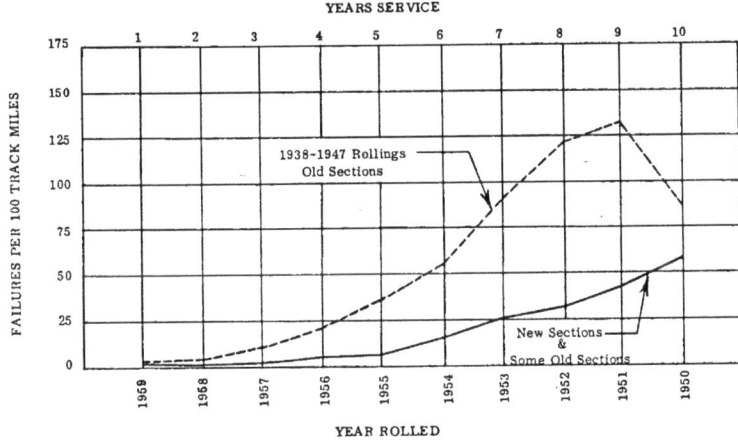

Fig. 4 - Control Cooled Rail Failures to December 31, 1960 Per 100 Track Miles – All Types
Excluding Engine Burn Fractures - Service and Detected.

TABLE 1 - SERVICE FAILURES FROM TRANSVERSE FISSURES AND DETECTED FAILURES FROM TRANSVERSE DEFECTS BY RAILROADS AND BY YEAR FAILED - ALL ROLLINGS BY ALL PROCESSES

Service Failures

Year Failed	1951	1952	1953	1954	1955	1956	1957	1958	1959	1960	Total
AT&SF	52	54	18	19	16	15	7	9	7	12	209
ACL	9	4	7	4	1	3	2	5	6	9	50
B&O	100	91	100	76	59	150	150	102	95	101	1024
B&OCT	2	0	2	4	5	12	6	9	6	6	52
Ban Aroos	0	0	0	1	0	0	0	0	0	1	2
B&LE	2	2	5	0	1	1	1	0	0	0	12
B&A	-	-	-	-	-	1	1	1	2	3	8
B&M	7	24	21	25	25	15	19	17	10	16	164
CP	75	47	47	31	39	26	20	16	20	10	337
C. of Ga.	24	14	12	10	4	5	0	0	1	0	70
C.&O. (Sys.)	29	43	21	16	39	19	52	30	22	13	284
C&EI	15	21	18	19	14	14	1	1	0	0	102
C&NW	85	74	72	62	73	38	38	37	29	27	535
CB&Q	59	28	23	17	12	8	7	5	10	7	176
CI&L	4	5	2	5	-	0	0	4	0	0	20
CMStP&P	42	45	44	30	21	21	11	16	12	6	248
CRI&P	56	46	49	27	22	21	13	8	8	13	263
C&S (n)	0	4	-	2	0	1	0	1	3	0	11
D&H	10	10	5	0	0	0	0	0	0	0	25
DL&W	3	5	-	-	-	-	-	-	-	(b)	8
D&RGW	14	-	-	-	3	4	3	5	0	(b)	32
Erie	19	20	19	9	10	15	3	10	9	13	127
FEC	1	7	0	9	4	0	1	0	0	0	12
GTW	9	7	16	6	4	9	1	8	2	0	62
GN	94	63	47	45	42	29	24	26	23	14	407
IC (Sys.)	55	28	33	25	10	15	5	6	6	5	188
JCL (NY&LB)	14	14	11	9	7	7	5	6	0	5	76
KCS	6	6	0	8	1	2	10	9	3	10	48
L&HR	1	1	0	0	0	0	0	2	0	0	2
L&NE	0	1	0	2	0	0	2	1	1	0	7
LV	12	20	8	2	8	2	8	3	2	3	68

Detected Failures

Year Failed	1951	1952	1953	1954	1955	1956	1957	1958	1959	1960	Total
AT&SF	1169	1088	970	908	1278	1120	1178	1423	917	1166	11217
ACL	758	741	853	851	836	594	648	565	603	590	7039
B&O	745	641	596	696	650	752	909	670	784	779	7222
B&OCT	7	2	0	3	1	1	2	0	0	1	17
Ban Aroos	7	11	24	22	16	33	23	18	14	22	190
B&LE	0	138	36	0	26	27	45	13	31	11	327
B&A	-	-	-	-	-	52	41	22	12	20	147
B&M	356	311	292	284	300	282	250	50	154	122	2401
CP	3800	4251	1504	1805	1375	1258	976	1210	1084	563	17826
C. of Ga.	243	301	0	367	570	426	694	685	637	859	4782
C.&O. (Sys.)	643	684	683	552	383	523	200	497	346	194	4715
C&EI	212	97	150	206	152	136	253	154	181	190	1731
C&NW	1216	817	836	861	778	807	675	838	1020	1037	8885
CB&Q	679	687	463	940	922	795	825	293	277	381	6262
CI&L	80	58	58	75	81	124	66	69	36	117	764
CMStP&P	763	733	982	923	810	893	864	749	668	679	8069
CRI&P	369	537	322	412	528	744	1076	228	239	339	4794
C&S (n)	55	182	327	308	133	140	121	5	-	-	1271
D&H	430	367	327	308	314	219	140	124	7	7	2243
DL&W	225	257	-	-	-	-	-	-	-	-	482
D&RGW	645	-	-	455	514	502	436	310	449	406	3717
Erie	475	459	435	469	270	317	275	180	161	272	3313
FEC	13	30	94	175	129	73	87	61	77	27	766
GTW	85	71	96	74	82	67	99	117	77	118	886
GN	1069	1525	932	1163	914	1362	1262	583	695	642	10147
IC (Sys.)	1226	1279	1088	908	830	826	782	593	602	521	8655
JCL (NY&LB)	93	30	15	58	83	67	56	42	33	25	502
KCS	11	56	89	58	47	54	71	56	90	87	619
L&HR	11	8	8	6	6	1	3	4	5	0	52
L&NE	12	12	0	14	15	0	15	8	24	25	125
LV	163	105	108	123	94	147	72	57	69	32	970

TABLE 1 (Continued)

Service Failures

Year Failed	1951	1952	1953	1954	1955	1956	1957	1958	1959	1960	Total
LI	0	0	0	0	12	-	9	4	-		25
L&N	63	40	40	29	29	27	30	24	25	23	330
Mo. Cent.	2	3	2	-	1	8	5	1	1	3	26
NSl.P&SSM	11	6	2	0	10	1	1	2	0	3	35
MKT	19	13	14	8	8	18	3	1	1	0	85
N.P Lines	72	51	42	23	11	6	11	16	9	18	259
NC&StL	6	5	4	1	1	2	0	(c)	(c)	(c)	19
NYC (Sys.) (d)	177	130	155	114	59	95	63	66	39(e)	28(e)	926
NYC&StL	16	13	10	9	7	6	4	7	4	6	82
NYNH&H	1	0	2	4	5	4	2	1	1	0	20
NYC&W	25	10	8	6	1	3	3	(f)	(f)	(f)	53
N&W	1	3	2	4	3	7	3	3	4	5	35
NP	11	1	0	-	1	9	4	7	3	0	36
PRR	103	85	122	76	67	129	95	51	46	48	822
P&LE	1	0	0	0	0	0	0	0	0	0	1
Reading	0	0	0	0	0	0	0	5	0	0	5
EF&P	0	0	0	0	1	0	0	2	0	0	2
Rutland	4	0	2	5	6	3	0	1	0	1	22
StL-SF	16	0	3	5	6	7	3	1	4	0	45
SAL	7	14	7	12	9	6	10	17	15	19	116
SP	75	76	69	63	55	38	23	13	20	23	455
Southern	106	70	63	41	32	20	24	36	19	30	441
TENO	88	79	55	28	16	1	0	4	17	31	319
T&P	8	7	5	6	3	2	5	4	2	5	47
UP	13	15	13	18	10	7	1	5	3	0	85
Virginian	0	1	0	0	0	3	7	0	0(h)	(h)	11
W. Md.	15	14	8	4	0	2	0	0	0	1	44
ALL ROADS	1639	1320	1207	913	767	837	693	600	493	506	8975

Detected Failures

Year Failed	1951	1952	1953	1954	1955	1956	1957	1958	1959	1960	Total
LI	18	25	15	14	49	-	13	19	-	14	167
L&N	843	925	711	592	696	589	606	760	646	815	7185
Mo. Cent.	89	55	46	-	36	36	43	45	28	44	422
NSl.P&SSM	116	138	81	632	48	68	67	59	94	61	1364
MKT	290	162	270	259	212	245	187	153	174	245	2197
N.P Lines	1308	1267	981	783	725	594	574	608	555	484	7879
NC&StL	43	43	11	3	6	10	7	(c)	(c)	(c)	123
NYC (Sys.) (d)	1344	1284	1151	1612	508	956	1152	841	584(e)	348(c)	9780
NYC&StL	84	77	60	56	39	41	32	11	56	33	487
NYNH&H	77	48	73	56	62	81	118	40	42	74	671
NYC&W	20	86	110	182	28	37	(f)	(f)	(f)	(f)	463
N&W	124	157	113	135	103	166	403	248	478	389	2316
NP	377	372	417	429	339	436	381	373	269	203	3596
PRR	1844	1322	1792	1574	1600	1880	1783	1810	1430	594	16029
P&LE	12	20	26	22	26	10	9	7	7	1	140
Reading	451	205	396	460	271	290	274	299	297	351	3294
EF&P	128	64	55	25	24	18	18	23	12	20	387
Rutland	28	0	0	0	0	0	0	0	0	0	28
StL-SF	377	410	398	346	309	568	1674	2145	287(g)	198	6712
SAL	324	318	344	344	300	253	259	257	230	317	2946
SP	960	970	1005	728	794	934	833	623	368	259	7504
Southern	1682	1490	1622	2157	1732	1442	1263	1579	1225	1208	15400
TENO	788	749	606	708	456	463	353	470	115	182	4890
T&P	114	90	70	115	58	57	49	35	77	60	725
UP	4142	4157	4632	5003	4074	4044	3113	3421	3640	3722	39948
Virginian	73	45	39	41	57	66	84	197	77(b)	(h)	679
W. Md.	133	156	153	59	105	84	35	33	86	59	903
ALL ROADS	31333	30178	26138	29376	24578	25718	25439	23848	20303	19460	256371

- No report received
(a) Includes FW&D failures
(b) Merged with the Erie Railroad forming the Erie-Lackawanna Railroad Company
(c) Merged with the L&N Railroad
(d) Includes IHB and CCC&StL Failures
(e) NYC (Western District did not report)
(f) Ceased operations 3-29-57
(g) Includes engine burn fractures
(h) Merged with the N&W Railroad

TABLE 2 - TONS OF RAILS AND TRACK MILES OF EACH YEAR'S ROLLINGS 1950 - 1959 INCL., REPORTED BY 54 RAILROADS

Year Rolled	OH CONTROL COOLED ONLY	
	TONS	TRACK MILES
1950	1,311,038	6,335.78
1951	1,229,261	5,908.58
1952	987,006	4,796.63
1953	1,207,782	5,683.69
1954	823,831	3,833.60
1955	883,125	4,085.60
1956	865,017	4,024.40
1957	814,723	3,480.30
1958	401,532	1,855.90
1959	458,260	2,131.10
TOTAL	8,981,575	42,135.58

TABLE 3 - SERVICE AND DETECTED FAILURES OF ALL TYPES EXCEPT ENGINE BURN FAILURES ACCUMULATED FROM DATE ROLLED TO DECEMBER 31, 1960, PER 100 AVERAGE TRACK MILES, CONTROL COOLED RAIL ONLY, IN ALL ROLLINGS, FROM ALL MILLS

Year Rolled	YEARS OF SERVICE									
	1	2	3	4	5	6	7	8	9	10
1950	3.1	6.0	9.6	14.1	19.6	25.9	35.8	45.6	54.9	57.7
1951	2.0	3.4	5.0	10.4	16.3	21.7	27.5	34.9	42.7	
1952	2.0	2.8	4.3	7.9	13.2	18.7	25.3	32.3		
1953	0.8	2.0	4.0	9.2	15.6	22.5	30.6			
1954	0.5	1.3	3.5	6.1	10.9	15.9				
1955	0.7	1.5	2.3	4.0	6.9					
1956	0.4	1.3	3.3	6.3						
1957	0.5	1.3	2.5							
1958	1.0	1.8								
1959	2.0									

TABLE 4 - TRACK MILES AND 1960 FAILURES, ALL TYPES, IN ROLLINGS 1950 TO 1959, INCL.
OPEN-HEARTH CONTROL-COOLED RAIL ONLY

ROAD	TRACK MILES BY MILL										1960 FAILURES	
	ALG	CARN	COLO	DOM	GARY	INLD	LACKA	ST.LTN	TENN	TOTAL	EBFs EXCL.	EBFs ONLY
AT&SF			1839		886	227				2752	21	5
ACL								241	894	1135	4	
B&O		378			412		45	583		1418	6	
B&OCT						29				29	1	
Bon Arbor								50		50		
B&LE		39								39	2	
Bos & Alb							45			45		
B&M	11							63		74		
CP	3464			794			30			4288	730	8
C. of Ga.									235	235	1	
C&O (Sys.)	11	4			753	577	233	69		1647	9	
C&EI					85	11				96	1	
C&NW					459	143	132			734	1	
CB&Q			746		781	151				1678	8	1
CI&L					79	37				116		
CStP&P					881	260				1141	3	1
CRI&P			160		587	184				931	1	
CCC&StL												
-P&E					272	52	11			335	128	2
C&S			374							374		
D&H								145		145		2
D&RGW			306							306	28	
Erie		217			297	20	63	6		603	3	
FEC		14						30	114	158	3	
GTW					164	53	27			244	5	
GN			491		589	228	200	11		1519	20	
IC					935	377			270	1582	27	3
KCS					182	23			12	217		
L&HR								18		18		
L&NE								16		16		
LV							224			224		
L&N					51				1560	1611	36	6
Mo. Cent.							71	3		74	1	
MStP&SSM	11				245	206	12			474	3	
MKT			53		87	20				160		
MP Lines			671		624	112			110	1517	5	
NYC - E					48	25	550			623	16	
NYC&StL		34			277	40	98			449	5	
NYNH&H		77						284		361	4	2
N&W		784						331		1115	305	4
NP			306		309	68	121			864	21	5
PRR		408			164	44		375		991	5	
P&LE		101								101	1	
Reading								297		297		
RF&P								137		137		
St. L-SF									750	750	2	
SAL								433	662	1095	3	2
SP			2234							2234	205	11
Southern					16	15		699	1302	2032	12	
T&NO			559						115	674	43	1
T&P			134						175	309		
UP			1559		591	87				2237	693	3
W. Md.		99						199		298	1	
TOTAL	3497	2155	9292	794	9774	2989	1862	3990	6199	40552	2363	56

NOTE: The following railroads did not report in 1961 and were omitted from this table. UtB, JC1. (NY&LB) and NYC - W
The NC&StL merged with the L&N Railroad
The Virginian merged with the N&W Railroad
The DL&W merged with the Erie Railroad

TABLE 5 - ACCUMULATED FAILURES AND FAILURES PER 100 TRACK MILES, IN ROLLINGS 1950 TO 1959 INCLUSIVE FROM DATE ROLLED TO DECEMBER 31, 1960, SERVICE AND DETECTED, BY MILL AND TYPE OF FAILURE

OH CONTROL COOLED RAIL ONLY

MILL	ACCUMULATED FAILURES TO DEC. 31, 1959 (EXCL. EBFs)										TRACK MILES	TRACK MILE YEARS	FAILURES PER 100 TRACK MI. YEARS
	TF VER U of I	CF & DF	VSH	HSH	OTHER HEAD	BROKEN	WEB IN JT.	WEB OTHER	BASE	ALL TYPES			
ALGOMA	9	12	361	8	412	40	980	20	329	2171	3484	21428	10.13
CARNEGIE (ET)	2	871	7	14	65	4	41	5	8	1017	2369	16111	6.31
COLORADO		2742	86	216	640	24	444	268	8	4428	9710	59278	7.47
DOMINION		3	70	4	69	3	969	6	35	1159	793	5747	20.17
GARY		300	45	18	275	37	76	154	7	912	9879	62601	1.46
INLAND	3	111	11	7	30	15	48	9	3	237	3088	20153	1.18
LACKAWANNA	2	45	41	5	97	10	91	12	20	323	2003	12624	2.56
STEELTON	10	254	28	11	40	5	29	5	1	383	4271	36306	1.05
TENNESSEE		52	42	31	54	15	91	33	3	321	6281	37605	.85
ALL MILLS	26	4390	691	314	1682	153	2769	512	414	10951	41877	271853	
FAILURES PER 100 TRACK MILE YEARS	.01	1.61	.25	.12	.62	.06	1.02	.19	.15	4.03			

Definition of symbols:

TF - Transverse Fissure   VSH - Vertical Split Head
CF - Compound Fissure   HSH - Horizontal Split Head
DF - Detail Fracture   EBF - Engine Burn Fracture

TABLE 5a – ACCUMULATED FAILURES AND FAILURES PER 100 TRACK MILES, IN ROLLINGS 1950 TO 1959, INCLUSIVE FROM DATE ROLLED TO DECEMBER 31, 1960, SERVICE AND DETECTED, BY RAIL SECTION AND TYPE OF FAILURE

OH CONTROL-COOLED RAIL ONLY

RAIL SECTION	ACCUMULATED FAILURES TO DEC. 31, 1960 (EXCL. EBFs)										TRACK MILES	TRACK MILE YEARS	FAILURES PER 100 TRACK MI. YEARS
	TF VER U of I	CF & DF	VSH	HSH	OTHER HEAD	BROKEN	WEB IN JT.	WEB OTHER	BASE	ALL TYPES			
106 RE			1							1	30	76	1.32
112 TR		5	2	2	1		1			11	1690	11035	0.10
115 RE	1	130	84	34	178	44	70	35	7	583	12552	84986	0.68
119 CF&I					1		1		1	3	589	1689	0.18
127 NYC (Mod)		4	7		32	1	6	3	10	63	741	3735	1.69
129 TR		7		3	2		4			16	362	2505	0.64
132 RE	7	1192	64	39	133	24	147	45	9	1660	10572	71699	2.32
133 RE		2581	21	59	496	3	25	47	4	3236	2568	16616	19.48
136 CF&I		3	2		3		1		3	12	1560	4782	0.25
140 RE	9	30	5	2	23	2	17	3		91	2415	14954	0.61
155 PS		18	3	4	6					31	303	2063	1.50
TOTAL	17	3970	189	143	875	74	272	133	34	5707	33382	214140	2.67
TOTAL PER 100 TRACK MILE YEARS	0.01	1.85	0.09	0.07	0.41	0.03	0.13	0.06	0.02	2.67			

Rail

TABLE 6 - ACCUMULATED FAILURES OF ALL TYPES FOR OH CONTROL COOLED RAIL, ONLY IN ROLLING 1950 - 1959, INCL., ACCUMULATED TO DECEMBER 31, 1960, SERVICE AND DETECTED, SEGREGATED BY ROADS AND MILLS.

ROADS	TF Ver U of I	CF & DF	VSH	HSH	Other Head	Broken	Web In Jt.	Web Other	Base	EBFo Excl. Accum. Total	EBFo Excl. 1960	EBFs Only Accum. Total	EBFs Only 1960
ALGOMA													
CP	9	12	361	8	411	40	980	20	329	2170	442	8	8
C&O (Sys.)	0	0	0	0	1	0	0	0	0	1	1	0	0
TOTAL	9	12	361	8	412	40	980	20	329	2171	443	8	8
CARNEGIE													
B&O	0	2	0	1	5	0	2	0	2	12	1	3	0
B&LE	0	0	0	2	1	0	0	0	0	3	2	0	0
Erie	0	0	0	0	1	1	6	0	0	8	3	7	0
NYC&StL	0	0	0	0	1	0	9	0	0	10	1	0	0
NYNH&H	2	0	0	0	0	0	0	0	0	2	1	2	2
N&W	0	855	4	7	51	2	18	5	6	948	261	26	3
PRR	0	14	0	3	6	0	4	0	0	27	1	37	0
P&LE	0	0	1	1	0	0	1	0	0	3	1	3	0
W., Md.	0	0	2	0	0	1	1	0	0	4	0	0	0
TOTAL	2	871	7	14	65	4	41	5	8	1017	271	78	5
COLORADO													
AT&SF	0	8	7	1	11	0	32	12	1	72	16	6	4
CB&Q	0	5	0	1	0	0	1	0	0	7	3	1	1
CRI&P	0	0	0	1	0	0	0	0	0	1	0	0	0
C&S	0	0	1	1	0	0	0	0	0	2	0	29	0
D&RGW	0	59	4	7	3	0	0	0	0	73	28	0	0
GN	0	0	1	0	5	0	0	0	0	6	0	0	0
MP Lines	0	0	5	1	0	0	3	0	0	9	3	7	0
NP	0	5	4	0	19	2	2	0	0	32	6	5	0
SP	0	169	41	128	187	18	351	175	4	1073	205	120	11
T&NO	0	8	5	13	27	1	29	35	0	118	37	1	1
UP	0	2488	18	63	398	3	26	46	3	3035	609	16	3
TOTAL	0	2742	86	216	640	24	444	268	8	4428	907	185	25
DOMINION													
CP	0	3	70	4	69	3	969	6	35	1159	286	0	0
TOTAL	0	3	70	4	69	3	969	6	35	1159	286	0	0
GARY													
AT&SF	0	0	1	1	3	0	8	1	0	14	3	1	1
B&O	0	1	3	0	2	1	5	0	0	12	2	12	0
C&O (Sys.)	0	14	6	1	4	1	8	0	1	35	4	3	0
C&EI	0	0	0	0	1	0	0	0	0	1	0	0	0
C&NW	0	0	0	0	2	0	0	0	1	3	0	0	0
CB&Q	0	3	1	3	1	0	2	0	0	10	3	0	0
CMStP&P	0	2	1	0	1	0	0	0	0	4	2	1	1
CRI&P	0	0	1	0	0	0	1	1	0	3	0	2	0
Erie	0	0	0	0	0	0	1	0	0	1	0	5	0
GTW	0	0	4	1	3	6	2	1	0	17	4	0	0
GN	0	55	2	2	78	17	10	0	1	165	11	0	0
IC	0	8	5	0	8	3	11	2	1	38	12	3	1
KCS	0	0	0	1	0	0	0	0	0	1	0	0	0
L&N	0	0	0	0	0	0	1	0	0	1	0	0	0
MStP&SStM	0	0	2	0	0	2	2	0	1	7	0	0	0
MKT	0	0	1	0	0	0	0	0	0	1	0	0	0
MP Lines	0	0	1	0	0	0	0	1	0	2	0	1	0
NYC (Sys.)	0	1	0	0	1	0	0	0	0	2	0	0	0
NYC&StL	0	5	5	1	10	1	11	143	0	176	132	6	2
NP	0	0	4	1	13	4	7	2	0	31	12	4	0
PRR	0	1	1	1	0	2	3	0	0	8	2	6	0
UP	0	210	7	6	148	0	4	3	2	380	75	7	0
TOTAL	0	300	45	18	275	37	76	154	7	912	262	51	5

TABLE 6 - CONTINUED

ROADS	TF Ver U of I	CF & DF	VSH	HSH	Other Head	Broken	Web In Jt.	Web Other	Base	EBFs Excl. Accum. Total	EBFs Excl. 1960	EBFs Only Accum. Total	EBFs Only 1960
INLAND													
AT&SF	0	0	0	0	0	0	4	2	0	6	2	1	0
B&OCT	0	0	0	0	0	0	2	0	0	2	1	1	0
C&O (Sys.)	2	21	1	1	3	3	7	2	1	41	4	3	0
C&EI	0	0	0	1	0	0	0	0	0	1	1	0	0
C&NW	0	1	0	0	0	0	0	1	0	2	1	0	0
CB&Q	0	4	0	0	2	0	2	0	0	8	2	0	0
CMStP&P	0	0	0	0	0	1	1	0	0	2	1	0	0
CRI&P	0	0	0	.1	1	2	1	0	0	5	1	0	0
Erie	0	0	0	0	0	0	0	0	0	0	0	1	0
GTW	0	0	0	0	2	0	3	0	0	5	0	0	0
GN	0	31	5	0	12	0	10	0	1	59	6	0	0
IC	0	2	2	1	6	0	6	1	1	19	4	0	0
MStP&SStM	0	1	3	1	0	8	8	0	0	21	2	0	0
MP Lines	0	1	0	0	0	0	1	0	0	2	1	0	0
NYC&SL	1	0	0	0	0	0	1	2	0	4	0	0	0
NP	0	1	0	0	1	1	2	0	0	5	0	0	0
PRR	0	1	0	0	0	0	0	0	0	1	1	0	0
UP	0	48	0	2	3	0	0	1	0	54	9	0	0
TOTAL	3	111	11	7	30	15	48	9	3	237	36	6	0
LACKAWANNA													
CP	0	0	2	0	12	0	14	0	7	35	3	0	0
C&O (Sys.)	2	28	0	0	2	0	19	2	0	53	0	1	0
C&NW	0	0	0	0	1	5	0	6	0	12	0	0	0
GTW	0	0	25	1	0	1	1	1	0	29	1	0	0
GN	0	6	3	2	39	0	0	0	1	51	2	0	0
Me Cent.	0	0	1	0	0	0	2	0	0	3	1	0	0
MStP&SStM	0	0	0	1	0	0	0	0	0	1	1	0	0
NYC (Sys.)	0	8	10	1	40	1	49	3	11	123	16	8	0
NYC&SL	0	3	0	0	1	3	3	0	0	10	0	0	0
NP	0	0	0	0	2	0	3	0	1	6	3	1	0
TOTAL	2	45	41	5	97	10	91	12	20	323	27	10	0
STEELTON													
ACL	0	0	0	1	0	0	1	0	0	2	0	0	0
B&O	7	6	4	1	9	0	5	0	0	32	3	11	0
Ban Aroos	0	0	1	1	3	0	0	0	0	5	0	0	0
B&M	0	0	0	0	0	0	0	0	0	0	0	5	0
C&O (Sys.)	0	0	0	0	4	0	0	0	0	4	0	0	0
D&H	0	3	2	1	0	0	0	1	0	7	0	2	2
GN	0	0	1	1	0	0	0	0	0	2	1	0	0
NYNH&H	1	1	4	3	3	0	4	3	0	19	3	22	0
N&W	2	237	3	1	14	0	11	1	1	270	44	8	1
PRR	0	5	4	0	5	1	3	0	0	18	1	23	0
Reading	0	2	0	0	0	0	2	0	0	4	0	0	0
RF&P	0	0	0	0	2	0	0	0	0	2	0	0	0
SAL	0	0	5	1	0	0	0	0	0	6	0	1	1
Southern	0	0	2	0	0	2	3	0	0	7	0	0	0
W. Md.	0	0	2	1	0	2	0	0	0	5	1	1	0
TOTAL	10	254	28	11	40	5	29	5	1	383	53	73	4
TENNESSEE													
ACL	0	0	1	2	4	1	6	0	0	14	4	5	0
C of Ga.	0	0	1	0	0	2	1	0	1	5	1	0	0
FEC	0	0	1	1	0	1	2	1	0	6	3	0	0
IC	0	12	2	3	8	0	0	0	0	25	11	2	2
L&N	0	35	19	14	20	4	15	19	1	127	36	18	6
MP Lines	0	0	1	0	0	0	1	0	0	2	1	2	0
StL-SF	0	0	2	3	2	1	3	1	0	12	2	0	0
SAL	0	0	4	1	1	1	0	0	0	7	3	1	1
Southern	0	3	10	6	18	5	16	0	1	59	12	9	0
T&NO	0	2	1	1	1	0	47	12	0	64	6	0	0
TOTAL	0	52	42	31	54	15	91	33	3	321	79	37	9
ALL MILLS	26	4390	691	314	1682	153	2769	512	414	10951	2363	448	56

TABLE 7

RAIL FAILURES IN THE WEB WITHIN THE JOINT BAR LIMITS FOUND IN 1960
ON RAIL OF 100' LB. AND ALL HEAVIER SECTIONS

Railroad	Rail Rolled Previous to 1937				Rail Rolled in 1937 and After				Joints Inspected with Defect Detecting Instruments
	Detected Failures		Service Failures		Detected Failures		Service Failures		
	Bolt Hole	Other	Bolt Hole	Other	Bolt Hole	Other	Bolt Hole	Other	
AT&SF	46	49	23	0	245	177	141	0	2,903,170
ACL	54	88	115	172	62	94	23	23	43,601
B&O	30	1	1335	88	4	0	34	17	–
B&OCT	1	0	26	12	0	3	2	5	9,180
Ban & Aroos	1	1	6	5	1	0	5	0	–
B&LE	0	0	0	0	0	0	1	0	–
Bos & Alb	19	18	3	1	43	41	0	0	34,750
B&M	84	12	88	3	3	3	4	0	384,192
CP	180	145	36	2	722	481	235	70	2,203,832
C of G	0	0	0	0	12	2	0	0	–
C&O (Sys.)	5	10	56	25	2	1	19	11	24,438
C&EI	0	0	20	0	0	0	4	0	0
C&NW	61	38	540	16	43	42	338	7	1,013,184
CB&Q	45	5	55	1	615	1199	57	89	533,807
CI&L	0	0	23	3	0	0	4	0	0
CMStP&P	337	55	130	9	74	35	20	0	356,523
CRI&P	62	1	97	59	9	1	84	58	–
CCC&StL	24	112	26	58	146	269	8	13	227,556
C&S	0	0	0	2	0	0	0	0	–
D&H	3	0	9	1	3	0	6	0	2,069
D&RGW	1	0	13	4	0	3	5	6	282,150
Erie	274	33	17	55	12	85	6	38	193,865
GTW	26	0	22	4	14	1	41	1	13,694
GN	118	0	17	0	165	0	45	0	128,276
IC	31	4	35	3	36	35	78	18	190,095
KCS	0	0	0	1	7	4	4	1	–
L&HR	4	0	18	0	2	0	7	0	9,600
L&NE	1	0	4	0	1	0	1	0	–
LV	0	0	33	10	0	0	0	0	8,737
LI	4	16	25	0	6	7	2	0	66,960
L&N	130	328	65	0	49	216	39	0	–
Me. Cent.	5	4	14	16	0	0	2	5	–
MStP&SStM	39	0	6	0	17	0	9	0	60,841
MKT	0	0	0	0	0	0	48	0	0
MP Lines	7	7	3	13	23	84	16	28	–
NYC (Sys.)	60	26	93	25	387	275	127	55	94,591
NYC&StL	0	0	19	0	2	1	146	1	–
NYNH&H	38	128	7	218	6	120	3	138	–
N&W	2	12	7	6	2	10	4	3	9,421
NP	1	0	253	12	0	0	217	77	–
PRR	955	348	2325	325	668	643	202	153	1,631,015
P&LE	0	0	0	0	10	56	2	5	10,550
Reading	3	0	0	0	3	0	0	0	–
RF&P	0	0	0	1	4	23	0	1	–
Rutland	0	0	4	0	0	0	0	0	0
StL – SF	3	15	125	6	1	0	65	0	44,739
SAL	0	0	14	2	0	0	20	7	–
Southern	64	118	0	0	301	870	0	0	46,711
SP	1	6	0	12	136	139	51	194	–
T&NO	115	56	9	4	58	37	8	1	157,877
T&P	122	26	5	0	174	34	11	2	229,398
UP	0	0	3	1	180	475	21	8	2,294,000
W. Md.	0	0	0	0	0	0	3	3	–
Totals	2956	1662	5724	1175	4248	5466	2168	1038	13,208,822

TABLE 8 – ACCUMULATED TRANSVERSE FISSURE FAILURES IN CONTROL COOLED RAIL AS VERIFIED BY LABORATORY INVESTIGATION, MILL AND YEAR ROLLED TO OCTOBER 1, 1961

Mills	1935	1936	1937	1938	1939	1940	1941	1942	1943	1944	1945	1946	1947	1948	1949	1950	1951	1952	1953	1954	1955	Total
Algoma					2b	2a	2b	1b	1a	1a		1a 2b	1a 1b	1b	9b	6b	1b					31
Carnegie (ET)			1c		1a				2c								2c					6
Colorado	*				1c		1c															2
Dominion						1b																1
Gary		7b	4b	1b											1b							13
Inland	1a		3a		3a	3a	8a		1a	7a 1b	1a	2a		1a			1a	3a		1a	1a	37
Lackawanna	5a							6a	2a				3a	1a	1a	1a	3a					22
Steelton		22a	13a	11a	10a	15a 1c	6a 1c	15a	8a	8a	1a	2a	2a	3a	1a	4a		4a				127
Tennessee		1b							2c	1a												4
TOTAL	6	30	21	12	17	22	18	22	16	18	2	7	7	6	12	11	7	7	0	1	1	243

Note- (a) TRANSVERSE FISSURE from hot torn steel. (b) TRANSVERSE FISSURE from shatter cracks due to improper cooling. (c) TRANSVERSE FISSURE from inclusion. Summary - 41 T.F.'s from shatter cracks, 11 T.F.'s from inclusions, 191 T.F.'s from hot torn steel.

* No CC rail rolled

TABLE 9

WELDED ENGINE BURNS AND FAILURES

Railroad	Engine Burns Welded Prior To 1960	Burns Welded In 1960	Failed Welded Engine Burns During 1960
AT&SF	90,642	15,457	1
B&O	1,310	131	1
C&O	41,074	3,954	9
C&NW	13,327	1,897	0
D&H	7,937	47	2
EJ&E	49,374	5,780	0
IC	21,001	11,432	0
PRR	301,315	20,670	19
RF&P	17,941	0	0
StL – SF	2,748	684	0
SAL	14,636	7,062	0
Southern (Sys.)	No Record	No Record	5
Southern (West. Div.)	145,591	Discontinued Keeping This Information	
SP	222	321	10
Total	707,118	67,435	47

Report on Assignment 5

Economic Value of Various Sizes of Rail

J. C. Jacobs (chairman, subcommittee), W. J. Cruse, J. A. Bunjer, O. E. Fort, John Ayer, Jr., S. H. Barlow, H. B. Berkshire, T. F. Burris, M. W. Clark, C. O. Conatser, G. H. Echols, D. T. Faries, C. E. R. Haight, K. K. Kessler, R. R. Lawton, F. R. Micheal, R. H. Patterson, H. F. Smith, A. P. Talbot, D. J. White, W. L. Young.

Your committee submits the following report as information. It is a continuation of Study "A" reflecting changes in the test mileage and computed to show averages after 17 years. The labor and material averages are computed to compensate for the decrease in track mileage.

Study A

RESULTS OF STUDY OF ILLINOIS CENTRAL RAILROAD NORTHWARD TRACK, MATTOON TO SAVOY, ILL., TEST SECTIONS OF 112-LB AND 131-LB RAIL

112-Lb Rail	131-Lb Rail

MP 152.24–172.00 laid in 1942 and 1943
Original test included:
  19.76 track miles
  18 turnouts
  1 railroad crossing
  22 public road crossings
  2 private grade crossings
  24-in joint bars
	MP 132.00–152.24 laid in 1944
	Original test included:
	  20.24 track miles
	  21 turnouts
	  3 railroad crossings
	  22 public road crossings
	  6 private grade crossings
	  36-in joint bars

Changes in rail mileage:
  1950—MP 152.09–152.24 laid in 115-lb,
    0.15 miles added to test.
  1952—MP 155.87–160.52 relaid in 132-lb,
    4.65 miles dropped from test.
  1953—MP 160.52–163.55 relaid in 132-lb,
    3.03 miles dropped from test.
  1954—MP 152.09–155.87 relaid in 132-lb,
    3.78 miles dropped from test.
  1956—MP 170.79–172.00 relaid in 132-lb,
    1.21 miles dropped from test.
  1957—MP 170.79–163.55 relaid in 132-lb,
    7.24 miles dropped from test.
    (completing removal of 112-lb rail)
	Changes in rail mileage:
	  1950—MP 152.09–152.24 laid in 115-lb,
	    0.15 miles dropped from test.
	  1960—MP 142.82–152.09 laid in 132-lb,
	    9.26 miles dropped from test.

Average annual traffic density—28,000,000 gross tons

COMPARISON OF THE TWO SECTIONS COST OF INVESTMENT—1944 PRICES

Item	Investment Charges per Mile	
	112-Lb	131-Lb
Gross cost—rail and other track material	$12,643	$14,413
Less estimated salvage	Cr. 4,284	Cr. 5,011
Net Cost—rail and other track material	$ 8,359	$ 9,402
Labor cost to lay	1,338	1,473
Total cost	$ 9,697	$10,875
Estimated life—based on 1960 conditions	15 years	20 years
Annual Cost—Rail and other track material	$ 557	$ 470
Labor to lay	89	74
Interest at 6 percent*	839	953
Total annual investment cost	$ 1,485	$ 1,497

* On gross outlay for labor and material.

MAINTENANCE LABOR AND MATERIAL PER MILE

	112-Lb					131-Lb			
Year	Miles Maintained	Man Hours	Cross Ties	Cu Yd Ballast	Year	Miles Maintained	Man Hours	Cross Ties	Cu Yd Ballast
1943	19.76	2,480	716	628					
1944	19.76	413	5	17	1944	20.24	2,606	1,065	647
1945	19.76	701	236	251	1945	20.24	131	0	30
1946	19.76	1,166	416	579	1946	20.24	370	5	114
1947	19.76	645	208	273	1947	20.24	748	172	301
1948	19.76	1,005	186	294	1948	20.24	245	38	185
1949	19.76	1,574	541	423	1949	20.24	670	10	116
1950	19.91	694	174	159	1950	20.09	1,642	402	323
1951	19.91	667	48	193	1951	20.09	614	59	52
1952	15.26	748	30	140	1952	20.09	1,144	62	304
1953	12.23	1,110	91	292	1953	20.09	1,089	139	120
1954	8.45	543	44	83	1954	20.09	392	0	21
1955	8.45	239	32	159	1955	20.09	631	91	64
1956	7.24	401	8	0	1956	20.09	535	10	52
1957	7.24	274	40	0	1957	20.09	394	0	0
					1958	20.09	242	78	214
					1959	20.09	53	0	0
					1960	10.82	97	26	0
Total 15 years	------	12,660	2,775	3,491	Total 17 years	------	11,603	2,157	2,543

AVERAGE OF 15 YEARS FOR 112-LB AND 17 YEARS FOR 131-LB RAIL

	112-Lb		131-Lb		Savings by Use of 131-Lb	
	Charge	Percent	Charge	Percent	Charge	Percent
Man-hours	844		682		162	
Cost at $1.25*	$1,055	55	$ 853	60	$ 202	42
Cross ties	185		127		58	
Cost at $3.32*	$ 614	33	$ 422	30	$ 192	40
Ballas (stone and slag) cu yd	233		150		83	
Cost at $1.00*	$ 233	12	$ 150	10	$ 83	18
Total maintenance	$1,902		$1,425		$ 477	
Percent		100		100		100
Investment charges	$1,485		$1,497		$ 12Cr.	
Total Cost	$3,387		$2,922		$ 465	
Percent						13.7

*Average prices 1943–1960.

SUMMARY

The greater savings realized through the use of 131-lb rail have been in labor and cross ties, partially due to the use of longer and heavier joint bars, larger tie plates and greater rail rigidity.

Similar maintenance standards have been practiced on both test sections.

Report on Assignment 6

Joint Bars: Design, Specifications, Service Tests, Including Insulated Joints and Compromise Joints

Embert Osland (chairman, subcommittee), John Ayer, J. M. Bentham, H. B. Berkshire,
R. M. Brown, J. A. Bunjer, R. E. Catlett, J. B. Clark, C. A. Colpitts, C. O.
Conatser, W. J. Cruse, V. E. Hall, C. J. Henry, S. R. Hursh, H. W. Jenkins,
J. S. Parsons, B. R. Perkins, G. L. P. Plow, R. C. Postels, R. B. Rhode, C. R.
Riley, V. R. Terrill.

This report, submitted as information, covers the rolling-load and physical-property tests of basic-oxygen-steel joint bars, rolling-load and bend tests of three compromise joints produced by thermit welding and a rolling-load test of an insulated joint using fiber-glass bars. The investigation is conducted by the Engineering Division research staff of the Association of American Railroads under the supervision of G. M. Magee, director of engineering research.

In order to establish data for consideration of basic-oxygen-steel joint bars for inclusion in AREA specifications, rolling-load tests are in progress at the Research Center, and physical-property tests were conducted by Professor R. E. Cramer at the University of Illinois. Six pairs of these joint bars of Joint Bar Co., K–4, 115 RE design have been submitted for testing by the Algoma Steel Company.

Rolling-load tests using the 36-in-stroke machines with a 44,400-lb wheel load have been performed on two rail joints with two pairs of these bars. One joint failed at 835,600 cycles and the other at 286,400 cycles, each from a crack in center of the bottom of the bar. Tests on two other joints using the next two pairs have exceeded 1,500,000 cycles to date.

The four bars from the failed joints were submitted for physical-property tests to Professor Cramer who reports in the following table the hardness and mechanical tests of the four basic-oxygen joint bars. Tensile specimens were taken near the middle of each bar from both the top and bottom of each bar. The two Brinell readings were taken on each end of the tensile specimens.

BRINELL AND MECHANICAL TESTS

Bar Number	1st Brin. Number	2nd Brin. Number	Yield Strength PSI	Tensile Strength PSI	% Elong.	% Red. Area
62A top.................	265	255	86,000	127,500	12.0	38.4
62A bottom.............	241	248	84,000	125,000	12.5	42.0
62A top.................	255	255	87,000	127,000	12.5	43.0
62A bottom.............	241	241	83,500	125,000	13.0	42.5
62B top.................	262	255	84,200	125,000	13.0	38.6
62B bottom.............	255	255	81,200	127,000	13.0	40.9
62B top.................	253	248	85,500	126,000	12.0	35.4
62B bottom.............	255	255	84,500	126,000	12.0	34.0

It can be noticed that three of the bars were slightly harder on the top than on the bottom. A chemical analysis of each of the bars will be performed and reported in the final report.

The frequent occurrence of service failures of compromise joint bars has led to the investigation of producing a compromise joint economically by means of a thermit weld. This has become feasible with the advent of a prefabricated mold and an improved quality of thermit welds. Three compromise joints of 115 RE to 132 RE rail sections have been produced by this method. Rolling load tests using the short stroke machine putting the head area in tension and the long stroke machine putting the weld area in reverse flexure are now in progress. A slow bend test has been performed at the University of Illinois. A final report will be made when the investigations are completed.

Research and testing on insulated rail joints has continued actively over the past year. Several design changes were made in the Vulcabond (AAR) joint that was tested in the laboratory and field, and several other types of insulated joints were tested in the laboratory. The Vulcabond joints are now installed on 30 railroads in quantities of 1 to 194 for periods up to two years. Their service performance is being followed. An extended abstract of Engineering Research Report 9, issued on March 1961, was published in Bulletin 559, bringing the results up to September 1, 1961. Since that data a fifth fiber-glass joint was tested with results similar to the other four. This joint had a larger cross section and gave slightly less deflection, but the deflection was still relatively great compared to the continuous type and the Vulcabond joint. It was run 115,000 cycles at ½ load and 6,000 cycles to failure at full load. A second 115-lb Vulcabond joint is being tested, and some design variations of the 132-lb joint are scheduled for test.

Report on Assignment 8

Causes of Shelly Spots and Head Checks in Rail: Methods for Their Prevention

L. S. Crane (chairman, subcommittee), W. D. Almy, John Ayer, Jr., B. Bristow, T. F. Burris, J. A. Bunjer, M. W. Clark, C. J. Code, W. J. Cruse, F. L. Etchison, D. T. Faries, O. E. Fort, C. J. Henry, C. C. Herrick, W. H. Hobbs, S. R. Hursh, T. B. Hutcheson, R. F. Lawson, Lee Mayfield, B. R. Meyers, J. G. Roney, J. S. Wearn, W. L. Young.

This is a progress report, presented as information.

The investigation this year has been conducted by (a) the AAR Engineering Division research staff, and (b) the University of Illinois. The AAR this year is providing all of the funds; the AISI has dropped its support of this investigation.

The work conducted by the AAR Engineering Division research staff is under the general direction of G. M. Magee, director of engineering research, and is covered by the report prepared by Kurt Kannowski, metallurgical engineer, included in this report as Appendix 8–a. It covers the inspection of service tests of fully heat-treated and alloy rail installations. There are five tests of fully heat-treated rail, three of high-silicon rail, one of chrome–vanadium and one of columbium-treated rail.

The heat treatment in some of the locations has shown considerable value in extending rail life under shelly conditions and in resisting head flow on the low side of curves. In order to obtain the full value of the application of heat-treated rail two of the service test curves were transposed this year, and a much longer service life may be expected.

The high-silicon and low-alloy rails are showing a noticeable resistance to wear and shelling.

The chrome–vanadium rail on the Duluth, Missabe & Iron Range Railway shows excellent performance and has had a service life twice as long as standard rail.

The work conducted by the University of Illinois is covered by the report prepared by Professor R. E. Cramer, which is included in this report as Appendix 8–b. This report covers (1) rolling-load tests to produce shelling in high-silicon chrome–vanadium rail, columbium-treated rails, basic-oxygen standard carbon rails, and flame-hardened rails produced by the UP process, and (2) end-quench hardenability curves determining the quenching characteristics of some of the low-alloy rail steels.

Appendix 8–a

Service Tests of Heat-Treated and Alloy-Steel Rail

Great Northern Railway Service Test of 115 RE Columbium-Treated Rail

In order to investigate the effect of small amounts of columbium in the improvement of the quality of rail steel, three ingots of standard analysis rail steel were treated with ¼ lb of columbium/ton in the first ingot, ½ lb of columbium/ton in the second ingot and 1 lb/ton in the last ingot at the Pueblo plant of the Colorado Fuel and Iron Corporation in cooperation with the Molybdenum Corporation of American in conjunction with the AAR Research Center.

This heat, No. 12143, was tapped on March 21, 1960. The heat had a clean tap and poured with a good stream. A good pouring temperature was observed. The ingot molds were clean and coated. All of the three test ingots were poured higher in order to obtain prolongs on the "A" rails.

The ladle analysis was 0.74 percent C, 0.80 percent Mn, 0.008 percent P, 0.017 percent S, and 0.14 Si. The analysis of the columbium content was 0.014–0.016 percent for the first ingot, 0.029–0.030 percent for the second ingot, and 0.044–0.051 percent for the third ingot.

The 6-ft prolongs from each of the "A" rails were drop tested on March 22, 1960. All three test rails took three blows each without failure. The test rails were then nicked and broken and found to be of sound and acceptable quality. The standard nick-and-break tests of the test ingots were examined, and a remarkable correlation between the grain refinement of the fractures and the amount of columbium was noted. The grain refinement increased with the amount of columbium.

Drop test results on the "A" rails from the three columbium-treated ingots were as follows:

Ingot No.	No. of Blows	Remarks
7A	3	Did not break
8A	3	Did not break
9A	3	Did not break

Results of the nick-and-break tests showed no pipe to be present either in 7A, 8A or 9A test pieces; and only one pipe was present in the whole heat, namely, test piece 3A.

Results of final inspection of the columbium rails are as follows:

(1) Rail 9A showed pipe in the finished rail.

(2) Rail 8A had a medium scab in the head located in the middle of the rail.

(3) Rails 8D and 8E had scabby web running for a distance of 5-ft from one end.

The balance of the rails was free of any surface or interior defects. Mechanical and physical tests on these rails were made by Professor Cramer of the University of Illinois and a commercial laboratory. They showed some slight improvements in these properties as compared with regular rail steel.

Professor Cramer has two test pieces each from ingot **7B**, **8B** and **9B**. He reports the following data on the rolling-load test for shelling and the mechanical tests.

Spec. No.	Section	Avg. Brin. Hard.	Yield Strength 0.2% Offset PSI	Tensile Strength PSI	Elong. in 2 in. %	Red of Area %	Charpy Value No Notch	Charpy Value Key Hole	Endur- ance Limit PSI
1209 A B	115 RE rail	267	72,500	132,800	10.0	19.2	133	4.2	
1210 A B	115 RE rail	267	75,900	134,000	8.8	19.7	151	4.9	
1211 A B	115 RE rail	269	77,600	136,300	9.3	20.1	148	3.4	65,000*

*Cycles for failure of 50,000-lb wheel load—2,051,000.

TABLE 1—REPORT OF PHYSICAL TESTS

To Molybdenum Corporation of America
Yield stretch by extensometer—0.2 percent offset

	Parent Steel	¼ Lb/Ton Columbium	½ Lb/Ton Columbium	1 Lb/Ton Columbium
Rail marked	12–143B6	B7–30K	B8–30P	B9–30–S
Diameter in inches	0.505	0.505	0.505	0.505
Area in square inches	0.2003	0.2003	0.2003	0.2003
Yield point, actual	13,700	13,900	13,800	15,300
Yield point, psi	68,400	69,390	68,800	76,380
Ultimate strength	26,300	26,600	26,500	27,200
Tensile strength, psi	131,300	132,800	132,300	135,800
Elongation in 2 in.	0.20	0.21	0.21	0.21
Elongation, percent	10.0	10.5	10.5	10.5
Reduction of area, percent	18.4	18.1	20.2	19.5
Brinell hardness	241	241	248	255
No-Notch Charpy Impact Test:	*Ft–Lb*	*Ft–Lb*	*Ft–Lb*	*Ft–Lb*
Room temperature	119 75 173	188 134 113	104 120 117	127 221 131
Average	122	145	114	160
Zero deg F	126 108 101	160 129 143	132 124 112	132 227 115
Average	112	144	123	158
−20 deg F	55 84 79	85 75 181	113 144 122	140 119 159
Average	73	114	126	139

TABLE 2—V-NOTCH CHARPY IMPACT TESTS

	Parent Steel Ft-Lb	¼ Lb/Ton Columbium Ft-Lb	½ Lb/Ton Columbium Ft-Lb	1 Lb/Ton Columbium Ft-Lb
Room temperature	8¼ 6.0	7.5 5.5	7.0 7½	4.0 3.0
Average	7⅛	6½	7¼	3½
Minus 20 deg F	6.0 7¼	4¼ 5¼	4½ 5½	4.0 4¾
Average	6⅝	4¾	5.0	4⅜
Minus 40 deg F	5½ 7¾	3¾ 2½	3¾ 4¾	4.0 4½
Average	6⅝	3⅛	4¼	4¼

It can be noted from Tables 1 and 2 that the impact values at lower temperatures show some improvements with the no-notch charpy impact tests. The notch sensitivity of V-notch charpy impact tests shows such a little difference that no significance can be given to these results.

Rolling-load tests were performed on the short-stroke machines at the AAR Research Center on flash-butt-welded rail joints consisting of rails from each of the different columbium analysis rails. All of the tests ran out at 2,000,000 cycles without a failure. The same type of specimen was also subjected to the standard rail drop test. All of the rail joints failed at the first blow which is consistent with the results obtained on previous drop tests with the same type of weld.

Based on the results of these tests the rails produced from the test ingots were placed in a service test on Curve 22 on the Great Northern Railway. This 4-deg curve had been laid with standard 115 RE rails in a service test comparing fully heat-treated rail. This curve is extensively described in the AREA Proceedings, Vol. 57, pages 837–850, and its history in regard to shelling is on record.

The rails were laid on Curve 22 on May 18 and 19 and are identified as follows, starting at the receiving end:

Rail Identification	Lb of Columbium/Ton	Rail Identification	Lb of Columbium/Ton
CF&I 3-60, 12143-A-7C	¼	CF&I 3-60, 12143-F-7C	¼
CF&I 3-60, 12143-D-8C	½	CF&I 3-60, 12143-A-8C	½
CF&I 3-60, 12143-A-6C	1	CF&I 3-60, 12143-D-9C	1
CF&I 3-60, 12143-F-9C	1	CF&I 3-60, 12143-C-9C	1
CF&I 3-60, 12143-E-9C	1	CF&I 3-60, 12143-C-8C	½
CF&I 3-60, 12143-E-8C	½	CF&I 3-60, 12143-D-7C	½
CF&I 3-60, 12143-F-8C	½	CF&I 3-60, 12143-C-7C	¼
CF&I 3-60, 12143-E-7C	¼		

The balance of the curve was laid with fully heat-treated rail that had been flash butt welded to 78-ft lengths. Contours were obtained on the test rails after the mill scale had worn off.

Duluth, Missabe & Iron Range Railway Service Test of Chrome-Vanadium Alloy Rail

The test installations covered by this report was laid in March 1954, using chrome-vanadium alloy rolled at the Gary Works of United States Steel Corp. in December 1953. The majority of the rail was taken from heat No. 976186, with an analysis of

0.64 C, 1.51 Mn, 0.018 P, 0.027 S, 0.41 Si, 0.06 Cu, 0.04 Ni, 1.11 Cr, 0.00 Mo and 0.16 V. Brinell hardness readings ranging from 352 to 375, with an average 10-point reading of 359.8, was obtained from a piece of new rail.

The purpose of these tests was to obtain a comparison between standard control-cooled rail and the chrome–vanadium alloy rail for use in curves. Four curves were selected, namely, a 2-deg curve at MP 3.93 and a 4-deg curve at MP 3.58 on the Missabe Division near Duluth, Minn., and a 4-deg curve at MP 32.2 and a compound 6-deg, 7-deg curve at MP 26.65 on the Iron Range Division near Two Harbors, Minn. These installations are described in detail in AREA Proceedings, Vol. 57, page 833.

All the curves selected have had a history of heavy rail renewals due to either shelling or excessive abrasion on the gage side of the head of the high rail. Each curve was laid with approximately half of its length being standard CC rail and the balance CV rail. In each case the same type of rail was laid on both the high and low sides of the curve.

One chrome–vanadium rail was removed from track after the passage of the first train due to a failure through a bolt hole at the joint. This rail was submitted for laboratory analysis, and it was found that the rail failed due to a fracture caused by a severe blow immediately adjacent to the bolt hole.

Three lengths of CV rail were removed from the area of the safety switch on the curve at MP 26.65 on June 9, 1955, at 24,000,000 gross tons, due to abrasion on the gage side. Previous to this, standard CC rail had been removed from this location after 8,000,000 to 10,000,000 gross tons of traffic.

Seven CC rails were removed from the side of curve 26.65 in October 1956 after 61,500,000 gross tons because of abrasion and shelling. The balance of the CC rail was removed from this curve in February 1958 after 95,200,000 gross tons. Twelve CV rails were also removed from the high side of this curve at the same time because of abrasion on the gage side. The balance of CV rail from the high side was removed in July 1960 at 135,000,000 gross tons, also because of abrasion. The rail on the low side of this curve is still in track at this time.

The standard CC rail was removed from the high side of the other three test installations at 90,000,000 to 100,000,000 gross tons of traffic because of shelling and abrasion. The majority of CC rail on the low side of these curves was removed at approximately 120,000,000 gross tons.

The majority of the CV rail, except for the high side of curve 26.65, is still in track and is not showing an excessive amount of wear. While it is not considered that these tests are completed until all the CV rail is removed from track, it now appears that it can be reasonably expected that about 50 percent more life from CV rail can be obtained in areas where rail renewal is due primarily to abrasion, and possibly two to three times more life where failure is due to excessive shelling or metal flow.

The traffic over these curves is primarily iron ore carried in closely coupled, short 50- and 70-ton ore cars on 33-in wheels. Steam locomotives were used exclusively in these areas during 1954 and 1955 with conversion to diesel beginning in 1956 and completed in 1958. Trains on the Missabe Division consisted of approximately 80 cars with steam operation and were increased to approximately 100 cars with the conversion to diesel. On the Iron Range Division, they consisted of approximately 100 cars with steam and have increased to approximately 140 cars with diesel.

The inspection of the Proctor Hill installation on October 5, 1961 at 142,000,000 gross tons of traffic showed that no substantial change has taken place over last year in the gage corner development.

Norfolk & Western Service Test of 132 RE Heat-Treated Rail
at Maher and Looney's Curve

These installations are on a 6-deg curve near Maher, W. Va., at M.P. 481 + 210 ft and on the 4-deg, 7-deg, 12-deg Looney's Curve at M.P. 455 + 582 ft, and are described extensively in the AREA Proceedings, Vol. 57, pages 834–835.

Sixty-six heat-treated rails were installed in both the high and low sides of the 6-deg curve at Maher on August 2 and 9, 1954. On September 21, 1959, after the rails were in service over five years and had carried 168,900,000 gross tons of traffic, they were transposed, high rails to the low side and low rails to the high side of the curve.

As reported previously 3 of the 66 original high-side rails were removed from service prior to the transposition due to the development of detailed fractures from shelling. Thus only 63 heat-treated test rails were relocated in the low side of the curve. Two low-side rails were not reinstalled in the high side during the transposition. These two rails were used elsewhere. By the time of this inspection (May 11, 1961) the rails in their transposed positions had carried an additional 44,500,000 gross tons of traffic, making a total of 213,400,000 gross tons.

Following is a tabulation of the service developments noted in the rails on May 6, 1959, the date of the last inspection prior to the transposition:

High Side of Curve

Head checks only .... 3 rails
Flaking only (light, medium and heavy) .... 50 rails
Black spots .... 4 rails
Small shells .... 6 rails
Removed from service .... 3 rails

Total .... 66 rails

Low Side of Curve

Mild crushing of head metal .... 2 rails

The service developments noted in the 63 high-side rails, of course, no longer are of note in the rails in their present transposed position. The mild crushing noted on the top of the head of the two low-side rails are still noticeable in their high-side positions; in fact this mild condition was also noted in a third rail at the time of this inspection.

The field-side corner of the head of the former low rails had developed a mild fin of plastically deformed metal which, of course, has considerable work to do in the new location as gage corner in the high-side rail. This fin has for the most part been ground away by the wheels, and the gage corner is assuming a contour normal in high-side rails.

At the time of the 1960 inspection 34 of the 64 rails in the high side were clear of gage corner service developments. Light flaking in 20 and medium flaking in 10 were noted in the other 30 high-side rails. By the time of this 1961 inspection, curve wear had changed these conditions so that 46 rails appeared free of gage corner service developments, and light flaking was noted in 17 rails and medium flaking in one. One rail had a heavy wheel burn at midrail.

There has been little protection from wear for the rails in this curve since the curve oilers have been ineffective.

The low-side rails appeared to be in excellent condition. Most of the rail ends have been built up by welding and all appear in good condition.

It now appears that the transposition will considerably extend the life of these fully heat-treated rails which, prior to the transposition, had given better than two times the life of previous regular rails in this curve.

Fifty-five heat-treated (oil-quenched) rails were laid in the high side and 54 in the low side of the compound Looney's curve on August 2 and 9, 1954. An inspection was made on May 11, 1961, at which time the test rails had carried approximately 118,-700,000 gross tons of traffic.

Following is a rough tabulation of the apparent head condition of the high-side rails as noted in May 1958, 1959, 1960 and 1961:

Service Development	May 1958	May 1959	May 1960	May 1961
Head checks	3	0	0	0
Light flaking	17	15	3	6
Medium flaking	1	1	3	0
Heavy flaking	0	0	0	0
Black spots	3	3	3	1
Shelling	7	13	23	22
Clear	24	23	23	24
Out of service	0	0	0	2

One rail was removed from service because of a detector-car indicated transverse development, and another rail broke under traffic on July 20, 1960. The failures were attributed to shelling in the high side of the curve. About half of the heat-treated rails in the low side of the curve have resisted heavy crushing while others show heavily deformed surface head metal.

There has been a noticeable increase in the progression of the service developments in both the high and low sides of this curve in the last two years. However, after 81 months, this should be expected in this location where non-heat-treated rails previously installed lasted only 15 to 18 months in the high side and 6 to 9 months in the low side of the curve.

Norfolk & Western Railway Service Test of 132 RE Heat-Treated Rail at Kermit, W. Va.

A final report on the service test of the 132 RE heat-treated rail at Kermit, W. Va., was given in 1957 and may be found in the Proceedings, Vol. 58, pages 1030–1032. Since 8 of the original 12 heat-treated rails on the high side and 10 of the original 11 in the low side of the curve are still in service, an inspection was made on May 11, 1960. The rails in this test, which originally consisted of 23 heat-treated and 24 non-heat-treated rails laid in both the high and low sides of the curve, were installed May 3, 1949.

The original test of fully heat-treated versus non-heat-treated rails in the 6-deg curve just west of Kermit was partially concluded as of May 23, 1960. As covered in the report of the May 11, 1960, inspection the remaining 7 of the original 12 fully heat-treated rails in the high side of the curve were removed on May 23, 1960, after more than 11 years of service, having carried 449,600,000 gross tons of traffic. They were replaced with new, regular, non-heat-treated rails, thereby, in effect, establishing a new comparative cycle between heat-treated and non-heat-treated rails in this service; the third set of comparative non-heat-treated rails were installed during the life of the original heat-treated high-side rails.

During the year, since installation on May 23, 1960, the new untreated high-side rails have carried only 3.8 million gross tons of traffic, less than one-tenth the average annual tonnage for the 11 preceding years. Consequently, the running surface of the head of the new rails is little more than polished.

Ten of the original 11 heat-treated rails are still in the low side of this test track. One rail was removed from service in October 1953 because of a bad engine-wheel burn. While two other rails, 87581–E11 and C10, were badly marked by wheel burns, they have remained in service.

These low-side heat-treated rails have now carried 453,400,000 gross tons of traffic and look as good as the adjacent fifth set of non-heat-treated comparative rails. Batter and crushing at both ends of each rail have been corrected by welding.

Pennsylvania Railroad Service Test of 140 RE High-Silicon Rail

This test installation of high-silicon 140 RE rail is located in the Pennsylvania Railroad Altoona District No. 4 track, east of Mifflin, Pa., between M.P. 152 and M.P. 153 + 2530. The test rails are installed in the 5-deg Mifflin Reverse Curve, in the 3-deg 15-min Stone House Curve, and in the 2-deg Casners Curve in alternating groups of three high-silicon and three standard high-carbon rails on both the high and low sides of the curves.

This installation has been previously reported in the AREA Proceedings, Vol. 58, page 962 and Vol. 60, pages 872 and 873.

During the May 1961 inspection it was noted that 161,000,000 gross tons of traffic had passed over it. Again it was observed that the gage corner developments related to shelling had not progressed extensively over those observed last year. The difference in the resistance to gage corner developments between the high-silicon rail and the standard blue-end rail is very slight at this point of the test. The high-silicon rail had some medium flaking while the standard blue-end rail had heavy flaking on one rail and medium flaking on the balance of the test rails. It was noted that gage corner defects were held to a minimum because of curve wear. The curve oiler was not operating.

Pennsylvania Railroad Service Test of 155 PS High-Silicon Rails

The 155 PS high-silicon rails were produced at Steelton in September 1953 and installed by the PRR. This installation is extensively described in the Proceedings, Vol. 53, page 1029.

The rails were laid October 5–8, 1953, in the No. 1 eastbound track in the 2-deg 37-min Bixler Curve, M.P. 164, east of Lewiston, Pa. Alternate groups of 5 high-silicon rails and 5 rails of standard analysis (blue ends) were installed in the high and low sides of the curve.

The previous 152 PS rails in this curve had shown light to medium flaking with some shelly spots in the high-side rails after 12 years of service.

At the time of the May 1958 inspection light to medium flaking was common to both the high-silicon and the standard rails in the portion of the curve at full elevation.

The test rails throughout the curve were transposed in conjunction with the regular rail program in January 1959. The high side rails were removed to the low side and vice versa. A total of 184,770,000 gross tons of traffic had been carried by the test rails to January 1, 1959. Some degree of flaking, ranging from very light to heavy, was noted in every high-side rail in the transposed portion of the curve. The degree of flaking was comparable in both high-silicon and standard analysis rails.

At the time of the May 1961 inspection it was reported that 236,777,000 gross tons had passed over the curve from October 1933 until January 1, 1961. This would place the tonnage passed over the curve since the transposition at 52,000,000 gross tons. During the inspection it was noted that some curve wear had taken place, but the condition of the gage corner developments had not changed appreciably from that of last year.

Great Northern Railway Service Test of 115 RE Heat-Treated Rails

This installation is extensively described in the AREA Proceedings, Vol. 57, pages 837–850. An inspection was made in May 1961. The 115 RE test rails were put into service in two comparable curves on February 7, 1961, east of Carlton, Minn. Reviewing the consist of the tests: 88 fully heat-treated rails were installed in both the high and low sides of the 4-deg Curve No. 20, and 88 non-heat-treated, end-hardened rails were placed in the high and low sides of the 4-deg Curve No. 22 for comparison. The heat-treated rails were from heat 89521 rolled at Steelton in November 1950 and were oil quenched and tempered. The comparative non-heat-treated rails were supplied by Illinois and were field end hardened.

The original non-heat-treated test rails in the high side of Curve No. 22 were replaced August 1, 1955, because of extensive shelling after carrying 193,000,000 gross tons of traffic. At least one rail had failed in service from a detailed fracture from shelling.

In the low side of the curve the original non-heat-treated rails were badly crushed and very rough. A considerable amount of plastic deformation of the head metal had taken place. Six of these non-heat-treated rails were removed from service on July 22, 1960, after carrying approximately 371,000,000 gross tons of traffic.

The remaining non-heat-treated test rails in the low side of Curve No. 22, as well as all of the original fully heat-treated test rails in both the high and low sides of Curve No. 20, were in service at the time of this May 1961 inspection and had carried 384,-065,777 gross tons of traffic.

Shelling and considerably more and heavier curve wear and flaking were in evidence in the second set of non-heat-treated rails in the high side of Curve No. 22 after only 191,065,777 gross tons of traffic than in the original heat-treated rails in the high side of Curve No. 20 at the time of this inspection. This was true in spite of the fact that the heat-treated rails had carried 193,000,000 more gross tons, or over two times as much traffic.

Generally speaking, little change had taken place in the majority of the rails since the 1960 inspection. Shelling spots were observed in 7 rails and flaking in 20 rails. Sixteen rails appeared clear of gage corner service developments.

The heat-treated rails in the low side of the curve again appeared to be in excellent condition. The resisistance to wear and plastic flow has been outstanding; however, the field corner of the head of a number of these low rails has become squared by the flow of metal toward the field side, and chipping of the surface metal had occurred at spots in some rails.

Contour tracings were made on rails throughout the test curves, as shown on Fig. 1. Compared with the 1960 contours, little change could be observed, undoubtedly because of the reduction in traffic, so the conditions noted following the 1960 inspection still apply. Briefly, these figures show that in the high side of the curves the wear in the

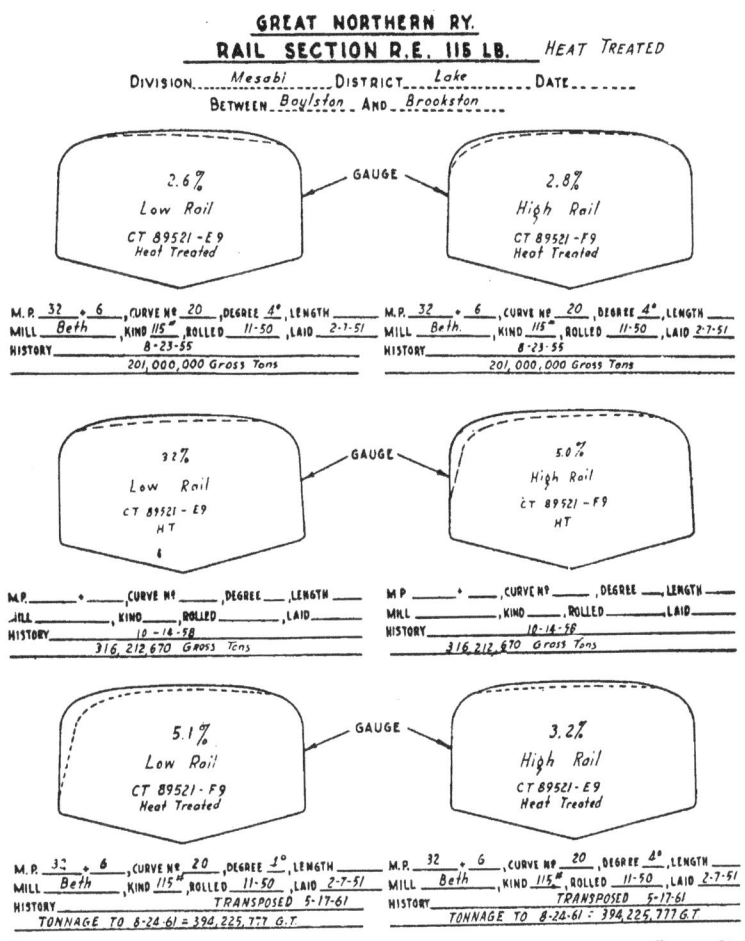

Fig. 1—Representative contours of the fully heat-treated rail on Curve 20.

second set of non-heat-treated rails in Curve No. 22 is equal to the wear in the original heat-treated rails in Curve No. 20, the heat-treated rails having carried more than twice the tonnage. In the low side of the curves the figures show the wear in the non-heat-treated rails to be as much as three times that in the heat-treated rails.

In the interest of getting the best possible rail life, the Great Northern decided to transpose the fully heat-treated rails in Curve No. 20. On May 17, 1961, the day following the inspection, the rails in the high side of the curve were moved to the low side, and the low-side rails were moved to the high side in direct transposition. Thus, after more than 384,000,000 gross tons of traffic, a new gage corner of the head of the heat-treated rails was put into service. The condition of the rails at the time of the transposition is shown on Figs. 2 to 4.

Six rails at the west end and 5 rails at the east end of the curve were not transposed and remained as originally installed.

The remaining low side and the second set of high side untreated rails in Curve No. 22 were replaced May 17, 1961, with flash welded fully heat-treated 78-ft rails. The condition of these rails at the time of the removal is shown on Figs. 5 and 6.

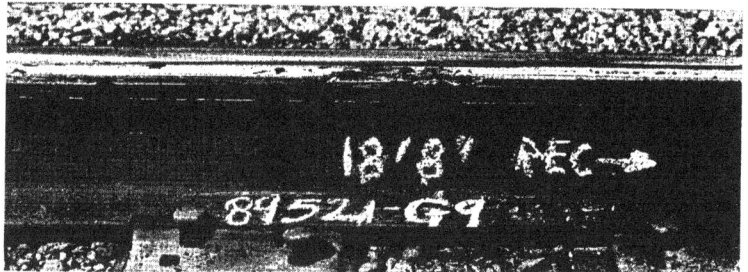

Fig. 2—Heavy shell on one fully heat-treated rail.

Fig. 3—Typical continuous flaking on some of the heat-treated rails.

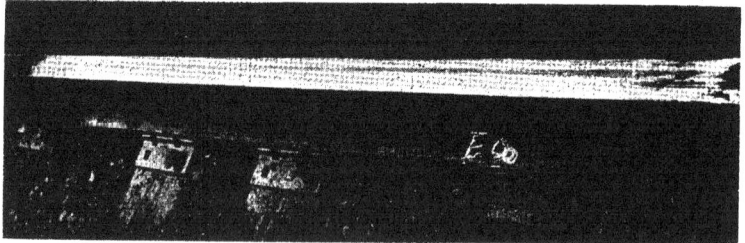

Fig. 4—Typical square field side of the low rail of the heat-treated rail.

Fig. 5 (above)—Typical high side of the second set of standard control-cooled rails. Heavy flaking and shelling.

Fig. 6—Typical low side of the second set of standard control-cooled rails.

Chesapeake & Ohio Railway Service Test of 132 RE Heat-Treated Rails

This test installation is described extensively in the AREA Proceedings, Vol. 57, page 833.

The 12 heat-treated (oil-quenched) and the 12 non-heat-treated, end-hardened 132 RE rails near Martha, W. Va., were inspected May 9, 1961. These rails were laid in the high and low sides of the northbound (loaded) track in a 3-deg 6-min curve on May 2, 1949. The test rails had carried approximately 321,000,000 gross tons of traffic to May 1, 1961.

Heat-treated rail 87581–D11 was the only test rail in the high side of the curve which continued to be free of gage corner service developments. Head checks and accompanying black areas along with small black spots appeared in the other five heat-treated rails.

Non-heat-treated high-side rail 87581–C37 was turned end for end on April 3, 1958; the field side thus became the gage side of the rail. The change was made on account of the shelling which had developed in the gage corner since the rail had been in service. This was the only change in the rails in this test installation to the time of this inspection.

Four of the other five non-heat-treated rails had developed a number of shells in the gage corner of the head, and flaking and black areas were noted in the remaining high-side non-heat-treated rail.

Both the heat-treated and non-heat-treated rails in the low side of the curve appeared in excellent condition, with more plastic flow noted in the non-treated rails. The first heat-treated rail in the low side of the curve, 87581–C14, was marked 7 ft from the receiving end by engine-wheel burns.

Comparatively speaking, progression of the service developments in these test rails had been slow through the 1959 inspection. However, an increase in the gage corner service developments, particularly in the non-heat-treated rails, was noted at the time of both the 1960 and 1961 inspections.

Wear measurements made by C&O personnel have continually indicated considerably more wear in the non-heat-treated than in the heat-treated rails. The measurements for the three previous years indicated the following amounts of wear (the figures for 1961 were not available at the time of this report):

	Heat-Treated			Non-Heat-Treated		
	1958	1959	1960	1958	1959	1960
Head Wear:						
High Rail	0.038	0.039	0.043	0.065	0.073	0.080
Low Rail	0.050	0.051	0.061	0.088	0.090	0.099
Gage Wear:						
High Rail	0.041	0.048	0.053	0.077	0.082	0.085

On June 12, 1961, after more than 12 years of service and after carrying approximately 325,000,000 gross tons of traffic, the test rails were transposed. It will be most interesting to see how well the rails take the transposition after this rather extensive amount of service in the original location. The condition of both the standard control-cooled and the fully heat-treated rails prior to the transposition is illustrated on Figs. 7 to 9.

Fig. 7—Standard control-cooled rail showing shelling and black spots.

Fig. 8—Heat-treated rail showing shelling and black spots.

Fig. 9—Low-side heat-treated rail.

Appendix 8–b

Shelly Rail Studies at the University of Illinois

By R. E. Cramer

Research Associate Professor, University of Illinois

Organization and Acknowledgment

The shelly rail investigation at this laboratory is financed by the Research Department of the Association of American Railroads. Jerry Crum, student test assistant, has worked on a part-time basis, and Marion Moore, mechanic, has repaired and operated the rolling-load machines.

Rolling-Load Tests to Produce Shelling in 136-Lb High-Silicon Chrome–Vanadium Rails

Rails 1202 and 1203 listed in Table 1 have high silicon, 0.66 percent, together with 0.20 percent chromium and 0.12 percent vanadium. These rails have Brinell hardnesses of 329 and 330, together with high strength. Two rolling-load tests were reported last year on rail 1202. They ran 1,682,000 cycles and 5,805,000 cycles, which are rather high averages. This year another test on a different rail (**Fig. 1**) from the same heat No. 03–092 rolled in 1959 ran 11,809,200 cycles and did not fail or show any signs of developing shelling. Close examination of the rail shown in **Fig. 1** reveals that the steel in the rail head did not flow to the gage side even at 11,809,200 cycles, which is very unusual for a rail of only 330 Brinell hardness. These tests seem to be higher than would be expected from the hardness and mechanical properties of these rails, but they are also very encouraging. This steel has only a small amount of alloy content compared to the chromium–vanadium alloy rails which have been previously tested. Cer-

Fig. 1—Chrome–vanadium rail after 11,809,200 cycles. No failure developed and the rail head did not flow or deform, only the mill scale on the side of the rail head has fallen off and a very thin fin has developed at the gage corner.

TABLE 1

MECHANICAL AND ROLLING LOAD TESTS, OCTOBER 1, 1960 - SEPT. 30, 1961

Spec. No.	Chemical Analysis C	Mn	Si	V	Size and Kind of Rail	Avg. Brin. Hard	Yield Strength .2% Offset psi	Tensile Strength psi	Elong. in 2 in %	Red. of Area %	Charpy No Notch	Value Key Hole	Endurance Limit psi	Cycles for Failure of 50,000 lb. Wheel Load
1202^A_B	.76 Cr. .20	.73	.66	.12	136 lb. High Si-Cr-V Colorado for S.P.R.R.	329	99,100	163,700	6.3	13.8	93	3.0	72,500	1,682,200 5,805,400
1203^A_B	.76 Cr. .20	.73	.66	.12	136 lb. High Si-Cr-V Colorado for S.P.R.R.	330	97,300	159,700	6.0	12.2	90	2.7	70,000	11,809,200 Stopped No Failure
1206^A_B	.68	.81	.15		112 lb. Algoma-Basic Oxygen Steel-Experimental	260	69,000	130,000	8.2	18.1	131	1.7	62,000	3,275,600* 3,120,000
1207^A_B	.68	.81	.15		112 lb. Algoma-Basic Oxygen Steel-Experimental	257	72,800	132,000	8.2	17.3	135	1.6	60,000	2,556,500 4,903,200
1208^A_B	.68	.81	.15		112 lb. Algoma-Basic Oxygen Steel-Experimental	261	70,800	130,700	7.3	15.9	118	2.0	60,500	2,984,300 1,800,000
1209^A_B	.74	.82	.14	Cb .015	115 lb. Columbium Rail Colorado for G.N.R.R.	267	T72,500 C69,400	132,800	10.0	19.2	133	4.2	62,000	2,458,100
1210^A_B	.74	.81	.14	.030	115 lb. Columbium Rail Colorado for G.N.R.R.	267	T75,900 C69,500	134,000	8.8	19.7	151	4.9	63,000	
1211^A_B	.74	.80	.14	.047	115 lb. Columbium Rail Colorado for G.N.R.R.	269	T77,600 C71,800	136,300	9.3	20.1	148	3.4	65,000	2,051,000 2,304,800
UPA1	.78	.84	.19		133 lb. Single Flame hard. by rising water UP,RR.	370								11,501,100
$DUPB^1_2$	.78	.84	.19		133 lb. Double Flame hard. by rising water UP,RR.	420								4,038,200 5,498,200
DUPC1	.78	.87	.17		133 lb. Double Flame hard. by receding water UP, RR.	400								3,297,800
UPD^1_2	.79	.87	.17		133 lb. Single Flame hard. by receding water UP, RR.	380								2,913,200 3,628,000

T - Tension   C - Compression.   *Web failure no shelling crack.

tainly this low chrome–vanadium, high-silicon alloy would not be expected to have the brittleness which was found to be objectionable in the 1 percent chrome–vanadium rails that failed in service tests on both the Pennsylvania Railroad and the Norfolk & Western Railway as reported by Kurt Kannowski of the AAR Laboratory (see AREA Proceedings, Vol. 62, pages 627–629).

Laboratory rail 1202 was from the C rail of ingot 1 and rail 1203 was from the C rail of ingot 11. This heat of rails was laid in the tracks of the Southern Pacific Company in the Techachapi Mountains where a heat of the 1 percent chrome–vanadium rails was also tested and proved to be too brittle for the service. It will be interesting to observe both the fourth laboratory rolling-load test specimen and the service these rails give in track.

Rolling-Load Tests of Basic-Oxygen 112-Lb Steel Rails

Last year mechanical tests only were reported on three basic-oxygen steel rails furnished by the Algoma rail mill. Six rolling-load tests have now been completed as shown in Table 1 on three of these rails, Nos. 1206, 1207 and 1208. Specimen A of rail 1206 ran 3,275,600 cycles and developed a head and web separation crack but no shelling crack. This may have been because it was of the 112-lb rail section which does not have a good fillet between the rail head and web. Rail 1206 B did develop shelling at 3,120,000 cycles as shown in Fig. 2. The two specimens from rail 1207 ran 2,556,500 and 4,903,200 cycles in the standard rolling-load test. Fig. 2 shows these specimen after the cracks developed. It will be noted that the gage corners show a large amount of plastic flow or deformation before the shelling cracks developed. Since these six tests average 3,106,500 cycles, it appears that there should be no hesitation in using basic-oxygen steel rails in track for service tests.

Rolling-Load Tests of 115-Lb Columbium Steel Rails

Last year the mechanical properties of three rails containing different amounts of columbium were reported, together with two rolling-load tests of rail 1211, both of which ran slightly over 2 million cycles. As shown in Table 1 of this report, a rolling-load test of rail 1209 A, which contained the lowest amount of columbium, ran 2,458,100 cycles. This rail deformed a large amount at the gage corner before the shelling crack developed as shown in the picture in Fig. 2 of rail 1209 A. So far, tests of the columbium rails have not indicated any appreciable benefits from the columbium content, and no further rolling-load tests of these rails are planned.

Rolling-Load Tests of Flame-Hardened Rails

In AREA Proceedings, Vol. 61, pages 877 and 878, a description is given of the Union Pacific Railroad's method of flame-hardening rails. At that time the rails were flame hardened by rising water and receding water, so two test specimens were prepared for rolling-load tests by each process. One each was single flame hardened and one each double flame hardened. Six tests are now completed, as shown in Table 1 and Fig. 3. The patterns of the flame-hardened areas and the Rockwell C hardness tests of these rails are shown in Fig. 4. The D in the middle two designations indicates double treatment. Three rails treated with rising water ran 11, 5 and 4 million cycles while the three rails treated with receding water ran about 3 million cycles. There are too few tests to draw definite conclusions, but it might be suspected that the treatment with rising water is the more desirable. Certainly all tests show an improvement of the flame-hardened rails over standard carbon steel rails in resisting shelling. It should be noted that these rails have surface Brinell hardness of about 400 which should also give them good abrasion resistance on curves.

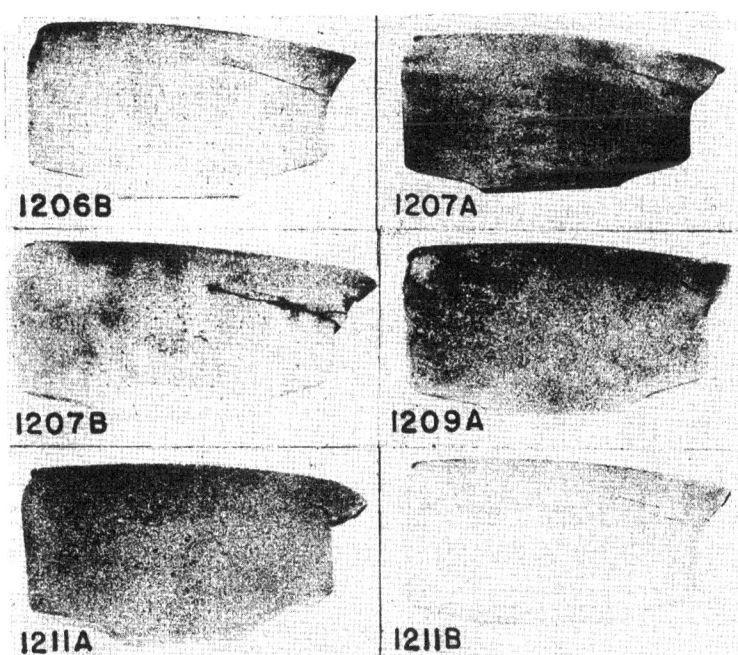

Fig. 2—Shelly failures produced in rolling-load tests.

Specimen Number	Size and Kind of Rail	Average Brinell Hardness	Cycles of 50,000-Lb Wheel Load
1206B	112 lb basic oxygen	260	3,120,000
1207A	112 lb basic oxygen	257	2,556,500
1207B	112 lb basic oxygen	257	4,903,200
1209A	115 lb 0.015% columbium	267	2,458,100
1211A	115 lb 0.047% columbium	269	2,051,000
1211B	115 lb 0.047% columbium	269	2,304,800

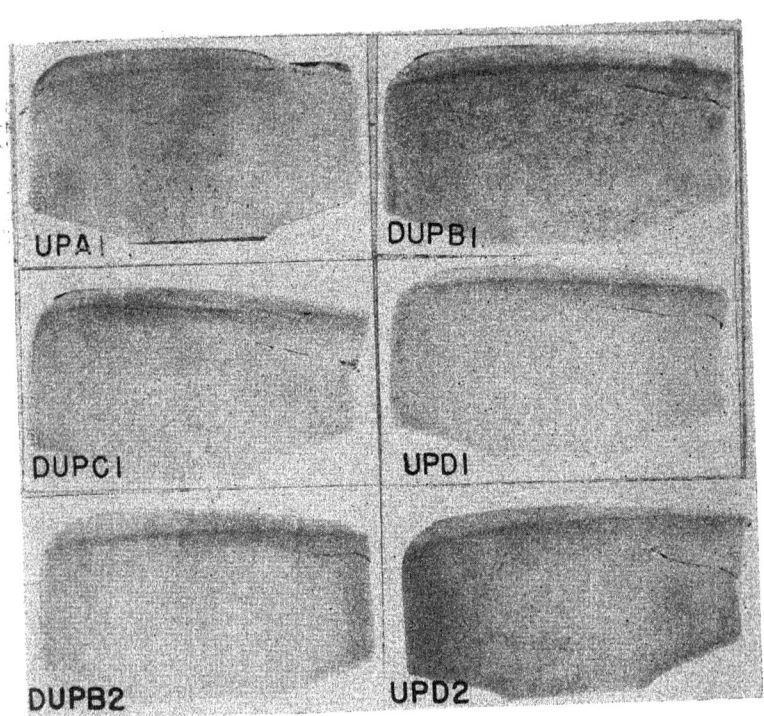

Fig. 3—Shelly failures produced in flame-hardened rails by rolling-load tests.

Specimen Number	Size and Kind of Rail	Average Brinell Hardness	Cycles of 50,000-Lb Wheel Load
UPA1	133 lb single flame hardened	370	11,501,000
DUPB1	133 lb double flame hardened	420	4,038,200
DUPC1	133 lb double flame hardened	400	3,297,800
UPD1	133 lb single flame hardened	380	2,913,200
DUPB2	133 lb double flame hardened	420	5,498,200
UPD2	133 lb single flame hardened	380	3,628,000

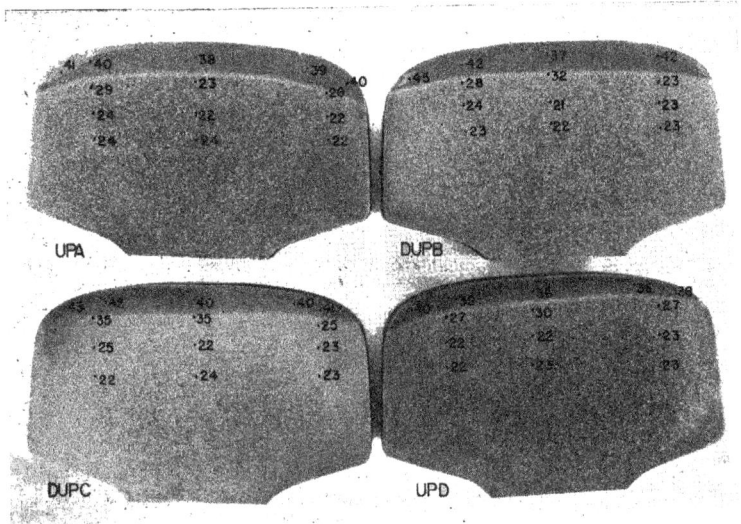

Fig. 4—Hardness patterns and Rockwell C hardness tests of Union
Pacific flame-hardened rails.

End-Quench Hardenability Tests of Four Rails Used in Rolling-Load Tests

Since it appears that there is likely to be future interest in the heat treatment of
rails either by full quenching or by flame hardening, a few preliminary end-quench
hardenability tests, sometimes called Jominy tests, were made. "Hardenability" of steel
is the property that determines the depth and distribution of hardness induced by
quenching. Hardenability should not be confused with hardness as such or with maximum hardness. The maximum attainable hardness of any steel depends on carbon content. The depth of penetration of hardness is often controlled by alloy content and
grain size. The end-quench hardenability tests measure both maximum hardness and
depth of penetration.

Briefly, the end-quench hardenability test consists of preparing specimens 1 in. in
diameter by 3 in long. In our tests these specimens were packed in carbon and heated
to 1450 F. They were then hung in a fixture, and a stream of water $\frac{1}{2}$ in. in diameter
was forced against the bottom end of the specimen for 10 min, while the remainder
was exposed to still air at room temperature.

After cooling, two parallel flats 180 deg apart and 0.015 in. deep are ground along
the entire length of the bar, and Rockwell C hardness measurements were taken at inter-
vals of $\frac{1}{32}$ in for the first $\frac{1}{2}$ in; $\frac{1}{16}$-in intervals for the next $\frac{1}{2}$ in; $\frac{1}{8}$-in intervals for
the second inch and $\frac{1}{4}$-in intervals for the remainder of the bar. These Rockwell C
hardness readings were then plotted as shown in Fig. 5. In this figure curves are given
for a basic-oxygen rail, No. 1206; a 0.047 percent columbium rail, No. 1211; a high-
silicon-vanadium rail, No. 1201; and a high-silicon chrome-vanadium rail, No. 1202.
There is a decided similarity among the curves, but there is some variation in the dis-

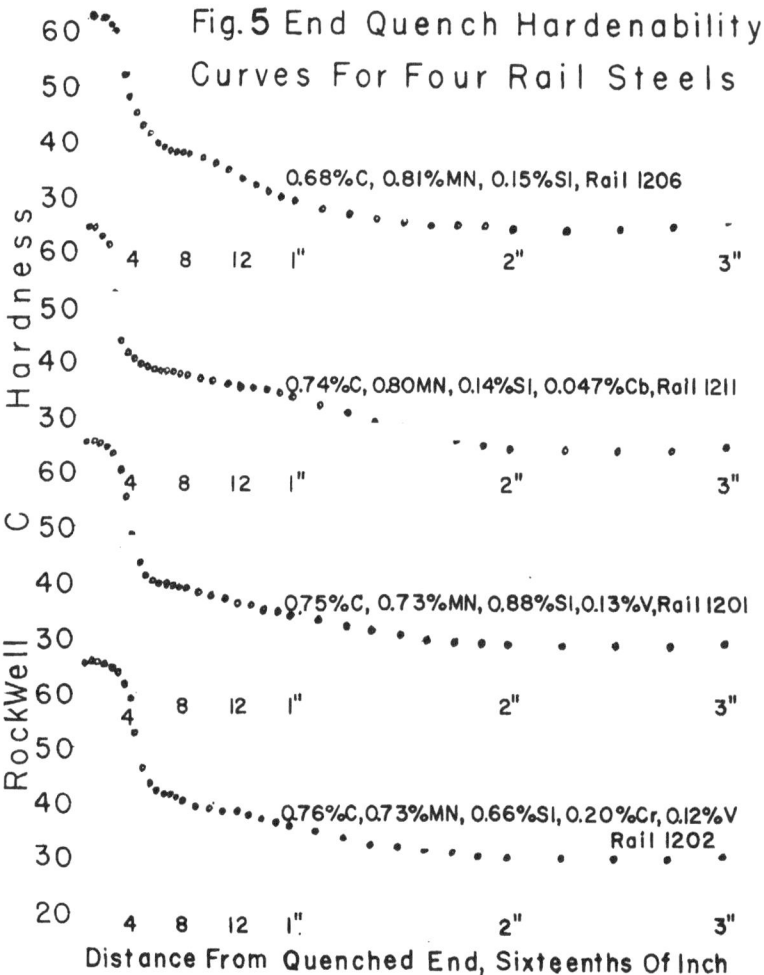

Fig. 5 End Quench Hardenability Curves For Four Rail Steels

0.68%C, 0.81%MN, 0.15%SI, Rail 1206

0.74%C, 0.80MN, 0.14%SI, 0.047%Cb, Rail 1211

0.75%C, 0.73%MN, 0.88%SI, 0.13%V, Rail 1201

0.76%C, 0.73%MN, 0.66%SI, 0.20%Cr, 0.12%V Rail 1202

Distance From Quenched End, Sixteenths Of Inch

Rockwell C Hardness

tance from the quenched end to a reading of 40 Rockwell C hardness. These rails were all very low in alloy content, so not much difference in hardenability would be expected. Duplicate tests were made on all specimens, and the results were almost identical, so it is believed that the small variations shown in the four curves are true differences in hardenability. Whether these variations would be significant in the heat treatment of full-size rails seems rather doubtful.

Summary

1. High-silicon chromium–vanadium rails have given extremely high rolling-load tests of almost 12 million cycles.

2. Basic-oxygen standard carbon-steel rails have given an average of 3,106,500 cycles in six tests.

3. Another rolling-load test of rails containing columbium is reported which ran 2,458,100 cycles.

4. Rolling-load tests are reported on six flame-hardened rails produced by the Union Pacific process. The tests ran from 11 million to 3 million cycles.

5. End-quench hardenability curves are given for four rail steels to furnish information on the quenching characteristics of low-alloy rail steels.

Report on Assignment 9

Standardization of Rail Sections

T. B. Hutcheson (chairman, subcommittee), W. D. Almy, S. H. Barlow, G. V. Begany, Jr., B. Bristow, J. A. Bunjer, T. F. Burris, C. J. Code, L. S. Crane, W. J. Cruse, O. E. Fort, C. E. R. Haight, C. J. Henry, C. C. Herrick, Ray McBrian, Embert Osland, B. R. Perkins, J. G. Roney, H. F. Smith, H. M. Williamson.

Engineering Aspects of Current Rail Sections*

The AAR Research Center staff was requested by the Rail committee to make a study of the CF&I rail sections compared to the existing AREA sections. The study has divided itself into four phases which will be discussed separately. These are:

1. Flexural stiffness and strength
2. Shape of head contour
3. Depth of head
4. Localized web and fillet stresses

Each of these four phases will be discussed in order.

Flexural Stiffness and Strength

As a background for discussion of the investigation Figs. 1 to 8, incl., have been prepared to give the detailed dimensions and physical properties of each of the rail sections included in this study. Fig. 1 shows the 100 RE section. Fig. 2 shows the 106 CF&I section with the 100 RE shown for comparison by the light dashed line. The differences in details of design are readily apparent from the figures and will not be discussed here but will be referred to later. Fig. 3 shows the 115 RE rail section and Fig. 4 shows the 119 CF&I section with the 115 RE indicated for comparison by the light dashed line.

* An abstract of Report No. ER-15 issued by the Research Department, AAR. Copies of the full report can be obtained from the director of engineering research, AAR, 3140 S. Federal St., Chicago 16.

Figs. 5 and 6 shows the same information for the 132 RE and 136 CF&I rail sections. Although not contemplated in the original study the New York Central redesigned its 127-lb section while this study was under way and asked that it be included. Fig. 7 shows the new design of the 136 NYC section with the 136 CF&I shown for comparison by the light dashed line. Fig. 8 shows the 140 RE section with the 136 CF&I shown for comparison with the light dashed line.

These eight figures show the moment of inertia of each rail section, which is indicative of its flexural stiffness as a beam and of its ability to distribute the wheel load lengthwise along the track to the supporting ties. It is also a measure of its ability to bridge over one or more ties that may not be directly supporting the rail or which for some reason do not offer the same resistance to rail depression as other ties. In other words, the property of rail stiffness contributes to uniformity of rail depression and smoothness of ride with the variabilities of tie supporting conditions as they exist even in well maintained track.

Fig. 9 shows graphically the moment of inertia for seven of the rail sections included in the study. The 136 NYC is not shown because its properties are so nearly the same as those for the 136 CF&I that the difference cannot be depicted in the graph. It will be noted that the stiffness of the various rail sections do not depart far from a straight line relationship relative to the weight.

The flexural strength of the rail is indicated by its section modulus. The bending moment developed in the rail by applied wheel loads divided by the section modulus of the head gives the maximum flexural stress developed in the head, and the bending moment divided by the section modulus of the base gives the maximum flexural stress developed in the base. Fig. 9 also shows the section modulus of the head and of the base for each of the seven rail sections.

It will be observed from Fig. 9 that the 100 RE is efficiently designed for its weight from the standpoint of stiffness and strength. The 106 CF&I is somewhat below the "line" for flexural stiffness and strength in the base. The 115 RE is on the line for flexural stiffness, somewhat below in flexural strength in the head, and somewhat above in flexural strength in the base. The 119 CF&I is close to the line in all three respects. The 132 RE is on the line in flexural stiffness, somewhat below in flexural strength in the head, and above the line in flexural strength in the base. The 136 CF&I is somewhat above the line for flexural stiffness and on the line in flexural strength for head and base. The 140 RE is somewhat below the line in flexural stiffness, somewhat above in flexural strength in the head, and somewhat below in flexural strength in the base.

Shape of Head Contour

In developing the designs of the 115 RE and 132 RE rail sections many contours were taken on the 112 RE and 131 RE sections in tangent track on several different railways. It was found from these contours that the effect of wheel wear is to form a uniform contour on the top of rail head regardless of the contour provided when the rail is rolled. If the rolled contour is flatter than the wear contour, as was the case with the 112 RE and 131 RE sections, there is a cold working initially on the gage corner of the rail, with resultant flow and formation of a lip on the gage side. Since this cold working results in severe plastic flow and the setting up of residual stresses in the gage corner location where shelly cracks develop, it seemed logical that the situation would be helped if the rail was initially rolled to a contour reducing this cold working, plastic flow and residual stresses to a minimum. From the analysis of the worn rail contours taken (see AREA Proceedings, Vol. 44, page 599, and Vol. 45, page 449) a top contour was designed for the 115 RE and 132 RE sections that was adopted in 1947 (see AREA Proceedings, Vol. 48, page 658) to give the best possible fit between the rolled rail contour and the contour to which it would become worn by the car wheels.

Rail contours taken on several railroads of 115 RE and 132 RE rail in tangent track have indicated that when the rail is to proper gage and cant the wear pattern is very uniform around the gage corner and on top of the rail head. Rail contours that have been taken on 119 CF&I and 136 CF&I sections, however, have indicated that there is a plastic flow of the rail from the top of the head into the gage corner to some extent (see AREA Proceedings, Vol. 59, page 981), resulting in a worn contour being formed which has the same shape as that provided initially in the 115 RE and 132 RE sections. As a result there is some plastic flow. This is to be expected, as observation of Fig. 9A showing a comparison of the gage corner profile for the 131 RE, the 132 RE and the 136 CF&I sections shows that the gage corner profile provided in the 136 CF&I sections is about an intermediate condition in relieving the wheel bearing on the gage corner between that in the 131 RE and in the 132 RE sections. The extent to which the cold flow in the 119 CF&I and 136 CF&I sections on the gage corner, as a result of this contour, will affect the development of shelling is not judged to be serious. However, it would appear that the top-of-rail contour provided in the 115 RE and 132 RE sections would be preferable to that provided in the 119 CF&I and 136 CF&I sections from the standpoint of prolonging the period before shelling cracks start to develop.

It will also be noted from Figs. 3, 4, 5 and 6 that the width of head of the 119 CF&I has been made $\frac{1}{16}$ in narrower than for the 115 RE, and the width of head of the 136 CF&I also $\frac{1}{16}$ in narrower than the 132 RE. Although this is a small change, nevertheless it would appear to be a step in the wrong direction because of the importance of wheel contact pressures on the development of failures within the rail head which are now the type of rail failures of most concern. Narrowing the rail head will tend to make the wheels hollow to a sharper radius, and as a result the width of the ellipse of contact bearing pressure will be somewhat decreased and the intensity of bearing pressures somewhat increased. However, taking into account the relatively small amount of new rail laid each year, it is very unlikely that the effects of this will be noted for many years to come. A further unfavorable condition attendant upon decreasing the width of rail head is the reduction in the amount of metal available to resist side wear on the high rail of curves.

This small reduction in the width of rail head, together with the increase in the length of the fillet radius on the bottom corners of the rail head, combine to reduce the flat portion of the fishing contact on the under side of the head of the 136 CF&I section to where only headfree-type joint bars could be satisfactorily used with it. This same condition exists in the 136 NYC section, whereas with both the 132 RE and the 140 RE sections there is sufficient flat fishing area to permit the use of either headfree or head-contact bars, in the event that a railroad does not want to be confined to the use of the headfree type only. This also results in a lesser bearing surface for supporting the rail on the filler blocks in turnout and crossing frogs.

Depth of Head

It will be noted that the depth of head for the CF&I sections has been increased compared to the corresponding RE sections by $\frac{3}{32}$ in for the 106 CF&I, $\frac{1}{8}$ in for the 119 CF&I, and $\frac{3}{16}$ in for the 136 CF&I. Increasing the depth of head, particularly if the additional metal is added on top of the rail as was done with the CF&I sections, tends to increase the moment of inertia and section modulus of the rail. It is not the most efficient way of accomplishing increase in stiffness and strength, as this can be done most effectively by having the metal in the head and web as far from the neutral axis as possible, as in the ordinary I-beam design. Nevertheless, the additional metal in the CF&I sections does add to the stiffness and strength of the rail as is indicated by the moment of inertia and section modulus, and previously discussed. Thus the additional metal is of some benefit in this respect. The additional thickness of head is also of benefit in reducing the localized stresses in the upper web and fillet area of the rail from the direct pressure of the applied wheel load. Another advantage of increasing the depth of the head is in preventing wheel flanges from striking the joint bars as vertical wear develops on top of the rail head. It is, however, not necessary to provide additional depth of head for this purpose since with the AREA designs of joint bars developed for the 115 RE and 132 RE rail sections, the top contour of the bar was shaped to provide adequate clearance for wheel flanges during the life of the rail.

A fourth advantage of increasing the depth of the head is that more metal is provided for vertical wear. Actually, however, very little rail is removed from track because the amount of vertical wear becomes excessive. In order to obtain factual data on this the chief engineers of AAR Member Roads were asked to have measurements made of the actual loss in height due to wear of rail being removed from main-line tangent track during the year 1957. Information on this was furnished by 20 railroads and the results were analyzed and presented in the AREA Proceedings, Vol. 60, page 970. The graph shown in that report shows that the rate of vertical rail head wear decreases as the average annual traffic density increases. This, in conjunction with the fact that the heavier rail sections will be used in locations of higher traffic density, makes it appear that a provision of $\frac{3}{16}$ in for vertical head wear will be ample to obtain the full usable life of the rail in most cases. Other considerations that make provision for more than $\frac{3}{16}$ in wear of doubtful value are: (1) the difficulty in replacing failed rails with rails that will have nearly enough the same amount of vertical wear to provide a satisfactory condition in main line track without too much joint impact due to the difference in height of abutting rail ends, and (2) the fact that with the trend towards increasing wheel loads, it appears most likely that incidence of head failures and rail surface failures (spalling) will dictate renewal of rail before the amount of vertical head wear would require renewal. This $\frac{3}{16}$-in provision for head wear would also allow for grinding 0.020 in from the rail at least twice during its life in main line location. This $\frac{3}{16}$-in value may therefore be kept in mind in reviewing the localized web and fillet measurements.

Localized Web and Fillet Stresses

Head and web fillet cracks at the rail ends within joint bar limits on the gage side of the low rail of curves, and of rails in wet tunnels and insufficiently drained road crossings and station platforms, have constituted an important number of rail failures in past years. In the design of the 115 RE and 132 RE rail sections adopted in 1947, the upper web was increased somewhat in thickness and the upper fillet radii were increased in length, in order to reduce the stress range resulting from the contact pressure of the wheel occurring anywhere on the top of the rail head from the gage side to the field side. The contact pressure of the wheel is distributed to the rail web as a compressive stress in the web area directly under the wheel load. This direct stress is combined with a bending of the head on the web if the wheel bears eccentrically on the head as is frequently the case, and this condition exists particularly on the high rail of curves where the wheel flange of the leading axle of the truck normally contacts the high rail in guiding the truck around the curve.

The same type of test was made to determine the localized stresses throughout the web, including the upper and lower fillet areas, that was made in the earlier tests when the 115 RE and 132 RE sections were designed and as affected by concentric and eccentric loading applied to the rail head. The reports of the earlier tests on these sections may be found in the AREA Proceedings, Vol. 47, page 449. The laboratory tests included in this report of the rail sections shown in Figs. 1 to 8, incl., were made with static applied loading in the laboratory at the Research Center, with the loading conditions carefully controlled to be the same for each rail section. In addition, measurements were made on seven of the eight rail sections included in this study without head wear, with $\frac{1}{16}$ in planed off the rail head to simulate vertical head wear of this amount, and with $\frac{3}{8}$ in planed from the top of the rail head to represent a very extreme condition of vertical head wear. Measurements with the $\frac{1}{16}$-in and $\frac{3}{8}$-in simulated vertical head wear were not made on the 136 NYC section as this would have involved delaying the report, and the stress measurements obtained with this section were so nearly the same as with the 136 CF&I section that the delay did not seem justified.

The stress measurements obtained for each rail section give a complete pattern of stress distribution throughout the rail web, including the fillet areas, not only with the new rail section, but by proration the stress distribution can be closely determined for any amount of vertical head wear up to the $\frac{3}{8}$ in maximum included.

This last phase of the investigation, which is presented in the current report for the first time, was conducted by the Engineering Research Division staff of the AAR Research Department, W. M. Keller, vice president, under the general direction of G. M. Magee, director of engineering research. E. E. Cress, principal research engineer, was in direct charge of the assignment, assisted by M. F. Smucker, assistant electrical engineer, who made the laboratory tests at the Research Center.

Significance of Web Stresses

Table 1 shows the maximum range of stresses as determined from the laboratory measurements in the upper fillet area at the location where head and web separations might develop. The significant range of stress in this fillet area is shown for each of the rail sections and for the full head, with $\frac{1}{16}$-in simulated wear, and with $\frac{3}{8}$-in simulated wear. It must be recognized that these stress ranges are with a simulated laboratory condition of loading. However, the stress ranges shown in Table 1 for the 115 RE and the 132 RE rail sections with full head are in good agreement with the maximum stresses measured under service conditions and previously reported in the Proceedings, Vol. 53, page 921.

Data on the fatigue strength of specimens cut from the web of rail steel with the as-rolled surface are shown in the diagram in Vol. 51, page 637, and also in the same volume on pages 643 to 645 incl. It will be noted that for a range of high compression to low tension as developed in the fillet area, the total range of endurance strength so determined is on the order of 75,000 psi. Comparing this endurance strength with the total range of stresses as determined in the laboratory tests and shown in Table 1, it is evident that there is ample reserve to prevent fatigue failure for all of the sections tested, even with the extreme amount of $\frac{3}{8}$ in removed from the head, except with the possible exception of the 100 RE which showed a total range of 77,000 psi with $\frac{3}{8}$-in simulated wear. Even here it is doubtful if a fatigue failure would result, except in the case of unusual corrosion environment, because for most of the life of the rail the range of stress would be below 75,000 psi, and it is further doubtful there would be more than $\frac{3}{16}$-in vertical wear except in a very few instances with the 100 RE section. In considering the possibility of the development of head and web separations account should also be taken of the large reduction made in stresses in this area in the change from steam to diesel locomotives, as shown in Vol. 53, page 921.

For the benefit of those who may be interested, the maximum stress in the upper fillet area was calculated for each of the rail sections tested, with the exception of the 136 NYC, by the method developed by C. J. Code, assistant chief engineer—staff, Pennsylvania Railroad, and published in AREA Proceedings, Vol. 48, page 987.

Summary

In this investigation of the engineering aspects of current rail sections the work divided itself into the following four phases:

(1) Flexural Stiffness and Strength—The additional stiffness as indicated by the moment of inertia and strength as indicated by the section modulus of the 106 CF&I, 119 CF&I and 136 CF&I sections are increased approximately in proportion to the additional amount of metal provided in these sections. Thus the justification for the additional investment would depend upon the benefits in track maintenance cost as related to the increased stiffness and strength of these rail sections, with due allowance being made for increased costs due to added inventory and other labor costs attendant upon a change in rail section on a particular railway.

(2) Shape of Head Contour—Rail contour measurements taken in the field indicate that the rolled contour of the 115 RE and 132 RE sections fit the worn wheel contour better than do the 119 CF&I and 136 CF&I sections. Thus it would be expected that somewhat better service performance would be obtained from the 115 RE and 132 RE sections insofar as the development of rail shelling is concerned. On the other hand, the contour of the 106 CF&I section fits the worn wheel contour better than the 100 RE contour, and would, therefore, be expected to be somewhat better in resisting the development of shelling.

The narrowing, even though slight, of the rail head in the CF&I sections would appear to be a step in the wrong direction, although probably not a serious one, because of its effect on decreasing the radius of hollowing of worn car wheels with resultant increase in wheel contact pressures and internal direct stresses and shearing stresses within the rail head. Narrowing the width of head also reduces the amount of metal available to resist side wear on curves. The relatively large radius used in the lower corners of the rail head on the CF&I sections substantially reduces the width of the flat fishing surface underneath the rail head and thus decreases the supporting area on fillers in turnout frogs and crossing frogs.

(3) Depth of Rail Head—The increased depth of rail head in the CF&I sections provides additional metal for vertical head wear. However, measurements of the actual vertical head wear as related to traffic carried and traffic densities indicate that the present RE sections have adequate depth of head to provide a full usable life of the rail, and it appears doubtful that the additional height in the CF&I sections would give any increased life as a result thereof.

(4) Localized Web and Fillet Stresses—Measurements under identical loading conditions with full head sections and sections with $\frac{3}{16}$-in and $\frac{3}{8}$-in simulated wear indicate that the upper fillet stresses in the web are reduced to some extent with the CF&I sections compared to the corresponding RE sections. However, even with $\frac{3}{8}$-in simulated

(Text continues on page 569)

TABLE 1. MAXIMUM RANGE OF MEASURED STRESS IN THE UPPER FILLET AREA

(40,000 Lb Load Statically Applied with 3/4 In Eccentricity
Each Side of the Center of the Rail Head.)

Stress in units of 1,000 lb Compression (-) Tension (+)

Condition of Rail	Range of Stress	Total Range	Ratio to End. Limit*
100 RE			
Full Head	-42 to +8	50	0.67
3/16 in Sim. Wear	-49 to +10	59	0.79
3/8 in Sim. Wear	-62 to +17	79	1.05
106 CF&I			
Full Head	-32 to +6	38	0.51
3/16 in Sim. Wear	-36 to +9	45	0.60
3/8 in Sim. Wear	-43 to +12	55	0.73
115 RE			
Full Head	-31 to +4	35	0.47
3/16 in Sim. Wear	-36 to +6	42	0.56
3/8 in Sim. Wear	-43 to +9	52	0.69
119 CF&I			
Full Head	-27 to +5	32	0.43
3/16 in Sim. Wear	-31 to +6	37	0.49
3/8 in Sim. Wear	-37 to +7	44	0.59
132 RE			
Full Head	-29 to +4	33	0.44
3/16 in Sim. Wear	-33 to +5	38	0.51
3/8 in Sim. Wear	-38 to +7	45	0.60
136 CF&I			
Full Head	-25 to +3	28	0.37
3/16 in Sim. Wear	-28 to +4	32	0.43
3/8 in Sim. Wear	-31 to +5	36	0.48
136 NYC			
Full Head	-23 to +2	25	0.33
140 RE			
Full Head	-23 to +2	25	0.33
3/16 in Sim. Wear	-26 to +3	29	0.39
3/8 in Sim. Wear	-31 to +5	36	0.48

*Endurance limit taken as -60 to +15. See AREA Proceedings, Vol. 51, page 637.

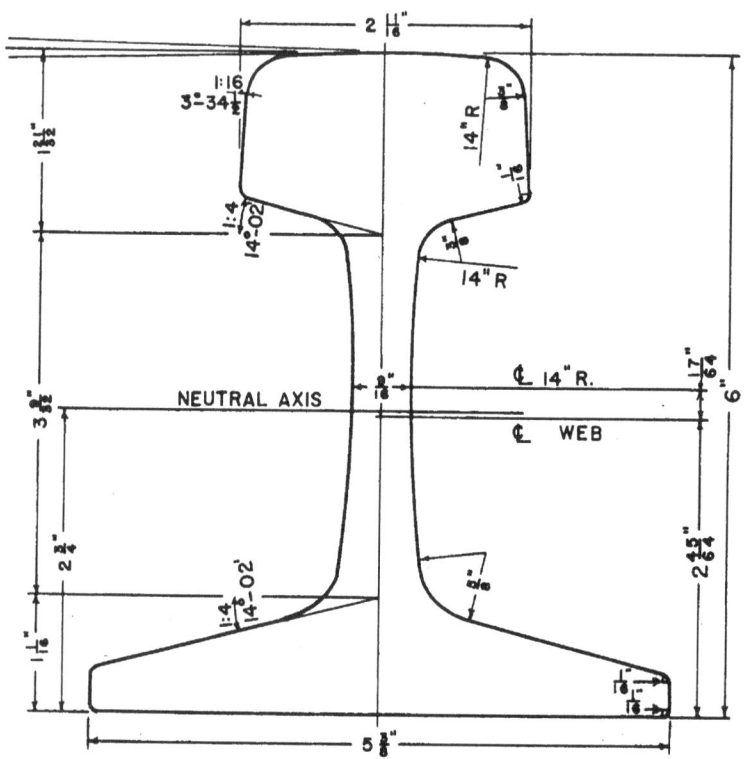

PROPERTIES OF SECTION

AREA:	Head	3.80 Sq In	38.2%	Moment of Inertia	49.00
	Web	2.25 " "	22.6%	Section Modulus, Head	15.10
	Base	3.90 " "	39.2%	" " Base	17.80
	Total	9.95 " "	100.0%	Ratio M. I. To Area	4.93
Weight per yard			101.5 lb	Ratio Sec. Mod. Head To Area	1.52

Gross Tons per mile of single Track   157.14

Fig. 1—100 RE rail section.

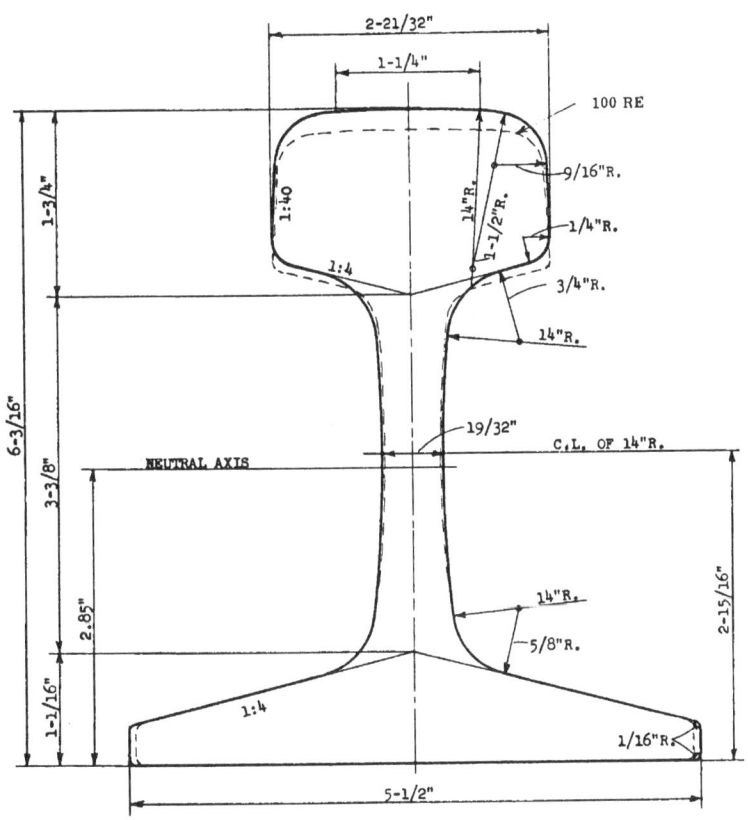

PROPERTIES OF SECTION

AREA: Head 4.00 **Sq** In   38.3%   Moment of Inertia   53.6
      Web   2.50 "   "   23.9%   Section Modulus, Head   16.1
      Base 3.95 "   "   37.8%   "   "   Base   18.8
      Total 10.45 "   "   100.0%   Ratio M. I. To Area   5.1
Weight per yard   106.6 lb   Ratio Sec. Mod. Head To Area   1.5

Gross Tons per mile of single Track   167.5

Fig. 2—106 CF&I rail section.

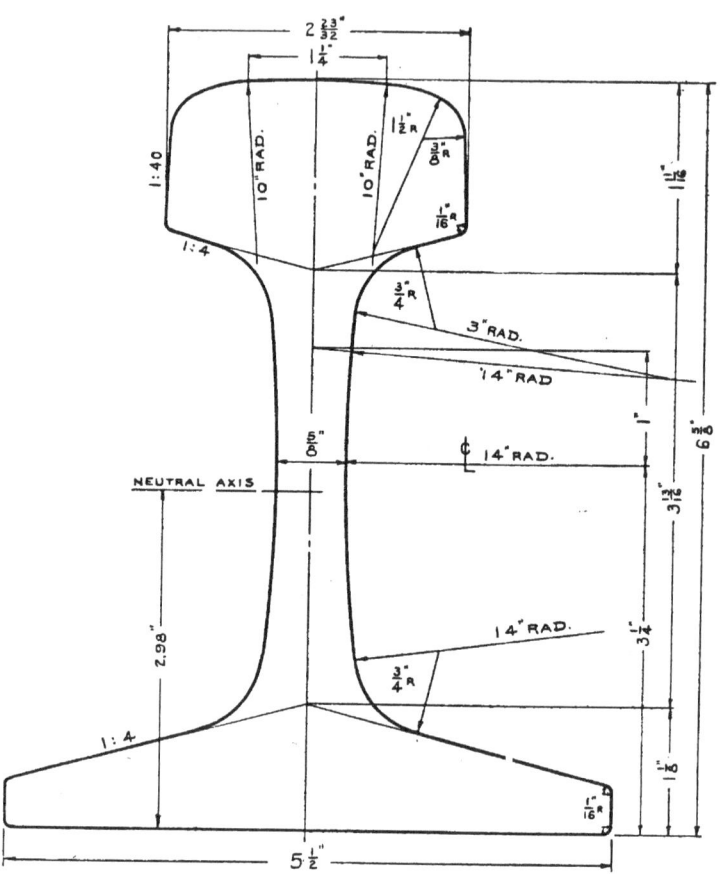

PROPERTIES OF SECTION

AREA:	Head	3.91 Sq In	34.8%	Moment of Inertia	65.6
	Web	3.05 " "	27.1%	Section Modulus, Head	18.0
	Base	4.29 " "	38.1%	" " Base	22.0
	Total	11.25 " "	100.0%	Ratio M.I. To Area	5.83
Weight per yard			114.75 lb	Ratio Sec. Mod. Head To Area	1.60

Gross Tons per mile of single Track 180.32

Fig. 3—115 RE rail section.

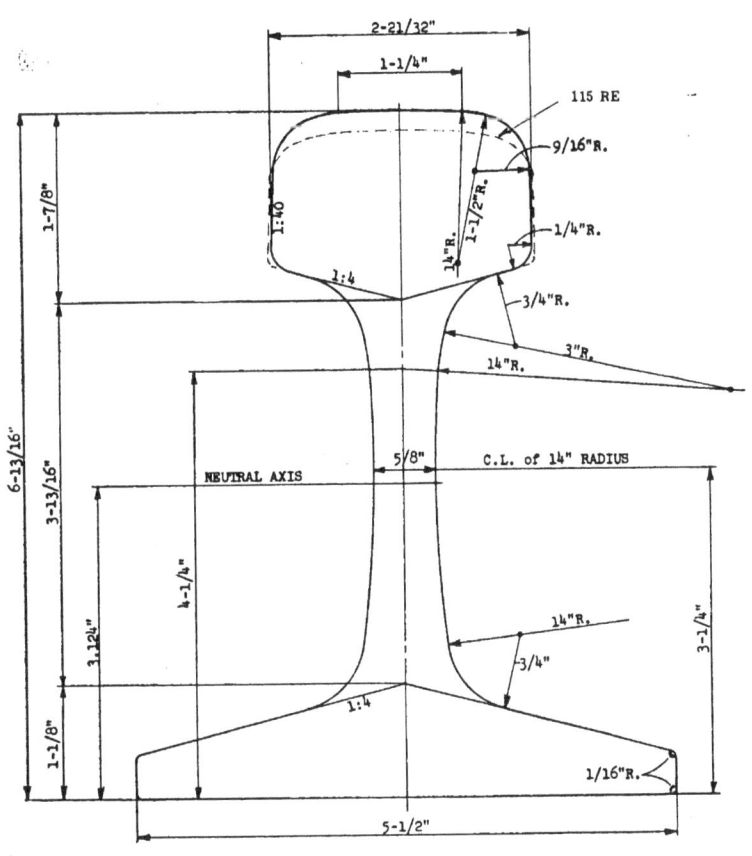

PROPERTIES OF SECTION

AREA:	Head	4.32 Sq	In	37.1%	Moment of Inertia	71.4
	Web	3.04 "	"	26.1%	Section Modulus, Head	19.4
	Base	4.29 "	"	36.8%	" " Base	22.9
	Total	11.65 "	"	100.0%	Ratio M. I. To Area	6.13
Weight per yard				118.8 lb	Ratio Sec. Mod. Head To Area	1.7

Gross Tons per mile of single Track 187.0

Fig. 4—119 CF&I rail section.

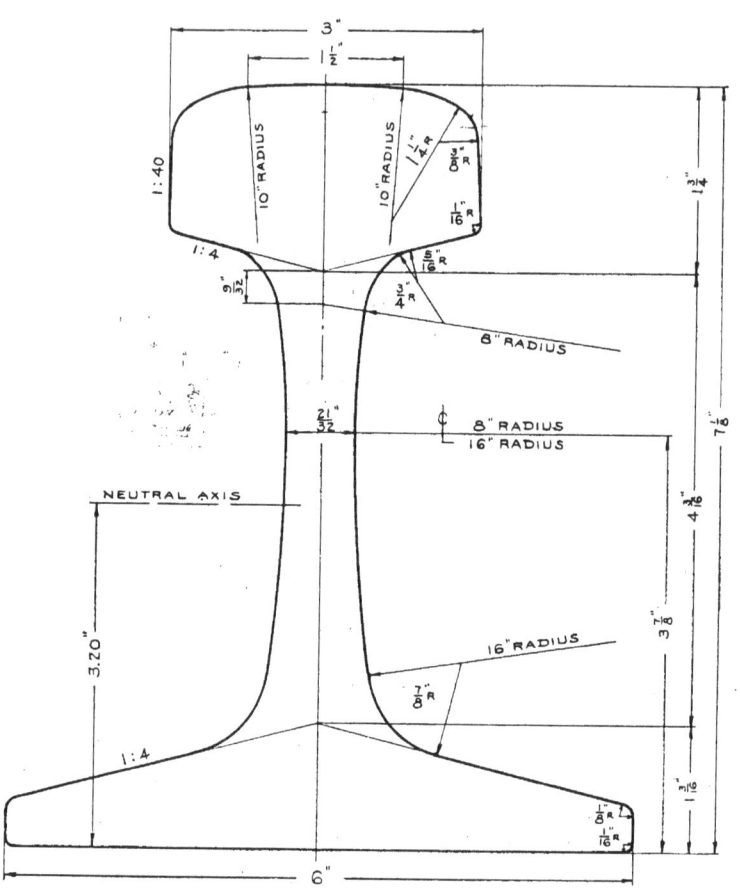

PROPERTIES OF SECTION

AREA:	Head	4.42 Sq In	34.1%		Moment of Inertia	88.2
	Web	3.66 " "	28.3%		Section Modulus, Head	22.5
	Base	4.87 " "	37.6%		" " Base	27.6
	Total	12.95 " "	100.0%		Ratio M. I. To Area	6.81
Weight per yard			132.1 lb		Ratio Sec. Mod, Head To Area	1.74

Gross Tons per mile of single Track   207.8

Fig. 5—132 RE rail section,

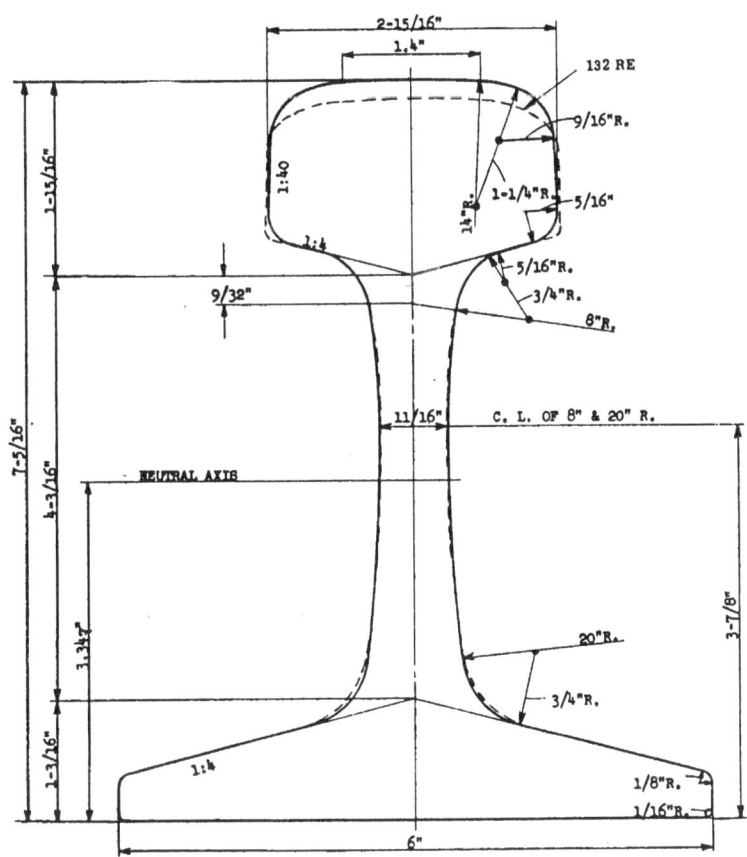

PROPERTIES OF SECTION

AREA: Head   4. 86 Sq In.   36. 4%   Moment of Inertia   94. 9
       Web   3. 62 "   "   27. 1%   Section Modulus, Head   23. 9
       Base   4. 87 "   "   36. 5%   "   "   Base   28. 3
       Total 13. 35 "   "   100. 0%   Ratio M. I. To Area   7. 1
Weight per yard   136. 2 lb   Ratio Sec. Mod. Head To Area   1. 8

Gross Tons per mile of single Track   213. 7

Fig. 6—136 CF&I rail section.

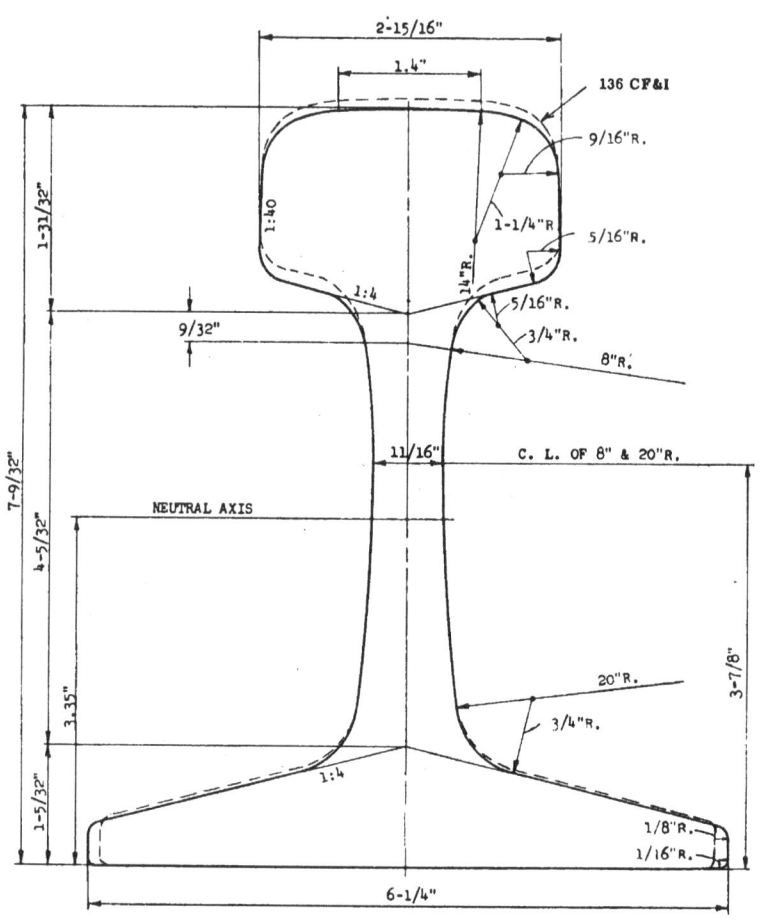

PROPERTIES OF SECTION

AREA:	Head	4.95 Sq In	37.0%	Moment of Inertia	93.9
	Web	3.63 " "	87.2%	Section Modulus, Head	23.9
	Base	4.78 " "	35.8%	" " Base	28.1
	Total	13.36 " "	100.0%	Ratio M.I. To Area	7.0
Weight per yard			136.3 lb	Ratio Sec. Mod. Head To Area	1.8

Gross Tons per mile of single Track   213.7

Fig. 7—136 NYC rail section.

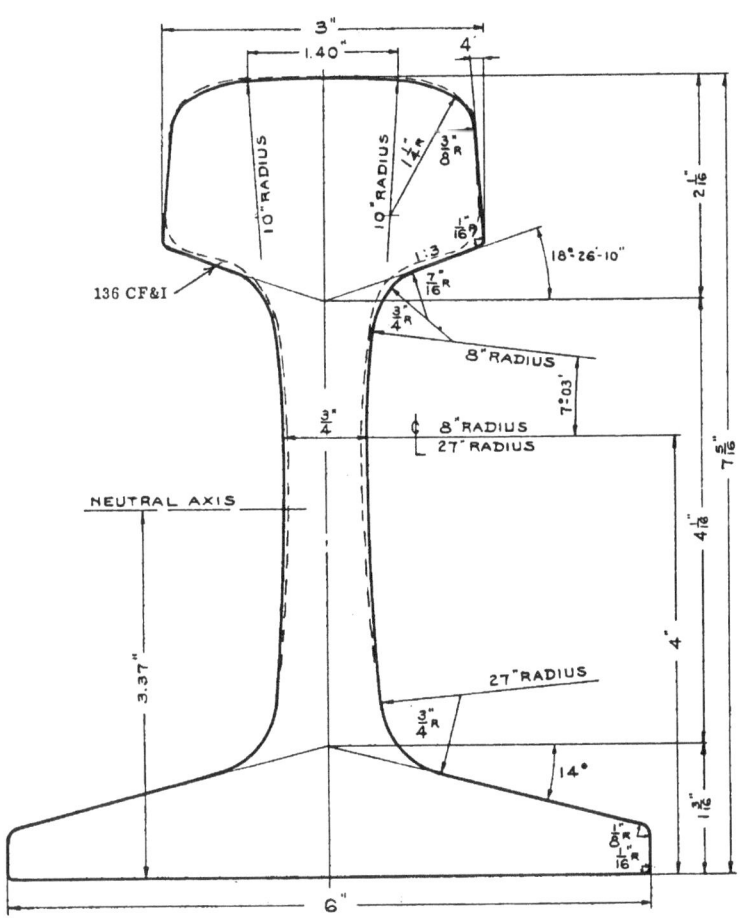

PROPERTIES OF SECTION

AREA:	Head	5.0	Sq	In.	37.0%	Moment of Inertia	97.0
	Web	3.9	"	"	28.0%	Section Modulus, Head	25.0
	Base	4.9	"	"	35.0%	" " Base	29.0
	Total	13.8	"	"	100.0%	Ratio M.I. To Area	7.03
Weight per yard					140.0 lb.	Ratio Sec. Mod. Head To Area	1.81

Gross Tons per mile of single Track   220.0

Fig. 8—140 RE rail section.

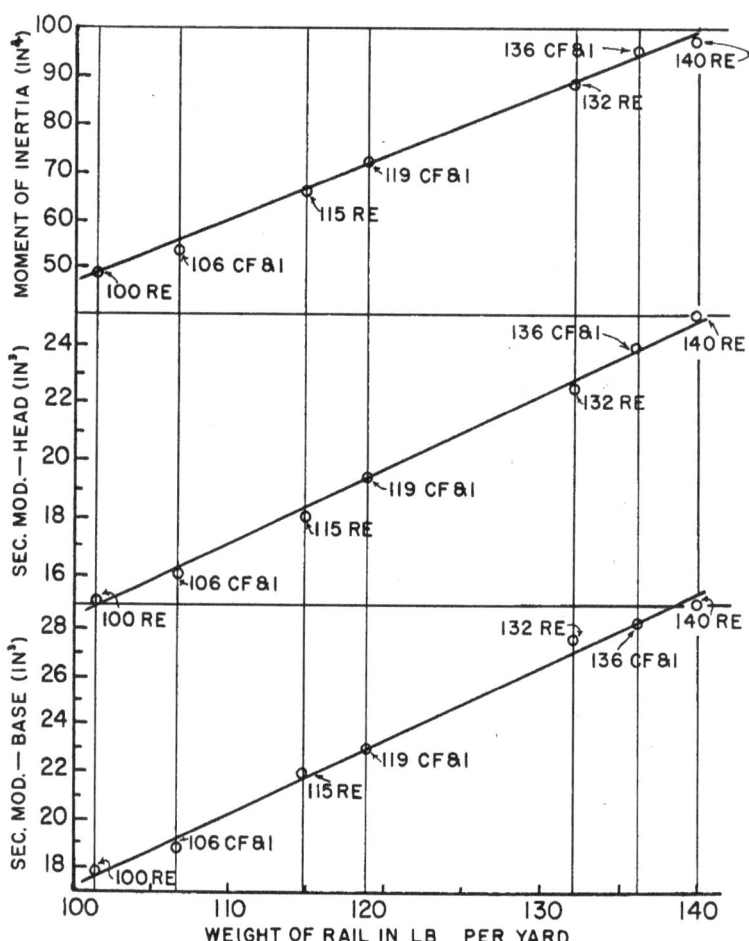

Fig. 9—Relation of flexural stiffness and strength of the various
rail sections.

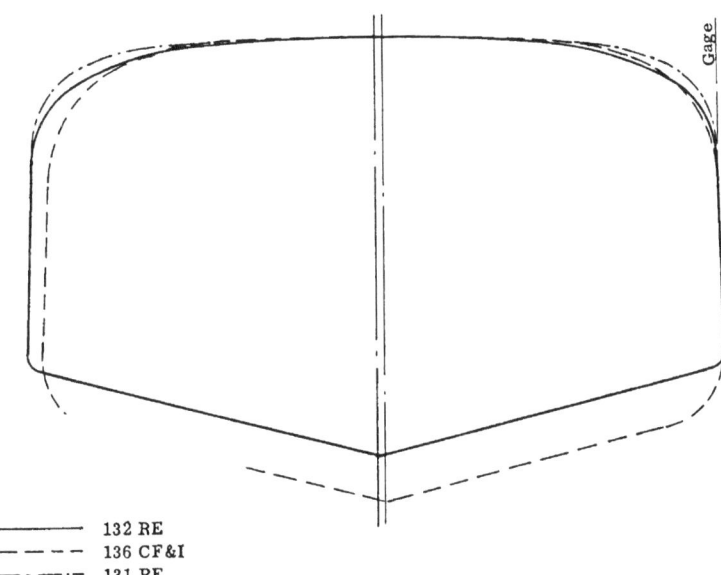

——————— 132 RE
— — — — 136 CF&I
—·——·— 131 RE

Fig. 9A—Comparison of bearing condition at the gage corner of the 131 RE,
132 RE, and 136 CF&I rail sections (enlarged about 1½ times).

wear, which is a very extreme amount of wear to be permitted from practical stand-
points, there is good reason to expect that the present AREA sections have low enough
stress levels under normal service conditions so their full service life will be obtained
without requiring renewal because of head and web separations. For those roads having
high traffic density and a large amount of curvature, the 140 RE section is available
for use if desired for these conditions.

If standardization of rail sections and frog and switch material is an objective to
be desired for railways in the United States, this objective can be attained only by
keeping rail sections unchanged unless there is a well defined need for a change. The
engineering aspects and service performance for 14 years or longer of the present AREA
rail sections do not indicate that such a need exists.

Report on Assignment 10

Service Performance and Economics of 78-Ft Rail, Specifications for 78-Ft Rail

Collaborating with Committee 5

A. P. Talbot (chairman, subcommittee), J. P. Barker, J. A. Bunjer, O. E. Fort, W. D. Almy, S. H. Barlow, Blair Blowers, R. E. Catlett, Jr., W. J. Cruse, G. H. Echols, V. E. Hall, J. C. Jacobs, H. W. Jenkins, R. R. Lawton, B. R. Meyers, F. R. Micheal, C. E. Morgan, R. H. Patterson, C. F. Parvin, R. C. Postels, R. B. Rhode, W. D. Simpson, V. R. Terrill, D. J. White, W. L. Young.

This is a progress report, submitted as information.

Service test installations of 78-ft rail under the sponsorship of this committee were made in 1950 on the Pennsylvania Railroad near Hamlet, Ind., and on the Chicago & North Western Railway near Calamus, Iowa. The rail anchorage pattern used on these installations did not adequately control the joint gap opening, and a further test installation was made on the Illinois Central Railroad, south of Monee, Ill., in 1956 to determine a rail-anchor pattern to maintain proper rail joint gap openings. Additional anchors were applied to the test stretch at Hamlet in 1954. Rail for all of these tests were fabricated by welding two 39-ft rails to provide the 78-ft rail.

A proposed table of expansion to be used in laying 78-ft rail, with anchorage pattern of alternate ties boxed, to provide uniform joint gaps is shown in Bulletin 563, February 1961, page 638. Other reports on these service tests may be found in the Proceedings Vol. 59, 1958, page 992; Vol. 60, 1959, page 972; and Vol. 61, 1960, page 883. No field measurements on these tests were made by the AAR staff during the past year.

Concerning the economics of 78-ft rail vs. 39-ft rail, actual cost data are available from only one railroad.

In 1950 the Pennsylvania installed three test stretches of 78-ft rail, each one mile in length, with a comparative adjacent mile of standard 39-ft rail of the same section. These installations were 155 PS at Ryde, Pa., installed October 1950, 140 PS at Coshocton, Ohio, installed August 1950, and 133 RE at Hamlet, Ind., installed September 1950. Maintenance cost data have been kept on these installations, and the maintenance cost to December 31, 1960, or a little more than 10 years of service, is as follows:

	155-Lb		140-Lb		133-Lb		Average	
	78-Ft	39-Ft	78-Ft	39-Ft	78-Ft	39-Ft	78-Ft	39-Ft
Million G.T. to 12/31/60	414		263		110		262	
Man-hours	2879	1985	4637	5474	3063	3629	3526	3696
Cost to 12/31/60	$6900—$4770		$11130—$13120		$7360—$8700		$8463—$8863	
Cost per mile per year	$ 690—$ 477		$ 1113—$ 1312		$ 736—$ 870		$ 846—$ 886	
Cost difference	$+213		$—199		$—134		$—40	

An average rate of $2.40 per hr was used to convert the man-hours to money in this tabulation.

The relatively high maintenance cost of the 155-lb 78-ft rail is thought to be due to soft subgrade through the 78-ft rail stretch.

The above indicates that, to date, there has been a modest saving by the use of 78-ft rail. There will be an increase in the saving when the need for reformed joint bars develops. Bars were renewed on the 140-lb rail in 1961 (not included in the cost above) and will be required on the other installations in the near future.

Batter measurements on the conventional joints in both the 78-ft and 39-ft rail show very little difference even though the joint gaps are generally wider in the 78-ft rail. All the rail in the test, both 78-ft and 39-ft was end hardened, and this has evidently retarded rail end batter to a very noticeable extent.

The AAR test installations should be observed at intervals by the research staff and results reported. The Pennsylvania expects to continue to keep maintenance costs on the three installations, and this information will be available to the committee.

The small difference in maintenance cost to date hardly justifies a final conclusion at this time.

Report of Committee 1—Roadway and Ballast

F. N. BEIGHLEY,
Chairman

L. J. DENO,
Vice Chairman

R. H. BEEDER, *Secretary*
G. B. HARRIS
W. P. ESHBAUGH
G. W. BECKER
G. D. MAYOR
K. W. SCHOENEBERG
A. E. LEWIS
R. D. WHITE

H. G. JOHNSON
R. D. BALDWIN
T. W. CREIGHTON
C. E. WEBB
W. T. ADAMS
R. A. ANDERSON
C. W. BAILEY
A. S. BARR
E. W. BAUMAN
C. R. BERGMAN
L. H. BOND
K. W. BRADLEY
J. G. CAMPBELL
H. W. CLARKE
T. T. CONNELLY
M. W. COX
B. H. CROSLAND
A. P. CROSLEY (E)
H. F. DAVENPORT
M. B. DAVIS
G. W. DEBLIN
J. W. DeMOYER
W. M. DOWDY
J. B. FARRIS
J. S. FELTON, JR.
J. K. FISHER
J. E. GRAY
M. B. HANSEN
F. W. HILLMAN (E)
H. O. IRELAND
R. J. KEMPER

H. W. LEGRO (E)
R. R. MANION
A. MANSON
P. G. MARTIN
W. C. McCORMICK
E. W. McCUSKEY
H. E. MOORE
PAUL McKAY (E)
W. G. MURPHY
F. R. NAYLOR
J. E. NEWBY
G. F. NIGH
S. J. OWENS
A. J. PACELLI
F. S. PATTON
G. W. PAYNE (E)
F. L. PECKOVER
W. F. PETTEYS
J. W. POULTER
C. W. REEVE*
L. G. REICHERT
E. L. ROBINSON
R. W. SCOTT
G. E. SHAW
W. M. SNOW
S. W. SWEET
O. W. TRIESCHMAN
STANTON WALKER
A. J. WEGMANN
E. L. WOODS

Committee

---

* Died October 1, 1961.
(E) Member Emeritus.
Members listed in bold face are the official representatives of the Engineering Division, AAR.

To the American Railway Engineering Association:

Your committee reports on the following subjects:

1. Revision of Manual.

    Progress report, including recommendations submitted for adoption ..... page 575

2. Physical properties of earth materials.

    (a) Roadbed. Load capacity. Relation to ballast. Allowable pressures.
    (b) Structural foundation beds, collaborating with Committees 6 and 8.

    Progress report applicable to both (a) and (b) covering recommendation for revision to Manual .... page 580

3. Natural waterways: Prevention of erosion.

    Your committee is reviewing present Manual material with a view of updating the material to reflect more recent data and research, and general current practices in determining size of waterway openings and prevention of erosion.

4. Culverts.

  (a) Erosion control for outlet structures.

    Research on this assignment was started in 1958 as a three-year proj-
ect. Budget curtailment in 1959, 1960 and 1961 forced cancellation.
It is felt the subject is of enough importance that in the near future
work will be resumed at Colorado State University as originally
planned.

5. Specifications for pipelines for conveying flammable and non-flammable
substances, collaborating with Committees 15 and 20.

  Progress report, including recommendations submitted for adoption ..... page 582

6. Roadway: Formation and protection:

  (a) Roadbed stabilization.

    No report.

  (b) Slope protection by use of additives.

    Your committee has submitted progress reports on this assignment
but is continuing its study to include the overall subject—Slope Protection—for Manual material.

  (c) Performance of filter material in subdrains.

    Second progress report has been submitted and appears in Vol. 63,
Bulletin 566

  (d) Gypsum treatment of subgrade to improve track drainage.

  Progress report, submitted as information .... page 586

7. Tunnels:

  (a) Ventilation, changes necessary for operation of diesel power.

    Your committee has submitted progress reports on this assignment and
is continuing its study to develop recommendations for Manual
material

  (b) Clearance; methods used to increase collaborating with Committee 28.

    Your committee has assembled a list of methods that have been used
in the past, including cost data. New and more recent methods are
being used which will be submitted as recommendations for the
Manual.

  (c) Methods of open cutting.

    This is a relatively new assignment to which the subcommittee has
not been able to devote its attention because of time necessary to
review present Manual material.

8. Fences.

  Progress report, including recommended revisions to the Manual .... page 589

9. Roadway Signs:

  (a) Reflectorized and luminous roadway signs collaborating with Committee 5 and 9 and with the Communication and Signal Section, AAR.

    Research has been discontinued temporarily due to curtailment of
research funds; however, the committee feels the subject should not
be dropped as new products are constantly being developed, some of
which might be adaptable for this purpose.

(b) Develop standard close clearance warning sign, collaborating with Committee 28.

Progress report, submitted for adoption .... page 591

10. Ballast:

    (a) Tests.

    (b) Special types of ballast.

    Progress report, submitted as information .... page 593

11. Chemical control of vegetation, collaborating with Committee 22 and with Communication and Signal Section, AAR.

    Progress report, submitted as information .... page 596

THE COMMITTEE ON ROADWAY AND BALLAST,

F. N. BEIGHLEY, *Chairman.*

AREA Bulletin 570, February 1962.

MEMOIR

Charles Wis'ar Reeve

Charles Wistar Reeve, treasurer, Young & Greenawalt Company, died October 1, 1961, in Chicago. He is survived by his wife and one son, Charles B. Reeve, now in the U. S. Army.

Mr. Reeve was born July 6, 1904, at Philadelphia, Pa. He graduated from Barringer High School, Newark, N. J., and held a B.S. degree in Engineering from Union College, Schenectady, N. Y. He was a member of Westminister Presbyterian Church, Albany, N. Y., and held membership in the B.P.O.E. Lodge at Oneonta, N. Y.

Mr. Reeve entered railroad service in 1928 as transitman with the Delaware & Hudson Railroad. He was promoted to the position of track supervisor in 1941 and to bridge and building supervisor in 1947. In 1954 he left railroad service to join Young & Greenawalt Company, manufacturers of corrugated pipe and metal products, with which he was associated until his death.

Mr. Reeve joined the American Railway Engineering Association in 1947, and from 1948 through 1954 was a member of Committee 22—Economics of Railway Labor. In 1956 he became an associate member of Committee 1, which membership he held until his death.

His interest, his untiring efforts and generous contributions to the work of Committee 1 will long be remembered by the membership. It is with a deep sense of appreciation of his association with us that Committee 1 records this tribute to his memory.

Report on Assignment 1

Revision of Manual

G. B. Harris (chairman, subcommittee), R. D. Baldwin, G. W. Becker, T. W. Creighton, W. P. Eshbaugh, H. G. Johnson, A. E. Lewis, G. D. Mayor, K. W. Schoeneberg, C. E. Webb, R. D. White.

Your committee has made a study of its Chapter in the Manual and submits the following recommendations for adoption:

Page 1-1-16

SPECIFICATION FOR RIPRAP STONE

Reapprove without change.

Pages 1-1-17 to 1-2-22.2, incl.

ROADBED STABILIZATION

Reapprove without change.

Pages 1-1-23 to 1-1-35, incl.

ROADWAY PROTECTION

Reapprove without change.

Page 1-1-36

METHODS OF PROTECTION AGAINST DRIFTING SAND

Reapprove without change.

Pages 1-1-65 to 1-1-67, incl.

METHODS OF OPENING SNOW BLOCKADES

Reapprove without change.

Pages 1-2-1 to 1-2-2, incl.

BALLAST—GENERAL

Reapprove without change.

Pages 1-2-2 to 1-2-4, incl.

SPECIFICATIONS FOR PREPARED STONE, SLAG AND GRAVEL BALLAST

Reapprove without change.

Page 1-2-5

SPECIFICATIONS FOR PIT-RUN GRAVEL BALLAST

Reapprove without change.

Page 1-2-5

SPECIFICATION FOR SUB-BALLAST

Reapprove without change.

Pages 1-2-6 to 1-2-12, incl.

BALLAST SECTIONS FOR SINGLE AND MULTIPLE TRACK ON TANGENT AND CURVES

Reapprove without change.

Pages 1-3-8 to 1-3-14, incl.

PREVENTION OF EROSION

Reapprove without change.

Pages 1-4-13 to 1-4-16, incl.

SPECIFICATIONS FOR BITUMINOUS-COATED CORRUGATED METAL PIPE AND ARCHES

Reapprove without change.

Pages 1-4-17 to 1-4-18, incl.

INSTALLATION OF PIPE CULVERTS

Reapprove without change.

Pages 1-6-20 to 1-6-23, incl.

METHODS OF PROTECTING THE ROADWAY AGAINST DRIFTING SNOW

Reapprove without change.

Pages 1-7-4 to 1-7-9, incl.

SPECIFICATIONS FOR ONE, TWO, THREE AND FOUR-TRACK OVERHEAD METAL WARNING AND METAL SIDE WARNINGS

Reapprove without change.

Pages 1-7-10 to 1-7-12, incl.

SPECIFICATIONS FOR ONE AND TWO-TRACK OVERHEAD WOOD WARNINGS AND WOOD SIDE WARNING

Reapprove without change.

GLOSSARY

Change all references in the Glossary attributed to Committee 2 (presently denoted by the italic numeral 2) to read *1*.

As the result of a special review of Chapter 1 of the Manual, headed by L. J. Deno, the following additional recommendations are submitted for adoption:

Pages 1-1-36.1 to 1-1-36.6, incl.

CONSTRUCTION AND PROTECTION OF ROADBED ACROSS RESERVOIR AREAS

Reapprove with the following revision:

Page 1-1-36.4—Revise title of Sec. D from "Construction of Embankment Protection" to "Specifications for Construction of Embankment Protection."

Pages 1-1-45 to 1-1-64, incl.

ROADWAY DRAINAGE

Reapprove with the following revisions:

Page 1-1-58—Revise Art. 6 to read:

"The proper treatment of soft spots and water pockets is important for economical operation under present high speed and heavy wheel loads. Higher speeds and heavier wheel loads, with resultant heavy impact, may develop soft spots in roadbeds which have adequately supported lighter loads at lower speeds.

"Soft spots and water pockets should be given prompt attention. The longer they exist the greater the resulting maintenance expense, as well as the greater expense involved in providing a permanent remedy."

Page 1-1-59—In the third paragraph of Art. 7—Treatment, change the word "must" to "may."

Pages 1-1-63 and 1-1-64, Art. 10—Maintenance Methods—Revise *"(a) Shimming and Bracing Rules"* to read *"(a) Recommended Shimming and Bracing Methods."*

In all the paragraphs of this article change the word "must" to "should" wherever it occurs. Also, in the first sentence of the third paragraph on page 1-1-64, change the words "is prohibited" to "should be avoided."

Pages 1-3-15 to 1-3-20, incl.

MEANS OF PROTECTING ROADWAY AND BRIDGES FROM WASHOUTS AND FLOODS

Reapprove with the following revisions:

Page 1-3-15—Revise Sec. B. Roadway, Art. 2. Permanent Protection, Par. (a), second sentence, by eliminating the words "which is occasionally the basis of damage suits and."

Page 1-3-17—Revise Sec. C. Bridges, Art. 2. Permanent Protection, Par. (a), by deleting in the third sentence the words "as well as to eliminate possible claims for damage to adjacent property insofar as possible."

Page 1-3-18—In the first paragraph, delete the entire third sentence which reads: "Care should be taken that this heading effect does not damage valuable land for which the railway company would be liable for damages."

Pages 1-4-19 to 1-4-24, incl.

JACKING CULVERT PIPE THROUGH FILLS

Reapprove with the following revisions:

Page 1-4-23, Art. 12. Digging Methods—In the fifth sentence of the first paragraph delete the word "safely." Also, delete the last three sentences of this paragraph, which read: "In very favorable soil the distance may be as much as 4 ft. Ordinarily it will be 1 or 2 ft. It is dangerous to carry excavation farther ahead of the pipe than is absolutely necessary."

Pages 1–4–33 to 1–4–35, incl.

METHODS OF INSTALLING CULVERTS INSIDE EXISTING CULVERTS

Reapprove with the following revisions:

Page 1–4–34, Art. 4. Backfilling—In the first sentence change the word "must" to 'should."'

Pages 1–4–35 to 1–4–37, incl.

CONDITIONS REQUIRING HEADWALLS, WINGWALLS, INVERTS AND APRONS, AND REQUISITES THEREFOR

Reapprove with the following revisions:

Page 1–4–35—In the first two paragraphs of Sec. A, change the word "must" to "should" wherever it appears.

Pages 1–7–1 to 1–7–4, incl.

ROADWAY SIGNS

Reapprove with the following revisions:

Page 1–7–1—Change the title of Sec. A to read: "A. Roadway Signs." Also, delete the entire paragraph following this heading and substitute: "Among the roadway signs in general use on the railways are the following:"

In Art. 1 (b) "No Trespass" signs, delete the words "dangerous or."

Page 1–7–2—In Art. 5 (b) Fire-Risk Signs, change the word "inflammable" to "flammable" in two places.

Pages 1–8–1 to 1–8–5, incl.

TUNNELS

Reapprove with the following revisions:

Page 1–8–1—Change the title of this document to read: "Specifications for Construction, Excavation and Temporary Lining of Tunnels."

In the second line under Sec. A. Alinement and Grade, chage the word "approved" to "recommended."

Change the title of Sec. B to read: "B. Construction, Excavation and Temporary Lining of Tunnels."

Report on Assignment 2

Physical Properties of Earth Materials

(a) Roadbed, Load Capacity, Relation to Ballast, Allowable Pressure
(b) Structural Foundation Beds, Collaborating with Committees 6 and 8

W. P. Eshbaugh (chairman, subcommittee), H. F. Davenport, M. B. Hansen, H. W. Legro, R. R. Manion, F. L. Peckover, J. W. Poulter, S. W. Sweet.

Your committee submits for adoption the following recommendations with respect to the Manual, which apply to both Assignments 2 (a) and 2 (b).

Pages 1–1–37 to 1–1–43, incl.

PHYSICAL PROPERTIES OF EARTH MATERIALS

Reapprove with the following revisions:

Page 1–1–37: Change last sentence of the first paragraph to read: "Any earth material or shale·not readily classified as rock is assumed to be soil for the purpose of this discussion, which takes into consideration roadbed and foundation bed soils, their properties, performance, and methods of exploration and test."

Page 1–1–37: Change second paragraph to read:

"The term soil or soils is loosely defined as sediments or other unconsolidated portions of the earth's crust produced through the chemical and physical weathering of rocks. Soil may also consist of organic matter from plants and animals that has accumulated over a long period of time, material carried or laid down by running water, windblown silts or sands, volcanic ashes, glacial material or any other residual or transported product of rock decay. Soil engineering is well established in engineering practice today. However, it is to be understood that complete knowledge of the properties and performance of soils is not yet available. The theories of soil mechanics can be used only with limitations. Some theories are based on the assumption that soils are plastic or elastic materials of known properties such as steel, whose physical and chemical properties can be controlled during the manufacturing process and whose performance can be predicted. This is far from the observed behavior in many cases. For instance, the same soil can perform very differently under various conditions of moisture and density."

Page 1–1–37: Add at the end of the third paragraph: "Following is a table of soil classification as proposed by the American Society for Testing and Materials:

Soil	Grain Size, Millimeters
Gravel	Larger than 2.0
Sand, coarse	2.0–0.42
Sand, fine	0.42–0.074
Silt	0.074–0.005
Clay	Smaller than 0.005
Colloidal clay	Smaller than 0.001

Page 1–1–37: Change first sentence of the fourth paragraph to read: "Sands and gravels often occur in nature in pure form, silts occasionally, usually as a result of wind-blown deposits, but clay soils are almost invariably found with some portion of sand or silt."

Page 1–1–37: In the fifth line of the fifth paragraph, add the word "relatively" before the word "nonplastic."

Page 1-1-38: Add the following to Par. 2: Cohesion and internal friction between the soil particles make up the shearing strength of a soil. Shearing strength is a property which enables soil to maintain equilibrium on a sloping surface of a cut or fill and which greatly influences the bearing capacity of a foundation soil."

Page 1-1-38: Delete the second sentence of Par. 3 and substitute the following: "The compressibility of a soil is a function of the decrease in volume of voids. If the voids are filled with water compression can occur only as a result of escape of water from the voids. The compressibility of sands in which the grains bear almost directly on adjacent grains is slight. It is much larger in clays where the minute scale-like grains are separated by water which is squeezed out and the volume of voids is decreased."

Page 1-1-38: Change Par. 5 to read.

"5. *Permeability* determines the rate of consolidation of soils. All soils contain continuous voids and are thus said to be permeable. However, there are large differences in the degree of permeability of various earth materials. Highly permeable soils such as sands will consolidate under loads very rapidly because excess water can escape from the voids quite readily. A saturated clay which has low permeability will require more time for the excess water to squeeze out and consequently will consolidate over a long period of time."

Page 1-1-38: Change Par. 6 to read:

"6. *Capillarity* in soils is a phenomenon which tends to draw or retain moisture above the water table. It is caused by the surface tension of water and its molecular attraction to the walls of the hair-fine tubes or spaces between the soil particles. The void spaces of sands are too large to act as capillaries except to a slight degree. Higher capillarity, such as clays possess, contributes to the properties of cohesion, compressibility and elasticity. Capillarity is an important contributing factor in the frost heaving of soils.

Page 1-1-39: In the third paragraph add the following sentence between the present first and second sentences: "Some success has been attained in stabilization of clay subgrade soils by the introduction of additives such as lime, portland cement and gypsum in relatively small quantities."

Page 1-1-40: Change first paragraph under Sec. D—Exploration and Tests, to read as follows:

"Before proceeding with plans for important structures, earthwork or corrective procedures, field soil investigations should be made, and appropriate laboratory tests should be run on soil samples obtained from the site. Field soil investigations may be divided into three classes all of which include ground-water data:

(1) Foundation investigations—to determine the capability of the subgrade soil to support a structure.

(2) Earthwork investigations—to examine sites of proposed cuts and embankments and to determine suitability of soils from potential borrow areas.

(3) Failure investigations—to determine the cause of instability or failure of roadbeds or foundations."

Page 1-1-41: In the sixth paragraph change the second sentence to read: "Certain basic tests should be performed on selected samples from all projects, namely, grain size and shape analyses on granular samples which may give an indication of relative stability and compressibility, and the liquid limit and plastic limit tests on cohesive samples which may give a measure of the potential cohesion, shearing resistance when wet, compressibility and dry strength."

Page 1–1–41: In the ninth paragraph change the last sentence to read: "For questionable clays, mineralogical analyses or swell tests will identify materials with unfavorably high swelling properties."

Page 1–1–42: In the last paragraph, second sentence, transpose the last two words, i.e., make them read "sample tube" instead of "tube sample."

Report on Assignment 5

Specifications for Pipelines for Conveying Flammable and Non-Flammable Substances

Collaborating with Committee 15 and 20

K. W. Schoeneberg (chairman, subcommittee), W. T. Adams, C. R. Bergman, L. H. Bond, J. G. Campbell, H. W. Clarke, G. B. Harris, H. G. Johnson, F. S. Patton, C. W. Reeve, R. W. Scott, A. J. Wegmann.

Last year the revised Specifications for Pipelines for Conveying Flammable and Non-Flammable Substances, Sec. A, for Flammable Substances, as presented by your committee, was adopted and published in the Manual. During this past year your committee has revised Sec. B, for Non-Flammable Substances, of these specifications, and now presents the revised version of this section for adoption and publication in the Manual, replacing the present specifications, Sec. B, for Non-Flammable Substances, found on pages 1–5–6 through 1–5–9.

SPECIFICATIONS FOR PIPELINES FOR CONVEYING FLAMMABLE AND NON-FLAMMABLE SUBSTANCES

B. FOR NON-FLAMMABLE SUBSTANCES

1. Scope

Pipelines included under these specifications are those installed to carry steam, water or any non-flammable substance which, from its nature or pressure, might cause damage if escaping on or in the vicinity of railway property.

2. General Requirements

a. Pipelines under railway tracks and across railway rights-of-way shall be encased in a larger pipe or conduit called the casing pipe as indicated in Fig. 2.

Casing pipe may be omitted under the following conditions:

(1) Under secondary or industry tracks as approved by the chief engineer of the railway company.

(2) In streets where the pipeline is of steel pipe material and where the stress in the pipe from internal pressure and external loads does not exceed 40 percent of the specified minimum yield strength.

(3) For non-pressure sewer crossings where the pipe strength is capable of withstanding railway loading.

Pipelines shall be installed under tracks by boring or jacking, if practicable.

Pipelines shall be located, where practicable, to cross tracks at approximately right angles thereto but preferably at not less than 45 deg, and shall not be placed within culverts nor under railway bridges where there is likelihood of restricting the area required for the purposes for which the bridges or culverts were built, or of endangering the foundations.

b. Pipelines laid longitudinally cn railway rights-of-way shall be located as far as practicable from any tracks or other important structures. If located within 15 ft of the center line of any track or where there is danger of damage from leakage to any bridge, building or other important structure, the carrier pipe shall be encased or of special design as approved by the chief engineer of the railway company.

c. Any replacement of a carrier pipe shall be considered a new installation, subject to the requirements of these specifications.

d. Where laws or orders of public authority prescribe a higher degree of protection than specified herein, then the higher degree of protection so prescribed shall supersede the applicable portions.

e. Pipelines and casing pipe shall be suitably insulated from underground conduits carrying electric wires on railway rights-of-way.

3. Carrier Pipe

Carrier line pipe and joints shall be of accepted material and construction as approved by the chief engineer of the railway company. Joints for carrier line pipe operating under pressure shall be mechanical or welded type.

The pipe shall be laid with sufficient slack so that it is not in tension.

4. Casing Pipe

Casing pipe and joints shall be of leakproof construction, capable of withstanding railway loading.

WALL THICKNESS FOR STEEL CASING PIPE FOR E 72 LOADING (INCLUDING IMPACT)

Minimum Thickness Inches	Diameter of Pipe Inches	Minimum Thickness Inches	Diameter of Pipe Inches
0.188	Under 12¾	0.375	24 and 26
0.219	12¾	0.406	28
0.250	14 and 16	0.438	30
0.281	18	0.500	32 and 34
0.312	20	0.563	36, 38 and 40
0.344	22	0.625	42

Steel pipe to have a minimum yield strength of 35,000 psi.

When casing is installed without benefit of a protective coating, and said casing is not cathodically protected, the wall thickness shown above shall be increased to the nearest standard size, which is a minimum of 0.063 in greater than the thickness shown except for diameters under 12¾ in.

Cast iron pipe may be used for a casing provided the method of installation is by open trench. Cast iron pipe shall conform to American Standards Association Specification A 21. The pipe shall be of the mechanical-joint type or plain-end pipe with compression-type couplings. The strength of cast iron pipe to sustain external loads shall

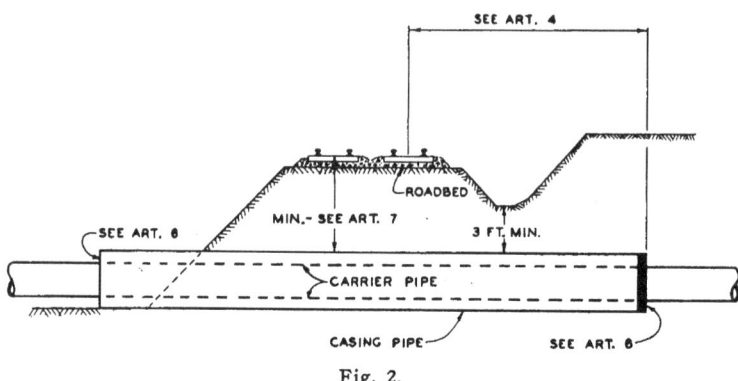

Fig. 2.

be computed in accordance with ASA A 21.1 "Manual for the Computation of Strength and Thickness of Cast Iron Pipe." Pit cast pipe may be used up to 20 in diameter. Centrifugal cast pipe shall be used for pipes over 20 in diameter.

For pressures under 100 psi in the carrier pipe, the casing pipe may be reinforced concrete pipe conforming to the AREA Specifications for Reinforced Concrete Culvert Pipe, Part 10, Chapter 8, or bituminous-coated corrugated metal pipe conforming to the AREA specification for such pipe, Part 4, this Chapter.

The inside diameter of the casing pipe shall be at least 2 in greater than the largest outside diameter of the carrier pipe, joints or couplings, for carrier pipe less than 6 in. in diameter; and at least 4 in greater for carrier pipe 6 in and over in diameter. It shall, in all cases, be great enough to allow the carrier pipe to be removed subsequently without disturbing the casing pipe or roadbed.

Casing pipe under railway tracks and across railway rights-of-way shall extend to the greater of the following distances, measured at right angles to center line of track:

    a. 2 ft beyond toe of slope.

    b. 3 ft beyond ditch.

    c. A minimum distance of 25 ft from center line of outside track when end of casing is below ground.

If additional tracks are constructed in the future, the casing shall be extended correspondingly.

5. Construction

Casing pipe shall be so constructed as to prevent leakage of any substance from the casing throughout its length except at ends. Casing shall be so installed as to prevent the formation of a waterway under the railway, with an even bearing throughout its length, and shall slope to one end (except for longitudinal occupancy).

Installations by open-trench methods shall comply with Installation of Pipe Culverts, Part 4, this Chapter.

Bored or jacked installations shall have a bored hole diameter essentially the same as the outside diameter of the pipe plus the thickness of the protective coating. If voids should develop or if the bored hole diameter is greater than the outside diameter of the pipe (including coating) by more than approximately 1 in, remedial measures as approved by the chief engineer of the railway company shall be taken. Boring operations shall not be stopped if such stoppage would be detrimental to the railway.

Tunneling operations shall be conducted as approved by the chief engineer of the railway company. If voids are caused by the tunneling operations, they shall be filled by pressure grouting or by other approved methods which will provide proper support.

Where casing and/or carrier pipe is cathodically protected, the chief engineer of the railway company shall be notified and suitable test made to insure that other railway structures and facilities are adequately protected from the cathodic current in accordance with the recommendations of Reports of Correlating Committee on Cathodic Protection, published in July 1951 by the National Association of Corrosion Engineers

6. Protection at Ends of Casing

Where the ends of the casing are below ground they shall be suitably protected against the entrance of foreign material, but shall not be tightly sealed. Where the ends of the casing are at or above ground surface and above high-water level they may be left open, provided drainage is afforded in such manner that leakage will be conducted away from railway tracks or structures.

7. Depth of Installation

a. Casing pipe under railway tracks and across railway rights-of-way shall be not less than 5½ ft from base of railway rail to top of casing at its closest point, except that under secondary or industry tracks this distance may be 4½ ft. On other portions of rights-of-way where casing is not directly beneath any track, the depth from ground surface or from bottom of ditches to top of casing shall not be less than 3 ft.

b. Carrier pipe installed under secondary or industry tracks without benefit of casing shall be not less than 4½ ft from base of railway rail to top of pipe at its closest point nor less than 3 ft from ground surface or from bottom of ditches.

c. Pipeline laid longitudinally on railway rights-of-way, 50 ft or less from center line of track, shall be buried not less than 4 ft from ground surface to top of pipe. Where pipeline is laid more than 50 ft from center line of track, minimum cover shall be at least 2 ft.

8. Shut-Off Valves

Accessible emergency shut-off valves shall be installed within effective distances each side of the railway as mutually agreed to by the chief engineer of the railway company and the pipeline company. Where pipelines are provided with automatic control stations at locations and within distances approved by the chief engineer of the railway company, no additional valves shall be required.

9. Approval of Plans

Plans for proposed installation shall be submitted to and meet the approval of the chief engineer of the railway company before construction is begun. Plans shall be drawn to scale showing the relation of the proposed pipeline to railway tracks, angle of crossing, location of valves, railway survey station, right-of-way lines and general layout of tracks and railway facilities. Plans should also show a cross section (or sections) from field survey, showing pipe in relation to actual profile of ground and tracks. If open cutting or tunneling is necessary, details of sheeting and method of supporting tracks or driving tunnel shall be shown.

In addition to the above, plans should contain the following data:

	Carrier Pipe	Casing Pipe
Contents to be handled	....	....
Outside diameter	....	....
Pipe material	....	....
Specification and grade	....	....
Wall thickness	....	....
Actual working pressure	....	....
Type of joint	....	....
Coating	....	....
Method of installation ....,..	....	....

Protection at ends of casing:
Both ends .... One end .... Type ....
Bury: Base of rail to top of casing .... ft........in
Bury: (Not beneath tracks) .... ft........in
Bury: (Roadway ditches) .... ft........in
Cathodic protection ....

10. Installation

The execution of the work on railway rights-of-way, including the supporting of tracks, shall be subject to the inspection and direction of the chief engineer of the railway company.

Report on Assignment 6

Roadway: Formation and Protection

(a) Roadbed Stabilization
(b) Slope Protection by Use of Additives
(c) Performance of Filter Materials in Subdrains
(d) Gypsum Treatment of Subgrade to Improve Track Drainage

A. E. Lewis (chairman, subcommittee), K. W. Bradley, E. A. Bauman, B. H. Crosland, J. B. Farris, J. E. Gray, H. O. Ireland, W. G. Murphy, J. E. Newby, A. J. Pacelli, G. W. Payne, J. W. Poulter.

Your committee submits as information the following report pertaining to Assignment 6 (d) only.

Lime Treatment of Subgrade

During April and May 1961, the Santa Fe graded and stabilized the subgrade with lime over a 4.1-mile industrial lead near Bay City, Tex. This site is on the flat coastal plain along the Gulf of Mexico, and the soils involved are rather heavy-textured clays. No granular material is available, except mud shell, within reasonable haul limits, and without some protection for the subgrade water pockets, even under limited operations, can easily develop. Shell from the Gulf is relatively expensive from commercial sources, and it appeared that stabilization of the subgrade soils with the addition of lime might be both beneficial and economical in this area. Much of the land has been used for rice crops. The water table is within 2 or 3 ft of the surface, varying somewhat with seasonal rainfall, but the area itself is in a rather high rainfall section of the state.

As noted above the soils are heavy silty clays except in one or two water-course areas where sandy silt soil appears. This latter material is limited in extent and covers probably not more than 25 percent of the project after grading. The grading project consisted of side ditching to possibly a 3-ft depth from the bottom of the ditch to the top of the subgrade.

Two samples were taken, one of each of the two soils described above, and run in the AAR laboratory with 3 and 5 percent lime added to check plasticity factors, maximum density, optimum moisture and increase of strength with the lime additive. A sample of the hydrated lime proposed tor. use was also received from the Santa Fe.

Lime in the area costs about $10 to $15 per ton at the plant. With the lime addition eliminating the need of approximately 6 in of shell sub-ballast, the economics appeared to be in favor of the lime stabilization.

Test results of the soil-lime mixtures were determined at the AAR lab as follows:

TABLE 1—UNCONFINED COMPRESSION TESTS

Moisture	0 Day PSI	2-Day Cure PSI	5-Day Cure PSI	7-Day Cure PSI
Raw clay	35			
Clay + 3% lime		83	172	202
Clay + 5% lime		89	103	128
Raw sandy silt	38			
Silt + 3% lime		79	145	153
Silt + 5% lime		107	105	132

TABLE 2—ATTERBERG LIMITS

Mixture	24-Hr Hydration			48-Hr Hydration		
	LL	PL	PI	LL	PL	PI
Raw clay	43.2	11.1	31.2	-------	-------	-------
Clay + 3% lime	36.0	26.9	9.1	38.5	27.6	10.9
Clay + 5% lime	42.3	28.3	14.0	45.3	32.2	12.1
Raw sandy silt	27.9	13.3	14.6	-------	-------	-------
Sandy silt + 3% lime	27.9	20.5	7.5	29.4	21.6	7.8
Sandy silt + 5% lime	26.9	21.8	5.1	28.1	24.7	3.4

TABLE 3—STANDARD PROCTOR MOISTURE-DENSITY TEST RESULTS

Mixture	Maximum Dry Density Lb/Cu Ft	Optimum Moisture Content Percent
Raw clay	103.1	19.7
Clay + 3% lime	98.1	22.5
Clay + 5% lime	94.8	21.3
Raw sandy silt	110.9	14.6
Sandy silt + 3% lime	106.9	15.4
Sandy silt + 5% lime	104.2	16.0

Additional tests on a sample of silty clay soil from a proposed project about 40 miles east were also run at this time. These do not include unconfined compressive strengths but the data as determined follow:

MISSOURI PACIFIC—LIME STABILIZATION TESTS

Silty Clay Soil Sample from Chocolate Bayou

Mixture	Moisture Content Data (Percent)			
	As Molded	3-Day Moist Cure	7-Day Moist Cure	7-Day Moist Cure + 8-Day Capillary
Raw soil_____	18.2	18.5	19.0	23.6
Soil + 3% lime_____	18.6	19.1	20.1	22.2
Soil + 5% lime_____	18.8	19.7	19.6	21.8

Mixture	Atterberg Limits					
	24-Hr Hydration			48-Hr Hydration		
	LL	PL	PI	LL	PL	PI
Raw soil_____	32.0	14.5	17.5	31.3	12.3	19.0
Soil + 3% lime_____	32.8	24.6	8.2	32.5	26.6	6.9
Soil + 5% lime_____	33.8	23.6	10.2	32.9	24.9	8.0

Mixture	Proctor Moisture Density—ASTM Std.	
	Optimum Moisture Content Percent	Maximum Dry Density Lb/Cu Ft
Raw soil_____	18.2	105.3
Soil + 3% lime_____	18.6	102.1
Soil + 5% lime_____	18.8	99.7

Procedures for lime stabilization are generally similar to those used for cement or asphalt work of a similar nature. All procedures involve the obtaining of an intimate mix of the materials, with the subgrade soils consisting of 90 percent or more of the mix.

A more detailed description of the work on the Bay City project follows.

The lime being used was pulverized hydrated material, mainly calcium hydroxide. The material was manufactured in Cleburne, Tex., and shipped to Wadsworth, Tex., in covered hoppers where it was unloaded by screw conveyors into a tank truck primarily designed for lime hauling, capacity about 25 tons. The truck was equipped with three spiral screws for unloading on the subgrade through three hoses in contact with the ground during spreading. The spread was fairly uniform over a width of about 8 ft. Gradation of the lime met suggested specifications of the Lime Association which show 100 percent passing a No. 20 sieve and a minimum of 85 percent passing a No. 100 sieve. This lime also had a high content through No. 200 sieve material. Fifty-five percent oxide content is the required minimum.

Two general soil types are involved as noted above. For the clay 5 percent of lime by weight for an 8-in compacted depth was added. For the silt and silty sand 4 percent lime for the 8-in depth was applied.

The compacted raw soil subgrade was first scarified and 4 in of the material bladed out in windrows on each side of the grade. Additional scarifying and/or disking of the remaining depth was beneficial. The lime was then added from the distributor truck, usually requiring two or three passes, depending on lime content over the 700- to 900-ft sections which were the common lengths for processing. Following the lime application the open grade was disked and the windrows brought in and disking continued. Water up to the optimum content was applied during this work. After a semi-uniform distribution of the lime had been obtained the section was smoothed to approximate grade and sealed with rubber-tire rolling for protection against weather during the hydration period.

The length of this period was determined by the type of soil involved. For as complete as possible pulverization in the clay, two to three days appeared needed. In the lighter material such as the silty sand encountered on about 25 percent of the project, one to two days were sufficient.

After the hydration period the section was again ripped open as before and the disking and blading continued until satisfactory pulverization had been obtained and lime was uniformly mixed throughout. Additional water is required to bring the mixture to the optimum content for efficient compaction. The lime effect was felt in the change of optimum upward for the lime admixture as illustrated by the optimum for the clay soil changing from 19.7 percent for the raw soil to 24.3 percent for the soil plus 5 percent lime.

After pulverization, mixing and sprinkling, the processed material was rolled with sheepsfoot rollers and rubber-tired rollers as required to obtain maximum compaction according to the standard. Final shaping was by blade.

There was a very noticeable and distinct change of soil characteristics in the processed area. Sprinkling for several days followed the final compaction to act as a further curing agent. On these sections there was little or no penetration of water—it acting only to preserve moisture already incorporated. There was also little or no rutting or mud under rubber-tired traffic, although the surface was not abrasion resistant.

An estimate of the cost of this work shows that the 8-in stabilization depth over a 20-ft width competes very favorably with the cost of 6 in of compacted shell base. From appearances and results sub-ballast was not needed, and a top ballast of screenings with the lime subgrade was sufficient to produce a stable track for the anticipated traffic.

During the hurricane in the fall of 1961 and the rains that followed, several areas of this project were under water from overflowing irrigation ditches. This resulted in flow across the grade. In several areas the current had removed the top ballast of limestone screenings, but as far as could be detected the supporting power of the subgrade was not affected.

Report on Assignment 8

Fences

H. G. Johnson (chairman, subcommittee), F. W. Hillman, G. F. Nigh, W. O. Treischman.

Your committee has reviewed the material on fencing in Chapter 1 of the Manual and submits for adoption the following recommendations in connection therewith:

Pages 1-6-1 to 1-6-3, incl.

SPECIFICATIONS FOR WOOD FENCE POSTS

Reapprove with the following revisions:

Page 1-6-2, Sec. D. Manufacture: Below Art. 1. General Requirements, add the following:

2. Tolerances

Posts shall not be more than 1 in shorter nor 3 in longer than the length specified. Posts shall not be more than ¼ in smaller nor more than 1 in larger than the diameters, widths and perimeters specified.

Page 1-6-3: Delete Sec. F. Workmanship, including Art. 1. Tolerances, and change the designations of present Secs. G and H to F and G, respectively.

Pages 1-6-3 to 1-6-6, incl.

SPECIFICATIONS FOR CONCRETE FENCE POSTS

Reapprove with the following revisions:

Page 1-6-5, Fig. 1—Plan of Concrete Fence Post: Revise second paragraph to read: "Fence wires shall be attached to post by molded holes through post, loops of galvanized wire projecting from the post, or other suitable device."

Revise third paragraph to read: "When conditions require the use of a stronger post, the greater strength can be economically obtained by increasing the size of reinforcing rather than the size of the post."

Pages 1-6-6 to 1-6-9, incl.

SPECIFICATIONS FOR METAL FENCE POSTS

Reapprove without change.

Pages 1-6-14 to 1-6-18, incl.

SPECIFICATIONS FOR ELECTRIC SHOCK FENCES

Reapprove with the following revisions:

Page 1-6-14, Art. 2. Construction: In the second line of the paragraph referring to Fig. 3, change the word "this" to "thus."

Page 1-6-17, Art. 3. Materials: Insert the word "Wire" in italics below the article heading.

Revise sentence under "Posts" to read: "Wood, steel, or concrete posts may be used."

Page 1-6-19

STOCK GUARDS

Reapprove without change.

Page 1-6-24

SPECIFICATIONS FOR WOOD SLAT PORTABLE SNOW FENCES

Reapprove without change.

Report on Assignment 9

Roadway Signs

(a) Reflectorized and Luminous Roadway Signs, Collaborating with Committees 5 and 9 and with the Communication and Signal Section, AAR

(b) Develop Standard Close Clearance Warning Sign, Collaborating with Committee 28

R. D. Baldwin (chairman, subcommittee), A. S. Barr, J. S. Felton, Jr., Paul McKay.

Under Assignment 9 (b) your committee submits for adoption and publication in the Manual the accompanying drawing (see next page) showing recommended design for "No Clearance" sign, which was selected for reasons of maximum simplicity, clarity, economy, and minimum obsolescence. The lettering is recommended to be black on white enamel or reflex-reflecting material.

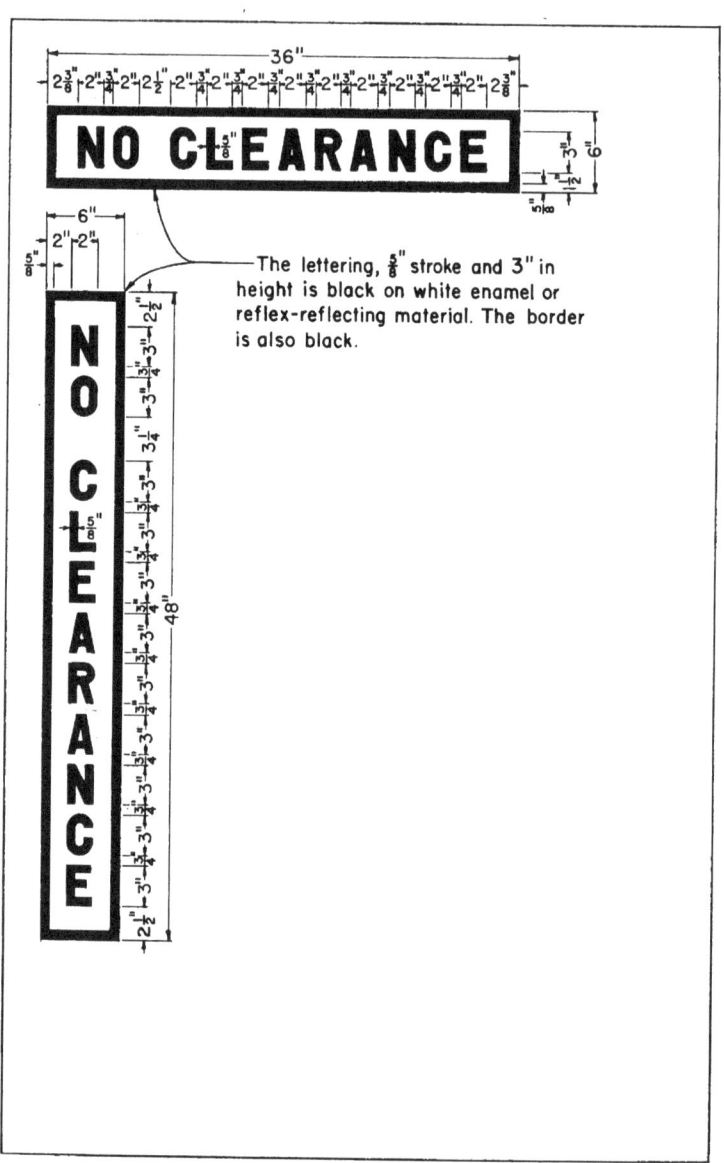

The lettering, $\frac{5}{8}$" stroke and 3" in height is black on white enamel or reflex-reflecting material. The border is also black.

Report on Assignment 10

Ballast

(a) Tests

(c) Special Types of Ballast

T. W. Creighton (chairman, subcommittee), E. W. Bauman, J. G. Campbell, J. E. Gray, W. C. McCormick, E. W. McCuskey, E. L. Robinson, Jr., Stanton Walker.

Under Assignment 10 (c) your committee presents a progress report on the 1961 condition of the asphalt-ballast and bridge-deck-treated sections on various railroads. The treatments were placed in 1959 and 1960 under a cooperative project between the Asphalt Institute, the Association of American Railroads, Research Department, and participating railroads. No additional work was done in 1961.

The 1959 applications were reported in some detail in the AREA Proceedings, Vol. 61, 1960, page 715. The progress report on these projects and the 1960 work was reported in abstract in the Proceedings, Vol. 62, page 707, and in more detail in Report No. ER-10 of the Research Department, AAR. The latter report is available on request to the AAR.

1959 BALLAST TREATMENTS

Santa Fe Railway

Daggett–Barstow, Calif.—Reports from the railroad, including maintenance costs for a full year, indicate that the project is performing very satisfactorily as is also the control section without treatment. Over the four miles included in both sections the spotting and lining labor required has been nominal, about equally divided between the control and treated areas.

Peach Springs—Truxton, Ariz.—Maintenance records on this section show requirements for spotting and lining on the treated area to be very low. On the untreated control section similar requirements are also low but are seven times greater. The amount involved, however, in both cases is too small to permit definite conclusions.

Suwanee–Marmon, N. M.—The treated section on the eastward track on this project became subject to severe pumping and poor ballast drainage early in 1961 and was taken out of test in March. The asphalt-treated ballast was bladed out from the end of the ties, and the track was raised on the material in the crib sections. It appeared that the ballast prior to treatment had become fouled with wind-blown fine sand and silt and that the asphalt coverage was insufficient completely to seal the surface from water penetration. For the two-mile treatment on the western track a number of cribs showed disturbance with relatively minor pumping. Here too, the asphalt quantity seemed deficient. In general, however, the treated area is in fair shape. Maintenance records show very nearly the same amount of spotting and lining work for both the treated area and untreated control section. Neither amount is excessive.

Lecompton–Topeka, Kans.—The general condition of this treatment is good. There is an exception for about one-half mile along the Kaw River where the roadway is subject to periodic movement towards the river. This area has required lining and spotting at times and subgrade grouting is being considered. A second exception is on a curve for a distance of one-quarter mile, approximately. Here it has been necessary to reline the curve and break the asphalt seal. It has not been definitely established whether this curve kicked out of line following the treatment or prior thereto. It has been the experience on other projects that the treatment has been of value in maintaining line,

It is at present the opinion of the viewers that this was a case where the treatment disclosed more clearly the off-standard alinement present at the time of application.

There is a minimum of disturbance in the cribs and at the tie ends over the remainder of the project. Some was indicated at joints where minor spotting has been performed. Coverage of asphalt and cover coat is good. Ballast section is full. Comparative costs between the treated and untreated control sections are not available.

Bucklin–Marceline, Mo.—After inspection in October 1961 this section of asphalt is classified as fair to good. There are several short sections of disturbance near road crossings with pumping ties, only several of which extend over an area of 25 to 50 ties. As with other jobs it is particularly apparent on this project that there is disturbance at both insulated and regular joints in the welded rail. It appears possible that some of this disturbance is started by rail or tie movements, breaking the seal in the cribs and permitting entry of water. This, however, has not been definitely established. The line is excellent as is also the adherence of asphalt and cover coat to ties and track fittings. The comparative cost between the treated and untreated sections is not yet available. The control section is also in good condition. Ballast section is full with slow lateral drainage.   .

Williamsfield–Dahinda, Ill.—This project was inspected in October and is reported as excellent. Very little disturbance in the cribs or at the tie ends was noted. Coverage of both ties and track fittings was very good, and no pumping ties were in evidence except at road crossings where this condition was minor. Line and surface has been maintained in excellent shape. At the time of treatment this section was shaped to permit drainage from the center of track to the sides, and this appears to be a very favorable factor in the excellent performance of the treatment. The control section, untreated, is also very good.

Victorville and Oro Grande, Calif.—These sections were treated in the vicinity of cement mills to prevent penetration of flue dust into the ballast. From reports by the railroad they are still performing very satisfactorily. One project has required a small amount of spotting and lining while the second project has required none. As this was a special treatment, no control sections could be set up for similar conditions, but reports from local maintenance officers indicate that this treatment has been very successful for the purpose intended.

Monica and Princeville, Ill.—These rail crossings were also special projects placed to hold line and surface under high-speed operations. Both are double track, and of these three out of four required reworking under the crossing proper, including a routine change out of frogs. In other areas the surface has required respotting, but from verbal reports from local maintenance officers the treatment has been fully successful in maintaining line. As is often found, drainage facilities under the crossing proper are not fully adequate. It has been suggested in any future work of this type that the area under the crossing proper receive no treatment, but that the approaches can be well benefitted by asphalt application.

1959 AND 1960 BRIDGE DECK TREATMENTS

Pittsburgh & West Virginia Railway

Bridge decks on this railway treated in 1959 and 1960 totaled about five miles of track. They were inspected in the spring of 1961 and all appeared to be in excellent shape. As was noted last year, adherence of the treatment and cover coat on the track fittings for the 1959 work was only fair. For the 1960 work this was excellent. In 1960 both new and old decks were treated. One bridge deck, approximately three months old and still oozing creosote, was sprayed with the hot asphalt. There had been some question at the time whether there was full compatability between the creosote and asphalt, but results in the 1961 inspection show no indications of any deterioration of the seal.

1960 BALLAST TREATMENT

Norfolk & Western Railway

This treatment was inspected in September 1961 and can be reported as being in excellent condition. There is little or no crib disturbance, and only at some of the joints is there a small pull-away of the asphalt treatment from the tie ends. Line and surface are very good. At insulated and regular joints some indication of slight disturbance can be ascertained, but this at present has in no way caused any deterioration of the treatment.

This job consists of continuous welded rail on tangents with shorter sections of 137-ft or 78-ft sections on curves or road crossings. This job received a rail-grinding treatment this year which was in no way connected with the asphalt treatment. The control section is also in very good condition.

Monon Railroad

This project also is excellent. It consists of both welded and conventional 39-ft rail, but the line and surface are excellent with a possible superiority in line on the welded rail. Treatment is tight practically throughout with no indication of tie movements or disturbed cribs. The control section is also excellent, but appears to be slightly inferior in line to that of the treated areas.

Texas & Pacific Railway

This project was not inspected in 1961, but reports from the railroad indicate that it has performed very satisfactorily. A very small amount of tie tamping was required. The control section, untreated, required some spotting but as yet direct comparison is not possible.

Chicago & North Western Railway

This project has also performed satisfactorily. It has improved drainage in the area through a passenger station and has maintained good asphalt coverage of the ties and track fittings. As with other projects it is noted that it has been successful in sealing cracks and splits in ties where conditions were not too severe, and even on those ties with rather open splits the asphalt has sealed the area sufficiently to delay ready absorption of water.

Louisville & Nashville Railroad

It was necessary to remove this section from test this year. The pocketed condition in soft subgrade at this location was not helped by the asphalt application, as pumping action caused a breaking up of the asphalt in the crib area as well as at the tie ends. The asphalt-penetrated ballast was tamped under the ties in raising the track. This section will be observed further to determine any benefits received from the asphalt-coated ballast under the ties.

1960 CHICAGO, ROCK ISLAND & PACIFIC RAILROAD BRIDGE DECKS

It was possible for the AAR to inspect only one bridge out of the many treated on this road. This bridge, however, was in very good shape with good adherence of the asphalt to the ties and track fittings. Further reports from the railroad indicated that results are good throughout the system with no adverse criticism. It was noted that the asphalt had been of value in filling checks and splits in the ties, and indications are good that the life of the bridge deck will be increased.

The Rock Island, in 1960, treated 6.87 miles of bridge decks, averaging 1.43 gal of asphalt per track foot.

CONCLUSIONS

At the time of this report the asphalt treatments of 1959 are over two years old, the 1960 applications one year or more. A pattern of performance is beginning to emerge, and within the next year or two it seems probable that cost and labor requirements can be reported in more detail. At this time a few general observations can be made based on the performance to date.

The asphalt ballast treatment is not a curative process. It will not correct soft track or fouled ballast conditions. The track must be up to a high standard at the time of treatment, including line, surface and, particularly, subgrade conditions. It was noted that those sections on which the surface was shaped to drain laterally were generally in the better condition. Also, sufficient asphalt should be used to insure a full seal.

The treatments, however, perform very satisfactorily under the required conditions. It appears that such treatment will be of value in maintaining these high standards over a greater length of time than similar open track. It is on this premise that the treatments may be justified economically. The following years will begin to develop the pattern.

Specifically, the asphalt treatments appear to be of appreciable assistance in maintaining line and surface. The coverage of the ties, plates and the waterproofing of open checks and splits can provide very definite extension of tie life. This also holds for bridge ties and timbers. The coverage of track fittings and base of rail should also be beneficial in reduction of corrosion, although the extent of such savings will be difficult to assess. Also to be noted is the possible increase in anchoring efficiency.

Report on Assignment 11

Chemical Control of Vegetation

Collaborating with Communication and Signal Section, AAR

C. E. Webb (chairman, subcommittee), C. W. Bailey, T. T. Connelly, J. W. DeMoyer, R. J. Kemper, S. J. Owens.

The following report, presented as information, concerns the results of application of granular and pelletized soil sterilants and dormant spray application with hormone-type chemicals for the control of brush. The work was instituted in 1958 as a cooperative test between the AAR Research Department, three railroads and various chemical companies. The AAR staff was to correlate and report the information recorded. Because of staff and budget curtailments the full plan could not be completed. This report includes only the conclusions derived from the limited work performed. The applications and earlier results have been tabulated and reported in the Proceedings, Vol. 60, 1959, Table 7, page 773, and Vol. 61, 1960, Tables 8 and 9, pages 808–820.

Two methods of application for the dry materials were used. The broadcast type, consisting of coverage of the soil area at a uniform rate, was done either by hand or by the use of a spray gun or seeder, with several plots receiving an aerial application. The second method was a basal or stem treatment through which one to three table spoonsful of the commercial product was placed by hand around each stem. Under this method the density of the growth determines the amount of chemical used per acre.

On the areas over which the broadcast applications were made the results have been highly irregular—in some cases very good, but in general poor except at high rates. Even with these rates the chemicals indicated appreciable selectivity as to the type of brush controlled. General indications, however, showed the possible value of the dry materials as an auxilliary agent in brush-control projects.

The basal method of application for the dry materials resulted in fair to good control of most species of brush. Because of the expense of application it appears that materials to be applied in this manner will probably be limited to a clean-up operation for root-suckering species and other resistant growth, such as red maple.

Dormant-spray applications have also been reported (in Table 8 referred to above). For this purpose the hormone-type chemical was mixed with oil. Both 2,4,-D—2, 4, 5,-T mixtures and 2, 4, 5,-T alone have been used. As these are sprayed in oil, material containing 6 lb of acid equivalent per gallon can be used, which at times may result in some economy.

The dormant-spray procedure has given many indications of good to excellent results. It has the advantage of reducing possible crop-damage claims to counter the possible increased cost of the oil solution. It also has the advantage of permitting better coverage of the ground area, reducing the stand of small brush and affecting the root and stump regrowth advantageously. It is now being used to a much greater extent as the policy of some railroads. The difference between results from use of the straight T and the mixture of D and T is still undetermined, but it has been demonstrated tentatively that 9 to 15 lb of acid equivalent are usually the minimums required for satisfactory results from either material.

Conclusions

General conclusions of the limited cooperative research can be stated as follows: The use of broadcast application of granular and pelletized material is not usually satisfactory as to results. The basal application can be effective but expensive. The principal use of these materials appear to be for spot or clean-up applications.

The dormant-spray procedure is a valuable tool with results that are equal and often superior to foliage applications. No one spray system, however, will answer all brush problems. It appears that a full program should consist, where necessary, of both foliage and dormant sprays and followed by spot clean up with dry materials. These requirements involve a knowledge of brush species and their reactions to the various chemicals.

New equipment has been developed which gives indications of value in procuring desired results. One item is an air-blast outfit which propels the stream from the spray nozzles into the brush at high velocity to increase coverage. Another item is an air ejector for granular or pelletized chemicals. This can be adapted for under-bridge work as well as for the roadbed section.

Report of Special Committee on Continuous Welded Rail

F. L. REES, *Chairman*

W. J. JONES,
Vice Chairman

R. E. DOVE, *Secretary*

A. H. GALBRAITH

M. S. REID

D. T. FARIES	R. A. HOSTETTER
C. R. MERRIMAN	S. R. HURSH
H. F. GILZOW	T. B. HUTCHESON
M. P. ANDERSON	H. W. JENKINS
S. H. BARLOW	A. C. JONES, JR.
C. N. BILLINGS	A. J. KUZAK
BLAIR BLOWERS	WOLTERS LEDYARD
E. J. BROWN	L. W. LEITZE
J. A. BUNJER	A. B. LEWIS
J. E. CAMPBELL	C. P. MARTINI
J. D. CASE	LEE MAYFIELD
C. O. CONATSER	A. S. McRAE
W. E. CORNELL	B. M. MONAGHAN
L. S. CRANE	C. E. MORGAN
F. W. CREEDLE	WM. NUETZEL
W. J. CRUSE	T. P. POLSON
A. G. ELLEFSON	S. H. POORE
O. E. FORT	B. R. PRUSAK
W. E. GARDNER	J. R. RYMER
R. G. GARLAND	E. F. SALISBURY
B. J. GORDON	WM. J. SAVAGE
E. P. HACKERT	R. W. SHAW
W. F. HANNA	T. C. SHEDD
J. B. HARTRANFT	H. W. SMITH
E. M. HODGES	C. W. WAGNER
J. L. HODGKINSON	C. E. WELLER

Committee

Members listed in bold face are the official representatives of the Engineering Division, AAR.

To the American Railway Engineering Association:

Your committee reports on the following subjects:

1. Fabrication.

    Awaiting further comments on its 1961 report on Basic Considerations in Fabrication of Continuous Welded Rail, your committee has not progressed further the development of specifications for the fabrication of such rail, but expects to do so in 1962.

2. Laying.

    Progress report, presented as information .... page 601

3. Fastenings.

    Your committee is giving consideration to comments and criticisms received regarding the specifications for the number and position of rail anchors on continuous welded rail, adopted at the 1961 convention, but these were received too late to permit offering any proposed revisions in the current report of the committee. Further consideration will be given to the matter in 1962.

4. Maintenance.

Due to a change in subcommittee chairman during the year there is no report, but the committee will continue its present investigation of practices on the various railroads with respect to the preparation of the roadbed before and after laying continuous welded rail, looking to a comprehensive report in 1963.

5. Economics.

Due to a change in subcommittee chairmanship during the year, the committee has no report. During the first part of 1962 the committee plans to send a questionnaire to those railroads which have laid continuous welded rail to develop the overall savings, if any, they have obtained through the use of this rail. It is the feeling of the committee that enough continuous welded rail has now been in service for a sufficient period to permit at least some roads to furnish figures having real significance.

6. Welding second-hand rail.

With insufficient material available, the committee has no report, but it is working on the development of general specifications for second-hand rail classified as suitable for welding and laying in secondary main and yard tracks.

THE SPECIAL COMMITTEE ON CONTINUOUS WELDED RAIL,

F. L. REES, *Chairman.*

AREA Bulletin 570, February 1962.

Report on **Assignment** 2

Laying

M. S. Reid (chairman, subcommittee), E. J. Brown, H. F. Gilzow, R. A. Hostetter, T. B. Hutcheson, H. W. Jenkins, A. C. Jones, Jr., W. J. Jones, A. B. Lewis, B. M. Monaghan, S. H. Poore, F. L. Rees, C. E. Weller

This is a progress report, submitted as information.

The total mileage of continuous welded rail as of the end of 1961, by years, is as follows:

Year	Track Miles	Year	Track Miles
1933	0.16	1949	33.05
1934	0.95	1950	50.25
1935	4.06	1951	37.25
1936	1.52	1952	40.00
1937	31.23	1953	80.00
1939	6.04	1954	87.00
1942	5.48	1955	266.50*
1943	6.29	1956	461.43*
1944	12.88	1957	550.12*
1945	4.81	1958	460.24*
1946	3.91	1959	1070.57*
1947	18.70	1960	1260.62*
1948	29.93	1961	1020.63*

5543.62

* 1955— 72   miles were electric-flash butt welded.
* 1956— 89.10   "   "   "   "   "   "
* 1957—159.65   "   "   "   "   "   "
* 1958—312.13   "   "   "   "   "   "
* 1959—691.92   "   "   "   "   "   "
* 1960—961.20   "   "   "   "   "   "
* 1961—926.50   "   "   "   "   "   "

The total for 1961 includes 3.83 track miles of yard tracks and 275.78 track miles of second-hand rail.

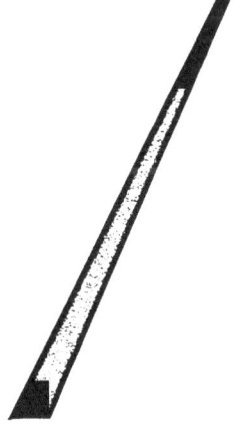

VEGETATION CONTROL
WITH
CHEMICALS

READE MANUFACTURING COMPANY, INC.

Jersey City—Chicago—Minneapolis—Kansas
City—Birmingham—Stockton

SERVING RAILROADS OF AMERICA FOR
MORE THAN FORTY YEARS

W
E
E
D

A
N
D

B
R
U
S
H

C
O
N
T
R
O
L

How GRS Syncroscan®
Paid Off Between
Buffalo and Cleveland

In 1956, the New York Central initiated its modern cTc program by placing a 163-mile stretch in service between Buffalo and Cleveland. The installation demanded a traffic control system with enormous capacity and speed. GRS Syncroscan was the answer. Syncroscan continuously checks—with electronic speed—the location of every train, the position of each switch and signal. Fast-acting relays control switches and signals with ample speed to keep up with operator's decisions.

By using Syncroscan—like the Central —you can handle many miles of busy multiple-track main line, including major interlockings, from a single office. And you can do it better, faster, and more economically than with your present method.

FOUR TRACKS CUT TO TWO The $6 million project produced $3 million in salvage.

FREIGHT TIME SAVED Train speed increased from 20-30 mph to 60 mph—freight time reduced from 7 hours to 3½.

GREATER FLEXIBILITY Two tracks give better operation—fast trains can now be run around slower trains.

EXCELLENT RETURN Estimated saving of 87% on the investment.

9100

SPEED SWING

MODEL 441

Developed and Built
for Railroad Maintenance

180° BOOM SWING

DOES ALL JOBS!

LAYING STANDARD RAIL

CUTS MAINTENANCE COSTS

12 FAST CHANGE ATTACHMENTS

- Forks
- 1¼ Cu. Yd. Bucket
- Tote Hook
- 18' Boom Extension
- Fork Tie Baler
- Track Cleaning Bucket
- Back Hoe
- Clamshell
- Back Filler Blade
- Pull Drag Bucket
- 4 Cu. Yd. Snow Bucket
- Pile Hammer

Optional Attachment
Flanged Wheels, Hydraulically Controlled

PETTIBONE MULLIKEN CORPORATION

RAILROAD **PMCO** DIVISION

141 W. JACKSON   CHICAGO 4, ILL.

*80 Years of Service
to the Railroad Industry*

Model N U Tie Cutter

HERE IS THE WINNING TEAM

The Woolery NU Tie Cutter and the Woolery Tie-end Remover preserve the line and surface of the track and at the same time reduce the cost of tie renewals. Ties can be removed without trenching, jacking up track or adzing tops of rail-cut ties. With this team you simply cut both ends of tie, pry out center piece, insert in its place the tie-end remover and out go the tie ends pushed by the double acting, double ended hydraulic cylinder of the Tie-end remover.

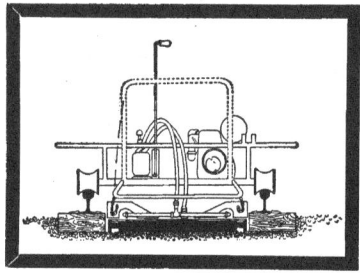 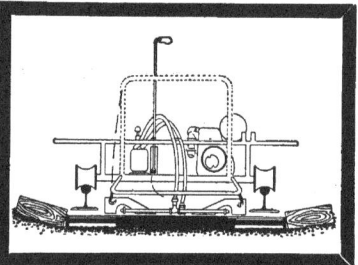

FOR HIGHEST EFFICIENCY USE TWO TIE CUTTERS WITH ONE TIE-END REMOVER

WOOLERY MACHINE COMPANY
MINNEAPOLIS, MINN.

for effective
weed control...

- Concentrated BORASCU®
- POLYBOR-CHLORATE®
- UREABOR®
- MONOBOR-CHLORATE®

These <u>borate weed killers</u> are proving best for roads in every way... *efficiency, safety, economy, convenience, easy application.*

Today's use of borates for maximum control of vegetation began years ago with our pioneer work in the field. Continued research has developed the group of herbicides, listed above, which most roads now favor for every phase of weed control. These four weed killers are nonselective. They are widely used for year-round maintenance of weed-free conditions about trestles, tie piles, yards, signals, switches, and rights of way. Find out how you, too, can do a better job on weeds... write today.

AGRICULTURAL SALES DEPARTMENT

U.S.BORAX

630 SHATTO PLACE · LOS ANGELES 5, CALIFORNIA

Tamping bars begin penetration of ballast in approximately a vertical position

As penetration continues they are directed at an angle right under the tie

And sidewise under the rail, thereby achieving maximum consolidation in the vital load-bearing zone

The result is a smoother, more densely consolidated tie-bed than can be accomplished with any other equipment

For Track that "stays put" longer use

Remove a tie that has been tamped with a JACKSON Track Maintainer and you will find the ballast so uniformly and thoroughly keyed together as to resemble a mosaic floor . . . perfect consolidation from the end of the tie, right under the rail and to the desired distance between the rails. This is the result of Jackson's exclusive, powerful vibratory action producing, in the case of the 1962 Track Maintainer, more than 6700 lbs. of downward force and over 2100 lbs. of horizontal force 4200 times per minute *in the ballast* for each of the eight tamping units. Track of this character is bound to "stay put" longer and be more economical to maintain.

JACKSON TRACK MAINTAINER

Write, wire or phone for details of the 8 highly important improvements incorporated in the 1962 model.

As always,
direct sales, leases
and service to
all U.S. railroads

R2-2A

JACKSON VIBRATORS, INC.
LUDINGTON, MICHIGAN

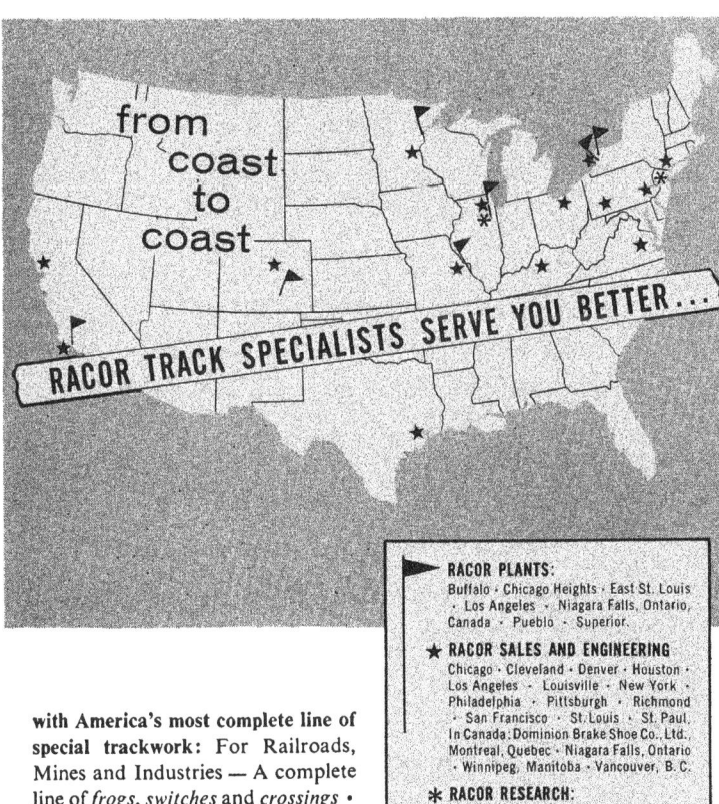

with America's most complete line of special trackwork: For Railroads, Mines and Industries — A complete line of *frogs, switches* and *crossings • Trackwork for installation in paved areas • Manganese steel guard rails • Automatic switch stands • Samson switch points • Snow-Blowers • Switch point guards • Rail and flange lubricators • Tie pads • Racor studs • Dual spike setters • Dual spike drivers • Mechanical car retarders.*

with America's most complete trackwork manufacturing facilities: Coast to coast to serve your needs.

RACOR PLANTS:
Buffalo · Chicago Heights · East St. Louis · Los Angeles · Niagara Falls, Ontario, Canada · Pueblo · Superior.

★ **RACOR SALES AND ENGINEERING**
Chicago · Cleveland · Denver · Houston · Los Angeles · Louisville · New York · Philadelphia · Pittsburgh · Richmond · San Francisco · St. Louis · St. Paul, In Canada: Dominion Brake Shoe Co., Ltd., Montreal, Quebec · Niagara Falls, Ontario · Winnipeg, Manitoba · Vancouver, B. C.

✳ **RACOR RESEARCH:**
Chicago · Mahwah, N. J.

with America's most complete trackwork engineering service: This lies in making available to our customers Racor's engineering experience—*practical* experience from years of designing and manufacturing . . . *advanced* experience solving tomorrow's trackwork problems today in Racor research laboratories.

Why not let us help *you* with your trackwork problems?

A-7126

RAILROAD PRODUCTS DIVISION
530 Fifth Avenue, New York 36, New York

E-x-t-e-n-d T-i-e L-i-f-e!
Hold Gage!
USE TIE PLATE
LOCK SPIKES

One-piece Design

LOCK SPIKES hold tie plates firmly in place on cross-ties and bridge timbers.

LOCK SPIKES are quickly and easily driven, or removed, *with standard track tools.*

Driven to refusal, the spread shank is compressed by the walls of the hole. Tie plates are held against horizontal and vertical movement under spring pressure. Play between the spike and the hole is eliminated—abrasion and seating of tie plates is overcome.

LOCK SPIKES hold their position in the tie, and redriving to tighten the plate is not required. They provide a quiet and strengthened track.

Annual cost of ties and maintenance expense is reduced by extending the life of ties and holding gage. Here is one answer to conservation of materials and labor. Write for free folder.

BERNUTH, LEMBCKE CO., INC.
420 Lexington Avenue, New York 17, N. Y.

Actual Size

FOR MORE ON-TRACK PRODUCTION TIME
MECHANIZE WITH A
MULTI-GANG*
PACKAGE UNIT

This versatile unit Tamps—Jacks—Lines—Removes and Inserts Ties—
Pulls Spikes—Drives Spikes—Bolts and Drills Rails—Transports Men,
Tools and Material.

For complete information contact:

Tamper **LIMITED, Railway Division**
160 St. Joseph St., LACHINE, Montreal 32, Canada

Tamper **INC.,** 53 Court St., P. O. Box 778
PLATTSBURGH, N. Y.

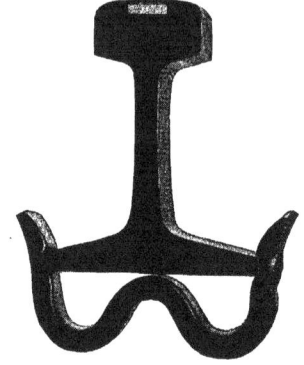

Get Spring-
through-Fall
weed control with

UROX® Liquid Weed Killer!

... as long as 8 to 18 months control after a single application

UROX Liquid Weed Killer kills fast . . . you can see weeds wilt and start to die within 12 hours, regardless of weather.
UROX Liquid Weed Killer handles easier . . . ideal for railroad spray trains. Won't clog strainers and nozzles . . . mixes with fuel, diesel, or ordinary weed oils.
UROX Weed Killer lasts longer . . . it builds up in soils. You use small "booster" treatments in subsequent years.
UROX Weed Killer saves you money . . . cumulative effectiveness means cumulative savings through the years.
Get the complete story now on money-saving, labor-saving UROX Weed Killers. Write or phone your nearest General Chemical Sales Office.

GENERAL CHEMICAL DIVISION
40 Rector Street, New York 6, N.Y.

AREA Publications—Price List

The following include some of the Association publications available from the secretary's office on order. Prices shown are for Members only:

Member
Price

Manual of Recommended Practice, complete in 2 volumes, including binders
(first copy) $15.00
Extra binders, each 4.50
Annual Supplements, each 1.00

Separate Chapters

6—Buildings 1.50
7—Wood Bridges and Trestles 1.00
9—Highways ·.. 0.50
11—Engineering and Valuation Records 1.25
13—Water, Oil and Sanitation Services 1.00
14—Yards and Terminals 1.00
15—Iron and Steel Structures 1.25
16—Economics of Railway Location and Operation 0.75
20—Contract Forms 1.25
22—Economics of Railway Labor 0.50
25—Waterways and Harbors 0.25
27—Maintenance of Way Work Equipment 0.50
28—Clearances 0.25
29—Waterproofing 0.25
Flexible-cover, loose-leaf binder for separate chapters, each 0.40

Portfolio of Trackwork Plans—119 plans, 8 sheets of specifications, 5 sheets
definitions of terms, complete with leatherette cover $12.50
Track Scale Pamphlet—109 pages, flexible cover 0.80
Federal Valuation of Railroads—87 pages, flexible cover 1.00
Instructions for Mixing and Placing Concrete—24 pages, flexible cover 0.40
Notes on Railroad Location and Construction Procedures from the School of
Experience—43 pages, flexible cover 0.50
Handbook of Instructions for the Care and Operation of Maintenance of Way
Equipment—149 pages, hard cover 0.85
Instructions for Care and Safe Operation of Welding and Grinding Equipment—23 pages, flexible cover 0.30
Specifications for Steel Railway Bridges (fixed spans)—70 pages, flexible
cover 0.75
Specifications for Movable Railway Bridges—73 pages, punched sheets 1.00

Chapters 1, 3, 4, 5, 8, and 17 are not available.

CONTINUOUS RAIL
—Quickly, Economically with the
NCG® RAIL WELDING SYSTEM

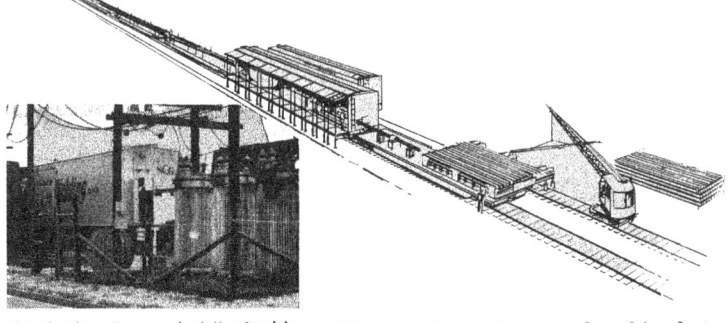

A typical transformer sub-station furnishing commercial power for NCG Automatic Rail Welding System.

When "Flashing" stops, the weld upset is sheared. The weld is then ground with abrasive belts to a smooth surface.

A pusher moves strings of welded rail onto flat cars ready for shipment.

Now small work crews do a big, fast job with the continuous, highly automated NCG Rail Welding System using commercial electricity. Time-wasting annealing and normalizing are eliminated. All operations are automatic, under push-button control.

The NCG Rail Welding System brings the highly desired advantages of continuous rail to many roads which previously deemed it beyond budget acceptance. It may be purchased or leased. Write for details now. NATIONAL CYLINDER GAS, DIVISION OF CHEMETRON CORPORATION 840 N. Michigan Ave., Chicago 11, Ill.

NCG®
NATIONAL CYLINDER GAS
Division of CHEMETRON *Corporation*

©1961 Chemetron Corporation

PROGRESS REPORT

Here are the up-to-date facts on the SPENO Ballast
Cleaning and the SPENO Rail Grinding Services.

BALLAST CLEANING

SPENO Engineering and Research has developed a superior screening arrangement so
that we are now using an improved Ballast
Cleaner with greater efficiency.

RAIL GRINDING

Our Rail Grinding Service has been so well
received we are now building a *THIRD* Rail
Grinding Train to take care of the increased
demand.

*SPENO is constantly developing means for
better service to make sure that the Railroads
receive everything they pay for — and more*

Just Ask the Railroads That have used us!

American Railway
Engineering Association—Bulletin

Vol. 63, No. 572   June—July 1962

PROCEEDINGS ISSUE

CONTENTS

Published by the American Railway Engineering Association, Monthly, January. February. March, November and December; Bi-Monthly, June–July, and September–October, at 2211 Fordem Avenue, Madison, Wis.; Editorial and Executive Offices, 59 Van Buren Street, Chicago 5, Ill.
Second class postage paid at Madison, Wis.
Accepted for mailing at special rate of postage for in .Section 1103, Act of October 3, 1917, authorized on June 29, 1918.
Subscription $10 per annum.

PROCEEDINGS

OF THE

SIXTY-FIRST ANNUAL CONVENTION

OF THE

American Railway Engineering Association

Engineering Division, Association of American Railroads

HELD AT

CONRAD HILTON HOTEL, CHICAGO
March 9 and 10, 1962

VOLUME 63

BULL. 572

OFFICERS, 1961-1962

R. H. BEEDER
President
Chief Engr. Sys.
A.T.&S.F. Ry.

C. J. CODE
1st Vice President
Asst. Chief Engr.—Staff
Penn. RR.

L. A. LOGGINS
2nd Vice President
Chief Engr.
S. P. Co., Tex. &
La. Lines

F. R. WOOLFORD
Past President
Chief Engr.
Western Pacific RR.

E. J. BROWN
Past President
Chief Engr.
Burlington Lines

A. B. HILLMAN
Treasurer
Ret. Chief Engr.
C.&W.I. RR.

NEAL D. HOWARD
Executive Secretary

604

DIRECTORS, 1961-1962

W. E. CORNELL
1959-62
Engr. of Track
N.Y.C.&St.L. RR.

F. L. ETCHISON
1959-62
Chief Engr.
Western Md. Ry.

W. J. CRUSE
1959-62
Engr. M. W.
Great Northern Ry.

C. R. RILEY
1959-62
Spec. Asst. to V. P.
B.&O. RR.

C. J. HENRY
1960-63
Chief Engr.
Penn. RR.

J. M. TRISSAL
1960-63
Vice Pres. & Chief Engr.
Ill. Cent. RR.

W. B. THROCK-MORTON
1960-63
Chief Engr.
C.R.I.&P. RR.

J. A. BUNJER
1960-63
Chief Engr.
Union Pacific RR.

J. H. BROWN
1961-64
Chief Engr.
St. L.-S. F. Ry.

J. E. EISEMANN
1961-64
Chief Engr., West.
Lines
A. T. & S. F. Ry.

W. H. HUFFMAN
1961-64
Asst. Ch. Engr.-Const.
C.&N.W. Ry.

F. R. SMITH
1961-64
Chief Engr.
Union. RR.

605

NUMERICAL INDEX TO COMMITTEE REPORTS

PROGRAM

Sixty-First Annual Meeting
Conrad Hilton Hotel, Chicago
March 9–10, 1962

Friday, March 9

Morning Session—Waldorf Room—9:00 to 12:00

Invocation, by Rev. Hurschel Allen, Associate Minister, First Presbyterian Church, Evanston, Ill.

Presidential Address—R. H. Beeder, Chief Engineer System, Atchison, Topeka and Santa Fe Railway.

Report of Executive Secretary—Neal D. Howard

Report of Treasurer—A. B. Hillman, Retired Chief Engineer, Chicago & Western Indiana and Belt Railway of Chicago.

Greetings from the National Railway Appliances Association, J. P. Kleinkort, (President), Railroad Manager, Railroad Products Division, American Brake Shoe Company.

Keynote Address—C. D. Buford, Vice President, Operations and Maintenance Department, AAR.

	Bulletin Numbers
Reports of Committees	
28—Clearances (10:45)	568
Address—What About Clearances for Piggy-back Operations? (Illustrated), by G. M. Magee, Director of Engineering Research, AAR	
11—Engineering and Valuation Records (11:20)	569
20—Contract Forms (11:32)	567
25—Waterways and Harbors (11:52)	568

Association Luncheon—Williford Room—12:00 Noon

Presentation of those at speaker's tables.

Post-luncheon showing of French National Railways' film "The Modern Track", which details all aspects of the French Railways' use of continuous welded rail.

Afternoon Session—Waldorf Room—1:25 to 5:15

14—Yards and Terminals (1:25)	567
Address—Latest Developments in Classification Yards, (illustrated) by A. V. Johnston, Chief Engineer, Canadian National Railways.	
Address—Piggyback Service Moves Ahead (illustrated), by D. C. Hastings, General Superintendent Terminals, Atlantic Coast Line Railroad.	
16—Economics of Railway Location and Operation (2:34)	567

607

Saturday, March 10

Waldorf Room—8:30 to 12:30

Report of the Tellers

March 10, 1962

We, the Committee of Tellers, appointed to canvass the ballots for officers and for members of the Nominating Committee, find the count of ballots as follows:

No. of Votes

For President

C. J. Code, Assistant Chief Engineer-Staff, Pennsylvania Railroad, Philadelphia 4, Pa. 1,589

For Senior Vice President*

L. A. Loggins, Chief Engineer, Southern Pacific Company, Texas and Louisiana Lines, Houston, Tex.

For Junior Vice President

T. F. Burris, Chief Engineer-System, Chesapeake & Ohio Railway, Huntington, W. Va. 1,575

For Directors (first four men elected)

W. L. Young, Chief Engineer, Norfolk & Western Railway, Roanoke, Va. ... 1,006

T. B. Hutcheson, Chief Engineer, Seaboard Air Line Railroad, Richmond, Va. 896

C. E. Defendorf, Chief Engineer, New York Central System, New York, N. Y. 887

John Ayer, Jr., Chief Engineer, Denver & Rio Grande Western Railroad, Denver, Colo. 844

C. Neufeld, Engineer of Bridges, Canadian Pacific Railway, Montreal, Que. 814

A. S. Krefting, Chief Engineer, Soo Line Railroad, Minneapolis, Minn. 687

D. V. Messman, Asst. to Chief Engineer, Southern Railway System, Washington, D. C. 597

D. T. Faries, Chief Engineer, Bessemer & Lake Erie Railroad, Greenville, Pa. 545

For Members of Nominating Committee (first five men elected)

A. L. Sams, Principal Assistant Engineer, Illinois Central Railroad, Chicago . 1,097

J. F. Beaver, Chief Engineer, Southern Railway System, Washington, D. C. . 975

B. B. Lewis, Professor of Railway Engineering, Purdue University, Lafayette, Ind. 855

J. J. Schmidt, Assistant Director Research, Denver & Rio Grande Western Railroad, Denver, Colo. 791

E. M. Hastings, Jr., Wire Crossing Engineer-System, Chesapeake & Ohio Railway, Richmond, Va. 788

(Continued on next page)

---

* Under the provisions of the Constitution, L. A. Loggins advances automatically from junior vice president to senior vice president.

S. E. TRACY, Superintendent Work Equipment, Chicago, Burlington & Quincy
Railroad, Chicago 781

F. A. HESS, Maintenance of Way Engineer, Indiana Harbor Belt Railroad,
Hammond, Ind. 740

F. N. BEIGHLEY, Roadway Engineer, St. Louis–San Francisco Railway, Springfield, Mo. 675

L. C. COLLISTER, Manager, Tie and Timber Treating Department–System,
Atchison, Topeka & Santa Fe Railway, Topeka, Kans. 606

P. D. BRENTLINGER, Forester, Pennsylvania Railroad, Philadelphia, Pa. 510

THE COMMITTEE OF TELLERS
J. E. WIGGINS, *Chairman*

R. A. BARDWELL	R. H. HOLDEN	C. ROSS
H. R. BECKMANN	W. R. HYMA	F. H. SMITH
T. W. BROWN	J. E. INMAN	J. J. STRELL
J. BUDZILENI	L. S. MARRIOTT	S. A. STUTES
K. J. DECAMP	C. A. MEADOWS	L. L. TAMELING
H. E. GRAHAM	R. W. MIDDLETON	T. TANTILLO
V. E. HALL	D. J. MOODY	D. C. TEAL
J. W. HAYES	C. F. MUELDER	J. F. WAGNER
J. HEHN	R. E. PEARSON	E. R. WILTZ
L. A. HICKS	H. L. READ	I. V. WILEY

Proceedings

Running Report of the 61st Annual Meeting of the American
Railway Engineering Association (Engineering Division, Association of American Railroads), March 9–10, 1962, Conrad
Hilton Hotel, Chicago, Including Abstracts of All Discussions, All Formal Action on Committee Presentations,
Specific Papers and Addresses Presented in Connection
with Committee Reports, and Other Official Business
of the Association

Opening Session, March 9, 1962

President R. H. Beeder*, Presiding

The opening session of the 61st Annual Meeting convened at 9:10 am.

PRESIDENT BEEDER: Gentlemen, will the meeting please come to order.

This is the 61st Annual Convention of the American Railway Engineering Association and the concurrent Annual Meeting of the Engineering Division of the Association
of American Railroads.

Having said this, I am almost tempted to call for the first committee report, skip-
ping a lot of the usual preliminaries because, as you know, with a 1½-day convention
instead of 2½ days, we are confronted with completing the work of this meeting,
which includes hearing from our 22 committees, in three-fifths the usual time.

We wanted to start on the right foot and have our invocation at this time, but
I do not believe Reverend Hurschel Allen is in the audience. Perhaps the snow is too
deep out in Evanston. So we will postpone the invocation and go ahead with our
program at this time.

Again we meet in new surroundings and under new conditions. From the spacious
ballroom of McCormick place, where we met last year, we meet today in somewhat
restricted but very hospitable quarters, I am sure, and appropriate and adequate, we
hope, to our restricted-attendance meeting.

As you know, there is no exhibit of the National Railway Appliances Association
in connection with our convention this year; but I am pleased to see many of our
railway supply friends here, and I hope you will avail yourself of their presence to
discuss any of your problems with them.

While, as you know, there are certain restrictions on our members as regards
attendance at this meeting, there are no restrictions whatever on members' wives, and
I am glad to note that a goodly number of them are here with us—some here in this
room at the present time. We welcome all of you ladies, hope that you will enjoy your
short stay in Chicago, and want you to know that we shall be happy to have you sit
in on any of our sessions to hear any particular addresses or committee presentations.

I particularly invite you to be present at the closing business session of our convention in this room at noon tomorrow, beginning about 12 o'clock, when the announcement will be made of the results of our annual election of officers and when the new
officers will be installed.

---

* Chief Engineer System, Atchison, Topeka & Santa Fe Railway.

Just one more thing by way of preliminaries. May I ask that all of you register your attendance, if only to get on the record and to swell the record of what we know will be the smallest registration at an annual meeting of this Association since its early days.

It is not without some significance that whereas official invitations to this convention were extended to only about 250 of our members, 132 railroad men had already registered their attendance by 4 pm yesterday, along with 129 supply men, making a total of 261, with the prospect that this 250 figure will be exceeded considerably before this convention is over.

Now, without further delay I would like to present to you those here with me at the speaker's table. Here is one place where we really cut corners. Instead of all of our officers and directors, past presidents and a number of special guests, as in past years, I have alongside me today only your two vice presidents, your executive secretary and your treasurer, whom I am sure I can present to you in about one-tenth the usual amount of time. To save still further time, may I ask that you withhold your applause until all have been introduced.

First, I would like to present to you your senior vice president, C. J. Code, assistant chief engineer—staff, Pennsylvania Railroad; your junior vice president, L. A. Loggins, chief engineer, Southern Pacific Company, Texas & Louisiana Lines; your treasurer, A. B. Hillman, retired chief engineer, Chicago & Western Indiana and Belt Railway of Chicago; and your executive secretary, Neal Howard. [Applause]

In taking this short cut this morning, I would not have you feel that we will completely overlook our officers and directors and those of our past presidents who could be with us at this convention. It is my plan to present all of them to you at the main speaker's table at our annual luncheon this noon; and, incidentally, if you have not already purchased your tickets for the annual luncheon may I ask that you excuse yourself from the room and do this immediately.

The first item of official business on our program is approval of the minutes of our 1961 Annual Meeting, which were published in the June–July Proceedings Issue of the AREA Bulletin, No. 565, a copy of which was furnished to each member. Spread out, as they are, over 250 pages in this Bulletin, it is obviously impossible to read these minutes here. However, assuming that many of you have at least reviewed these minutes, I will entertain a motion that they be approved.

VOICE: I so move.

[The motion was duly seconded, was put to a vote, and carried.]

Address of President R. H. Beeder

Members of the American Railway Engineering Association and guests: In accordance with the provisions of our Constitution the next order of business gives me the privilege of summarizing some of the activities of our Association during the past year and commenting on the effect that those events may have on our future.

Actually, all of us are kept up to date on the affairs of our Association by that fine bi-monthly publication the *AREA News,* which is so ably edited by Executive Secretary N. D. Howard and his staff. This publication is certainly a tremendously effective instrument of communication within our Association. Long may it live!

In my opinion, the past year has certainly been a "year of decision" for our officers and directors. I suppose that every Board of Direction of our Association since its founding 63 years ago could have made the same statement and still have been making a reasonable appraisal. But few have been confronted with the general desire of railroad managements, due to recent economic conditions, to reduce the time and expense of their men in attending annual meetings.

These circumstances have resulted in our present 1½-day, restricted-attendance Annual Meeting which we convene here today. Of course, as an Association, the AREA can act in any manner we desire. We can act unilaterally and have 2½-day Annual Meetings every year, or follow any other independently selected arrangement. In my opinion, such actions would not be realistic, because AREA is a vital and extremely important segment of the entire railway and associated industries. It would not serve our best interests to move off on a tangent when the railway industry as a whole needs our intensely productive and dedicated Association. You and I are exactly the same men here this morning that we will be Monday morning when we report back to our managements. Perhaps some of us will report back sooner if we have an emergency. I am sure that our work here in AREA, as well as our day-to-day assignments in industry, should both be carried on with the same respect to the aims and goals of our respective managements.

I am hopeful that the conditions under which our railroads have been working will turn toward the better. Implementing some of the recommendations of the 15-man Presidential Commission which reported on work rules last week should certainly be a long-overdue, forward step in the removal of some of the shackles from our industry. Attainment of the objectives prescribed in the railroads' "Magna Carta Program" will also improve our ability to serve the transportation needs of our country.

As you know, there was some previous discussion within the railway industry as a whole to eliminate in 1962 most, if not all, of the annual meetings of the railway groups, including our AREA, in the interests of economy. This we could not do as we must progress highly important material for our Manual of Recommended Practice. The very essence of our Association is the tremendously effective work of our technical committees, and their results must be progressed at least on an annual basis if we are to keep anywhere near abreast of advancing technology. Even though our Annual Meeting here during the next 1½ days will be condensed and will be principally a business meeting, your Board of Direction has acted to provide as many special features as possible within the time limits imposed. I am sure you will find much of interest and benefit in all of the presentations.

Our recent economic climate has also resulted in some other unfavorable aspects in the affairs of our Association. One of the most serious of these is the further curtailment of the research work of the AREA which is carried out by the Engineering Division research staff of the Association of American Railroads. Our technical committees proposed an expenditure of $516,950 for research projects and technical services, but the total budget allotments for these items during 1961 could only be maintained at a level of approximately $222,000, or much less than half.

While as retiring president as of the end of this convention, it is not my province to discuss with you future plans of the Board of Direction, there is one matter respecting one aspect of the future plans of the Association already considered by your present Board, concerning which it is appropriate for me to advise you. I refer to plans for a meeting or meetings of the Association in 1963.

You have already read, I am sure, the article which appeared in the Convention Issue of the *AREA News,* advising that decision has been reached by those involved to carry through a plan suggested by the Association of American Railroads to bring together in 1963, within a 2-week period (and possibly at 4-year intervals thereafter), the annual meetings of major railway groups, with a concurrent combined exhibit throughout this period by all of the related railway supply organizations. In 1963, such combined meetings will be held within the period Wednesday, October 9, through Wednesday, October 16, in Chicago, with a concurrent combined exhibit at McCormick Place.

As now contemplated, the meetings of the AREA, the Mechanical Division, AAR, the Purchases and Stores Division, AAR, and the Communication and Signal Section, AAR—and possibly one or more smaller groups—will be held the latter half of the first week, and the meetings of the Roadmasters' Association, Bridge and Building Association, Coordinated Mechanical Associations—and possibly one or more other groups— will be held in the first half of the following week.

I have said that the AREA would be represented in this joint enterprise, but not that any decision has already been reached to this effect, or as to the character of its participation, because, unquestionably, our Association must actively take part in this industry-wide display of strength and technological developments in the railway industry.

This plan for coordinated meetings and exhibits in 1963 presents our Association with problems, not the least of which is the fact, as you no doubt know, that our Constitution stipulates specifically that AREA conventions shall be held in March. In view of this, and to permit the Board to consider shifting the Association's Annual Convention from March to October 1963, and, in fact, to the fall—or to any other month of the year—in future years, the Board at its meeting yesterday approved submitting to the membership amendments to the Association's Constitution which, in the future, would place in the hands of the Board the responsibility of setting the dates for the Annual Convention in each year in the light of existing circumstances and conditions. These circumstances and conditions, which are becoming increasingly variable, include general economic conditions, the availability of adequate hotel accommodations, the availability of suitable exhibit facilities, and the proposed periodic repetition of combined meetings and exhibits as planned in 1963. Passage of these amendments would also open the door to future realistic negotiations with the Roadmasters' and Bridge and Building Associations, as has been suggested many times, looking to the possibility of consecutive meetings (not concurrent meetings) of the AREA, Roadmasters' and Bridge and Building Associations in the fall of each year, with combined exhibits of their related railway supply organizations every two years—if such can be mutually agreed upon as in the overall interest of all concerned.

Thus, as now planned, a letter ballot covering the proposed amendments to the Constitution will accompany the next issue of the *AREA News* for your careful consideration and action. I believe that approval of these amendments will give your Board of Direction badly needed flexibility of decision in setting the dates of our Annual Conventions, which is highly important under present and prospective conditions. Personally, I am in full accord with these proposed amendments and will vote for them.

These Annual Meetings of our Association take a terrific amount of planning and work by our Committee on Convention Arrangements and Executive Secretary Neal Howard and his staff. The planning for this meeting has been and will be particularly tough, because of the new location and the lack of a previous 1½-day pattern to steer by. I want to thank all of you and a perfectly splendid Board of Directors who have worked together so harmoniously. Thank you. [Applause]

PRESIDENT BEEDER [continuing]: What we were to have done a little earlier in the program we shall do now. We shall have the invocation by the Reverend Hurschel Allen, associate minister, First Presbyterian Church, Evanston, Ill. Reverend Allen.

Invocation

REVEREND HURSCHEL ALLEN: Let us bow together as we pray.

Eternal God, we acknowledge Thee as the author and sustainer of all life. We praise Thee as the One in whom each of us lives and moves and has his being. Grant that we may so consciously accept the breath of this and every new day, as a trust from Thee, that we may daily live as responsible and disciplined stewards—stewards of time and energy, stewards of things and our very existence.

As on this World Day of Prayer we are aware of our movement into another lenten season, grant to us a double blessing. We pray, O God, that we may gain strength and confidence through the experience of sharing with people in all areas of the world in the fellowship of prayer. At the same time, increase our awareness of Thy presence, that our personal experience during this season may be one of intensified individual relationship with Thee.

Bless, O God, this convention. Bless each participant here, that this may be a time of renewal and encouragement, a time of growth in fellowship with others who share our interests, our concerns, our purposes. Help us to see our purposes always in accord with Thy eternal purpose.

In the Name of the Master, Amen.

PRESIDENT BEEDER: Thank you, Reverend Allen. It has been a pleasure and privilege for us to have you start off two of our conventions in the past three years, and we appreciate your very appropriate petition in our behalf. We know that you have met with us this morning at some inconvenience to yourself, so we want you to know that you are excused, although if your time permits we will be happy to have you remain with us as long as you may desire.

Thank you again, Reverend Allen.

As you all know, this Association has an executive secretary, and the next official order of business is to hear from him. Mr. Howard, I shall be pleased if you will present your report at this time.

Report of Executive Secretary

Mr. President, Members of the American Railway Engineering Association, and guests:

I am glad the invocation preceded my report because I feel a little stronger in having the Reverend Hurschel Allen appear ahead of me. I wouldn't have you think that Mr. Allen was late because he got up late, or for any other reason. I owe it to him to tell you that, as he pointed out to me in the rear of the room a few minutes ago, my letter to him said that this meeting would begin at 9:30, and he was here at 9:25. [Laughter]

As with our conventions as a whole, I hope we get our starting time back to something that is standard again. Mr. Woolford started his convention at 8:30, Mr. Brown at 9:30, and now we are starting at 9, and I mixed it up. I apologize to you as well as to Mr. Allen.

Confronted with a number of disturbing influences which reflected themselves adversely on Association activity, 1961, by comparison with recent years, was not an exceptionally good year for our Association. But neither can it be said that it was a bad year, and your secretary is pleased to report that, to the present time, the effect of none of these disturbing influences has been very serious; that the membership of

the Association remains relatively high; and that the activities and production of committees continue at a high level. Furthermore, that largely as a result of practical
expedients adopted, and the practice of rigid economy generally, as will be reported
by your treasurer, the Association is financially sound.

But it must be said that some of the unfavorable and disturbing influences referred
to—economic and otherwise—have already had an adverse effect on the total membership of the Association, and that unless these influences disappear or are largely overcome, they will have not only a more serious effect on Association membership, but an
adverse effect upon future Association activity, its financial soundness, and its ability
to serve its members and the railroad industry.

As of February 1, 1962, the membership of the Association stood at 3347, a net
loss of 60 members compared with the membership of 3407 one year earlier. Thus,
after an unbroken record of growth in membership from 1944 through 1957, the membership has slipped backward in two of the last four years—1958 and 1961—and might
have slipped backward in the other two years had losses in those two years not been
overcome by special circumstances or special member interest and effort.

The net loss of membership during the year resulted primarily from a combination
of the enrollment of fewer new members and the loss of a larger than usual number
of members through resignations and being dropped because of non-payment of dues.

Unmistakably, the fewer new member enrollments in 1961 reflected the relatively
poor economic condition which continued to prevail in the railroad industry during
the year, the continued tightening up of organizations within many railroads, the restricted recruitment by the railroads of additional technically trained college graduates,
and the result of several major railroad consolidations in recent years. Unmistakably,
too, some of these same factors reflected themselves in the larger number of members
who resigned or who were dropped as of the end of 1961. To the extent that such was
true, the situation was beyond the immediate control of the Association. But to the
extent that the causes of fewer enrollments and larger defections may have been the
result of more circumscribed and restricted membership and Association activity, whether
resulting from outside influences or self imposed, both the Association and the railroad
industry should be seriously concerned about the further loss they stand to sustain in
the future unless these unfavorable influences are offset or removed. The Association
must constantly seek to enhance its efficiency and usefulness, and must question seriously
any proposals, which, in the interest of real or fancied immediate benefits, would weaken
its structure and long-term effectiveness.

Time will not permit me to comment on those aspects of your secretary's report
with respect to the activities of committees during the year, Association publications,
research work, and the stepped-down plans for the 1962 convention of restricted length
and attendance. But with respect to the last named—our 1962 convention, your
secretary's report says:

"Thus, the Association's work for 1961 is being climaxed by an annual convention
in March 1962, which will accomplish the official business planned, and which will,
unquestionably, prove of interest and benefit to those who attend. But it must be said
that this stepped-down type of convention has been accepted enthusiastically by few,
if any, members, and is a keen disappointment to many. What effect this may have
upon the various aspects of the well being of the Association—membership, membership interest, productivity, and finances—especially if these restricted conventions are to
be repeated periodically, is a matter for conjecture, but one which should be a matter
of serious concern to the officers of the Association, to the membership as a whole, and

to the railroad industry which the Association has served so effectively for the past 63 years.

"If there are to be ill effects from the restricted-scope 1962 convention, they can be overcome only by the continued loyalty of members, the full cooperation of the Association of American Railroads and of individual railroads, stepped-up committee activity, and the production of even more valuable Association reports and publications —all of which are to be hoped for in the year ahead."

To make reference to the importance of stepped-up committee activity and the production of even more valuable Association reports and publications is particularly appropriate here this morning, because I am well aware that I am addressing, in large part, our committee officers and subcommittee chairmen, upon whom the Association must rely to bring about this stepped-up committee interest and activity.

To these more formal remarks I would like to add briefly that it has been a pleasure and privilege for me and the members of your secretary's staff to work for and with all of you during the past year, and that we look forward to that pleasure and privilege—in the interest of the Association—in the year ahead.

Thank you. [Applause]

PRESIDENT BEEDER: Thank you, Mr Howard. It is a matter of regret to all of us that you were unable to say, as you have for so many years in the past, "The Association has had a favorable year in every respect." On the other hand, as you did say, the Association did not have a bad year, and it held its own or moved ahead on many fronts. To me, this is a highly creditable showing under the conditions which have existed, and I want to take this opportunity to thank our officers and directors, our committee chairmen and members, our secretary and his staff, and all those who have collaborated with our Association to make this possible.

Our next order of business is the report of our treasurer, A. B. Hillman, who has now been our treasurer for the past 10 years. Mr. Hillman, will you please present your report at this time.

Treasurer's Report

Mr. President, members and guests: Last year I reported that 1961 would confront the Association with new and difficult financial problems, and that the Board of Direction had taken bold, if not drastic measures, to meet this situation. How well the situation was met, I am happy to say, is evident by reference to the financial statement for the year, which indicates an excess of receipts over disbursements of $10,051.53—and that the Association ended 1961 in a sound financial condition.

I will not attempt to detail the various items of receipts and disbursements during the past year as they are all set forth in the financial statement for the year, which appears in the March Bulletin. I would point out however, that total receipts, while some $2000 lower than in 1960, were some $5000 higher than estimated. Our lower receipts resulted in large part from lower membership receipts of $1300, and no receipts from the AAR for research report printing, which in 1960 amounted to $5,832. Offsetting these losses were higher receipts in 1961 over 1960 in the items of sales of publications, interest on investments, Manual, and convention registration fees.

Total disbursements were some $14,900 under those of 1960, and approximately $3700 lower than anticipated—due to efficient management and strict economies—principally the elimination of the bound volume of the Proceedings.

It is well that the year 1961, through savings and rigid economies, was a favorable one financially for the Association, as the year ahead will present a difficult financial problem. This is because the Manual must be reprinted in its entirety in order to replenish a depleted stock, and the 1962 Supplement thereto will be even larger than the large 1961 Supplement. Therefore, even further drastic economy measures to offset loss of revenues, to augment income, and to hold disbursements to the minimum, will not prevent the Association from incurring a substantial deficit in 1962, with disbursements exceeding receipts by at least $13,000. Should a large supplement to the Portfolio of Trackwork Plans be issued in 1962, as seems likely, then this presently anticipated deficit will be considerably larger.

But, while it is expected that the Association will incur a large deficit in the year ahead, this can be overcome in the next few years, I am sure, if we can retain our present membership level, continue to have the interest of our members and that of the railroads, and continue the widespread sale of Association publications. Thank you. [Applause]

PRESIDENT BEEDER: Thank you, Mr. Hillman. We greatly appreciate your continued valuable service to this Association as its treasurer. It is highly gratifying that due to outstanding economies adopted, along with a combination of other circumstances, our Association had a highly favorable year financially; but from what you have said, we are certainly going to go off the deep end in 1962, especially if we have to print a big supplement to the Portfolio of Trackwork Plans in addition to reprinting the Manual. Thank you again, Mr. Hillman.

Gentlemen, you have heard the reports of our executive secretary and treasurer. A motion is in order that these reports be accepted. Do I hear such a motion?

VOICE: I so move.

[The motion was duly seconded, was put to a vote, and carried.]

PRESIDENT BEEDER: There is one other feature of our usual program that we did not feel we could dispense with, and that is giving recognition to and hearing a few words from the president of the National Railway Appliances Association, J. P. Kleinkort. Joe, we would like to have a word or two from you.

Greetings from NRAA

J. P. KLEINKORT: Mr. President, officers, members and guests:

I bring you greetings from all 154 members of the National Railway Appliances Association. Our association feels honored to have their president given this opportunity, since it provides us with a chance to voice our appreciation for the close cooperation between AREA and NRAA existing today as it has in the past.

The period of change which has confronted both of our organizations during the past two years has been made a great deal easier for all of us through a free exchange of experience and needs. There have been many letters written, meetings held, and numerous discussions about conventions and exhibits. Our principal concern is exhibits in cooperation with AREA. We want exhibits, and we believe you need them for a number of reasons:

First of all, to show our improved products to as many potential users as possible, and at the lowest cost.

Second, through an exchange of ideas, to get maximum benefit from these products, as well as suggestions for their improvement that will lead to more economical and varied use.

Third, to help determine the volume of products required to satisfy the need.

Fourth, to instruct user personnel in proper operation and maintenance.

Fifth, to permit your personnel to show you what they want and need to lower maintenance costs and speed up the work.

Sixth, to expand sales to spread development cost and provide funds for more development.

Much of the reduction in your maintenance costs is due to equipment and materials you found at our exhibits, particularly during the last 10 years. We must join hands, as we are doing in maintaining productive conventions and exhibits, to continue the outstanding maintenance achievements of the past 10 years. The NRAA and its members are ready and willing to do their part.

At this time I also would like to thank your Association for the opportunity extended to us to permit qualified members in the supply industry to join the AREA. I do hope that if there are any members here of our Association who have not availed themselves of this opportunity, and looking forward to the expenses the AREA is going to have in the coming year, it might be a very helpful thing to avail yourselves of this opportunity and help the AREA at the same time.

At this time I would also like to commend the officers and board of direction of your Association for doing a fine job in the past year.

Thank you. [Applause]

PRESIDENT BEEDER: Thank you, Joe. You haven't had to plan and arrange for an exhibit in connection with this year's convention, but I know you have had your hands full looking ahead to 1963. It has been a pleasure working with you and the other officers of your Association. We appreciate the courtesies and cooperation which your members extend to our members and to our Association, and we look forward to the fullest possible cooperation between our two groups in the years ahead.

Thank you again, Joe.

Until a few days ago I had a beautiful introduction prepared for the man we had expected would be our next speaker—our keynote speaker—but I will now turn that over to your next president for use another year, since our keynote speaker, Curtis D. Buford, vice president, Operations and Maintenance Department, AAR, had to cancel his appearance here this morning to participate in a special AAR Member Road meeting called in Washington today to consider the report, recently released, of the Presidential Commission on Railroad Work Rules. This is a keen disappointment to all of us, but I am pleased to have a brief statement from Mr. Buford which I would like to read to you:

"To the Members of the American Railway Engineering Association—AAR Engineering Division:

"First, I would like to convey to you all my sincere regrets for having had to cancel my plan to be with you at this time. It has always been a privilege and a real pleasure to be at your annual meetings and to participate in your beneficial programs.

"A special AAR Member Road meeting has been called in Washington today, and I know you will understand that I must be in attendance.

"Let me express to you in this way my best wishes for a most successful meeting. I know you have an interesting and productive program lined up, and I am confident that your Association will continue its good work on behalf of the railroad industry. We still have our problems, but I personally feel that your officers and committees have satisfactorily solved many of them and are making real progress on others. Your cooperation is always needed, and is indeed appreciated.

"I hope that the shortened pattern you have adopted for this meeting works out successfully. Your president, Board of Direction, and executive secretary have worked industriously on this new procedure; and while it is an innovation, I feel sure it will result in a highly productive meeting.

"I also would like to congratulate your group, as well as the members of the National Railway Appliances Association, for the ingenuity and aggressiveness they have shown in working out the preliminary details for the combined railway meetings and exhibit at McCormick Place in October 1963. I am confident that an all-out effort will be made by all of the participating organizations to insure the success of that undertaking.

"Let me again convey to you my regrets at not being able to be with you today, and my sincere wishes for a very successful meeting. I am sure your deliberations will produce many worth-while results.

"May I remind you of these four provisions of the so-called 'Magna Carta' program which the industry is now endeavoring to have enacted into law: 'Freedom from subsidized competition, from outmoded regulation, from discriminatory taxation, and freedom to provide a diversified transportation service,' and respectfully suggest that this program merits the full support of each of you, both as individuals and as members of your Association."

PRESIDENT BEEDER (continuing): Mr. Buford, for the record I want to say that we miss you here this morning, and that we appreciate your message of regret and good wishes, and look forward to your being with us another year.

Gentlemen, with all of the preliminaries of our convention out of the way, it is time for us to get down to the bedrock work of this convention, which includes the reports of our various committees and action upon their many Manual recommendations.

This year, with official invitations to this convention confined to the officers and subcommittee chairmen of our various committees, I cannot have the pleasure of inviting the entire personnel of committees to the speaker's platform as the various committees present their reports; but it can be expected, I am sure, that our committee officers and their subcommittee chairmen are here in large measure; and as I invite the different committees to present their reports, I hope all of these men will come promptly to and take their places at the speaker's table—the chairman of the committee occupying the chair at my immediate left, where a microphone has been provided for his use.

Discussion on Clearances

[For report, see Bulletin 568, pp. 337–340]

PRESIDENT BEEDER: The lead-off report of this convention will be that of our Committee 28—Clearances, of which J. G. Greenlee, clearance engineer, Pennsylvania Railroad, is chairman. I would appreciate it if Mr. Greenlee and the other members of his committee present will come to the platform promptly.

While these gentlemen are coming to the speaker's table, I would like to extend to all of you present the privilege of the floor, and invite your comments on or criticism of any of the report presentations within the limited amount of time which can be made available for discussion.

While it may well be said that most of the Manual recommendations to be presented to you for adoption have been thoroughly threshed out in committee discussions, with little more to be said respecting them, we would particularly appreciate any pertinent comments with respect to any of these recommendations as they are brought to our attention by our various committees.

Mr. Greenlee, I yield this microphone to you so that you can proceed with the presentation of you committee's report.

CHAIRMAN J. G. GREENLEE: Mr. President, members of the Association and guests:

Committee 28—Clearances, has nine assignments. Under two of these assignments we shall make recommendations for publication in the Manual. For the remaining assignments we shall give progress or status reports. In addition to this, we shall present a very interesting and timely special feature concerning tests of piggyback cars.

Assignment 1—Revision of the Manual.

CHAIRMAN GREENLEE: The chairman of Subcommittee 1 is B. Bristow, principal assistant engineer, Chicago, Rock Island & Pacific Railroad. Mr. Bristow has not yet shown up so I shall give you a resumé of his report.

Your committee submits for adoption the following recommendations with respect to the Manual: That page 28–1–1, General Information, be reapproved with the following revisions: Change Par. 2 to read: "On curved track the clearances shall be increased to allow for the overhang and tilting of a car 85 ft long, 65 ft between centers of trucks, and 15 ft 1 in high." Also change the Note to read: "Note: Pars. 1, 2, 3 and 4 apply to Figs. 1 to 8 incl., Par. 5 applies to Figs. 1 to 8 incl., and also to Fig. 1, Part 3, this Chapter."

I move that these changes be accepted for publication in the Manual.

[The motion was duly seconded, was put to a vote, and carried.]

Assignment 2—Clearances as Affected by Girders Projecting Above Top of Track Rails, Structures, Third Rail, Signal and Train Control Equipment, Collaborating with Committee 18, with Communication and Signal Section, AAR, and with Mechanical and Operating-Transportation Divisions, AAR.

CHAIRMAN GREENLEE: This assignment is completed for the time being. However, it has been a very important assignment and may be revived in the near future. C. W. Hamilton, engineer of design, Wabash Railroad, St. Louis, is subcommittee chairman.

Assignment 3—Review Clearance Diagrams for Recommended Practice, Collaborating with AREA Committees Concerned and the AAR Joint Committee on Clearances.

CHAIRMAN GREENLEE: Under this assignment the committee is giving further study to proposed Fig. 9—Clearance Diagram for Overhead Bridges and Other Structures not Otherwise Provided for, which is being held in abeyance. R. L. Williams, office manager, Engineering Department, Illinois Central Railroad, and vice chairman of Committee 28, is chairman of the subcommittee.

Assignment 4—Compilation of the Railroad Clearance Requirements of the Various States.

CHAIRMAN GREENLEE: Subcommittee Chairman J. F. Smith, inspector, Illinois Central Railroad, is not present. Under this assignment the committee has submitted clearance dimensions for the State of New York, which are published in Bulletin 568, December 1961. These statutes became effective April 20, 1961 and, being new, were not shown on the clearance chart dated October 1, 1961, published in Bulletin 561. Those wishing to add these dimensions to their present charts are referred to Bulletin 568.

Assignment 5—Clearance Allowances to Provide for Vertical and Horizontal Movement of Equipment Due to Lateral Play, Wear and Spring Deflection, Collaborating with the Mechanical Division, AAR.

CHAIRMAN GREENLEE: Subcommittee Chairman E. E. Mills, draftsman, Pennsylvania Railroad, Chicago, will present the report on this assignment.

E. E. MILLS: Mr. President, Mr. Chairman, members of the Association and guests.

Your committee submits a report of progress in the collection of data relative to the static and dynamic behavior of flat-car trailer carriers and three-level auto carriers.

Tests were run, May 24, 1961, on the Burlington Railroad main line from Chicago to Galesburg, Ill., at speeds up to 70 mph; on May 25, 1961, on the branch line from Galesburg to Barstow, Ill., and return, at speeds up to 63 mph; and from Galesburg to Chicago in a passenger train, at speeds up to 83 mph on May 26, 1961.

On May 30, 1961, a similar car equipped with a three-level auto rack and loaded with 12 new automobiles was run from Chicago to St. Paul, Minn., at speeds up to 70 mph, but predominantly around 50 mph because of the length of the train.

Because these tests were conducted on a single type of car, they are considered not entirely conclusive, and similar tests on different types of cars are to be conducted as soon as proper arrangements are concluded.

The data secured from the Lackawanna freight car tests conducted in 1959 are being evaluated and these, together with the data secured from trailer and auto carrying cars already secured and yet to be secured, will be the subject of a further report.

Assignment 6—Compilation in Table Form of Offsets for Overhanging Loads on Curves.

CHAIRMAN GREENLEE: Subcommittee Chairman W. P. Kobat, draftsman, Atchison, Topeka & Santa Fe Railway, Chicago, will present this report.

W. P. KOBAT: Your committee now submits for adoption and publication in the Manual the two tabulations showing offsets for overhanging loads on curves that were presented as information with last years report (Proceedings, Vol. 62, 1961, page 412).

Sheet 1 gives the offsets for a 1-deg curve at the middle of car or load for distances between truck centers from 20 to 65 ft, with illustrations. Sheet 2 gives the end offsets for a 1-deg curve for end overhang from 6 to 40 ft.

The purpose of the tabulations is to save time in calculating clearances for excessive-dimensioned shipments on curves.

MR. KOBAT: Mr. President, I move that the tables be adopted.

[The motion was duly seconded, was put to a vote, and carried.]

Assignment 7—Methods of Measuring High and Wide Shipments, Collaborating with Mechanical Division, AAR.

CHAIRMAN GREENLEE: This assignment is also completed and will be replaced with a new assignment. R. A. Skooglun, chief draftsman, Northern Pacific Railway, St. Paul, Minn., is subcommittee chairman and was unable to be with us today.

Assignment 8—Review Present Methods of Presenting Published Clearance Information to Determine How This Can Be Simplified and/or Standardized.

CHAIRMAN GREENLEE: Samples of published clearances of various roads have been forwarded to this committee. These samples are being analyzed and evaluated. On completion of this study the committee will recommend a simplified and standardized form. J. A. Crawford, assistant engineer, Chesapeake & Ohio Railway, Richmond, Va., is subcommittee chairman.

Assignment 9—Review Clearance Records of Various Roads, Looking to Developing a Standardized Method For Charting All Obstructions.

CHAIRMAN GREENLEE: Sample clearance charts have been received from various roads. These charts are being studied and analyzed. Their best features will be combined, and a sample chart will be submitted to the full committee for further study and review. M. E. Vosseller, draftsman, Central Railroad of New Jersey, Jersey City, N. J., is subcommittee chairman.

CHAIRMAN GREENLEE: We will now present a special feature—an address entitled, "What About Clearances For Piggyback Operation"? by G. M. Magee, director of engineering research, Association of American Railroads.

What About Clearances for Piggyback Operation?

By G. M. MAGEE
Director of Engineering Research, AAR

The trailer-on-flat-car and auto-carrier types of railroad business have been expanding in recent years and have led to the development of new types of cars. Naturally the question has been raised in the minds of many chief engineers and other engineering personnel as to what the effects of these new types of cars may have on clearances, and perhaps also on operating speeds on curves. The trailer-on-flat-car offers a somewhat different problem than the conventional type of freight car because there is a double spring support of the trailer body. For example, in rounding curves at other than the equilibrium speed an ordinary freight car would lean inward or outward to an extent governed by the characteristics of its springing arrangement. With the trailer-on-flat-car, however, in addition to the lean due to the springing system under the flat car, there is also the tilting of the trailer body on its own tires and springs. There is also the question whether with two separate masses and two separate springing arrangements there may be a forced vibration condition set up of the trailer body that might be different from that of the car body with only a single spring arrangement.

To obtain further knowledge on this subject an arrangement was made with the Burlington Railroad to run a series of tests on its lines in May of 1961. The plans made contemplated the use of a flat car with two loaded trailers supported thereon. The flat car was a conventional 85-ft-long TTX type car, and two 40-ft Burlington trailers were used loaded with ties which were found to give the maximum permitted weight on the trailers together with maximum height of center of gravity of combined loading of 90 in. The arrangements made contemplated three test runs, one to be with the trailers-on-flat-car operated in a highspeed freight train between Chicago and Galesburg, Ill.; the second, with the trailers-on-flat-car operated in a special train on a branch line from Galesburg to Barstow, Ill., and return; and the third, with the trailers-on-flat-car in a passenger train from Galesburg to Chicago. The Burlington dynamometer car was operated next to the flat car in order to provide a car for the recording instruments. In general, the instrumentation consisted of Brush oscillographs which made a direct-writing record of the various movements on the trailer and flat car that it was desired to measure. Vertical accelerations and displacements and lateral displacements were measured at the car-floor level and lateral displacement at an upper level on an upright post attached to the car floor at a height corresponding to that of the upper-level accelerometer in the trailer. Vertical and lateral accelerations and displacements were also measured on the trailer floor and lateral accelerations and displacements at the upper level. In addition one vertical gyroscope was placed on the car floor and another on the floor of one trailer to measure the inclination of the floor from the horizontal.

In the operation from Chicago to Galesburg in the regular freight train it was not practical to take many photographs or movies. However, on the special test train turn around between Galesburg and Barstow movies were taken which will serve to give a better idea of the manner in which the tests were conducted.

[The movie was then shown, with the following commentary by Mr. Magee]:

The test train consisted of a diesel unit followed by a spacer car, then the flat car with the two trailers, then the instrument car, a baggage car and the caboose. We see here a panorama of the special test train standing in the station at Galesburg awaiting clearance for starting on the test run. This shows the flat car with the two trailers secured with the conventional tie-down method. The red car is the dynamometer or instrument car. The deflection of the springs was also measured on each side of one truck of the flat car from which could be determined at any instant the actual load transmitted from the car to each truck frame and the position of the resultant load with relation to the center of the track. This shows the method of fastening the rear wheels, and here is the trailer hitch used at the front part of the trailer for securing the trailer to the flat car.

Now as I mentioned previously there was a high-level accelerometer mast attached to one end of the flat car and this is a panorama showing the bipod arrangement with one pipe going up as a mast and one other serving as a brace. Then at the top part of this mast an accelerometer was placed for measuring the lateral displacements, and here we see a close-up of the top part of the mast with the Statham accelerometer in position. This roll gyroscope was used to measure the variation of the level of the car floor from a true horizontal position. This gyro is especially designed to minimize drift; it gives measurements of roll angle to an accuracy of about 15 min.

The various recording instruments were placed in the dynamometer car. This is a two-element Brush oscillograph which recorded the deflections of the truck springs. Then two other six-element Brush oscillographs were used to record the accelerations and displacements at the various locations and the roll from the two gyros. The boxes that have the red light in them are amplifier units that take the signals from the accelerometers and amplify them enough so they will actuate the recording elements in the Brush oscillographs. These pictures were taken by E. E. Mills of the Pennsylvania Railroad, a member of the AREA Clearance committee and subcommittee chairman on this assignment. We want to express our appreciation to him for his interest in taking the time not only to take the pictures but also to develop, edit and prepare this film for your viewing.

This is another view in the dynamometer car taken while the train was standing still. Here we have quite an interesting picture looking forward from the dynamometer car and showing the movement of the trailer body while rounding a curve on the branch line turn around run. You will notice that the trailer is quite steady. Here you get a similar view showing the motion of the trailer with the flat car passing through a cross-over. No pictures were taken on the return trip from Galesburg to Chicago. To supplement the running tests, a static lean test was made in the Burlington shops at Aurora with the wheels blocked up on one side of the flat cars by different amounts and the amount of lean of the trailer determined for each amount. These pictures show the car and trailer with the equivalent of 6 in superelevation with the car standing still. The amount of lean in this particular view looks to be more than it actually was because the picture itself was taken at a little bit of an angle. Here is shown a level bar being used to measure the tilt or inclination of the car floor in the static tests; the same procedure was used to measure the tilt of the trailer floor. Now by having static measurements with 4 in and 6 in equivalent unbalanced elevation it was possible to correlate these measurements with the measured displacements with the train in operation with corresponding amounts of unbalance.

Upon completion of the tests with the trailer-on-flat-car a test run was made with a three-level auto-carrier car. This test was run on a regular high-speed freight between Chicago and St. Paul on the Burlington. Here we see the auto carrier with 12 large automobiles and the dynamometer car waiting to be coupled into the train for making the run. Pictures taken here from the instrument car looking ahead towards the auto carrier car give a good idea of the relative motions that occur; you will note particularly that there was considerable rocking of each automobile body on its own springs that was different from the general movement of the rack on the carrier car. Insofar as clearances are concerned, it is of course obvious that the clearance problem, if any, would be with the rack itself, as the auto bodies are much narrower than the flat car and in themselves would not offer any problem insofar as side clearance is concerned.

These tests were conducted in collaboration with the Burlington Railroad under the sponsorship of the AAR Joint Committee on Relation Between Track and Equipment and AREA Committee 28—Clearances. The determination of clearance requirements is a complicated subject. It involves play between the track and wheel gage; play between component parts in the freight car; weight, height of center of gravity, and spring characteristics of the car; effect of irregularities in track alinement and surface, as related to speed, on car roll and oscillations; amount of overhang and effect of unbalance on curves; and the physical dimensions of the car itself.

These tests were necessarily limited in scope to one type of trailer car and did not include many tests at speeds resulting in higher amounts of unbalance on curves which would be required to establish maximum limits for extreme conditions. However, it may be stated that in these tests with the type of equipment used there was no unsafe condition noted in tests with the trailer-on-flat-car on main-line track up to a top speed of 83 mph on tangent and $4\frac{1}{2}$ in unbalance on curves; on branch line up to a top speed of 63 mph on tangent and almost 4 in unbalance on curves; and with the three-level auto carrier up to a speed of 70 mph on tangent and up to 1 in unbalance on curves.

Because of the nature of the tests, it seemed advisable to restrict this report to railroad personnel. Copies of the report have therefore been furnished to chief engineering officers and chief mechanical officers of Member Roads and to the particular committees concerned, and no other distribution will be made.

In evaluating the data obtained in these tests, it should be kept in mind that the cars and trailers used were in very good condition. Trailers handled on flat cars should be in good condition and it should be known that the trailer springs are functioning satisfactorily and that tires are properly inflated. It is believed that the report contains data of interest in the design and operation of this type of equipment and it is hoped that similar tests can be made on other types of trailer cars at a later date.

(Text continued on page 631)

Fig. 1—Truck trailers on flat car in special train, branch line run.

Fig. 2—Instrument car and trailers-on-flat-cars in regular freight train, main line run.

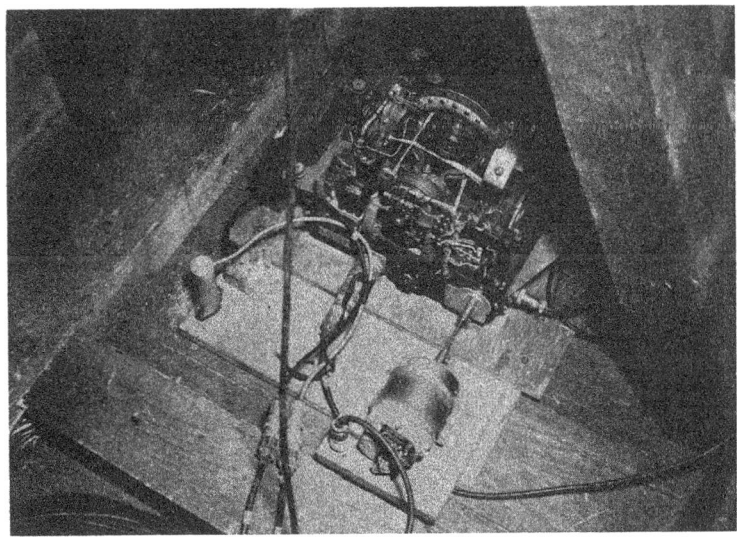

Fig. 3—Trailer gyroscope and lower accelerometers.

Fig. 4—Truck of trailer car, showing spring deflection potentiometer.

Fig. 5—Upper accelerometer mast for flat car.

Fig. 6—Recording oscillographs and amplifiers in instrument car.

Fig. 7—Triple-deck auto carrier in regular freight train, main line.

Fig. 8—Static lean test, 6-in oak shims under one side of car.

Fig. 9—Level for measuring car angle in static lean tests.

In closing I would like to express our appreciation to E. J. Brown, chief engineer, and J. D. Rezner, mechanical assistant to the vice president, of the Burlington Railroad and members of their staff for their active participation in conducting the tests. From our staff, Randon Ferguson, electrical engineer, was in direct charge of the tests; M. F. Smucker, assistant electrical engineer, supervised the testing, and Ralph Schinke, stress analyst, assisted in preparing the report. [Applause]

CHAIRMAN GREENLEE: On behalf of myself and the entire committee, as well as the Association, I would like to thank Mr. Magee for his splendid talk.

Mr. President, this concludes the report of Committee 28—Clearances.

PRESIDENT BEEDER: Thank you, Mr. Greenlee. Your committee is carrying on valuable work on behalf of the Association and the railroads in these days of higher speeds and not infrequent heavier and higher loadings, and we appreciate the work which it is doing.

With piggyback operations on the increase on practically all of our railroads, we particularly appreciate the studies that your committee is making through Mr. Magee and members of his staff, and I want to add my thanks to Mr. Magee for his interesting presentation here on the data collected during piggyback runs on the Burlington. We hope that further test runs can be made, as necessary, to complete the information which we will need in order to handle this new type of business most expeditiously with maximum safety.

Mr. Greenlee, you are excused, with our thanks to you and your committee. [Applause]

Discussion on Engineering and Valuation Records

[For report, see Bulletin 569, pp. 427–456]

PRESIDENT BEEDER: Our next report will be that of our Committee 11—Engineering and Valuation Records, of which L. W. Howard, land and tax commissioner, Chicago & Western Indiana Railroad, and Belt Railway of Chicago, is chairman. I shall be glad if Chairman Howard and the members of his committee will come to the platform at this time and present their report.

Assignment 1—Revision of Manual.

CHAIRMAN L. W. HOWARD: Mr. President, members and guests:

Your Committee 11 has completed its assignment on the subject of revision of the Manual for the current revision period. The subcommittee chairman, W. A. Krauska, assistant engineer, Missouri Pacific Railroad Company, could not be here today, so I shall ask E. W. Smith, assistant to chief engineer, St. Louis–San Francisco Railway, to make the report on Assignment 1. Mr. Smith is no stranger to this report, as his Subcommittee 3 reviewed Part 4 of the Manual and the report incorporates its recommendations in connection therewith.

E. W. SMITH: Your committee offers the following recommendations with respect to Part 2, Chapter 11 of the Manual:

Pages 11–2–1 to 11–2–4, incl., Construction Reports and Property Records—Relation to Current Problems. Reapprove with revisions shown on pages 430 and 431 of Bulletin 569.

Pages 11–2–4 to 11–2–7 incl., Authority for Expenditure. Reapprove with revisions shown on page 431 of Bulletin 569.

Pages 11–2–8 and 11–2–9, incl., Detailed Estimate. Reapprove with revisions shown on pages 431 and 432 of Bulletin 569.

Pages 11–2–10 to 11–2–21 incl., Register of Authorities for Expenditure; Time Roll-Labor-Monthly; Track Foreman's Daily Material Report; Foreman's Monthly Material Report—Bridges; Foreman's Bridge Section Tool Report; Foreman's Daily Report of Work Train Performance; Conductor's Daily Report of Work Train Performance. Reapprove all these documents without change.

Pages 11–2–22 to 11–2–29 incl., Roadway and Structures—Records and Reports, Reapprove with revision shown on page 432 of Bulletin 569.

Pages 11–2–30 to 11–2–48 incl., Bridge Construction Reports; Monthly Track Material Report; Report of Quantities in Completed Work; Cost of Property Retired; Roadway Completion Report; Equipment Completion Report; Record of Ballast Changes; Progress Profile. Reapprove all these documents without change.

Mr. President, I so move.

[The motion was duly seconded, was put to a vote, and carried.]

MR. SMITH: Your committee offers the following recommendations with respect to Part 3: Pages 11–3–1 to 11–3–19 incl., Cost Accounting Methods, Statistical Records and Forms for Analyzing Expenditures for Assistance in Controlling Expenditures. Reapprove with revisions shown on pages 433 and 434 of Bulletin 569.

Mr. President, I so move.

[The motion was duly seconded, was put to a vote, and carried.]

MR. SMITH: Your committee offers the following recommendations with respect to Part 4: Pages 11–4–1 to 11–4–9 incl., General and Text. Reapprove with revisions shown on pages 434, 435 and 436 of Bulletin 569.

Pages 11–4–10 to 11–4–24, Graphical Symbols. Reapprove with the revisions shown on page 436 of Bulletin 569.

Pages 11–4–25 and 11–4–26, Methods of Folding Drawings; pages 11–4–27 to 11–4–31, incl., Specifications for Preparation of Maps and Profiles. Reapprove without change.

Mr. President, I so move.

[The motion was duly seconded, was put to a vote, and carried.]

Assignment 2—Bibliography on Subjects Pertaining to Engineering and Valuation Records.

CHAIRMAN HOWARD: The report on Assignment 2 will be given by Subcommittee Chairman J. Bert Byars, assistant to chief engineer, Denver & Rio Grande Western Railroad.

J. BERT BYARS: The bibliography report made each year by Committee 11 is for the use and benefit of all members of the AREA and other railroad personnel.

I should like to encourage all of you and your associates to take advantage of the material compiled in these reports. This material pertains to railroad accounting, depreciation, valuation, and other items related thereto. Knowledge of these subjects is becoming more vital each year in the operation and maintenance of railroad properties. Valuation is used for depreciation purposes and tax bases as well as setting up tariff rates.

A great deal of time and money is spent each year by the railroad industry in compiling and maintaining property records, and making the necessary reports to the Interstate Commerce Commission. These vital statistics, as they may be called, require the services of many capable experienced men, and the use of the finished product is of great importance to the railroad supervisory personnel.

The ever-constant emphasis on economies in railroad operation requires each officer and supervisor to have a knowledge of the effect of valuation and accounting procedures. I believe that a study of the bibliography report of Committee 11 each year will prove beneficial in this respect.

Assignment 3—Office and Drafting Practices.

CHAIRMAN HOWARD: This subcommittee has submitted a progress report which appears on page 446 of Bulletin 569. Subcommittee Chairman E. W. Smith, assistant to chief engineer, Frisco, has already been up here to report on Assignment 1, but I wish to thank him for the energetic way he has attacked his assignment during his first year as subcommittee chairman.

Assignment 4—Use of Statistics in Railway Engineering.

Assignment 5—Construction Reports and Property Records.

CHAIRMAN HOWARD: Assignments 4 and 5 have been combined into a single assignment for 1962, with co-chairmen. I would like to have these co-chairmen rise and be recognized. They are W. J. Pease, assistant general auditor, Illinois Central Railroad, and H. R. Williams, valuation engineer, Union Pacific Railroad.

Assignment 6—Valuation and Depreciation.

The report on Assignment 6 is published on pages 447–449 of Bulletin 569 and includes, in addition to current statistics regarding the work of the Interstate Commerce Commission, a valuable summary of the report of the Committee on Valuation of the National Association of Railroad and Utilities Commissioners. The chairman of this subcommittee, C. R. Dolan, assistant to engineer of capital expenditures, Missouri Pacific Railroad, had a prior commitment in Washington, D. C., so in his absence I shall present his report as a progress report and express to him the thanks of the committee.

Assignment 7—Revisions and Interpretations of ICC Accounting Classifications.

CHAIRMAN HOWARD: The report on Assignment 7 appearing in Bulletin 569 covers the status of current ICC subjects being studied by the industry and the Commission, and revisions in the ICC Uniform System of Accounts. The chairman of the subcommittee is M. M. Gerber, accounting engineer, Baltimore & Ohio Chicago Terminal Railroad.

B. H. MOORE [AAR]: Mr. Gerber, has your subcommittee knowledge of the fact that the new accounting classification was released by the Interstate Commerce Commission during January and that some changes have been made from those previously published?

M. M. GERBER: Yes, there were some changes, among which were changes in accounting for rail replaced by rail of lighter weight. Formerly the heavier rail was retired at current prices. The new accounting provides for retirement of the excess weight at cost. The new classification can be obtained from the Superintendent of Documents, U. S. Government Printing Office, Washington, D. C., at 65 cents per copy.

CHAIRMAN HOWARD: Mr. Gerber, I presume if these changes were made after publication of your report this year, you will report them in next year's report, will you not?

MR. GERBER: I shall do so.

Assignment 8—Instructions for Making Engineer Field Checks and Their Application to Completion Reports.

Assignment 9—Simplification of Annual Reports on Form 588 to the Interstate Commerce Commission, and Underlying Completion Reports.

CHAIRMAN HOWARD: Subcommittees 8 and 9 are holding up work awaiting disposition of Valuation Order 30 by the Interstate Commerce Commission. C. F. Olson, valuation engineer, Great Northern Railway, is chairman of Subcommittee 8. H. N. Halper, retired valuation engineer, Erie–Lackawanna, was chairman of Subcommittee 9 until this year, and Committee 11 would like to express its appreciation for his valued services. F. A. Roberts, his successor as valuation engineer of the Erie–Lackawanna, will be chairman of Subcommittee 9 for 1962.

CHAIRMAN HOWARD [continuing]: The committee's report for 1962 includes memoirs for three of our members—Charles B. Martin, vice chairman-designate at the time of his death, James H. Roach, and Robert C. Watkins, and it is my sad duty to add to those memoirs one reporting the death on January 18, 1962, of George J. Harris. The committee requests its inclusion in the Proceedings of the Association.

MEMOIR

George J. Harris

It is with deep sorrow that Committee 11 records the death on January 18, 1962, of George J. Harris, who became a member of the committee in 1959.

Mr. Harris was born in Washington, D. C., on January 14, 1906, and after graduating from Eastern High School, he entered the service of the Southern Railway in 1923. After serving in various capacities in the operating and accounting departments for a period of 21 years he became auditor of construction in 1954. In 1959 he joined the staff of the Association of American Railroads as assistant to the vice president, Finance, Accounting, Taxation and Valuation Department.

While Mr. Harris was a member of the committee but a short time, he brought to it a wealth of knowledge in capital expenditure accounting, and depreciation and tax procedures. He participated actively in the affairs of the committee and his early passing is a source of regret to his many friends on the committee.

He served in the U. S. Army (1943–1945) during World War II.

Interment was in Arlington National Cemetery, Va. Surviving him are his wife, Beulah, and a son, George, Jr., of Bethesda, Md.

B. H. MOORE,
W. S. GATES, JR.,
Committee on Memoir.

CHAIRMAN HOWARD [continuing]: At the conclusion of this Annual Meeting I am stepping down to the status of committee member, and I should like at this time to express my appreciation to all the members of Committee 11 who have placed in me their trust to serve as their chairman for the past three years. The committee has responded to the demands placed upon them in an unselfish and pleasant manner, and the pleasure of serving as their chairman has been all mine.

I also wish to extend to the officers of the Association and the executive secretary and his staff thanks for all the help and courtesies rendered to me. I heartily recommend that all members of the Association seek service on its committees as chairmen, subcommittee chairmen, or working committee members. I can guarantee that their efforts spent in such matters will return untold dividends.

Mr. President, at this time I should like to present the new vice chairman of Committee 11, H. R. Williams, valuation engineer of the Union Pacific Railroad, who will please rise and be recognized. [Applause]

Mr. President, I should now like to present to you Milton C. Wolf, valuation engineer, Northern Pacific Railway, the new chairman of Committee 11. The work of your committee is in good hands. [Applause]

PRESIDENT BEEDER: Thank you, Mr. Howard, and your committee, for the important work which it has carried on during the last three years under your direction We especially appreciate the detailed attention which your committee has given to the review of its material in our Manual during the past two years in order that it might be completely up to date.

We welcome Mr. Williams as the new vice chairman of Committee 11 and Mr. Wolf as the new chairman, knowing that the past good work of the committee will continue under their direction.

Mr. Wolf, if you will step to the podium I should like to present you with your official chairman's gavel. I am sure you will use it wisely and effectively in the years immediately ahead. [Applause]

Thank you again, Chairman Howard. Your committee is now excused with the thanks of the Association.

Discussion on Contract Forms

[For report, see Bulletin 567, pp. 167–175]

PRESIDENT BEEDER: Our next committee to report is Committee 20—Contract Forms, the chairman of which is D. F. Lyons, office engineer, Chicago South Shore & South Bend Railroad, Michigan City, Ind. Will Mr. Lyons and the other members of his committee please come to the platform and present their report. Again may I repeat my earlier invitation for comments and discussion from the floor.

Mr. Lyons, you may proceed with your presentation.

CHAIRMAN D. F. LYONS: Mr. President, fellow members and guests.

Committee 20 has six assignments. We are reporting on three of them and have a brief comment to make on a fourth. You will find our report covered in Bulletin 567 beginning on page 167.

Assignment 1—Revision of Manual.

CHAIRMAN LYONS: W. D. Kirkpatrick, chairman, Branch Line Committee, Missouri Pacific Railroad, is chairman of the subcommittee handling Assignment 1—Revision of Manual. However, due to the size of the task, it has been handled by divisions of the entire committee.

J. L. Perrier, division engineer, Chicago & North Western Railway, who is chairman of Subcommittee 1 (d) and vice chairman of Committee 20, will report at this time.

J. L. PERRIER: Mr. President, fellow members and guests:

Last year Committee 20 divided its membership into seven subcommittees to review Chapter 20 of the Manual, and several of these committees presented their reports at the 1961 convention. The following report completes the review

We recommend reapproval without change of the following Construction Agreements in Part 1, Chapter 20 of the Manual:

> Form of Bond.
> Form of Cost-Plus Percentage Construction Contract.
> Form of Construction Contract for Minor Projects.

Mr. President, I so move.

[The motion was duly seconded, was put to a vote, and carried.]

MR. PERRIER: We recommend reapproval without change of the following Agreements Covering Passenger and Freight Facilities in Part 2, Chapter 20 of the Manual:

> Form of Agreement for the Organization and Operation of a Joint Passenger Terminal Project.
> Form of Agreement for Cab Stand and Baggage Transfer Privileges.
> Form of Agreement for Joint Use of Passenger Station Facilities.
> Form of Agreement for Joint Use of Freight Terminal Facilities.

Mr. President, I so move.

[The motion was duly seconded, was put to a vote, and carried.]

MR. PERRIER: We recommend reapproval of Form of Agreement for Wire or Cable Line Crossings, pages 20–3–27 to 20–3–30 incl., with one minor revision as published on page 169 of Bulletin 567, Vol. 63.

Mr. President, I so move.

[The motion was duly seconded, was put to a vote, and carried.]

MR. PERRIER: We recommend reapproval of Part 4, Chapter 20, Agreements Covering Track, with revisions to the following Agreements:

Form of Agreement for Trackage Rights
Form of Agreement for Industry Track
Form of Agreement for Crossing of Railways at Grade.
Form of License for Private Road Crossing.
Form of Agreement for Operation of Commissary and Boarding Outfits.
Form of Agreement for Furnishing Water from Railway Water Systems to Employees and Others.
Form of Agreement for Purchase of Water.
Form of Agreement for Placing Snow or Sand Fences Off the Railway Company's Property.
Form of Lease Covering the Use of Railway Tracks for Storage of Tank Cars Containing Liquefied Petroleum Gases, Anhydrous Ammonia and Other Flammable or Dangerous Materials.

The revisions are shown in detail on pages 169–173 inclusive of Bulletin 567, Vol. 63.

Mr. President, I so move.

[The motion was duly seconded, was put to a vote, and carried.]

MR. PERRIER: It is recommended that the following Agreements be transferred from Part 4 to Part 7—Miscellaneous Agreements.

Form of Agreement for Operation of Commissary Outfits.
Form of Agreement for Furnishing Water
Form of Agreement for Purchase of Water

The work of reviewing these parts of the Manual and preparing the report was done under the direction of the following subcommittee chairmen:

W. D. Kirkpatrick, chairman, Branch Line Committee, Missouri Pacific Railroad; Clarence Young, assistant engineer, Baltimore & Ohio Railroad; E. M. Hastings, Jr., wire crossing engineer—system, Chesapeake & Ohio Railway; and J. L. Perrier, division engineer, Chicago & North Western Railway.

This completes our review of the Manual.

Assignment 3—Form of Lease Covering Right to Strip-Mine on Railway Miscellaneous Physical Property.

CHAIRMAN LYONS: The report on Assignment 3 will be presented by E. M. Hastings, Jr., wire crossing engineer—system, Chesapeake & Ohio Railway. Mr. Hastings.

E. M. HASTINGS: Mr. President, Assignment 3 covers the preparation of a form of lease covering right to strip-mine on railway miscellaneous physical property. A draft of this form was published in Bulletin 560 on page 351 and was presented as information at the last convention. No adverse comments have been received.

I move that the aforementioned form be adopted by the Association and published in the Manual.

[The motion was duly seconded, was put to a vote, and carried.]

Assignment 4—Form of Agreement to Cover Disposal of Surplus Railway Property.

CHAIRMAN LYONS: The report on Assignment 4 will be presented by Subcommittee Chairman E. W. Smith, assistant to chief engineer, St. Louis–San Francisco Railway.

E. W. SMITH: Mr. President, members and guests. Assignment 4 covers the preparation of a form of agreement to cover disposal of surplus railway property. The committee has prepared a draft which appears on page 173 of Bulletin 567 as information. We invite your comments and suggestions.

Assignment 7—Bibliography on Subjects Pertaining to Contract Forms.

CHAIRMAN LYONS: K. J. Silvey, area engineer, Pennsylvania Railroad, is chairman of the subcommittee handling Assignment 7. He will be glad to receive any references to subjects that any of you may think may be of value for bibliography purposes.

I thank the subcommittee chairmen and members for their labors of the past year.

I now wish to introduce to you the new secretary and new subcommittee chairmen for the coming year. Please withhold your applause until all are named.

Our new secretary, Donald G. West, general industrial agent, Detroit, Toledo & Ironton Railroad.

Subcommittee 1 chairman, C. L. Gatton, assistant engineer, Louisville & Nashville Railroad.

For a new assignment this year, Subcommittee 3 chairman, R. C. Heckel, assistant engineer, grade crossings, New York Central.

Subcommittee 5 chairman, F. B. Mallas, division engineer, Northern Pacific. [Applause]

Mr. President, this concludes the report of Committee 20.

PRESIDENT BEEDER: Thank you, Mr. Lyons. You have done a good job in carrying forward the work of your committee during your first year as its chairman; and, as I said with respect to the work of Committee 11 earlier, we appreciate the extensive work done by your committee in reviewing its many Manual documents in order that, with or without revisions, they might be completely up to date in our Manual as to be reprinted later this year.

Your committee is now excused with the thanks of the Association. [Applause]

Discussion on Waterways and Harbors

[For report, see Bulletin 568, pp. 341–350]

PRESIDENT BEEDER: We shall hear next from our Committee 25—Waterways and Harbors, the chairman of which is J. F. Piper, assistant chief engineer—construction, Pennsylvania Railroad.

Mr. Piper, will you and the members of your committee please come to the platform and present your report.

CHAIRMAN J. F. PIPER: Mr. President, members of the Association and guests:

Your committee's work this year has been rather sparse as measured by the printed report, but a definite start has been accomplished in two fields: (1) the use of hydraulic models for the study and resolution of waterway problems, and (2) The study of railway terminal facilities for roll-on, roll-off, lift-on and lift-off operations.

As a matter of interest, this committee met last fall in Baltimore coincidental with the biannual meeting of the Permanent International Association of Navigation Congresses, in which a substantial number of the committee hold membership.

Assignment 1—Revision of Manual.

CHAIRMAN PIPER: Your Committee recommends that Part 1—Public Improvements, Their Costs and Benefits, and Part 5—Dredging Specifications, of Chapter 25, the current Manual, be reapproved in their present form and given a current date.

Mr. President, I so move.

[The motion was duly seconded, was put to a vote, and carried.]

Assignment 3—Bibliography Relating to Benefits and Costs of Inland Waterway Projects Involving Navigation.

CHAIRMAN PIPER: Subcommittee 3 has been annotating the bibliography relating to benefits and costs of inland waterway projects previously published, and this work with six additional references has been published in the current Bulletin. M. A. Michel, assistant to chief engineer, Pittsburgh & Lake Erie, is chairman of the Subcommittee.

Assignment 7—Relative Merits and Economics of Construction Materials Used in Waterfront Facilities.

CHAIRMAN PIPER: Subcommittee 7 presents two reports as information: (1) Azobé as a construction material and its comparison with greenheart in waterfront facilities; and (2) Service performance records of greenheart in docks and harbors of the United Kingdom. Both of these papers have been published in Bulletin 568. The subcommittee chairman is Shu-t'ien Li, professor, Department of Civil Engineering, South Dakota School of Mines and Technology.

Mr. President, I would like to introduce to the Association F. J. Olsen, incoming chairman of Committee 25. [Applause] He is resident engineer of the Baltimore & Ohio Chicago Terminal Railroad.

PRESIDENT BEEDER: Thank you, Mr. Piper. Your committee is one of the smaller committees of our Association, with the smallest chapter in our Manual. Possibly that is because of the highly specialized nature of its work, which is definitely of greater interest to some railroads than others; but we continue to believe that there is an important place in our Association for your committee, handling the engineering and construction phases of waterway and waterfront problems, without encroaching upon the interests and responsibilities of the General Waterways Committee of the AAR.

So, we hope that the work of your committee will continue aggressively under your new chairman, Mr. Olsen.

Mr. Olsen, if you will step to the podium I would like to present you with this chairman's gavel to assist you in the conduct of your committee's work in the three years immediately ahead. [Applause]

This completes our morning session; but before I recess the meeting for our Annual Luncheon I would like to remind you that our afternoon session will convene in this room immediately after the luncheon, with another interesting and intensive program of reports and special features.

The meeting is now recessed for the Annual Luncheon, which will be held in the Williford Room, directly across the corridor from this room.

[The meeting recessed at 11:20 am.]

Annual Luncheon
Williford Room
Friday, March 9, 1962

[The Annual Luncheon was held in the Williford Room, beginning at 12:00 noon, with a total of 572 members and guests in attendance. At the main speaker's table were seated the officers and directors of the Association and several special guests. At a long table immediately in front of the main speaker's table were seated the chairmen of the Association's 22 standing and special committees. In greeting those at the Luncheon, President Beeder said:]

PRESIDENT BEEDER: Members of the American Railway Engineering Association and guests—we appreciate your participating in our Annual Luncheon. Like the smaller size of our business sessions this year, the size of this Luncheon gathering is much smaller than usual, but for those of us who are here, we hope it will be just as enjoyable as in the past.

It may add to your enjoyment if I tell you now that, to permit more convention session time, we have no formal program for this Luncheon, and no speaker.

With this brief comment, I hope you will now relax and enjoy your luncheon.

[After the Luncheon, President Beeder introduced those at the speaker's table and then those at the chairmen's table. Following the introduction of the last committee chairman, President Beeder said:]

PRESIDENT BEEDER: To all of you committee charimen I want to express my personal thanks for your interest in behalf of our Association. To those seven of you who are completing your 3-year term of office and are stepping down, I wish to express our special appreciation for the service you have rendered. While your guiding hand in the work of your committees will be missed, your influence for good will continue with your continued membership on your committees, and we are confident that ·the work of your committees will go forward under the men who have been selected to succeed you.

When I told you at the beginning of our Luncheon that we had no formal program to follow our Luncheon, I was more correct, I expect, than some of you may have thought. In the first place, unlike previous years, I cannot at this time announce to you the results of our annual election of officers since, under our Constitution, the polls cannot close until noon of the second day of our convention. However, these results will be made known to you at our Closing Business Session tomorrow noon, when our new officers will be installed. I hope many of you will be present.

In the second place, in spite of the fact that our convention program shows this Luncheon ending up with a showing of the French National Railway's film, "The Modern Track", we have decided, in the interest of better viewing and your greater convenience, to shift that showing, immediately, to our meeting room across the hall—preceeding the reconvening of our afternoon session. So, with the hope that all of you had a pleasant luncheon together, and will enjoy seeing this film—which is all about rail welding on the French Railways—the Luncheon is now adjourned back to the Waldorf Room.

Afternoon Session, March 9, 1962

[The meeting reconvened at 1:30 pm, President Beeder presiding.]

PRESIDENT BEEDER: Thank you for moving back to this room a little early. The film you are about to see has been made available to us through the courtesy of the French National Railways and Past President Ray McBrian, director of research, Denver & Rio Grande Western Railroad, who brought it back from his recent trip to Europe. Since it will be readily evident to you, it is needless for me to say, I am sure, that it is narrated by an Englishman.

If you are now ready, Mr. Operator, the signal is "GO".

[The film was shown.]

PRESIDENT BEEDER: Gentlemen, that's how they do it in France. Our thanks to the French National Railways and to Past President McBrian for making this film available to us.

Gentlemen, I shall now officially convene our afternoon session.

Discussion on Yards and Terminals

[For report, see Bulletin 567, pp. 235-258]

PRESIDENT BEEDER: The first committee to report is Committee 14—Yards and Terminals, the chairman of which is A. S. Krefting, chief engineer, Soo Line, at Minneapolis, Minn.

If Mr. Krefting, along with his vice chairman, secretary and subcommittee chairmen, will come to the speaker's table, I shall be pleased to turn this microphone over to him.

While the committee is coming to the platform may I remind you that you have the privilege of the floor, to the extent that time will permit, to make comments or ask questions.

Mr. Krefting, you may proceed.

CHAIRMAN A. S. KREFTING: Mr. President, members of the Association and guests: Committee 14 during the past year has completed its review of Chapter 14 of the Manual, and has studied seven other assigned subjects. Reports on four of these subjects will be made at this time. These reports have been printed in Bulletin 567, pages 235–258, incl. Discussion from the floor is invited at the conclusion of each report.

The subcommittee chairmen have worked hard on their reports, and I am glad that they are here to present their reports at this time.

Assignment 1—Revision of Manual.

CHAIRMAN KREFTING: The report on Revision of Manual, exclusive of the revisions to Part 5 covering Scales, will be presented by Subcommittee Chairman F. E. Austerman, chief engineer, Chicago Union Station Company.

F. E. AUSTERMAN: Mr. President, Vice President Code, Chairman Krefting and members of the Association:

During the past two years Committee 14 has completely reviewed the material in Chapter 14 of the Manual. A number of changes in Parts 1, 2 and 3 of the Chapter were approved at the convention a year ago.

We have one additional minor correction to recommend in Part 2—Passenger Terminals, a number of modifications to recommend in Part 4—Locomotive Terminals, and we rewrote the section on "Stores Facilities, Including Reclamation, Scrap and Material Yards" to bring it into line with current practice.

Mr. President, I move that Part 2—Passenger Terminals, and Part 4—Locomotive Terminals, of Chapter 14 of the Manual, be reapproved with the changes shown in detail in Bulletin 567, and that the material on "Stores Facilities, Including Reclamation, Scrap and Material Yards", contained in Part 4 of the Manual, be replaced with the revised material printed in Bulletin 567.

[The motion was duly seconded, was put to a vote, and carried.]

MR. AUSTERMAN: Committee 14 also recommends the addition to the Glossary of four terms commonly used in yard and terminal work.

Mr. President, I move that the definitions for the four terms shown at the bottom of page 339 of Bulletin 567 be added to the Glossary.

[The motion was duly seconded, was put to a vote, and carried.]

Assignment 3—Scales Used in Railway Service, Collaborating with Committee 18.

CHAIRMAN KREFTING: The report on Assignment 3, including revisions to the Part of Chapter 14 of the Manual pertaining to scales, will be made by a member of the subcommittee, W. P. Buchanan, supervisor of scale inspectors, Pennsylvania Railroad.

W. P. BUCHANAN: Mr. President and members:

Our committee during the year made a complete review of Part 5 of Chapter 14 of the Manual which relates to scales. We feel that a number of relatively minor revisions should be made in this part of the Manual to bring it up to date.

Mr. President, I move that the rules and specifications in Part 5 of Chapter 14 of the Manual listed on pages 240 to 243, incl., Bulletin 567, be reapproved with the changes described in detail.

[The motion was duly seconded, was put to a vote, and carried.]

MR. BUCHANAN: Our committee has for a number of years continued a study of two-draft motion weighing. During the past two years representatives of our committee made tests on two such scales which are in railroad service. The results of these tests, together with records furnished by individual railroads on several other scales used for two-draft uncoupled motion weighing, are published in our report this year. The report also contains our comments as to the advantages and general requirements for scales to be used for two-draft uncoupled motion weighing.

In addition to our study on uncoupled motion weighing, the committee during the year took a leading part in arranging for a modification of the rule in the National Code of Rules for Weighing and Re-weighing Freight Cars, which prohibited weighing coupled cars in motion. The committee felt that the rule tended to hinder the possible development of scale equipment which might make coupled motion weighing feasible and was, therefore, undesirable.

At the recommendation of Committee 14, Sec. C of Rule 3 of the aforementioned code was revised to read: "Cars may be weighed in motion either coupled or uncoupled, only upon a weighing system, including scales, approach and leaving trackage, properly designed for weighing in this manner and in charge of a competent weighmaster."

The vice president—Operations and Maintenance Department, of the AAR advised chief operating offices of this change in a letter dated January 10, 1962.

The committee does not feel that sufficient data are available definitely to establish that any equipment already developed will consistently weigh coupled cars in motion with sufficient accuracy. The action taken by our committee in recommending the change in the rule was intended solely to make it possible to weigh coupled cars in motion if and when suitable equipment to produce sufficiently accurate weights becomes available.

Mr. President, this report is submitted as information.

PRESIDENT BEEDER: Thank you, Mr. Buchanan. The report will be so received.

Assignment 4—One-Spot Car Repair Facilities, Collaborating with Committee 6.

CHAIRMAN KREFTING: The report on Assignment 4 will be made by Subcommittee Chairman F. S. King, district engineer, Pennsylvania Railroad.

F. S. KING: Mr. Chairman, this report is for information only.

Many railroads, in an effort to improve their operation and at the same time reduce expenses, have installed one-spot repair track facilities. The basic principle of this system of car repairs is the moving of cars to the repair facility or facilities, rather than men carrying tools and materials to cars requiring repairs. The rate of repairing cars requiring the light type of repairs ranges fom 10 to 20 cars per spot per 8-hr trick.

Railroads contacted have reported actual annual savings in car repair costs on a single-spot, single-track operation from $20,000 to $50,000, compared to the old-style rip-track operation. Other benefits result from less delay to loaded cars, reduced switching costs, and better working conditions for shop employes.

Our subcommittee report on this assignment appears in Bulletin 567, November 1961. We recommend that the subject be discontinued.

Assignment 5—Applications of Car Retarders at Locations in Yards Other Than on a Hump.

CHAIRMAN KREFTING: The report on Assignment 5 will be made by Subcommittee Chairman B. E. Buterbaugh, special engineer, St. Louis–San Francisco Railway.

B. E. BUTERBAUGH: The report of your committee on this subject was published in Bulletin 567, November 1961, page 254.

Several railroads are testing the retarders described in the report, making installations on pull-down ends of classification tracks, receiving and departure tracks, and on car repair and car-cleaning tracks. Preliminary reports indicate satisfaction with their use under proper conditions, with some savings being realized.

This report is presented as information with recommendation that the subject be discontinued.

Assignment 7—Waterfront Terminals.

CHAIRMAN KREFTING: The report on Assignment 7 will be made by Subcommittee Chairman J. J. Tibbits, office engineer, Erie–Lackawanna Railroad.

J. J. TIBBITS: Mr. President, Mr. Chairman, members and guests:

Under Assignment 7 your committee has submitted a report on car float facilities as one phase of waterfront terminals. The report appears on pages 255 through 258 of Bulletin 567.

The report covers in a descriptive manner the principal kinds of car float and car ferry facilities in use by railroads in the United States.

No recommendations in regard to specific types of transfer devices, fender construction or the design of supporting yards have been offered in the report. The determination of such features must be governed solely by a thorough engineering study of the characteristics of the waterway to be crossed, the topography and subsoil conditions of the terminal sites on each shore, and the type and volume of traffic to be handled.

This is a progress report, submitted as information.

CHAIRMAN KREFTING: I am glad to be able to present to you at this time two special features which, while they have not been directly arranged for by this committee, do apply to its field of study.

The first speaker, A. V. Johnston, needs no introduction to this group, as he is an active member of this Association and served as a director during the years 1957 to 1960. Mr. Johnston, in his capacity as chief engineer of the Canadian National Railways, has supervised the design and construction of several modern yards recently completed and others now under construction. Our committee last June had the pleasure of inspecting the fine yard recently completed at Montreal.

Mr. Johnston will address you on the subject, "Latest Developments in Classification Yards". Mr. Johnston.

Latest Developments in Classification Yards

By A. V. JOHNSTON
Chief Engineer, Canadian National Railways

Mr. President, members and guests:

In considering my approach to the subject of "Latest Developments in Classification Yards", it occurred to me that the material could be divided into two main categories. The first would include new developments and innovations which are being built into yards at the present time; the second area of discussion would cover new features and concepts which are being planned or considered for future yards.

Some gravity classification yards, both manual and automatic, have been built and others are under construction in the British Isles, in Western Europe and other parts of the world, but in the time available to me I propose to deal almost exclusively with yard developments on the North American Continent, which I think will be of paramount interest to this group.

Although most new developments are not exclusive to any one particular yard or railroad, I am more familiar with our own installations and having convenient access to records, statistics and results, my remarks, in general, will refer to and be illustrated by details of major new yards recently completed or still under construction on the CNR.

The Canadian National is in the midst of a vast yard improvement program involving the construction of four major fully automatic classification yards. Two of these yards are now in operation at Moncton, N. B., and Montreal, Que., and two others are under construction. Symington Yard near Winnipeg, Man., will be ready for operation in the late summer of 1962 and a new yard on the outskirts of Toronto, Ont., will be ready in 1965. All four hump yards are designed to attain the same humping rate of 5 cars per min. All of the yards have essentially the same track profile from crest of hump to clearance point in order that the desired humping rate may be readily sustained. Since some of the hump leads fan out to as many as 84 classification tracks, the hump grades are of utmost importance in controlling the speed and separation of cars.

Cars leave the crest of hump on a steep "kick-away" grade of 6 percent so that speeds of up to 12 mph are attained in the first 125 ft. A measuring section of 1¾-percent grade is then traversed before the car enters a master retarder on another 6-percent grade. With distance from crest to clearance as great as 1450 ft, a second set of retarders is necessary for the final adjustment of speed and separation so that each car reaches its destination at a safe coupling speed.

All hump yards are equipped with automatic sensing devices which supply the computer with the actual position of the last stationary car on each classification track. The grades on the body tracks in these yards are 0.1 percent or less (predominantly 0.08 percent) as a further aid to control the speed and separation of cars. These non-accelerating body track grades permit closer control over the exit speeds from the group retarders so that cars are more accurately positioned on the classification track. The general trend is toward flat classification tracks, although we have found that in areas where heavy and frequent snowfall is prevalent, very flat grades may be a disadvantage.

The three major hump yards at Montreal, Winnipeg and Toronto are designed for dual-hump operation, a system whereby two separate trains can be humped side by side. Moncton Yard is a single-hump operation and is considerably smaller than the other three. In the dual-hump operation, the main classification yard can be used as a single

Slide 1.

yard or as two separate yards when required. When operated as a single yard, the automatic switching system is capable of routing cars into the entire yard. A second push-button machine is equipped to route cars into half of the yard on occasions when the main classification yard is used as a double yard.

Slide 1 is a view of the retarder control machine located in the next to top floor of the retarder tower in Montreal Yard. You will notice there were two operators on duty and the yard was being operated as a dual-humping double yard at the time this picture was taken. Slide 2 shows the Montreal Yard plan layout.

This yard is referred to in the promotional literature as the largest automatic freight classification yard in the world, and perhaps this is true as I have no evidence to the contrary. However, what I think is more important and unique is that this yard has dual main humps and also a second single hump used for sorting local and wayfreight traffic. Cars destined for local delivery in the Montreal area are first classified over the dual humps into a local receiving yard, which consists of 6 tracks in the main classification yard. From these 6 center tracks of the main yard, cars are pushed over the hump for final classification in the local yard, containing 40 classification tracks. I do not propose to describe all the features of the Montreal Yard, many of which are standard, but for those of you who may be interested, there is a quite complete description of this yard in the February 1962 issue of the International Railway Journal.

The main dual-hump area is encircled and marked "A" to the right of the screen [slide 2], the local hump area marked "C" is pretty much in the center, the car repair area marked "B" is in the upper center, the diesel shop and locomotive servicing area marked "D" is below and to the left of the local hump yard and the car-cleaning area marked "E" is in the upper left. Slide 3 is an enlargement of the dual-hump area which

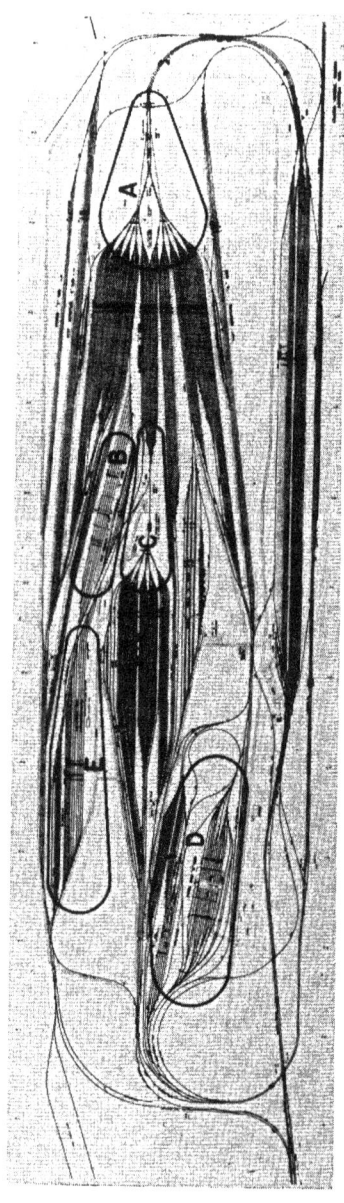

Slide 2.

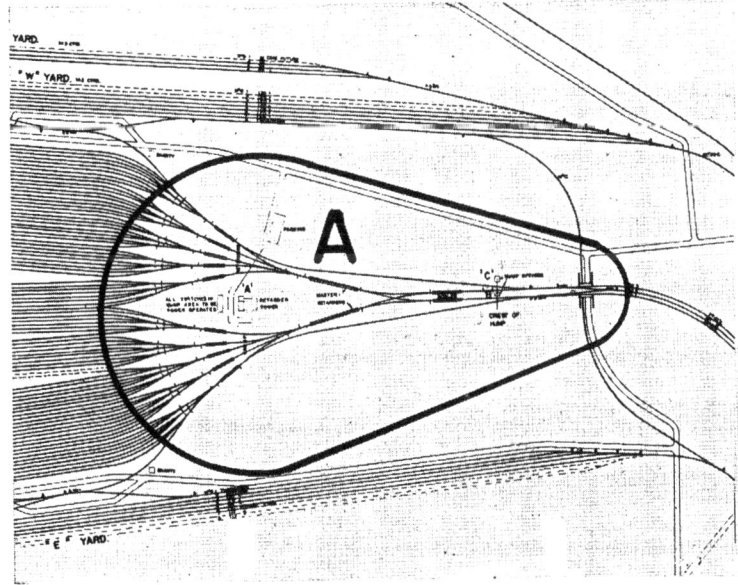

Slide 3.

shows the arrangement in more detail. Slide 4 shows an enlargement of the car-repair area above and the local hump area below marked "C". Slide 5 shows the diesel shop and locomotive servicing area and Slide 6 shows details of the car-cleaning area. Slide 7 [see Slide 2] shows again the overall yard plan.

The reason for the tandem hump arrangement is that Metropolitan Montreal, with a population of about 2 million people, is served by only two railways, so far as freight handling facilities are concerned. For this reason each railway has a very large number of industrial classifications and transfer runs originating in a major yard such as this. It was found impossible to provide sufficient classification tracks for all our require-ments connected to the dual main humps, and for this reason the concept of the tandem local hump evolved. There are 84 classification tracks connected to the main dual humps through 10 group retarders. The local yard consists of 40 classification tracks and 5 retarder groups. You will note the parallel arrangement of receiving, classification and departure tracks, which is the current trend, rather than a tandem arrangement of these facilities. However, the final decision in this respect at any specific location depends upon the dimensions of the available land.

Slide 8 shows a model of Montreal Yard. This model is made of molded fiberglass to accurate scales. The model is 6 ft wide by 25 ft long; the horizontal scale is 1 in equals 50 ft and the vertical is 1 in equals 10 ft. It cost less than $10,000 and was built primarily for training yard staffs. It has been most effective in this respect and has contributed a great deal to smoothing out the initial operation of the yard.

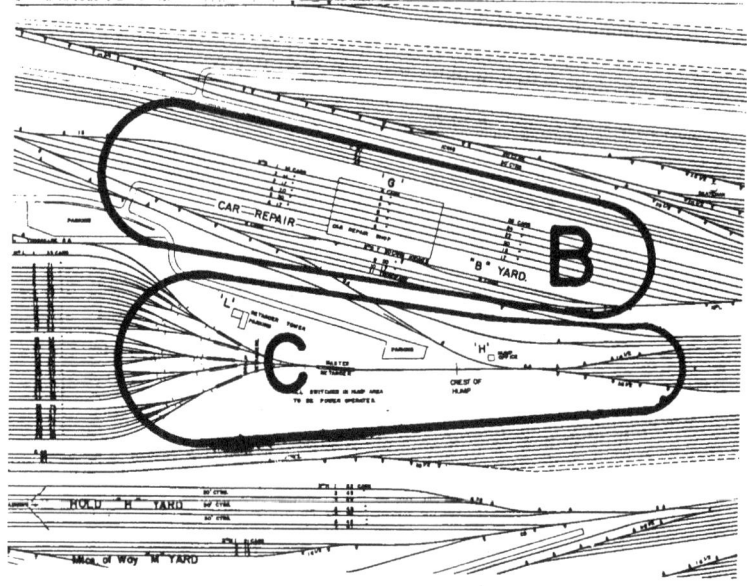

Slide 4.

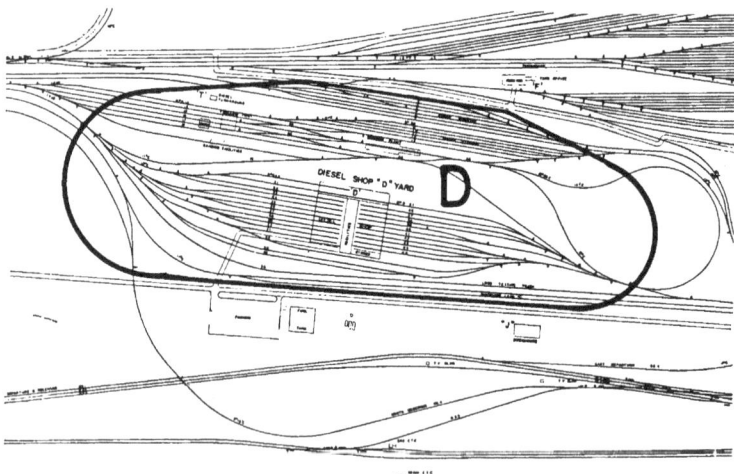

Slide 5.

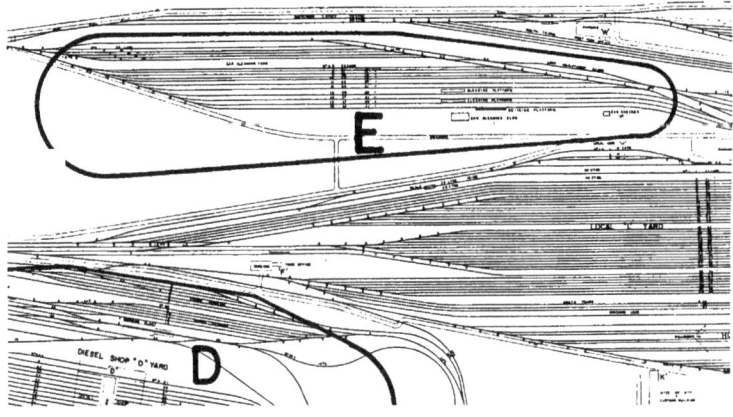

Slide 6.

Slide 9 shows the modern yard office building which houses the CTC control machine for all yard entrances and exits. It also houses the television closed-circuit monitors, the integrated data-processing equipment, special punched-card and teleprinter equipment, which I shall describe later, the yard model room, the clerical staff and the usual facilities for yard and road crews.

Advance consist lists are prepared from information received by teleprinter or transceiver for all inbound trains. The inbound clerk then adds the classification track number into which each car is to be humped and the sequence number. Slide 10 shows a keypunch operator punching the yard cards from one of these lists.

Slide 11 shows a close-up view of the advance consist on the left and several punched yard cards on the right which have been prepared from it. You will note that the classification track number for each car has been entered on the advance consist list in the second column from the right, and this track number is also shown on each card. Some of the track numbers are followed by the letter "L", which designates a track in the local yard.

In order to transmit hump lists and switch lists from the main yard office to the hump and departure towers concerned, a key-punch, intercoupler-teletype system has been developed for all Canadian National hump yards. This highly versatile machine will convert 12-impulse code-punched card language to the 5-impulse code required for a teleprinter to produce a typewritten hard copy, and these typewritten lists can be produced simultaneously at any number of locations where the information is required.

Slide 12 shows a card punch machine, an intercoupler and a teleprinter in operation. This intercoupler arrangement can also be used to transmit copies of switch and hump lists from the yard to city railway offices to provide customers with information on car locations. The development and use of the intercoupler has permitted the complete elimination of pneumatic tube systems from all four yards, with substantial savings in first cost, operation and maintenance.

Slide 8.

Slide 9.

Slide 10.

Slide 13 shows a battery of two teleprinters located in the hump building. The machine on the left is receiving a typewritten hard copy of the hump list for the train consist which we have been tracing through the yard. The teleprinter on the right is simultaneously producing car labels to correspond with the cars on the hump list. The production of these car labels is an ingenious by-product of the transmission of the hump list from the main yard office. A minor adjustment to the mechanism of the tele-printer, which is producing these car labels, permits it to advance nine lines each time the hump list advances one line so that duplicate information is typed on each simul-taneously. The Canadian National system is divided into 99 blocks and each block is subdivided into a series of zones. The block and zone number, printed in large numbers or letters in the spaces provided on the car label, easily identify the destination of the car. The large figures and letters on the card can be read by train and switch crews or other employees at a distance of 5 to 10 ft from the car, and they have immediate information for switching or routing at intermediate terminals, and placing at final destination.

Slide 14 shows a close-up of the punched yard cards on the left which are fed into the intercoupler mechanism at the yard office. In the center is the hump list for the train and on the right are some typical car labels. These car labels show, on the line at the top of the card, the car initials and number, the contents, the consignee, destina-tion, block number and zone number. The sole clerical effort involved in producing these car labels is copying the block and zone code from its special position at the top right-hand side of the punch card to the larger spaces provided in the body of the card.

(Text continued on page 655)

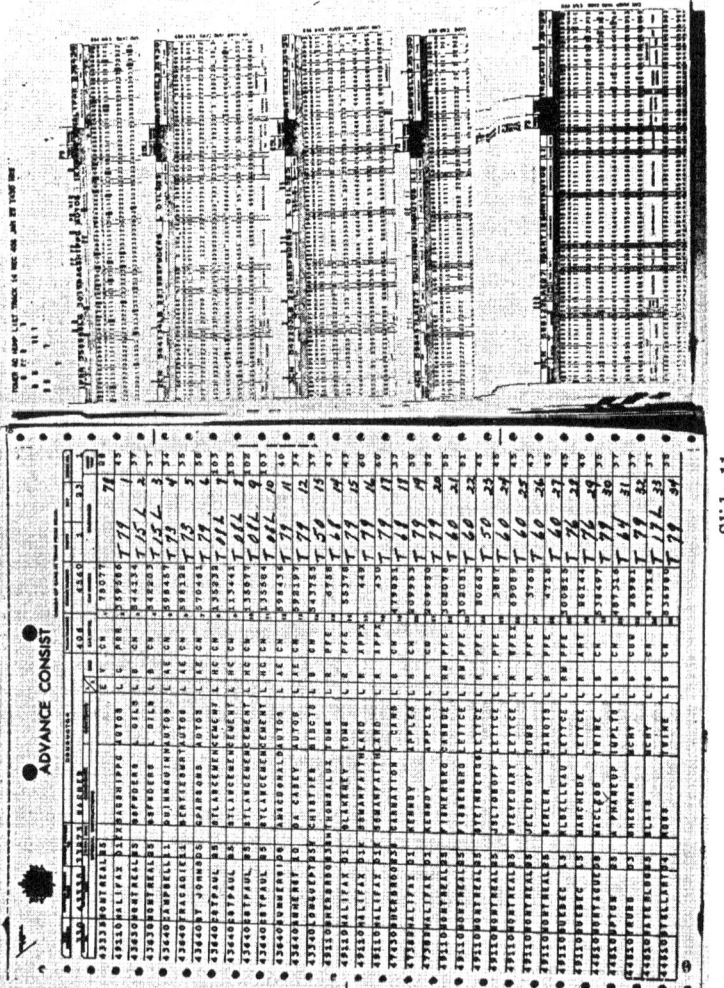

Slide 11.

Slide 12.

Slide 13.

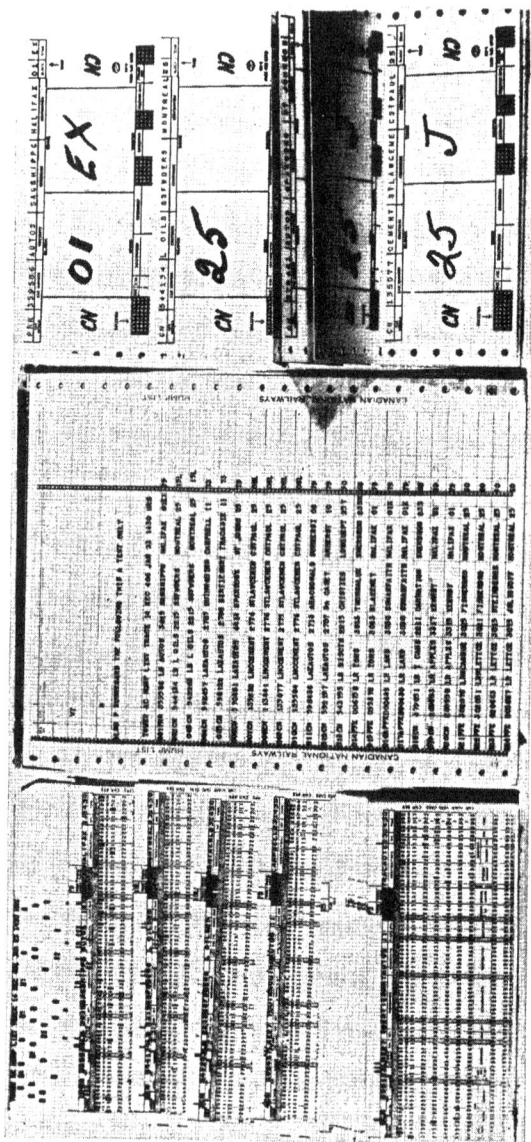

Slide 14.

The labeler then takes these cards and as the cars pass by him over the hump, he staples the corresponding label on each of the cars where it remains until the car reaches its destination.

Yard Lighting

In all four hump yards a system of mercury-vapor lights on high wooden poles has been adopted rather than the conventional method of using banks of floodlights on high towers.

In general the lighting arrangement consists of three or four luminaires, with a power consumption of 1000 w each, spaced at about 400 ft centers throughout the entire yard. Special consideration was paid to the illumination of classification tracks, retarder areas, leads, icing tracks and service tracks.

Inert Retarders

In all four yards, inert retarders are being used instead of skates and skate men. The retarders are located at the end of the body track grade just previous to a 0.3-percent decelerating grade. Inert retarders installed in the two yards which are now in operation are operating successfully and perform satisfactorily, within the limits of their design capacity, in preventing cars from fouling the leads or running through exit switches.

Slide 15 [not reproduced herein] shows one type of inert retarder with a retarder shoe on the inside of each running rail which bear against the inside of the car wheels. The mechanism of these retarders is practically all above the top of the ties, and very little clearance is required beneath the running rails for its installation and operation.

Slide 16 [not reproduced herein] shows a retarder of a different design. This model has retarding shoes on each side of the running rail which bear against both the inside and outside of the car wheel simultaneously. You will note that this retarder is considerably longer than the previous model and is mounted on one rail only, with a guard rail opposite. However, this type of retarder also comes in shorter lengths, and multiple units can be installed on the track either in tandem or parallel to each other. The energy-absorbing mechanism of this retarder is located principally below the top of the ties and requires greater clearance under the base of the running rail.

Communications

Two-way radio for yard-train crews throughout the yard and for carmen in the receiving and departure yards, is being used instead of paging speakers. The two-way radio system has proved to be most satisfactory in the existing yards, and improved installations, using all-transistor units, are planned for the yards at Winnipeg and Toronto.

All yard locomotives are equipped with four-channel radios for transmitting and receiving. Only one channel is open to the locomotive crew at any one time. Each radio unit has a locking device under the control of the locomotive foreman, and a change in the channel on any unit can be made only by him or his authorized representative. The four-channel radio units segregate radio communication in various parts of the yards but allow interchangeability of locomotives by simply unlocking and resetting the channel control.

Car Repair Facilities

In a large automatic classification yard it is essential to concentrate car repairs in one area, and because of climatic conditions we have found it necessary in our yards to have all car-repair work done under cover. The car-repair facilities at Montreal Yard are located adjacent to the local hump as shown in Slide 17 [see slide 4] with access from several tracks of the main classification yard. Cars for repairs are placed at the southerly end of the car repair building by a yard locomotive, and they are moved into the car repair building with a trackmobile as required. The repaired cars are taken out of the northerly end of the car repair building by the trackmobile 'and are then picked up when ready by a yard locomotive and returned to the main hump for reclassification.

Slide 18 [not reproduced] shows the trackmobile moving six cars into the southerly end of the car repair building. There are six tracks through this building and each has a working capacity of six cars under cover.

Slide 19 [not reproduced] shows the southerly end of the building and the general type of design.

Slide 20 [not reproduced] shows the easterly side of the building which contains the office space, woodworking shop and similar supporting facilities. This building is heated by infrared overhead units fired by natural gas. Although the air temperature within the building may be only 30 to 40 deg, depending upon the outside temperature and the frequency with which the doors are opened, working conditions are quite comfortable, and efficiency has been improved because of the overhead infrared heating system. All types of car repairs from the lightest to the heaviest can be done within this building, 'and all cars which are found in need of repairs within the yard are handled at this one central point.

Spot Car Cleaning

The actual number of tracks and platforms varies depending on local conditions. At Symington Yard, the 10-car platform is bounded by 2 long tracks on which some 45 cars await cleaning on each track, and a similar number of cleaned cars are ready for rehumping beyond the platform. A system of car pullers is used to position the cars at the platform so that a continuous flow of cars is maintained past the working point. A second platform is provided for refrigerator cars.

At Montreal Yard two platforms serving four tracks are used with a third platform for de-icing refrigerator cars.

Even with the arrangement of one platform, the efficiency and speed of car cleaning has been greatly improved by bringing the cars to a stationary, centralized, and well-equipped platform compared to the former system of cleaning, which consisted of special equipment mounted on trucks and taken to the stationary cars spotted on the cleaning tracks.

Buildings

I have shown you slides of a few buildings to illustrate specific functions of the yard and now I would like to point out some of the architectural details of other buildings in Montreal Yard.

Slide 21 [not reproduced] shows a general view of a three-story dormitory, operated by the YMCA for the accommodation of train and engine road crews.

Slide 22 [not reproduced] shows a close-up view of the wing of this dormitory, the lower level of which contains a cafeteria; the upper level is a lounge for employees off duty.

Slide 23 [not reproduced] shows the reception desk used for room assignment and general administration of the building.

Slide 26.

Slide 24 [not reproduced] is an interior view of the cafeteria. This cafeteria is open 24 hr a day and serves meals to all railway employees within the yard.

Slide 25 ([not reproduced] is a partial view of the lounge area.

Slide 26 shows the exterior and general architectural shape of the retarder tower. This building houses the complete electronic equipment for the main classification yards, including the master computer, communications facilities, the retarder control console located in the next to top floor of the building and the yardmaster's tower at the top. The two expanded floors and the next two floors within the tower itself contain all of this equipment. At the present time, we are considering the advantages of locating all the equipment on one floor at ground level in future retarder control towers. The tower shaft space would be unused from the ground floor level to the retarder control room except for wiring and stairways. This would permit the installation of a rather simple type of man-lift for the limited number of employees who would be required to go up to the two top floors.

Slide 27 is an architectural sketch of the-type of building envisaged along these lines for the new yard at Toronto.

Future Developments

The modern concept of an automatic classification yard became possible in the mid-1950's with the increasing availability of electronic computers. Further major developments appear to have remained on a plateau from that time except for refinements in the system, such as the use of one master computer to control the entire operation compared to several computers, each of which was required to control a specific group of

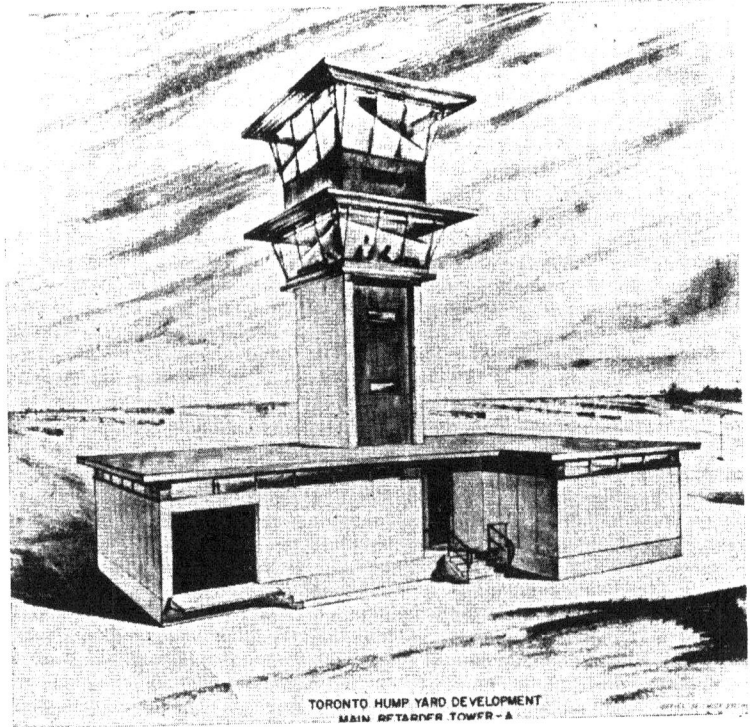

TORONTO HUMP YARD DEVELOPMENT
MAIN RETARDER TOWER - A

Slide 27.

retarders in the initial installations. The control data produced by the master computer must be applied through a system based on application relays and other equipment with physical moving parts.

Slide 28 shows two roll-out racks, the top one of which is the master speed-control panel and the lower one the car space-motion detector in a modern installation. You will note the printed circuits and the plug-in type of equipment in these racks.

Slide 29 shows a track diagram panel of the hump and retarder area which is located in the retarder relay room. The progress of all cars over the dual humps can be followed on this track diagram, and, when necessary, a car movement can be simulated over either hump and into any classification track to locate the source of possible trouble. This simulated operation does not produce any interference with the actual operation of the humps.

Slide 30 shows a bank of plug-in-type application relays. The equipment shown here is only a small portion of that actually required in connection with the operation

Slide **28.**

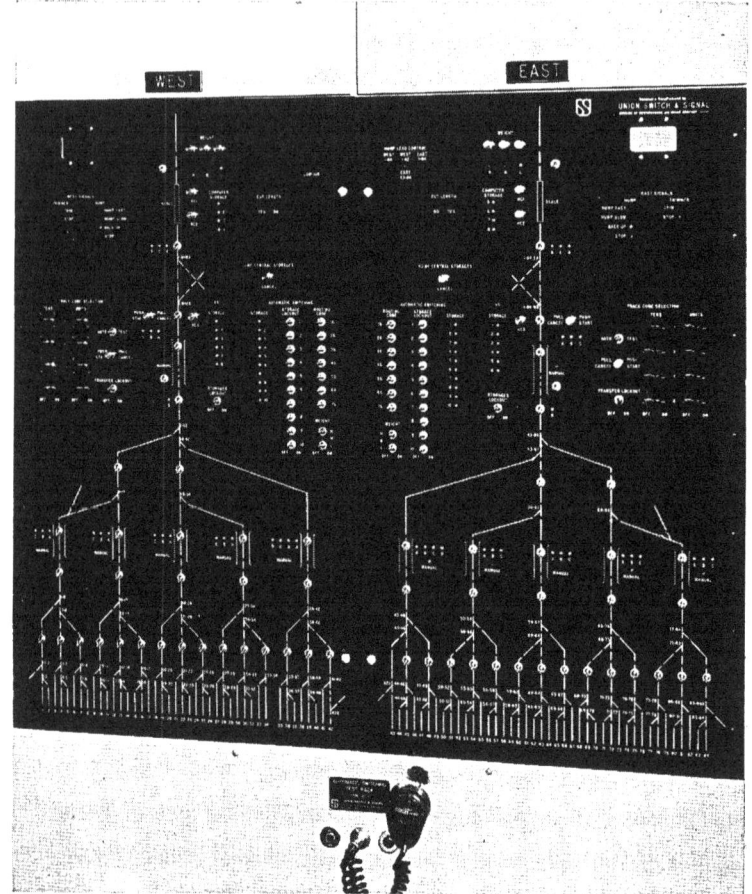

Slide 29.

of a dual hump automatic classification system. I believe the next major advance in this field will be the use of solid-state elements such as transistors, semi-conductor diodes and ferrite cores which will reduce the sheer volume of equipment required as dramatically as the application of the transistor to produce the pocket-size radio. The application of solid-state elements will eliminate all moving parts except those actually required. to change the signal aspects, actuate the retarders, and throw the switches.

One of the most important requirements for the future is more accurate information on the rolling qualities of individual cars. It has been said to me that the rollability of

Slide 30.

a freight car is less predictable than the flight path of a guided missile. A number of things have been tried to improve the travel distance of a hard-rolling car, including guard rails to straighten the trucks as soon as they reach tangent track, the development of rail-head lubricators for application to the curves below the group retarders, and I believe that at least one railroad is attempting to simulate the rail conditions experienced during rainy weather when it is generally conceded freight cars roll with much greater consistency.

In our fourth yard, to be built at Toronto, the design of which is still fluid, we are making every effort to reduce the amount of reverse curvature between the group retarders and the classification tracks. This is being done by the best arrangement of tracks and switches to obtain optimum rollability.

I understand a suggestion has been advanced in connection with the construction of another hump yard to construct an offset in the hump track of approximately 7 ft just below the hump crest and introduce sharp reverse curvature consisting of two 14-deg curves which would permit more accurate electronic measurement of the effect of curve resistance on the rolling characteristics of each individual car. This information would then become part of the computer equation for improved predictability of car progress through the curved track area.

I believe an important feature in the design of new automatic classification yards will be provision for handling piggyback and container shipments. With the rapid increase in this type of traffic, it not only has to be humped but provision must be made for quick handling of local cars to unloading points. In the design of the Toronto Yard, we have selected an area, adjacent to the main humps, which is connected directly to the top end of the receiving and departure tracks, for loading and unloading trailers and containers as this traffic develops. Incoming cars can be put over the hump or taken directly to the unloading area. Outbound loaded cars can be taken directly to the departure tracks, if expedient, or can be put over the hump for wider classification.

At the present time, the Canadian National is making studies on the feasibility of developing hump yards for smaller terminals. In these smaller terminals the emphasis will not be on a high rate of humping but rather on the largest possible number of classifications. Consequently, the studies have concentrated on track layouts involving about 12 tracks per group. With such a system the distance from the crest to clearance point will be increased, but fewer retarders and less total retardation are necessary than with systems using 8 or 9 tracks per group. Further economies may be possible in these smaller yards by using an all-manual rather than automatic speed control. For example, very substantial improvements over present flat switching operations are possible with a manually controlled, two-group, 24-track yard to serve industrial transfer and way-freight operations.

Thank you for your kind attention. [Applause]

CHAIRMAN KREFTING: Mr. Johnston, I wish to extend the thanks of the Association to you for your interesting and informative address. The pictures of yards recently constructed on the Canadian National Railways were also excellent, and we are very grateful.

The next speaker also is known to many of you. D. C. Hastings, general superintendent terminals, Atlantic Coast Line Railroad and the vice chairman of this committee, will address you on the subject, "Piggyback Service Moves Ahead." Mr. Hastings.

Piggyback Service Moves Ahead
By D. C. HASTINGS
General Superintendent Terminals, Atlantic Coast Line Railroad

At a time when the most important and valuable transportation industry in America is fighting so desperately for its very existence—an industry hamstrung by archaic rules and restricted by obsolete laws and regulations—it is a bit refreshing and encouraging to look at one of the expanding phases of railroading—Piggyback Service. Regardless of the attempts of some companies to identify this fast-growing method of transportation as "Trailer-on-Flat-Car-Service" and others as "Trailer-Train Service", its most widely used name is "Piggyback." This name is also used to cover the various methods of containerization and the movement of automobiles on flat cars equipped with multi-level loading devices.

Excluding the operations of the United States Army military railway service during the Civil War, the Long Island was possibly the first railroad to provide piggyback service on a commercial basis. In December 1884, the Long Island announced it would carry loaded farm wagons to the East River for $4 per wagon, such charge to cover two horses and a driver. In January 1885, at Albertson on the Locust Valley Branch, the first all-piggyback train was dispatched. There were flats for the wagons, box cars for the horses and a coach for the teamsters. The nine ensuing years saw this form of transportation die out, and with a few exceptions, it was not revived until May 9, 1926, when the Chicago North Shore & Milwaukee Railroad initiated a piggyback service of trailers loaded on flat cars. From that year until 1954 the service remained in an incubator state barely alive and unable to exist by itself as a mode of transportation.

Finally the babe was able to be released and in 1955 the railroads had an awkward, unsettled and uncertain adolescent on its hands that gave promise of growing faster than any form of transportation service the industry ever offered. This has been the case—from 168,150 cars of loaded trailers in 1955 to 591,246 in 1961, the service has shown an increase of 423,096 cars or 252 percent in just six years.

The child has outgrown its equipment, its terminals and its rates, and today we have a six year old still bursting at every seam and giving promise of being a real transportation giant if fed properly with more equipment, larger and more flexible terminals and proper rates.

The railroads today are offering piggyback service in connection with the use of highway trailers in five basic plans. Each of these plans is designed to fill the need of either a motor carrier, a railroad, a shipper, or a combination of these needs.

Plan I—Railroads transport trailers owned by motor common carriers on a division of the truck rate—actually, in practice, a flat charge per trailer based on weight and distance regardless of the commodity. The trucker solicits and bills all freight at truck rates, takes the trailers to and picks them up from the railroad piggyback terminal, and performs any required road-haul before or after the rail movement. The railroad has no direct contact with the shipper in this plan and simply substitutes for the trucker on part or all of the road haul.

Plan II—Railroads transport their own trailers under their own truck-competitive tariffs. Under this all-rail plan the railroad deals directly with the shipper, furnishes all of the equipment, and provides pick-up and delivery between shipper plants and railroad terminals, either by railroad-owned tractors or by contract with local draymen. The pick-up and delivery is usually confined to established territories contiguous to rail terminals.

Plan III—Railroads transport trailers owned or leased by shippers on a flat rate per mile. The shipper delivers his trailers to the rail terminal, the railroad puts them on flat cars, ties them down, transports them to destination and unloads them. The responsibility for final delivery rests with the shipper or his agent.

Plan IV—Railroads transport trailers owned or leased by shippers on flat cars also owned or leased by shippers on a flat rate per mile or a flat charge per car, regardless of whether the trailers are loaded or empty. The railroad merely performs the rail-haul movement.

Plan V—Railroads transport their own trailers or common carrier truck trailers on joint rates which include the entire movement by both rail and highway. Plan V is similar to Plan I, but is actually a joint operation which, in effect, extends the territory of each participating carrier into that served by the other, permits each to handle shipments originating in or destined to the other's territory, and allows each to sell for the other.

The phase of piggyback service involving the movement of containers may fall under either Plan II or III, such as the Southern Railway's containerized freight and the New York Central's Flexivan Service.

New automobiles move via piggyback under Plan I or V where the highway trailer for transporting new cars owned by a common carrier is loaded onto a flat car.

New automobiles on bi- or tri-level auto racks on flat cars, while not an actual plan of piggyback service, is usually considered as part of that type of operation since the autos are transported on a per car rate which covers the rail transportation from one auto loading terminal to another. The new automobiles are assembled at the loading point and are loaded onto the multi-level flats over an adjustable ramp which is used for each level. They are then moved to destination by rail where they are unloaded over a similar adjustable ramp. The autos are assembled in groups according to consignee on the parking area and are finally delivered to the dealer by highway trailer or under their own power.

The success of the piggyback service offered today stems from its ability to combine the flexibility and service features of highway transportation with the speed, dependability and economic advantages of rail transportation to form a transportation service that is most attractive to any shipper. Just the other day I heard a very prominent shipper say that this is one of the most attractive things that the railroads have offered in a century.

Door-to-door service at attractive rates and on good dependable schedules is the dream of the shipping public and is the type of service that must be offered today if the transportation company expects to remain in business. Piggyback is designed by the railroads to answer this need of the shipping public. A trailer is sent to the warehouse of our customer where he loads his products right at his front door. This he can do with his regular warehouse freight-handling equipment to meet his own specifications and schedules. When he has his shipment ready, he advises the railroad and the trailer is picked up by a tractor and brought via highway to the rail terminal where it can be parked to be loaded later or loaded immediately onto a flat car.

The trailer is loaded onto a flat car by backing up a ramp, and is tied down in the usual way. These ramps vary from large concrete and asphalt multiple-track structures at busy terminals to simple wooden ramps serving only one track at outlying stations. Some roads have made ramps out of old flat cars by removing the trucks from one end. The ease with which piggyback service can be established depending on the volume to be handled, and the simplicity of construction at small stations is another reason why the operation is expanding so rapidly. Having reached final destination, our piggy-back trailer is then unloaded from the flat car and moves on out from the others to be delivered to the consignee who, when he opens the doors finds a shipment ready for immediate unloading and damage free.

The consignee can then with his own warehouse equipment accomplish the unload-ing of the trailer, and the door-to-door shipment by piggyback has met the modern need of the shipping public.

To accomplish this operation, we find many types of equipment and terminal facil-ities [see accompanying photographs], all basically the same, but some having special features to meet particular needs or demands of the service.

Yes, we can truly say that piggyback service is moving ahead. The future looks bright—the potential is terrific—the opportunities are unlimited—however, the volume will directly depend on our ability to meet the demands of the shippers with equipment and develop expedited methods of terminal operation. Our present large terminals are too slow for the volume of trailer traffic that they are being called upon to handle. It is in this last area that we as members of the American Railway Engineering Asso-ciation have our greatest challenge. Are you ready to go to work? Our Association and our respective managements are both looking to us for the solution. [Applause]

Flat cars with tri-level automobile racks spotted at adjustable
unloading ramp.

Trailers parked at terminal.

Trailer being loaded onto a flat car.

Having reached final destination, piggyback trailer is unloaded from flat car.

TTX standard 85-ft flat car, which is made by several car-building companies.

A.

An 85-ft TTX car equipped with tri-level auto racks.

An 85-ft Pullman-Standard car equipped for both containers and trailers.

An 85-ft TTX car showing the new deck design.

ACF trailer hitch, which is operated by either air or electric
impact wrenches.

Fruit Growers Express refrigerated trailers now in regular service
between Florida and the East, via the Southern, Seaboard Air Line, and
Atlantic Coast Line.

Fruehauf–General American system of roll on-roll off or drive on-drive off, utilizing the G–85 car.

Southern's containerized freight.

Southern Pacific Clejan type cars.

New York Central's Flexivan yard at the Bronx, New York City.

Unloading terminal for new cars on the ACL at Jacksonville, Fla.

Unloading terminal for new cars at Oakland, Calif., on the Southern Pacific.

Container unloading area, Southern Railway at Alexandria, Va.

Multiple track ramp on the Burlington.

Piggyback hold yard at Oakland on the Southern Pacific.

Solid piggyback train on the Atlantic Coast Line.

MR. KREFTING: Mr. Hastings, your address has covered a timely subject in which we are all interested. It has been interesting to follow the development in piggyback service through your fine talk and slides. Please accept the thanks of the Association.

Mr. President, there is one other matter which should be taken care of before our committee is dismissed. This is my last year as chaiman of Committee 14. I would like to present the incoming chairman, D. C. Hastings, general superintendent terminals, Atlantic Coast Line Railroad, and the new vice chairman, H. J. McNally, chief engineer, New York Improvements, Pennsylvania Railroad. [Applause]

PRESIDENT BEEDER: Thank you, Mr. Krefting; and thank you both, Mr. Johnston and Mr. Hastings. Time will not permit me to comment on your interesting presentations other than to say that each of you has made a great contribution to our convention program.

Mr. Krefting, you have done a splendid job as chairman of Committee 14 for the past three years, which is clearly evident in the 1960, 1961 and 1962 reports of your committee. But it is good to know, as you complete your term as chairman, that Committee 14 has a lot of able men to carry on.

We welcome as the new chairman of your committee, Mr. Hastings, and your new vice chairman, Mr. McNally. If Mr. Hastings will please step up to the podium I would like to present him with a chairman's gavel as the symbol of his authority for the next three years. [Applause]

Mr. Krefting, you are now excused, with our thanks to you and the members of your committee. [Applause]

Discussion on Economics of Railway Location and Operation
[For report, see Bulletin 567, pp. 137–160]

PRESIDENT BEEDER: Our next committee to report is Committee 16—Economics of Railway Location and Operation, the chairman of which is C. L. Towle, assistant vice president and chief engineer, Detroit, Toledo & Ironton Railroad, Dearborn, Mich. Mr. Towle, if you and the members of your committee will come to the platform I shall be pleased to turn the microphone over to you. Mr. Towle, you may proceed.

CHAIRMAN C. L. TOWLE: Mr. President, members of the Association and guests: Your Committee 16 is reporting on four of its eight assignments. These are presented on pages 39 through 51 of Bulletin 566 and pages 137 through 160 of Bulletin 567. One of these is a final report and two are progress reports submitted as information. One progress report, that of Subcommittee 1, will be presented for adoption as Manual material. Your committee invites your comments regarding its presentations, and will be glad to reply to any questions raised thereon.

Assignment 1—Revision of Manual.

CHAIRMAN TOWLE: A. S. Lang, director of data systems, New York Central System, chairman of Subcommittee 1, is unable to be present, and in his absence the report on Assignment 1 will be presented by George Rugge, assistant engineer, Atchison, Topeka & Santa Fe Railway, a member of the subcommittee.

GEORGE RUGGE: Mr. President and gentlemen:

Your committee has submitted $7\frac{1}{2}$ pages of new material to replace that now appearing on various pages in Chapter 16 of the AREA Manual. In addition, your committee has recommended deletion in its entirety of existing Part 4 of Chapter 16.

The bulk of the new material deals with various considerations involved in line location and replaces material in Part 1 of the present Manual chapter. It enunciates the basic capital return criterion in slightly greater detail than in the present version.

It also includes as new material a summary of the basic principles of location laid down by Wellington.

Under the heading of Special Considerations, the new material includes a series of subjects, such as temporary construction and compensation for curvature, which are widely scattered throughout the present chapter. It also includes new sections on helper districts and balanced profiles.

A second part of the new material submitted deals with train resistance. This is to replace and supplement material now covered in Part 3 of the chapter. The new material explains the mechanism of train resistance, presents the Davis Formula, and gives references on other train-resistance considerations.

Because of page numbering conflicts, both the material on location and that on train resistance cannot practicably be inserted in the Manual at this time. This problem will be rectified when additional material now being prepared is available to replace the remaining material in the old (that is, the present) chapter. In the meanwhile, a note will be inserted in the chapter Table of Contents calling attention to the existence of new (and approved) material in the Proceedings.

A third part of the new material involves only minor changes in wording in two otherwise unchanged sections of the chapter.

Mr. President, I move that this report be adopted and inserted in the Manual (as soon as practicable) to replace material now appearing on designated pages in Part 1, 2 and 3 of Chapter 16, and that Part 4 be deleted in its entirety.

[The motion was duly seconded, was put to a vote, and carried.]

Assignment 3—Determination of Maintenance of Way Expense Variation with Various Traffic Volumes and Effect of Using Such Variations, in Terms of Equated Mileage or Other Factors, for Allocation of Available Funds to Maintenance of Way, Collaborating with Committees 11 and 22.

CHAIRMAN TOWLE: Subcommittee 3 is currently awaiting completion and analysis of various studies under way on several major railroads; and when this information is available it will be correlated with regression analysis approach and with other data currently under consideration by this subcommittee.

Responsible for this assignment are L. E. Ward, industrial engineer, Pennsylvania Railroad, and J. W. Bolstad, chief engineer, Ft. Dodge, Des Moines & Southern.

Assignment 4—Potential Applications of Electronic Computers to Railway Engineering and Maintenance Problems in Research, Design, Inventory, Etc., Collaborating with Committees 11 and 30.

CHAIRMAN TOWLE: In the absence of Ferdinand Wascoe, chairman of Subcommittee 4, the report on Assignment 4 will be delivered by L. P. Diamond, assistant engineer, research, Chesapeake & Ohio Railway, vice chairman of the subcommittee.

L. P. DIAMOND: Mr. President and members of the Association:

In AREA Bulletins 566 and 567, Progress Reports were published on two phases of Assignment 4 by your Subcommittee.

The report appearing in Bulletin 566, entitled, "The Digital Terrain Model Approach to Railroad Route Location", by Paul O. Roberts, assistant professor of civil engineering, Massachusetts Institute of Technology, was submitted by your committee to describe in some detail one way that computers can be used by railway engineers to determine the best solution rather than a possible solution to a given problem at minimum cost.

The Digital Terrain Model System is a system of integrated electronic computer programs which aids the location engineer in selecting the route for which costs of construction are the minimum. The report describes the data inputs necessary for the operation of DTM, the computations made and the outputs of the system. With these outputs and estimates of costs, the location engineer can prepare construction cost summaries for each trial alinement for use in making the final location decision.

The report further analyzes the engineering design cost implications of the DTM approach, with the conclusion that the integrated computer program approach taken under this system produces significant savings over conventional computer techniques when four trial line computations are exceeded and, furthermore, that it would be impractical to perform the same number of trial lines by hand.

In summary, the significant thing about a production approach to location engineering such as the DTM system is not only that the cost of application is generally lower than for other engineering techniques, but that it also provides a practical means of finding a better line location.

In Bulletin 567 your committee reported on progress in compilation of information on the use of electronic computers by railroads for engineering and maintenance problems. In response to a questionnaire submitted late in 1959 to 69 railroads, 67 replies were received up to June 1961. Seven of the 67 indicated that they were utilizing computers on enginering and maintenance problems. Thirteen, who at the time were not using computers, anticipated using them in the future. Eleven different types of problems were being analyzed by the 7 roads using computers, while it was anticipated that 17 more different types of problems would probably be "fair game" for computer solution. The majority of roads using or expecting to use computers for these problems indicated a willingness to share their knowledge.

Recognizing that computer usage by railroad engineering departments is changing rapidly, your committee will submit another questionnaire shortly to bring this compilation up to date.

This report is presented as a progress report for the information of the Association, with the recommendation that the subject be continued.

The committee especially wishes to express its appreciation to Professor Roberts of MIT for his courtesy in submitting a fine paper on "The Digital Terrain Model Approach to Railroad Route Location."

Assignment 5—Methods of Reducing Time of Freight Cars Between Loading and Unloading Points, Collaborating with Car Service Division, AAR, Communication and Signal Section, AAR, Operating-Transportation Division, AAR, and American Association of Railroad Superintendents.

CHAIRMAN TOWLE: Subcommittee 5 has completed a plan of approach covering various phases of the problems, and proposes, with the necessary permission to circulate a questionnaire to operating officials of American railroads to gain further information for use in this study. The subcommittee chairman is W. J. Dixon, director of industrial engineering, Baltimore & Ohio.

Assignment 6—Features of Economic and Engineering Interest in the Study, Design, Construction and Operation of New Railway Line Projects, or Major Line Relocations, Proposed, in Progress, or Recently Completed.

CHAIRMAN TOWLE: In the absence of T. D. Wofford, staff engineer, Illinois Central, chairman of Subcommittee 6, the report on Assignment 6 will be presented by H. A. Lind, assistant to chief engineer, Chicago, Burlington & Quincy Railroad, vice chairman of the subcommittee.

H. A. LIND: Mr. President, Mr. Towle, and members of AREA:

Your committee is pleased to present in its report information on two outstanding railroad construction projects. One of these projects involves a rather unusual piece of construction to connect existing lines of the Pacific Great Eastern Railway terminating at Vancouver and Squamish and to extend this railroad northward to the Peace River. This report was very ably prepared by J. C. Martin, assistant system supervisor, terminal operation, Canadian National Railways.

The other project is the Santa Fe relocation of main line in Arizona, involving the construction of 44 miles of double-track between Williams and Crookton. Many of you have seen movies on the construction of the line, and it is believed you will find the report an interesting supplement. Your committee is indebted to the Santa Fe for the information and to Mr. Inman, assistant engineer on that road, for his efforts in putting the information into report form.

It is the aim of Subcommittee 6 to present reports of special interest to railroad engineers. Many of the projects which we intend to cover are written up in trade journals, but the magazine articles are often lacking in technical content, especially economic information, that engineers want. We believe that everyone interested in railroad location and operation will benefit from the reports on this assignment.

The report to this convention is the second since initiation of the assignment. Your committee already has several additional reports in stages of development and will be able to continue worthwhile activity into the foreseeable future. There are a number of significant and interesting line changes and new construction projects under way or in the planning stages. Reports on these will be developed wherever possible.

This progress report is submitted as information, with the recommendation that the subject be continued.

Assignment 7—Engineering, Maintenance and Operating Benefits to Be Derived from Increased Joint Use of Railway Facilities, Collaborating with Committees 11, 14, and 20.

CHAIRMAN TOWLE: Subcommittee 7 under the chairmanship of J. W. Barriger, president, Pittsburgh & Lake Erie Railroad, has submitted, for the present, a final report on its assignment in Bulletin 567. Chairman Barriger submits the following message on the subject for your consideration:

"We do wish again to emphasize the importance of continuous attention to this channel as a means of promoting economy in maintenance and operating expenses.

"We especially recall the report of the Senate Subcommittee on Surface Transportation and its indictment of the railroad industry for its lack of interest 'in such matters as joint use of railroad facilities in order to eliminate waste, such as multiple terminals and yards that require expensive interchange operations; reductions in duplications of freight and passenger services by pooling and joint operations; abandonment or consolidation of non-paying branch and secondary lines . . . ' "

I wish at this time to thank Chairman Barriger and the members of his subcommittee for their fine work on this assignment, with the recommendation that the assignment be concluded until current developments in our industry are implemented and assessed.

Assignment 8—Innovations in Railway Operations.

CHAIRMAN TOWLE: Subcommittee 8 is currently engaged in assembling data for a progress report covering various aspects and ramifications of containerization. The subcommittee chairman is A. L. Sams, principal assistant engineer, Illinois Central.

Assignment 11—Review of Developments in New Methods and Modes of Transport.

CHAIRMAN TOWLE: Subcommittee 11 is gathering data and studying material for possible future reports on pipeline operations and air-flow vehicles. Its chairman is F. J. Richter, publisher, Modern Railroads.

Mr. President, this concludes our presentation of the report of Committee 16.

PRESIDENT BEEDER: Thank you, Mr. Towle, for the continuing good work of your committee in bringing its Manual chapter up to date, and for the valuable reports which your committee brings to this Association. The opportunities for effective, economy-producing studies by your committee are almost unlimited, and we shall look to your committee for further valuable reports in the future.

You are now excused with the thanks of the Association. [Applause]

Discussion on Cooperative Relations with Universities

[For report, see Bulletin 569, pp. 403–425]

PRESIDENT BEEDER: The third of the eight committees to report this afternoon is our Committee 24—Cooperative Relations with Universities, the chairman of which is W. W. Hay, professor of railway civil engineering at the University of Illinois.

Professor Hay, I would appreciate it if you and the other members of your committee present would take your places here at the speaker's table and proceed immediately with the presentation of your report.

CHAIRMAN W. W. HAY: President Beeder, members and guests:

The report of Committee 24 is found in Bulletin 569, pages 403 to 425, incl. The committee is currently working on seven subjects.

Assignment 1—Stimulate Greater Appreciation on the Part of Railway Management of (a) the Importance of Bringing into the Service Selected Graduates of Colleges and Universities, and (b) the Necessity for Providing Adequate Means for Recruiting Such Graduates and Retaining Them in the Service.

CHAIRMAN HAY: At the 1961 convention Subcommittee 1 reported the preparation of a questionnaire which would provide a profile of the present-day railway engineer. This questionnaire has been revised to permit punch card recording of the data.

Within a few weeks questionnaire forms will be sent to the chief engineers of Class I Member Roads, with a request for their cooperation in obtaining responses, not only from their own departmental staff but from all graduate engineers employed by their company. Special attention to non-engineering employment areas will be appreciated.

The problems faced by railroads in recruiting young graduate engineers indicate the necessity of establishing the characteristics of those to whom railroad employment has proven attractive, and determine what thoughts they have concerning their own education as compared to the technical requirements of the industry. The maximum benefit from this questionnaire can only be obtained through the maximum and enthusiastic support of the engineering officers receiving the questionnaire.

This is presented as a report of progress.

J. F. Davison, assistant to chief engineer, Canadian National Railways, is the able chairman for Assignment 1. He is also the AREA representative in the American Society for Engineering Education. Mr. Davison was unable to attend the annual meeting of ASEE held last June in Lexington, Ky., so T. D. Wofford, Jr., staff engineer, Illinois Central Railroad, went in his stead. Since Mr. Wofford is unable to be here today, I shall ask Mr. Davison to read a brief report that Mr. Wofford has prepared.

J. F. Davison (reading Mr. Wofford's comments): Mr. President, members and guests:

The 1961 annual meeting of the American Society for Engineering Education was held on the campus of the University of Kentucky, at Lexington, Ky. I attended the meeting as the official representative of the American Railway Engineering Association and of Committee 24. The Illinois Central Railroad is also an industrial member of ASEE, and I served as our company representative.

Several thousand people from education, industry and government assemble once a year at these meetings. The topics of discussion involve problems of engineering education, professional development of technical personnel and research activities of common interest to employers and educational institutions.

During a meeting of the Civil Engineering Division, there was considerable discussion about the teaching of transportation engineering which is of concern to the railroad industry. One of the outstanding features of the session was an address by Professor Hay, chairman of Committee 24.

Meetings of the Relations with Industry Division are the focal point of contact between educators and representatives of industry at the ASEE meetings. There was considerable discussion at the Lexington meeting about proposals to revise the content of engineering curricula in the various universities and what had been accomplished to date.

The most active industries participating in the ASEE meetings are from a relatively small number of companies that have a great amount of interest in research and development. These firms naturally have a large proportion of engineers and scientists in their organizations. I suspect that the influence of these companies in the ASEE is out of proportion to the number of engineers they employ insofar as the whole engineering profession is concerned. This leads me to believe that other segments of industry, including railroads, and other organizations that employ large numbers of engineers, need to make their influence felt more than they do now.

Insofar as I could determine, I was the only railroad representative present at the meeting in Lexington. Several professorial members of Committee 24 were present, including Professor Lewis from Purdue, Dean Kerekes from Michigan School of Mines, and Professor Hay from the University of Illinois. A more extensive participation by individual railroads in the ASEE would be beneficial in many ways.

Assignment 2—Stimulate Among Colleges and University Students a Greater Interest in the Science of Transportation and Its Importance in the National Economic Structure by (a) Cooperating with and Contributing to the Activities of Student Organizations in Colleges and Universities, and (b) Presenting to Students and Their Counselors a Positive Approach to the Attractive and Interesting Features of the Railroad Industry and the Advantages of Choosing Railroading as a Career.

Chairman Hay: The entire membership of Committee 24 has been active in contributing to Assignment 2. Members have been speakers to Student Chapters of ASCE at various universities and high schools. Arrangements have been made for inspection trips to modern railway yards. Open-houses have been held at universities, and equipment has been made available to universities for the inspection of the students.

The program of student affiliates is again active this year. During the 1960-1961 school year there were 52 student affiliates enrolled with the AREA. Several of these have taken railroad jobs and applied for AREA junior membership.

This subcommittee has been assigned the task of making a list of speakers to be available to various colleges and universities. Any information regarding speakers in the railway industry will be received with appreciation by Subcommittee 2.

B. B. Lewis, professor of railway engineering, Purdue University, is giving this assignment a well-directed chairmanship. Professor Lewis, will you please stand?

Assignment 3—The Cooperative System of Education, Including Summer Employment in Railway Service.

CHAIRMAN HAY: In fulfilling Assignment 3 in 1961, the committee followed essentially the same procedures used in preceding years. A questionnaire was sent to the chief engineering and maintenance offices of the railroads in late February, requesting information concerning their summer employment needs. The replies to the questionnaire were returned to the subcommittee chairman; the information was tabulated, reproduced in the AREA secretary's office, and sent to some 125 engineering colleges in late March.

A similar program is being followed in 1962. Because of the heavy correspondence involved in answering the many applications for summer employment, the committee is attempting this year to spread this load by having applications sent to several officials along the railroad rather than to any one individual. It also endeavored to simplify the questionnaire.

Committee 24 has indicated by a general expression of opinion among the membership that it considers this project to be worth while.

The committee has become aware of the fact that a number of young engineers have learned from experience obtained through summer employment with a railroad that there are satisfactory careers to be had in the railroad field. Consequently, as stated above, the project is going forward again this year, and the committee requests your continued cooperation.

The able chairman for Assignment 3 is W. A. Oliver, professor of civil engineering, University of Illinois. Professor Oliver, will you stand? Thank you.

Assignment 4—What Constitutes a Desirable Curriculum for Students to Pursue in Preparation for a Career in Railroad Engineering?

CHAIRMAN HAY: T. D. Wofford, Jr., staff engineer, Illinois Central Railroad, is chairman of Subcommittee 4. I shall present Mr. Wofford's report in his absence.

CHAIRMAN HAY (reading Mr. Wofford's comments): This subcommittee is very pleased to present its report on Assignment 4, concerning "What Constitutes a Desirable Curriculum for Students to Pursue in Preparation for a Career in Railroad Engineering?"

Your subcommittee has found this to be a very live subject. One of the most interesting discoveries was the tremendous range of viewpoints that people within the railroad· industry have in connection with this subject. It seems that there are almost as many opinions as to what a student should study in college as there are people with different degrees of experience and responsibility in the industry. Engineers who have a great deal of technical responsibility favor a strong technical education. Those officers who have primarily administrative responsibilities often favor a somewhat broader curriculum that cultivates interest beyond strictly technical fields. The findings of your subcommittee seem to bear out the theory that the education of an engineer is a never-ending process that must continue long after the college diploma is received.

In making the studies necessary for this report, your subcommittee found that the college staffs themselves are giving much study to the subject. College curriculums are undergoing intensive analysis by educators at this time. In fact, there seems to be a considerable difference of opinion among educators as to what the ideal college curriculum should include. A few schools have already taken steps radically to revise their engineering curriculum. Most schools are making changes slowly, but it appears that new concepts of educating engineers are going to be widely adopted in the future.

The current report is submitted as a final report with the recommendation that the assignment be reactivated within the next five years. A review of the assignment will help the railroad industry keep informed on future trends in engineering education and to bring its influence to bear wherever desirable.

CHAIRMAN HAY (continuing): Mr. Wofford and his committee have given this subject a thorough study.

Assignment 5—Ways in Which Railroads Can Cooperate with Universities in Developing Research, Including the Revising of "Suggested Topics for Theses on Railroad Subjects."

CHAIRMAN HAY: The report on Assignment 5 will be given by its chairman, H. E. Hurst, division engineer, The Milwaukee Railroad.

H. E. HURST: Mr. President, members, and guests:

The report on Assignment 5 of this committee is presented on pages 413 to 416, incl., of Bulletin 569. In the report, we have sought to direct your attention to the ever-increasing importance of research and graduate-study activity in our institutions of higher learning. Our report complements to a large extent the address by Professor K. B. Woods of Purdue University which he presented to the Association last year.

The increasing emphasis on graduate programs is a direct result of the rapid advancement of the sciences and scientific developments which are influencing the entire engineering profession.

Your committee, in addition to showing the importance of this work, is seeking and working towards the promotion of interest in railroad problems. In doing so, your attention is invited to the significance and importance of promoting the interest of college faculty members in the railroad transportation field. One inescapable conclusion is that interest and motivation will be enhanced by making monetary funds available for such work. While millions are being spent in research and graduate studies, we feel that a relatively small investment by the railroads will promote a greatly increased interest in their problems. It is in this direction that we are turning our attention for the coming year.

In reviewing the report of this assignment, you will note that the "Suggested Topics For Study And Research On Railroad Subjects" have been revised, and these listings will be further up-dated from time to time. In this task, I am sure all suggestions will receive careful consideration.

This is a progress report offered for your information.

CHAIRMAN HAY: Thank you, Mr. Hurst. You have presented a well prepared and challenging report.

Assignment 6—Procedures for Orienting and Developing Newly Employed Engineering Personnel.

CHAIRMAN HAY: Work was progressed on Assignment 6 under the chairmanship of C. E. R. Haight, chief engineer, Delaware & Hudson Railroad.

Assignment 7—Stimulating an Interest by College and University Staff Members in Current Railroad Problems and Practices, Including AREA Membership.

CHAIRMAN HAY: Work was also progressed on Assignment 7 under the chairmanship of Frank Kerekes, dean of the faculty, Michigan College of Mining and Technology.

CHAIRMAN HAY (continuing): With deep regret I must announce the resignation, for reasons of health, of H. E. Kirby from the vice chairmanship of Committee 24. Mr. Kirby has been of great help to me in conducting the work of this committee. I wish here to express my deep appreciation for his service and cooperation, and to extend to him the thanks and best wishes of Committee 24.

Committee 24 is fortunate to have an able and experienced man to assume the vice chairmanship. It gives me pleasure and satisfaction to introduce our new vice chairman, J. F. Davison, assistant to chief engineer, Canadian National Railways. Mr. Davison. [Applause]

President Beeder, this concludes the report of Committee 24.

PRESIDENT BEEDER: Thank you, Professor Hay. Your committee, both through the professorial members on its roster and the informative reports which it develops, continues to give our Association and the railroad industry valuable contacts with college campuses and their student bodies. We particularly appreciate your interest, and that of the other professors on your committee, in the work of Committee 24.

We are sorry that Herman Kirby, through retirement from railroad service, had to relinquish the vice chairmanship of your committee, because he has been a powerful worker in our Association; but we welcome his successor, Mr. Davison, who has likewise demonstrated his interest and ability.

Thank you again, Professor Hay. Your committee is excused with the thanks of the Association. [Applause]

The remainder of our Convention session this afternoon will be devoted to structures and structural materials, as we hear successively the reports of our structural committees.

Discussion on Wood Bridges and Trestles

[For report, see Bulletin 569, pp. 451–456]

PRESIDENT BEEDER: The first of the structural committees to report is Committee 7—Wood Bridges and Trestles, the chairman of which is K. L. DeBlois, structural engineer, New York Central System, Chicago. Again I can invite to the platform only the chairman, vice chairman and subcommittee chairmen of this committee. I will be pleased if they will come promptly to the platform. Mr. DeBlois, you may proceed.

CHAIRMAN K. L. DEBLOIS: Mr. President, members and guests of the American Railway Engineering Association:

The report of Committee 7 will consist of brief reports covering work accomplished during the past year. These reports, except for the one on Assignment 6, can be found in Bulletin 569, commencing on page 451. The committee invites your comments and questions at the conclusion of each presentation.

Assignment 1—Revision of Manual.

CHAIRMAN DEBLOIS: Our first report is on Assignment 1 and will be presented by Subcommittee Chairman D. V. Sartore, assistant engineer of bridges, Burlington Railroad. Mr. Sartore.

D. V. SARTORE: Mr. President, members and guests:

In accordance with the secretary's request that all Manual material be reviewed prior to reprinting of the Manual in 1962, your committee has continued its review of Chapter 7 with the results shown on pages 452 through 455 of Bulletin 569. Your committee felt it proper to divide this material into three categories for presentation to the Association.

The first consists of documents to be reapproved with revisions where the revisions are of an editorial nature or were suggested as a result of a special Manual review.

Mr. President, I move that the recommended action for the following documents: "Glossary", "Instructions for Inspection of Timber Trestle Railway Bridges", "Methods of Fireproofing Wood Bridges and Trestles", "Plans for Open-Deck Pile and Framed Trestles, Multiple-Story Trestles and Ballast Deck Pile and Framed Trestles", and "Recommended Practice for Design of Wood Culverts", as shown on pages 452 through 455 of Bulletin 569, be adopted.

[The motion was duly seconded, was put to a vote, and carried.]

MR. SARTORE: The second category consists of a document rewritten to take into consideration constructive comments brought to the attention of your committee.

Mr. President, I move that the recommended action for the document, "Use of Guard Rails and Guard Timbers", as shown on pages 454 and 455 of Bulletin 569, be adopted.

[The motion was duly seconded, was put to a vote, and carried.]

MR. SARTORE: The third category consists of a document that was rewritten for the purpose of deleting out-dated material.

Mr. President, I move that the recommended action for the document, "Recommended Practice for Overhead Wood Highway Bridges", as shown on page 455 of Bulletin 569, be adopted.

[The motion was duly seconded, was put to a vote, and carried.]

Assignment 6—Applications of Synthetic Resins and Adhesives to Wood Bridges and Trestles, Collaborating with Committees 8 and 15.

CHAIRMAN DEBLOIS: Our next report is on Assignment 6. It will be presented by Subcommittee Chairman L. R. Kubacki, area engineer—structures, Pennsylvania Railroad.

L. R. KUBACKI: Mr. President, members and guests:

Last year your committee presented a preliminary report on this assignment. At present this preliminary report is being revised and the revised version will be published in the near future. The revised report, in addition to changes in the formulations, will include some general instructions for the use of epoxy resins.

Regarding experimental work, we have obtained knowledge of the physical properties of epoxy resins from known field applications and small-scale laboratory tests, and have started some field performance tests.

The laboratory and field tests that have been or are being conducted are as follows:

(1) At the Research Center preliminary tests were performed using two epoxy formulations as waterproofing coatings over concrete to determine whether the formulations would pass Committee 29 acceptance tests for waterproofing concrete surfaces. A total of nine specimens were used. Six were coated and three were uncoated control specimens.

Three specimens were coated with the liquid portion of formulation 1101-32. (This will be designated formulation P1 in the current revision.) Three specimens were coated with formulation 632-1555.

The results of the tests indicated that formulation P1 is a satisfactory coating and formulation 632-1555 an unsatisfactory coating. Formulation 632-1555 was removed from the list of applications and formulation P1 now appears in its place in the current revision.

(2) On January 10, 1962, 40 steel plates were installed on the Huey Long Bridge, New Orleans, La. Thirty-six of these plates were coated with an epoxy resin and four were left uncoated. All plates were of A7 steel, 7 in by ½ in by 10 in. The plates were numbered from 101 to 140, incl.

The plates numbered 101 to 106, incl., were coated with an epoxy formulated according to formulation P1. This is a one-coat application. Plates numbered 107 to 112, incl., were coated with formulation C1. This is also a one-coat application. Plates numbered 113 to 118, incl., were coated with formulation 285–1 for the first coat and 285–6 for the second coat. Plates numbered 119 to 124, incl., were coated with formulation 285–22 for the first coat and 632–1255 for the second coat. Plates numbered 125 to 130, incl., were coated with an epoxy tar known as "Dearclad 765."

Plates numbered 131 to 136, incl., were coated with an epoxy tar coating known as formulation 242–2. Plates numbered 137 to 140, incl., were left uncoated.

The coated plates were sand blasted prior to coating. Plates numbered 137 to 140 were not sand blasted and were installed with the mill scale intact.

(3) Under the supervision of the AAR Research Center, the top flanges of two beam spans installed recently by the Chesapeake & Ohio have been coated with an epoxy in lieu of being painted. The beam spans are open deck with ties placed on top flanges of beams. Before the epoxy was applied the top flanges of one span were sandblasted and the other span was cleaned in conventional manner.

Two formulations were used. One was the formulation C1 shown in the current revision. This was a one-coat application. The other coating was a two-coat application; the first coat consisted of formulation 285–22 and the second coat was formulation 632–1255. This application was made in the summer of 1961.

(4) Under the supervision of the Research Center and through the cooperation of the Chesapeake & Ohio, a test installation was made using epoxy adhesive for installing a treated timber crossing adjacent to treated timber crossing installed with spiral dowel fasteners.

Formulation 991–67 was used with a primer coat of formulation P1. This installation was made in the summer of 1961.

(5) Through the cooperation of the Research Council for Riveted and Bolted Structural Joints, a pilot investigation is being conducted at the University of Illinois to investigate the use of epoxies to prevent slippage of bolted joints. A progress report may be available this month.

In addition to these items, the investigation into the physical properties of epoxy resins is continuing. Also, performance tests are being conducted in conjunction with other laboratory tests performed at the Research Center.

I would like to mention that we are always open for suggestions for new applications and formulations. Also, if any of the railroads have an application where they would like to try an epoxy, please contact this committee for recommended formulation or other assistance. The more field applications we have the better we are able to evaluate the various formulations.

This report is submitted as information.

Assignment 7—Repeated Loading of Timber Structures.

CHAIRMAN DEBLOIS: Our next report is on Assignment 7. C. V. Lund, assistant to chief engineer, Milwaukee Road, is chairman of this subcommittee.

Laboratory tests were completed during the past year on 24 glued laminated Douglas fir bridge stringers in static and repeated loading. These tests were conducted principally to explore the relationship of the position of wheel loads to strength in horizontal shear, and constitute part of a more comprehensive program of tests on sawn and glued laminated stringers being undertaken in cooperation with the American Institute of Timber Construction, the National Lumber Manufacturers Association, the American Wood Preservers Institute and the Forests Products Laboratory, United States Department of Agriculture. A second series of tests using southern pine stringers is in progress.

Assignment 10—Rules for Rating Existing Wood Bridges and Trestles.

CHAIRMAN DEBLOIS: Our last report is on Assignment 10. W. A. Genereux, assistant engineer, Canadian Pacific Railroad, is chairman of this subcommittee; and since he could not be with us today, he has asked me to present the report.

The report, appearing on page 456 of Bulletin 569, sets forth two corrections to the tentative rating rules presented as information at the 1961 convention and published in the Proceedings, Vol. 62, 1961, pages 539–542. The changes noted in Bulletin 569 were approved by letter ballot of the committee.

Mr. President, I move that these rating rules, with the revisions, be adopted and published in the Manual.

[The motion was duly seconded, was put to a vote, and carried.]

CHAIRMAN DEBLOIS: Mr. President, this concludes the reports on our assignments. [Vice President C. J. Code assumed the Chair.]

VICE PRESIDENT C. J. CODE: Thank you, Mr. DeBlois. We appreciate the careful review which your committee has made of its chapter of the Manual and the recommendations brought to this convention. Your committee has a number of highly important assignments, and we hope you will make real progress on them in the year ahead.

Your committee is now excused with the thanks of the Association. [Applause]

Discussion on Masonry
[For report, see Bulletin 568, pp. 275–314]

VICE PRESIDENT CODE: Our next structural committee to report is Committee 8—Masonry. The chairman of this committee, completing his first year as chairman, is D. H. Dowe, assistant engineer of bridges, Seaboard Air Line Railroad, at Richmond, Va. Mr. Dowe, if you and the other members of your committee will come to the platform, we will be pleased to have your report at this time.

CHAIRMAN D. H. DOWE: Mr. Chairman, members of the Association and Guests:

Your Committee on Masonry has reported on six assignments in Bulletin 568, pages 274 to 314, incl. Four of these assignments include Manual revisions, and for each of these assignments, I shall call upon the chairman of the subcommittee, if present, to give the report.

Assignment 1—Revision of Manual.

CHAIRMAN DOWE: E. P. Wright, special engineer, Illinois Central Railroad, Chicago, who was subcommittee chairman at the time the report on this assignment was prepared, recently resigned from Committee 8, and in his absence I shall make the report on Assignment 1 as shown on pages 276 and 277 of Bulletin 568.

During the past year a special review was made of Chapters 8 and 29 of the Manual to correct certain minor inconsistencies that occurred throughout these chapters. All of these changes appear in the report on Assignment 1.

Mr. Chairman, I move that the revisions as shown under Assignment 1, pages 276 and 277 of Bulletin 568, be adopted.

[The motion was duly seconded, was put to a vote, and carried.]

Assignment 2—Design of Masonry Structures, Collaborating with Committees 1, 5, 6, 7, 15, 28 and 30.

CHAIRMAN DOWE: Subcommittee Chairman F. A. Kempe, Jr., assistant bridge engineer, Northern Pacific Railway, St. Paul, Minn., will report on Assignment 2.

F. A. KEMPE: Mr. Chairman, members and guests:

Your committee has reviewed the specifications in Parts 2, 7, 8 and 10 of Chapter 8 of the Manual, and we recommend their reapproval with the revisions published in Bulletin 568, pages 276–278, incl.

Mr. Chairman, I move that the Specifications for Design of Plain and Reinforced Concrete Members, pages 8-2-1 through 8-2-27 of the Manual, be reapproved with the four revisions recommended in Bulletin 568.

[The motion was duly seconded, was put a vote, and carried.]

MR. KEMPE: Mr. Chairman, I move that the Specifications for Reinforced Concrete Arches, pages 8-7-1 through 8-7-7 of the Manual, be reapproved with the four revisions recommended in Bulletin 568.

[The motion was duly seconded, was put to a vote, and carried.]

MR. KEMPE: Mr. Chairman, I move that the Specifications for Reinforced Concrete Rigid-Frame Bridges of One Span, pages 8-8-1 through 8-8-6 of the Manual, be reapproved with the one revision recommended in Bulletin 568.

[The motion was duly seconded, was put to a vote, and carried.]

MR. KEMPE: Mr. Chairman, I move that Specifications for the Placement of Concrete Culvert Pipe, pages 8-10-1 through 8-10-5 of the Manual, be reapproved with the three revisions recommended in Bulletin 568.

[The motion was duly seconded, was put to a vote, and carried.]

MR. KEMPE: Your committee has rewritten the Specifications for Reinforced Concrete Culvert Pipe, pages 8-10-6 through 8-10-14 of the Manual, as published on page 280 of Bulletin 568. The new specification provides for the use of ASTM Specifications, Designation C 76.

Mr. Chairman, I move that Specifications for Reinforced Concrete Culvert Pipe be adopted as written in Bulletin 568.

[The motion was duly seconded, was put to a vote, and carried.]

MR. KEMPE: Considering the advent of prestressing into the concrete industry, your committee reviewed the Specifications for Design of Concrete Transmission Poles, pages 8-12-1 through 8-12-6 of the Manual, and recommends that the subject matter be withdrawn.

Mr. Chairman, I move that the Specifications for Design of Concrete Transmission Poles be withdrawn, as recommended in Bulletin 568.

[The motion was duly seconded, was put to a vote, and carried.]

MR. KEMPE: Mr. Chairman, this concludes the report on Assignment 2.

Assignment 4—Deterioration and Repair of Masonry Structures.

CHAIRMAN DOWE: A brief progress report on Assignment 4 is presented on page 280 of Bulletin 568. The subcommittee chairman- is J. M. Gilmore, engineer bridges and buildings—maintenance, New York Central.

Assignment 6—Prestressed Concrete for Railway Structures, Collaborating with Committee 6.

CHAIRMAN DOWE: A brief progress report on Assignment 6, prepared by Subcommittee Chairman J. R. Williams, assistant engineer of bridges, Chicago, Rock Island & Pacific, may be found on page 281 of Bulletin 568.

Assignment 7—Quality of Concrete and Mortars, Collaborating with Committee 6.

CHAIRMAN DOWE: The report on Assignment 7 will be made by Subcommittee Chairman J. W. Dolson, assistant bridge engineer, Missouri Pacific Railroad, St. Louis, Mo. Mr. Dolson.

J. W. DOLSON: Mr. Chairman and gentlemen:

Your committee has completed that portion of its assignment relating to review of Part 1—Specifications for Concrete and Reinforced Concrete Railroad Bridges. and Other Structures. The reviewed and completely rewritten version appears in Bulletin 568, pages 281 to 312, incl.

Part 1 has been expanded in scope and content and revised editorially to affect practically all sections. These changes are too voluminous to be covered in any detail by this report, but have been reviewed and unanimously approved by the membership of Committee 8.

Mr. Chairman, I move that revised Part 1 be approved by the Convention.

[The motion was duly seconded, was put to a vote, and carried.]

Assignment 8—Waterproofing for Railway Structures, Collaborating with Committees 6, 7, and 15.

CHAIRMAN DOWE: Subcommittee Chairman R. J. Brueske, assistant division engineer, Chicago, Milwaukee, St. Paul and Pacific Railroad, La Crosse, Wis., will report on Assignment 8.

R. J. BRUESKE: Mr. President, fellow members, and guests:

During the past year, your Committee reviewed the specifications in Chapter 29—Waterproofing, and recommends the changes listed on pages 313 and 314 of Bulletin No. 568. Included in the proposed changes is the temporary withdrawal of the Specifications for Bituminous Emulsions for Dampproofing, Part 3, pending further study and complete revision. The materials presently specified are no longer commercially available.

Mr. President, I move the adoption of the specification changes for Chapter 29—Waterproofing, as listed on pages 277, 313 and 314 of Bulletin 568.

[The motion was duly seconded, was put to a vote, and carried.]

CHAIRMAN DOWE: Mr. Chairman, this concludes the reports on our assignments. As a special feature of Committee 8, I am pleased to present Harold R. Hutchens, general manager, Concrete Products Division, Carter–Waters Corporation, Kansas City, Mo., who is an associate member of Committee 8. He will tell us about "The Use of Lightweight Concrete in Railroad Work."

Use of Lightweight Concrete in Railroad Work
By HAROLD R. HUTCHENS
General Manager, Concrete Products Division, Carter–Waters Corp.

We hear much today about lightweight structural concrete and lightweight aggregate. One reason for this is a greater demand for a lightweight structural concrete due to the rapid increase in the use of precast and prestressed concrete structures. Another reason is because of the increased knowledge of the structural properties of the material and its general acceptance into recognized codes and specifications. As proof of this please look over, if you haven't done so already, the new proposed ACI Building Code published in the February issue of the ACI Journal, and note the new reference and specifications for lightweight concrete. Still another reason is the increasing scarcity of good quality natural aggregates in some parts of the country.

Based on an individual's experience, lightweight concrete could mean any number of things, for example: Gaseous or foam concrete, no-fines concrete, or concrete made with lightweight aggregates. Even narrowing the field of concrete made with lightweight aggregates, (an aggregate weighing less than 70 lb per cu ft) could mean many things.

A concrete spectrum of weights from 10 to 120 lb per cu ft could be developed, and such a spectrum is shown in the accompanying illustration. This whole lightweight-

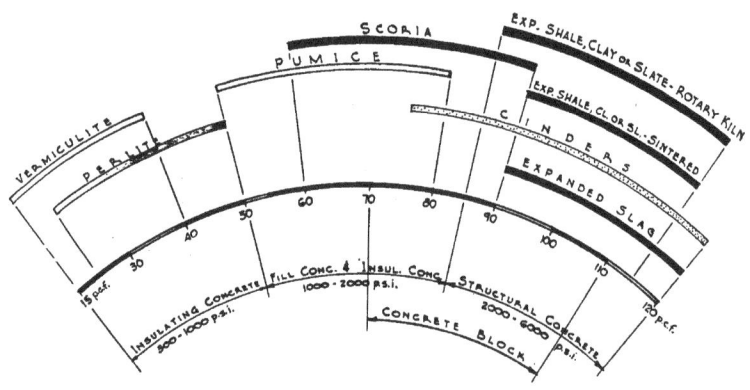

SPECTRUM OF LIGHTWEIGHT CONCRETES

aggregate industry represents some 250 plants and is still growing. It produces some 15,000,000 tons of aggregate per year under 130 different trade names.

The following are a few comments on these various concretes:

(1) *Vermiculite and Perlite:* Super lightweight—used for roof fills, fireproofing and insulation.

(2) *Pumice and Scoria:* Used mostly for concrete block and fill concrete. Has generally poor concrete structural properties.

(3) *Cinders:* Availability is diminishing. Used mostly for concrete block and fill concrete. Has poor structural concrete properties.

(4) *Sintered Shales and Expanded Slag:* Used mostly for block. Has fair concrete properties with maximum strength of about 4000 psi with excessive cement.

(5) *Rotary Kiln Shales (Clay or Slate):* Concrete produced from these lightweight aggregates has good to excellent structural properties. Its structural uses range from low-strength insulating concrete to high-strength concrete for prestressed concrete members. There is no difficulty in getting strengths from 2000 to 6 or 7000 psi with reasonable cement factors.

As food for thought, imagine the results of blending these lightweight aggregates with sands to produce a "not-so-light" lightweight concrete with considerably improved properties, or using excessive air-content to produce super lightweight concrete. The spectrum goes from complicated to crazy when this is done. This points to some exciting future potentials of the structural lightweight concrete industry.

My remarks from this point on will be directed toward lightweight concrete made with lightweight aggregates produced by the rotary kiln process. This industry is represented by the Expanded Shale, Clay and Slate Institute with headquarters in Washington, D. C. This organization along with its members are ready and willing to answer any of your questions and supply you with information regarding this aggregate.

To make the picture a little more confusing, there are at least 25 different trade names for aggregates produced by the rotary kiln process. For example, there are Haydite, Basalite, Materialite, Galite, Idealite, Norlite, Kenlite, Ridgelite, Shale-lite, etc., all essentially the same material.

During my association with the lightweight concrete industry, I have been approached more than a few times by engineers who have never designed with structural lightweight concrete. Their questions regarding design indicate that they are apprehensive because this material is so "new" and "so entirely different"—so they think. Basically it is still concrete; cement-and-water paste is the binder; expanded shale aggregate particles instead of sand and gravel or crushed stone are the filler. For those properties that depend mainly on the quality of the paste, such as compressive strength, durability and permeability, lightweight concrete and heavyweight concrete are similar. For properties that depend mainly on the aggregate, such as weight, modulus of elasticity, tensile strength, creep and shrinkage there is a divergence between lightweight concrete and heavyweight concrete. The basic principles of concrete design with lightweight concrete are still as valid as ever—it merely behooves the engineer to acquaint himself with the proper properties of structural lightweight concrete and use them logically and intelligently.

As examples of illogical and unintelligent evaluation of properties, let me cite two which have in the past led to unwarranted rejections of lightweight concrete. One property of the aggregate different from heavyweight aggregate is its greater absorption, hence newly hardened concrete will have a greater percentage of uncombined water within it. This, the uninformed argue, must make the concrete less durable and more permeable, i.e., less water tight. On the contrary, lightweight concrete is found by test to be durable provided that the aggregate is furnished and mixed in a less-than-saturated condition with the cellular spaces in the aggregate not completely filled even after mixing; and especially after an initial drying-out period after normal moist curing, the expansion that accompanies freezing can be accommodated without breaking up the aggregate. A quality paste will completely coat each aggregate particle and prevent moisture from building up in the aggregate, once the concrete is hardened, even under a head of water. And the initial free water in the aggregate is given up slowly to provide "internal curing", more completely hydrating the cement and improving the paste's qualities.

That this is not just a haphazard guess can be seen in the Expanded Shale Institute's Report on the Selma, a World War I ship with a lightweight concrete hull that had lain partially submerged in salt water for 34 years, at which time an investigation showed the concrete to be dry at a depth of ¼ in and that reinforcing bars with only ⅝ in of cover were as sound as the day they were tied and placed in the forms. The many bridge decks of lightweight concrete throughout the country certainly give testimony to durability and wearability.

Another property which frightens the fearful away is the creep and shrinkage in lightweight concrete. It is generally believed that these values are substantially greater than for heavyweight concrete.

About 2½ years ago, producer members of the Expanded Shale, Clay & Slate Institute instigated a comprehensive research program on creep and shrinkage. This research program includes between 20 and 25 member aggregates together with several heavyweight aggregate concretes. This research program is being conducted by the U. S. Bureau of Standards. Many of you saw the tests in progress at a Committee 8 meeting in Washington over a year ago. These tests include concentrically loaded specimens of each concrete and are being extended to include eccentrically loaded specimens and full-sized members. When these tests are completed information on creep and shrinkage loss values will be available for almost all rotary-kiln-produced lightweight aggregates as well as comparisons with several heavyweight aggregates. These tests will run for at least 10 years and longer if space at the Bureau is available. It is expected that information on long-time deflection will be forthcoming from these tests. This is something about which very little is known for either heavyweight or lightweight concrete.

The results of these tests have not yet been released publically; however, these facts are now apparent:

(1) Some lightweight concretes have less creep and shrinkage than some heavyweight concretes and that for those that did have higher values the divergence is not nearly as great as is generally assumed.

(2) The range in creep and shrinkage values for the heavyweight concrete tested is about the same as for the lightweight concretes tested.

(3) The average lightweight concrete in this investigation at the end of one year exhibited about 23 percent greater creep and shrinkage losses than normal weight concretes made from high-quality dense aggregates. These are average values at the end of one year of testing and should not be construed as final results.

Even when confronted with a higher shrinkage or creep strain it must be remembered that this material is more elastic, that is, it has a lower modulus of elasticity. The combination of a slightly higher shrinkage strain with a lower E value may actually result in lower shrinkage stresses. It has been reported by contractors that lightweight slabs placed on a day that would invariably cause plastic or shrinkage cracks in heavyweight concrete, came through free of cracks. Lower shrinkage stresses and the ability to hold moisture within the aggregate and release it slowly to the concrete may account for this better performance, which is just another plus value for lightweight concrete.

What is the one feature about lightweight structural concrete that makes it desirable? Why the increased fuss and bother about it in the trade journals? Why is the question of its use cropping up more and more in government agencies and in new specifications and building codes written by specifying bodies that up until recently had nothing whatever to do with such a material? Very simply, structural lightweight concrete has been proven to have the same compressive strength and other properties as the concrete most people are familiar with, using roughly the same cement factor, but which weighs only two-thirds as much. The ratio of strength to weight is improved by 50 percent.

This improvement in strength-weight ratio has accounted for the rapid growth in the use of prestressed concrete and will be equally potent in pushing along the increased use of lightweight structural concrete. How does this effect the railroad bridge and building engineer? The railroad bridge engineer will realize immediately that the 30 to 35 percent decrease in dead load will not appreciably effect his design since the dead load is usually such a small proportion of the total design load. In the case of prestressed and precast concrete bridges, however, this 30 to 35 percent saving in weight does become important. Many of these structures will be erected under traffic, and the ease and speed with which the members can be handled is important. The lighter the member the quicker and easier it can be seated in its final position. It may also be possible to handle the members with lighter equipment. Remember that a bridge girder made with heavyweight concrete and weighing 30,000 lb will weigh only 20,000 lb using lightweight concrete. Consider also the ease of handling and driving a prestressed concrete pile compared to a heavyweight concrete pile weighing one and one-half times as much. Experience has shown that a prestressed lightweight-concrete pile drives equally as well as a heavyweight concrete pile. Some railroad bridge engineers are considering the possibility of casting prestressed concrete bridges in two pieces rather than the proposed standard three or four pieces per span. Some have already experimented with the assembly of a complete span at an off-the-site location and transporting the completed span to the bridge site. Consider the effect of the saving in weight on either of these two techniques. It is evident therefore that the larger the unit the more important the ratio of strength to weight becomes. In passing, consider other uses of lightweight concrete in the railroad industry in which this saving in weight is important due to ease of handling:

(3) Prestressed concrete cross ties.

(2) Signal foundations and structures.

(3) Precast cribbing.

(4) Prestressed concrete beams and slabs and precast slabs for freight terminals and other railroad building structures.

The strengths we are talking about today for prestressed concrete work are the high strengths of 4000 psi in 24 hr and 5000 and 6000 psi at 28 days. These strengths have been consistently achieved in most of the more than 50 prestressing plants that are using expanded-shale concrete. It has been indicated previously that the cement content to obtain the above strengths are consistent with that required for heavyweight concrete for roughly the same maximum size of aggregate. This brings us to a brief discussion of design and control of lightweight concrete mixes.

In August 1959 the American Concrete Institute published an ACI Standard entitled "Recommended Practice For Selecting Proportions for Structural Lightweight Concrete." This ACI standard is essentially the end result of research work done by the Portland

Cement Association and others. This standard develops a method of designing a light-weight concrete mix based on trial mixes and the development of a "specific gravity factor." This method has simplified to a great extent the problem of how to design a lightweight mix, which has plagued the industry for so long. From experience I can testify that the method recommended in this ACI standard will work, and with the necessary precautions and controls applied, the high-strength concrete required for prestressed concrete can easily be obtained day after day and year after year.

To summarize, these facts are important in the design of lightweight concrete structures:

(1) Select a good lightweight aggregate which will produce high-strength concrete with a normal cement factor.

(2) Determine from the producer of these aggregates or concrete using these aggregates the properties of the lightweight concrete. Realize that after these properties are known, the design, whether it be for ordinary reinforced con-crete or for prestressed concrete, is precisely the same as for heavyweight concrete or for any other structural material.

(3) The main advantage of the use of lightweight concrete is simply the fact that it weighs two-thirds as much as heavyweight concrete. There are other advantages, such as a built-in curing device and the possibility that the lower E value will cause less shrinkage cracks. My own experience with pre-stressed lightweight concrete design and in the production of prestressed con-crete members indicates that there is no reason to fear the use of this material in railroad structures.

In spite of this confidence in the material, we are insisting that tests be made and, in fact, are furnishing prestressed concrete bridge girders to the AAR laboratories for load tests. These tests will be conducted alongside heavyweight girders of the same size. It is hoped that these tests will get under way within the next 30 to 60 days. We have every confidence that they will pass the test with flying colors; however, it is our responsibility to prove without question the merits of lightweight concrete rather than to expect the railroad design engineers to take our word for it and thus assume our responsibility. [Applause]

CHAIRMAN DOWE: Thank you, Mr. Hutchens, for a most enlightening and inter-esting talk.

Mr. Vice President, this concludes the presentation of Committee 8.

VICE PRESIDENT CODE: Thank you, Mr. Dowe. It is evident that the work of your committee has continued to move forward smoothly and aggressively during your first year as chairman, and we particularly appreciate the large amount of attention given by your committee in bringing its Manual chapter up to date. It is also evident that during the past year your committee has readily absorbed the work of our former Committee 29 on Waterproofing.

We thank you too, Mr. Hutchens, for your informative presentation on "The Use of Lightweight Concrete in Railroad Work."

Mr. Dowe, you are now excused, with our thanks to you and your Committee. [Applause].

Discussion on Impact and Bridge Stresses

[For report, see Bulletin 567, pp. 161–166]

[President Beeder resumed the Chair.]

PRESIDENT BEEDER: We will hear next from our Committee 30—Impact and Bridge Stresses. The chairman of this committee, who until early in 1961 was design engineer, Erie–Lackawanna Railroad, is D. W. Musser, now associated with Trygve, Hoff & Associates, with headquarters in Cleveland, Ohio. Mr. Musser, if you and the other members of your committee will come to the platform we shall be pleased to hear your report.

CHAIRMAN D. W. MUSSER: Mr. President, members of the Association and guests:

Committee 30—Impact and Bridge Stresses, is reporting on three of its ten assignments. These are presented in Bulletin 567, pages 161 through 166. The three are all progress reports and are presented as information. Your comments regarding these reports are invited.

I shall also comment briefly on a few of our other assignments.

Assignment 2—Steel Truss Spans.

CHAIRMAN MUSSER: Although no work was done explicitly for this committee, the AAR research staff, under the direction of E. J. Ruble, made a service test on the 1409-ft-long cantilever truss span of the Pittsburgh & Lake Erie Railroad's Ohio River Bridge at Beaver, Pa. The investigation was conducted principally to determine the magnitude of stress reversal in certain members of the anchor arm and to determine if heavier ore cars could be permitted on the structure.

Assignment 4—Longitudinal Forces in Bridge Structures.

CHAIRMAN MUSSER: No tests were conducted and no reports written on this subject this year. We do, however, anticipate some work on this subject next year, either as a separate report or at least in connection with another report. Based on preliminary work done on this subject in connection with other assignments, it has become increasingly evident that the longitudinal forces that exist in a fixed structure are a very small percentage of the total applied load.

Assignment 5—Distribution of Live Load in Bridge Floors: (a) Floors Consisting of Transverse Beams, (b) Floors Consisting of Longitudinal Beams.

CHAIRMAN MUSSER: Assignment 5 is the first of the three on which we reported in Bulletin 567.

In February 1961 the AAR published report ER-5—The Lateral and Longitudinal Distribution of Loading in Steel Railway Bridges. Bulletin 566 contains a summary of this report. This study was conducted at the University of Illinois by W. W. Sanders, Jr., and W. H. Munse under the sponsorship of Committee 30.

The study of this problem began several years ago while we were making our impact studies on steel girder bridges. In reviewing the test results for those studies it was noted that the loads in the floor system of bridges were much smaller than our present design criteria specify and that the loads were transferred to the floor members more uniformly than anticipated. It was felt, therefore, that our present design criteria were too conservative and that if a more accurate design formula could be developed it would mean considerable savings to the railroad industry.

With this in mind the study was undertaken at the University of Illinois to develop a more realistic specification for the distribution of live load in the floor systems of steel railway bridges.

The first portion of the study consisted of a review of the existing methods of analysis for bridge floor systems and then the development of methods of analysis accurately to determine the distribution of wheel loads to the floor. The second objective was to obtain a number of solutions of simulated bridge floors using these methods of analysis. These solutions included evaluations of the two general types of bridges now in use and were compared with the results of a number of actual bridge tests conducted by the AAR. The third objective was to derive for design office use a method of computing the liveload distribution which takes into account the principal factors affecting the distribution.

Two basic bridge floor systems were studied. The first—transverse floorbeams supported by heavy longitudinal edge girders or trusses, and the second—longitudinal beams connected by a series of transverse diaphragms. In addition results were obtained indicating the effect on the behavior of the bridges of (1) size and spacing of floorbeams or stringers, (2) type of floor covering, (3) diaphragms, (4) ballast, and (5) number of tracks.

An analytical study of steel railway bridge floors offers numerous problems. In most cases the floor systems are indeterminate to a number of degrees. Also, the structural design is such that variations are many, even in a bridge of a given general type. Therefore, to study the various types of bridge floors it was necessary to idealize the structures. For this study the idealized structure was assumed to be a gridwork of beams supporting a slab or plate, simply supported on two edges and free on the other two.

There are several mathematical proceedures available to analyze this gridwork of beams; however, only two were used in this study: (1) a moment-distribution method and (2) the orthotropic-plate theory. The first of these two procedures assumes no interaction between the slab and the beams, whereas the second method assumes full composite action.

The results of this study have been very good. The highly theoretical equations developed give results which agree very well with recorded results. The theoretical equations are, of course, quite long and it was felt that the proceedures were too complicated for use in the design office.

We have, therefore, prepared design procedures which are simpler to use, but which, of course, are not as accurate as the theoretical analysis. The committee is presently studying these equations and in the near future hopes to approve them for submission to Committee 15 for their use in reviewing their specification.

It might be well to point out that bridges with properly designed slabs over 6 in thick and with ballast, yield the best results insofar as load distribution is concerned. Bridges with metal floor plates yield results similar to existing specifications.

Assignment 6—Concrete Structures.

CHAIRMAN MUSSER: A report on tests conducted by the AAR research staff on a Santa Fe Railway prestressed concrete bridge near Colorado Springs, Colo., appears in AAR Engineering Report ER–18 and in Bulletin 567. The bridge has two 71-ft 6-in spans with each span consisting of four modified "T" girders. These bridges carry the Santa Fe Railway over the entrance roads to the Air Force Academy and are the largest prestressed concrete railway spans in America today. A few of the more pertinent results of this test were:

(1) Load distribution to the girders, as measured by strains in the bottom of the girders at midspan, was approximately equal.

(2) The maximum total impact, as calculated from recorded strains on the bottom of the girders at midspan, was 13.5 percent and was recorded at speeds of 41 and 55 mph. AREA impact for this span is 34.2 percent using the nominal axle loads on the test locomotive.

(3) The maximum recorded impact, as determined from horizontal shearing strain, was 52.7 percent.

(4) The vertical strain distribution indicated composite action between the slab and girders.

A report on tests conducted by the AAR Research Staff on a Florida East Coast Railroad precast prestressed concrete bridge near Pompano Beach, Fla., was reviewed by the committee and approved for publication. The test spans consist of six beams per track and are 30 ft 6 in long. One of the spans had shear keys and both spans were tested before and after transverse tensioning of the beams. A few of the more pertinent results of the test were:

(1) Both shear keys and transverse tensioning are effective in distributing the track loads to the beams.

(2) The test trains operated at speeds up to 60 mph, yet no impact value exceeded the calculated AREA value of 54 percent. The greatest impact values were 38 percent for the locomotive and 52 percent under the cars.

Assignment 7—Timber Structures.

CHAIRMAN MUSSER: Under this assignment the AAR research staff prepared and released Report ER–1 "Investigation of 60-Ft Glued Laminated Beams on the Weyerhaeuser Timber Company Railroad." This investigation is of particular significance in that the beams are the largest glued laminated beams in the United States carrying regular diesel locomotives, and are of dimensions that could conceivably reflect future design practice. The test span is a deck-type structure consisting of four 52-in laminated girders about 60 ft long. A few of the more pertinent results of this test were:

(1) Measurements of the shear strains indicate an effective transfer of stress across the glue lines.

(2) Distribution of the live load to the four beams, based on flexural strains, was reasonably consistent with the eccentricity of the track.

Assignment 9—Use of Electronic Computers for Railroad Bridge Problems.

CHAIRMAN MUSSER: Seventeen additional moment and shear tables for special heavy railroad cars were calculated with the moving load program and distributed to the chief engineers of Member Roads. We believe that we have pretty well covered the field of special loads with these tables. However, if there are any additional loads that any of you would like to see published, please contact E. R. Andrlik, our subcommittee chairman, or the chairman of this committee.

CHAIRMAN MUSSER: This completes the resumé of the current work of Committee 30. I would now like to introduce the subcommittee chairmen who are responsible for the committee's activities. As I give their names I ask that each chairman stand and be recognized.

E. S. Birkenwald, engineer of bridges, Western Lines, Southern Railway System, Subcommittee on Steel Truss Spans.

A. T. Granger, dean of engineering, University of Tennessee, Subcommittee on Viaduct Columns.

J. A. Erskine, assistant bridge and building engineer, Gulf, Mobile & Ohio Railroad, Subcommittee on Longitudinal Forces in Bridge Structures.

Dr. N. M. Newmark, head, Department of Civil Engineering, University of Illinois, Subcommittee on Distribution of Live Load in Bridge Floors.

P. L. Montgomery, division engineer, New York, Chicago & St. Louis Railroad, Subcommittee on Concrete Structures.

C. V. Lund, assistant to chief engineer, Chicago, Milwaukee, St. Paul & Pacific Railroad, Subcommittee on Timber Structures.

James Michalos, chairman, Department of Civil Engineering, New York University, Subcommittee on Vibrational Characteristics of Bridges.

E. R. Andrlik, bridge designer, Atchison, Topeka & Santa Fe Railway, Subcommittee on Electronic Computers.

A. R. Harris, engineer of bridges, Chicago & North Western Railway, Subcommittee on Continuous Structures.

N. E. Ekrem, assistant bridge engineer, Great Northern Railway, Subcommittee on Composite Structures.

———————

CHAIRMAN MUSSER (continuing): Thank you, gentlemen. Since I am at this time completing my term as chairman of Committee 30, I would like to express my appreciation for the assistance you have given me during the past three years. My thanks, also, are extended to the members of the committee, the AAR research staff and to Executive Secretary N. D. Howard for their cooperation and help during this period. I would also like to express my appreciation to Trygve Hoff and Associates, the consulting engineering firm where I have been employed this past year, for allowing me to attend this Convention and our committee meetings and for the use of the secretarial personnel for all of my necessary correspondence.

At this time I would like to introduce the new vice chairman of Committee 30: N. E. Ekrem, assistant bridge engineer of the Great Northern Railway.

I would now like to introduce J. W. Davidson, bridge engineer of the Burlington Railroad, who will become the new chairman of Committee 30. Mr. Davidson has been an able assistant during my term of office, and I am certain he will do an excellent job of administering the affairs of the committee.

Mr. Beeder, this concludes the report of Committee 30—Impact and Bridge Stresses.

PRESIDENT BEEDER: Mr. Musser, your report will be received as information. Thank you.

Your committee continues to give valuable assistance to our other structural committees, and its work during the past three years under your direction has been no exception. We greatly appreciate the informed direction which you have given to your committee during your term as chairman.

We are pleased to welcome Mr. Davidson as the new chairman of your committee and Mr. Ekrem as the new vice chairman, assured that they will make a good team in carrying forward the work of Committee 30.

Mr. Davidson, if you will please step to the rostrum I would like to give you this chairman's gavel as the symbol of your authority for the next three years. [Applause]

Mr. Musser, you and your committee are now excused with the thanks of the Association.

Discussion on Iron and Steel Structures

[For report, see Bulletin 569, pp. 379–401]

PRESIDENT BEEDER: Continuing our consideration of structural matters, we will hear next from our Committee 15—Iron and Steel Structures, the chairman of which, completing his first year in office, is C. Neufeld, engineer of bridges, Canadian Pacific Railway. Mr. Neufeld, will you and the other official members of your committee who are present please come to the platform and present your report. Mr. Neufeld.

CHAIRMAN C. NEUFELD: Mr. President and members of the Association:

Before proceeding with the presentation of our reports, Committee 15 wishes to express its sorrow at the death of one of its members. Henry C. Tammen, partner in the firm of Howard, Needles, Tammen & Bergendoff, and a member of this committee since 1933, passed away suddenly on July 6, 1961. His Memoir is included in Bulletin 569, along with our annual report.

In compliance with the regulations of the Association, all material being presented to the convention for adoption and publication in the Manual has received the endorsement of the Committee by letter ballots, in the form of an affirmative vote of at least two-thirds of the voting membership.

Assignment 1—Revision of Manual.

CHAIRMAN NEUFELD: Subcommittee Chairman E. S. Birkenwald, engineer of bridges, Southern Railway, will present our report on Assignment 1.

E. S. BIRKENWALD: Mr. President, your committee submits for adoption Manual material which requires either revision or reapproval, found in Bulletin 569, pages 382 to 387, incl. In addition, Manual material on pages 387 to 398, incl., of this Bulletin is presented now as information, to be considered one year hence for adoption.

Revisions to the Specifications for Steel Railway Bridges have been made for clarification, except for Art. 1, Sec. B, page 15-1-32, and Art. 2 (c), Sec. B, page 15-1-33, which, when adopted, will permit the use of the basic-oxygen process in the manufacture of structural and high-strength steel. This process is relatively new, permitting greater efficiency in the production of steel.

Revisions to the Specifications for Movable Railway Bridges have been made to conform with modern practice, except for the revision to the "Foreword", which has been made for clarification.

Revisions to the Instructions for the Maintenance Inspection of Steel Bridges have been made so as to meet the Association's requirements for recommended practice.

Mr. President, I move that the revisions to the Specifications for Steel Railway Bridges, the Specifications for Movable Bridges, and the Instructions for the Maintenance Inspection of Steel Bridges, published on pages 382 to 386, incl., Bulletin 569, be adopted for publication in the Manual, and that, with these revisions, these parts of Chapter 15 of the Manual be reapproved.

[The motion was duly seconded, was put to a vote, and carried.]

MR. BIRKENWALD: On pages 386 and 387 are listed the specifications and recommended practices in Chapter 15 of the Manual which have been reviewed by your committee, which recommends reapproval without change. These documents are:

Specifications for Steel Railway Turntables.
Specifications for the Design of Rigid-Frame Steel Bridges.
Specifications for the Design of Continuous Steel Railway Bridges.
Specifications for the Erection of Steel Railway Bridges.

Instructions for the Mill Inspection of Structural Steel.

Instructions for the Inspection of the Fabrication of Steel Bridges.

Instructions for the Inspection of Bridge Erection.

Classification of Railway Bridges.

Rules for Rating Existing Iron and Steel Bridges.

Methods of Strengthening Existing Bridges.

Fusion Welding.

Track Anchorage on Bridges and Similar Structures.

Specifications for Assembly of Structural Joints Using High Strength Steel Bolts in Steel Railway Bridges.

Mr. President, I move that these documents in Chapter 15 of the Manual be reapproved without change.

[The motion was duly seconded, was put to a vote, and carried.]

MR. BIRKENWALD: On pages 387 to 390, incl., your committee presents as information proposed revisions to the Specifications for Steel Railway Bridges, necessitated by the introduction of ASTM A 36 steel by the steel industry. Since there is only $1 per ton differential between A 36 steel and the A 7 steel now utilized by the Association in its Specifications, it can be anticipated that savings resulting from the higher unit stresses which can be used for A 36 steel will dictate the use of A 36 over A 7 steel. As a matter of fact, there are known instances where A 36 steel has been substituted by the manufacturer when A 7 steel has been ordered.

Likewise, your committee presents as information, on pages 390 to 398, incl., proposed Specifications for Structural Joints Using High-Strength Bolts in Steel Railway Bridges, which it hopes the Association will substitute one year hence for the present Specifications for Assembly of Structural Joints Using High-Strength Steel Bolts in Steel Railway Bridges.

The proposed specifications recognize two types of joints, friction or bearing; permit the use of high-strength steel bolts without hardened washers under certain conditions; and provide a Commentary which outlines the basis for the more important provisions of the specifications.

Mr. President, this concludes the report on Assignment 1.

Assignment 2—Composite Steel and Concrete Spans; Non-Ferrous Metal Bridges, Collaborating with Committees 8 and 30.

CHAIRMAN NEUFELD: James Michalos, professor of structural engineering and chairman of the Department of Civil Engineering, New York University, is chairman of the Subcommittee on Assignment 2. Due to Prof. Michalos' absence I shall read the report on his assignment.

Your committee submits, as information, Specifications for Composite Steel and Concrete Spans, as published in Bulletin 569, pages 398 and 399, to be considered for adoption one year hence. The specifications give the criteria to be applied in the design of simple-span bridges consisting of steel beams and concrete slab work integrally.

Assignment 4—Stress Distribution in Bridge Frames, (a) Floorbeam Hangers, (c) Truss Bridge Research Project.

CHAIRMAN NEUFELD: E. T. Franzen, engineer of bridges, Chicago, Rock Island and Pacific Railroad, will present our report on Assignment 4.

E. T. FRANZEN: Mr. President and members:

The work on Assignment 4 (a)—Floorbeam Hangers, was essentially completed with revision of the Manual as recommended and approved last year. Further minor revision will be made to clarify unit stresses for the design and rating of floorbeam hangers.

An amendment to the material of Chapter 15 on Strengthening of Existing Bridges has been submitted as information, which points out methods of reducing or eliminating the possibility of fatigue cracking in floorbeam hangers.

Assignment 4 (c) pertains to the Truss Bridge Research Project at Northwestern University.

During the past year investigation has been made of the ultimate carrying capacity of a truss span with a damaged end post. It was found that the carrying capacity was reduced as the amount of damage increased; however, the calculated axial stress in the damaged end post under maximum load for the condition of greatest damage was equal to only 52 percent of the yield stress. From this it was apparent that the load must have been redistributed to the remainder of the bridge. The damaged end post has been removed and replaced with a new end post. A similar series of tests will be conducted on another end post of a different design.

Assignment 10—Effect of Continuous Welded Rail on Bridges, Collaborating with the Special Committee on Continuous Welded Rail.

CHAIRMAN NEUFELD: J. C. King, assistant engineer of bridges and structures, Canadian National Railways, will present the report on Assignment 10.

J. C. KING: Mr. President and members:

A progress report published in Bulletin 569, pages 400 and 401, is submitted as information. It is a brief review of the practice presently in use by a number of roads, but before the committee can prepare a final report it would appreciate receiving suggestions and experience records from other railroads so that these can be studied.

CHAIRMAN NEUFELD: The following subcommittee chairmen are responsible for assignments on which we have status or brief progress reports only. Will they please stand as I call their names:

A. R. Harris, engineer of bridges, Chicago & North Western Railway, is chairman of Subcommittee 5—Design of Steel Bridge Details.

J. G. Clark, of Clark, Daily and Dietz, Consulting Engineers, is chairman of Subcommittee 7—Bibliography and Technical Explanation of Various Requirements in AREA Specifications Relating to Iron and Steel Structures.

R. W. Gustafson, bridge engineer, Great Northern Railway, is chairman of Subcommittee 8 on Specifications for the Design of Structural Plate Pipe with Diameters Greater than 15 Ft.

Three subcommittee chairmen were unable to be present: J. E. South, system engineer—structures, Pennsylvania Railroad, chairman of Subcommittee 3—Corrosion of Deck Plates; R. C. Baker, engineer of structures, Chicago & Eastern Illinois Railroad, chairman of Subcommittee 6—Preparation and Painting of Steel Surfaces; and Ellis E. Paul, partner, Howard, Needles, Tammen & Bergendoff, who will be the new chairman of the Subcommittee on Composite Steel and Concrete Spans; Non-Ferrous Metal Bridges, Collaborating with Committees 8 and 30.

Mr. President, I would now like to invite comments or questions from the floor concerning our report. If there are no comments or questions, this concludes the report of Committee 15. [Applause]

PRESIDENT BEEDER: Thank you, Mr. Neufeld and your committee, for the highly important work which you are continuing in keeping our Association and the railroads completely up to date with respect to new developments in steel structures design and construction. As you no doubt know, the specifications which have been developed by your committee are widely used in structural courses in many engineering colleges of the country, and as a guide for bridge engineers on many railroads throughout the world. All of this has reflected great credit upon our Association.

Thank you again, Mr. Neufeld. Your committee is now excused with the thanks of the Association. [Applause]

Discussion on Buildings

[For report, see Bulletin 567, pp. 191–234]

PRESIDENT BEEDER: The last of our structural committees to be heard from today, and the last feature on this afternoon's program, is the report of our Committee 6— Buildings, which will be supplemented by a four-part symposium on Air Rights Developments on the Railroads.

The chairman of Committee 6, completing his first year as chairman, is K. E. Hornung, architect, The Milwaukee Road, Chicago. Mr. Hornung, if you and the other members of your committee will come to the platform we shall be very pleased to hear your report. Mr. Hornung, you may proceed.

Assignment 1—Revision of Manual.

CHAIRMAN K. E. HORNUNG: Mr. President, members and guests of the Association:

Committee 6—Buildings, has been most active this past year, our major activity being the work on Assignment 1—Revision of Manual.

As those of you who make reference to Chapter 6 of the Manual know, a considerable portion consists of complete specifications for building construction and materials. Such specifications are implements to which reference is made frequently for building construction, and therefore require current revision. To carry out this task, the entire membership of Committee 6 was assigned to work on Assignment 1. Our goal to review all Manual material will be completed by this fall.

I should like to present W. G. Harding, architect of the Wabash Railroad, who has given so much of his time and talent as chairman of Subcommittee 1 on this Manual review. Mr. Harding.

W. G. HARDING: Mr. President, members of the Association and guests:

Your committee has diligently pursued, during the past year, its assignment of bringing up to date and modernizing the specifications of Chapter 6, to the end that with the proposed reprinting of the Manual this year, the majority of the specifications will have been reviewed.

The report on Assignment 1, covering additions, revisions or deletions to the Manual material, is printed on pages 191 through 234 of Bulletin 567, and is too voluminuos to review in detail in the allotted time. All material therein presented has been approved by letter ballot of the committee.

Mr. President, I move the adoption of the foregoing additions, revisions and deletions to the Manual.

[The motion was duly seconded.]

PRESIDENT BEEDER: There is a considerable amount of work covered here in a great many pages, and it shows a lot of effort on your part in reviewing these matters. Are there any comments or questions?

[The motion was put to a vote and carried.]

Assignment 2—Specifications for Railway Buildings.

CHAIRMAN HORNUNG: Subcommittee 2, under the direction of Chairman H. T. Seal, engineer of buildings, Chesapeake & Ohio, is studying and preparing specifications for newly developed materials and construction methods.

Assignment 4—Wind Loading for Railway Building Structures.

CHAIRMAN HORNUNG: Subcommittee 4, with G. A. Morison as chairman, who is assistant engineer, Canadian Pacific Railway, is continuing the study of available data on wind loading on railway buildings.

Assignment 8—Infra-Red Ray Heating, Collaborating with Committee 18.

CHAIRMAN HORNUNG: Subcommittee 8, headed by Chairman J. W. Gwyn, assistant engineer, Missouri Pacific, consists of an enthusiastic group of capable engineers with the most interesting and exciting subject of infra-red ray heating. In this study we are collaborating with Committee 18.

In this relatively new method of heating by radiation, many applications to the railroad industry are being explored. Data are being obtained from actual installations, and in the near future will be made available to the Association.

Imagine, if you will, a method of heating diesel locomotives to maintain the temperature of this equipment at 60 deg, with outside temperatures of —30 deg and ambient temperatures inside the enginehouse well below zero, where protection of the locomotive or similar equipment during layover is the only motive for heating. Infra-red ray appears to offer such protection at a fraction of the operating cost of conventional building heating systems.

We are enthused with the data being obtained this winter, and know the Association will find it most interesting.

Mr. President, this concludes the reports on the work of Committee 6.

PRESIDENT BEEDER: Thank you, Mr. Hornung.

CHAIRMAN HORNUNG: It is now the privilege of Committee 6 to present a special feature—a symposium on Air Rights Developments on the Railroads.

Webster defines "symposium" as "collection of opinions on a subject." For this collection of opinions we are most fortunate to have been able to prevail upon four gentlemen who are familiar with this subject. While they are coming forward to the speaker's table I should like to identify them.

J. M. TRISSAL, vice president and chief engineer, Illinois Central Railroad.

H. J. McNally, chief engineer, New York Improvements, Pennsylvania Railroad.

C. E. Defendorf, chief engineer, New York Central System.

F. E. Austerman, chief engineer, Chicago Union Station Company.

This is a most timely subject, for the utilization of air rights over railroad property provides a means of additional income.

Our first presentation will be by J. M. Trissal. Mr. Trissal.

Symposium—Air Rights Developments on the Railroads

On the Illinois Central

By J. M. TRISSAL
Vice President and Chief Engineer, Illinois Central Railroad

When your president invited me to appear before this august body to take part in a symposium, I thought that I had better have a clear understanding of what I was getting into. Imagine my surprise when Webster gave these definitions of a symposium in this order:

1. A drinking party,
2. A feast preceeded by drinks, and finally,
3. A conference at which a particular subject is discussed.

The last one relieved my mind no end.

What are Air Rights?

I like to think of air rights as three-dimensioned subdivisions of land and space so as to make them adaptable to different uses at various elevations. Since a three-dimensional subdivision involves many engineering problems not encountered in the usual subdivision of land, the engineering department on the Illinois Central handles all air-rights negotiations.

Our sales and option contracts—developed jointly with our law department—include, among other items, the following salient points:

1. Plan showing permissible location of columns and caissons. This is very important as length of transfer girder spans over tracks will have a great influence on construction cost and hence the selling price of the air rights.
2. Elevation of top of caissons and low steel in air-rights plane.
3. Method of conveyance—more about this later.
4. Provision for a construction contract covering staging of construction operations—more about this later.
5. Control of assignment of contract to prevent speculation.
6. Guarantee of completion of building and recapture provision on default.
7. Construction of viaducts surrounding block to be at no expense to railroad.
8. Rearrangement of railroad facilities and provision for adequate lighting and ventilation to be at purchaser's expense.
9. Selling price to be net to railroad.
10. Provision for approval of Illinois Commerce Commission. We must satisfy the Commission that we are receiving a price which is adequate and that the construction will not interfere with our common carrier function.

One question often asked is, how do we set a price on air rights? Our price is based on two considerations. First, price of comparable naked land less extra cost of air-rights construction, taking into account saving in basement excavation and ability to get two decks of parking below the viaduct level without loss of rentable space; and second, what the traffic will bear.

Fig. 1 shows our air-rights area north of Randolph Street in Chicago. Existing and future viaducts are shown. The four blocks to the right of the Outer Drive contain 16 acres within building block lines. The 12 blocks to the left of the Outer Drive contain 32 acres within block lines, exclusive of the Prudential Building. Blocks 5, 6, 7, 8, 11

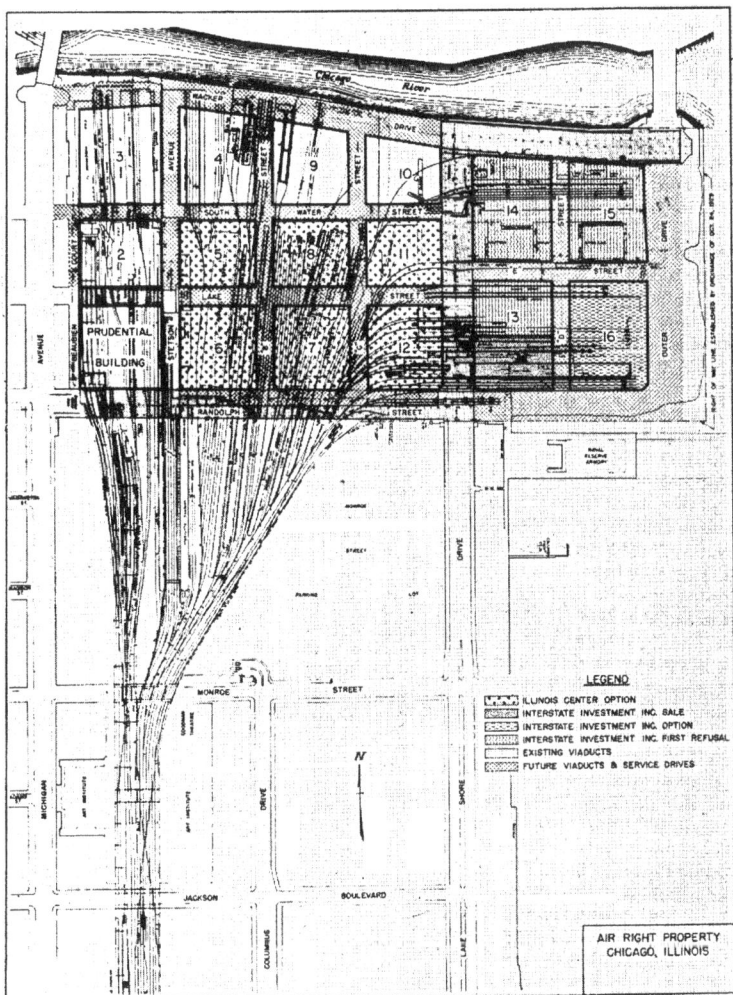

Fig. 1

Fig. 2.

Fig. 3.

and 12 are under option to the Illinois Center Corporation. The south half of Block 13 is covered by a sales contract to Interstate Investments, which also has an option to purchase Block 16 and a right of refusal to the north half of Block 13 and Blocks 14 and 15. Within the last fortnight we reached an agreement with another developer for an option on Blocks 2, 3, 4, 9 and 10. In these last two cases, the option price is subject to escalation based on the cost-of-living index. In all cases options, which are for a fixed period, are only granted for a substantial consideration.

Fig. 2 is an artist's conception of the Interstate development, a 39-story building containing 940 apartments as well as a swimming pool, restaurants and ancillary commercial development. We expect to pass title to the air rights within 30 days.

Fig. 3 is an artist's conception of the Illinois Center development as now planned. The initial stage will be the four high-rise apartment buildings shown in the foreground. Each building will contain 1200 apartments.

Fig. 4 shows our property between 11th Place and 29th St. containing 167 acres which is susceptible to air-rights development. The areas lying south and north of 23rd St. viaduct are covered by purchase contracts. A total of about 9 acres is involved. You will note that they are adjacent to McCormick Place. A hotel is planned for the area south of 23rd St. and apartments for the area north of 23rd St.

Fig. 5 is an artist's conception of McCormick Inn to be located south of 23rd St.

Fig. 6 illustrates how we subdivide space-vertically. We convey fee title to the space occupied by caissons and columns and all space above the air-rights level which, except under viaducts, is elevation 25. This results in many individual lots, each with its legal description. In the Prudential deed there are over 500 separate lots, each with its legal description. You will note that there is space for two levels under the building below the viaduct level which is not included in computing the cubic content allowed under the zoning ordinance. Basement area is largely above instead of below ground. Based on the use of columns 3 ft in width, our basic spans are 18 ft for 1 track and 31 ft for two tracks.

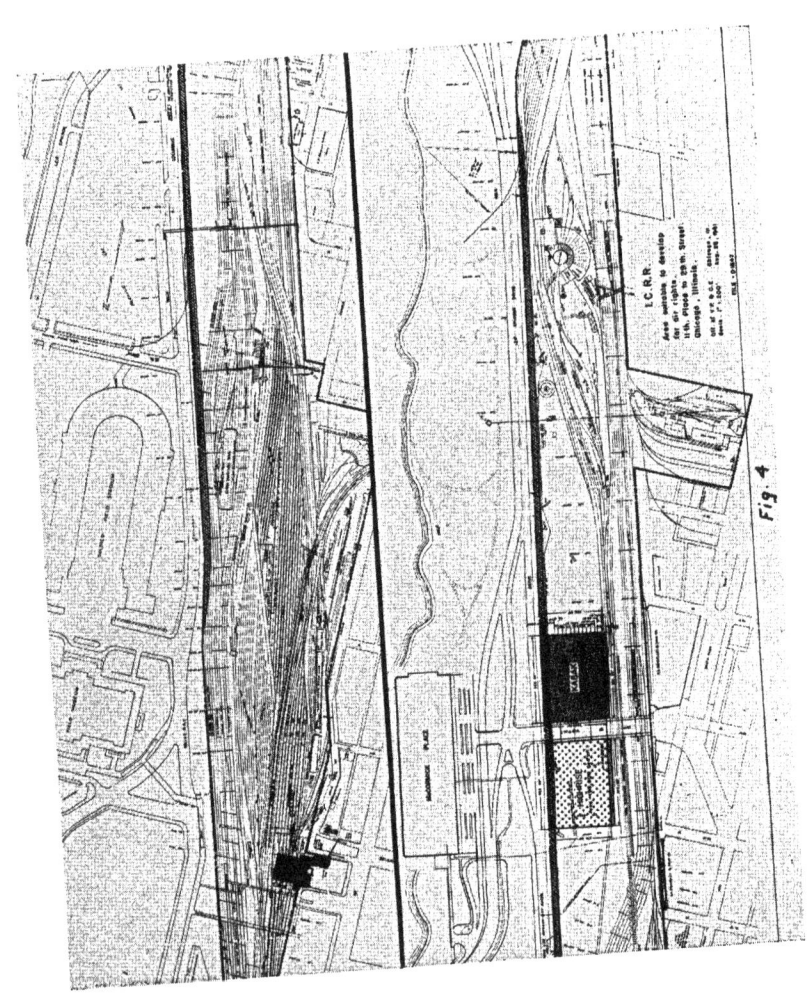

Fig. 4

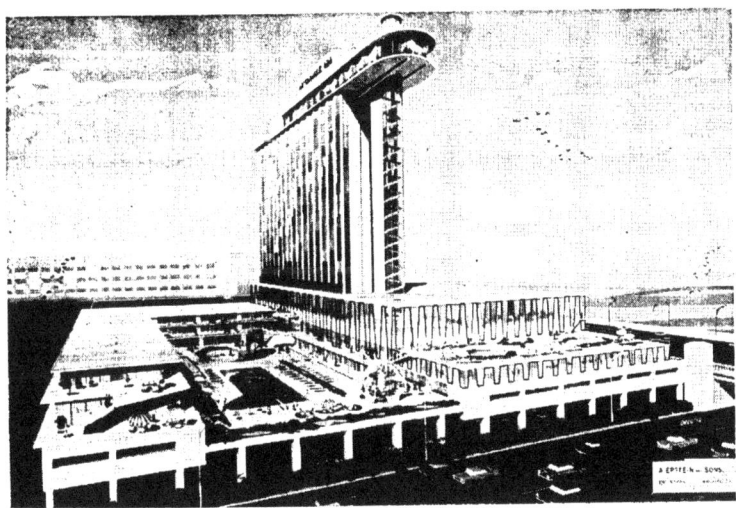

Fig. 5.

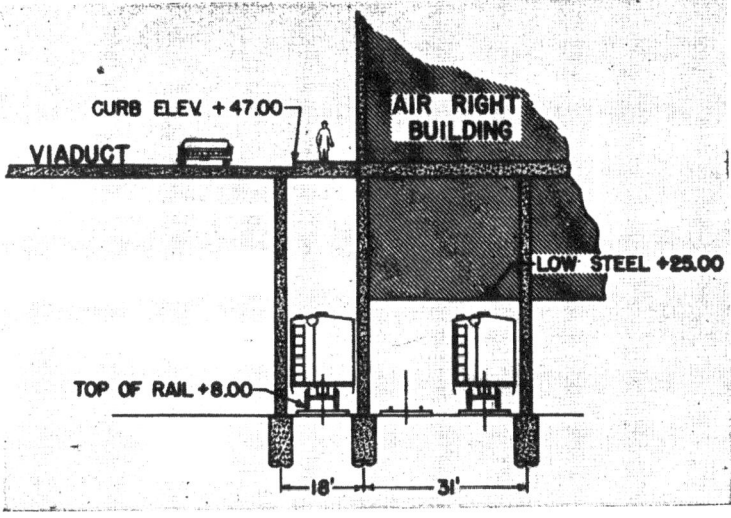

Fig. 6.

The next four figures cover the staging of construction of the Prudential Building which was unusually complicated due to the necessity of maintaining our electrified suburban operation under the building all during construction.

Fig. 7 shows track facilities as they existed at the start of construction.

Fig. 8 shows the first stage, which was to construct some of the caissons. Tracks in service and tracks out of service are indicated.

Fig. 9 shows completed caissons as well as those being constructed. Suburban tracks in service are shown as well as tracks being constructed but not yet in service.

Fig. 10 shows construction moving westward. After completion of the caisson construction, the same procedure is followed for columns and floor slabs.

In closing, I would like to mention one other facet of air rights—that is, selling air-rights space to toll roads when they cross over our right-of-way. We have two cases in Chicago and five in west Kentucky where this has been done. To date we have not sold any space for highways which are not part of a toll road system, but we are working on it.

Thank you. [Applause]

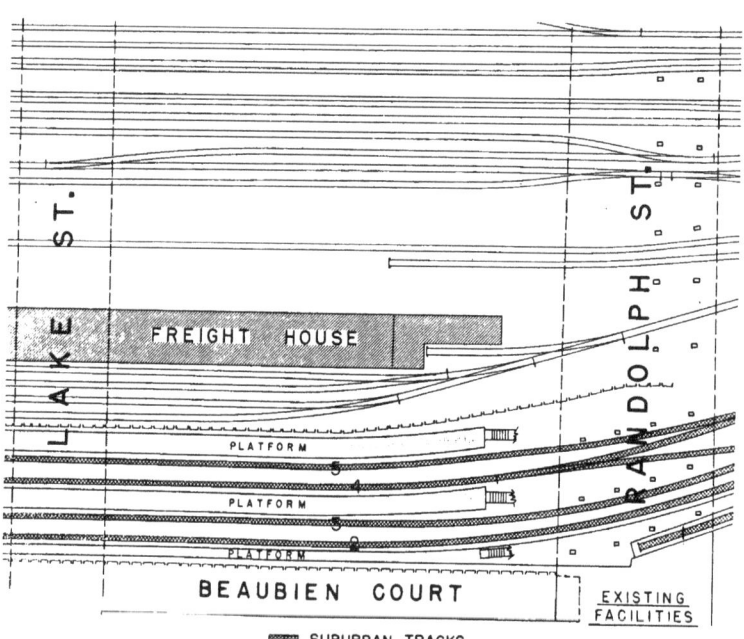

Fig. 7.

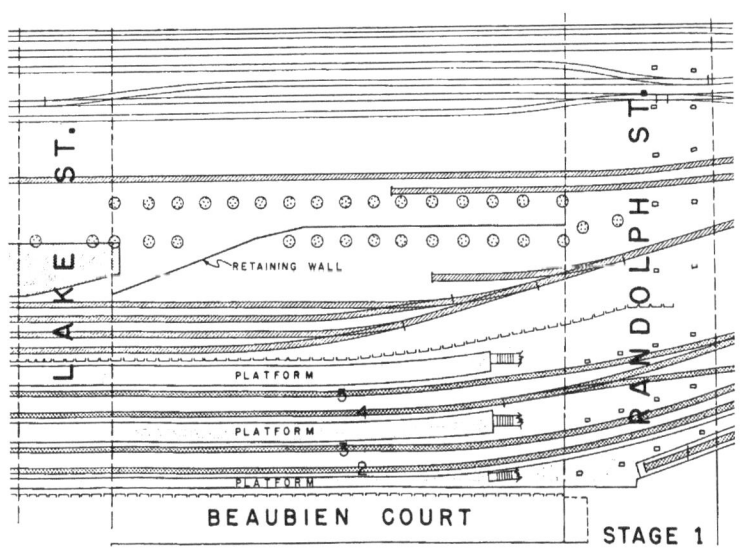

BEAUBIEN COURT

STAGE 1

⊗ CAISSONS BEING CONSTRUCTED
▨▨▨ SUBURBAN TRACKS IN SERVICE
▨▨▨ FREIGHT TRACKS OUT OF SERVICE

Fig. 8.

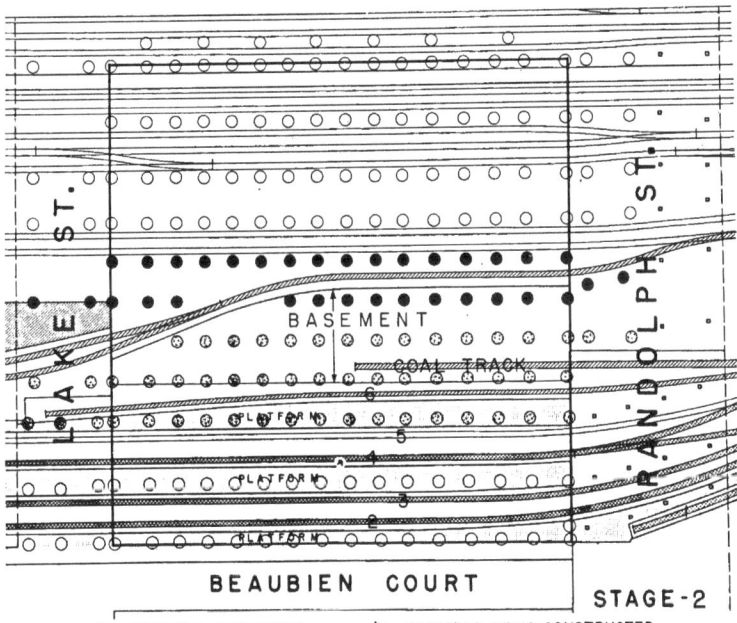

BEAUBIEN COURT

STAGE-2

● CAISSONS COMPLETED   ⊗ CAISSONS BEING CONSTRUCTED
▨▨▨ SUBURBAN TRACKS IN SERVICE
▨▨▨ TRACKS BEING CONSTRUCTED NOT IN SERVICE

Fig. 9.

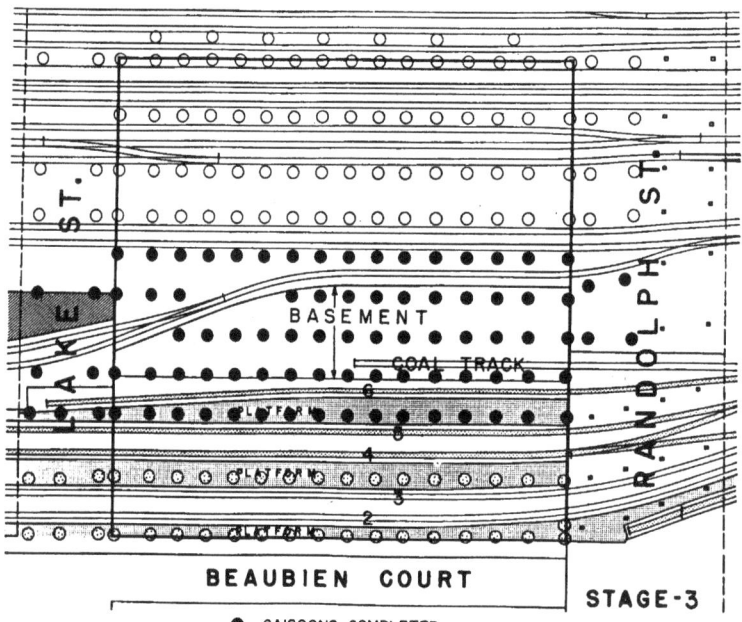

BEAUBIEN COURT

STAGE-3

● CAISSONS COMPLETED
⊕ CAISSONS BEING CONSTRUCTED
▭ SUBURBAN TRACKS IN SERVICE

Symposium—Air Rights Developments on the Railroads

The Engineers Stake in Planning the Development of Air Rights

By H. J. McNALLY

Chief Engineer, New York Improvements, Pennsylvania Railroad

The present-day changes in our urban communities and the growth of our country's utility systems increase the potential value and uses of our railroad properties. Our industrial-growth increases, the demand for utility services, the elimination of open land in our urban areas and the increase in "high-rise" multiple-story construction in our cities has increased the number of potential buyers in the real estate market looking at railroad properties. Most of the railroad companies we work for have property in the cities. In many instances this property is located in the heart of the most valuable sections of those communities.

The development of the railroad "Air Rights" has been in process for many years in a limited sense. The potential for the future is growing more and more. This is not only true in the glamor spots such as the New York Central's Grand Central area in New York City, the Illinois Central's lake-front property here in Chicago, the Pennsylvania Railroad's Pennsylvania Station area in New York City and other similar spots around the country, but the potential use of the railroad right-of-way has also grown and is growing in all urban areas and, to a limited extent, elsewhere. For example, in many of our cities it is virtually impossible for an electric power company to acquire a new right-of-way. In the case of planning the construction of a building over railroad right-of-way there are cases where foundation costs allocated to the number of square feet being developed may have been $4 to $5, or more, per square foot when the planning was limited to three or four floors of development. However, when you get into the high-rise category it is a possibility that foundation costs may be reduced to $1.50 to $2, or less, per square foot, thus making it competitive with construction cost in adjacent areas.

Now this is where the railroad's engineering people fit into the picture. They have the knowledge as to what is necessary to maintain railroad uses of the property and, with a little dreaming on their part, can conceive of ways and means by which the property can also be used for other purposes. We must assist our income-producing counterparts who handle our real estate matters. Many of us are looked upon by our management as the "spenders", the fellows always asking for money for rail, ties, machinery, etc. If we work toward developing the use of our property we may be able to temper this opinion a little.

In many of our cities the railroads' properties are the last large plots available on a single purchase, or lease basis, in the downtown area. Where city planning codes are restrictive, quite often the railroads' right-of-ways are zoned in the least restrictive catagories, thus increasing the potential value to a larger part of the real estate market.

The railroad engineer, knowing what is essential and what relative value exists in the various parts of the railroad structures and functions, can look at a potential project in a different light than the real estate developer. There may be a good chance of relocating parts of the railroad facilities or a change in operations, thus freeing areas for construction purposes or permitting construction at a reduced cost. This, of course, with the thought of producing a return for the railroad from a piece of property instead of its use continuing as an expense.

In and around passenger stations, in our yard layouts, etc., we may occupy more space than required. The construction cost necessary to reduce the size of the area or areas used may not be warranted by any savings in the cost of operations alone. However, when construction cost in air rights is considered, along with the possible savings in operating costs, such savings may warrant the change in operations and may also make the development of the air rights attractive to a buyer.

This thought is not restricted to in-town areas only but applies anywhere on the right-of-way. As an example, you may have a communication pole line which you have considered replacing by leasing circuits. Any savings found may be marginal or may not warrant a change. This same property may have been considered for use by an electric utility pole line. The corrective induction interference equipment necessary and other factors may not permit efficient use of this property by the utility. However, when both thoughts are considered together you may eliminate a maintenance expense you now bear and at the same time produce an additional return for your company. All utility companies have plans for growth. It is incumbent upon us to investigate how the use of our railroad right-of-way can fit into these plans and see that its use is included. This thought does not apply to the air rights only, of course, but also applies to the sub-surface use of the land.

Many times the railroad ownership on each side of the operating right-of-way does not provide sufficient area for a particular development. However, the right to connect the pieces by building over or under the railroad may increase the uses such property has and increase its value. Under these circumstances joint use of some of the facilities for access, drainage purposes, etc., may be necessary.

The railroad engineering people do have a stake in the development of the railroad property, and more specifically the air rights, and should extend themselves toward that end.

[Mr. McNally then presented slides showing various air-rights developments on the Pennsylvania Railroad in Pittsburgh and Philadelphia, none of which are reproduced herein, and concluded with the following remarks:]

Now the project I am personally most interested in at the present time is the redevelopment of Pennsylvania Station in New York City. Here we plan to confine all of the station operations below street level, except access and loading and unloading of trucks. Many of the present functions will be altered or changed. Above ground Madison Square Garden Center Incorporated will develop a complex on the eight acres plus, including a multi-storied office building on the 7th Ave. end of the two city blocks involved. On the 8th Ave. end a circular-shaped building is planned which will house the main arena with a 20,000 to 24,000-seat capacity, a smaller 4000-seat arena, an ice-skating rink, bowling alleys and other functions now provided to the people of the city by the present Madison Square Garden. The physical work incident to this project is not started to date, but we are all working toward the end that it will be soon. [Applause]

Recent view of the 7th Ave. end of the Pennsylvania Railroad station
in New York City.

Cencept showing the "after" appearance of the Pennsylvania Railroad
station in New York City when the station facilities are moved below
ground.

Symposium—Air Rights Developments on the Railroads
On the New York Central

By C. E. DEFENDORF
Chief Engineer, New York Central System

The New York Central has been interested in the development of air rights for a good many years. In fact, the first building in this category was a postoffice structure built in 1909 over two levels of terminal platforms and tracks at Grand Central in New York. Since that time we have been increasingly interested in the development of air rights because of the income derived. During the past 6 or 7 years, however, we have become much more active and are now involved in air rights studies and actual construction in Boston, Cleveland, and particularly, New York City. To indicate the various locations involved in the New York City area, I would like to present a slide [Slide 1] showing diagrammatically the many locations involved.

The most interesting area in air rights, of course, is the Grand Central Terminal section where in excess of 20 office buildings and hotels have been built over two levels of terminal tracks since 1909. During the past few years 5 of the original multi-story air-rights structures have been demolished and replaced by larger and higher structures; most times, of course, for a different type of occupancy. At present a 12-story luxury apartment building is being torn down to be replaced by a 52-story office building. I have a slide showing the Grand Central express level tracks with the air-rights structures outlined. [Slide 2]

It is interesting to note that there are, or will be, when the present construction is completed, over 16,700,000 sq ft of office and hotel space over this area. The most outstanding structure of the group is the Pan-Am Building, being built by Grand Central Terminal Building, Inc., at a cost of $100,000,000. [Slide 3]

This 2,400,000-sq ft structure is to be 59 stories high, and is being built over the main portion of the tracks and platforms. It is necessary, of course, to maintain terminal operation during all phases of construction.

The 808 ft high tower of the new structure is supported on 99 new columns which are carried down to bed rock. Each of these columns rests on a steel grillage and a vibration pad made up of steel, lead, and asbestos. The function of these pads is to prevent vibration due to train operation from being transmitted up into the new building. I have a slide showing a grillage being placed between tracks. [Slide 4]

From street level to grillage the lower portion of these 99 columns were 60 ft long and had to be dropped through upper and lower deck openings to the grillage below. [Slide 5]

These columns also have to be isolated from the track structure throughout their length so as to prevent vibration transmission.

The lower 10-story base portion of the Pan-Am Building is carried on 200 columns from the original structure. These columns have been spliced above the train deck area and also rest on vibration pads. I am happy to say that the structure is up to the 44th floor and is expected to be ready for occupancy by December of this year.

It is the responsibility of the railroad to see to it that the design of the structure, including the wind bracing, and all phases of the construction procedure are analyzed so as to permit the railroad to operate safely and to maintain train schedules; and to see that passengers are in no danger while in the terminal area (this is an important factor, since over 95,000 passengers use the facilities each work day).

(Text continued on page 721)

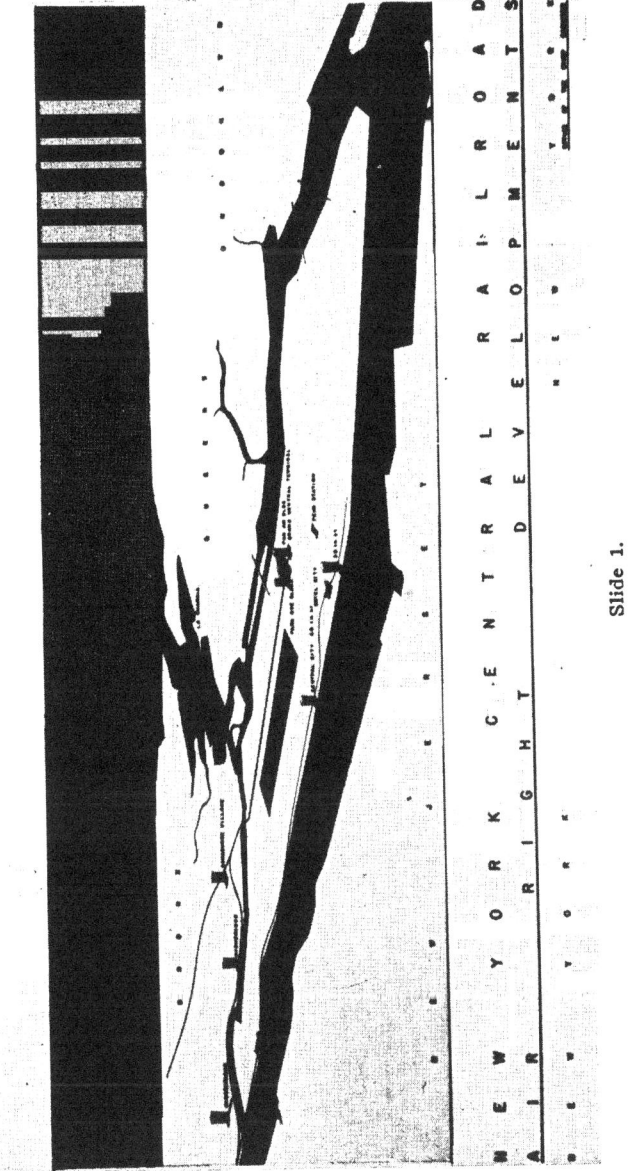

Slide 1.

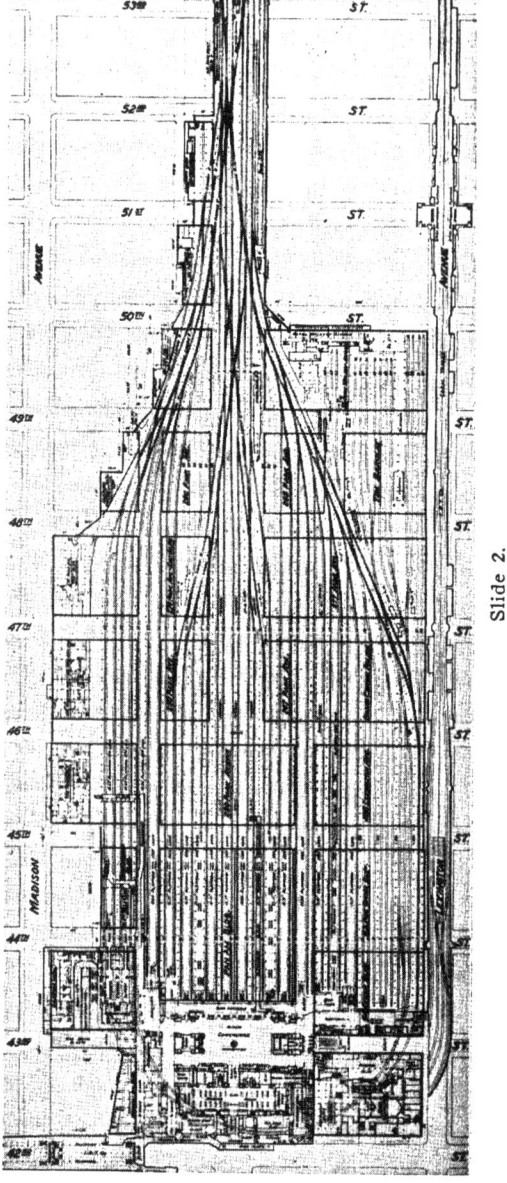

Slide 2.

Slide 3.

Slide 4.

Slide 5.

Because of the time limit, this talk is merely an outline. However, I would be happy to review with anyone interested our experience, which I am sure would be of considerable help to any of you who are ready to embark on air-rights projects. [Applause]

Symposium—Air Rights Developments on the Railroads
Chicago Union Station Area
By F. E. AUSTERMAN
Chief Engineer, Chicago Union Station Company

Railroads have been erecting buildings and structures over their tracks for a century. At the Chicago Union Station, air-rights buildings have been under study since 1926, with two buildings completed by 1930—the Daily News building, and the U. S Post Office. Foundation conditions are essentially the same at these locations. The Daily News' 25-story building—now known as the Riverside Plaza—is built on caissons resting on the limestone strata at an elevation of —100 city datum. The low section of the building and the plaza is built on spread footings resting on hard pan at elevation —55 city datum. The spacing of columns varies from 45 to 100 ft spanning the tracks and 20 ft along the tracks.

Caissons for the Post Office rest on limestone rock at elevation —85 city datum. Columns are spaced at 45 ft centers spanning the tracks and about 29 ft centers along the tracks.

The ventilation systems installed in the two buildings exemplify two different methods of removing the fumes. Designed primarily for steam locomotives, both systems proved just adequate when the conversion to diesel locomotives was made.

The Daily News building was constructed on an air-rights block of 104,000 sq ft between Madison and Washington Sts., two blocks north of the Union Station. The 25-story portion covers 20 percent, with a plaza along the river covering about 30 percent. The remainder of the building is low section building to 7-stories high.

The track area covers 62 percent of the block, with 4500 lin ft of smoke slot over all trackage. Included in the track area are four double-slip switches and five turnouts with smoke slots fitted over them. The smoke slot and concrete slab ceiling are suspended by ½ in by 4 in steel bars encased in concrete. After more than 30 years there has been no appreciable deterioration of these steel supports. The cost of maintenance has been nominal.

Chambers from 2 to 10 ft in height cover the entire track area connecting to a high-rise stack with 220 sq ft of area to the roof of the building (335 ft above the track level). There is some exhaust fan draft ventilation exhausted into this shaft, but basically the ventilation is natural high-rise draft with heated space around the shaft. The basic ventilating system was designed by Holabird & Roche, architects.

The Post Office covers 14 station and mail tracks over an area of 280,000 sq ft between Van Buren St. and Harrison St., one block south of the Union Station. Smoke slots are constructed over every track and have a total length of two miles. The smoke slots open into a ventilating chamber 1 to 8 ft high, which in turn connects to 8 shafts about 40 sq ft each in area. These shafts go to the roof of the pylons at elevation +220. Fifteen huge exhaust fans draw the fumes from the chamber up these shafts. The system operates well on most days, but on heavy humid days the gases are not drawn completely upward and frequently come back down through other smoke slots.

The smoke slots and concrete ceiling are suspended by ¾-in diameter steel rods. These deteriorated rapidly in the early days until covered by pitch, lead sleeves and an asphaltic fabric tape. The cost of maintenance of the entire system, which includes a large drainage system with heated down spouts, has been high.

Erwin S. Wolfson, New York City builder, proposes to erect a group of office buildings in the three blocks north and south of the Station between the aforementioned air-rights buildings. Plans include a private two-lane street along the river to be named Riverside Drive.

Plans for the first building, a 20-story office building, is under study by Architects Skidmore, Owings & Merrill. Over one-half of the 90,000-sq-ft block will be plaza. The street level will be ramped up from Madison and Monroe Sts., with a service entrance from Canal St. Escalators will handle the people to the second level where elevators are planned.

Plans for ventilating the more than 100,000 sq ft of track area will generally follow the Daily News building system. [Applause]

MR. AUSTERMAN: Mr. President, the symposium members asked me to tell you that if there are any questions they will be pleased to try to answer them.

CHAIRMAN HORNUNG: Gentlemen, are there any questions you would like to ask the four members of our symposium?

J. H. BROWN (Frisco): I would like an explanation on the subject of clearance. I believe Mr. Trissal said 31 ft for two tracks plus a 3-ft column. Is that in conformity with the Illinois clearances, or are there special provisions made?

MR. TRISSAL: We have a special order in the State of Illinois in the area north of Monroe Street. We are allowed 13-ft track centers, 7½-ft side clearance, and 16½-ft vertical clearance. We have an air-rights level now of 25 and a track level of 8, so we get 17 ft vertical clearance instead of the 16½ ft allowed under the special order of .the Illinois Commerce Commission.

PRESIDENT BEEDER: 'Thank you, Mr. Trissal. Are there any other questions? If not, thank you, Mr. Hornung. In the light of developments in recent years it is evident that your Chapter 6 required quite a going over, and we appreciate the large amount of attention which your committee has given to this matter during the past two years. With all of this work now behind you, I hope you will be able to make real progress on your other assignments during the coming year.

Gentlemen, we have just heard a most interesting and enlightening symposium on a matter which is engaging the attention, profitably, of an increasing number of railroads, with much more of the same to come, I am sure.

Accordingly, I know I express your feeling when I say to Mr. Trissal, Mr. McNally, Mr. Defendorf and Mr. Austerman that we are very much indebted to them for their discussions of air rights developments on their railroads.

Mr. Hornung, you are now excused, with our thanks to you and your committee. [Applause]

PRESIDENT BEEDER (continuing): You will be interested to know that the convention registration up until about 4 pm today included 335 railroad men and 295 guests, a total of 630. As was to be expected, this is a far cry from the total registration of 1946 at the close of the first day of our convention last year, but under the circumstances it is gratifying, and is really larger than was expected.

The meeting is now adjourned, to reconvene tomorrow morning in this room promptly at 8:30 am. So, get to bed early, and let's get off to a prompt 8:30 start in the morning on the interesting program still ahead of us.

[The meeting adjourned at 5:45 pm]

Morning Session, March 10, 1962

[The meeting reconvened at 8:30 am, President Beeder presiding.]

PRESIDENT BEEDER: The meeting will please come to order.

This is an early start; and while I realize that our entire group has not assembled as yet, it is essential that we begin on time if we are to complete today's program on time, which we expect will be about 12:30 pm.

Discussion on Water, Oil and Sanitation Services
[For report see Bulletin 567, pp. 99–135]

PRESIDENT BEEDER: The first committee to report this morning is our Committee 13—Water, Oil and Sanitation Services, the chairman of which is D. C. Teal, superintendent water supply, system, Chesapeake & Ohio Railway, at Richmond, Va., who is completing his three-year term as chairman. Mr. Teal, we are pleased to have your committee back on our platform again, and we would appreciate your coming up here promptly.

While the committee is coming to the platform, may I remind you that you have the privilege of the floor to ask questions or make comments. I hope you will take advantage of this to the extent that you can contribute something and time will permit.

Mr. Teal, will you proceed, please.

CHAIRMAN D. C. TEAL: Mr. President, members of the Association and guests:

The report of Committee 13 may be found in Bulletin 567, pages 99 to 135, incl. Our nine assignments cover a variety of subjects. Subcommittee 1 has been especially active in their review of Manual material, and their report contemplates extensive revisions that will have to be voted on by the members of the Association here present. Subcommittees 2, 3, 5 and 6 have published informative reports, a brief summary of which will be presented to you by their respective chairmen. You are invited to interrupt at any time with questions or comments.

Assignment 1—Revision of Manual.

CHAIRMAN TEAL: The report on Assignment 1 will be presented by E. C. Harris, engineer of tests, Missouri Pacific Railroad, St. Louis, Mo., chairman of the subcommittee. Mr. Harris.

E. C. HARRIS: Mr. President and members of the Association:

In accordance with its assignment, the committee has reviewed Chapter 13 of the Manual, giving concerted study to Parts 4, 5 and 6 which concern the chemistry of water conditioning. Our recommendations with respect to these parts will follow discussion of our work on other parts of the chapter.

There are eight documents which the committee recommends for approval without change:

In Part 1:

    (1) Pages 13-1-1 and 13-1-2 under "General Principles of Water Supply Service", that document entitled "Supply".

Under Part 3—Specifications for Equipment, Materials and Methods Used in Water Supply:

    (2) Page 13-3-1, "Specifications for Cast Iron Pipe and Special Castings."

    (3) Page 13-3-1, "Specifications for Hydrants and Valves."

    (4) Page 13-3-17, "Specifications for Welded Steel Tanks for Water or Oil Storage."

    (5) Page 13-3-2, "Specifications for Laying Cast Iron Pipe."

Under Part 8—Sanitation:

    (6) Page 13-8-1, "Water for Drinking Purposes."

    (7) Page 13-8-2, "Railway Sewage Disposal Facilities."

    (8) Page 13-8-21, "Sterilization of New and Repaired Water Wells."

Mr. President, I move that the Association reapprove the eight aforementioned documents without change.

[The motion was duly seconded, was put to a vote, and carried.]

MR. HARRIS: In a number of documents in Chapter 13 minor revisions and limited deletions of certain sentences are recommended to bring the wording of the subject matter up to date to conform with present-day requirements. Documents falling within this classification follow:

    (1) Pages 13-2-1 to 13-2-15, "Pumping Plants", in which 12 sentences are recommended for revisions, 4 sentences for deletion.

    (2) Page 13-8-24, "Waste Disposal". Revise one sentence.

    (3) Page 13-9-3, "Diesel Locomotive Fuel and Water Service". Revise two sentences.

Mr. President, I move that the three aforementioned documents, with minor revisions and deletions as indicated, be reapproved by the Association.

[The motion was duly seconded, was put to a vote, and carried.]

MR. HARRIS: Due to changes in the times which have brought about different and improved methods of operations, a number of the documents of Chapter 13 have become outdated and obsolete and are recommended for deletion from the Manual by the committee. The particular documents recommended for such deletions follow:

(1) Pages 13–7–1 to 13–7–2—Part 7: "Characteristics of Water." This pertains only to the steam locomotive and is recommended for deletion in its entirety.
(2) Pages 13–3–6 and 13–3–7, "Specifications for Wood Water Tank."
(3) Pages 13–3–8 to 13–3–10, "Specifications for Tank Hoops."
(4) Pages 13–3–11 to 13–3–13, "Specifications for Steel Structures for Wood Water Tanks."
(5) Pages 13–3–14 to 13–3–16, "Specifications for Timber Substructure for Water Tanks."
(6) Pages 13–M–1 to 13–M–6, "Water Service Records."
(7) Pages 13–M–7 to 13–M–9, "Water Service Organization."

Mr. President, I move that the seven aforementioned documents be deleted from the Manual for the reasons specified.

[The motion was duly seconded, was put to a vote, and carried.]

MR. HARRIS: As previously mentioned, your committee has placed special emphasis on the study of Parts 4, 5 and 6 of Chapter 13 which cover the chemical treatment of water for various purposes, specifications for chemicals used in the conditioning of water, and chemical procedures used in the testing of water. Most of these sections have been rewritten by your committee for the purpose of bringing the subject matter up to date to conform with present-day practices of other industries in the field of water conditioning.

A brief description of these revised sections is herewith outlined to the Association for vote of approval.

Pages 13–4–1 to 13–4–6. "Water Treatment." The section entitled "General" was rewritten to present a more concise and detailed description of various methods of conditioning water, and also to eliminate any reference, both direct and indirect, to the steam locomotive.

Pages 13–4–7 to 13–4–16. "Treatment of Water for Cooling Purposes", was revised to eliminate any recommended practices which have become outmoded, and also to eliminate specific terms which are no longer consistent with accepted standards.

Pages 13–5–1 to 13–5–11, "Specifications for Chemicals Used in Water Treatment." This section was rewritten to include only references to specifications for chemicals published by the American Water Works Association which is the recognized authority in this field.

Pages 13–6–1 to 13–6–10, "Water Analysis and Interpretation of Results", was rewritten to include a system of expressing results of water analysis in the terms which are used by most modern industrial laboratories, and to show only those references on laboratory procedures published by the American Public Health Association and the American Society for Testing and Materials.

For the benefit of those who are familiar with only the railroads' system of reporting results of water analysis in terms of grains per U. S. gallon, this system of units was retained in the committee's revised version of pages 13–6–4 to 13–6–6, "Rapid Field Tests."

Mr. President, I now move that revised Parts 4, 5 and 6, as presented in the foregoing discussion, be approved with the revisions indicated.

[The motion was duly seconded, was put to a vote, and carried.]

Assignment 2—Prevention of Corrosion in Hot and Cold Water Systems.

CHAIRMAN TEAL: The committee's report on Assignment 2 will be presented to you by Subcommittee Chairman J. J. Dwyer, chief chemist—system, Chesapeake & Ohio Railway, Huntington, W. Va.

J. J. DWYER: Mr. President, members of AREA, and guests:

The title of this report is Corrosion Prevention in Potable Hot Water Systems. It appears on page 115 of Bulletin 567 for November 1961. While it was written primarily for the use of railroads, it contains information of interest to each of you present here today. The reason is that the hot-water heater and hot-water piping in your own home constitute a potable hot water system such as described in this report.

We are continually updating the information in this report, and in light of recent comprehensive research, notably that of Obrecht and Quill at Michigan State University, we have re-evaluated the previous suggestion of red brass for temperatures above 140 F, and now advise Admiralty tube or metal (approximate composition: 71 percent copper, 1 percent tin, 28 percent zinc, 0.06 percent arsenic) at a cost of about 5 percent over copper, or for the most severe service (both high temperatures and high velocities) 90/10 cupro-nickel alloy, at about 65 percent over cost of copper.

The hot-water corrosion problem is summarized briefly as follows:

Causes

1. Corrosive dissolved gases, such as carbon dioxide, oxygen, hydrogen sulfide, ammonia.
2. Dissolved copper.
3. High temperature.
4. High velocity, turbulence.
5. Galvanic couples.
6. Soft water.

Remedies

1. Use deaerators where practicable.
2. Use air release valves where practicable.
3. Do *not* use pneumatic tanks in hot-water systems.
4. Use type K copper tube for temperatures up to 140 F.
5. Confine temperatures to the range 130–140 F.
6. Where higher temperatures are required, use 90/10 cupro-nickel alloy.
7. Confine velocities to 5 fps, preferably 4 fps.
8. Insulate galvanic couples.
9. Use cathodic protection in hot water tanks when practicable.
10. Use appropriate chemical treatment. This may be:
    (a) Carbonate balance system.
    (b) Sodium silicate.
    (c) Sodium metaphosphate.
    (d) Mixture of (b) and (c).
    (e) pH adjustment.
    (f) Dealkalization.

This report is presented at this time as information. It is currently being revised and rewritten for submission as Manual material a year hence.

Assignment 3—Design, Construction and Operation of Coach Servicing Facilities to Comply with Regulations of U. S. Public Health Service.

CHAIRMAN TEAL: Our report on Assignment 3 will be presented to you by Subcommittee Chairman C. F. Muelder, assistant to engineer of buildings, Chicago, Burlington & Quincy Railroad, Chicago. Mr. Muelder.

C. F. MUELDER: Mr. President: No new developments concerning design or construction of coach servicing facilities have come to the attention of the committee. However, through the efforts expended on the Hydrant Research Project sponsored by the committee, a certain post-type hydrant, formerly available commercially, is again being manufactured. Replacement parts for these hydrants are also now available.

Another function of this committee is to advise of any changes or new regulations of the U.S. Public Health Service or other governmental bodies as they may pertain to the scope of water, oil or sanitation services. Complete details of this report are found in the November 1961 Bulletin.

We call your attention to several significant items:

The revision of the 1946 Drinking Water Standards, U. S. Public Health Service, was made available September 1, 1961. The principal changes were the revision of chemical limits and the addition of such items as detergents, fluorides, arsenic, cyanides, iron manganese and nitrates. Also, a section on radioactivity has been introduced. It is interesting to note that several of the items added are man-made chemical pollutants and are indicative of the increasing problems and need for effective protection of our water resources.

A new program for grading railroad dining cars has been placed in effect. Cars are graded A, B, or C according to the defects found at the time of inspection. Grades have no expiration date and remain current until reinspection indicates otherwise. Grade A is given only to those cars on which there are no defects. Grades B and C are provisional ratings given according to the type of defects noted.

Greater emphasis is now being placed by the U.S. Public Health Service on inspection of sanitary drinking water facilities of diners, coaches, sleeping cars, etc. U.S. Public Health inspectors are now equipped with new modern field test equipment, enabling them to determine more quickly the sanitary qualities of water samples taken on trains. Any equipment found to have unsatisfactory water is to be taken out of service until conditions are corrected. This may require draining, cleaning and flushing, sanitizing, or shopping the car, depending on the defects noted. After correction, this equipment can be placed back in service, notifying the U.S. Public Health Service that the defects have been corrected.

Originally, inspections of water hoses, buckets, hydrants, and ice houses were of prime importance. This new approach not only requires that the water sources be adequate and the handling of watering equipment be properly cared for, but the distribution and storage on railroad equipment must be maintained in such a manner as to provide proper sanitary water as it is dispensed and consumed. Control of the water itself precludes the necessity of extensive inspections by the U.S. Public Health Service in various railroad yards. It places greater responsibility on railroad personnel to keep up local standards. It is our expectation that this emphasis will require our industry to police itself more thoroughly than ever in order to avoid the problems that arise when equipment is taken out of service.

Your committee will continue to keep abreast of new regulations, using its influence where possible to promote the best interests of the railroad industry, keeping you informed of changes in governmental policy or regulations as they may occur.

Assignment 4—Cathodic Protection of Pipelines and Steel Storage Tanks, Collaborating with Committee 18.

CHAIRMAN TEAL: Your committee notes progress in its study on Assignment 4 but has not developed enough information during the past year to justify a formal report. The proper handling of this assignment also includes collaboration with and approval by the recently reactivated Committee 18—Electricity, all of which takes extra time.

Assignment 5—Economic Justification of and Methods for Conditioning Water for Use in Steam Generators and Engine Cooling Systems of Diesel Locomotives.

CHAIRMAN TEAL: Our report on Assignment 5 will be presented to you by Subcommittee Chairman R. A. Bardwell, engineer of tests, Chicago & Eastern Illinois Railroad, Danville, Ill. Mr. Bardwell.

R. A. BARDWELL: Committee 13 in past years has presented a number of subcommittee reports on techniques and chemicals recommended for treatment of diesel locomotive waters. The report this year evaluates the economic benefits from various methods and types of water treatment.

The extent of the savings realized from the use of various methods of water treatment for diesel locomotives has been gathered from various railroads. The figures show that treatment is economically justified in all cases, with the best methods giving the largest savings.

Methods to calculate individual savings are offered in this report. The report indicates that an approximate return of $13 can be realized for each $1 judiciously spent for cooling-water treatment, and at least $2.60 for each $1 spent for steam-generator water treatment. Net annual material savings can average almost $500 for each diesel cooling system and over $1000 for each steam generator under the conditions reported. These figures, exclusive of labor, are felt to be conservative. To assure continuity of operations, the best water treatment methods available are recommended to effect these savings.

Assignment 6—Railway Waste Disposal.

CHAIRMAN TEAL: Assignment 6 has been on our docket for a number of years and is one that we feel should be continued indefinitely because of changing conditions on the railroads, increasing interest on the part of the U.S. Public Health Service which is concerned with these matters, and to keep the Association informed as to latest developments in the now very active stream pollution control field.

The chairman of the subcommittee, T. A. Tennyson, engineer of tests and sanitation, St. Louis–Southwestern Railway, Pine Bluff, Ark., will present the report on this subject. Mr. Tennyson.

T. A. TENNYSON: In handling waste disposal plans and problems it is important to know the regulatory agencies involved and how to establish contact with them. Since locations and titles of officials are not the same for all states, your committee has compiled and published a list of state and interstate agencies which have the function of water pollution control, along with their addresses and the titles of the officials in charge. This report is submitted as information.

CHAIRMAN TEAL: Mr. President, as retiring chairman I would like to express publicly my thanks to the members of Committee 13, its officers and the chairmen of its subcommittees, and to the AREA office staff for the support they have given me, for the work they have done to help me carry on the work of this committee, and for the privilege and pleasure of being associated with them.

As my last official act I want to present to you the new chairman and vice chairman of Committee 13. Your new chairman is E. C. Harris, engineer of tests, Missouri Pacific Railroad. Will you please stand to be recognized, Mr. Harris? [Applause]

Your new vice chairman of Committee 13 is T. A. Tennyson, Jr., engineer of tests and sanitation, St. Louis–Southwestern Railway. Mr. Tennyson, will you please stand to be recognized? [Applause]

Mr. President, this concludes the report of Committee 13.

PRESIDENT BEEDER: Thank you, Mr. Teal. Your committee, under your direction for the past three years, has presented many interesting and valuable reports with respect to the wide range of subjects coming under its jurisdiction, all of which is appreciated. We particularly appreciate the effort of your committee during the past year in reviewing and updating its chapter of the Manual.

We are pleased to welcome Mr. Harris as the new chairman of Committee 13 and Mr. Tennyson as the new vice chairman. Mr. Harris, I congratulate you upon your advancement to chairman, and would like to present you with this chairman's gavel. [Applause]

Mr. Teal, you and your committee are excused, with the thanks of the Association

Discussion on Electricity
[For report see Bulletin 569, pp. 457–484]

PRESIDENT BEEDER: Moving on to the next feature of our program, I would now like to invite to the speaker's table the members of our newest committee, Committee 18—Electricity, which, you will recall, was re-established early in 1961 to take over the fixed-property work of the former Electrical Section, Engineering and Mechanical Divisions, AAR.

The chairman of this committee is P. B. Burley, superintendent communications and electrical engineer, Illinois Central Railroad, Chicago. Mr. Burley, we will be pleased if you and the other members of your committee will come to the platform promptly and make your presentation.

CHAIRMAN P. B. BURLEY: Mr. President, members of the Association and guests:

The report of Committee 18 will be found starting on page 457 of Bulletin 569, Vol. 63. This year Committee 18 recommends revisions in the Manual and offers two reports as information.

Our work program suffered greatly from the death of Robert C. Welsh, Jr., in October 1961. Mr. Welsh was chairman of Subcommittees 4 and 8—Power Supply, Motors and Controls. A Memoir to him appears in our report

Assignment 1—Revision of Manual.

CHAIRMAN BURLEY: Moving to the details of our report, Subcommittee Chairman W. O. Muller, electrical engineer—fixed property, Missouri Pacific Railroad, will present the report on Assignment 1.

W. O. MULLER: Mr. President, your committee submits the following recommendations with respect to the AAR Electrical Manual:

Section 14—Safety. Delete in its entirety.

Section 11—Electric Heating. Revise Chapter 2, Parts 4, 5, 6, 8 and 10 as shown in detail in our report in Bulletin 569, January 1962.

Mr. President, I move that these Manual recommendations be adopted by this Association as submitted.

[The motion was duly seconded, was put to a vote, and carried.]

Assignment 10—Wire, Cable and Insulating Materials, Collaborating with Mechanical Division, AAR.

CHAIRMAN BURLEY: F. T. Snider, foreman, Office of Electrical Engineer, Pennsylvania Railroad, will present the report on Assignment 10.

F. T. SNIDER: Mr. President and members:

This year one new wire and cable standard was published which is of interest to all Member Roads. This is the IPCEA-NEMA Standards Publication on Thermoplastic-Insulated Wire and Cable. Copies of this specification are available from either organization.

Major revisions were also made during 1961 to the IPCEA-NEMA Standards Publication for Rubber-Insulated Wire and Cable for the Transmission and Distribution of Electrical Energy, IPCEA No. S–19–81, Third Edition, NEMA No. WC3–1959.

On pages 460 through 463 of Bulletin 569 we have presented an article on the use of aluminum conductors for insulated power cables. This article can be summed up by the first sentence in the summary: "Aluminum's place in the wire and cable field will be decided largely by the overall economics of the installation."

Mr. President, this report is submitted as information.

Assignment 13—Railway Electrification, Collaborating with Mechanical Division, AAR.

CHAIRMAN BURLEY: Subcommittee Chairman L. B. Curtis, assistant engineer, Pennsylvania Railroad, will present a report on Railway Electrification. Mr. Curtis.

L. B. CURTIS: Mr. President, members, and guests:

It has been some time since a report on railroad electrification has been made to this body. The work of this subcommittee is somewhat unique in that it applies directly to only a few of the railroads in AREA. Yet the subcommittee believes that in the not too distant future many more American railroads will become involved, as is true in so many foreign countries.

The report starts on page 464 of Bulletin 569, January 1962. The first assignment on the use of carbon shoes on electric locomotive pantographs (we are particularly concerned with this subject as it affects the contact wire), the use of aluminum in electrification, and the possibilities of the use of prestressed concrete poles.

The next assignment has to do with the thermit welding of aluminum, principally for grounding purposes. This problem has not been solved as yet, but progress is being made.

The assignment that follows is quite comprehensive. The first part has to do with keeping American railroads informed of the new developments in electrification in the United States and abroad. Most improvements are in foreign countries where electrification is flourishing. The most prominent items are the increased use of 25-kv commercial frequency electrification and the development and use of semi-conductor rectifiers to convert this a-c power to direct current on the locomotive.

Since first cost of electrification is one of the major deterrents in this country, study is continuing on methods of reducing this cost. A thorough report on 600-v third-rail systems, particularly applicable to commuter service, is made. A brief report is given of the use of semi-conductor rectifiers in this country.

Finally, a table is presented giving the latest figures on railroad electrification mileage and type throughout the world. Our attention has been called by Mr. Trissel to an error in this table on page 484 in regard to the Illinois Central mileage. The correct figures are 38 route miles and 130 track miles.

Your subcommittee hopes that it can keep alive for American railroads the advantages possible with electrification in main-track areas where electrification can be justified economically.

This is a progress report submitted as information.

D. F. LYONS [CSS & SB]: Mr. President, may I make a comment at this time?

On Table 1, page 484, I note what appears to be an omission. The third longest electric railroad in the United States—The BA&P, 134 route miles—does not appear on that list.

MR. CURTIS: The reason is that it is a captive road owned by the Anaconda Copper Mining Company and is operated entirely as a subsidiary of Anaconda. The American railroads in our list are all Class I Roads.

CHAIRMAN BURLEY: I would like to take this opportunity to present others on this committee who are here with me.

J. J. Schmidt, our vice chairman, assistant director of research, Denver & Rio Grande Western.

E. B. Hager, assistant engineer, Illinois Central Railroad, chairman of Subcommittee 9—Electrolysis and Electrolytic Corrosion.

B. D. Allison, electrical engineer, Chicago & North Western Railway, chairman of our new Subcommittee 11—Electric Heating.

E. M. Hastings, Jr., wire crossing engineer—system, Chesapeake & Ohio Railway, chairman of Subcommittee 15, Relations with Public Utilities.

There are two others who unfortunately could not be with us. They are: T. F. Jelnick, electrical engineer line property, Burlington Lines, who is now handling Subcommittee 4 and 8—Power Supply and Motors and Controls, and W. R. Preece, Jr., assistant electrical engineer, Baltimore & Ohio Railroad, chairman of Subcommittee 5—Illumination.

Mr. President, this concludes the report of Committee 18.

PRESIDENT BEEDER: Thank you, Mr. Burley. I hope that you and the other members of your committee, many of whom are relatively new members in our Association, are beginning to feel completely at home with us and find our methods and procedures quite acceptable.

Your committee has made a good start in its 1961 work, and we look forward to reports on your other assignments at our next convention. Thank you again, Mr. Burley. You and your committee are now excused with the thanks of the Association. [Applause]

Discussion on Highways
[For report see Bulletin 567, pp. 117–189]

PRESIDENT BEEDER: Next on our program is the report of our Committee 9—Highways, the chairman of which is J. M. Trissal, vice president and chief engineer, Illinois Central Railroad, and a director of our Association. Mr. Trissal, if you, along with your vice chairman, secretary and subcommittee chairmen will come to the speaker's table, I shall be glad to turn the microphone over to you. Mr. Trissal.

CHAIRMAN J. M. TRISSAL: Mr. President, members of the American Railway Engineering Association and guests:

The report of Committee 9 is published in the November 1961 Bulletin 567 starting on page 177. Your committee is reporting on four of its five assignments. Work is progressing on the other assignment.

Assignment 1—Revision of Manual.

CHAIRMAN TRISSAL: Subcommittee Chairman E. R. Englert, cost control engineer, Louisville & Nashville Railroad, will present the report on Assignment 1.

E. R. ENGLERT: In 1961, the Board Committee on Manual requested that all committees review and present to the Association all material in their chapters of the Manual for reapproval, with or without change, no later than the 1962 convention. Your committee has completed its review of Chapter 9 and our recommendations were published in Bulletin 567 on pages 178 to 185, incl.

Mr. President, I move that these recommendations be adopted.

[The motion was duly seconded, was put to a vote, and carried.]

Assignment 3—Merits of Various Types of Highway–Railway Grade Crossing Protection, Collaborating with Communication and Signal Section, AAR.

CHAIRMAN TRISSAL: In the absence of Subcommittee Chairman J. A. Jorlett, structural engineer, New York Improvements, Pennsylvania Railroad, who I understand is trying to find his house off the coast of New Jersey, the report will be given by Vice Chairman R. W. Mauer, assistant engineer, Atchison, Topeka & Santa Fe Railway.

R. W. MAUER: The Armour Research Foundation of the Illinois Institute of Technology, with funds provided through the Research Department of the Association of American Railroads, has produced the final report on an "Analysis of Railroad Crossings and Accident Data for the State of Ohio During the 10-Year Period, 1949 through 1958." The Ohio Department of Highways and 12 railroad companies operating in the State of Ohio assisted invaluably by providing information concerning the physical characteristics, volume of vehicular traffic, accident records and number and speeds of trains for 7416 crossings.

The principal objectives of the report were:

1. To determine relationships between accident rates and characteristics of the selected highway grade crossings, and
2. To obtain risk factors for these crossings. (The risk factor for a given crossing is defined as the expected accident rate over a period of 10 years.)

The program was carried out in three well defined phases: (1) the data were examined and summary tables constructed; (2) correlations between accidents and crossing characteristics were investigated; and (3) a regression surface was fitted to observational data to provide risk factors.

Prediction equations were derived for each of the following three crossing categories:

1. Crossings with painted crossbucks
2. Crossings with automatic flashing lights
3. Crossings with automatic flashing lights and gates.

The calculations required were performed on Armour Research Foundations's Univac 1105 digital computer and auxiliary equipment. A further study had been planned to determine the effect of changed protection on accident rates, but because of financial limitations, it was not possible to accomplish this study.

Your committee recognizes that there are weaknesses in the report which would require careful evaluation of all factors before this method of determining risk factors is used. Accordingly, Committee 9 will not release this report.

This is a progress report submitted as information. Your committee recommends that the subject be continued.

Assignment 5—Recommended Method of Developing Annual Maintenance Cost of the Various Types of Highway–Railway Grade Crossing Protection, Collaborating with Communication and Signal Section, AAR.

CHAIRMAN TRISSAL: Subcommittee Chairman F. C. Cunningham, division engineer, Chesapeake & Ohio Railway, will present the report on Assignment 5.

F. C. CUNNINGHAM: Your committee has been assembling information from various railroads on their experience in regard to the maintenance cost of the various types of highway-railway grade crossing devices. The Communication and Signal Section, AAR, is progressing a concurrent study of this subject and has initiated a one-year actual cost record study on a number of railroads which they expect to complete by July 1, 1962.

Committee 9 is holding its report in abeyance pending the outcome of this cost study. An evaluation of the results of this study, along with the information the committee now has, will form the basis of future recommendations.

This is a progress report submitted as information. Your committee recommends that the subject be continued.

Assignment 6—Methods of Providing Additional Advance Warning to Highway Traffic Approaching a Highway–Railway Grade Crossing.

CHAIRMAN TRISSAL: Subcommittee Chairman Raymond Dejaiffe, chief engineer, Toledo Terminal Railroad, will present the report on Assignment 6.

RAYMOND DEJAIFFE: Your committee reports on two installations of special advance warning signs connected with track circuits and used in connection with automatic protection to reduce gate breakage; and two installations of octagonal highway stop signs used at highway-railway crossings, which resulted in greatly improved accident records.

Also included is a report on the experimental use of "rumbler strips" or rough pavement ahead of existing dangerous highway intersections to alert the vehicle drivers to the danger. Consideration is being given to the use of similar rumbler strips on the approaches to highway-railway grade crossings.

It should be noted that all the above installations were made by different governmental agencies at no cost to the railroads.

This report is submitted as information and your committee recommends that the subject be continued.

CHAIRMAN TRISSAL: This year's convention marks the end of my term as chairman of Committee 9, and I want to express my sincere thanks to the members of the committee, and especially the subcommittee chairmen, for their splendid help and cooperation during the last three years. I also want to thank Secretary Neal Howard and his staff for their excellent assistance, guidance and cooperation.

Our new vice chairman will be Raymond Dejaiffe, chief engineer, Toledo Terminal Railroad. Mr. Dejaiffe's selection as vice chairman is in recognition of his loyalty to the committee and the splendid work he has done both as a member and as a subcommittee chairman. Mr. Dejaiffe. [Applause]

As my last official act, I take great pleasure in presenting my successor as chairman of Committee 9, R. W. Mauer, assistant engineer, Atchison, Topeka & Santa Fe Railway. As vice chairman during the past three years and secretary the preceding three years, he has demonstrated his interest in committee work and his loyalty to our committee. He will make our organization an outstanding chairman. Mr. Mauer. [Applause]

Mr. President, this concludes the report of Committee 9.

PRESIDENT BEEDER: Thank you, Mr. Trissal. Your committee has presented another series of valuable reports on a subject of great importance to the railroads, and we appreciate the further attention that it has given to is Manual chapter in order that the documents therein may be in step with best practices and procedures.

Our thanks to you, Mr. Trissal, for your interested and effective guidance of the work of Committee 9 for the past three years. As you give up your responsibilities, we are pleased to welcome as chairman, Mr. Mauer, and as vice chairman, Mr. Dejaiffe, knowing that they will be effective in progressing the work of your committee.

Mr. Mauer, it is with pleasure that I present you with a chairman's gavel as the symbol of your authority for the next three years. [Applause]

[Vice President L. A. Loggins assumed the chair.]

Discussion on Economics of Railway Labor
[For report, see Bulletin 568, pp. 259-273]

VICE PRESIDENT LOGGINS: We will now hear from one of our more "popular" committees, possibly because it has the word "economics" in its title, combined with the fact that it gets around each year to see interesting maintenance-of-way operations. I refer to our Committee 22—Economics of Railway Labor, the current chairman of which is J. E. Eisemann, chief engineer, Western Lines, of the Santa Fe System. Jack, if you and the other members of your committee will come to the platform, I shall be glad to turn this microphone over to you. Mr. Eisemann.

J. E. EISEMANN: Mr. Chairman, members and guests:

The activities of Committee 22—Economics of Railway Labor, are published in Bulletin 568 on pages 259 through 273. Initially the subjects for study were developed by our Subcommittee A—Recommendations for Further Study and Research. J. S. Snyder, assistant regional engineer, Southwestern Region, Pennsylvania Railroad, who is chairman of this subcommittee, will report on his efforts for the coming year.

J. S. SNYDER: Committee 22 has taken under consideration two new studies for 1962. One of these is Labor Economies to be Derived from Cropping Rail in Track versus Building Up Rail Ends by Welding. Howard Seeley, engineer maintenance of way, Detroit, Toledo & Ironton, has taken the chairmanship of the subcommittee for this study.

The other new subject is Labor Economies Inherent to Various Methods of Taking Up Track. John Stang, industry planning analyst, New York Central, has accepted the chairmanship of the subcommittee.

Assignment 1—Revision of Manual.

CHAIRMAN EISEMANN: W. W. Hay, professor of railway civil engineering, University of Illinois, has a report to present concerning revision of Manual which is submitted for your consideration, acceptance and adoption.

PROFESSOR W. W. HAY: Mr. Chairman, members and guests:

Subcommittee 1 presents for action by this convention the following proposed changes in Committee 22's Manual material. These proposals have received the necessary two-thirds approval of the committee by written ballot.

The material on pages 22–1–1, 22–1–2 and 22–1–3, Recruiting, Training and Welfare, has been rewritten into a more condensed version under the heading, Recruiting and Training, as set forth on pages 260 and 261 of Bulletin 568. This revision, too lengthy to be read in detail, deletes certain features not generally applicable, emphasizes the demands placed on recruiting and training by highly organized and mechanized gangs, and shortens the presentation of the remainder by eliminating reference to benefit associations and insurance.

Also, on page 22–1–3 in the material entitled Annual Inspection and Prize Awards, the title has been changed to Annual Inspection for Checking Progress and Planning Future Programs, and the first two paragraphs have been deleted and replaced with the following:

"A well-devised plan for an annual inspection provides a desirable means for checking progress and making future programs. This can be combined with a properly directed competition to increase the interest and activities of maintenance-of-way forces."

Mr. Vice President, I move the adoption of these revisions.

[The motion was duly seconded, was put to a vote, and carried.]

PROFESSOR HAY: Pages 22–1–4 and 22–1–5, Outfit Cars for Housing. Reapprove with the following revisions:

Under Sec. A. Design, item 4, delete the words: ". . . both as a matter of convenice and safety."

Page 22–1–5. Delete item 4 of Sec. B and replace with:

"4. Inspection should be made at sufficiently frequent intervals to ensure the proper physical and sanitary conditions of the equipment."

Mr. Vice President, I move the adoption of these revisions.

[The motion was duly seconded, was put to a vote, and carried.]

PROFESSOR HAY: Pages 22–2–1 and 22–2–2. Programming Work and Diversion of Traffic. Reapprove with the addition of a new section on the use of radio, as printed on page 262 of Bulletin 568.

Mr. Vice President, I move the reapproval of this material with the addition of the new section.

[The motion was duly seconded, was put to a vote, and carried.]

PROFESSOR HAY: Page 22–2–3, Weight of Rail. Reapprove with the following revision:

Delete Par. (e) and replace with:

"(e) For lines of high traffic density, the saving in track labor following the installation of heavier rail sections may reach 40 percent of the total expenditure for this item."

Page 22–2–3, Vegetation Control. Reapprove with the following revision:

Delete the words "fire hazard and" from the third line of the second paragraph.

Mr. Vice President, I move the reapproval of these documents with the revisions mentioned.

[The motion was duly seconded, was put to a vote, and carried.]

PROFESSOR HAY: Page 22–2–5, Mechanical Equipment. Reapprove with the revisions given on page 262 of Bulletin 568.

Mr. Vice President, I so move.

[The motion was duly seconded, was put to a vote, and carried.]

PROFESSOR HAY: Pages 22–3–1 to 22–3–4, incl., Programming work. Reapprove with the following revision:

Page 22–3–1. In the second paragraph, third from the last line of that paragraph, delete the phrase "it is necessary" and substitute the phrase "it is suggested."

Page 22–3–5, Mechanical Equipment. Reapprove with the revisions given on page 263 of Bulletin 568.

Mr. Vice President, I so move.

[The motion was duly seconded, was put to a vote, and carried.]

Assignment 2—Analysis of Operations of Railways that Have Substantially Reduced the Cost of Labor Required in Maintenance of Way Work.

CHAIRMAN EISEMANN: Assignment 2 will be reported on by Subcommittee Chairman E. J. Sierleja, regional industrial engineer, Pennsylvania Railroad.

E. J. SIERLEJA: This report is the 20th of a series on an assignment that has been reassigned annually since 1935. The report may be found beginning on page 259 of Bulletin 568.

This year's analysis was of the track rehabilitation process being used at the time of the committee's inspection by the Baltimore & Ohio Railroad on its Buffalo Division at Bradford, Pa.

I wish to take this opportunity to extend our thanks to the B&O for its cooperation and hospitality during our field trip.

This report is submitted as information.

Assignment 3—Labor Economies to be Derived from Work Measurement Standards for Comparison of Work Performance Among Various Gangs or Divisions.

CHAIRMAN EISEMANN: The subcommittee chairman for Assignment 3 is H. J. Fast, coordinator of work study—system, Canadian National Railways. I would like Mr. Fast to stand and be recognized. The study is continuing and we are reporting progress.

Assignment 4—Labor Economies to be Realized Through the Use of a Combined Surfacing-Timbering Gang vs. Separate Gangs for Surfacing and Timbering.

CHAIRMAN EISEMANN: H. W. Kellogg, assistant chief engineer, Chesapeake & Ohio Railway is chairman of Subcommittee 4, and I would like to have him stand and be recognized at this time. This study was completed this year and a final report has been published as information.

Assignment 5—Labor Economies to be Effected Through Use of Power Tools and Mechanized Equipment by Bridge and Building Gangs.

CHAIRMAN EISEMANN: W. E. Chapman, chief engineer of maintenance, Central of Georgia, at Savannah, Ga., is responsible for the report on Assignment 5, I would like to have Mr. Chapman stand and be recognized. This subject has been under study for some time, and this particular effort will close it out. The final report is published as information.

*Assignment 7—Labor Economies in Track Maintenance to be Derived
Through the Use of Combination On-Off-Track Equipment vs. On-Track
Equipment Only.*

CHAIRMAN EISEMANN: This subject was studied and report prepared under the direction of Subcommittee Chairman T. L. Kanan, assistant engineer of track, Colorado & Southern Railway Company, Denver. Mr. Kanan, will you kindly stand and be recognized.

The subject matter has been under study for two years. Progress is being made.

*Assignment 8—Labor Economies to Be Derived from the Welding,
Distributing, Laying and Maintenance of Continuous Welded Rail, Collaborating with the Special Committee on Continuous Welded Rail.*

CHAIRMAN EISEMANN: W. J. Jones, engineer maintenance of way and structures—system, Southern Pacific Company, is chairman of Subcommittee 8. Mr. Jones, will you please rise and be recognized. This is also a continuing subject, and progress is reported.

CHAIRMAN EISEMANN (continuing): I wish to record at this time that Committee 22 lost a hard-working, well liked member in the death last fall of A. H. Stimson, supervisor—maintenance of way material and equipment, Pennsylvania Railroad. A memoir has been prepared and will be published in the Proceedings.

MEMOIR
Alfred Hopper Stimson

Alfred Hopper Stimson, supervisor maintenance of way material and equipment, Pennsylvania Railroad, died suddenly at Titusville, Pa., October 12, 1961.

Mr. Stimson was born on January 11, 1907, at Cincinnati, Ohio, and received his higher education at the Virginia Military Institute, graduating with a degree of Bachelor of Science in Civil Engineering.

He entered the service of the Pennsylvania Railroad at Newark, Del., July 1, 1927, as an assistant on the engineering corps. In 1934 he was promoted to supervisor of track at Cincinnati, and was promoted to division engineer at Philadelphia in 1940.

He served as division engineer at Fort Wayne, Ind., and Pittsburgh, Pa., until his promotion to supervisor maintenance of way materials and equipment at Buffalo, N. Y., in November 1955.

He is survived by his wife, Catherine T. Stimson; two daughters, Mrs. Gilbert (Hannah) Morcroft, and Elizabeth Anne Stimson; and a son, Alfred H. Stimson, Jr.

Mr. Stimson joined the American Railway Engineering Association in 1955 and in the following year became a member of Committee 22.

Mr. Stimson gave generously of his time to committee work and was very interested in the activities of the Association. It is with sorrow and regret that we here record his passing.

J. E. EISEMANN
J. S. SNYDER
D. E. RUDISILL
Committee on Memoir

CHAIRMAN EISEMANN: Mr. Vice President, this concludes the presentation of Committee 22.

VICE PRESDENT LOGGINS: Thank you, Mr. Eisemann, and your committee for your work during the past year and for the interesting and informative reports presented to this convention. Nothing is more important in railway engineering and maintenance than to increase the efficiency of our maintenance operations, and we look to your committee for guidance and recommendations in this regard.

Thank you again, Mr. Eisemann, and the members of your committee. You are now excused with the thanks of the Association. [Applause]

[President Beeder resumed the Chair.]

Discussion on Maintenance of Way Work Equipment
[For report see Bulletin 568, pp. 351-378]

PRESIDENT BEEDER: Our next report deals with the power tools and machines which, to a large, extent, have made possible the economies in the use of railway labor reported on by Committee 22. I refer to our Committee 27—Maintenance of Way Work Equipment, the chairman of which is R. S. Radspinner, supervisor, roadway machines and equipment, Chesapeake & Ohio Railway, at Saginaw, Mich. Mr. Radspinner, if you and the other members of your committee will come to the platform promptly, I shall be glad to turn this podium over to you. Mr. Radspinner.

CHAIRMAN RADSPINNER: Mr. President, members and guests:

The reports of this committee will be found in Bulletin 568, pages 351 to 378, incl. We are reporting on eight assignments, three of which are progress reports and three that are continuing subjects; two are final. Our presentation will be confined to a brief summary; but the committee feels, because of the ever increasing interest in work equipment, that much of value can be gained by those who read the complete report in the Bulletin. This committee will be happy to answer any questions from the floor after each subcommittee chairman completes his report, or at the end of our complete presentation.

Assignment 1—Revision of Manual.

CHAIRMAN RADSPINNER: The changes recommended in Chapter 27 of the Manual this year include the complete rewriting of the material on Wire Rope Used with Work Equipment, and minor word changes resulting from the special Manual review suggested by the AAR. Paul Martin, methods engineer—system, New York Central Railroad, and W. T. Hammond, regional engineer, Pennsylvania Railroad, handled a great deal of the work on the special review.

The specific revisions are as follows:

Care and Operation of Maintenance of Way Equipment: Several word changes were made as the result of the special review mentioned above.

Wire Rope Used with Work Equipment: Rewritten to incorporate improvements in both material and techniques. Data on sizes and types brought up to date.

Motor Cars, Push Cars and Trailers: Word changes resulting from the special review are recommended.

Gentlemen, these recommendations were approved by letter ballot of this committee and are submitted for adoption.

Mr. President, I move that the recommendations presented be included in the Manual.

[The motion was duly seconded, was put to a vote, and carried.]

Assignment 1 (a)—Revision of Handbook of Instructions for Care and Operation of Maintenance of Way Equipment.

CHAIRMAN RADSPINNER: The report on Assignment 1 (a) was prepared under the direction of Subcommittee Chairman R. E. Berggren, supervisor, maintenance of way equipment, Illinois Central Railroad. Mr. Berggren was unable to attend this meeting, so G. E. Roberts, assistant engineer, Santa Fe, will present his report at this time.

G. E. ROBERTS: Instructions for the care and operation of three machines are given in this report—a four-tool spot tamper, an eight-tool multi-purpose tamper and a tie spacer. All these machines have sufficient general distribution and use to justify their inclusion in the Handbook. Therefore, it is the recommendation that this material be included in the next revision of the Handbook.

Assignment 2—Improvements to Be Made to Existing Work Equipment.

CHAIRMAN RADSPINNER: Our report on Assignment 2 includes changes recommended in work equipment now available and used by our railroads. G. L. Zipperian, supervisor of work equipment, Great Northern Railway, is the subcommittee chairman but he is unable to be with us today, so I shall read his report.

Your committee submits a progress report as information. It is a continuation of the reports submitted by this committee in previous years. It pertains to changes in existing work equipment that the committee has found to be both desirable and practical.

This report covers the improvements which we suggest be made to eight machines. The suggestions originated with members and others directly connected with the operation and maintenance of the machines.

These machines are:

Spike pullers, mechanical
Tie adzer
Tie coaters
Truck cranes, on-and-off track operation
Tie tampers, two types of production machines
Under-track positioning and control carriage for plow
Utility crane and tie inserter
Power track wrench

The recommended improvements in the report were submitted to the manufacturers of the equipment for their consideration. They expressd gratitude for the suggestions and advised that they would cooperate, where possible, in effecting the desired improvements. Many of the suggestions are now being incorporated in production-model machines.

Assignment 3—Standardization of Parts and Accessories for Work Equipment.

CHAIRMAN RADSPINNER: Assignment 3 is a continuing subject under which we shall study various components or accessories used in the mechanical, hydraulic, pneumatic or electric systems of work equipment. We are now working in hydraulics, covering the standard hydraulic tank and its fittings. We propose suggesting (the first of a series) specifications that can be used in ordering equipment which will provide us with tanks well constructed, and easy to operate and maintain.

Assignment 4—Study of Average Annual Costs of Repairing and Serv-icing Various Widely Used Multi-Tool Power Tampers.

CHAIRMAN RADSPINNER: Our committee has recommended that Assignment 4 be dropped because the realistic information needed for an accurate and satisfactory report on the subject is not available.

Assignment 5—Set-offs and Set-off Attachments for Work Equipment.

CHAIRMAN RADSPINNER: The report on Assignment 5 is a final report and will be presented by Subcommittee Chairman R. W. Bailey, engineer of scales and work equipment, Chicago & North Western Railway.

R. W. BAILEY: Mr. President, members and guests:

Various types of portable set-offs were studied, and a general type similar to that being furnished by manufacturers was described and illustrated. This is a basic unit consisting of various beam supports connected to vertical posts connected to ground supporting bases. These units are light in weight and their parts are easily handled by two men. Their use eliminates the need for the crib type of set-offs.

The report further deals with refinements in and extensive designs of complete hydraulic units for the quick removal of major units or combinations of units used by a gang, as well as a portable turnout for this same purpose.

This is a final report. I recommend that it be accepted as information.

Assignment 6—Procurement and Stocking of Parts and Materials for Repair of Work Equipment.

CHAIRMAN RADSPINNER: Under Assignment 6 information is being assembled from railroads to study the various methods now being used. Recommendations as to good practice will be developed and reported.

Assignment 7—Vehicles and Equipment for Routine Track Inspection.

CHAIRMAN RADSPINNER: A final report on vehicles and equipment for routine track inspection, which covers the subject very completely, is published in Bulletin 568. It includes good pictures of the equipment and tools described. H. D. Hahn, assistant chief engineer, Chicago & Illinois Midland Railway, subcommittee chairman, will make the report.

H. D. HAHN: Mr. President, members of the Association, and guests:

This report on Assignment 7 can be found on pages 369 to 378, incl., of Bulletin 568. It is a final report, submitted as information.

Although the standard conventional modes of transportation are still actively used by track inspectors and track inspection parties, there has been a trend in recent years toward the use of highway-rail vehicles in this work. These vehicles can be furnished as a complete unit, or the highway-rail conversion equipment can be purchased separately and installed at the railroad's shop or at the factory. There are a wide range of makes and models of passenger cars and trucks which can be equipped for highway-rail operation, and the conversion equipment presently available from two manufacturers is described in detail in this report. A third manufacturer offers highway-rail vehicles on a complete unit basis only, and these units are described in this report, along with the optional equipment which can be installed on almost any make or model of vehicle.

Other vehicles which are briefly referred to in this report are the larger bus-type vehicles which operate exclusively on the rails, and the on-track vehicles which measure the quality of track and record track characteristics. The former is used solely for the transportation of track inspection parties while the latter type, which can be either self-propelled or pulled by a passenger train, is used to determine the condition of the track from recorded data.

In addition to the optional equipment which can be installed on a highway-rail vehicle, a description of other units of inspection equipment is included in this report. The items described include a track inspection machine, two different track gage indicators, a track recorder, and the combination track level and gage. Several railroads have indicated that they are equipping certain inspection crews with a small complement of power tools to expedite on-the-spot repairs of track defects where possible. The future may bring a change in the vehicles and equipment used by track inspection personnel as evidenced by the present-day modifications which many railroads have made in conjunction with force reductions and the lengthening of assigned territories.

Assignment 8—Equipment for the Control and Performance of Jacking in Track Surfacing Operations.

CHAIRMAN RADSPINNER: Much of the equipment being studied under Assignment 8 is still under development. We propose to cover all tamping power jacks and auxiliary wire, optical, and electronic devices now on the market and in use for surfacing operations.

CHAIRMAN RADSPINNER (continuing): Mr. President, this concludes the presentation of the report of Committee 27.

PRESIDENT BEEDER: Thank you, Mr. Radspinner, and your committee, for another group of informative reports on this all-important matter of power tools and machines with which to accomplish our work most effectively and economically.

As has been said before, our Association looks to your committee to keep us up to date on all important developments in this equipment, and as to the best practices in the maintenance and repair of this equipment. With developments coming so fast, this places a real responsibility on your committee, but we are confident that it will continue to meet that responsibility.

Your committee is now excused, with our thanks. [Applause]

Discussion on Ties and Wood Preservation
[For report see Bulletin 568, pp. 315–336]

PRESIDENT BEEDER: Beginning the home stretch of our convention program, we will hear next from our Committee 3—Ties and Wood Preservation. The chairman of this committee is R. B. Radkey, engineer of ties and treatment, Illinois Central Railroad, Chicago. If Mr. Radkey and the members of his committee will come to the platform, we shall be glad to hear their report at this time.

CHAIRMAN R. B. RADKEY: Mr. President, members and guests:

Committee 3—Ties and Wood Preservation, is a new committee this year, resulting from the consolidation of old Committee 3—Ties, and old Committee 17—Wood Preservation. We are responsible for both Chapter 3 and Chapter 17 of the Manual. Details of our report have been printed in Bulletins 565, June–July; 566, September–October, and 568, December. Several subcommittees have prepared significant data, but we shall limit our presentation to the high points.

Assignment 1—Revision of Manual.

CHAIRMAN RADKEY: The report on Assignment 1 will be presented by Subcommittee Chairman C. S. Burt, assistant to vice president, purchases and stores, Illinois Central Railroad.

C. S. Burt: Mr. President and members of the Association:

This committee has been engaged in a critical review and study of Chapter 3—Ties, of the Manual. The recommendations for changes therein, now presented for your consideration and approval, are intended to bring the chapter fully up to date, so that the methods and devices described therein will be abreast of the best current practice throughout the industry.

Accordingly, the following documents are offered for reapproval without change:

Installation and Keeping Records of Test Sections.
Marking Ties for Service Records.
Traffic Unit for Use in Comparing Tie Life.
Best Practices for Tie Renewals.
Fundamentals to Be Considered in Designs of Substitute Ties.

Mr. President, I so move.

[The motion was duly seconded, was put to a vote, and carried.]

Mr. Burt: The following documents are offered for reapproval with minor changes:

Explanation of Cross Tie Design.
Size of Holes Bored For Spikes.
Economic Comparisons of Ties.
Conservation of Timber Supply.

In connection with Conservation of Timber Supply, Item (8) of Par. 2 as printed on page 321 of Bulletin 568 should be changed to read: "(8) improved track maintenance."

Mr. President, I move the reapproval of these documents with the minor changes described in the report.

[The motion was duly seconded, was put to a vote, and carried.]

Mr. Burt: It is believed that splitting and checking are the major causes of failure of oak and other species of hardwood cross ties. Also, it is generally recognized that certain types of anti-splitting devices provide a worthwhile measure of protection and assistance in reducing these defects to the minimum.

In recent years the types of anti-splitting devices have become somewhat standardized. The same is true with respect to the location of the device in the tie, and the time and method of application. Dowels, possibly the most effective means of retarding splitting and checking, have recently come into more general use for this purpose.

Therefore, the committee recommends the elimination of Specifications for Devices to Control the Splitting of Wood Ties, adopted in 1944, and elimination of the Application of Anti-Splitting Devices, also adopted in 1944, as these two documents now appear in the Manual, and the substitution therefor of the rewritten versions of these two documents as printed on pages 316 to 318 of Bulletin 568.

In this connection, a typographical error is pointed out for correction in the new Specifications for Devices to Control the Splitting of Wood Ties. The last line under Art. 2, Dowels, should be changed to read: ". . . A 107, with a minimum of 0.2 of 1 percent copper."

Mr. President, I move the adoption of the revised documents just mentioned as they now appear in the committee's report, including correction of the typographical error.

[The motion was duly seconded, was put to a vote, and carried.]

MR. BURT: The use of untreated switch ties is no longer considered an economical practice. Accordingly, we are now recommending that Specifications for Switch Ties be revised to eliminate all references to the use of untreated switch ties. It is also recommended that all references to hewed switch ties be eliminated, as such ties are no longer available.

Mr. President, I move the adoption of the revisions to the Specifications for Switch Ties, as set forth in the report.

[The motion was duly seconded, was put to a vote, and carried.]

MR. BURT: During recent years the procurement of cross ties, loading on line, the receipt of ties at treating plant, handling through the plant, treatment, distribution of treated ties to the line of road, and stacking of treated ties on line of road have changed so greatly that it becomes necessary to revise considerably that portion of Chapter 3 dealing with The Handling of Ties From the Tree Into the Track.

Mr. President, I move the adoption of the revised document just mentioned, as presented in the report.

[The motion was duly seconded, was put to a vote, and carried.]

MR. BURT: Mr. President, this completes the report of Subcommittee 1.

Assignment 2—Cross and Switch Ties.

CHAIRMAN RADKEY: The work on Assignment 2 is under the direction of Subcommittee Chairman H. F. Kanute, engineer layout and design, St. Louis–San Francisco Railway. This subcommittee reports that a specification for dried ties is not practical and that present tests on concrete ties do not yet indicate need of a Manual specification. There is variation on different railroads in the use of secondhand ties and we hope to report on this subject next year. We will also consider the possibility of changes in timber tie design and spacing.

Assignment 3—Wood Preservatives.

CHAIRMAN RADKEY: The work on Assignment 3 is under Subcommittee Chairman W. W. Barger, chief inspector, Tie and Timber Treating Department, Atchison, Topeka & Santa Fe Railway. Under this assignment all preservative specifications in Chapter 17 of the Manual have been reapproved or revised and brought up to date. Study will continue on a new creosote-coal tar solution specification and a new creosote specification for use in marine piles.

Conditions precluded the inspection of a treating facility this past year. However, the committee hopes to resume this function when practical in the future.

Assignment 4—Conditioning and Preservative Treatment of Forest Products.

CHAIRMAN RADKEY: The report on Assignment 4 will be presented by Subcommittee Chairman L. C. Collister, manager, Tie and Timber Treating Department, System, Atchison, Topeka & Santa Fe Railway.

L. C. COLLISTER: Mr. President, your committee submits for adoption the following recommendation with respect to Chapter 17 of the Manual:

Page 17–4–9: To the table of specific requirements for preservative treatment of lumber, timbers and bridge ties, add oil-borne preservatives for fir, hemlock, larch and oak, as follows:

	Above Ground	Ground Contact
Oil-borne preservatives		
Pentachlorophenol	0.30	0.4
Copper Naphthenate (Copper metal)	0.05	0.1

Mr. President, I move the adoption of this recommendation.

[The motion was duly seconded, was put to a vote, and carried.]

MR. COLLISTER: We also reported, as information, on the treatment of salvaged ties as observed during an inspection trip on the Baltimore & Ohio Railroad.

For further information we presented information on a test of fire-retardant treatments for timber trestles using an emulsion of borax/boric acid, penta, petroleum oil, diesel oil and marasperse.

Our report as printed in Bulletin 568 also includes a definition of "Refusal Point" and "Refusal Treatment."

Furthermore, we reviewed the retentions for TC–TD ties, and it is the recommendation of the committee that this instruction be dropped.

This concludes the report of Subcommittee 4

C. J. CODE: [Pennsylvania Railroad]: At the risk of delaying the proceedings, I would like to ask a couple of questions about this report. I was very much interested in the report on the fire tests of timbers with special preservative treatment, and I wanted to ask whether a comparative test or control test was made with ordinary creosote treatment. One reason why I ask the question is that I ran some tests several years ago myself in which we took segments of new ties, both creosote-treated and untreated, put a handful of waste on them, poured fuel oil on the waste, set it afire, and when the fire in the waste went out neither the ties with coatings nor the ties without coatings had caught fire.

We did the same thing standing the segment of tie on end and setting a fire alongside it, and still we couldn't set fire to the tie even though it had no special protection against fire. The only ties we could set afire were some old ties that had splits in them so that the fire could get down inside them.

PRESIDENT BEEDER: Did you try using a fusee? [Laughter]

MR. COLLISTER: As a matter of fact, we didn't burn a full-scale control replica as such, but last June we had a full-scale replica of our California-type bridge under test with a protective coating on it that we thought was going to stand up all right, but after 20 min we found that the protective coating was not giving us the protection we thought was necessary; we let the fire burn, and in four hours the bridge collapsed. It burned to completion.

This material was treated with a creosote-petroleum mixture. We don't use straight creosote treatment on our bridge material. In the earlier stages of our test work we found to our sorrow that we needed something in addition to preservative treatment, because one day a brand-new, completely installed bridge just south of Topeka was being cleaned up, and an acetylene torch touched off a fire. Before it could be extinguished the bridge had burned to completion.

MR. CODE: That satisfies me entirely. We don't have any tumbleweed available to start fires with, either.

PRESIDENT BEEDER: Come and see us. [Laughter]

Assignment 5—Service Records.

CHAIRMAN RADKEY: Assignment 5 is under the direction of Subcommittee Chairman W. L. Kahler, general inspector, forest products and treatment, Missouri Pacific Railroad. Mr. Kahler is not with us today. If he were here he would say that the statistics on tie renewals and costs per mile of track as furnished by the AAR show an average of 43 ties per mile in the year 1960, and 56 ties per mile for the five-year average of 1956 to 1960, incl. The 43 ties per mile in 1960 is another all-time low.

The termite stake test, a research project being conducted in cooperation with the AAR at the University of Florida at Gainesville, Fla., is beginning to show results. The data show that only the specimens treated with coal tar-creosote have resisted all decay and termite attack. Three previous progress reports have been made since 1959. An inspection was made in February this year and results will be published later in 1962.

A service test of 313 creosote-treated fence posts installed by the New York Central Railroad in right-of-way fence in the vicinity of Rome, N. Y., in 1930, is described in our report, giving the results of a November 1960 inspection.

Assignment 6—Methods of Prolonging Service Life of Ties.

CHAIRMAN RADKEY: Subcommittee 6 is headed by P. D. Brentlinger, forester, Pennsylvania Railroad.

A study was started in 1958 on the behavior of cross ties protected against splitting by various devices and methods. To date the procedural method has been published by the committee. Data recorded to date cover the behavior of ties and anti-splitting devices before, during and after yard seasoning. Conclusions cannot be drawn before ties are observed in service; too many decisions regarding the use or non-use of anti-splitting devices have been based on observations before ties were subjected to use in track. Future reports will deal with behavior of the ties in track.

There has been called to the attention of the committee the failure of six red gum and one yellow pine tie on the Santa Fe treated by the one-step seasoning and treating method after eight year's service.

Assignment 7—Substitutes for Wood Ties.

CHAIRMAN RADKEY: Subcommittee 7 is headed by M. J. Hubbard, assistant chief engineer—system, Chesapeake & Ohio Railway.

The Association of American Railroads Engineering Division research staff has prepared a comprehensive report covering their investigation of prestressed concrete ties. The investigation was started in 1957, even though it was realized that the supply of wood ties was ample at present and in the foreseeable future, as it was felt desirable to have a few test installations on several railroads for service performance.

[Mr. Radkey then gave highlights of the summary of the AAR report as printed on pages 330 to 335, incl., of Bulletin 568, concluding as follows:]

CHAIRMAN RADKEY: Based on an assumption of 50 years' service life for concrete ties and 25 years for treated wood ties, a slight annual savings is shown from the installation of concrete ties in new track, but when installed as renewals in existing tracks, practically no savings over wood ties are shown.

The original installations of concrete ties have not been in service long enough to determine any maintenance trend.

Assignment 8—Making Charcoal from Used Ties.

CHAIRMAN RADKEY: Subcommittee 8 is headed by G. A. Williams, regional engineer, Pennsylvania Railroad.

For the past several years, inquiries have been made of various charcoal producers in regard to the possibility of making charcoal from used ties. However, overall results to date are not promising. We shall present a final report on this assignment next year.

CHAIRMAN RADKEY (continuing): We contemplate making a few changes in committee organization for more efficient functioning. Consolidation of the two committees has created a few unforeseen problems. We are looking forward to our coming year's work.

I wish to thank Vice Chairman W. E. Fuhr, the subcommittee chairmen and members of Committee 3 for their generous response and cooperation in the past year's work.

F. R. WOOLFORD (Western Pacific): Mr. Radkey, I noticed in your report that you stated that tie renewals averaged in the vicinity of 43 ties per mile and then in the concrete tie investigation a tie life of 25 years for wood ties was used. How do you correlate these statements? If we are renewing 43 ties per mile we are talking about a tie life of around 70 years.

CHAIRMAN RADKEY: The 43 ties per mile is the average for the year 1960, Mr. Woolford. If you go back and take the number installed over, say, a period of 35 or 40 years, the answer will be different. We feel we are definitely at a low point in our tie installations.

MR. WOOLFORD: Has your committee brought up anything that would indicate the average life of ties in the past 15 or 20 years? What is the life of a wood, tie?

CHAIRMAN RADKEY: That is a good question. What is the average life of a man?

MR. WOOLFORD: You can answer the question better than that. [Laughter] I can't take that answer.

PRESIDENT BEEDER: In the statistics that were used to compare the two types of ties, I think a life of 25 years was used for the timber tie, was it not, Mr. Radkey?

CHAIRMAN RADKEY: That is correct. We have test-section data that can establish a 25-year life. One major railroad in this country today feels that they are now achieving a 50-year average life with timber ties. Most of us feel that the answer lies somewhere in between. How you can come up with something that will not be a blind guess is another question.

MR. WOOLFORD: Don't you think a 35-year life would be a little more reasonable in the study?

CHAIRMAN RADKEY: For a timber tie? Yes, I do.

C. J. CODE (Pennsylvania): In connection with this discussion, I think you have to distinguish between main-track ties and all ties. The 35-year figure represents the average of all ties. You are including a lot of side tracks and minor branch lines where you probably do get 50 or 60 years out of creosoted ties, whereas in main track you may have a tie life as low as 15 years.

I think the 25-year figure used for wood ties in the cost comparison with concrete ties probably represents a fair average for main-track ties. On the other hand, I would like to ask what the basis is for assuming a 50-year life for a concrete tie.

CHAIRMAN RADKEY: That is also a good question. I don't know. That is the figure —is Mr. Magee here? [Laughter] Would you care to answer that, please?

G. M. MAGEE (AAR): I would like to point out that in this study we were a little reluctant to make an economic comparison. We did it because we were more or less requested to do so by the Tie committee. They felt that they would like to have something in the way of economics to go on.

The 50-year life is an assumption, and we don't know any more than you do whether concrete ties are going to last 50 years or not. I don't think I shall be here to know.

However, from the data that are available today resulting from tests of reinforced concrete and prestressed concrete, and exposure tests, along with the advances in air entrainment, we selected 50 years as being, in our best judgment, a figure that should be attainable insofar as exposure is concerned.

One fact we have to take into account certainly is derailments. As the gentleman from England said last year when he spoke here, we don't think we ought to design a track for derailments; but, on the other hand, we know that we do have them. Certainly the life of those ties subjected to derailments will be shortened. They may be reusable or they may be completely destroyed. That is something else we don't know.

In this study we used a figure of 25 years for wood ties because that figure was indicated in one of the Tie committee past reports, as referred to in the AAR report.

W. E. CORNELL (Nickel Plate): In view of the discussion I have just heard, it would seem to me that the economic comparison is somewhat out of order in this report.

PRESIDENT BEEDER: In what way, Mr. Cornell?

MR. CORNELL: Because it doesn't mean anything. We don't know anything about the life of concrete ties, and we are very unsure and very doubtful about the life of wood ties.

I agree with Mr. Woolford about the 35 years for wood ties, but we use an unrealistic figure of 25 years for the purpose of comparison; and to people who may not be as informed as Mr. Woolford, it would give a very improper picture.

CHAIRMAN RADKEY: Mr. Cornell, the report on prestressed concrete ties is presented for information. It is not going into the Manual. It was ably prepared by the people ·at the laboratory, and it would do us all good to read the report in detail.

The thing that they were up against is that it is hard to come up with any kind of an answer unless an assumption is made. You can make the same calculations they made in their report, and vary the answer with whatever assumption you think is appropriate.

The second thing I would like to comment on is this: We are very prone to talk about the average service life of a tie. I would suggest that on your railroads you don't worry too much about what the book shows as the average life, but that you base your tie renewals on the actual physical condition of the ties on the railroad. [Laughter and Applause]

MR. WOOLFORD: Mr. Radkey, I must differ with you there to a certain extent. We have our management to answer to on the railroad. Management wants to compare the life of our ties with the life that other railroads are getting.

I also disagree with Mr. Code in that we are only getting 15 years of life out of our ties in main track. If we are, we certainly are incurring some deferred maintenance, considering the few ties we are putting in today.

MR. CODE: Under certain special conditions only, Mr. Woolford.

MR. WOOLFORD: I have found no ties on my railroad that are lasting only 15 years. Maybe the Pennsylvania has some, but I think that is a very short life for a wood tie. I think from our practice and the number of ties we are putting in, we must be getting a 35-year average life out of our wood ties if they are properly protected and properly taken care of and properly put in.

I don't think we can consider damage from derailments, and so on, as affecting the life of a tie. We have to consider only ties in normal use in track, properly treated, protected with pads and tie plates, and their ends protected from splits.

CHAIRMAN RADKEY: I know that on our property we have realized a 25-year average life for ties we installed 25 years ago, and longer ago than that. We think in 25 years we have made some improvements. We certainly are satisfied that those improvements are going to be reflected in longer life of ties.

I have probably talked too much, Mr. President.

PRESIDENT BEEDER: Thank you, Mr. Radkey. It is pretty difficult to assign an average life to ties, because there are so many intangibles and differences among them, and in seasoning and treating methods, etc. Improved treatment, of course, will make a big difference in average tie life. All those things enter into the estimate. Thank you, Mr. Woolford.

Mr. Radkey, you have certainly stimulated an interesting discussion this morning. We thank you and your committee for additional informative and interesting reports this year. It seems a little strange to hear your committee reporting on such matters as wood preservatives and fire-retardant treatments for bridge trimbers, along with all aspects of cross ties; but, unquestionably, the assumption by your committee during the past year of the work of our former Committee 17—Wood Preservation, was a desirable move, and from all appearances it has proved to be a happy union. We hope your committee will keep up its good work in the year ahead.

Your committee is now excused with the thanks of the Association. [Applause]

Discussion on Roadway and Ballast
[For report see Bulletin 570, pp. 573–597]

PRESIDENT BEEDER: The next committee to report to us is our Committee 1—Roadway and Ballast, another one of our more popular committees and of which I am proud to be a member.

The chairman of this committee is Frank Beighley, roadway engineer, St. Louis–San Francisco Railway, at Springfield, Mo. Frank, if you will bring your "gang" up here I shall be delighted to turn this microphone over to you.

CHAIRMAN F. N. BEIGHLEY: Mr. President, members of the Association and guests:

During the past year Committee 1 held two meetings during which each of our 11 subcommittees reported on its activities, consisting mostly of reviewing Manual material. However, progress was made on each individual assignment.

We are submitting this year reports on only 8 of the 11, the other 3 not having been progressed to the point where reports can be submitted. The reports will be found in Bulletin 570, Part 1, pages 573 to 597, incl. As the individual reports are presented, we invite your comments, criticism, or any questions you may care to ask regarding any or all of them.

Before we have the presentation of the various subcommittee reports, we wish to pay tribute to one of Committee 1's valued associate members who passed away during the past year, whose Memoir will be given by George Harris, assistant engineer of the Chesapeake & Ohio Railway.

G. B. HARRIS: Mr. President, Mr. Chairman, members and guests:

It is with deep regret that I announce to you the death on October 1, 1961, in Chicago, of one of our most beloved and valuable members, Charles Wistar Reeve, treasurer of the Young and Greenawalt Company. Mr. Reeve first entered railroad service in 1928 with the Delaware & Hudson Railroad and left railroad service in 1954 to join Young and Greenawalt Company, manufacturers of corrugated pipe and metal products, with which company he was associated until his death. He joined this Association in 1947 and from 1948 through 1954 was a member of Committee 22—Economics of Railway Labor. In 1956 he became an associate member of this committee, which membership he held until his death. His untiring efforts, his assistance to various committee chairmen, and his work on the committee will long be remembered by all of us.

It is with a deep sense of appreciation for his work with this Association, and especially Committee 1, that we submit a Memoir in his honor.

Assignment 1—Revision of Manual.

CHAIRMAN BEIGHLEY: The report on Assignment 1 will be given by Subcommittee Chairman G. B. Harris, assistant engineer, Chesapeake & Ohio.

G. B. HARRIS: Your committee for the past two years has been making a study of its chapter in the Manual. This year Subcommittee 1 was divided into two sections, one to study the Manual material for possible editorial revision, and the other for technical revision. Mr. Deno has prepared a report on the recommended editorial revisions of the Manual which he will present later.

At this time I should like to call your attention to the recommendations of Committee 1 found on pages 576 and 577 of Bulletin 570. Your committee recommends that the portions of the Manual listed on these two pages be reapproved without change.

Mr. President, I so move.

[The motion was duly seconded, was put to a vote, and carried.]

MR. HARRIS: The only change recommended in the Glossary by the committee is that those definitions presently denoted by the italic numeral 2 be changed to read 1. Perhaps all of you present are not familiar with the fact that Committee 1 is a combination of what was formerly Committees 1 and 2. This is an editorial matter, and I do not believe it requires a vote of this convention.

L. J. Deno, staff engineer—maintenance, Chicago & North Western, will present the second part of Subcommittee 1's report. Mr. Deno.

L. J. DENO: Mr. President, members and guests:

During the past year your committee conducted a special review of Chapter 1 of the Manual. As a result a number of revisions to the chapter are recommended, primarily to correct inconsistencies in the language of specifications compared to that in documents of information and suggested practices. All of the recommended revisions are detailed on pages 578 and 579 of Vol. 63, Bulletin 570, Part 1, dated February 1962, and will be subsequently referred to only by Manual page number.

Recommended revisions to Part 1—Roadway, are:

Pages 1-1-36.1 to 1-1-36.6, incl., "Construction and Protection of Roadbed Across Reservoir Areas." Reapprove with revision on page 1-1-36.4.

Pages 1-1-45 to 1-1-64, incl., "Roadway Drainage." Reapprove with revisions on pages 1-1-58, 1-1-63 and 1-1-64.

Mr. President, I move that these documents of Part 1 be reapproved with the revisions mentioned.

[The motion was duly seconded, was put to a vote, and carried.]

MR. DENO: Recommended revisions to Part 3—Natural Waterways, are:

Pages 1-3-15 to 1-3-20, incl., "Means of Protecting Roadbed and Bridges From Washouts and Floods." Reapprove with revisions on pages 1-3-15, 1-3-17 and 1-3-18.

Mr. President, I move that this document of Part 3 be reapproved with the revisions mentioned.

[The motion was duly seconded, was put to a vote, and carried.]

MR. DENO: Recommended revisions to Part 4—Culverts, are:

Pages 1-4-19 to 1-4-24, incl., "Jacking Culvert Pipe Through Fills." Reapprove with revisions on page 1-4-23.

Pages 1-4-33 to 1-4-35, incl., "Methods of Installing Culverts Inside Existing Culverts." Reapprove with revisions on page 1-4-34.

Pages 1-4-35 to 1-4-37, incl., "Conditions Requiring Headwalls, Wingwalls, Inverts and Aprons, and Requisites Therefor." Reapprove with revisions to page 1-4-35.

Mr. President, I move that these documents of Part 4 be reapproved with the revisions mentioned.

[The motion was duly seconded, was put to a vote, and carried.]

MR. DENO: Recommended revisions to Part 7—Signs, are:

Pages 1-7-1 to 1-7-4, incl., "Roadway Signs." Reapprove with revisions to pages 1-7-1 and 1-7-2.

Mr. President, I move that this document of Part 7 be reapproved with the revisions mentioned.

[The motion was duly seconded, was put to a vote, and carried.]

MR. DENO: Recommended revisions to Part 8—Tunnels, are:

Pages 1-8-1 to 1-8-5, incl., "Tunnels." Reapprove with revisions to page 1-8-1.

Mr. President, I move that this document of Part 8 be reapproved with the revisions mentioned.

[The motion was duly seconded, was put to a vote, and carried.]

Assignment 2—Physical Properties of Earth Materials.

CHAIRMAN BEIGHLEY: The report on Assignment 2 will be presented by Subcommittee Chairman W. P. Eshbaugh, chief engineer, Genesee & Wyoming Railroad.

W. P. Eshbaugh: Your committee submits for adoption recommendations with respect to "Physical Properties of Earth Materials", Manual page 1-1-37 to 1-1-43, incl. Mr. President, I move that the aforementioned document be reapproved with the revisions set forth on pages 580 through 582 of Bulletin 570.

[The motion was duly seconded, was put to a vote, and carried.]

Assignment 3—Natural Waterways: Prevention of Erosion.

CHAIRMAN BEIGHLEY: The chairman of Subcommittee 3 is G. W. Becker, special engineer—drainage, Chicago, Rock Island & Pacific Railroad. The subcommittee is reviewing present Manual material with a view to updating it to reflect more recent data and research, and general practice in determining size of waterway openings and prevention of erosion.

Assignment 4—Culverts, (a) Erosion Control for Outlet Structures.

CHAIRMAN BEIGHLEY: G. D. Mayor, division engineer, Chesapeake & Ohio Railway, is chairman of Subcommittee 4. "Erosion Control for Outlet Structures" is an assignment on which research was started in 1958 under a grant to the Colorado State University as a three-year project. Because of budget curtailment in 1959, 1960 and 1961 the project has been discontinued. It is hoped that funds will be made available for resumption of this research in the near future.

Assignment 5—Specifications for Pipelines for Conveying Flammable and Nonflammable Substances.

CHAIRMAN BEIGHLEY: The chairman of Subcommittee 5 is K. W. Schoeneberg, chief engineer, Akron, Canton & Youngstown Railroad, who will give the report.

K. W. SCHOENEBERG: President Beeder, Chairman Beighley, members and guests:

At last years convention, the revised Specifications for Pipelines for Conveying Flammable and Non-Flammable Substances, Sec. A, for Flammable Substances, as presented by your committee, were adopted and have been published in the Manual.

During this past year, your committee has revised Sec. B of these specifications, for Non-Flammable Substances, and now these specifications in finalized form as contained in Bulletin 570, February, 1962, pages 582 to 586, incl., are presented by your committee for adoption and publication in the Manual, replacing the present Specification, Sec. B, for Non-Flammable Substances, now appearing on Manual pages 1–5–6 through 1–5–9.

Mr. President, I move the adoption of the new specifications as contained in Bulletin 570, pages 582 to 586, incl., to replace the like portion of the existing document.

[The motion was duly seconded.]

PRESIDENT BEEDER: This is a very important subject, and a lot of work has gone into it. Are there any comments?

F. R. WOOLFORD (Western Pacific): I should like to ask a question about Art. 7. Depth of Installation. What was the basis for using 5½ ft as the minimum depth from base of rail to top of casing pipe.

MR. SCHOENEBERG: That figure has been carried for years, and the committee felt that it should be continued. You will note that this distance may be 4½ ft under secondary or industry tracks.

MR. WOOLFORD: That's right, but don't you think that the 5½-ft dimension should be restudied and that we ought to be a little more realistic? I think you will find that few carriers use that dimension; we certainly get considerable criticism from people wanting to put pipes under our tracks when we hold to the 5½-ft minimum.

MR. SCHOENEBERG: Mr. Woolford, we certainly do get a lot of criticism, as you say. I have a thick file on it. These items are continually under study, and we are now in the process of collaborating with the American Water Works Association. They are now starting a revision of their specifications and we are going to work with them; and possibly, as reported by your committee last year, we still haven't come to a realistic, finalized form of Sec. A, for flammable substances as yet. We are still working with the API and the pipeline people on that part of the specification.

MR. WOOLFORD: I would like to ask that that one item be restudied.

MR. SCHOENEBERG: It certainly will, sir, as well as the rest of them. As you and I both stated, we are getting questions constantly from all sources on various things just like that. We have some people who say, "We will go to any depth." Others say they want to stay closer to the surface, and then ask why they have to go so deep. Possibly, future years will see a more realistic revision of our specifications.

PRESIDENT BEEDER: These specifications will change, I am sure, Mr. Woolford, but the present revision is bringing us much closer to the optimum solution, I think, than we have ever been before.

Thank you for your question. Is there any further discussion?

[The motion was put to a vote and carried.]

Assignment 6—Roadway: Formation and Protection.

CHAIRMAN BEIGHLEY: The chairman of Subcommittee 6 is A. W. Lewis, engineer of construction, Seaboard Air Line Railroad. Mr. Lewis could not be present at this convention, so his report will be read by G. B. Harris, assistant engineer, Chesapeake & Ohio.

G. B. HARRIS: Your committee this year presents a report on the lime treatment of a railroad subgrade in Texas—a new industrial lead. It has now been in service about 9 months with every indication that the treatment is satisfactory. The report describes the project in detail, and it will not be necessary to comment further on it now.

The use of lime for the treatment of clay subgrades is gaining momentum in the highway and airport fields. It is particularly adapted to heavy clay materials in flat, wet and slowly drained areas and for railroads also in those areas where suitable materials for sub-ballast are not available or are costly. In certain of such projects it appears possible that the use of lime will reduce or eliminate the requirements of sub-ballast and this consideration can be balanced against treatment costs.

The lime, either slaked or unslaked, works in several ways to improve clay soils. First it reduces plasticity to a large degree, secondly it makes the clay material readily pulverizeable, and thirdly with time a certain strength is built up in the soil. These factors also make the compacted lime-treated subgrade relatively waterproof.

Lime requirements are usually 5 percent by weight or less of the soil. A 6-in depth is common but it can be more or less. Lime treatment is relatively inexpensive if the lime is available at less than $15 per ton f.o.b. plant. It will not produce the strength of soil cement but is less expensive and with heavy plastic clays is much more readily worked.

This report is presented as information.

Assignment 7—Tunnels, (a) Ventilation, (b) Clearance, Methods Used to Increase, (c) Methods of Open Cutting.

CHAIRMAN BEIGHLEY: R. D. White, division engineer, St. Louis–San Francisco Railway, is chairman of Subcommittee 7. Under Assignment 7 (a) progress reports have been submitted, and it is hoped that recommendations for Manual material can be developed in the near future. Under Assignment 7 (b) the committee is investigating numerous data and hopes to develop new methods for increasing tunnel clearance as Manual material. Under Assignment 7 (c) sufficient data on methods of open cutting have not yet been collected to permit presenting a report at this time.

Assignment 8—Fences.

CHAIRMAN BEIGHLEY: Subcommittee Chairman H. G. Johnson, assistant engineer, Milwaukee Road, will present the report on Assignment 8.

H. G. JOHNSON: Your committee has reviewed the material on fencing in Chapter 1 of the Manual and our recommendations in connection therewith are printed on pages 589 and 590 of Bulletin 570.

Mr. President, I move that these recommendations be adopted.

[The motion was duly seconded, was put to a vote, and carried.]

Assignment 9—Roadway Signs, (a) Reflectorized and Luminous Roadway Signs, (b) Develop Standard Close Clearance Warning Sign.

CHAIRMAN BEIGHLEY: Subcommittee Chairman R. D. Baldwin, district engineer, Pennsylvania Railroad, will give the report on Assignment 9.

R. D. BALDWIN: Mr. President and gentlemen:

Under Assignment 9 (b) your committee submits for adoption and publication in the Manual the "No Clearance" sign illustrated on page 592 of Bulletin 570. This sign was selected for reasons of maximum simplicity, clarity, economy, and minimum obsolescence.

Mr. President, I move that this sign be adopted.

[The motion was duly seconded, was put to a vote, and carried.]

Assignment 10—Ballast: (a) Tests, (c) Special Types of Ballast.

CHAIRMAN BEIGHLEY: The chairman of Subcommittee 10 is T. W. Creighton, regional engineer, Canadian Pacific Railway, who will now present his report.

T. W. CREIGHTON: Under assignment 10 (c)—Special Types of Ballast, your committee presents a progress report on the conditions in 1961 of the sections of asphalt-treated ballast and bridge decks on various railroads which were installed in 1959 and 1960 under a cooperative project among the Asphalt Institute, the AAR Research Department, and the participating railroads. The report appears on pages 593 to 596 of Bulletin 570. No additional work was done in 1961. The tests are being continued.

Assignment 11—Chemical Control of Vegetation, Collaborating with Communication and Signal Section, AAR.

CHAIRMAN BEIGHLEY: Subcommittee Chairman C. E. Webb, engineer of tests, Southern Railway System, will now report on Assignment 11.

C. E. WEBB: The report appearing in Bulletin 570 covers the results of a cooperative test started in 1958 among the AAR Research Department, three railroads, and several chemical companies. This project was concerned with the application of granular and pelletized soil sterilants and dormant spray application with hormone-type chemicals for the control of brush. Drastic budget curtailment did not permit completion of the intended scope of the original project nor the further exploration of some promising avenues of approach.

In summary, the results of broadcast application of granular and pelletized materials were very erratic and in general poor. Basal or stem treatment with dry materials resulted in fair to good control of most species. The dormant spray procedure has given many indications of good to excellent results that are equal and often superior to foliage applications. This method has the advantage of better coverage of the ground area, with crop damage minimized. It is now being used to a much greater extent as the policy of some railroads.

CHAIRMAN BEIGHLEY: Mr. President, this concludes the report of Committee 1.

PRESIDENT BEEDER: Thank you, Mr. Beighley. After hearing the report of Committee 1 I think it can be said with all modesty that the committee has been "on the ball" during the past year, both in reviewing and updating its Manual chapter and in progressing its other assignments.

Frank, we appreciate your able direction of this work, and we all look forward to working with you toward further progress in the year ahead.

You are now excused with the thanks of the Association. [Applause]

Discussion on Track

[For report see Bulletin 570, pp. 485–495]

PRESIDENT BEEDER: Our next committee will, I am sure, not only bring us several interesting reports, but will also confront us with several major decisions as an Association. I am referring to our Committee 5—Track, the chairman of which is Stuart Poore, office engineer, Chesapeake & Ohio Railway, at Richmond, Va.

Mr. Poore, we are ready to hear the report of your committee and any recommendations you have to make to this Association.

CHAIRMAN S. H. POORE: Mr. President, members of the Association, and guests: Your Committee 5 has a brief report this year; however, the significance of it is far-reaching.

Assignment 1—Revision of Manual.

CHAIRMAN POORE: At this time it is my pleasure to introduce to you R. J. Hollingsworth, engineer roadway, track and equipment, Baltimore & Ohio Railroad, chairman of Subcommittee 1, who will present the report of his committee.

R. J. Hollingsworth: Mr. President and Gentlemen:

Your committee submits for adoption the following recommendations with respect to Chapter 5 of the Manual:

Page 5-2-5 covers a spike for use at rail joints and as an alternate for general use. A survey of the major spike manufacturers developed that production of this spike has been negligible.

Page 5-2-6 covers the ⅝ by 6-in and $\frac{9}{16}$ by 5½-in track spikes now in general use.

Mr. President, I move that the spike design on page 5-2-5 be deleted from the Manual and that the spikes on page 5-2-6 be reapproved without change.

[The motion was duly seconded, was put to a vote, and carried.]

Mr. Hollingsworth: Pages 5-3-1 to 5-3-4, incl., cover "Spirals." I move that these pages be reapproved with certain editorial changes as indicated on page 487 of Bulletin 570, which were made as a result of a special Manual review.

Mr. President, I so move.

[The motion was duly seconded, was put to a vote, and carried.]

Mr. Hollingsworth: Pages 5-3-5 to 5-3-9 cover "String Lining of Curves by the Chord Method."

Mr. President, I move that these pages be reapproved with certain editorial changes as indicated on page 487 of Bulletin 570.

[The motion was duly seconded, was put to a vote, and carried.]

Mr. Hollingsworth: Pages 5-3-9 to 5-3-11 cover "Elevations and Speeds for Curves."

Mr. President, I move that this document be reapproved with certain editorial changes as indicated on page 487 of AREA Bulletin 570.

[The motion was duly seconded, was put to a vote, and carried.]

Mr. Hollingsworth: I move that the material on page 5-3-13 covering "Vertical Curves" be reapproved without change.

[The motion was duly seconded, was put to a vote, and carried.]

Mr. Hollingsworth: I move that the material on page 5-3-14 covering "Permanent Monuments" be reapproved with certain editorial changes as indicated on page 487 of Bulletin 570.

[The motion was duly seconded, was put to a vote, and carried.]

Mr. Hollingsworth: I move that Part 4—Track Construction, pages 5-4-1 through 5-4-4, be reapproved with certain editorial changes as indicated on page 488 of Bulletin 570.

[The motion was duly seconded, was put to a vote, and carried.]

Mr. Hollingsworth: Pages 5-5-1 to 5-5-4 cover "Specifications for Laying Rail" and "Temperature Expansion for Laying Rails."

I move that the rail temperature expansion table offered the convention in 1961, Bulletin 563, page 647, Part B, be substituted for the expansion table now carried on pages 5-5-2 and 5-5-4, and that these documents be reapproved with certain editorial changes as indicated on page 488 of Bulletin 570.

[The motion was duly seconded, was put to a vote, and carried.]

Mr. Hollingsworth: I move that the material on pages 5-5-4 and 5-5-4.1 be retitled "Rail Creepage—Number and Position of Rail Anchors for Jointed Track and Where Temperature Expansion is Provided", to differentiate from anchorage recommended for continuous welded rail, and that certain editorial changes be made as indicated on page 489 of Bulletin 570.

[The motion was duly seconded, was put to a vote, and carried.]

Mr. HOLLINGSWORTH: I move that page 5-5-6 covering "Track Bolt Tension Practice" be reapproved without change.

[The motion was duly seconded, was put to a vote, and carried.]

Mr. HOLLINGSWORTH: Page 5-5-7 covers "Gage." I move that Par. (c) be changed to read:

"Wide gage due to worn rail, within the permissible limits of wear, should be corrected by closing in or by interchanging the low and high rail", and that certain editorial changes be made as indicated on page 489 of Bulletin 570.

[The motion was duly seconded, was put to a vote, and carried.]

Mr. HOLLINGSWORTH: The subject, "Lubrication of Rail on Curves", is covered on page 5-5-10.

I move that this section be reapproved with the substitution of the word "fuel" for "coal."

[The motion was duly seconded, was put to a vote, and carried.]

Mr. POORE: Thank you, Mr. Hollingsworth. With these revisions the Track Chapter of the Manual will be more valuable and, we believe, better organized.

Assignment 2—Track Tools, Collaborating with Committee 1 and with Purchases and Stores Division, AAR.

CHAIRMAN POORE: C. E. Peterson, assistant engineer, Atchison, Topeka & Santa Fe Railway, is chairman of Subcommittee 2 and will make the report.

C. E. PETERSON: Mr. President and members of the Association:

Your committee submits the following recommendations with respect to specifications and plans for track tools appearing in the Manual, together with specifications and plans for steel drive spikes:

Reapprove without change: Specifications for Track Tools; Specifications for Ash and Hickory Handles for Track Tools; Recommended Limits of Wear for Tools to Be Reclaimed; Specifications for Steel Drive Spikes, and Plans for Drive Spikes.

Mr. President, I so move.

[The motion was duly seconded, was put to a vote, and carried.]

Mr. PETERSON: The revisions recommended in the Plans for Track Tools are as follows:

Page 5-6-9: Insert "Metal Specifications" above the headings for the list of plans. For Plan 33-61, Drive Spike Extractor Socket Wrench, under "Hardness" change "See Plan" to "300-350."

Page 5-6-25: Plan 30-53, AREA Spot Board, change "cast iron plate" to "malleable iron," for use with gage stop. Also change "C-9-4 x 1⅝ in 7¼ channel" to "4 in 7.25 lb channel."

Mr. President, I move the adoption of these revisions.

[The motion was duly seconded, was put to a vote, and carried.]

Mr. PETERSON: Reapprove without change the balance of the Plans for Track Tools, pages 5-6-10 to 5-6-26, incl.

Mr. President, I so move.

[The motion was duly seconded, was put to a vote, and carried.]

Mr. PETERSON: Your committee also submits, as information, a progress report on the claw bar and snap-on rachet track wrench.

This completes the report on Assignment 2.

CHAIRMAN POORE: Thank you, Mr. Peterson.

Gentlemen, it would seem that in this day of mechanization hand tools would be going out of style and use; but in view of the interest in this subject by many railroads, it appears that there is still an important place for these tools in railroad maintenance.

Assignment 3—Standardization of Trackwork Plans, Collaborating with Communication and Signal Section, AAR.

CHAIRMAN POORE: You will recall that at the March 1960 convention this committee suggested, and the Association approved, the preparation of a series of standard plans designed to reduce, if possible, the multiplicity of design and alternates now appearing in the Portfolio of Trackwork Plans. Subcommittee 3 is now prepared to offer to you certain plans. C. J. McConaughy, track designer, Southern Pacific Company, is the subcommittee chairman. Mr. McConaughy.

C. J. McCONAUGHY: Mr. President and gentlemen:

Your Subcommittee 3 has completed its study on Standardization of Trackwork Plans and received the necessary two-thirds vote of approval for submitting them to this convention.

Bulletin 570, Vol. 63, Part 1 covers the presentation of the proposed standard plans. The plans themselves are printed in Part 2 of Bulletin 570. Also included in Part 2 are the proposed revisions of present Plans 111–55, 221–55 and 223–55. Revisions to their companion sheets are outlined in Part 1.

Your committee submits for adoption the proposed revisions to Plan 221–55 covering removal of sharp edges, and an addition to Appendix A of the Specifications for Special Trackwork, also covering the removal of sharp edges.

Mr. President, I move that the proposed revisions be adopted.

[The motion was duly seconded, was put to a vote, and carried.]

MR. McCONAUGHY: Your committee submits for adoption revisions to Plans 111–55, 113–55, 115–55, 117–55, 121–55, 123–55, 125–55, 127–55 and 223–55 covering increasing the shoulder lengths on the various flat plates on both the field and gage sides, as printed in the Bulletin.

Mr. President, I move that the proposed revisions be adopted.

[The motion was duly seconded, was put to a vote, and carried.]

MR. McCONAUGHY: It is the intention of this committee to apply the revisions just approved to the proposed new trackwork plans.

Your committee has developed plans for five turnouts—Nos. 6, 8, 10, 15 and 20—with subsidiary plans giving complete details of component parts. Part 1 of the Bulletin covers the recommendations and Part 2 contains the plans themselves.

In order to have these plans available for this convention it was necessary that the work of preparing the drawings be distributed among various groups. This accounts for the variation in positioning the various details, as well as some editorial errors. The minor variations and editorial discrepancies will be corrected before the plans are issued in the 1962 Supplement to the Trackwork Portfolio.

Mr. President, I move that the proposed new Plans, as printed in Bulletin 570, Part 2, be adopted.

[The motion was duly seconded, was put to a vote, and carried.]

Assignment 4—Prevention of Damage Resulting from Brine Drippings on Track and Structure, Collaborating with Committee 15 and Mechanical Division, AAR.

CHAIRMAN POORE: Because of the lack of funds for the necessary research work, no report on Assignment 4 is offered this year. L. W. Leitze, engineer of track, Great

Northern Railway is chairman of the Subcommittee. Mr. Leitze, will you rise and be recognized.

Assignment 5—Design of Tie Plates and the Use of Rubber or Composition Plates Under Insulated Joints, Collaborating with Committees 3 and 4 and the Communications and Signal Section, AAR.

CHAIRMAN POORE: Subcommittee 5 has no report to present at this time. Certain inspections are planned for 1962. L. A. Pelton, district engineer, Pennsylvania Railroad, is chairman of this subcommittee. Mr. Pelton, will you stand and be recognized.

Assignment 6—Hold-Down Fastenings for Tie Plates, Including Pads Under Plates; Their Effect on Tie Wear, Collaborating with Committee 3.

CHAIRMAN POORE: Subcommittee 6 is chairmanned by N. C. Kieffer, Jr., division engineer in charge of construction, Louisville & Nashville Railroad. A brief progress report on this assignment was published in Bulletin 570. Mr. Kieffer, will you please stand to be recognized.

Assignment 7—Effect of Lubrication in Preventing Frozen Rail Joints and Retarding Corrosion of Rail and Fastenings.

CHAIRMAN POORE: The chairman of Subcommittee 7 is R. G. Garland, assistant division engineer, Atchison, Topeka & Santa Fe Railway. Under this assignment service tests have been set up on the Chicago & North Western. The tests have not progressed to the extent that results can be reported at this time. Mr. Garland, please stand to be recognized.

Assignment 8—Laying Rail Tight with Frozen Joints.

CHAIRMAN POORE: Test installations of tight rail are in service; and although no money was available in 1961 and none is in sight for 1962, we hope to report results some time soon. This subcommittee is headed by V. M. Schwing, engineer of track, Bessemer & Lake Erie Railroad. Mr. Schwing, will you please stand?

Assignment 9—Critical Review of the Subject of Speed on Curves as Affected by Present Day Equipment, Collaborating with AAR Joint Committee on Relation Between Track and Equipment.

CHAIRMAN POORE: Subcommittee 9 has been hampered in its work during the last two years because of the fact that certain railway equipment has not been available to them. However, there is printed on page 67 of Bulletin 566 a study on "Speed of Trains Through Turnouts", which your committee proposes to use in revising the table of turnout speeds now appearing in the Manual. L. H. Jentoff, assistant chief engineer, maintenance of way, Erie–Lackawanna, is chairman of this subcommittee but unfortunately is not present.

Assignment 10—Methods of Heat Treatment, Including Flame Hardening, of Bolted Rail Frogs and Split Switches, Together with Methods of Repair by Welding; Explosive Hardening of Manganese Steel Trackwork.

CHAIRMAN POORE: Under Assignment 10 we have had a service test on the Milwaukee Road for some time. Service data on this installation are being accumulated, and we hope to bring in a conclusive report on this subject soon. J. M. Salmon, Jr., chief engineer, Clinchfield Railroad, is the chairman of this Subcommittee. Mr. Salmon, will you please stand?

CHAIRMAN POORE (continuing): Mr. President, I would like to announce certain changes in committee personnel. J. B. Wilson, chief engineer, Georgia Railroad, has resigned as vice chairman of Committee 5. We have been fortunate in that Mr. Salmon has volunteered at my request to accept the appointment to succeed Mr. Wilson.

Mr. President, this concludes the report of your Track committee.

F. R. WOOLFORD (Western Pacific): Mr. President, May I say a word? I think Mr. Poore is to be highly complimented for his pushing through the work on standardization of trackwork plans. You will remember that the subject was quite thoroughly discussed on the floor when I was president.

PRESIDENT BEEDER: I recall that very clearly, Mr. Woolford.

MR. WOOLFORD: I hope in completing this work that all of the railroad members of AREA will use the new plans. We have been trying for many years to standardize turnouts. The committee has now come up with what I think are very good plans. We have reduced our turnouts to a minimum, and I think it behooves us as an organization to follow them.

PRESIDENT BEEDER: I think we shall see further continued progress in standardization, which you have advocated for so long, Mr. Woolford.

CHAIRMAN POORE: I have a pleasant job to do at this time. Some of you, I am sure, have heard of the Chesapeake & Ohio track inspection car. In some respects it was designed to cause roadmasters and track engineers to quake in their boots. Its major objective, however, is to inform management of track conditions. In either case, it is a wondrous machine.

The man who will tell you about this device has been in railroad maintenance for many years and is widely known in railroad circles. He is the chief engineer system of the Chesapeake & Ohio Railway.

It is my pleasure to introduce your nominee for the office of junior vice president (I feel certain I am correct in referring to him as vice president-elect) of this Association, and my boss, T. Fred Burris, who will speak on the subject, "The New C & O Track Inspection Car". Mr. Burris.

The New C&O Track Inspection Car

By T. FRED BURRIS
Chief Engineer System, Chesapeake & Ohio Railway

Considering the large annual expenditure on track maintenance, it has been our thought that an instrument to measure or determine the effectiveness of the expenditure would be useful to Management. In the past, the need for track work has been variously interpreted by different maintenance officers. No one man can reliably evaluate an entire railroad system. Usually one portion is evaluated by one man and another portion by another and so on, and the chief engineer had no way of knowing if his maintenance officers were viewing their respective territories with the same maintenance standards.

The C&O first built a track inspection car in 1935 which we felt would look at all parts of the line with the same eye and provide an appraisal free from personalities. This car was basically a mechanical device and after years of operation was no longer able to produce reliable information. Plans were made to rebuild the equipment using electronic devices wherever possible to measure and record the findings.

Among the most important features of the design of the new car is that accurate measurements can be made at all speeds through 100 mph. This will permit reliable evaluation of the entire system in a reasonable time at realistic train speeds.

At the end of the roadway maintenance season in the fall the inspection car is run, and the information received helps to determine how effective budgeted maintenance programs were in accomplishing their objective. Early spring inspection serves to confirm seasonal maintenance programs as well as to indicate areas of future maintenance activity.

Accordingly, the development of the new car (the RI-2) was directed toward providing an important aid in efficient and economic quality control of track on the C&O System.

The overall system of the RI-2 includes means for measurement of such track characteristics as curvature, cross level or superelevation of rails, surface, and joint condition. These measurements are continuously recorded on tape along with landmarks and other notes indicating location, as well as speed of the car. The recording and control facilities are compactly located near the center of the car. Facilities are provided for unobstructed visual observation of the railroad through an observation deck seating 31 persons. The car is placed as the last car in a train with the observation end towards the rear. At the other end of the car are kitchen, office, conference, sleeping and toilet facilities. Two men can operate all the measuring and recording facilities.

The operation and facilities of the RI-2 which I have just summarized are illustrated in more detail in the slides which follow. The first slide [Slide 1] shows an exterior view of the RI-2. This car is a converted passenger car 78 ft 8 in long over platform I beams, 13 ft 7 in high, and 10 ft wide. The observation end of the car, with an inclined seating arrangement, has track and floodlights for night observation and glass panels arranged for unobstructed vision. The rear truck supports practically all the primary measuring devices on the car. This is a six wheel truck, 5 ft 6 in between wheel centers. The distance between this truck center and the forward truck center is 55 ft 11 in.

Slide 1.

A large box mounted on the rear wheel journal box cover contains a two-axis vertical gyroscope. The purpose of the gyroscope is to measure cross level of rails through its inner axis and track surface through its outer axis. A diagram showing its operation will be shown later. For protection against shock and vibration the gyroscope rests on shock mounts at each of the four corners of the base with lateral shock absorbed through a cradle mounting cushioned in sponge rubber. A device mounted on the center wheel journal box cover measures the extent of low joints at the rail ends. It is actuated by a flexible wire cable attached to a reference beam which, in turn, is attached to the journal box covers of the two end wheels of the truck. A later slide will show its mode of operation.

The recording tape is driven at a speed proportioned to train speeds by a tachometer which is attached to the truck from under the car body.

Another measuring device is mounted near the center of the forward truck under the car body. This equipment measures track curvature by measuring the movement of the truck around the center line of the car as the car negotiates a curve.

The measuring device is called a RVDT which is short for "rotary variable differential transformer." It translates the mechanical motions just described into an electrical voltage proportioned to the mechanical motion.

The next slide [Slide 2] is a diagrammatic representation of the manner of measuring curvature by the RI–2. As the truck enters a curve it rotates around its center pin. The flexible cable mounted on the truck frame is wrapped around the wheel of the RVDT and causes the wheel to turn proportionately to the truck rotation. The

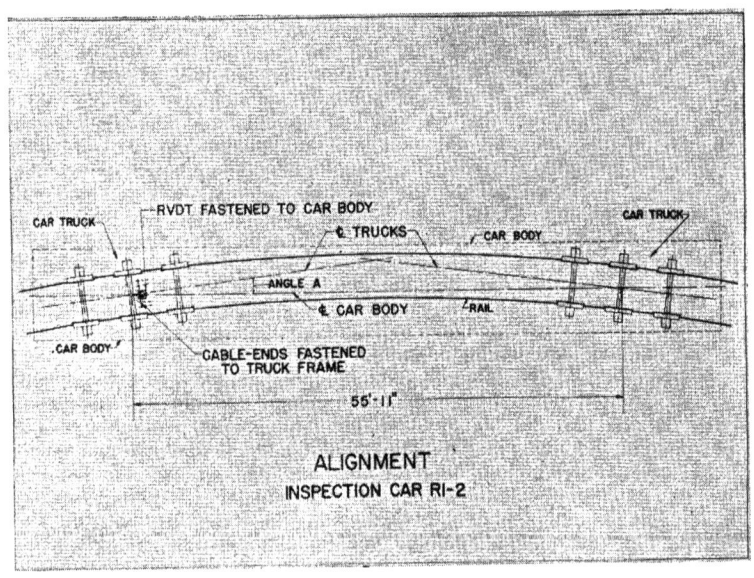

Slide 2.

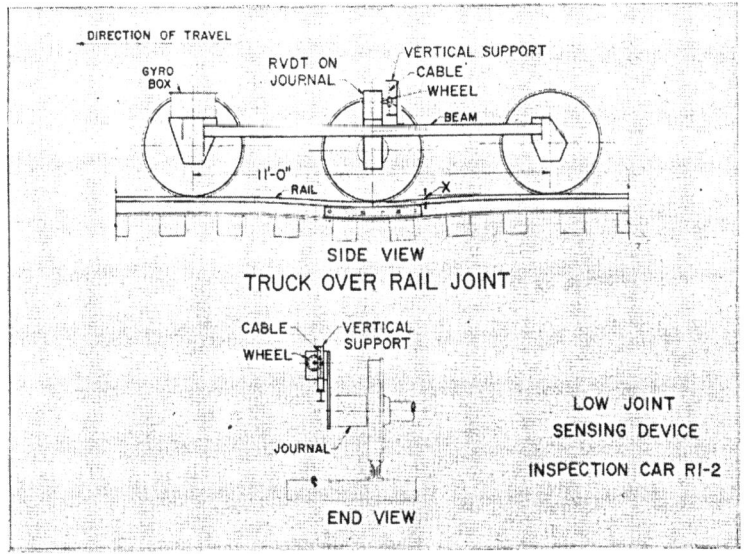

SIDE VIEW
TRUCK OVER RAIL JOINT

LOW JOINT
SENSING DEVICE
INSPECTION CAR RI-2

END VIEW

Slide 3.

angle between the center line of the car and the center line of the truck is translated into curvature.

The next slide [Slide 3] is a diagrammatic representation of the manner of measuring low joints. If the joint is low as the center wheel passes over the rail ends, it will descend below the level of the wheels on the leaving and receiving rails. In so doing, it will cause the RVDT wheel to move a distance proportional to the drop of the wheel below the reference beam. This generates a voltage proportional to this movement which is recorded and translated into distance. The movement of the center wheel upward is also detected and recorded.

The next slide [Slide 4] indicates the manner of measuring cross level and surface with a gyroscope. The gyroscope is a rugged two-axis type which can measure the generalized properties of pitch and roll by independent departures of each of its axes from the reference. The roll axis measures rail superelevation since one side of the truck is higher than the other by the amount of superelevation. The roll axis of the gyroscope moves away from its reference proportional to the distance one side of the track is higher than the other. This gyro movement is translated into inches of superelevation and recorded on the tape and measured. The pitch axis of the gyroscope, like the roll axis, departs from its reference independently when there are changes in track gradient.

All of the electrical signals I mentioned before are transmitted by shielded wiring to a console on the inside of the car in the center of the car. It is arranged to permit observers to watch and the operator to work. On the left-hand side of the console is a Visicorder. This is an optional oscillograph that changes the electrical voltages trans-

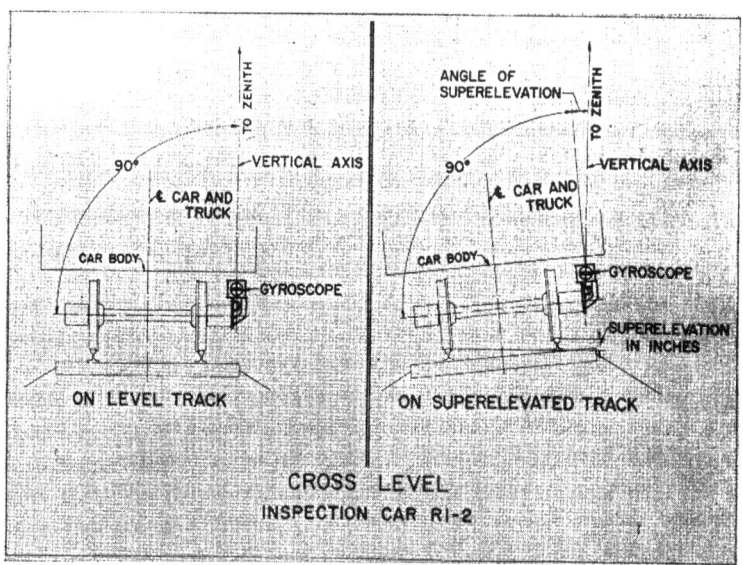

Slide 4.

mitted by the measuring devices into movements of tiny mirrors mounted on sensitive galvanometers. A very bright light shines on these mirrors. The movement of the mirrors is reflected by the light through a lens system on to a photosensitive paper tape. This light leaves marks of the recordings on the paper.

The next slide [Slide 5] shows the paper tape emerging from the Visicorder. The speed of the paper tape is proportional to the speed of the car. To observe track characteristics in greater detail the scale can be expanded by adjustment. The scale is normally 13.2 in of tape per mile of track. As the paper tape moves along the length of the console table a regular 20-w fluorescent light positioned above the paper exposes the paper to obtain positive print of proper contrast.

The next slide [Slide 6] shows the control panel for all measuring devices on the car. The switches turn the instruments on and off and colored pilot lights indicate whether the current is turned on into each of the indicated instruments. The electrical supply to the console is 115-v 60-cycle current through an amplidyne which is converted for desired use. The plus-and-minus-marked lights on the right and left top of the console are adjustable indications of low joints in excess of acceptable deviations.

The observation deck is illustrated in the next slide [Slide 7] looking from the center of the car towards the rear. The sloping deck permits unobstructed vision through the rear glass panels or the side windows. There are three movable chairs close to the rear glass panels for ranking officers. There is a microphone on the extreme right rear car post for public address purposes.

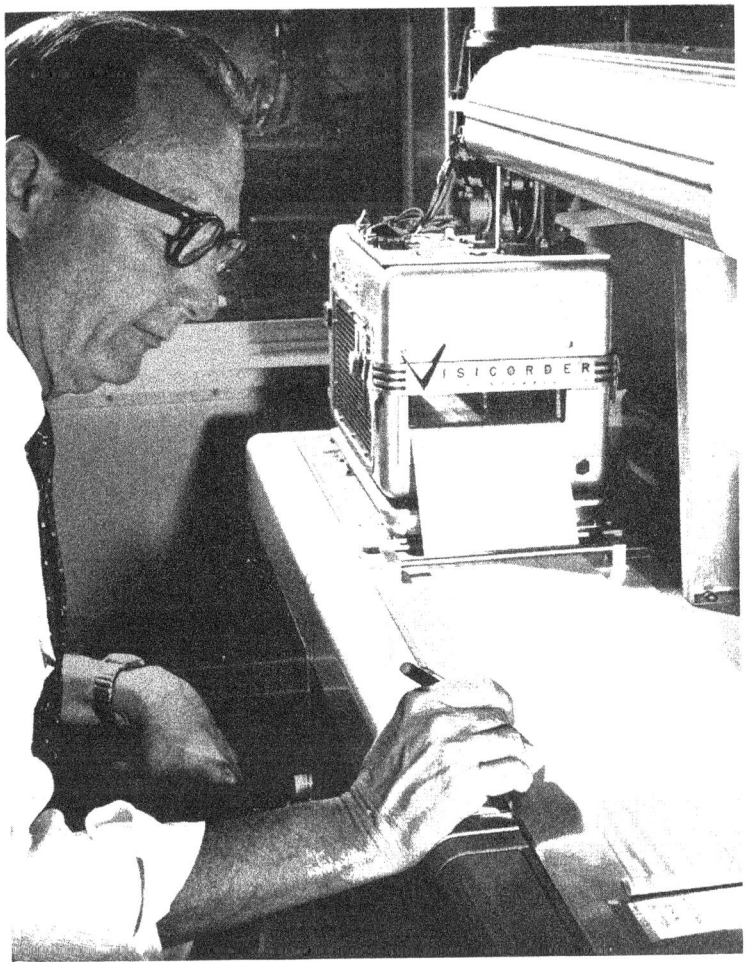

Slide 5.

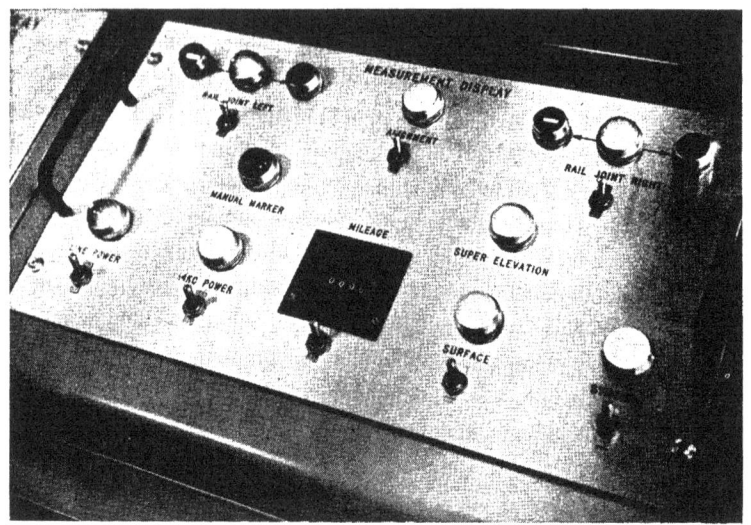

Slide 6.

Slide 7.

One of the operators of the car sits at the baywindow in the center of the car. He is a local man familiar with the territory. He calls out mile posts or landmarks, as he passes them, to the operator seated on the console. The console operator marks the name of the landmark on the record tape next to a mark on the tape which is made by the window man pressing a button next to the window.

As one enters the car from the forward vestibule the first room is a kitchen equipped to serve light meals. The next room towards the center of the car is an office equipped with a telephone for plugging in at terminals, a desk for paper work and examination of record tapes, and two berths which fold into the bulkheads when not in use.

The next room is a conference room equipped to seat four, to store various supplies and clothing, and to sleep two in separate berths which fold to the bulkheads when not in use.

On the record tape, starting from the bottom, the first trace is speed-time. The distance the trace moves up from its zero point at the bottom of the paper is proportional to speed. The pips shown on this trace are made at 10 second interval.

The next trace up is a quarter-mile marker. Each mile is a long mark, the quarter miles are shorter. Its purpose is to aid relation of paper record to locations on track. Going up the line the next two traces indicate the low joints of the left and right rails. The trace marks in the upward direction indicate a low joint while the trace marks in the downward direction indicate a center wheel bounce of upward movement. Proceeding upward the next trace is the superelevation or cross level. The straight portion of the line, which is on its calibrated zero mark, indicates tangent track. As the car enters the spiral of a curve the trace departs from its reference base, indicating an inclined line up or down as the curve is to the right or left, until it reaches the circular portion of the curve. This is the straight portion of the line either above or below the zero position. The amount of the displacement is calibrated to give readings of superelevation in inches. Wiggles in this trace indicate irregularities of superelevation.

Next up the line is the curvature trace, coordinated with the superelevation trace. The curvature trace marks are interpreted in a similar fashion to the superelevation trace except that the displacement from the reference is interpreted in degrees of curvatures, and the irregularities in the trace are interpreted as alinement faults if they are excessive. Near the top of the paper is the surface trace. Changes in track surface or gradient are indicated by departures of the trace from its reference mark. Irregularities in the line indicate surface irregularities.

This photosensitive paper record tape, which is 6 in wide, has reference marks automatically imprinted as a background as the Visicorder operates. The lighter lines are 0.1 in apart and the heavier lines are $\frac{1}{2}$ in apart. Thus, visual observation of the paper tape allows measurement of the track characteristics by tape inspection. Each trace is calibrated in reference to the linear displacements of the measuring instruments on the car. Also, each trace is located at a particular place on the paper tape by adjustment of the galvanometer in the Visicorder.

The traces of the superelevation, alinement, surface and joints are examined, and when they depart from the predetermined standard a maintenance error is noted. The other traces are used to find where the trace marks are in relation to the actual track and analyze the record with respect to speed.

The record tape is graded by engineering personnel in the office according to the standard briefly described as follows and illustrated in the next slide [Slide 8]:

If the record on the joints traces exceed the reference line by the equivalent of $\frac{3}{8}$ in then one demerit is counted for each such excess. Similarly, if there are excesses from

EXAMPLE OF GRADING METHOD
ROADWAY INSPECTION CAR, RI-2

D - E - M - E - R - I - T - S

TRACK SUPERVISOR'S DISTRICT	TRACK	MILES TESTED	LOW JOINTS	ALIGNMENT	CROSS LEVEL	SURFACE	TOTAL DEMERITS	(1) TOTAL DEMERITS PER MILE TESTED	TRACK AND ROADWAY EXPENSE	EQUATED MILES	(2) EXPENSE PER EQUATED MILE	INSPECTION INDEX
(A) MILE BY MILE EXAMPLE	# 2											
MP-0 TO MP-1		1	3	17	16	0	36					
MP-1 TO MP-2		1	7	23	22	2	54					
MP-2 TO MP-3		1	11	19	17	1	48					
ETC-												
(B) SUPERVISOR'S DISTRICT												
PIEDMONT	96L	95-0	369	427	413	295	1474	15-55	$200407	107-50	$89599	294
MOUNTAIN	9GL	93-4	288	312	300	192	1072	11-48	188827	10053	188080	826
ALLEGHANY	81	76-2	174	201	194	126	495	6-12	280984	280-13	C73048	221
ETC-												

THESE FIGURES INCLUDE ONLY DIRECT LABOR AND MATERIAL COSTS
FOR TIES, RAIL, AND BALLAST FOR PAST YEAR FOR ENTIRE DISTRICT

FOR ENTIRE TRACK SUPERVISOR'S DISTRICT

Slide 8.

the reference line of the alinement, cross level, and surface traces by 1 deg, ½ in, and ¼ in, respectively, within approximately two rail lengths, then a demerit is counted for each such excess. Demerits are accumulated for all traces combined for each mile inspected. In order to make due allowance for the type of territory and the amount of maintenance effort and money expended on the district, these total demerits are converted to total demerits per mile tested. Then the track and roadway expense on the district tested is converted into a figure of dollars per equated mile of the tested district. Multiplication of the total demerits per mile tested by the track and roadway expense per equated mile yields the inspection index, which characterizes the district inspected and is shown in the last column of the table. It can be seen that this procedure makes an assessment of how well the maintenance money available is spent in terms of accomplished results. For example, although the Alleghany district exhibits the fewest total demerits per mile of the three hypothetical districts shown, the Mountain district has the lowest inspection index. The lower the inspection index, the more effective the money is spent in maintaining track.

All of the grading is now performed manually. The left hand side of the console in the inspection car has been left empty to permit future electronic grading. The electrical voltages generated by the measuring instruments can be fed into a computer at the same time they are permitted to make the traces on the oscillograph. Circuitry for this operation is being planned which would permit electronic interpretation of the inspection at the same time the record is being made.

This is a modern development in track inspection and quality control that can be exercised on track maintenance. Data from the RI-2, as it accumulates, can also be used to set quality standards which recognize differences in territories, and budgetary and other requirements. Since the inspection car is principally an information processing device, it can be used as a rolling laboratory and a research vehicle for improved track maintenance and construction. [Applause]

CHAIRMAN POORE: Thank you, Mr. Burris, on behalf of the Track committee.

PRESIDENT BEEDER: Thank you very much, Mr. Poore and your subcommittee chairmen, for your Manual review work and, again, several interesting reports. It is to be regretted that work on some of your projects has had to be deferred due to limitation of research funds or other causes, but it is evident that you made real progress during the past year in the revision of a number of our track plans, and in the development of plans for the five standard turnouts, which have been presented to our Association today.

We welcome Mr. Salmon as the new vice chairman of your committee.

And we thank you, Mr. Burris, for your well-presented talk on your new track inspection car. Next to the space capsule, "Friendship 7", I suspect it is the last word in electronics. Your new car certainly appears to be a vast improvement over the one I saw a number of years ago. Did I understand that you still have a few changes to make in the gyroscopes?

MR. BURRIS: No changes, Mr. President, except that they are going to be enlarged.

PRESIDENT BEEDER: Has the completion date been set as yet?

MR. BURRIS: I have some men coming from England the latter part of this month to look at it. I am very much in hope that it will be in condition to operate thereafter.

PRESIDENT BEEDER: Thank you, Mr. Burris, for your very interesting presentation.

Mr. Poore, you are now excused, with our thanks to you and your committee.
[Applause]

Discussion on Rail

[For report see Bulletin 570, pp. 497–571]

PRESIDENT BEEDER: The next report is that of Committee 4—Rail. We should be pleased to have the members of the committee come to the platform at this time.

The chairman of Committee 4 is W. J. Cruse, engineer maintenance of way, Great Northern Railway, and a director of our Association. Mr. Cruse, please proceed.

CHAIRMAN W. J. CRUSE: Mr. President, members and guests:

The report of Committee 4 appears on pages 497 to 571, incl., Bulletin 570, Part 1. In the interest of time we shall present only one formal report, which has to do with our Manual material, and touch briefly on two other assignments.

During the year we lost one member by death and nine other members, three due to resignations and retirements; three were dropped because of the three-year rule for retired members, and three for other reasons. R. P. Winton, retired testing engineer maintenance of way, Norfolk & Western Railway, and a Member Emeritus of Committee 4, died on January 9, 1962. A Memoir has been prepared which will appear in the Proceedings issue.

MEMOIR

Robert Prince Winton

The Rail committee regrets to report the passing of Robert Prince Winton, retired testing engineer maintenance of way of the Norfolk & Western Railway, who died on January 9, 1962, in Roanoke, Va., after an illness of several months. Surviving are his wife, Mrs. Virginia Bedinger Winton; two daughters, Mrs. R. F. Bondurant, Roanoke, and Mrs. B. F. Weaver, Pebble Beach, Calif.; his stepmother, brother and sister, and five grandchildren.

Mr. Winton was born in New Haven, Conn., June 18, 1889, and graduated in electrical engineering from the Sheffield Scientific School of Yale University in 1909. After working five years with General Electric Company and the New York, New Haven & Hartford Railroad, he returned to Yale where he taught electrical engineering in the Sheffield Scientific School for two years.

He entered the service of the Norfolk & Western on November 1, 1922, as catenary engineer. He was appointed welding engineer in 1932 and testing engineer maintenance of way in 1947, which position he held until the time of his retirement in 1957. Prior to his appointment as welding engineer, he began to study the possibilities of welding with respect to reclaiming track materials.

Known as the "Professor" because of his teaching career, he was a prolific originator of home-made gadgets, among which are a machine for welding that attracted attention throughout the railroad industry; a device to hold track frogs during welding to prevent heat from warping the frog; a planer type of grinder to smooth welded portions of frogs; a reversing machine for cleaning the inside of track nuts; and a jig for properly locating holes to be drilled in switch rods. His genius helped the Norfolk & Western to pioneer with better methods of reclaiming many thousands of tons of roadway material.

Mr. Winton pioneered in the application of the oxyacetylene pressure process of butt welding rails, and had the reputation of getting the best possible welds by this process.

Joining the AREA in 1926, he served on Committee 4—Rail, from 1940 to the time of his retirement and was elected a Member Emeritus in 1960. He was for several years chairman of the Subcommittee on Rail End Batter; Causes and Remedies, was a member of the Joint Contact Committee and was active in the various studies of the Rail committee on the shelly rail problem. There were few matters pertaining to rail which escaped his attention, and his opinions and conclusions were always sought and respected.

Assignment 1—Revision of Manual.

CHAIRMAN CRUSE: The report on Assignment 1 will be presented by Subcommittee Chairman J. A. Bunjer, chief engineer, Union Pacific Railroad.

[Mr. Bunjer then read the report on Assignment 1 as printed on pages 498 to 503, incl., of Bulletin 570, concluding as follows:]

MR. BUNJER: Mr. President, I move that these recommendations be adopted.

[The motion was duly seconded, was put to a vote, and carried.]

Assignment 3—Rail Failure Statistics, Covering (a) All Failures; (b) Transverse Fissures; (c) Performance of Control-Cooled Rail.

CHAIRMAN CRUSE: The report on Assignment 3 will be presented by Subcommittee Chairman C. J. Code, assistant chief engineer—staff, Pennsylvania Railroad.

C. J. CODE: The report on Assignment 3 occupies 16 pages of Bulletin 570 and contains a mass of data which can scarcely be absorbed at one reading. In fact, if you read it three times you will still find things in it you did not see before. All the tables and charts contain valuable and authentic information which is useful, in fact essential, in studying the many factors which influence rail failure. The performance of control-cooled rail continues to be excellent, and the new sections which were introduced in 1947, and some more recently, show an outstanding improvement, particularly with respect to web failures which they were designed to prevent. One type of defect which has not responded noticeably to remedial measure is the detail fracture, commonly associated with shelly spots and heavy wheel loading.

The current report and earlier reports contain no table which shows all failures in rail of all ages. In the next 10 or 15 years, if wheel loads are increased without commensurate increase in wheel diameter, it will be essential to show the effect of such increased loads on rail failures. The effect may not show up to its fullest extent in the first 10 years of rail life and hence may not be reflected to the fullest extent in a report such as Table 5, which includes only rail failures in new rail in its first 10 years of service.

Consequently, beginning with 1962, you will be asked to furnish a report, which I am sure you all prepare for your own information, showing failure of all ages of rail, by rail section and by types of failure, in a form somewhat similar to Table 5. The results will be included in reports beginning with the 1964 Bulletin.

Attention is called to the fact that, in spite of the title, Table 5-a lists only the failures in the new sections.

MR. CODE (continuing): This has nothing to do with Subcommittee 3, but my attention was called this morning to an error in the report on Service Test of 140 lb High-Silicon Rail on the Pennsylvania Railroad, which goes to show that people do read the reports.

The last sentence of this item says, "The curve oiler was not operating." Our supervisor of track read the report and said, "There isn't any curve oiler there."

I wondered about this when I read the report but I didn't bother to check it, so our apologies go to Supervisor of Track Johnson and to District Engineer King.

Assignment 9—Standardization of Rail Sections.

CHAIRMAN CRUSE: The report on Assignment 9 will be presented by Subcommittee Chairman T. B. Hutcheson, chief engineer, Seaboard Air Line Railroad.

T. B. HUTCHESON: The report on Assignment 9 begins on page 553 of Bulletin 570, Part 1, and is in the form of a comprehensive review of the AAR Research Center's Report ER-15, entitled Engineering Aspects of Current Rail Sections. The report is submitted as information.

During the past year Subcommittee 9 secured information from rail mills as to sections currently rolled for United States, Canadian and Mexican railroads. It studied that information along with a poll taken by the subcommittee in 1959 as to sections then currently in use, and the Research Center's Report ER-15. Following such study, the subcommittee recommended to the Rail committee that the 136-lb CF&I section be designated 136-lb RE and included in the Manual as an AREA recommended section, and that the 133-lb RE section be eliminated as an AREA recommended section, and further that appropriate changes be made in recommended joint bars to conform with the recommendations regarding rail sections. These changes are included in the report on Assignment 1—Revision of the Manual.

CHAIRMAN CRUSE: Mr. President, I would like to take a moment to recognize those chairmen who were unable to present reports here today. They are:

K. K. Kessler, chief inspector, Baltimore & Ohio Railroad, Chairman of Subcommittee 4—Rail End Batter, Cause and Remedies.

J. C. Jacobs, engineer maintenance of way, Illinois Central Railroad, Chairman of Subcommittee 5—Economic Value of Various Sizes of Rail.

Embert Osland, office engineer, Atchison, Topeka & Santa Fe Railway, Chairman of Subcommittee 6—Joint Bars: Design, Specifications, Service Tests, Including Insulated Joints.

L. S. Crane, assistant chief mechanical officer, Southern Railway, Chairman of Subcommittee 8—Causes of Shelly Spots and Head Checks in Rail: Methods For Their Prevention.

A. P. Talbot, assistant engineer, Pennsylvania Railroad, Chairman of Subcommittee 10—Service Performance and Economics of 78-ft Rail, collaborating with Committee 5; Specifications for 78-ft Rail.

Mr. President, this concludes our report.

PRESIDENT BEEDER: Thank you, Mr. Cruse. You are extending a long record of good chairmen of our Committee 4, which is evident in the reports of your committee again this year. Our thanks, too, to Mr. Magee and the members of his staff, for the large amount of valuable assistance that they give to your committee, as well as to many others of our committees.

On the basis of the vote here today, I think it is in order to say that we welcome into our family of AREA rail sections the new 136-RE section, and say a fond farewell to our former 133-RE section.

Without time for further comments, may I merely say that you and your committee are now excused, with our thanks. [Applause]

Discussion on Continuous Welded Rail
[For report see Bulletin 570, pp. 599–601]

PRESIDENT BEEDER: The final report on our convention program is that of our Special Committee on Continuous Welded Rail, the chairman of which is F. L. Rees, division engineer, Santa Fe, at Fort Madison, Iowa. Mr. Rees has had a pretty rugged year as the result of an automobile accident, but we are happy that he came through okay and is here with us this morning. Mr. Rees, I shall be pleased to turn this microphone over to you if you and the other members of your committee will come promptly to the speaker's table.

Assignment 1—Fabrication.

CHAIRMAN F. L. REES: The Subcommittee on Fabrication has encountered many difficulties in attempting to develop specifications for the fabrication of welded rail. Last year a preliminary draft of the specifications was published in the Bulletin, and the membership at the convention was asked for pertinent comments regarding the proposal— but none were received. However, at the last meeting of the whole committee the discussion of these specifications absorbed most of the meeting. Discussion centered around the tolerances allowable and the economics involved in correcting the rails during welding production to meet the allowable tolerances. It was finally decided that probably the proposed specifications were more strict than mill tolerances and that welds could not be economically produced adhering to the proposed specifications without first reviewing mill tolerances on rail. The subcommittee will collaborate with the Rail committee and will later meet with the Joint Contact Committee to discuss mill tolerances. When welding secondhand rail the tolerances proposed could be more easily complied with.

The chairman of this subcommittee is A. H. Galbraith, welding engineer, Santa Fe.

Assignment 2—Laying.

CHAIRMAN REES: M. S. Reid is chairman of the subcommittee on laying welded rail. He is assistant chief engineer–maintenance of the Chicago & North Western Railway at Chicago.

This committee is continually looking for new methods and machines for laying welded rail. The committee also receives and compiles data on all welded rail laid each year, and these data appear on page 601 of Bulletin 570.

The Special Committee on Continuous Welded Rail has been in existence for almost 12 years and at our last meeting someone asked, "What is welded rail?" It is possible to stand so close to the trees that you can't see the forest. Immediately we had three or four different definitions—5 rails long, 6 rail long, 78 ft long—but the committee was directed to submit a definition for welded rail so everybody would know what it was. I wonder what it is we have been studying for the past decade.

Assignment 3—Fastenings.

CHAIRMAN REES: D. T. Faries, chief engineer, Bessemer & Lake Erie at Greenville, Pa., is the Chairman of the Subcommittee on Fastenings.

Last year this committee submitted a Manual recommendation on anchorage, which was adopted. After further review it was decided that some minor revisions are required, and these will be submitted in 1963. This committee is also collaborating with Subcommittee 10 of the Committee on Iron and Steel Structures concerning the anchorage of welded rail on long open-deck bridges.

Assignment 4—Maintenance.

CHAIRMAN REES: Subcommittee 4 evidently is the boomers' committee because every time the report is due there is a new subcommittee chairman. This assignment has more possibilities for Manual material than the other assignments. At present there is a questionnaire out requesting information on preparation of the roadbed, that is, the kind of surfacing and when it is done, when the ties are installed and how many per panel, and the ballast size, kind, and section used. The new chairman of this subcommittee is C. R. Merriman, engineer maintenance of way and structures, Chicago South Shore & South Bend Railroad at Michigan City, Ind.

Assignment 5—Economics.

CHAIRMAN REES: If you are not able to decide whether welded rail is economically justified on your particular road, I direct your attention to our Assignment 5, under which we expect to provide comparative costs in usuable terms. The argument has been advanced that welded rail has not been in use long enough to tell whether there are economies to be gained. This is true to some extent, yet in reviewing the compilations of Subcommittee 2 we find that 87 miles were laid in 1954, 266 miles in 1955 and 1070 miles in 1959. Gentlemen, someone has decided that there are economies that justify welded rail. That is the problem of this Subcommittee on Economics. They are still digging for the answers, and if you have any cost data on control sections of jointed rail adjacent to welded rail or if you have compiled any other economic data on welded rail, you are invited to submit the data to this subcommittee, whose chairman is T. C. Shedd, editor, Modern Railroads.

Assignment 6—Welding Second-Hand Rail.

CHAIRMAN REES: The last committee is a newly organized group especially formed to investigate the welding of second-hand rail, which is a more or less recent development. The mileage welded is still very small but this committee is seeking information on the problems and specifications for welding second-hand rail for branch lines, secondary mains and yard tracks. This subcommittee was formerly headed by H. F. Gilzow, assistant purchasing agent, Frisco, but is now chaired by J. F. Beaver, chief engineer, Southern Railway System, Washington, D. C.

CHAIRMAN REES (continuing): This completes my tenure as chairman of this committee. It has been a gratifying experience as well as a rewarding one. It has added to my knowledge of welded rail to find that even though men from other roads might not use the same procedures in handling welded rail as my home road, yet we all obtain similar results.

The subcommittee chairmen, as I have said before, are the backbone of any committee and I truthfully believe that this committee has the best. Neal Howard has kept me busy answering his letters and sometimes they come in so fast I overlook answering one now and then, but Neal has always stood ready to lend a helping hand when needed and I thank you for that, Mr. Howard.

My successor as chairman is W. J. (Jack) Jones, engineer maintenance of way and structures—system of the Southern Pacific at San Francisco. His right hand man in the position of vice chairman will be D. T. Faries, chief engineer, Bessemer & Lake Erie, who is also chairman of Subcommittee 3. This concludes the report of the Special Committee on Continuous Welded Rail.

PRESIDENT BEEDER: Thank you, Mr. Rees. Between your own disability and the "boomers" among your subcommittee chairmen, it is evident that your committee has been seriously handicapped in its work for the past year. However, we do appreciate the extent to which it has been possible to progress the work on at least some of your assignments, and we hope that, as now reorganized, the committee will make real progress on its assignments in 1962.

We welcome Mr. Jones as the new chairman of the committee and Mr. Faries as the new vice chairman, assured that they will push work on all of the committee's assignments in the year ahead.

Mr. Jones, if you will step to the podium, I would like to present you with this chairman's gavel to assist you in directing the work of your committee during the year ahead. [Applause]

Thank you, Mr. Rees, and all the members of your committee. You are now excused with the thanks of the Association. [Applause]

PRESIDENT BEEDER (continuing): As information, at high noon today the registration stood at a total of 648—348 railroad men and an even 300 nonrailroad men.

Closing Business Session

PRESIDENT BEEDER: Having completed the formal presentations of our technical committees, we are now ready to begin the Closing Business Session of this Convention, during which will be announced the results of our annual election of officers and the installation of our new officers for the ensuing year.

Before convening the Closing Business Session, I should like to say that it has been a great honor, privilege and rewarding experience to have been your president during the past year. Also, I want to extend my thanks to everyone who has contributed to the work of the Association during my term as president, and especially to everyone who participated in our Convention program.

This may not have been as long a Convention, timewise, as in the past, but it was certainly no less intensive and valuable. I wonder if you realize that during the last day and a half you have, among other things, acted on 359 Manual documents, which will result in the largest annual Manual Supplement ever produced by our Association. So, if nothing else can be said for this 1962 convention, it must be said that it accomplished important results, which could have been accomplished in no other way.

There are so many to whom I am indebted for seeing this Convention through successfully that I cannot begin to name them all; but I should feel particularly neglectful if I did not express my appreciation to Charlie Coverley and all the members of his Committee on Convention Arrangements, and to our secretary, Neal Howard, and his staff. Also, to the ladies who have assisted Mrs. Beeder in the social functions for the ladies during our Convention, and to the many railway supply men who were here, for their cooperation and courtesies throughout our meeting.

Again, my thanks to all of you.

I now call to order the closing business session of this convention. Is there any other business to come before us?

PAST PRESIDENT G. M. O'ROURKE: Mr. President, may I have the privilege of the floor?

PRESIDENT BEEDER: Mr. O'Rourke, you may. Please come to the platform.

MR. O'ROURKE: Ladies and gentlemen, it may surprise some of you to know that I am going to try to say some nice things about your president in the next few minutes.

President Beeder, a year ago I was pleased to escort you up to this launching pad and watch you being shot into orbit. Having observed the accomplishments of the past year, in compressing into one and a half days the business of this Convention that tried sorely the time and efforts of your predecessors to handle satisfactorily in two and a half days, I must say that you have been flying high. [Laughter]

The members realize that this constriction was necessary in the light of economic circumstances, as you described in your introductory comments at the beginning of this Convention; and I speak for them, I believe, when I say that we know you are leaving your high office to join the Past Presidents' Club with the knowledge of a job well done. The past presidents welcome you, and we are proud of you.

Yours has been a performance that will take a lot of doing to equal or surpass. Allow a warm personal friend, if you please, to congratulate you. You have done a magnificent job, and Mrs. O'Rourke would want to join me in telling you and Mrs. Beeder of our admiration and affection for you. May God bless you!

And now, in recognition of your many accomplishments, it is my pleasure to hand you this plaque, which reads as follows:

"The American Railway Engineering Association records its grateful appreciation to R. H. Beeder for his able administration of the affairs of the Association during his term as President."

We hope you will find a place to hang this so that you can look at it every day. [Applause]

PRESIDENT BEEDER: Thank you, Past President George Martin O'Rourke. I am most happy to join your Club. I shall cherish and value this wonderful plaque and the kind thoughts that you have expressed in presenting it to me. I already have wall space selected for it about 5 ft from the end of my desk so that when I look at the message conveyed by this plaque I can remember the many men whom I have come to know and respect through our work in this Association.

While thanking you, George, I want to thank again each member of the Board of Direction for his counsel and advice during the past year, and especially those members who are retiring from the Board, having completed their terms of office.

The close of this convention completes the service on the Board of Past President Frank Woolford, chief engineer of the Western Pacific Railroad, under the provision of the Constitution that past presidents remain on the Board for only two years following the completion of their term of office as president. I have a feeling that Frank wishes there were no such stipulation in the Constitution, and we might well wish the same; but that's what the Constitution says.

We are all deeply indebted to Mr. Woolford for his valuable service to our Association, both in an official and unofficial capacity; and although he will not be on our Board, I am sure we will want to call upon him in the future for advice on important matters as occasions arise. I shall be pleased if Mr. Woolford will stand to be recognized. [Applause] I am sure that not only will we want to continue to call on him, but he will want to continue to call on us from the floor and elsewhere when important matters arise that need discussion.

MR. WOOLFORD: Thank you, Mr. Beeder. May I reiterate what I said before—that I certainly enjoyed my tour of office with this Association, both on the Board of Direction and as president of the Association, and also as a past president.

PRESIDENT BEEDER: Thank you very much, Mr. Woolford. We greatly appreciate everything you have done.

The term of office of four of our directors also terminates with the close of this Annual Meeting. These directors are: W. E. Cornell, engineer of track, Nickel Plate Railroad; F. L. Etchison, chief engineer, Western Maryland Railway; W. J. Cruse, engineer maintenance of way, Great Northern Railway, and C. R. Riley, special assistant to vice president—operations and maintenance, Baltimore & Ohio Railroad.

All of these men who are retiring from the Board have rendered valuable service to our Association both as working members and in their official capacities on the Board; and, speaking for the members of the Association as a whole, I want them to know that their service has been greatly appreciated.

To the extent that our retiring directors are present in the room, I shall be glad if they will stand and permit us to show them our appreciation. [Applause]

Is there any other business to come before this meeting?

J. E. WIGGINS, JR. (Southern): Mr. Beeder, I have the official report of the Tellers Committee as a result of their count of the ballots in this year's election of officers. May I present it to you?

PRESIDENT BEEDER: I will be glad to have you come forward, Mr. Wiggins. This is a very important piece of unfinished business, and I shall be pleased to read the report to this meeting. As I read the names of those elected, I would appreciate your withholding your applause until all names have been read.

First, I will read the names of the four directors who have been elected:

W. L. Young, chief engineer, Norfolk & Western Railway, Roanoke, Va.

T. B. Hutcheson, chief engineer, Seaboard Air Line Railroad, Richmond, Va.

C. E. Defendorf, chief engineer, New York Central System, New York City.

John Ayer, Jr., chief engineer, Denver & Rio Grande Western Railway, Denver, Colo. [Applause]

I would now like to read the names of the five members elected to the Nominating Committee who, together with the five most recent living past presidents of our Association, will constitute the Nominating Committee as a whole for the coming 1963 election.

A. L. Sams, principal assistant engineer, Illinois Central Railroad, Chicago.

J. F. Beaver, chief engineer, Southern Railway System, Washington, D. C.

B. B. Lewis, professor of railway engineering, Purdue University, Lafayette, Ind.

J. J. Schmidt, assistant director—research, Denver & Rio Grande Western Railroad, Denver.

E. M. Hastings, Jr., wire crossing engineer—system, Chesapeake & Ohio Railway, Richmond, Va.

These five gentlemen are the newly elected members of your Nominating Committee. [Applause]

Having read to you the Tellers' report, it is now my pleasure and privilege to present to you the new directors and officers whom you have elected for the ensuing year. I will appreciate it if our newly elected directors will come to the speaker's table and stand at my left. Three of them are here.

Gentlemen, I congratulate you upon your election as directors of this Association, and welcome you to the Board of Direction. In your election you have been highly honored. Your election is not only recognition of your past interest in and service to this Association, but an expression of the confidence of the membership in each of you to assume leadership in the direction of the affairs of our Association. I am sure that each of you will live up to this confidence, and to the responsibilities which you assume as directors. You may be seated. [Applause]

Our senior vice president is L. A. Loggins, chief engineer, Southern Pacific Company, Texas & Louisiana Lines, at Houston, Tex., who under the Constitution automatically advances to this position from that of junior vice president. Mr. Loggins, I shall be pleased if you will come to the platform and take your place at my right. [Applause]

Your newly elected junior vice president is T. F. Burris, chief engineer—system, Chesapeake & Ohio Railway, who returns to the Board of Direction following a three-year term which ended with our 1961 convention. Mr. Burris, will you please come to the platform and stand here next to Mr. Loggins? [Applause]

Mr. Loggins, I congratulate you on your further elevation to high office in this Association, with the further obligations and opportunities which this affords you for still greater service to the Association.

Mr. Burris, I congratulate you upon your election as junior vice president, and welcome you back on the Board. Again, it can be said, we have two strong vice presidents.

You gentlemen may now be seated. [Applause]

As your president for the year ahead you have elected C. J. Code, assistant chief engineer—staff, Pennsylvania Railroad. To accord Mr. Code the special recognition due him, I have asked Past Presidents Woolford and Brown to escort him to the platform, and I would appreciate their doing so at this time. [Applause]

Mr. Code, I congratulate you upon your election to the highest office in this Association, and I now proclaim you president. This honor which has come to you is richly deserved, and it is with pleasure and complete confidence that I turn over the responsibilities of president to you.

In so doing, I have an important thing I want to give to you. It is this solid gold Emblem of the Association, which bears the engraved words on the back, "C. J. Code, President, 1962–1963." I am sure you will wear this Emblem as a badge of distinction to yourself and to the Association which it represents.

I now turn the podium over to President-Elect Code. [Applause]

PRESIDENT-ELECT C. J. CODE: Thank you, Mr. Beeder.

I am accepting with a mixture of pride, humbleness and some misgivings the office which has been bestowed on me today—pride and humbleness when I think of the many distinguished engineers who have preceded me in this office, and misgivings when I think of the responsibilities of the position under the conditions that confront our Association.

The American Railway Engineering Association has meant a great deal to me in the past 20 years—actually longer than that, since I had a Manual and looked with respect on its recommended practices long before I became a member. It has meant a great deal to me; and I want it to continue to mean a great deal to its membership, present and future, and particularly to the younger engineers who are just starting to benefit from membership.

I want you to know that, whatever difficulties may lie ahead, we are going to put forth every possible effort to keep the AREA the strong, active, outstanding Association that it has been since its inception.

Thank you. [Applause]

Is there any further business to come before this meeting? If not, before adjourning it I should like to remind all members of the Board of Direction, including the retiring members and newly elected members, that they will have a joint luncheon with the members of our Arrangements Committee immediately following the adjournment of this meeting, in Private Dining Room 2. This will be followed immediately by the post-convention meeting of the Board of Direction, which will be held in Private Dining Room 1.

If there is no further business to come before this meeting, I shall now declare the Sixty-First Annual Meeting of the American Railway Engineering Association, and concurrent Annual Meeting of the Engineering Division, Association of American Railroads, adjourned·

[The meeting adjourned sine die at 12:45 pm]

MEMOIR

George Joseph Ray

Died March 5, 1962

George Joseph Ray, eminent railroad engineer and administrator, civic leader, Christian gentleman, and the 22nd president of the American Railway Engineering Association (1924–1925), died in retirement at his home in Summit, N. J., on March 5, 1962, in his 86th year; after a prolonged period of ill health. He is survived by his wife, the former Ethel Wray Pearce, and by three daughters—Mrs. Benjamin H. Palmer, Jr., of Norwich, Conn.; Mrs. Graeme Jackson Pearce of Summit, N. J.; and Mrs. Everett Fay of Berkeley Heights, N. J.

Mr. Ray, who spent practically all of his professional career on the Delaware, Lackawanna & Western, was born at Metamora, Ill., on March 24, 1876. He received

George Joseph Ray

his higher education at the University of Illinois, from which he was graduated in 1898 with the degree of B.S., and from which, in 1910, he was given the degree of C.E. In 1916 he received the award of Doctor of Science from Lafayette College.

Mr. Ray entered railroad service in 1898 as a rodman on the Illinois Central Railroad, advancing progressively to assistant engineer, supervisor of track, and roadmaster, until March 1, 1903, when he left the Illinois Central to begin his career on the Lackawanna, which was to see him involved in or in responsible charge of many outstanding physical improvements on that road, and the development of track maintenance standards which were the envy of many railroads throughout the country.

Mr. Ray went to the Lackawanna as division engineer, serving in this capacity until January 1, 1909, when he was advanced to chief engineer. From February 1, 1919, to March 1, 1920, during the period of Federal control of the railroads, he was engineering assistant to the Regional Director, Eastern Region, United States Railroad Administration, returning to the Lackawanna as chief engineer on the latter date. On

January 1, 1934, in recognition of his managerial abilities, Mr. Ray was advanced to vice president—operations of his railroad, which position he held until his retirement on March 31, 1946.

As chief engineer of the Lackawanna during its heyday as an anthracite coal carrier, Mr. Ray made many contributions to the development and betterment of his railroad. Among the more notable improvement projects carried out under his direction were the following: The construction of the Lackawanna's New Jersey cut-off, 28.45 miles long, from Lake Hopatcong, N. J., to Slateford Junction, Pa., which was completed late in 1911; the construction of a new double-track line from Clark's Summit, Pa., to Hallstead, 39.6 miles long. which was completed late in 1915; the elimination of grade crossings through the populous suburban territory of The Oranges (New Jersey) and on the Montclair Branch to Montclair, N. J., a task complicated by the necessity of maintaining heavy commuter service throughout construction; and the electrification of the New Jersey lines between Hoboken and Dover, via Morristown, including branches to Montclair and Gladstone, involving 70 miles of road, with 160 miles of tracks.

The New Jersey cut-off was a double-track line, which was cut through the hills of New Jersey to provide a shorter route, improve grades, and reduce curvature. A feature of this new line was its Pequest fill, 3 miles long, from 70 to 110 ft high, and containing 6,625,000 cu yd of material—a feature which was particularly noteworthy, having been carried out before the advent of the heavy grading equipment available today. Other outstanding features of the line were the construction of a mutiple-arch concrete viaduct over Paulin's Kill, 1100 ft long, with seven spans and a height of 117 ft over the kill; and the construction of a multiple-arch concrete viaduct over the Delaware River, 1450 ft in length, with five arches of 150-ft span, two of 120 ft, and two of 50 ft, and a height of 64 ft over the river.

Outstanding features of the Clark's Summit–Hallstead line, which, like the New Jersey cut-off, was designed to provide a shorter route, improved grades, and reduced curvature, were again heavy grading operations, and the construction of a double-track, multiple-arch concrete viaduct over Tunkhannock Creek Valley. This viaduct, 2375 ft in length and 240 ft high over the creek proper, consists of ten clear spans of 180 ft each, and two filled spans of 100 ft each, and is still probably the largest reinforced concrete railroad bridge in the world. Both for its size and architectural beauty, and because built when reinforced concrete was largely in its infancy and on trial, this structure has received world renown.

While deeply involved in many construction projects, Mr. Ray never lost sight of the problems and expenses of maintenance. In fact, every detail of the projects carried out under his direction were with permanence and minimum future maintenance costs in mind. Accordingly, he developed high standards in all elements of the track structure and in the maintenance of the track itself. With this in mind, he was an ardent advocate of maintenance budgets geared to operating conditions over a period of years, unaltered by short swings in traffic; also of an arrangement which permitted maintenance expenditures in those months of the year which would be the most productive, regardless of traffic fluctuations and earnings. Giving expression to this belief in his presidential address to the AREA in 1925, Mr. Ray said: "The management that cuts down maintenance of way expenses at every indication of a falling off of business and again puts on forces when business picks up is working at a disadvantage and, as a result, uneconomically maintains the property."

Throughout practically all of his career with the Lackawanna Mr. Ray was an ardent worker in and supporter of the AREA. He joined the Association in 1906, and through succeeding years was a member of the following committees: 13—Water Service, 1907–1908; 5—Track, 1909–1918, being its vice chairman 1912–1916, and its chairman 1917–1918; 4—Rail, 1917–1937, and its chairman 1919–1924; 26—Standardization, 1920–1924; 24—Cooperative Relations with Universities, 1924–1933; 25—Rivers and Harbors, 1934–1936; and the Committee on Stresses in Railroad Track, 1917–1940. Mr. Ray was a Director of the Association 1914–1916, junior vice president 1922–1923, senior vice president 1923–1924, and president 1924–1925. He became a Life Member in 1941. In addition to his activity in the AREA, Mr. Ray was a member of the American Society of Civil Engineers and also of the Newcomen Society of England and America.

Beyond his attributes as an engineer and Association member, Mr. Ray had many other talents and interests. Varying with age, he was an excellent and enthusiastic sportsman, especially in hunting, fishing, and golf. In addition, he was an historian and had a wide knowledge of geology. Possible his keenest hobby was horticulture, in connection with which he maintained a private green house and grew flowers which he displayed at a number of national shows. This hobby was reflected generally over his railroad, which throughout his administration had some of the most attractively shrubbed and landscaped station grounds in the country.

Along with it all, Mr. Ray was modest and unassuming, had the confidence and respect of all, and was held in high regard by both labor and supervision. Reflecting no doubt the feeling of many who knew him best, one of them has said, "He was one of the outstanding men in railroad history. If the industry had a 'Hall of Fame', he would be an early selection."

G. A. Phillips, *Chairman*
G. D. Brooke
C. H. Mottier
Committee on Memoir

MEMOIR

Ralph Budd

Died February 2, 1962

Ralph Budd, Honorary Member of the American Railway Engineering Association, died at his home in Santa Barbara, Calif., on February 2, 1962, at age 82. Chief engineer of a railroad at 28 and a railroad president at 40, his career was marked by many outstanding achievements in railroad engineering and operation, and awards in recognition thereof. It was through his foresight and initiative that the American railroads entered the era of diesel power, and, as an early advocate of cooperative research through the agency of the Association of American Railroads, he was influential in the creation of the position of research engineer in the Association's Engineering Division.

Ralph Budd was born near Waterloo, Iowa, on August 20, 1879, the son of Charles Wesley and Mary Ann (Warner) Budd. He was graduated from Highland Park Col-

Ralph Budd

lege at Des Moines, Iowa, in 1897 with the degree of B.S. in Civil Engineering, but continued his studies in engineering at the same school for the next two years. He then began his railway career as a rodman on the Chicago Great Western Railway, advancing to instrumentman and draftsman before leaving that property to become a roadmaster on the Chicago, Rock Island & Pacific Railway. He was promoted to division engineer in 1903 and continued in that position for two years, when he returned to the Great Western as a division engineer.

It was during his service with the Rock Island that Mr. Budd was brought in contact with John F. Stevens, who became a vice president of that road in 1904—a contact which had a profound influence on the early years of his career. Mr. Stevens became chief engineer of the Panama Canal Commission in 1905, and with one of the most urgent problems confronting him, that of finding a capable engineer to take over the rebuilding of the Panama Railroad, he chose for this position the young division engineer he had known on the Rock Island and asked him to come to the Zone in 1906 with the title of engineer of construction and maintenance. Mr. Budd became chief engineer of the line in 1907, about the time that Mr. Stevens left the Canal

Zone after the project was turned over to the U. S. Corps of Engineers under the direction of Colonel G. W. Goethals. Two years later, when Mr. Stevens was elected president of the Spokane, Portland & Seattle Railway, the Oregon Trunk Railway, and the Spokane & Inland Empire Railroad, Mr. Budd accepted appointment as chief engineer of these properties—which were owned jointly by the Great Northern and Northern Pacific Railways.

The extension and development projects being carried out at that time on these railroads, all of them subsidiaries of the "Hill" system, constituted key steps in the ensuing "battle of the giants" (James J. Hill vs. Edward H. Harriman). So it was only natural that the young chief engineer should come to the attention of Mr. Hill and that the latter should call him to St. Paul in 1913 to become chief engineer and assistant to the president of the Great Northern Railway. Promotions to vice president and executive vice president followed in rapid succession, and in 1919, when he was only 40, he was advanced to president.

Mr. Budd's most outstanding service to that railroad lay in the fulfillment of a long-felt need, namely, a new crossing of the Cascade Range in western Washington that would obviate the troublesome operating problems of the existing line and eliminate several miles of snow sheds which had been a source of heavy capital outlay and constant maintenance expense. The answer to this was the construction of a new 43-mile line through the Cascades, with a tunnel 7.70 miles long, but a line which was 7.68 miles shorter, and with its summit elevation 501 ft lower than that of the old line. This work was begun in December 1925 and the new line was ready for operation in January 1929. Later in 1929 the Great Northern inaugurated a new transcontinental train, the Empire Builder, on a schedule which reduced the running time between Chicago and the Pacific Northwest by a full business day.

By that time Mr. Budd's reputation as a railway officer of distinction had extended beyond the borders of the American continent and during the summer of 1930, on the invitation of the Soviet Government, he made a survey and study of the Russian railroads, the Trans-Siberian Railroad, and the Chinese Eastern Railway, and presented recommendations for their rehabilitation and modernization.

Meanwhile, Mr. Budd had organized a highway bus subsidiary—the Northland Transportation Company, out of which grew the Greyhound Corporation, and his last contribution to the Great Northern was the extension of the company's line in 1931 into California, where it connects with the Western Pacific at Bieber.

In 1932 Mr. Budd became president and chairman of the Executive Committee of the Chicago, Burlington & Quincy Railroad, as well as of its several subsidiaries in Colorado, Wyoming and Texas. He was by no means a stranger to the Burlington, because, as an officer of the Great Northern, which, with the Northern Pacific, was a joint owner of the Burlington, he had served as a director of the Burlington since 1916. With the broader opportunities afforded by a railway system serving an area of greater population and industrial development, Mr. Budd's genius in his chosen profession came to full flower. Under his leadership the Burlington introduced the "Zephyrs", the first of which was the streamlined, light-weight, articulated train powered with a diesel engine, that made the historic "dawn to dusk" run from Denver to Chicago on May 26, 1934, in 10 hours and 4 minutes.

The date is significant—a day in the depth of the depression of the 30's, for it marked the beginning of the renaissance during which the railroads adopted many measures designed to increase their effectiveness. Chief among these were a new concept with respect to train speed and the transition from steam to diesel power. That the

Burlington was in the vanguard of this movement is shown by the inauguration of the Twin Cities' Zephyr service on April 21, 1935, and the introduction of the Advance Zephyrs (Denver to Chicago) on May 31, 1936. Under the leadership of its president, the Burlington was among the first carriers to extend diesel power to freight trains and, eventually, to all service. Also, it is of interest to note that after the idea of the dome car had been suggested to Mr. Budd, it was first embodied experimentally in a passenger train car undergoing repairs in the Burlington's shops at Aurora, Ill.

In 1940, during his presidency of the Burlington, Budd was appointed by President Roosevelt to the position of transportation commissioner of the Advisory Commission to the Council of National Defense, and following the entrance of the United States into World War II, he was commissioned colonel to serve as a liaison officer between the U. S. Army and the railroads of the Central Western Region.

Mr. Budd retired from the presidency of the Burlington in 1949 at the age of 70, and shortly thereafter became chairman of the Chicago Transit Authority, the public body which manages and operates the city's public transportation system. He retired from this position in 1954.

Mr. Budd received wide recognition for his achievements in engineering and transportation. In addition to being an Honorary Member of the AREA, he was elected to honorary membership by the American Society of Civil Engineers, the American Society of Mechanical Engineers, the Western Society of Engineers, and the American Railway Bridge and Building Association. Among awards conferred on him were the John Fritz Medal of ASCE, the Award of Merit of the American Institute of Consulting Engineers, the Washington Award of WSE, and the Henderson Award of Franklin Institute. The respect he commanded in the business world is attested by the directorships which he held in banks, insurance companies and other corporations, while his public spirit was demonstrated by his identity with the trustees of such institutions as the Newberry Library, the Museum of Science and Industry, the Zoological Society, and the Citizens' Traffic Safety Council, all of Chicago, and the James J. Hill Reference Library of St. Paul and Carleton College.

Mr. Budd was married in 1901 to Georgia Marshall who survives him, as do his two sons, John M. Budd, president of the Great Northern Railway, and Robert W. Budd of Charlottesville, Va., and a daughter, Mrs. Victor Hunt, of Mexico City, Mexico.

W. S. Lacher, *Chairman*
H. C. Murphy
E. J. Brown
Committee on Memoir

Report of The Executive Secretary

March 1, 1962

To the Members:

Nineteen hundred sixty-one, by comparison with recent years, was not a bad year for the American Railway Engineering Association. But neither can it be said that it was an exceptionally good year because the Association was confronted with a number of disturbing influences which reflected themselves adversely on Association activity. Principal among these were the relatively unfavorable economic climate in the railroad industry generally, and on some railroads in particular, accompanied by an attitude of restraint from some quarters toward the activities of various railroad groups, in the interest of economy. These had anything but a buoyant effect upon our Association throughout much of the year. Likewise, with less than former financial support from the railroads to function as the Engineering Division of the Association of American Railroads, to support research work proposed by committees, and to publish research reports prepared for AREA committees of the Engineering Division research staff, several aspects of Association and committee work were handicapped.

Some of the unfavorable and disturbing influences referred to—economic and otherwise—have already had an adverse effect on the total membership of the Association, and with certainty it can be said that unless these influences disappear or are largely overcome, they will have not only a more serious effect on Association membership, but an adverse effect upon future Association activity, its financial soundness, and its ability to serve its members and the railroad industry.

But your secretary is pleased to report that, to the present time, the effect of none of these influences has been very serious; that the membership of the Association remains relatively high; and that the activities and production of committees continues at a high level. Furthermore, that largely as the result of practical expedients adopted, and the practice of rigid economy generally, the Association is financially sound, as is indicated later in this report.

The 1961 Convention

Considerable impetus to the Association's activities during the year was given by the highly successful annual convention of the Association and concurrent exhibit of the National Railway Appliances Association held March 7-9 at McCormick Place, Chicago s new lakefront Exposition Center—an arrangement which resulted in a new peak of interest, coordination and efficiency. Even those who earlier had expressed some doubt as to the wisdom of tying the convention sessions and exhibit together under one roof, with housing headquarters two miles away at the Conrad Hilton Hotel, readily conceded later that the arrangement worked to near perfection, especially in view of the splendid facilities afforded, the good judgment of members in dividing their time between the convention sessions and the exhibits, and the shuttle bus service provided by the suppliers between the hotel and convention-exhibit hall throughout the convention hours each day.

Exactly how many attended the convention and exhibit is not known because there was no exhibit registration as such, and convention registration was not compulsory. But that the attendance was large—a near record—is seen in the fact that, in spite of the registration fee of $2.00, 1367 railroad men and 982 non-railroad men—a total of 2349—registered their attendance over the 3½-day period—a registration that was exceeded at only the 1955 convention (an exhibit year) when total registration reached 2375.

Committees of the Board of Direction
1961–1962

Executive Committee

R. H. Beeder (Chairman), E. J. Brown, F. R. Woolford, C. J. Code, L. A. Loggins

Assignments

C. J. Henry (Chairman), W. J. Cruse, J. A. Bunjer, J. M. Trissal, J. H. Brown

Personnel

F. L. Etchison (Chairman), C. R. Riley, C. J. Henry, J. E. Eisemann

Publications

W. B. Throckmorton (Chairman), W. E. Cornell, J. E. Eisemann, J. H. Brown

Manual

F. R. Woolford (Chairman), F. L. Etchison, L. A. Loggins, W. H. Huffman, F. R. Smith

Membership

C. R. Riley (Chairman), J. A. Bunjer, W. E. Cornell, W. H. Huffman, F. R. Smith

Finance

J. M. Trissal (Chairman), E. J. Brown, W. B. Throckmorton

Research

W. J. Cruse (Chairman), Ray McBrian, C. J. Code

Whereas in previous years AREA conventions have served concurrently as the annual meetings of the Construction and Maintenance Section of the Engineering Division, AAR, this year the convention served concurrently as the annual meeting of the Engineering Division by virtue of the fact that on January 1, 1961, the AREA took on the status of the Engineering Division as a whole.

All of the sessions on the first and third days were held in the spacious Banquet Room at the Exposition Center, but both the morning and afternoon sessions on the second day were held in the 573-seat theater-style Recital Hall, while the Banquet Room was being set up for, and was the scene of, the Annual Association Luncheon.

The pattern for the 1961 convention was much the same as in recent years and included the presentations of all of the Association's standing and special committees, which themselves included or were interspersed by 17 special features in the form of addresses, illustrated papers or motion pictures. All together the committees presented progress or final reports on 126 of their assignments, which included recommendations respecting 180 different documents in the Association's Manual of Recommended Practice—all of which were approved.

The principal social function of the convention was again the Annual Association Luncheon on the second day, which was participated in by 1085 members and guests. The principal features of the Luncheon program were an address by H. C. Murphy, president, Burlington Lines, on "The Railroads Look to the Future", and announcement of the result of the annual election of officers of the Association.

Not overlooked at the convention were the 275 women who accompanied their husbands and separately registered their attendance. Each day they enjoyed attractive reception and registration facilities at the Conrad Hilton Hotel, and, for the most part, participated in two especially pleasant functions arranged for their enjoyment. One of these was the Ladies Reception and Tea in the Williford Room at the Conrad Hilton; the other was the Women's Luncheon, which was held concurrent with the Annual Association Luncheon, in the V.I.P. Room at McCormick Place.

MEMBERSHIP

The continuing economic problems facing the railroad industry, with resulting problems presented to the Association, have finally substantiated the concern of your secretary in recent years for further membership growth—even sustaining the membership level of the Association. As of February 1, 1962, the membership of the Association stood at 3347, a net loss of 60 members compared with the membership of 3407 one year earlier. Thus, after an unbroken record of growth in membership from 1944 through 1957, the membership has slipped backward in two of the last four years— 1958 and 1961—and might have slipped backward in the other two years had losses in those two years not been overcome by special circumstances or special member interest and effort.

In 1959, a special effort by a number of chief engineering and maintenance officers to encourage membership among qualified non-members on their respective roads brought in nearly 100 new members, to put that year over the top. In 1960, the joining of 34 men from the Electrical Departments of the railroads and from related industries in order to participate in the Association's new Committee 18—Electricity, was a major factor in keeping the membership in the black in that year. But in 1961, even a combination of special effort by a number of chief engineering and maintenance officers on their roads, and the enrollment of an unusually large number of Associates under a more liberal interpretation of the restrictive "sales participation" clause of the Constitution, were not enough to offset the abnormally high losses through resignations and the dropping of members for non-payment of dues.

More specifically, the net loss of 60 members during 1961 resulted from the enrollment of only 161 new members (compared with 198 in 1960), the reinstatement of 30 former members (4 more than in 1960), and a decrease of 2 Junior Members (compared with a decrease of 10 in 1960), in combination with a loss of 249 members through deaths, resignations and being dropped because of non-payment of dues (compared with a total of 180 in these later categories in 1960).

MEMBERSHIP

(February 1, 1961, to February 1, 1962)

Members on the rolls February 1, 1961 3407
New Members 161
Reinstatements 30
Loss in Junior Membership .... —2

3596

Deceased 31
Resigned 69
Dropped 149

249
Net loss 60
Membership February 1, 1962 3347

MEMBERSHIP CLASSIFICATION AS OF FEBRUARY 1

	1956	1957	1958	1959	1960	1961	1962
Life	465	470	469	482	481	474	490
Member	2414	2478	2524	2491	2527	2554	2467
Associate	261	258	268	251	264	288	301
Junior	163	144	101	86	101	91	89
Totals	3303	3350	3362	3310	3373	3407	3347

Unmistakably the fewer new member enrollments in 1961 reflected the relatively poor economic conditions which continued to prevail in the railroad industry during the year, the continued tightening up of organizations within many railroads, the restricted recruitment of additional technically trained college graduates, and the result of several major railroad consolidations in recent years. Unmistakably, too, some of these same factors reflected themselves in the larger number of members who resigned or who were dropped as of the end of 1961. To the extent that such was true, the situation was beyond the control of the Association. But to the extent that the causes of fewer enrollments and larger defections may have been the result of more circumscribed and restricted membership and Association activity, whether resulting from outside influences or self imposed, both the Association and the railroad industry should be made aware of the loss they have already sustained, and should be seriously concerned about the further loss they stand to sustain in the future unless these unfavorable influences are offset or removed. The Association must constantly seek to enhance its efficiency and usefulness, and must question seriously any proposals which, in the interest of real or fancied immediate benefits, would weaken its structure and long-term effectiveness.

During the year ended February 1, 1962, there were a total of 31 deaths among the membership, as indicated in the roster of deceased members at the end of this report. Happily, this list includes no past officers of the Association, only one past Director—R. E. Dougherty, retired vice president, New York Central System—and only one past committee chairman—J. C. Mock, retired signal-electrical engineer, New York Central Railroad, who was chairman of former Committee 10—Signals and Interlocking. However, as will be noted, the list includes the names of many who contributed much to the work of the Association.

While coming after February 1, 1962, and, therefore, not included in the foregoing figures or referred-to list, it must be noted here, with a sense of great loss, the death on February 2, at his home in Santa Barbara, Calif., of one of the Honorary Members of the Association, Ralph Budd, an outstanding railroad engineer and retired president of the Burlington Lines.

Student Affiliates

Not included in the foregoing membership figures are the Student Affiliates which the Association began to enroll late in 1960 on college campuses—a relationship which was explained in the secretary's report for that year. Suffice it here to say that as of February 1, 1962, the Association had 47 Student Affiliates on 20 different campuses, compared with 42 such affiliates one year earlier.

As was expected, there was and will continue to be a large turnover in Student Affiliates as new undergraduates become interested in affiliation with the Association and upperclassmen and graduate students complete their studies.

Without exception, there has been nothing but appreciation on the part of Student Affiliates and faculty members for the service rendered in this campus effort, and while the benefits derived by the railroad industry and the Association are largely intangible, several Student Affiliates have already entered railroad service and have become Junior Members of the Association. But entirely aside from any direct return which the Association may realize from this effort, it can have the distinct satisfaction of making a substantial contribution to the education and training of a sizable number of young men.

ACTIVITIES OF COMMITTEES

Personnel of Committees

The continued interest of members in serving on committees, and of their railroads being represented on committees, is clearly reflected in the record of committee personnei during 1961. Specifically, throughout the year there were 1160 members (including 61 Members Emeritus) regularly assigned to 1251 places on the Association's 22 standing and special committees. This compares with 1128 who occupied 1247 places on the Association's 23 committees during the previous year.

In addition to their regular members during 1961, practically all committees again carried a number of "guest" members on their rosters—members assigned during the year on a guest basis, awaiting regular assignment with the official roster change to become effective with the close of the 1962 convention.

There were no special restrictions on the number of members permitted on committees in 1961, but again, to meet the desire of the Association of American Railroads for relatively small AAR committees, there was continued in 1961 the plan adopted by the Board of Direction in 1960, which provides that the chairmen of committees, the vice chairmen, the secretaries, and all subcommittee chairmen, alone make up the official Engineering Division representatives on the committees. At the same time, all of these men, together with all of the other members on the committees, constituted the AREA committees as a whole.

To set apart as a group the official Engineering Division representatives on any committee, for convenience and record purposes, the names of these representatives, showing their official capacities on committees, were grouped at the head of the list of personnel of the committee as a whole, as presented in the Committee Assignments Pamphlet and the Bulletins, and were set in bold-face type.

Reflecting the continued unfavorable economic conditions in the railroad industry generally, and on a few of the larger railroads of the country in particular—and in spite of an expectation of improved conditions in the year ahead—the number of members assigned to committees for 1962, effective with the official roster changes at the end of the 1962 convention, will be down some from 1961. Specifically, 1137 members have been assigned 1227 places on committees for 1962, which compares with the 1160 members who served in 1251 places during 1961.

Work of Committees

The work of committees during 1961 continued to follow much the same pattern as in previous years, their different subcommittees carrying out their own studies and investigations independently, or with the cooperation of the research staff of the Association of American Railroads, looking to the preparation of progress or final reports for information; of revising material appearing in the AREA Manual of Recommended Practice, the AAR Electrical Manual, and the Portfolio of Trackwork Plans; of developing new Manual and Portfolio material; and of carrying out special projects related to their assignments. That the work accomplished by committees was again substantial is seen in the fact that they produced one or more reports on 102 of their 174 assignments (not including Assignments A), 12 of which were final reports. Furthermore, continuing the practice established by the Board in 1958, all committees presented brief "progress" or "status" statements with respect to assignments on which they made no formal report.

Outstanding in the work of committees during the year was their continued review of material in their respective Manual chapters—this year with a two-fold purpose in mind. One purpose was to complete the routine review of material in their chapters prior to the scheduled reprinting of the Manual late in 1962 to replenish the supply on hand for sale. The other purpose was to remove from some documents of "recommended practice" in the Manual compulsive words, expressions or statements which had inadvertently crept into them in past years, contrary to purpose and intent. As a result of their overall Manual work, all of the 18 committees having jurisdiction over AREA Manual chapters included in their reports recommendations affecting Manual documents—recommendations which range from reapproval of these documents without change, to the complete rewriting of some documents. The magnitude of this phase of committee work is reflected in the following statement of Classification of Material Produced by Committees. With approval of all of these recommendations at the 1962 convention, there will necessarily result another large AREA Manual Supplement in 1962, which is certain to exceed in size the 672-page, record-size Supplement issued in 1961.

Another outstanding phase of committee work was that of Committee 5—Track, in the development and presentation for adoption, of new plans for 5 "standard" turnouts, incorporating recommendations previously submitted to and approved at the 1960 convention—each of these plans being accompanied by 2 or 3 other sheets covering details of the switch, frog, and guard rails basically needed for manufacturing purposes. In addition, the committee also developed, and is submitting for adoption, revisions in 10 other plans.

Classification of Material Produced by Committees

The work of committees during the year was again so diversified and extensive that it is impossible to do other than to refer to it in general terms in a report of this character. But there is presented in the following a general categorical classification of the results of this work, as published in the Bulletins of the Association, and to be presented to the 1962 convention:

Recommendations pertaining to the development, revision, deletion or reapproval without change, of 359 different specifications and recommended practices for inclusion in the AREA Manual and the AAR Electrical Manual; 57 reports on current developments in engineering practice and design; 12 reports dealing with economy in the use of labor and the recruiting and training of employees; 3 reports on new and improved power tools, machines, material and equipment; 4 reports involving statistics; 4 economic and analytical studies; 4 reports on relations with public authorities; and 3 bibliographies.

The work of committees affecting the AREA Manual included the presentation of 1 specification for adoption; the rewriting or revision of 45 specifications (with or without reapproval); the reapproval of 33 specifications without change; the deletion of 13 specifications; the presentation of 2 tentative specifications; the presentation of 4 recommended practices for adoption; the revision of 101 recommended practices, with or without reapproval; the reapproval of 89 recommended practices without change; the deletion of 15 recommended practices; the adoption of 1 agreement form; the rewriting or revision of 9 agreement forms; the reapproval of 8 agreement forms without change; the presentation of 1 tentative agreement form; and a number of revisions in the definitions of terms included in the Glossary.

Recommendations covering the AAR Electrical Manual, which were exclusively the work of Committee 18—Electricity, covered the revision of 5 specifications and the deletion of 1 document of recommended practice.

In addition to their work on the two Manuals, the reports of committees presented instructions with regard to 3 machines for inclusion in the Handbook of Instructions for Care and Operation of Maintenance of Way Equipment; 18 trackwork plans for adoption and inclusion in the Portfolio of Trackwork Plans; and revisions in 10 current trackwork plans.

With so much Manual work to be completed in 1961, looking to reprinting the AREA Manual late in 1962, some committees took on no new assignments in 1961, and the total number of new assignments given to committees was only 16, not including the 7 new assignments given to recreated Committee 18—Electricity. During 1962, the committees as a whole will work on 173 assignments, 20 of which are new.

Committee Meetings

Influenced by various circumstances and conditions, the 22 technical committees of the Association held fewer meetings in 1961 than during either 1960 or 1959, and, if specific attendance figures were available, they would undoubtedly show that the attendance at many committee meetings was somewhat smaller in 1961 than in the two previous years. Neither of these situations can be considered favorable, to the extent that they may have affected committee work adversely. In the interest of economy and minimum off-the-job time of members, the large majority of meetings were again held in Chicago or at points central to the larger number of committee members, and were confined to one day.

Specifically, 64 full committee meetings were held during the year ended March 1, 1962. This compares with 70 meetings held during the year ended March 1, 1961, and 73 meetings held during the year ended March 1, 1960. Of the 64 meetings held during the 1961 Association year, 44 were in Chicago (including 14 meetings held during the 1961 convention); 2 meetings each were held in Cleveland, Ohio, St. Louis, Mo., Washington, D. C., and New York City; and 12 were held in as many other cities.

Dictated by the scope of their work and other considerations, 5 committees each held 4 meetings; 10 committees each held 3 meetings; and 9 committees each held only 2 meetings. Combined with their meetings, 9 committees made 2 inspection trips to see facilities, structures, procedures or operations directly related to their work.

ASSOCIATION PUBLICATIONS

Again in 1961 the Association made widespread distribution of its publications beyond its own membership; continued or adopted several new policies with respect to its publications; and, through its committees, completed a monumental job of reviewing and updating its Manual of Recommended Practice, which resulted in the largest Supplement to the Manual ever issued by the Association.

During 1961 the Association continued the practice begun in 1960 of charging member holders of the Manual a flat fee of $1.00 for the Annual Supplement, as a means of offsetting the sizable cost of rendering this special service to that segment of the membership involved. Also, it continued the new policy, started late in 1960, of presenting in the Bulletin only edited and condensed versions of Engineering Division research reports prepared by the AAR research staff on behalf of committees; and carried through its plan, decided upon late in 1960, to discontinue the annual bound volume of the Proceedings.

The new policy regarding the publishing of Engineering Division research reports in the Bulletin, and the reasons therefore, are set forth in this report under comments on "Research Work". Suffice it to say further that, under the circumstances, this new policy has been received with few complaints from members, all of whom can secure complete copies of these reports in multilith form from the AAR Research Center, if they desire, and that it has resulted in substantial economies to the Association.

Likewise, discontinuance of the annual bound volume of the Proceedings was with no loss to the membership in published material, resulted in minimum inconvenience to members, and effected large savings to the Association. In place of the Proceedings, all of the Bulletins of the Association from the September–October 1960 issue through the June–July 1961 issue, except the March Year Book Issue, were punched for binding and all members who requested it were furnished, without charge, a two-post, hardcover, book-type binder, similar in every respect to the book casing on past volumes of the Proceedings, in which to house as a unit their copies of the Bulletin. Thus, without the past duplication of material in the Bulletin and Proceedings, members were able to assemble in neat bound form all of the Bulletins for the year September 1960 through July 1961—including the complete proceedings of the 1961 annual meeting, which appeared in the June–July issue.

Under the procedure adopted, members who expressed their desire for a 1961 binder on the card form submitted to them, and all new members who make similar requests in the future, will be sent a binder each year as soon as possible after the publication of the June–July Bulletin.

It is regretted that these binders cannot be issued simultaneously with the first Bulletin to be included in them, and thus permit housing in the binder the subsequent Bulletins as issued, but this cannot be done since only at the end of the publication year is it possible to know the aggregate number of pages in the Bulletins, and thus the exact size binder to produce a neat binding job. Thus, it will continue to be necessary for members to preserve their copies of the Bulletin carefully during the year for binding each fall. This is important because the Association may not be in a position to replace missing issues of the Bulletin, and, in any event, will have to charge for them.

Manual Supplement—Recruiting Brochure

As indicated earlier in this report under Work of Committees, the Association in 1961, as the result of recommendations developed by committees and adopted at the 1961 convention, necessarily produced a record-size Annual Supplement, containing 672 pages (336 sheets). By early August, copies of this Supplement had been sent to all those with standing orders for Supplements, and to all members who had paid the $1.00 fixed fee for the 1961 Supplement—a total of nearly 1500.

Also, late in October and early November, the secretary's office made the sixth annual distribution to colleges of the United States and Canada of the Association's Engineer Recruiting Brochure "The Railroad Field—A Challenge and Opportunity for Engineering Graduates". On the basis of earlier requests, this distribution included approximately 3350 copies to some 120 schools. At the same time, the chief engineering and maintenance officers of the Class I railroads of the United States and Canada were again apprised of the availability of the brochure—at 25 cents per copy—as a valuable aid to their respective roads in their individual recruiting efforts.

RESEARCH WORK

Due to a general reduction in the overall AAR budget, the extent of the research activities of the AREA to be carried out by or through the Engineering Division research staff of the AAR, was again greatly curtailed, total expenditures for these activities in 1961 amounting to only $263,100, compared with $398,400 in 1960. Of the total amount authorized for Engineering Division research in 1961, $41,100 was for AAR Detector Car Development and Leasing Service, leaving a total of only $222,000 for specific research projects and general technical services. This total expenditure for Engineering Division research projects and technical services—$222,000—was far less than $516,950 proposed by committees for research projects and for general technical services, and also than the necessarily reduced budget of $503,150 approved by the AREA Board of Direction for these purposes and recommended to the vice president, research, AAR.

To bring about the reduced 1961 budget, many proposed projects at the bottom of the priority list established jointly by the AREA Board and the director of engineering research were again entirely eliminated, cuts were made in the amounts requested for the remaining projects wherever possible, while still permitting some headway during the year; and participation in the work of Research Councils and contract research to be carried out by outside agencies and institutions was again practically eliminated.

Furthermore, for the first time in 10 years, the research budget did not include the usual appropriation in the amount of $5,000 to $7,000 to help defray the cost of publishing Engineering Division research reports in the AREA Bulletins. Unable to assume this added expense, the Association, by Board action, presented, for the most part, only edited and condensed versions of these reports, along with any conclusions and recommendations, in the Bulletins during 1961.

Any disadvantage or loss in this new arrangement to those who desired the full text of research reports was offset to some extent by the fact that the AAR Research Department had been producing copies of its complete reports in typewritten, multilith form, and has been sending copies of such to the chief engineering and maintnance officers of AAR Member Roads and to the members of sponsoring AREA committees. Also, copies of these reports have been made available to others, on request, from the AAR Research Center.

In most respects the total approved Engineering Division research budget for 1962 is a carbon copy of the reduced budget, or actual research expenditures, for 1961, this budget amounting to $262,900. Of this amount $41,100 is again for AAR Detector Car Development and Leasing Service, leaving a total of only $221,800 to progress research projects sponsored primarily by AREA committees and for general technical services. This authorized total expenditure for Engineering Division research and technical services is again far less than the amount proposed by committees for these purposes ($393,550), and also the necessarily reduced proposed budget of $323,600 approved by the AREA Board of Direction for these purposes, and recommended to the vice president, research, AAR.

Details of the authorized Engineering Division budget for 1962, exclusive of the authorized expenditures for Detector Car Development and Leasing Service, are presented in the accompanying tabulation, which shows for the different projects the expenditures authorized, compared with estimated expenditures for projects in 1961. In this tabulation it will be noted that the names of a considerable number of projects have been changed to simplify and identify or enlarge their scope. It will also be noted that one project has been discontinued and two new projects added. Missing entirely from this list are a considerable number of projects proposed by committees. Also, as will be evident to committees, the authorized expenditures for a number of the projects are less than initially requested.

Thus, necessarily, 1962 will be another year of delay and deferment in Engineering Division research and technical services, compared with the hopes and desires of AREA committees and the Board of Direction. Furthermore, again the 1962 budget provides no funds to help defray the cost of publishing AAR research reports in the AREA Bulletins. Thus, the Association plans again in 1962 to present only edited and condensed versions or summaries of these reports, for the most part, in the Bulletin, with any recommendations and conclusions.

TOTAL ALLOTMENTS FOR RESEARCH PROJECTS, ENGINEERING DIVISION, AAR,
EXCLUSIVE OF DETECTOR CAR DEVELOPMENT AND LEASING SERVICE

1943–1962

1943	$ 98,445	1953	364,100
1944	109,050	1954	351,307
1945	138,110	1955	351,653
1946	159,510	1956	365,050
1947	234,428	1957	476,845
1948	291,840	1958	563,709
1949	372,457	1959	353,800
1950	294,045	1960	350,300
1951	354,770	1961	222,000
1952	381,400	1962	221,800

SUMMARY OF PROJECTS INCLUDED IN 1962 APPROVED ENGINEERING DIVISION RESEARCH
BUDGET (EXCLUSIVE OF DETECTOR CAR DEVELOPMENT AND LEASING SERVICE),
SHOWING EXPENDITURES AUTHORIZED FOR EACH PROJECT, COMPARED
WITH AMOUNTS EXPENDED FOR PROJECTS IN 1961

	1961 Estimated Expenditure	1962 Approved Budget
Administration		
Research Office	$ 35,000	$ 35,000
Total	$ 35,000	$ 35,000
Committee 1—Roadway and Ballast		
Roadbed Stabilization—General	$ 15,000	$ 15,000
—Isotope Studies		
—Slip Plane Location with Plastic Tubes		
Vegetation Control by Ultrasonic Vibration	3,000	....
Total	$ 18,000	$ 15,000
Committee 3—Ties and Wood Preservation		
Development of Prestressed Concrete Ties and Fastenings	$ 5,000	$ 5,000
°Termite Control Investigation	400	400
Total	$ 5,400	$ 5,400

° Indicates change in name of project.

	1961 Estimated Expenditure	1962 Approved Budget
Committee 4—Rail		
Investigation of Failures in Control Cooled Rail	$ 3,000	$ 3,000
Rail Failure Statistics	5,000	5,000
°Insulated Rail Joint Development	6,000	3,000
Shelly Spots and Head Checks	15,500	20,050
°°Metallurgical Investigation of Basic Oxygen Steel for Rail and Joint Bars	....	2,250
Total	$ 29,500	$ 33,300
Committee 5—Track		
Corrosion Protection of Track and Structures from Brine Drippings	$	$ 800
Prestressed Concrete Crossing Frog Support	2,000	800
Explosive Hardening of Manganese Frogs	1,000	2,000
°Welding Heat Treated Carbon Steel Frogs and Switches	2,000	2,000
°°Riding Qualities of Equipment Through High Speed Turnouts	....	1,300
Specification Development for Tie Plate Fastenings and Tie Pads for Wood and Concrete Ties	10,000	8,600
Design of Spirals	4,000	3,600
Total	$ 19,000	$ 19,100
Committee 7—Wood Bridges and Trestles		
°Application of Synthetic Resins and Adhesives	10,000	$ 7,500
°Strength of Timber Stringers	6,600	9,000
°Non-Destructive Testing of Wood	2,000	2,000
Total	$ 18,600	$ 18,500
Committee 8—Masonry		
°Bearing Pads for Bridges	$ 2,000	$ 2,000
Total	$ 2,000	$ 2,000
Committee 15—Iron and Steel Structures		
°Truss Bridge Research	$ 5,000	$ 5,000
Total	$ 5,000	$ 5,000
Committee 16—Economics of Railway Location and Operation		
°Feasibility of Determining Track Maintenance Require- ments by Digital Computer Analysis	$ 1,000	$ 5,000
Total	$ 1,000	$ 5,000
Committee 30—Impact and Bridge Stresses		
°Steel Bridges	$ 7,500	$ 12,000
°Concrete Bridges	12,000	9,000
°Timber Bridges	6,000	4,000
Total	$ 25,500	$ 25,000

° Indicates change in name of project.
°° New project.

	1961 Estimated Expenditure	1962 Approved Budget
Special Committee on Continuous Welded Rail		
Butt-Welding of Rails:	$ 9,000	$ 9,000
Total	$ 9,000	$ 9,000
Joint Committee on Relation Between Track and Equipment		
°Relation of Wheel Load to Wheel Diameter	)	)
°Clearance Requirements	$ 22,000)	$ 18,500)
Dynamic Action of Piggyback Cars in Regard to Clearance, Stability and Ride Qualities	)	)
Total	$ 22,000	$ 18,500
Board Committee on Research		
Long-Range Weather Forecasting	$ 1,000	$ 1,000
Total	$ 1,000	$ 1,000
Electrical Laboratory and Instrumentation	$ 12,000	$ 12,000
Total	$ 12,000	$ 12,000
Technical Services		
General Technical Services	$ 18,000	$ 18,000
Total	$ 18,000	$ 18,000
Grand Total	$222,000	$221,800

° Indicates change in name of project.

FINANCES

The report of the Treasurer, Financial Statement, General Balance Sheet, and Statement of Cash Receipts and Disbursements for the calendar year 1961, all of which are presented herein, indicate that the Association continues in a sound financial condition, with total Receipts for the year exceeding Disbursements by $10,051, but with an inventory of saleable publications practically exhausted. A comparison of Receipts and Disbursements for the past two years is presented below:

	1960	*1961*
Receipts	$ 81,138.71	$83,461.73
Disbursements	83,978.29	73,410.20
	$—2,839.50	$10,051.53

The above, while presenting actual figures, does not present a true comparison, due to an unexpected and necessarily different method of handling payment for the 1961 Annual Luncheon. In 1960, and previous years, this payment was made immediately and directly from funds realized through the sale of Annual Luncheon tickets, and was kept out of the Association account, but in 1961, due to a change in hotel policy, funds realized through the sale of tickets, in the amount of $4325, had to be deposited in the bank account of the Association, and charges incurred for the Annual Luncheon were paid upon presentation of invoice to cover, at a later date. Therefore, a true comparison of 1960–61 Receipts and Disbursements would indicate:

	1960	1961
Receipts	$ 81,138.79	$79,136.73
Disbursements	83,978.29	69,085.20
	$—2,839.50	$10,051.53

COMPARISON OF RECEIPTS AND DISBURSEMENTS FOR A 20-YEAR PERIOD

	Receipts	Disbursements	Net Gain
1942	31,500.00	26,692.00	4,808,00
1943	28,736.00	23,809.00	4,927.00
1944	30,492.00	26,534.00	3,958.00
1945	32,305.00	29,305.00	3,000.00
1946	28,836.00	34,583.00	5,747.00*
1947	46,993.00	46,989.00	4.00
1948	57,741.00	53,062.00	4,679.00
1949	62,081.00	57,075.00	5,005.00
1950	59,752.00	51,795.00	7,957.00
1951	69,045.00	62,369.00	6,676.00
1952	77,514.00	76,964.00	550.00
1953	73,033.07	82,067.86	9,034.79*
1954	85,748.99	68,003.03	17,745.96*
1955	80,177.21	73,923.18	6,254.03
1956	79,531.11	70,336.17	9,014.04
1957	85,429.31	89,830.57	4,401.26*
1958	81,454.56	77,348.92	4,105.64
1959	80,407.16	80,297.48	109.68
1960	81,138.79	83,978.29	2,839.50*
1961	83,461.73	73,410.20	10,051.53

* Deficit.

Reviewing the financial picture briefly on the above basis, 1961 Receipts were actually only some $2000 lower than those of 1960, despite lower Membership receipts in 1961 of $1300, and no receipts from the AAR for Research Report Printing, which in 1960 amounted to $5832. Offsetting these losses were higher receipts, 1961 over 1960, in the following items and amounts: Sales of Publications of some $1000; Interest on Investments of $1000; Manual $1000; and Convention Registration Fees of $2400.

Disbursements for 1961 were considerably below those of 1960, and actually some $3700 under anticipated disbursements for the year. This was due to economies in printing through the discontinuance of the former printing of duplicate pages of the material appearing in the Bulletin, for later inclusion in the bound volume of the Proceedings; to the slightly smaller number of pages contained in the Bulletins; and to other economies effected in printing—which together, in themselves, resulted in a saving of some $3000 in the single item of Bulletin and Proceedings. Also, a saving of $1500 was realized in Miscellaneous Stationery and Printing, as a result of fewer orders for reprints than expected, and some $800 in Track Plans, due to an anticipated small Supplement to the Portfolio which did not materialize in 1961. Offsetting these savings to some extent was an over-expenditure of some $2000 for the Manual Supplement, the size of which could not be ascertained accurately when the 1961 Budget was prepared. All other disbursements, with minor over-under expenditures, very closely approximated disbursements anticipated for the year.

It is well that the year 1961, through savings and rigid economies, was a favorable one financially for the Association, as the year ahead will present a difficult financial problem. This is because the Manual must be reprinted in its entirety in order to replenish a depleted stock, and the 1962 Supplement thereto will be even larger than the large 1961 Supplement. Therefore, even further drastic economy measures to offset loss of revenues, to augment income, and to hold disbursements to the minimum will not prevent the Association from incurring a substantial loss in 1962, with disbursements exceeding receipts by at least $13,000. Should a large Supplement to the Portfolio of Trackwork Plans be issued in 1962, as seems likely, then this presently anticipated deficit will be considerably larger.

LOOKING AHEAD AT 1962

While completing their work for 1961, committees, the Board of Direction, and the secretary's office were necessarily looking ahead to plans for 1962—especially for the annual convention. One year earlier, stepped-up plans were being made for the full-scale 1961 convention, concurrent with an exhibit of the National Railway Appliances Association, at McCormick Place, Chicago. Now, in keeping with the feeling of some influences outside the Association that large annual meetings of railroad groups should be held at intervals less frequent than every year, committees, the Board, and the secretary's office were faced with the necessity of stepped-down plans for the 1962 convention, under a convention pattern established by the Board which calls for 2½-day full-program conventions every second year, and 1½-day, restricted-scope conventions in alternate years, beginning in 1962.

Under the plan adopted for the 1962 convention, official invitations to attend have been extended to the officers and directors of the Association, to the chairmen, vice chairmen and secretaries of committees, and to the current and incoming subcommittee chairmen of the different committees—a total of approximately 250. However, no restriction has been placed on the attendance of any other railroad members of the Association or railroad guests who, with the permission of, or upon instructions from, their respective railroads, desire to attend. Likewise, there has been no attendance restriction on non-railroad members of the Association or guests from the railway supply industry. Furthermore, the 1962 convention has been scheduled for Friday and Saturday, March 9 and 10, rather than early in the week, as in past years.

With the convention confined to 1½ days, its program will necessarily be highly condensed, but it will include the presentation of, and action on, all of the Manual recommendations of committees, supplemented by brief reports on at least the more important committee assignments, and by a limited number of special features.

Thus, the Association's work for 1961 will be climaxed by an annual convention in March 1962, which will accomplish the official business planned, and, unquestionably, prove of interest and benefit to those who attend. But it must be said that this stepped-down type of convention has been accepted enthusiastically by few, if any, members, and is a keen disappointment to many. What effect this may have upon the various aspects of the well being of the Association—membership, membership interest, productivity, and finances—especially if these restricted conventions are to be repeated periodically, is a matter for conjecture, but one which should be a matter of serious concern to the officers of the Association, to the membership as a whole, and to the railroad industry which the Association has served so effectively for the past 63 years.

If there are to be ill effects from the restricted-scope 1962 convention, they can be overcome only by the continued loyalty of members, the full cooperation of the

Association of American Railroads and of individual railroads, stepped-up committee activity, and the production of even more valuable Association publications—all of which are to be hoped for in the years ahead.

Respectfully submitted,

NEAL D. HOWARD,
Executive Secretary.

Deceased Members

F. L. BEAL
Retired Designing Engineer, St. Louis Southwestern Railway, Tyler, Texas

E. CHRISTIANSEN
Retired Engineer–Architect, Chicago, Rock Island & Pacific Railroad, Chicago, Ill

R. E. DOUGHERTY
Retired Vice President, New York Central System, New York, N. Y

R. L. GEIS
Division Engineer, New York Central System, Springfield, Ohio

W. E. GUIGNON
Retired Assistant Engineer, Pennsylvania Railroad, Philadelphia, Pa.

C. C. JACOBSON
Superintendent of Scales, Illinois Central Railroad, Centralia, Ill.

B. H. JONES
Retired Assistant Engineer, New York Central System, Pleasantville, N. Y.

J. H. KNOWLES
Retired Division Engineer, Southern Pacific Lines, San Antonio, Texas

C. B. MARTIN
Chief Valuation Engineer, New York Central System, New York, N. Y.

R. MATHER
Retired Engineer of Construction, Baltimore & Ohio Railroad, Akron, Ohio

J. S. McCAULEY
Division Engineer, Southern Pacific Company, Bakersfield, Calif.

R. L. McDANIEL
District Engineer, Atchison, Topeka & Santa Fe Railway, Los Angeles, Calif.

M. S. MILLER
Retired Special Engineer, Reading Company, Nassau, N. Y.

J. C. MOCK
Retired Signal—Electrical Engineer, Michigan Central Railroad, Detroit, Mich.

L. R. MORGAN
New York Central System, Syracuse, N Y

F. J. PITCHER
Retired Assistant to Chief Engineer, New York, New Haven & Hartford Railroad,
New Haven, 15, Conn.

B. H. PRATER
Retired Engineering Consultant, Union Pacific Railroad, Salt Lake City, Utah

R. W. PUTNAM
Retired Assistant Engineer Maintenance of Way & Structures, Southern Pacific Company,
Palo Alto, Calif.

C. W. REEVE
Treasurer, Young & Greenawalt Company, East Chicago, Ind.

J. H. ROACH
Retired Chief Valuation Engineer, New York Central System, Bronxville, N. Y.

C. ROBERTS
General Industrial Agent, Atlantic Coast Line Railroad, Atlanta, 3, Ga.

C. T. SCHWALB
Chief Engineer, Mississippi Export Railway, Moss Point, Miss.

F. C. SQUIRE
Retired Member, U. S. Railroad Retirement Board, Winnetka, Ill.

A. H. STIMSON
Supervisor—Maintenance of Way Material and Equipment, Pennsylvania Railroad, Buffalo, N. Y.

H. C. TAMMEN
Consulting Engineer, Howard, Needles, Tamman & Bergendoff, Short Hills, N. J.

W. F. TURNER
Retired Division Engineer, Southern Pacific Company, Sacramento, Calif.

R. C. WATKINS
Engineer (Valuation), Chesapeake & Ohio Railway, Richmond, Va.

R. C. WELSH
General Electrician, Pennsylvania Railroad, Philadelphia, Pa.

R. P. WINTON
Retired Testing Engineer Maintenance of Way, Norfolk & Western Railway, Roanoke, Va.

A. M. ZABRISKIE
Retired Chief Engineer, Central Railroad of New Jersey, Jersey City, N. J.

J. P. ZEARLEY
Supervisor—Methods and Cost Control, Southwestern Region, Pennsylvania Railroad, Indianapolis, Ind.

FINANCIAL STATEMENT FOR CALENDAR YEAR ENDING DECEMBER 31, 1961

Balance on Hand January 1, 1961 .... $147,040.39

RECEIPTS

Membership Account

Entrance Fees	$ 1,760.00	
Dues	42,302.90	$44,062.90

Sale of Publications

Proceedings	2,526.59	
Bulletins	2,444.09	
Manuals	9,948.09	
Specifications	2,058.77	
Track Plans	1,415.95	18,393.49

Advertising

Publications	5,978.16

Interest Account

Interest on Investments	4,585.11

Miscellaneous

Annual Convention Luncheon	4,325.00	
Registration Fees	4,469.50	
Ladies Luncheon	708.00	
Other	939.57	10,442.07

Total	$83,461.73

DISBURSEMENTS

Salaries	$28,259.54
Bulletins and Proceedings	18,499.52
Stationery and Printing	3,072.56
Rent	1,140.00
Postage	2,201.00
Supplies	475.98
Audit	400.00
Pensions	300.00
Social Security and Unemployment Taxes	1,255.01
Manuals	5,480.60
Track Plans	198.00
Committee and Officers Expenses	236.36
Annual Meeting Expenses	8,889.61
News Letter	2,008.50
Servicing Student Affiliates	105.50
Miscellaneous	888.02

Total	$73,410.20
Excess Disbursements over Receipts	10,051.53
Loss from Sale of Bonds	112.25

Balance on hand December 31, 1961 .... $156,979.67

REPORT OF THE TREASURER

To THE MEMBERS:

Balance on hand January 1, 1961 .... $147,040.39
  Receipts during 1961 ....$ 83,461.73
  Paid out on Audited Vouchers 73,410.20

Excess of Receipts over Disbursements 10,051.53   10,051.53
Balance on Hand December 31, 1961
  Consisting of Bonds at cost 153,403.98
  Cash in Northern Trust Company Bank 3,550.69
  Petty Cash 25.00 $156.979.67

We have made an examination of the accounts of the American Railway Engineering Association for the year ended December 31, 1961, and found them to be in accordance with the foregoing statement.

C. A. BICK,
P. D. MITCHELL,
Auditors.

GENERAL BALANCE SHEET

	1961	1960
ASSETS:		
Cash in Northern Trust Co. Bank ....$	3,550.60	$ 1,820.00 Cr. Bal.
Petty Cash	25.00	25.00
Due from members	116.50	16.50
Due from sale of publications	49.55	36.65
Due from sale of advertising	1,128.20	928.80
Prepaid postage	27.76	35.50
Furniture and fixtures	1,075.00	1,194.00
INVENTORIES:		
Publications (estimated)	500.00	500.00
Manuals	2,170.10	4,305.00
Track Plans	2,178.40	2,830.00
Binders, index and chapter	118.00	35.00
Paper stock	...	880.00
Investments (cost)	153,403.98	148,835.39
Interest accrued on investments	958.51	840.73
Totals	165,301.69	158,642.57
LIABILITIES:		
Members dues paid in advance	400.50	312.00
Surplus	164,901.19	158,330.57
Totals ....$	165,301.69	$158,642.57

STATEMENT OF CASH RECEIPTS DISBURSEMENTS YEAR 1961

Cash in Bank, January 1, 1961 ....Cr.—$1,820.00

RECEIPTS:

From members, sales of publications, interest, etc. 83,461.73

$ 81,641.73

DISBURSEMENTS:

Audited vouchers $73,410.20
Bond purchase 4,680.84   78,091.04

Cash in Bank December 31, 1961 .... $ 3,550.69

American Railway Engineering Association

CONSTITUTION

Revised to October 30, 1958

Article I

NAME, OBJECT AND LOCATION

1. Name

The name of this Association shall be the AMERICAN RAILWAY ENGINEERING ASSOCIATION.

2. Object

The object of the Association shall be the advancement of knowledge pertaining to the scientific and economic location, construction, operation and maintenance of railways.

3. Means to be Used

The means to be used for this purpose shall be:

(a) The investigation of matters pertaining to the object of the Association through Study and Research Committees.

(b) Meeting for the presentation and discussion of papers, and for action on the recommendations of committees.

(c) The publication of papers, reports and discussions.

4. Conclusions

The conclusions adopted by the Association shall be recommendatory.

5. Location

The office of the Association shall be located in Chicago, Ill.

Article II

MEMBERSHIP

1. Classes

The membership of this Association shall be divided into five classes: Members, Life Members, Honorary Members, Associates and Junior Members.

2. Qualifications

A. GENERAL

(a) An applicant to be eligible for membership in any class other than that of Junior Member shall be not less than 25 years of age.

(b) To be eligible for membership in any class, or for retention of membership as a Member, an Associate or a Junior Member, a person shall not be engaged directly or primarily in the sale to the railways of appliances, supplies, patents or patented services.

(c) The right to membership shall not be terminated by retirement from active service.

(d) In determining the eligibility for membership in any class, graduation in engineering from a school of recognized standing shall be considered as equivalent to three years of active practice, and satisfactory completion of each year of work in such school, without graduation, shall be considered as equivalent to one-half year of active practice.

(e) In determining the eligibility for Member under Section B (a) of this Article, each year of practical experience in engineering, or in science related thereto, prior to employment on a railway, if such experience were of the same specialized character as the current work of the applicant, shall be considered as equivalent to one year of railway service.

B. MEMBER

A Member shall be:

(a) An engineer or officer in the service of a railway corporation that is a common carrier, who has had not less than five years' experience in the location, construction, operation or maintenance of railways.

(b) A dean, professor, assistant professor, or equivalent in engineering in a university or college of recognized standing, or an instructor or equivalent in such university or college, who, with an engineering degree, has had at least two years' experience in teaching engineering.

(c) An engineer or member of a public board, commission or other official agency who, in the discharge of his regular duties, deals with railway problems.

(d) An editor of a trade or technical magazine who, in the discharge of his regular duties, deals with railway problems, and who has had the equivalent of five years' engineering or railway experience.

(e) A consulting engineer, engaged in private practice, or an engineer in his employ or in the employ of a consulting engineering organization, who has had the equivalent of five years' engineering experience.

C. LIFE MEMBER

A Life Member shall be a Member or an Associate who has paid dues for 35 years, or who has been retired under a recognized retirement plan and has paid dues for not less than 25 years.

D. HONORARY MEMBER

(a) An Honorary Member shall be a person of acknowledged eminence in railway engineering or management.

(b) The number of Honorary Members shall be limited to ten.

E. ASSOCIATE

An Associate shall be:

(a) An engineer of a railway which is essentially an adjunct of an industry, or which is used primarily to transport the products and materials of an industry to and from a railway which is a common carrier.

(b) A person qualified by training and experience to cooperate with Members in the object of this Association, but who is not qualified to become a Member.

F. JUNIOR MEMBER

(a) A Junior Member shall be not less than 21 years of age and shall be an engineering employee of a railway corporation who has had not less than three years of experience in the location, construction, operation or maintenance of railways.

(b) His membership in this classification in the Association shall terminate at the end of the calendar year in which he becomes 30 years of age.

(c) He may make application for membership other than as a Junior Member at any time when he becomes eligible to do so.

3. Transfers

The Board of Direction shall transfer from one class of membership to another, or may remove from membership, any person whose qualifications so change as to warrant such action.

4. Rights

(a) Members, and Life Members who were formerly Members, shall have all the rights and privileges of the Association. Life Members who were formerly Associates shall continue to have all the rights and privileges of Associates.

(b) Honorary Members shall have all the rights and privileges of the Association except those of holding elective office, provided, however, that Members or Life Members who are elected Honorary Members shall retain all the rights and privileges of the Association.

(c) Associates and Junior Members shall have all the rights and privileges of the Association except those of voting and holding elective office.

Article III

ADMISSION, RESIGNATION, EXPULSION AND REINSTATEMENT

1. Charter Membership

The Charter Membership of this Association consists of all persons elected to membership before March 15, 1900.

2. Application for Membership

(a) A person desirous of membership in this Association shall make application upon the form provided by the Board of Direction. In the event that Junior Membership is desired, the applicant shall so state.

(b) The applicant shall give the names of at least three Members of this Association to whom personally known. Each of these Members shall be requested by the Executive Secretary of the Association to certify to a personal knowledge of the applicant with an opinion of the applicant's qualifications for membership.

(c) If an applicant is not personally known to as many as three Members of this Association, the names of well-known persons engaged in railway or allied professional work to whom he is personally known shall be substituted, as necessary, to provide a total of at least three references. Each of these persons shall be requested by the Executive Secretary of the Association to certify to a personal knowledge of the applicant, with an opinion of the applicant's qualifications for membership.

(d) No further action shall be taken upon the application until replies have been received from at least three of the persons named by the applicant as references.

3. Election to Membership

(a) Upon completion of the application in accordance with Section 2 of this Article the Board of Direction through its Membership Committee shall consider the application and make such investigation as it may consider desirable or necessary.

(b) Upon completion of such consideration and investigation, each member of the Board of Direction shall be supplied with the required information, together with the recommendation of the Membership Committee as to the class of membership, if any, to which the applicant is eligible, and the admission of the applicant shall be canvassed by ballot among the members of the Board of Direction.

(c) In the event that an application has been made under the provisions of Section 2, Paragraphs (a) and (b) of this Article, a two-thirds affirmative vote of the entire Board of Direction shall be required for election.

(d) In the event that an application has been made under the provisions of Section 2, Paragraphs (a) and (c) of this Article, a unanimous affirmative vote of the entire Board of Direction shall be required for election.

4. Subscription to the Constitution

An applicant for any class of membership in this Association shall declare his willingness to abide by the Constitution of the Association in his application for membership.

5. Honorary Member

A proposal for Honorary Membership shall be endorsed by ten or more Members of the Association and a copy furnished each member of the Board of Direction. The nominee shall be declared an Honorary Member upon receiving a unanimous vote of the entire Board of Direction.

6. Resignation

The Board of Direction shall accept the resignation, tendered in writing, of any person holding membership in the Association whose obligations to the Association have been fulfilled.

7. Expulsion

Charges of misconduct on the part of anyone holding membership in this Association, if in writing and signed by ten or more Members, may be submitted to the Board of Direction for examination and action. If, in the opinion of the Board action is warranted, the person complained of shall be served with a copy of such charges and shall be given an opportunity to answer them to the Board of Direction. After such opportunity has been given, the Board of Direction shall take final action. A two-thirds affirmative vote of the entire Board of Direction shall be required for expulsion.

8. Reinstatement

(a) A person having been a Member, an Associate or a Junior Member of this Association and having resigned such membership while in good standing may be reinstated by a two-thirds affirmative vote of the entire Board of Direction.

(b) A person having been a Member, an Associate or a Junior Member of this Association and having forfeited membership under the provisions of Article IV, Section 3, may, upon such conditions as may be fixed by the Board, be reinstated by a two-thirds affirmative vote of the entire Board of Direction.

ARTICLE IV

DUES

1. Entrance Fee

(a) An entrance fee of $10 shall be payable to the Association with each application for membership other than Junior Membership. This sum shall be returned to an applicant not elected.

(b) No entrance fee shall be required for Junior Membership, except that a Junior Member, in transferring to another class of membership, shall pay the entrance fee prescribed for other classes of Membership.

2. Annual Dues

(a) The annual dues for each Member and each Associate shall be $15.

(b) The annual dues for each Junior Member shall be $5.

(c) Life Members and Honorary Members shall be exempt from the payment of dues. Life Members desiring to continue to receive the Bulletins and Proceedings of the Association may do so by paying a subscription fee prescribed by the Board of Direction

3. Arrears

A person whose dues are not paid before April 1 of the current year shall be notified by the Executive Secretary. If the dues are still unpaid on July 1, further notice shall be given, informing the person that he is not in good standing in the Association. If the dues remain unpaid by October 1, the person shall be notified that he will no longer receive the publications of the Association. If the dues are not paid by December 31, the person shall forfeit membership without further action or notice, except as provided for in Section 4 of this Article.

4. Remission of Dues

The Board of Direction may extend the time of payment of dues, and may remit the dues of any Member, Associate or Junior Member who, for good reason, is unable to pay them.

Article V

OFFICERS

1. Officers

(a) The officers of the Association shall be a President, two Vice Presidents, twelve Directors, an Executive Secretary and a Treasurer.

(b) The President, the Vice Presidents and the Directors, together with the two latest living Past Presidents continuing to be Members, shall constitute the Board of Direction, in which the government of the Association shall be vested; they shall act as the trustees and have the custody of all property belonging to the Association. The President, the Vice Presidents and the Directors shall be Members.

(c) The Executive Secretary and the Treasurer shall be appointed by the Board of Direction.

2. Term of Office

The term of office of the President shall be one year, of the Vice Presidents two years and of the Directors three years. The term of each shall begin at the close of the annual convention at which elected and continue until a successor is qualified. All other officers and employees shall hold office or position at the pleasure of the Board of Direction.

3. Officers Elected Annually

(a) There shall be elected at each annual convention a President, one Vice President and four Directors.

(b) The candidates for President and for Vice President shall be selected from the members or past members of the Board of Direction.

4. Conditions of Re-election of Officers

A President shall be ineligible for re-election, except as provided for in Section 5 (e) of this Article. Vice Presidents and Directors shall be ineligible for re-election to the same office, except as provided for in Section 5 (e) of this Article, until, at least one full term has elapsed after the end of their respective terms.

5. Vacancies in Offices

(a) If a vacancy should occur in the office of President, as set forth in Section 6 of this Article, the senior Vice President shall immediately and automatically become President for the unexpired term.

(b) If a vacancy should occur in the office of the senior Vice President, due to advancement under Section 5 (a) of this Article, or for reasons set forth in Section 6 of this Article, the junior Vice President shall automatically become senior Vice President for the unexpired term.

(c) If a vacancy should occur in the office of the junior Vice President, due to advancement under Section 5 (b) of this Article, or for reasons set forth in Section 6 of this Article, the Board of Direction shall by the affirmative vote of two-thirds of its entire membership, select a junior Vice President from the members or past members of the Board of Direction.

(d) A vacancy in the office of Director, due to advancement of a Director to junior Vice President under Section 5 (c) of this Article, or for reasons set forth in Section 6 of this Article, shall be filled by the Board of Direction by the affirmative vote of two-thirds of its entire membership.

(e) An incumbent in any office for an unexpired term shall be eligible for re-election to the office held; provided, however, that anyone selected to fill a vacancy as Director shall be eligible for election to that office. excepting that such appointee filling out an unexpired term of two years or more shall be considered as coming within the provisions of Section 4 of this Article.

6. Vacation of Office

(a) In the event of the death of an elected officer, or his resignation from office, or if he should cease to be a Member of the Association as provided in Section 2 (B), Article II; Section 6 or 7, Article III; or Section 3, Article IV, the office shall be considered as vacated.

(b) In the event of the disability of an officer or neglect in the performance of duty by an officer, the Board of Direction, by the affirmative vote of two-thirds of its entire membership shall have the power to declare the office vacant.

Article VI

NOMINATION AND ELECTION OF OFFICERS

1. Nominating Committee

(a) There shall be a Nominating Committee composed of the five latest living Past Presidents of the Association, who are Members, and five Members who are not officers.

(b) The five Members who are not Past Presidents shall be elected annually for a term of one year, when the officers of the Association are elected.

(c) The senior Past President who is a member of the committee shall be the chairman of the committee. In the absence of the senior Past President from a meeting of the committee the Past President next in seniority present shall act as chairman.

2. Method of Nominating

(a) Prior to December 1 of each year the chairman shall call a meeting of the committee at a convenient place, at which nominees for the several elective offices shall be selected as follows:

Office to be Filled	Number of Candidates to be named by the Nominating Committee	Number of Candidates to be elected at the Annual Election of Officers
President	1	1
Vice President	1	1
Directors	8	4
Nominating Committee	10	5

(b) The chairman of the Nominating Committee shall send the names of the nominees to the President and Executive Secretary not later than December 15 of the same year, and the Executive Secretary shall report the names of these nominees to the members of the Association not later than January 1 following.

(c) At any time between January 1 and February 1 any ten or more Members may send to the Executive Secretary additional nominations for any elective office for the ensuing year signed by such Members.

(d) If any person nominated shall be found by the Board of Direction to be ineligible for the office for which nominated, or should a nominee decline such nomination, his name shall be withdrawn. The Board of Direction may fill any vacancies that may occur in the list of nominees up to the time the ballots are sent out.

3. Ballots Issued

Not less than thirty days prior to each annual convention, the Executive Secretary shall issue a ballot to each voting Member of record who has paid his dues to or beyond December 31 of the previous year, listing the several candidates to be voted upon. When there is more than one candidate for any office, the names shall be arranged on the ballot in the order that shall be determined by lot by the Nominating Committee. The ballot shall be accompanied by a statement giving for each candidate his record of membership and activities in this Association.

4. Substitution of Names

Members may remove names from the printed ballot list and may substitute the name or names of any other person or persons eligible for any office, but the number of names voted for each office on the ballot must not exceed the number to be elected at that time to such office.

5. Ballots

(a) Ballots shall be placed in an envelope, sealed and endorsed with the name of the voter, and mailed to or deposited with the Executive Secretary at any time previous to the closure of the polls.

(b) A voter may withdraw his ballot, and cast another, at any time before the polls close.

(c) Ballots received in unendorsed envelopes, or from persons not qualified to vote, shall not be counted.

(d) The ballots and envelopes shall be preserved for not less than ten days after the vote is canvassed.

6. Closure of Polls

The polls shall be closed at 12 o'clock noon on the second day of the annual convention, and the ballots shall be counted by tellers appointed by the presiding officer.

7. Election

(a) The persons who shall receive the highest number of votes for the offices for which they are candidates shall be declared elected.

(b) In case of a tie between two or more candidates for the same office, the Members present at the annual convention shall elect the officer by ballot from the candidates so tied.

(c) The presiding officer shall announce at the convention the names of the officers elected in accordance with this Article.

Article VII

MANAGEMENT

1. President

The President shall have general supervision of the affairs of the Association, shall preside at meetings of the Association and of the Board of Direction, and, by virtue of his office, shall be a member of all committees, except the Nominating Committee.

2. Vice Presidents

The Vice Presidents, in order of seniority, shall preside at meetings in the absence of the President.

3. Treasurer

The Treasurer shall pay all bills of the Association when properly certified by the Executive Secretary and approved by the Finance Committee. He shall make an annual report as to the financial condition of the Association and such other reports as may be called for by the Board of Direction.

4. Executive Secretary

The Executive Secretary, under the direction of the President and Board of Direction shall be the Executive Officer of the Association and shall attend the meetings of the Association and of the Board of Direction, prepare the business therefor, and record the proceedings thereof. The Executive Secretary shall see that all money due the Association is collected, is credited to the proper accounts, and is deposited in the designated depository of the Association, with receipt to the Treasurer therefor. He shall personally certify to the accuracy of all bills and vouchers on which money is to be paid. He shall invest all funds of the Association not needed for current disbursements, as shall be recommended by the Finance Committee and approved by the Board of Direction, with notification to the Treasurer of such investments. The Executive Secretary shall conduct the correspondence of the Association, make an annual report to the Association, and perform such other duties as the Board of Direction may prescribe.

5. Auditing of Accounts

The financial accounts of the Association shall be audited annually by an accountant or accountants approved by and under the direction of the Finance Committee.

6. Board of Direction and Executive Committee

(a) The Board of Direction shall manage the affairs of the Association, and shall have full power to control and regulate all matters not otherwise provided for in the Constitution.

(b) The Board of Direction shall meet within thirty days after each annual convention, and at such other times as the President may direct. Special meetings shall be called on request, in writing, of five members of the Board of Direction.

(c) Seven members of the Board of Direction shall constitute a quorum.

(d) At the first meeting of the Board of Direction after the annual convention, the President shall appoint from the membership of the Board, subject to ratification by the Board, four members to serve with him as an Executive Committee which shall possess and may exercise during intervals between meetings of the Board, all of the powers of the Board on matters which in the judgment of a majority of the Executive Committee cannot properly be delayed until the next meeting of the Board. Actions of the Executive Committee shall be reported to the Board of Direction at the next meeting of the Board. The President shall be chairman of the Executive Committee. Actions of the Executive Committee shall be authorized by a concurring majority of its full membership. Members of the Executive Committee shall serve until their successors are appointed or until the Executive Committee is dissolved by action of a majority of the full membership of the Board of Direction. Following dissolution of the Executive Committee it may be re-created at any time by action of a majority of the full membership of the Board of Direction. If not so re-created prior to the next annual convention, the Executive Committee shall be reconstituted in the normal manner at the first meeting of the Board of Direction following the convention.

7. Administrative Committees

At the first meeting of the Board of Direction after the annual convention, the following Administrative Committees, each consisting of not less than three members, shall be appointed by the President. The personnel of these committees shall be subject to approval by the Board of Direction.

Assignments
Finance
Manual
Membership
Personnel
Publications
Research

Other special Administrative Committees may be appointed by the President at any time, and reappointed annually, if necessary, their personnel being subject to approval by the Board of Direction.

Membership on Administrative Committees shall be restricted to members of the Board of Direction, except that one or two members of the Administrative Committee on Research may be past members of the Board of Direction.

8. Study and Research Committees

The Board of Direction may establish continuing or special Study and Research Committees to investigate, consider, and report upon subjects appropriate to the object of the Association, as set forth in Art. I.

9. Duties of Administrative Committees

(a) Assignments

The Assignments Committee shall review and pass upon the recommendations of Association Study and Research Committees for subjects to be investigated, considered and reported on by these committees during the ensuing Association year, and shall report thereon to the Board of Direction for its approval. The Assignments Committee shall have authority to assign additional subjects or change the scope of any existing subjects at any time during the year, reporting its action thereon to the Board at its next regular meeting.

(b) Finance

The Finance Committee shall have immediate supervision of the accounts and financial affairs of the Association; shall approve all bills before payment, and shall make recommendations to the Board of Direction as to the investment of funds and other financial matters. The Finance Committee shall not have the power to incur debts or other obligations binding the Association, nor authorize the payment of money other than the amounts necessary to meet ordinary current expenses of the Association, except by authority of the Board of Direction.

(c) Manual

The Manual Committee, with the assistance of the Publications Committee, shall have general supervision over the Manual.

(d) Membership

The Membership Committee shall investigate applicants for membership and shall make recommendations to the Board of Direction with reference thereto.

(e) Personnel

The Personnel Committee shall review and pass upon applications of members for appointment to Study and Research Committees, and shall also appoint the chairman and vice chairman of such committees and make a report thereon to the Board of Direction for its approval. Should an unexpected vacancy in chairmanship or vice chairmanship of any such committee occur, the Personnel Committee shall have authority to fill such vacancy immediately, reporting its action thereon to the Board at its next regular meeting.

(f) Publications

The Publications Committee shall have general supervision over the publications of the Association. The Publications Committee shall not have the power to incur debts or other obligations binding the Association, nor authorize the payment of money except by authority of the Board of Direction.

(g) Research

The Research Committee shall encourage and coordinate the research activities of the Association, in the course of accomplishment of which it shall review and pass upon the recommendations of Study and Research Committees for research projects and shall report thereon to the Board of Direction, recommending for approval specific projects initiated by these committees or by the Research Committee and recommending allotments of funds for these projects in the research budget of the Association of American Railroads or from other sources compatible therewith; shall collaborate closely with the research staff of the Association of American Railroads; and when called upon by the Vice President—Research or the Vice President—Operations and Maintenance of that association, members of the Research Committee shall engage in the activities of advisory committees or groups of that organization and shall report from time to time to the Board of Direction on those activities.

10. Special Committees

The Board of Direction may appoint special committees to examine into and report upon any subject connected with the objects of this Association.

11. Discussion by Non-Members

The Board of Direction may invite discussions of reports from persons not members of the Association.

12. Sanction of Act of Board of Direction

An act of the Board of Direction which shall have received the expressed or implied sanction of the membership at the next annual convention of the Association shall be deemed to be the act of the Association.

Article VIII
MEETINGS

1. Annual Convention

(a) The Annual Convention of the Association shall be held in the City of Chicago, Ill., or in such other city as may be determined by the affirmative vote of two-thirds of the entire membership of the Board of Direction. The convention shall open on the second Tuesday in the month of March, or on the third Tuesday if the month of March has five Tuesdays, excepting that some other opening day in March may be designated by the affirmative vote of two-thirds of the entire membership of the Board of Direction

(b) The Executive Secretary shall notify all members of the Association of the time and place of the annual convention at least 30 days in advance thereof.

(c) The order of business at the annual convention of the Association shall be:

Reading of the minutes of the last meeting
Address of the President
Reports of the Executive Secretary and the Treasurer
Reports of committees
Unfinished business
New business
Installation of officers
Adjournment

(d) This order of business may be changed by a majority vote of Members present.

(e) The proceedings shall be governed by "Robert's Rules of Order" except as otherwise herein provided.

(f) Discussions shall be limited to Members and to those others invited by the presiding officer to speak.

2. Special Meetings

Special meetings of the Associations may be called by the Board of Directions on its own initiative, and may be so called by the Board of Direction upon written request of 100 Members. The request shall state the purpose of such meeting.

The call for such special meeting shall be issued not less than ten days in advance of the proposed date of such meeting and shall state the purpose and place of the meeting. No other business shall be taken up at such meeting.

3. Quorum

Twenty-five Members shall constitute a quorum at all meetings of the Association.

Article IX
AMENDMENT

1. Amendment

Proposed amendment of this Constitution shall be made in writing, shall be signed by not less than ten Members, and shall be acted upon in the following manner:

The amendment shall be presented to the Executive Secretary, who shall send a copy to each member of the Board of Direction as soon as received. If a majority of the entire Board of Direction so votes, the matter shall be submitted to the Association by letter ballot.

Sixty days after the date of issue of the letter ballot, the Board of Direction shall canvass the ballots which have been received, and if two-thirds of such ballots are in the affirmative the amendment shall be declared adopted and shall become effective immediately. The result of the letter ballot shall be announced to members of the Association.

Information and Rules for the Guidance of Committees

The following information and rules for the guidance of committees are designed to obtain the maximum benefits from the efforts of the members who make up the personnel of such committees. They are designed to effect a continuity of effort in committee work throughout the entire year, under a plan whereby the personnel of the committees and their respective assignments for investigation and report are set up and made public on or before the beginning of the calendar year, thus enabling the work to be continued without interruption, although the new personnel and subject assignments do not become officially effective until the beginning of the "Association Year," which starts with the close of the annual meeting.

The rules also take into account the fact that the publication of the committee reports must be spread out over a period of four months (November through February), to facilitate printing and to give members of the Association a reasonable length of time in which to study such reports in advance of the annual meeting.

SUBJECT ASSIGNMENTS

Reassigned Annually

The assignments for investigation and report of each committee shall be reviewed annually. To this end, each committee shall review suggestions for new subjects submitted by its Subcommittee A, by other members of the committee, or by others, and such suggestions as receive the approval of the committee shall be submitted by the committee chairman to the executive secretary of the Association not later than October 1. Each suggestion shall be accompanied by brief explanation of the purpose and scope of each proposed assignment, or change in the wording of present assignments. At the same time, the committee chairman shall submit the committee's recommendations covering the withdrawal or continuation of current assignments, with a brief statement of the reason or reasons therefor.

The recommendations received from the various committees shall be assembled and forwarded to the Board Committee on Assignments, which has the responsibility of authorizing the subject assignments to the various committees. Deviations from assignments thus authorized may be made during the course of the year only upon authority of the Board Committee on Assignments. However, this is not to be construed as preventing any committee from proposing additional urgent assignments at any time during the year, upon which it feels work should be begun promptly.

Scope of Assignments of Committees 22 and 27

The scope of assignments of Committee 22 will encompass studies relating to the economics of various types of work equipment as used by the labor forces to which they are assigned, including the labor savings that may be effected, production, and quality of work.

The scope of assignments of Committee 27 will encompass studies involving the mechanical features, operating characteristics, development and maintenance of work equipment, and fuels, lubricants, etc., necessary for its operation; also pertaining to such labor aspects as the selection and training of equipment operators, maintainers and repair forces.

Either or both committees may include in their considerations and reports factors of design or operation that affect productivity or quality of work, and such economic aspects as first cost, obsolescence, life, depreciation, and maintenance and repair costs as may be necessary to the comprehensive development of their respective assignments.

In the case of an overlapping assignment, the assignment should normally be handled by the committee principally affected in the light of the foregoing paragraphs, with the other committee collaborating.

COMMITTEE PERSONNEL

Reorganized Annually

The personnel of each committee shall be reorganized annually. It is desirable that 10 percent of the membership be changed each year. Members who do not attend meetings of the committee, who do not render service by correspondence, or who do not return letter ballots will be dropped. To this end the chairman of the committee shall submit to the secretary's office not later than October 1 the current Committee Member Activity Record Chart, filled out in full regarding each member, and showing in the appropriate columns which members he recommends be dropped because of delinquency in service to the committee, or for other reasons, and those members he recommends be continued on the committee. The chart, at the bottom, should also list any members he recommends for appointment to the committee, whether previously carried as "guest" members or not.

The recommendations received from the various committees shall be assembled and forwarded to the Board Committee on Personnel, which has the duty of appointing the committee personnel.

No additions to the personnel of committees will be made during the year following the official closing of committee rosters, October 1, except as provided for in the rules applying to "Guests."

Members who desire appointment to a committee should make application through the committee chairman or the executive secretary on the prescribed form.

As soon as committee chairmen have been advised by the secretary's office of the approved assignments and personnel of their committees—which should be shortly after the regular meeting of the Board of Direction in November—they should begin reorganizing their subcommittees as necessary, for the coming year's work, and, by December 15, must advise the secretary's office of the names of all subcommittee chairmen for the coming year, for listing among the official AAR representatives on committees in the forthcoming Committee Assignments pamphlet.

Chairmen, Vice Chairmen and Subcommittee Chairmen

Chairmen, vice chairmen and subcommittee chairmen must hold the grade of Member in the Association, and be in active service of their respective companies or organizations (not retired). These officers of a committee and of its subcommittees, along with the secretary of the committee, if any, constitute the official representatives of the Engineering Division, AAR, on the committee (except to the extent that any one or more of them may not be in the employ of an AAR Member Road), and their names will be set in bold-face type at the top of all published listings of the committee personnel.

The term of chairman and vice chairman shall be three years in each position, and will normally start at the beginning of the Association year, at the close of an annual meeting. However, the term of office of vice chairman will be shorter if he is appointed to fill a vacancy in the position of vice chairman. Chairmen completing their three-year term shall recommend to the Board Committee on Personnel a nominee for the chairmanship and a nominee for the vice chairmanship, with assurance of acceptances from such nominees if appointed by the Board Committee. The term of office of subcommittee chairmen may be more than three years.

In the event of a vacancy in the office of chairman, the office shall be filled by the vice chairman, subject to the approval of the Board Committee on Personnel. The three-year term of office of the chairman so approved, or of a new appointee shall be considered as having started as of the end of the immediately preceding convention if the appointment is made prior to the time the committee's report is due in the secretary's office, and as becoming effective as of the end of the next convention if the appointment is made after the committee's report is due in the secretary's office.

In the event of a vacancy in the office of vice chairman, it shall be the duty of the Board Committee on Personnel to fill the vacancy. The term of office of the vice chairman so appointed shall be considered as having started as of the end of the immediately preceding convention if the appointment is made prior to the time the committee's report is due in the secretary's office, and as becoming effective as of the end of the next convention if the appointment is made after the committee's report is due in the secretary's office.

Committee Secretary

Any chairman may appoint a secretary with duties usually encompassed by such office.

Size of Committees*

The total membership of any committee shall be limited to 70, including Members, Associates, and Junior Members, but not counting retired members, even though gainfully employed.†

In determining the membership of a committee, railroads having no more than 50 Association members may have not more than 2 members on any committee; railroads having 51 to 100 members may have not more than 3 members on any committee; railroads having more than 100 members may have not more than 4 members on any committee.

No college, university or other institution of learning shall have more than 2 members on any committee, and no manufacturer or supply company or other organization shall have more than 1 Member or Associate member on any committee.

Retired Members

Members who have retired from active service under normal retirement procedure, regardless of whether they undertake other employment (other than sales to the railroads), may serve on committees and subcommittees a maximum of three years following retirement, but cannot hold the office of chairman, vice chairman, or subcommittee chairman, and have no voting rights on technical matters. Their presence on the committee roster shall not be counted in the application of the rules affecting the total

* In applying any of the rules under the headings: Size of Committees and Associate Members, see paragraph under heading "Retired Members," and third last paragraph under heading "Member Emeritus."
† A temporary exception to this rule was authorized by the Board of Direction in 1960 to permit Committee 8—Masonry, to absorb members from Committee 29—Waterproofing, which was discontinued in 1961.

number of members permitted on committees, the number of associates permitted on a committee, or the rules having bearing upon the number of members on committee permitted from any railroad, supply company, or other organization. Following termination of their service on committees, retired members may continue to attend committee meetings as "visitors" subject to the approval of the committee chairman involved.

Associate Members

No company will be permitted to have more than one Associate member on any committee, and company representation shall not necessarily be continuing. However, in the event that a railroad member on a committee becomes associated with a manufacturer or supply company (in other than a sales capacity) after retirement from railroad service on pension, and thus automatically becomes an Associate member, he shall not be deprived of membership on the committee during the period of three years, following his retirement from railroad service. As regards the voting rights of Associate Members on committees, see "Voting in Committees."

The membership of Associates on a committee shall not exceed 10 percent of the total membership of the committee, except as may be occasioned by the exception provided in the preceding paragraph or the exceptions set forth under "Retired Members" and "Member Emeritus."*

Member Emeritus

This class of committee membership was established in 1953 in order to permit recognition of long-sustained meritorious service of committee members to committees, following their retirement.

To be eligible for this honor, or to retain this honor, a member must be in good standing in the Association as a Member, Honorary Member, Associate, or Life Member, and must have:

(a) Retired under normal retirement procedure from active service in the company with which he has been connected.

(b) Served on the committee at least 10 years. (Executive secretary's office can furnish service record on any retired committee member.)

(c) Rendered outstanding service to the committee over a period of years.

(d) Been proposed by at least five committee members in writing and voted the honor by a two-thirds affirmative letter ballot of all members of the committee, including Associates, retired members and Junior Members—the letter ballots to be returnable to the executive secretary's office within 60 days. (Secretary's office can furnish sample type letter ballot).

The number of such members permitted on any committee will be limited to five.

Furthermore, his election as Member Emeritus must be affirmed by the Board Committee on Personnel through the executive secretary's office.

Having been elected as Member Emeritus, the member's name will continue to appear on the roster of the committee, and he will have all rights and privileges of committee members except that of voting on technical matters (i.e., can serve on subcommittees, should he desire, in order that the committee might benefit from his knowledge and experience). Likewise, his name will continue to be shown in the printed roster of the

---

* An exception to this general 10-percent rule was authorized by the Board of Direction late in 1960 with respect to Committee 18—Electricity, which was recreated in 1961.

committee appearing in the Bulletins of the Association, and in the Assignments Pamphlets, in each case suitably designated as Member Emeritus. However, the names of Members Emeritus will not be designated by an "E" or otherwise in the alphabetical listing, railroad listing, Honorary Member listing, or Life Member listing in the March Bulletin.

Members Emeritus will not be counted in the application of the rules affecting the total number of members permitted on committees, the number of associates permitted on a committee, the rules having bearing upon the number of members on committees permitted from any railroad, supply company, or other organization, or the number of years that a retired member may serve on a committee. Any Emeritus title will terminate with the death of the recipient, or in the event of the termination of his membership in the Association for other reasons.

Nothing in these rules will prevent extending the honor of Member Emeritus to a retired committee member who may have taken up, or who subsequently takes up, other employment following his official retirement.

Tangible evidence of this honor will be given to those so named in the form of a pocket card, similar in form to a railroad pass, signed and sent out by the committee chairmen.

"Guests" and "Visitors"

The previously stated rule under Committee Personnel Reorganized Annually, that "no additions to the personnel of committees will be made during the year following the official closing of committee rosters, October 1, except as provided for under the rules applying to Guests," does not preclude the attendance at committee meetings of other members of the Association, and non-members of the Association, as "Visitors," with the approval of committee chairmen.

If there are vacancies on a committee roster after the official closing of committee rosters on October 1, (i.e., less than 70), or if vacancies occur during the following year, or are definitely in prospect at the end of that year, Association members (including Junior members), with the approval of committee chairmen and the Board Committee on Personnel, can be appointed as "guests" of that committee. As such, they may attend committee meetings and participate in the committee's activities, unofficially, looking to becoming regularly assigned members at the beginning of the next Association year (March).

"Guests" must always be designated as such on the rosters maintained by the committees and the secretary's office, but their names will not appear in published committee or subcommittee reports. Creation of this class of committee affiliation is not intended to increase the size of any committee beyond the 70 maximum set by the Board, but rather to make it possible to add to "short" rosters between official roster changes.

Furthermore, one need not be either a "regular member" or a "guest" of a committee to attend its meetings from time to time. With the approval of the committee chairman, who must be consulted as regards any specific meeting, any AREA member (including Junior Members), or any non-member may sit in on the meeting as a "visitor", listen to all deliberations and participate in discussions.

Service on More Than One Committee

No member of the Association shall serve on more than one committee, except that a member may serve on two committees if one or both of the committees are among the following: Committee 3—Ties and Wood Preservation; Committee 7—Wood Bridges and Trestles; Committee 18—Electricity; Committee 20—Contract Forms; Committee 24—Cooperative Relations with Universities; Committee 25—Waterways and Harbors; Committee 28—Clearances; Committee 30—Impact and Bridge Stresses; and the Special Committee on Continuous Welded Rail.

COMMITTEE ORGANIZATION AND PROCEDURE

Organizing the Committees

The new assignments and personnel of committees shall become effective at the close of the annual meeting in March. However, chairmen should begin reorganizing their committees late in the preceeding year, as soon as advised officially of the approved assignments and personnel of their committees—usually in the latter half of November. This is important, because the secetary's office must have, by December 15, the names of all subcommittee chairmen for the coming year for listing among the official AAR representatives on committees in the forthcoming Committee Assignments pamphlet.

Upon receipt of the new Committee Assignment pamphlet, it is the duty of the committee chairman—if he has not already done so—to notify new members promptly of their appointment and to advise old members of their reappointment or release. It is also his duty to complete the reorganization of his subcommittees without delay. In the Association year in which his term as chairman expires, the chairman should call on his successor for advice and assistance in the reorganization of the committee.

Subcommittees

In general, the committees are organized to conduct their work by the appointment of one subcommittee for each subject assignment. If deemed advisable, any subject may be subdivided into several parts and a separate sub-subcommittee assigned to each part.

Subcommittee chairmen should make a report on the status of their work at each committee meeting. If they cannot be present at any meeting, they should submit such report to the chairman in writing, to be read to the meeting, or should arrange for some member of their subcommittee, or of the AAR research staff to report for the subcommittee. This rule should be followed even though the subcommittee has little or nothing to report at any particular meeting.

Organization Charts

The chairman shall furnish the executive secretary of the Association two copies of the organization chart (schedule of subcommittee assignments and personnel) of his committee, and shall advise him currently of any subsequent revisions thereof. This chart may be in the form regularly used by committees, but should not be in the form of a blueprint, on which it is difficult to make corrections. White prints are acceptable. These charts should be in the hands of the executive secretary by February 1, and should be prepared with the greatest care to insure the accuracy of initials and names.

The names of "guest" members on committees, if any, (not "visitors") should appear on the charts, but should be clearly designated as such. These names may be arranged either alphabetically among the members or grouped at the bottom of the chart as desired by the various committees. Names of "visitors" should not appear on or be subsequently added to these charts. Charts should also list (a) names of all committee members who are collaborators with other AREA committees and other organizations, and (b) separately, the names of all non-members of the committee who are collaborators from other AREA committees and other organizations.

Handbook for Committee Chairmen

For the assistance and guidance of committee chairmen in the conduct of their committee work, the Association has published a small mimeographed "Handbook for Committee Chairmen", which contains the following material:

Procedures that Can Be Adopted by Committee Chairmen to Stimulate the Most Effective Committee Work.

Procedures Designed to Expedite the Conduct of Committee Meetings, Stimulate Greater Interest in Them, and Produce the Most Effective Results.

Report of a Well Conducted Committee Meeting.

Copies of this handbook are available to committee chairmen from the executive secretary's office.

Voting in Committees

Voting in committees and subcommittees on all Association matters shall be the prerogative of active Members only, except that retired Members, Associates, and Juniors may vote on matters of a social nature or on ballots for Member Emeritus of the committee.

Action on all matters considered by committees or subcommittees shall be determined by a simple majority vote, oral or otherwise, except in the case of voting on the approval of Manual recommendations for inclusion in the committee's report, which requires a two-thirds affirmative vote, by letter ballot, as set forth on page 21.

COMMITTEE AND SUBCOMMITTEE MEETINGS

Location and Number*

Most committees find it possible to conduct their work effectively with a maximum of three meetings each year. While these meetings can be held at any time to fit in best with the work of each committee, the trend in recent years has been for committees to hold their first (organization) meeting each year in January or February in order to get an early start on their new year's work, and not wait until after the annual convention in March. However, concurrent or overlapping committee meetings are undesirable from the standpoint that they may draw several key personnel from the same railroad at the same time, or otherwise adversely affect the attendance at the committee meetings. Accordingly, before committees decide definitely on meeting dates and send out notices thereof, they should clear these dates with the secretary's office.

Subcommittee meetings can likewise be held whenever desired, either independent of full committee meetings or in conjunction therewith. The latter plan has the advantage of minimizing travel time and possibly total time away from members' offices. Where subcommittee meetings are held in conjunction with general committee meetings, they may be held immediately before or after such meetings, or during such meetings if desirable, in recesses specifically called by the committee chairman for this purpose.

During 1960 the Board of Direction issued the following summary of committee meeting rules—which continue in effect—to give emphasis to the desirability of mini-

---

* Conference Rooms 707, 1218, and 2316 at AREA headquarters in Chicago, which will accommodate 28, 30, and 45 people, respectively, are available for committee meetings to the extent they have not been already committed for other use. The Conference Room at the AAR Research Center, 3140 South Federal Street, Chicago, which will accommodate up to 50, is likewise available for committee meetings. Arrangements for the use of rooms 707, 1218, and 2316 should be made through the secretary's office. For the use of the Conference Room at the Research Center arrangements should be made through the office of G. M. Magee, director of engineering research, at the Research Center.

mizing the expense and off-the-job time of committee members in attending committee meetings:

Limit the number of meetings held by committees to the absolute minimum consistent with carrying out their work.

Require the holdings of such meetings, to the fullest extent possible, at points most convenient to the majority of members.

Restrict the length of meetings to one day whenever possible.

Eliminate committee inspection trips except those essential to the work of committees.

Schedule meetings so as to minimize the off-the-job time of members.

The Association has no funds to defray the cost of meeting rooms. Therefore, committees should hold meetings where no charge is made for such rooms, or should work out some other arrangement agreeable to members of the committee.

Notices and Minutes

Committee chairmen shall send out, or arrange to have sent out, well in advance of meetings, copies of notices of all committee meetings to both committee members and collaborators. Two copies of all such notices should be sent to the secretary's office as early as possible for publication of meeting dates and places in the AREA News. In this latter regard, and especially if mailing of official notices is to be delayed, chairmen should give the secretary's office advance information about meetings, if possible. It should be kept in mind that the deadline for material for any issue of the News is the twentieth day of the month immediately preceding the date of issue.

Meeting notices, generally, should include or be accompanied by an agenda, preferably in timetable order, for the benefit of any members who may not be able to be in attendance the full time of any meeting. They should also include as much information as possible relative to any inspection trips or other features planned.

Minutes of all committee meetings should be prepared as soon as possible following meetings, and copies should be sent to all committee members and collaborators—with two copies to the secretary's office.

Reporting on Inspection Trips

In order that highlights of all committee inspection trips may be published in the News, committee chairmen should send detailed information concerning such trips to the secretary's office, or arrange to have such information sent, as soon as possible after the completion of such trips, keeping in mind that the deadline for news to appear in any issue of the News is the twentieth day of the month preceding the date of issue.

Included among details furnished should be the name of the host or hosts (companies or company representatives) on the occasion, the facilities or operations observed, and separately, the number of members and guests who participated.

COLLABORATION

Between AREA Committees and with AAR Committees

Subjects, the nature of which clearly indicates the possibility of overlapping interest of two or more AREA committees, or the interest of committees of other groups with which the Association has agreed to collaborate, carry an appended clause reading: "collaborating with ...." It is the duty of the chairmen of subcommittees having an assignment carrying this instruction to take the initiative in effecting such collaboration—by arranging for the appointment of a representative of the other interested group, should such be mutually decided as desirable, or by setting up an arrangement whereby the collaborating group will review and criticize any reports submitted to it. If a representative or collaborator is appointed, he should be kept fully advised of all activity of the subcommittee involved. Regardless of whether the assignment specifically mentions collaboration, committees should be on the alert to obtain the advice and assistance of other AREA committees or interested groups in dealing with any subject that imposes any questions of possible overlapping interest or responsibility.

The reports of subcommittees involving collaboration should be submitted to collaborators or collaborating groups whether they are for information only or involve specifications or recommended practice, and should be submitted as far in advance of filing date as possible. If they cannot be submitted prior to the committee's filing date for any reason, they should be submitted as soon thereafter as possible, and in any event prior to the annual meeting, so that if the collaborators, or the groups they represent, desire to comment thereon or to take exception thereto in any respect, such can be done in writing to the committee chairman or subcommittee chairman involved prior to the annual meeting, or in written or oral form at the annual meeting.

A committee undertaking revision of its Manual chapter should request collaboration of any committee that participated in the original development and adoption of the material under revision. The executive secretary of the Association will provide information concerning such previous collaboration.

If an AREA committee or subcommittee is asked to collaborate with another AREA committee, or with committees of any of the sections or divisions of the AAR, it shall appoint a representative to implement this collaboration, if such is mutually decided as desirable, or it should agree, without a specific collaborator, to review and criticize any reports submitted to it. Committee members appointed to collaborate with any other AREA or AAR committees should report currently to their own committees on any matters of interest resulting from the collaboration.

The names of all collaborators, whether to or from a committee, should be shown separately on the committee's organization chart, as set forth under "Organization Chart."

With Other Organizations

Many AREA committees appoint from their membership representatives to serve as collaborators on committees of the American Standards Association, the American Society for Testing Materials, the American Concrete Institute, or other outside organizations, these representatives acting either directly for the AREA committees or in behalf of the Association of American Railroads which may hold membership in the organizations involved. In all such cases, representation in these other organizations, either initially or otherwise, is handled through the AREA executive secretary's office. Thus AREA committee nominations for representatives on these outside committees, or for changes in representatives, are made through the executive secretary's office, which transmits the nominations to the organizations, secures their acceptance, notifies those interested, and makes official record thereof.

Beyond this point the representatives carry on their collaboration independent of the executive secretary's office, but each AREA committee should keep on its organization chart a record of all of the organizations with which it collaborates, and the names of its collaborators, as set forth under "Organization Chart".

Committee members appointed to collaborate with other organizations should report currently to their own committees on any matters of interest resulting from the collaboration.

WORK OF THE COMMITTEES

Objectives

The objectives of the Association are advanced through the work of the committees in two ways—(1) the development of useful information pertinent to their assignments to be presented to the Association "as information," and (2) the formulation of recommended practices to be submitted for adoption and publication in the Manual or in the Portfolio of Trackwork Plans.

Planning the Work

In pursuing the work on any assignment, the first step is necessarily one of fact finding, including (a) a study of available literature on the subject, particularly reports of previous investigations, (b) a compilation of current practice, especially recent changes in practice, and (c) resort to original tests or experimentation, after a canvass of all other sources of information indicates that research work is necessary.

Collection of Data

Committees are privileged to obtain data or information in any proper way. If desired, the executive secretary will *mail* circulars of inquiry, or questionnaires, prepared by committees. Where sufficient information can be secured from members of the committee, they alone should receive letters of inquiry or questionnaires. Where a broader representation of railroads is necessary or desirable, such letters of inquiry or questionnaires may be sent to the appropriate officer within the engineering and maintenance of way departments of selected additional roads or of all AAR Member Roads.

Only in special cases should communications of any kind be sent to officers in other than the Engineering and Maintenance of Way Departments (presidents or chief executive officers, chief operating officers, chief mechanical officers, etc.), and then only over the signature of, or with the explicit permission of, the heads of the appropriate AAR department, division or section, such to be arranged for through the executive secretary's office.

Circulars of inquiry or questionnaires should be brief and concise; the questions contained therein should be specific and pertinent, and not of such general or involved character as to preclude the possibility of obtaining satisfactory and prompt response; should specify to whom answers are to be sent; and should be furnished in duplicate so that a copy can be retained by persons replying.

Research

It is primarily the responsibility of Subcommittee A of each committee to bring together recommendations for further study and research on the part of the committee, based upon suggestions received from other members of the Association, or as the result of its own observations within or without the railroad industry. Any recommendations for assignments in the following year which call for research appropriations should be processed with the committee early in the Association year, beginning with the close of the annual meeting in March, in order that any proposal for research approved by the committee can be in the hands of the director of engineering research, AAR, with copy to the executive secretary, AREA, by July 1. These recommendations must be accompanied by a supporting statement setting forth: (a) the nature of the information sought; (b) how the railroads are adversely affected by the lack of this information; (c) the estimated cost of the investigation; (d) the estimated time to complete the work; (e) the basis for assuming that the investigation will produce the data desired; and (f) an estimate of the savings to be realized or other advantages to accrue from the successful completion of the investigation. A request for funds to continue or complete an investigation shall include also a statement of the results obtained to date.

Maintaining Manual Up to Date

Each committee shall critically review the material in its chapter of the Manual at such intervals as to insure that it is kept up to date. It shall resubmit all Manual material for revision or reapproval at intervals of not more than 10 years. This rule, however, is not intended to encourage the reapproval of documents only at 10-year intervals. On the contrary, and especially since each document in the Manual carries a reapproval line under its heading, committees are urged to recommend the reapproval of documents each time that revisions (major or minor) are proposed, using some such wording as "Reapprove with the following revisions". If such reapproval is not requested specifically when revisions are recommended, the document will continue to carry its previous adoption or reapproval line.

However, since two or more sheets must be issued in a Supplement every time a document is reapproved without revisions, to correct the document date and the contents page or pages, it is recommended that, in the interest of avoiding unnecessary printing costs, documents which do not require revisions should not be offered for reapproval at intervals of less than 8 or 10 years.

Group Revisions in Specific Years

While it is a healthy situation for committees to be constantly on the alert to improve their respective documents in the Manual, and while some revisions in Manual material will be of a character that will require that they be made at the earliest possible date, many changes will be of an editorial or less important character and will not demand that they be made immediately.

Accordingly, in the interest of economy, committees should, so far as possible, group their revisions in any specific document, or anywhere in their respective chapters, looking to submitting them as a group at intervals of two or three years or more, rather than separately year after year—thus avoiding the necessity for reissuing the same Manual pages, including contents pages, in successive years, to the greatest extent possible.

NATURE AND PREPARATION OF REPORTS

Form of Report

It is important that committee reports be prepared in accordance with the following instructions pertaining thereto, and the Style Standards for committee reports, as detailed on following pages in this pamphlet.

At Least Brief Progress or Status
Statements to Be Made on All Assignments

Committees should pursue their investigations on all assignments, but are expected to present detailed progress or final reports for publication only on assignments with respect to which pertinent information has been developed. However, with respect to all other assignments each year, the report of a committee shall include for each a brief statement as to the progress which has been made on it during the year, or in the event that no progress has been made, a brief statement as to the status of the project. These brief statements, which are intended to be informative to those interested in any specific subject in a year when no detailed report can be made, may be confined to as few as one or two sentences, which may be entirely adequate. In such cases, the statement should appear directly under the title of the assignment as listed in the chairman's introductory statement to the committee report as a whole, instead of the words used previously, "No report", or "Progress in study, but no report". On the other hand, if these "progress" or "status" statements, in order to be adequate to the situation, necessarily exceed 70 words, they should be presented in the usual manner as a subcommittee report, with assignment heading and subcommittee personnel. In such cases, the chairman's introductory statement should show, for the assignments in question, "Brief progress statement", and the page number, or "Brief status statement", and the page number, as the case may be.

Reports on Assignment A should not be submitted for publication.

Reports on Assignment A

In the case of Assignment A—Recommendations for further study and research, two reports on recommendations shall be made to the committee each year; (1) early in the Association year with respect to any proposed new assignments involving appropriations for the conduct of research work, as set forth in detail under "Research", on page 17; and (2) late in the summer or early fall, covering recommendations with respect to new assignments for study which do not call for research appropriations. This latter report should also include recommendations as to whether any existing assignments can be, or should be, discontinued. Neither of the reports on Assignment A will be presented in the Bulletins of the Association, or orally at conventions.

Information Reports or Recommended Practice

Whether the report on any particular assignment should take the form of "information" or a "recommended practice," depends largely on the nature of the assignment. Some assignments will be fulfilled completely by the presentation of information; others call for information in support of appended recommendations that are submitted for adoption. In still other cases, the primary objective is a comprehensive statement of recommended practices, but the development of these recommended practices may entail investigation or research work, the results of which are of such importance as to warrant their presentation as information prior to the submission of the recommendations. In some cases, it may be advisable to submit material in the form of recommended practice, but as information only, with a view to inviting suggestions and criticism that may serve as the basis for revisions prior to the resubmission of the material for adoption at a later date. This, however, is not mandatory.

When the work has been completed on any assignment, the committee should request of the Board Committee on Assignments that the assignment be discontinued. Its last report on such an assignment should be designated as "final report" only when the committee does not contemplate further study of the subject in the near or foreseeable future; otherwise, the report should be designated as a "progress report", with the recommendation that the subject be discontinued until there are further developments.

Writing of Committee Reports*

Many progress or final reports, whether based on research or other investigation. best lend themselves to written presentation in orderly sequence or chronological arrangement, ending with any conclusions or recommendations which may have been arrived at. However, in most cases, and especially in the case of long reports, to conserve the time of members who may or may not be interested in the details of the study involved, it is recommended that reports be introduced with a brief highlight summary statement of the background, purpose and extent of the study, as may be desirable, and including a synopsis of any conclusions, recommendations or other results—this latter material to supplement a more detailed presentation elsewhere in the report.

Reports of information, supplementing previous reports of progress, should make reference to the previous reports by Proceedings volumes, year and page number, and may include a brief review of material previously presented, but should avoid extended repetition of such material.

Use of Trade Names

Committee reports which are based upon physical research or field tests carried out by or through the research staff of the Engineering Division, AAR, may use trade names or manufacturers' names in referring to products, machines, devices or processes under test, in accordance with rules in effect with the AAR Engineering Division research staff.

No other committee reports, however, shall contain the trade names of products, machines, devices or processes, nor the names of manufacturers, in either text or cut captions, unless in each instance approval is secured from the Board Committee on Publications prior to the publication of the reports. To seek such approval, a committee must submit five copies of the report in question to the executive secretary's office, for transmission to the members of the Board Committee, six weeks prior to the scheduled filing date of the report. If time does not permit a ruling upon the request of the committee prior to the publication date of the report in question, the report of the committee must either be altered to eliminate the trade names or terms involved, or be withdrawn, at the discretion of the committee which prepared it.

Trade or manufacturers' names are not to be used anywhere in the Manual of Recommended Practice, the Portfolio of Trackwork Plans, the Handbook of Instructions for Care and Operation of Maintenance of Way Equipment, or other comparable publications of the Association.

Illustrations in Committee Reports

Committees may use illustrations within their reports, both photographs and line drawings, to the extent necessary to enhance the value of their reports, or to preclude detailed descriptions or the presentation of detailed data which would otherwise be required. For the physical requirements of such illustrations, see "Illustrations" under Style Standards. No illustrations, within themselves, shall show trade or manufacturers' names; neither shall the captions for such illustrations use trade or manufacturers' names, without prior approval on the part of the Board Committee on Publications, as is set forth under "Use of Trade Names".

---

* See also Style Standards for Committee Reports.

Nature of Manual Material*

Material adopted by the Association for publication in the Manual shall be considered Recommended Practice, but shall not be binding on the members. Recommended Practice, as defined in the Manual, is as follows: "A material, device, design, plan, specification, principle or practice recommended to the railways for use as required, either exactly as presented or with such modifications as may be necessary or desirable to meet the needs of individual railways, but in either event, with a view to promoting efficiency and economy in the location, construction, operation or maintenance of railways. It is not intended to imply that other practices may not be equally acceptable."

In specifications, the word "shall", and in some cases "must", is permissible, and may, in fact, be essential. In Manual material other than specifications, however, the words "shall" or "must", or equally compulsive words or expressions, must not be used. Rather, use "should", in the sense of "is preferred", or "is desirable", or use some equally non-compulsive words or expressions. Frequently, statements of preference or desirability in documents of recommended practice should be qualified by the words 'if practicable", or "where practicable", or "where warranted", or the like.

Avoid specific requirements as to limits of wear, frequency of inspection, amount of supervision, and the like.

Avoid the use of the word "approved" in all Manual material in the sense that the AREA puts an official stamp of approval on anything—which is not the case. Liksewise, avoid the use of the word "essential" in Manual material.

Avoid the use of the word "Standard" when referring to any AREA specifications or other Manual recommendations, or to the AREA Trackwork Plans, unless and until the Association should adopt certain so-called "standard plans", designated as such, to differentiate them from present recommended plans and alternates.

Printing of Manual Material*

Material offered for adoption and publication in the Manual, except as noted herein, should be submitted in full, regardless of its publication in previous years, unless the material in question appeared in substantially identical form not more than one year before being submitted for adoption. Such material shall appear in the report of the committee that is published not less than 30 days before the annual meeting at which it is to be presented. Recommended revisions of Manual material, if extensive, shall include only the proposed material, which shall be printed in full in the report of the committee. Manual material recommended for reapproval, or for deletion, shall be presented by title and page reference only. Likewise, plans, specifications or other documents of other organizations recommended for adoption by the AREA shall be presented by title and serial designation only, e.g., current ASTM specifications, designation D 17.

When entirely new material is offered for inclusion in the Manual, the committee sponsoring it should state specifically in its report the exact location the material is to have in the Manual.

Letter Ballot Required of Committee*

Any action recommended by a committee with respect to the adoption, revision, reapproval or withdrawal of Manual material must have received prior endorsement by the committee in the form of an affirmative vote of two-thirds of the voting membership of the committee, such vote to be taken by letter ballot. Associates, Junior

---

* Same applies to Portfolio of Trackwork Plans.

members, Members Emeritus, and retired members on a committee are not entitled to vote. Letter ballot of a committee on Manual material shall be taken only after approval at a regular meeting of the committee, by majority vote, to submit the material to letter ballot.

It is imperative that committee members promptly consider and vote on all letter ballots, seeking the advice of other committee members or specifically qualified officers on their own roads if in doubt as to whether to vote for or against a proposal.

If a member votes in the negative on any Manual proposal, it is encumbent upon him to state the reason or reasons therefor.

PUBLICATION OF REPORTS

Dates for Filing Complete Committee Reports

To insure the orderly publication of the reports in the four winter Bulletins of the Association—November–February, incl.—in accordance with a predetermined schedule, it is necessary that chairmen file complete reports with the executive secretary of the Association on or before the dates specified in the Committee Assignments for Study and Research pamphlet.

Reports to be published in the September–October issue of the Bulletin shall be submitted in the same manner by committee chairmen, or by members of the AAR research staff in their behalf, as other reports, on a schedule worked out with the secretary's office.

The manuscript of the report must be furnished in duplicate, preferably double spaced. Piecemeal filing of reports by subcommittee chairmen is permissible only under special arrangement (in writing) with the executive secretary of the Association.

The regular annual reports of committees—to appear in the winter Bulletins of the Association—must in each case include an introductory statement, or committee chairman's report, embodying the personnel and list of assignments of the committee, as set forth under Style Standards for Committee Reports, pages 25 to 28, incl.

Portrait Photographs of Committee Chairmen

During his first year as chairman, each chairman must furnish the secretary's office a portrait photograph of himself to be used with the reports of his committee as published in the Bulletins while he is chairman. If this has not been done prior to the filing of the committee's report, it must be done at that time. The photograph furnished need be of no special size, but should be black and white, clear, and an acceptable likeness of the chairman. These photographs will be returned to chairmen upon request.

PRESENTATION OF REPORTS AT ANNUAL MEETINGS

Presentation of Reports

Reports offered as information should be presented by title or by a brief highlight outline of the contents. Material submitted for adoption and publication in the Manual* may be presented by reading the title and subtitles, but the presiding officer may, upon request, authorize the reading of specific portions of the material being offered.

---

* Same applies to Portfolio of Trackwork Plans.

Since both the degree of effectiveness with which a report is received by those assembled in annual convention, and the accuracy with which it can be reported in the Proceedings, depend upon the clarity with which the oral presentation is made to the meeting, it is desirable that committee members write out and read their presentations, and that they speak directly and distinctly into the microphone at the rostrum, raising or lowering the microphone as may be necessary to that end. In the event that written presentations are read, a copy of such presentations should be given to the executive secretary or to the convention reporter before the speaker leaves the rostrum.

Visual Presentations

The use of illustrations in the form of slides, motion pictures, etc., as a part of or in conjunction with committee presentations, whether reports or special features, shall be governed by the following rule:

Films** produced by supply companies, manufacturers, and supply organizations depicting their products or services in any form are not to be used in connection with committee presentations, either supplementing committee reports or as special features, at annual meetings, and the use of trade association films is not encouraged. However, under special conditions, where a committee desires to use a trade association film in connection with its presentation, the matter must be referred to the Board Committee on Publications for approval, through the executive secretary's office, by January 1 of the year inquestion, in order that a ruling may be secured prior to the publication of the convention program in the AREA News. Trade association films to be considered under this rule must be of an educational, rather than of a sales-promotion type, must make no direct or indirect comparisons with other products or services, and may make reference to the associations which produced them in only an innocuous way.

Oral Discussions

Comments on or criticisms of any report may be offered from the floor. When necessary to insure accuracy, or upon request, the speaker's remarks will be submitted to him in writing before publication in the Proceedings, for the correction of diction, misstatements, and errors of reporting, but not for the elimination of remarks.

Written Discussions

Written discussions of published reports will be transmitted to the chairman of the interested committee who will read or present them by title or in abstract at the convention. Written discussions will be published in the Proceedings as a part of the discussion of the committee reports.

Action on Reports

No formal action is to be taken by the convention on material submitted as information, whether in the form of a progress or final report.

Action on material submitted for adoption and publication in the Manual will be one of the following:

(a) Adoption as a whole as presented.

** Wherever the word "film" is used, it applies as well to slides and any other form of visual presentation.

(b) Affirmative action on the amendment of a part or parts of the material presented, followed by adoption as a whole as amended.

(c) Adoption of a part, complete in itself, and referring the remainder back to the committee for further consideration.

(d) Recommittal with or without instructions.

Note.—An amendment which affects underlying principles, if adopted, shall of itself constitute a recommittal of such part of the report as the committee considers affected.

The Chair will decline to entertain amendments which in his opinion are primarily a matter of editing.

MISCELLANEOUS

Memoirs

The Association has developed a complete set of rules with respect to memoirs in committee reports or elsewhere in its publications, covering the scope, preparation and presentation of such memoirs. Copy of these rules, as well as the Association service record of any deceased member, can be secured from the executive secretary's office.

Letter Ballot of Membership

When and as required between annual meetings, recommendations for the adoption, deletion, revision or reapproval of Manual material shall be submitted to letter ballot of the Members of the Association under the following limitations:

(a) That the letter ballot shall be taken only after the Board of Direction has recognized the necessity for such emergency action, and

(b) That the propositions submitted by the committee shall have the approval of a special committee of the Board of Direction appointed by the President for that purpose, both as to the substance of the material offered and also as to the circumstances attending the consideration of the material by the committee.

The Board of Direction, acting under the provisions of paragraphs 6 (a) and 12 of Article VII of the AREA constitution, has the authority to amend, delete or revise Manual material at any time, subject to later confirmation or rejection by the membership, submission to the membership to be effected either by means of a letter ballot immediately following such Board action, or by a motion presented at the annual meeting.

Review by Association of American Railroads*

All material adopted for publication in the Manual and all recommendations for the revision or withdrawal of Manual material shall be referred to the vice president, Operations and Maintenance Department, Association of American Railroads, for review, before distribution is made thereof to holders or purchasers of the Manual, or parts thereof.

Publication and Distribution of Annual Supplement to Manual*

Revisions of or additions to the Manual authorized by action at each convention will be published annually in the form of loose-leaf sheets which will be made available

---

* Same applies to Portfolio of Trackwork Plans.

to all holders of the Manual at established prices. These supplemental sheets will be accompanied by instructions for insertion of the new sheets and the withdrawal of sheets that have been superseded, as well as those sheets that have been withdrawn by action of the Association.

In order that committee members who have purchased individual Chapters of the Manual in connection with their committee work may keep these separate Chapters up to date, the secretary's office will make available to them annually, through their committee chairmen, those supplement sheets required to this end.

Publication of Abstracts by Technical Journals

The following rules will govern the releasing of material for publication in technical journals:

Committee reports to be presented at an annual meeting will not be released for publication until after presentation to the annual meeting. Special articles, contributed by members and others, on which no action by the Association is necessary, will be released for publication in technical journals only after issuance in a Bulletin; provided, application therefor is made in writing and proper credit is given the Association, authors or committees presenting such material.

Advance Report of Committee 3—Ties and Wood Preservation

Report on Assignment 5

Service Records

W. L. Kahler (chairman, subcommittee), A. B. Baker, W. Buehler, C. M. Burpee, C. E. DeGeer, F. J. Fudge, W. E.. Fuhr, H. M Harlow, R. P. Hughes, R. B. Radkey, A. P. Richards, J. T. Slocomb.

Tie Renewals and Cost per Mile of Maintained Track

The annual statistics compiled by the Bureau of Railway Economics, AAR, providing information on cross tie renewals and cost data for 1961, are presented herewith in Tables A and B.

The 1961 figures for the Class I Roads of the United States as a whole compared with 1960 are as follows:

Year	Total New Wooden Ties Renewed	Renewals per Mile
1960	13,655,783	43
1961	10,999,747	35
5-year average, 1957 to 1961, incl.		48

The average cost in 1960 was $3.73 and in 1961, $3.87. As noted on the tables, these figures represent storekeepers average cost of ties charged out; they are not the actual costs or prices paid for the ties purchased during the period.

Table A

CROSS TIE STATISTICS (EXCLUDING SWITCH & BRIDGE) FOR CLASS I RAILROADS IN THE UNITED STATES AND LARGE CANADIAN RAILROADS

Calendar year ended December 31, 1961.

Road	New wooden ties reinforced		Cross ties laid in replacement								Track maintained by reporting railroad					Repaired gross ties-miles (thousands) &				New wooden cross ties replacement average				
			New wooden ties treated		Total all new wooden ties laid		Ties other than wood (%)																		
	Number	Average age cost	Number	Average cost	Number	Average age cost	New & second-hand (%) & cross-ties transposed	Total ties applied		Total miles	Total cross ties	Cross ties per mile	Total	Per mile of track	Per cent tie renewal to all ties in main-raised track	Number laid per mile of main-raised track	Renewal cost per mile of main-raised track	Renewal cost per 1,000 repaired gross ties miles						
1	2	3	4	5	6	7	8	9	10	11	12	13	14	15	16	17									

NORTHWESTERN REGION
Chicago & North Western
Chicago Great Western
Chicago, Milwaukee, St. Paul & Pac.
Duluth, Missabe & Iron Range
Duluth Winnipeg & Pacific
Great Northern
Green Bay & Western
Lake Superior & Ishpeming
Minneapolis, Northfield & Southern
Northern Pacific
Soo Line R.R. Co.
Spokane International
Spokane Portland & Seattle

Total

CENTRAL WESTERN REGION
Atchison Topeka & Santa Fe
Chicago Burlington & Quincy
Chicago Rock Island & Pacific
Colorado & Southern
Colorado & Wyoming
Denver & Rio Grande Western
Fort Worth & Denver
Northwestern Pacific
Pacific Electric
Southern Pacific Co. (Incl. T&NO)
Toledo Peoria & Western
Union Pacific
Western Pacific

Total

SOUTHWESTERN REGION
Kansas City Southern
Louisiana & Arkansas
Kansas City Oklahoma & Gulf
Missouri-Kansas-Texas
Missouri Pacific
Quanah Acme & Pacific
St. Louis San Francisco
St. Louis-San Francisco & Texas
St. Louis Southwestern Lines
Texas & Pacific
Texas Mexican

Total

Grand Total: United States

CANADIAN ROADS
Canadian National
Canadian Pacific
Ontario & Northland

a/ Gross ton-miles as shown in Columns 2, 4 and 6 represents storekeepers average cost of ties charged out of service, plus three times gross ton-miles of incoming and tenders in passenger service.

b/ Not reported.

Note: "Average cost" as shown in Columns 2, 4 and 6 represents storekeepers average cost of ties purchased during the period, and used. They are not the actual costs or prices paid for the ties purchased during the period.

Association of American Railroads, Bureau of Railway Economics, Washington, D.C.
from Annual Reports of Class I Railroads to the Interstate Commerce Commission.

May 1962

Note: All figures are exclusive of bridge and switch ties

Road	Numbers of new wood cross tie renewals per mile of maintained track						Aggregate cost of new wood cross tie renewals per mile of maintained track						Per cent new wood cross tie renewals to all ties in tracks					
	1957	1958	1959	1960	1961	5 year average	1957	1958	1959	1960	1961	5 year average	1957	1958	1959	1960	1961	5 year average
NEW ENGLAND REGION																		
Bangor & Aroostook																		
Boston & Maine																		
Canadian Pacific (lines in Me.)																		
Central Vermont																		
Maine Central																		
New York Connecting																		
New York, New Haven & Hartford																		
Rutland																		
Total																		
GREAT LAKES REGION																		
Ann Arbor																		
Delaware & Hudson																		
Detroit & Toledo Shore Line																		
Erie-Lackawanna																		
Grand Trunk Western																		
Lehigh & Hudson River																		
Lehigh & New England																		
Lehigh Valley																		
Monongahela																		
New York Central																		
New York Chicago & St. Louis																		
New York, Susquehanna & Western																		
Pittsburgh & Lake Erie																		
Pittsburgh & West Virginia																		
Wabash																		
Total																		
CENTRAL EASTERN REGION																		
Akron, Canton & Youngstown																		
Baltimore & Ohio																		
Bessemer & Lake Erie																		
Central R.R. of New Jersey																		
Chicago & Eastern Illinois																		
Chicago & Illinois Midland																		
Detroit & Toledo																		
Elgin, Joliet & Eastern																		
Illinois Terminal																		
Long Island																		
Missouri-Illinois																		
Monon																		
Pennsylvania																		
Penna.-Reading Seashore Lines																		
Reading																		
Staten Island Rapid Transit																		
Western Maryland																		
Total																		
POCAHONTAS REGION																		
Chesapeake & Ohio																		
Norfolk & Western																		
Richmond, Fred'burg & Potomac																		
Total																		
SOUTHERN REGION																		
Alabama Great Southern																		
Atlanta & Tennessee North & Northern																		
Atlanta & St Andrews Bay																		
Atlantic & West Point																		
Atlantic Coast Line																		
Carolina & Northwestern																		
Central of Georgia																		
Cincinnati, New Orleans & Tex.Pac.																		
Clinchfield																		
Florida East Coast																		
Georgia																		
Georgia & Florida																		
Gulf, Mobile & Ohio																		
Illinois Central																		
Louisville & Nashville																		
New Orleans & Northeastern																		
Norfolk Southern																		
Piedmont & Northern																		
Savannah & Atlanta																		
Seaboard Air Line																		
Southern Ry.																		
Tennessee Central																		
Western Ry. of Alabama																		
Total																		

Table 8

NUMBER AND AGGREGATE COST OF NEW WOOD CROSS TIE REMOVALS PER MILE OF MAINTAINED TRACK AND RATIO OF NEW WOOD CROSS TIES TO TOTAL CROSS TIES IN MAINTAINED TRACK

Class I roads in the United States and large Canadian roads, by years, and for the average of the five years 1957 to 1961, inclusive

Note: All figures are exclusive of bridge and switch ties

Road	Number of new wood cross tie removals per mile of maintained track						Aggregate cost of new wood cross tie removals per mile of maintained track						Per cent new wood cross tie removals to all ties in tracks					
	1957	1958	1959	1960	1961	5 year average	1957	1958	1959	1960	1961	5 year average	1957	1958	1959	1960	1961	5 year average
NORTHWESTERN REGION:																		
Chicago & North Western																		
Chicago Great Western																		
Chicago, Milwaukee, St. Paul & Pacific																		
Duluth, Missabe & Iron Range																		
Duluth, Winnipeg & Pacific																		
Great Northern																		
Green Bay & Western																		
Lake Superior & Ishpeming																		
Minneapolis, Northfield & Southern																		
Northern Pacific																		
Soo Line R.R. Co.																		
Spokane International																		
Spokane, Portland & Seattle																		
Total																		
CENTRAL WESTERN REGION:																		
Atchison, Topeka & Santa Fe																		
Chicago, Burlington & Quincy																		
Chicago, Rock Island & Pacific																		
Colorado & Southern																		
Colorado & Wyoming																		
Denver & Rio Grande Western																		
Fort Worth & Denver																		
Northwestern Pacific																		
Pacific Electric																		
Southern Pacific Co. (incl. Tab O.)																		
Toledo, Peoria & Western																		
Union Pacific																		
Western Pacific																		
Total																		
SOUTHWESTERN REGION:																		
Kansas City Southern																		
Kansas, Oklahoma & Gulf																		
Louisiana & Arkansas																		
Missouri-Kansas-Texas																		
Missouri Pacific																		
Quanah, Acme & Pacific																		
St. Louis-San Francisco																		
St. Louis-San Francisco & Texas																		
St. Louis Southwestern																		
Texas & Pacific																		
Texas Mexican																		
Total																		
Grand total - United States																		
CANADIAN ROADS:																		
Canadian National																		
Canadian Pacific																		
Ontario Northland																		

a. Not Class I prior to January 1, 1958. Data for prior year not available.

b. Not Class I prior to January 1, 1961. Data for prior years not available.

Association of American Railroads, Bureau of Railway Economics, Washington, D.C., from Annual Reports of Class I railroads to the Interstate Commerce Commission.

Compiled by

May 1962

INDEX OF PROCEEDINGS, Vol, 63, 1962

A

Accounting, ICC, classifications, revisions and interpretations, 449, 634

Agreement forms, bibliography, 175, 638

—bond, reapproved, 168,636

—cab stand and baggage transfer privileges, reapproved, 168, 636

—commissary and boarding outfits, revisions, 171, 636

—construction contract, cost-plus, reapproved, 168, 636

—minor projects, reapproved, 168, 636

—crossings at grade, revisions, 170, 636

—fences snow and sand, off railway property, revisions, 172, 636

—industry track, revisions, 170, 636

—joint freight terminal facilities, reapproved, 169, 636

—joint passenger station facilities, reapproved, 169, 636

—joint passenger terminal project, organization and operation of, reapproved, 169, 636

—private road crossings, revisions, 171, 636

—strip mine on railway miscellaneous physical property, lease, 173, 637

—surplus railway property, disposal of, 173, 637

—tank cars, containing liquefied petroleum gases, anhydrous ammonia and flammable or dangerous materials, storage of on railway tracks, reapproved, 173, 636

—trackage rights, revisions, 169, 636

—water, furnishing from railway systems to employees and others, revisions, 172, 636

—purchase of, revisions, 172, 636

—wire or cable line crossings, revisions, 169, 636

Aluminum, bonding, Thermit welding of, 465, 730

Annual meeting, closing session, 772

—invocation, 615

—opening session, 611

—program of, 607

Anti-splitting devices for ties, revised, 316, 741

Arches, reinforced concrete, revisions, 276, 687

Atchison, Topeka & Sante Fe, lime treatment of subgrade, 586, 751

—main line relocation in Arizona, 154, 678

—prestressed concrete girders, field investigation, 13

—speeds of trains through turnouts test, 67

—treated ballast test, 593, 752

Atlantic Coast Line, prestressed concrete tie test, 330, 745

Austerman, F. E., address, air rights developments, Chicago Union Station area, 721

B

Ballast, gravel, stone, pit-run gravel, slag, specifications, reapproved, 576, 749

—sections, single and multiple track, on tangent and curves, reapproved, 570, 749

—sub-, specifications, reapproved, 576, 749

—treated, progress reports on, 593, 752

Baltimore & Ohio, analysis of maintenance of way operations, 263, 736

Barriers, for dead-end streets, revisions, 184, 732

Beams, glued-laminated, tests on, 164, 697

Beeder, R. H., president's address, 612

Bibliography (See Agreement Forms)

—(See Engineering and Valuation Records)

—(See Waterways and Harbors)

Boilers, locomotives, washouts, water changes and blowdown of, deleted, 113, 724

Bolts, high-strength steel, assembly of structural joints, steel railway bridges, revised, 390, 699

—track (See Track Bolts)

Bonding, new types, Thermit welding of aluminum, 465, 730

Bridge floors, distribution of live load in, 163, 695

Bridge stringers, glued laminated Douglas fir, progress report, 452, 684

Bridges, classification of, reapproved, 387, 699

—composite, steel and concrete, specifications, 398, 700

—continuous welded rail on, effect, 400, 701

—erection, instructions for inspection of, reapproved, 386, 699

—floorbeam hangers, stress distribution in, 399, 700

—fusion welding, reapproved, 387, 699

—movable, railway, specifications, revisions, 383, 699

—non-ferrous metal, progress report, 379, 699

—prestressed-concrete, impact tests, 163, 696

—protecting from washouts and floods, means of, revisions, 578, 749

—steel railway, fixed spans, specifications, revisions, 382, 387, 699

—continuous, design, specifications, reapproved, 386, 699

—erection, specifications, reapproved, 386, 699

—fabrication, instructions for inspection, 386, 699

—lateral and longitudinal distribution of loading, 13

—maintenance inspection of, instructions for, revisions, 385, 699

—rating, rules for, reapproved, 387, 699

—rigid-frame, specifications, reapproved, 386, 699

—strengthening existing, methods, reapproved, 387, 699

Model N U Tie Cutter

HERE IS THE WINNING TEAM

The Woolery NU Tie Cutter and the Woolery Tie-end Remover preserve the line and surface of the track and at the same time reduce the cost of tie renewals. Ties can be removed without trenching, jacking up track or adzing tops of rail-cut ties. With this team you simply cut both ends of tie, pry out center piece, insert in its place the tie-end remover and out go the tie ends pushed by the double acting, double ended hydraulic cylinder of the Tie-end remover.

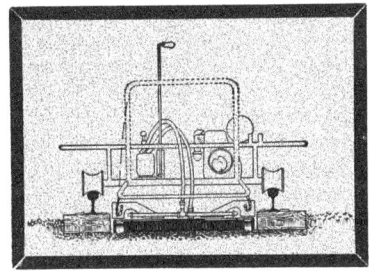 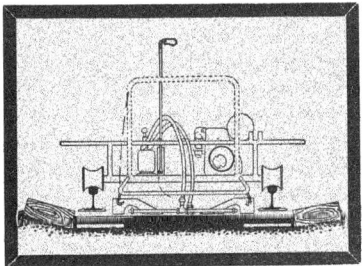

FOR HIGHEST EFFICIENCY USE TWO TIE CUTTERS WITH ONE TIE-END REMOVER

WOOLERY MACHINE COMPANY
MINNEAPOLIS, MINN.

for effective
weed control...

- Concentrated **BORASCU®**
- **POLYBOR-CHLORATE®**
- **UREABOR®**
- **MONOBOR-CHLORATE®**

These <u>borate weed killers</u> are proving best for roads in every way... *efficiency, safety, economy, convenience, easy application.*

Today's use of borates for maximum control of vegetation began years ago with our pioneer work in the field. Continued research has developed the group of herbicides, listed above, which most roads now favor for every phase of weed control. These four weed killers are nonselective. They are widely used for year-round maintenance of weed-free conditions about trestles, tie piles, yards, signals, switches, and rights of way. Find out how you, too, can do a better job on weeds... write today.

AGRICULTURAL SALES DEPARTMENT

U.S.BORAX®

630 SHATTO PLACE · LOS ANGELES 5, CALIFORNIA

AUTOJACK
ELECTROMATI

The only completely automatic track surfacing machine on the market

Proven in operation by North Ameri
leading railroads. Complete and a
matic control of surface and cross l
through tangent and curve terri
regardless of height of lift.

- Combination of Autojack and Electro
  equals or improves production of Ele
  matic alone.

- Precision of lift and uniformity of compa
  controlled automatically.

- All variations in lift, level and run-out
  trolled from operator's panel.

- Beam "sighting" for utmost precision.

- Front buggy self-propelled ahead of ta

TAMPER INC. 53 Court St., Plattsburgh,
SALES AND SERVICE: 2147 University Avenue
St. Paul 14, Minnesota
Phone: 645-5055

PROGRESS REPORT

Here are the up-to-date facts on the SPENO Ballast Cleaning and the SPENO Rail Grinding Services.

BALLAST CLEANING

SPENO Engineering and Research has developed a superior screening arrangement so that we are now using an improved Ballast Cleaner with greater efficiency.

RAIL GRINDING

Our Rail Grinding Service has been so well received we are now building a *THIRD* Rail Grinding Train to take care of the increased demand

SPENO is constantly developing means for better service to make sure that the Railroads receive everything they pay for — and more

Just Ask the Railroads That have used us!

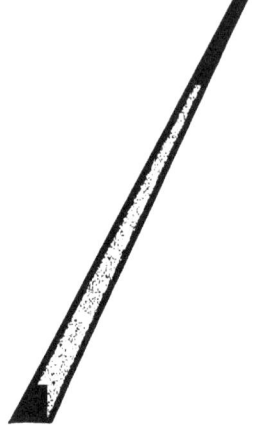

VEGETATION CONTROL
WITH
CHEMICALS

READE MANUFACTURING COMPANY, INC.

Jersey City—Chicago—Minneapolis—Kansas
City—Birmingham—Stockton

SERVING RAILROADS OF AMERICA FOR
MORE THAN FORTY YEARS

WEED AND BRUSH CONTROL

MODEL 441

Developed and Built
for Railroad Maintenance

180° BOOM SWING

DOES ALL JOBS!

LAYING STANDARD RAIL

CUTS MAINTENANCE COSTS

12 FAST CHANGE ATTACHMENTS

- Forks
- 1¼ Cu. Yd. Bucket
- Tote Hook
- 18' Boom Extension
- Fork Tie Baler
- Track Cleaning Bucket

- Back Hoe
- Clamshell
- Back Filler Blade
- Pull Drag Bucket
- 4 Cu. Yd. Snow Bucket
- Pile Hammer

Optional Attachment
Flanged Wheels, Hydraulically Controlled

PETTIBONE MULLIKEN CORPORATION

RAILROAD **PMCO** DIVISION
141 W. JACKSON   CHICAGO 4, ILL.

*80 Years of Service
to the Railroad Industry*

MATISA opens fixed location plant in Birmingham.

Welded Rail Shipments Now Faster, Cost Less

Continuing the "Story of Welded Rail" as pioneered by Matisa, new chapters are constantly being added.

Refined techniques to increase the safety of the "already safest" rail weld—to increase production speed of the "already fastest" rail weld and to decrease the cost of the "already least expensive" rail weld are constantly improving delivery, efficiency and cost features of Matisa *Thoroweld* Continuous Welded Rail.

The latest addition to the Matisa service is this new plant location in the Birmingham switching district. Added to the Chicago switching district plant in Argo, Matisa rail welds are now available to small as well as large railroads.

MATISA RAILWELD, INC.
1020 Washington Ave., Chicago Heights, Ill.

the most efficient use of hydraulics ever applied to a tamper!

TAMPING UNIT DOUBLE CLUTCHES ELIMINATED BY REVERSIBLE HYDRAULIC MOTORS!

HERE IS the world's highest tamping efficiency.

This machine retains the unbeatable Matisa principle of vibration-compaction tamping with the machine load always on tamped track, but now has many *PLUS* features.

For details, write for the New *Matisa* Speedtamper brochure.

MATISA EQUIPMENT CORPORATION

1020 Washington Avenue — Chicago Heights, Illinois

Never leaves you in the dark

Wayside signs of "Scotchlite" Reflective Sheeting light up, bold and bright, in headlamp beams. Trainmen see them far down the track, even in bad weather. You get improved train control, greater operational safety . . . yet RTC signs cost you less! No costly painting or field maintenance. Long-lasting signs for your entire system can be mass-produced in one central sign shop. Crossbucks, too. See how much money *your* road can save with RTC. See your 3M Representative.

Scotchlite®
BRAND

REFLECTIVE SHEETING

3M Reflective Products Division
MINNESOTA MINING & MANUFACTURING CO.

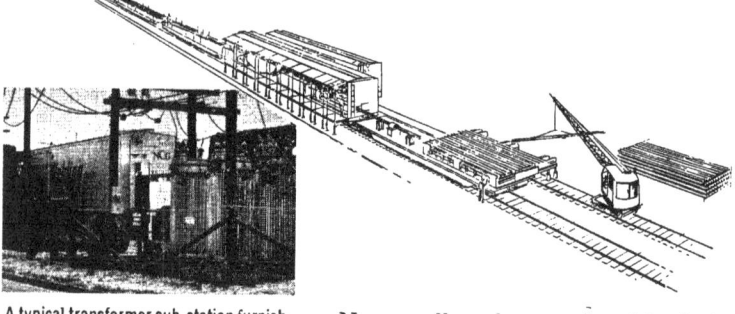

＿aw ＿＿wor ＿＿ines

Designed, Tested, and Proven
on America's Railroads

MONTGOMERY   **ALABAMA**

5 GOOD REASONS WHY THE '62 JACKSON TRACK MAINTAINER LEADS THE PARADE:

1 IT PUTS UP BETTER TRACK—LONGER LASTING TRACK due to the unique vibratory action that keys the ballast under the tie and under rail so thoroughly as to resemble a mosaic floor. And that's no applesauce. Removal of a tamped tie will prove it conclusively.

2 IT HAS TAMPING POWER GALORE. Over 7100 lbs. of vertical vibratory force and over 2100 lbs. of horizontal force IN THE BALLAST.

3 IT'S FAST. Powerful motors and improved tamper suspension insure rapid penetration. Fast double-acting workhead rams and fast, positive braking insure FAST INDEXING.

4 FULLY PUSH-BUTTON CONTROLLED for ease and speed of operation, with protection for tamping motors and generator against short circuits, single phasing and low voltage.

5 It's backed by the most competent and cooperative field organization in the railroad world . . . one that goes way out to insure maximum usefulness from all JACKSON tamping equipment. Let us give you the complete details.

JACKSON VIBRATORS, INC.
LUDINGTON, MICHIGAN, U.S A

R2-4A

Get any coating job done...
ANYWHERE
with

GRACO
HYDRA-SPRAY

The custom-built assembly shown above and to the right is an all-purpose rig designed to give *maximum* flexibility in coating and painting work. It was designed for field application of paints, lacquers, vinyls, cutback asphalts, creosotes, heavy oils and greases.

It uses the *economical* Graco Hydra-Spray Process, and proves once again, you get the job done faster and better with Graco than with any other coating system.

If speed of coating application, and material savings are important to you, write today for all the details of the Graco Hydra-Spray Process.

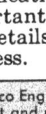

FREE! Graco Engineers are prepared to help you in the design of your paint and material spray assemblies. Your Graco Railway Representative will be glad to explain the many benefits of this service. Write or call him . . . *today!*

GRACO **RAILWAY DEPARTMENT**

GRAY COMPANY, INC.
MINNEAPOLIS 13, MINNESOTA

JOHN P. McADAMS, Eastern Sales Representative
2304 Wilson Boulevard, Arlington, Virginia

CHICAGO—(Broadview, Ill.)
R. D. Worley
3030 South 25th Ave.

PHILADELPHIA
The A. R. Kidd Co.
1036 Suburban Station Bldg.

NEW YORK—Newark, New Jersey
R. A. Corley
744 Broad Street

CLEVELAND
M. H. Frank Company, Inc.
1202 Marshall Building

LOUISVILLE
T. F. & H. H. Going
6308 Limewood Circle

SAN FRANCISCO
The Barnes Supply Company
Rm 504, 74 Montgomery Street

HOUSTON
Houston Railroad Supply Co.
1610 Dumble Street

ST. LOUIS
The Carriers Supply Company
818 Olive Street

TWIN CITIES—St. Paul, Minn.
The Daniel L. O'Brien Supply Compa.₩
Endicott-On-Fourth Bldg.

WASHINGTON—Arlington, Va.
Southeastern Railway Supply, Inc.
2304 Wilson Blvd.

MONTREAL—Ontario, Canada
International Equipment Co., Ltd.
360 St. James Street West

*You can
start like
this . . . .*⟩

*and end
up like
this . . . .*⟩

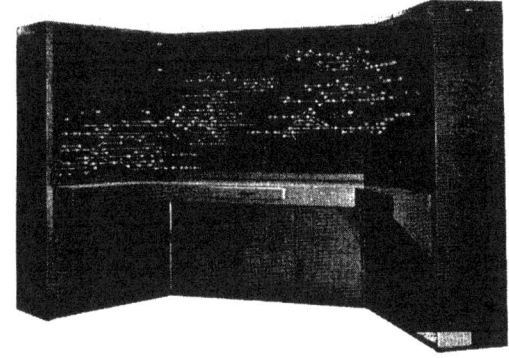

by using

SYNCROSTEP® REMOTE CONTROL

GRS Syncrostep is a packaged system—a control office stepper and application units and a field stepper and application units. You can start a large interlocking consolidation with one package and a small control panel. You can easily add on by additional packaged systems and expand your control panel at the same time. By this means, you create minimum disturbance to existing facilities.

E-x-t-e-n-d T-i-e L-i-f-e!
Hold Gage!
USE TIE PLATE
LOCK SPIKES

One-piece Design

LOCK SPIKES hold tie plates firmly in place on cross-ties and bridge timbers.

LOCK SPIKES are quickly and easily driven, or removed, *with standard track tools.*

Driven to refusal, the spread shank is compressed by the walls of the hole. Tie plates are held against horizontal and vertical movement under spring pressure. Play between the spike and the hole is eliminated—abrasion and seating of tie plates is overcome.

LOCK SPIKES hold their position in the tie, and redriving to tighten the plate is not required. They provide a quiet and strengthened track.

Annual cost of ties and maintenance expense is reduced by extending the life of ties and holding gage. Here is one answer to conservation of materials and labor. Write for free folder.

BERNUTH, LEMBCKE CO., INC.
420 Lexington Avenue, New York 17, N. Y.

Actual Size

AREA Publications—Price List

The following include some of the Association publications available from the secretary's office on order. Prices shown are for Members only:

Member
Price

Manual of Recommended Practice, complete in 2 volumes, including binders
(first copy) $15.00
Extra binders, each 4.50
Annual Supplements, each 1.00

Separate Chapters

6—Buildings 1.50
7—Wood Bridges and Trestles 1.00
9—Highways 0.50
11—Engineering and Valuation Records 1.25
13—Water, Oil and Sanitation Services 1.00
14—Yards and Terminals 1.00
15—Iron and Steel Structures 1.25
16—Economics of Railway Location and Operation 0.75
20—Contract Forms 1.25
22—Economics of Railway·Labor 0.50
25—Waterways and Harbors 0.25
27—Maintenance of Way Work Equipment 0.50
28—Clearances 0.25
29—Waterproofing 0.25
Flexible-cover, loose-leaf binder for separate chapters, each 0.40

Portfolio of Trackwork Plans—119 plans, 8 sheets of specifications, 5 sheets
definitions of terms, complete with leatherette cover $12.50
Track Scale Pamphlet—109 pages, flexible cover 0.80
Federal Valuation of Railroads—87 pages, flexible cover 1.00
Instructions for Mixing and Placing Concrete—24 pages, flexible cover 0.40
Notes on Railroad Location and Construction Procedures from the School of
Experience—43 pages, flexible cover 0.50
Handbook of Instructions for the Care and Operation of Maintenance of Way
Equipment—149 pages, hard cover 0.85
Instructions for Care and Safe Operation of Welding and Grinding Equipment—23 pages, flexible cover 0.30
Specifications for Steel Railway Bridges (fixed spans)—70 pages, flexible
cover 0.75
Specifications for Movable Railway Bridges—73 pages, punched sheets 1.00

Chapters 1, 3, 4, 5, 8, and 17 are not available.

from
coast
to
coast

RACOR TRACK SPECIALISTS SERVE YOU BETTER...

with America's most complete line of special trackwork: For Railroads, Mines and Industries — A complete line of *frogs*, *switches* and *crossings* · *Trackwork for installation in paved areas* · *Manganese steel guard rails* · *Automatic switch stands* · *Snow-Blowers* · *Switch point guards* · *Rail and flange lubricators* · *Tie pads* · *Racor studs* · *Dual spike setters* · *Dual spike drivers* · *Mechanical car retarders.*

with America's most complete trackwork manufacturing facilities: Coast to coast to serve your needs.

with America's most complete trackwork engineering service: This lies in making available to our customers Racor's engineering experience— *practical* experience from years of designing and manufacturing . . . *advanced* experience solving tomorrow's trackwork problems today in Racor research laboratories.

Why not let us help *you* with your trackwork problems?

Aeroquip Automatic Fueling Unit Stops Spillage, Speeds Refueling

Eliminate costly fuel spillage and overflow with the Aeroquip Automatic Fueling Unit. Fully automatic, it feeds fuel fast, shuts itself off when tank is full. It saves one gallon of fuel out of every fifty you buy, and eliminates oil pollution and contamination of drainage systems. Write for complete information.

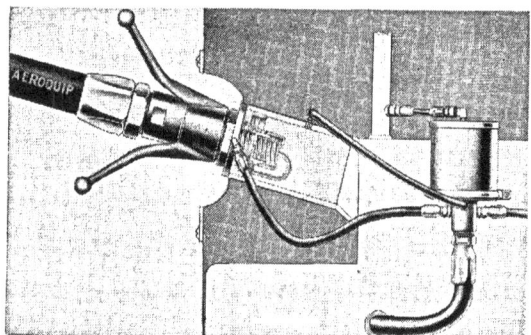

AEROQUIP CORPORATION • JACKSON, MICHIGAN
INDUSTRIAL DIVISION

INDUSTRIAL PLANTS: VAN WERT, O.; BURBANK, CALIF.; DALLAS, TEX.; PORTLAND, ORE.; CRANBURY, N.J.; ATLANTA, GA.
In Canada: Aeroquip (Canada) Ltd., Toronto 19, Ontario
In Germany: Aeroquip G.m.b.H., Baden Baden-Oos
AEROQUIP PRODUCTS ARE PROTECTED BY PATENTS IN U.S.A., CANADA AND ABROAD

AEROQUIP PRODUCTS ARE DESIGNED FOR BETTER RAILROADING

| FLEXMASTER Pipe Coupling | Diesel Manifold Clamp | Air brake, hot water, lube and fuel oil lines |

FLEXMASTER is an Aeroquip Trademark

CPSIA information can be obtained
at www.ICGtesting.com
Printed in the USA
BVHW080920211118
533723BV00010B/158/P

9 780260 451996